Fracture Mechanics of Ceramics

Volume 10
Fracture Fundamentals,
High-Temperature Deformation,
Damage, and Design

Volume 1	Concepts, Flaws, and Fractography
Volume 2	Microstructure, Materials, and Applications
Volume 3	Flaws and Testing
Volume 4	Crack Growth and Microstructure
Volume 5	Surface Flaws, Statistics, and Microcracking
Volume 6	Measurements, Transformations, and High-Temperature Fracture
Volume 7	Composites, Impact, Statistics, and High-Temperature Phenomena
Volume 8	Microstructure, Methods, Design, and Fatigue
Volume 9	Composites, R-Curve Behavior, and Fatigue
Volume 10	Fracture Fundamentals, High-Temperature Deformation, Damage, and Design

Fracture Mechanics of Ceramics

Volume 10
Fracture Fundamentals, High-Temperature Deformation, Damage, and Design

Edited by R. C. Bradt
University of Nevada-Reno
Reno, Nevada

D. P. H. Hasselman
Virginia Polytechnic Institute and State University
Blacksburg, Virginia

D. Munz
University of Karlsruhe
Karlsruhe, Germany

M. Sakai
Toyohashi University of Technology
Toyohashi, Japan

and V. Ya. Shevchenko
High Tech Ceramics Scientific Research Centre
Moscow, Russia, CIS

PLENUM PRESS • NEW YORK AND LONDON

ISBN 0-306-44203-5
Library of Congress Catalog Card Number 83-641076

Second part of the proceedings of the Fifth International Symposium on the Fracture Mechanics of Ceramics, held July 15–17, 1991, in Nagoya, Japan

A Division of Plenum Publishing Corporation
233 Spring Street, New York, N.Y. 10013

Printed in the United States of America

DEDICATION

JUNN NAKAYAMA
1930-1991

Junn Nakayama was a pioneer of the quantitative study of the fracture of ceramics. A physics graduate of Gakushuin University, he joined the research laboratory of the Asahi Glass Co., Ltd. in 1953. Within a decade he had developed the now-famous work-of-fracture test (*Japan Journal of Applied Physics*, **3**[7]442(1964)). He had already applied it to the quantitative analysis of the thermal shock damage of fireclay refractories (*Bulletin of the American Ceramic Society*, **45**[7]666(1966)) when others were just becoming cognizant of this simple, yet elegant technique. Those two studies were landmark efforts in establishing the framework for the microstructural design of improved thermal shock damage resistant refractories. The principles were subsequently extended to structural ceramics, as well as to the confirmation of the energy balance theories for thermal shock damage advanced by Hasselman.

Junn Nakayama spent his entire career with the Asahi Glass Co., becoming Director of the Research Center in Yokohama and subsequently Managing Director of the Asahi Glass Foundation in Tokyo. At this Fifth Fracture Mechanics of Ceramics Symposium in July of 1991, Junn Nakayama attended the opening ceremonies at the Japan Fine Ceramics Center in Nagoya, but was not able to participate in the meeting. When introduced to the attendees, he received a standing ovation. It is with deep regret to report to you that Junn Nakayama passed away on December 14, 1991. He will always be missed, but will never be forgotten.

PREFACE

These volumes, 9 and 10, of Fracture Mechanics of Ceramics constitute the proceedings of an international symposium on the fracture mechanics of ceramic materials held at the Japan Fine Ceramics Center, Nagoya, Japan on July 15, 16, 17, 1991. These proceedings constitute the fifth pair of volumes of a continuing series of conferences. Volumes 1 and 2 were from the 1973 symposium, volumes 3 and 4 from a 1977 symposium, and volumes 5 and 6 from a 1981 symposium all of which were held at The Pennsylvania State University. Volumes 7 and 8 are from the 1985 symposium which was held at the Virginia Polytechnic Institute and State University.

The theme of this conference, as for the previous four, focused on the mechanical behavior of ceramic materials in terms of the characteristics of cracks, particularly the roles which they assume in the fracture processes and mechanisms. The 82 contributed papers by over 150 authors and co-authors represent the current state of that field. They address many of the theoretical and practical problems of interest to those scientists and engineers concerned with brittle fracture.

The program chairmen gratefully acknowledge the financial assistance for the Symposium which was provided by the Japan Society for the Promotion of Science, the Inoue Foundation for Science, The Asahi Glass Foundation, the Nippon Sheet Glass Foundation for Materials Science, The Daiko Foundation, the Japan Fine Ceramics Center, and the Nagoya Convention & Visitors Bureau. Without their support, the magnitude and quality of this conference simply would not have been possible.

Unfortunately, the numerous individuals who contributed to the success of the conference cannot all be listed here. However, the program chairmen would especially like to recognize the contributions of Prof. N. Soga (President of the Conference, Kyoto University), Dr. O. Kamigaito (President of the Conference, Toyota C. R. D. Labs.), Dr. H. Awaji (JFCC), and Dr. H. Takahashi (Toyota C. R. D. Labs.) in planning and organization of the conference; Ms. E. Deguchi for her conscientious and efficient organization of the registration; Mr. N. Nakagomi for his patience and help in finally bringing these proceedings to press.

R.C. Bradt
Reno, USA

D.P.H. Hasselman
Blacksburg, USA

D. Munz
Karlsruhe, Germany

M. Sakai
Toyohashi, Japan

V.Y. Shevchenko
Moscow, USSR

July, 1991

CONTENTS

DISSIPATIVE PROCESSES ACCOMPANYING FRACTURE

J.T. Dickinson, S.C. Langford, and L.C. Jensen

Department of Physics
Washington State University
Pullman, WA 99164-2814 USA

INTRODUCTION

Failure by crack growth in materials is a serious problem in our technological society, reaching into our daily lives in our transportation vehicles, our structures, and all forms of machines. Exposure to natural stresses such as wind and gravity, mechanical forces and accelerations, and to radiation and chemicals is frequently unavoidable and often leads to crack formation, crack growth, and ultimately to catastrophic failure. As we push our technology to higher and lower temperatures, higher levels of stress, and wider ranges of environments, demands on materials become continually more intense. The role of fracture mechanics in the design and development of new materials has been crucial and has served as a guide to probing failure on microscopic levels. In this paper, I would like to explore some fundamental aspects of fracture which may stimulate and encourage extension of mechanical analyses to deeper levels of materials structure. After a brief introduction, this will include discussion of:

- fluctuations associated with a moving crack (chaotic aspects of fracture)
- resulting roughness of fracture surfaces (e.g., fractal geometry features)
- dissipative processes in fracture of inorganic crystals and glass
- time and spatially resolved measurements of model fiber-matrix debonding and pullout.

If we examine the forces which hold materials together, we find that they center on electrostatics. These forces depend on the electron distributions associated with the atoms in the solid, and depend on the type of bonding; .e.g. ionic, covalent, metallic and Van der Waals bonds, which result in net charge transfer or interactions due to net dipoles or induced dipoles. When a defect in a semi-brittle material experiences stress, we know that the forces between atoms distribute themselves in such a way that particular bonds are vulnerable -- localized strain increases. Interatomic potentials are many-body in nature; as strain rises between atoms, electron clouds in nearby atoms respond. Critical regions of strain exist where debonding and rebonding may occur, essentially in equilibrium. Fluctuations due to thermal motion and possibly due to stress (e.g., reflected stress waves in the bulk) can influence the rates of bond breaking / bond making. One could think of these fluctuations in terms of phonons bombarding and emitted from the bonds.

At critical strains, critical bonds can rupture irreversibly — this simply means the average separation of the atoms, $\langle r \rangle$, has exceeded distances where bond reformation cannot occur, therefore the crackfront has locally advanced. Alternatively, the flaw may initiate collective, concerted motion of atoms associated with slip or twinning. The desirable toughening associated with plastic deformation is of continued interest, because it has served

Fracture Mechanics of Ceramics, Vol. 10
Edited by R.C. Bradt *et al.*, Plenum Press, New York, 1992

us so well in metallurgy and could greatly benefit ceramic systems. We know that atomic structure plays a critical role in the differences between brittle vs ductile behavior; good examples are the essential role of impurities such as Hg and Tl in the embrittlement of aluminum and the recently proposed role of mobile oxygen atoms in the increased flow of fused silica.[1] These are effects arising from subtle changes in the local electronic structure, resulting in significant changes in the response to stress. On the atomic scale, the full theoretical description of these processes are still being developed using embedded atom techniques as well as more sophisticated many-body analyses. Nevertheless, it is the opinion of many that atomic scale studies of deformation and fracture will aid attempts to improve materials and their performance, particularly if connections to the macroscopic approaches can be made in useful ways.

Since our observations are generally experimental in nature, we summarize our laboratory techniques in the following section.

EXPERIMENTAL METHODS

The majority of our previous work[2] has been performed in vacuum at pressures ranging from 10^{-7} to 10^{-9} torr. For many studies, 20-24 samples are loaded into a carousel sample holder and sequentially rotated into position for testing. The load force is monitored to determine the rate of loading, the force at fracture, and the time of fracture. Samples are often instrumented with strain gauges and conducting arrays of thin film "wires" to provide more details on deformation and crack growth on sub-microsecond time scales.

Charged particles are often detected with a Channeltron Electron Multiplier (CEM), which produces fast (10 ns) pulses with high detection efficiencies for both electrons and positive ions. Excited neutrals with sufficient internal energy can also ionize upon collision with the CEM and be detected. By varying the CEM bias voltage and the voltages on nearby grids or barriers, the charge of the detected particles can be determined unambiguously.

Photons are usually detected with a photomultiplier tube, which also produces fast (30 ns) pulses with reasonable detection efficiencies, especially in the blue and ultraviolet. Photomultiplier tubes can be readily cooled (in vacuum, with liquid nitrogen) greatly reducing background noise levels. One of our photon detectors is housed within mm of the working area of a Applied Test Systems Universal Testing Machine, which in turn is light shielded as well as electronically shielded for time resolved phE and electrical transient measurements. Spectra of the phE can be obtained with a reasonable resolution spectrometer in a time resolved fashion using a gatable, image-intensified diode array and an optical multichannel analyzer (EG&G OMA-III).

The identifications of atomic and molecular species is undertaken with a variety of mass spectroscopies. Time-of-flight mass spectroscopy is of value for very short lived or highly peaked emissions. More sustained emissions are more readily studied with a quadrupole mass spectrometer (UTI-QMS), which can be tuned to a single mass peak. The time evolution (on a microsecond time scale) of a particular mass emission can be determined from the observed signals. Under the appropriate conditions, both these tools can be applied to studies of neutral emission (with ionizer) and positive or negative ion emission (without ionizer). As shown below, combinations of *two* quadrupole mass spectrometers offer unique capabilities for time resolved neutral emission experiments.

As mentioned above, interfacial failure often results in highly charged surfaces. One charge or the other will predominate on each surface, due to the contact potential difference between the dissimilar materials making up the interface. However, local variations in surface defect density can result in charged patches on both surfaces. Even one excess charge in a million surface sites yields very strong electric fields at the surface. A coil antenna placed near the sample is used to couple to the changing **B** fields associated with electromagnetic emissions often associated with electrical discharges within the crack and on the fracture surfaces. (We call the detected emission RE--for Radiofrequency Emission). Care must be taken to distinguish between near field vs far field radiation, the former accompanying *any* dE/dt, whereas the latter in this context is associated with electrical breakdown events.

GENERAL FEATURES OF FE/ELECTRICAL TRANSIENT STUDIES

In investigating the links between particle emissions/electrical signals and failure mechanisms, it is important to review some of the observations to date on these phenomena. As mentioned above, FE includes the emission of electrons, ±ions, neutral species in both ground states and in excited states, and visible photons. These emissions are often especially intense when failure occurs at or near a material interface. In general, the emissions are very sensitive to the locus of fracture, and velocity of crack growth, and electronic structure of the region near the crack tip.

The general features of the emissions accompanying cleavage or fracture (**cohesive failure**) are the following:[2,3]

(a) To a first approximation, the emission begins and is most intense *during* fracture.
(b) The kinetic energies of the electrons and ions are quite materials specific; if the materials are not prone to the creation of intense patches of electrical charge on the freshly created surfaces, the kinetic energy of the charged particles appears to be quite low. If the material is piezoelectric or if high densities of charged defects are produced during fracture (which is often the case in interfacial failure, for example), then the kinetic energies can be quite high, as high as several keV.[4,5]
(c) FE continues *after* fracture, suggesting that the fracture surfaces are in an activated state which relaxes, at least in part, via non-radiative processes. The "time constants" of this "after-emission", which vary from ms to several seconds, are also strongly material specific and can depend on the type of loading and strain rate preceding the fracture.
(d) In the few cases where thermal stimulation of the *fracture surface* has been performed, Electron Emission (EE) glow curves similar to those obtained following irradiation with x-rays or energetic electrons were observed. EE, when observed, could be correlated with the activation of known defect chemistry (e.g., V_K and V_F-center annihilation).

The electron and photon emission from fracture of inorganic dielectrics has been attributed to non-adiabatic processes involving fundamental excitations (e.g., excitons), or creation and recombination of point-like defects and charge carriers (electrons, holes). These mechanisms are consistent with a number of the above observations, and are consistent with bond breaking being the initial cause of the participating excitations.

For systems involving **interfacial failure,** the following model is appropriate:

(a) During crack propagation, charge separation occurs on the freshly created fracture surfaces.
(b) The rapid migration of charge carriers (usually electrons) along the surface leads to collisional excitations of the surface. This migration is most intense during fracture due to the high **E** fields. This stimulation of the fracture surface also is responsible for the after-emission from the fracture surfaces by producing charge carriers and traps suitable for luminescence and electron emission.
(c) In some systems, gas emission or gas in the environment (e.g. N_2) may be sufficient to precipitate microdischarging during fracture, yielding charged particles (generally electrons and positive ions) as well as photon emission (line spectra), and far field, long wavelength electromagnetic radiation.
(d) The static charge on the fracture surfaces leads to acceleration of the emitted electrons, modifying their energy distributions. A large portion of the EE is pulled back to the surface resulting in a self-bombardment process. This results in the emission of negative and positive ions (NIE and PIE) and excited neutrals (NE*) via electron stimulated desorption (ESD).

Different types of photon emission (phE) have been observed. phE can occur *prior* to fracture in a number of systems. In some cases, this light is due to microfracture events that precede failure. Moving dislocations can generate defects which recombine to yield light in some materials. Following fracture, some materials "glow" with characteristic decaying signals, much like phosphorescence. These emissions result from the recombination of charged trapped at surface defects, either by tunneling or by thermally activated motion of one charge carrier to a recombination centers. In addition, where charge separation is intense, light from microdischarges is quite evident. Typically, these events yield sharp spectral lines characteristic of the gases present in the crack tip, discernable with spectroscopy.

Neutral emission accompanying fracture often is due to the release of occluded gases from the freshly exposed surface. These gases may be trapped at interfaces, grain boundaries, voids, and similar structures. Of more fundamental interest is NE derived from bond scissions, sometimes referred to as mechanochemically derived NE. For instance, recent observations of NE bursts from the fracture of brittle alkali halides,[6] single crystal Ge,[7] as well as soda lime glass,[8] have been interpreted in terms of the energetic relaxation of plastic deformation produced during failure. Current work on calcite is showing similar behavior, where the major emission is in the form of CO_2 bursts after fracture, indicating that delayed, relaxation of deformation causes local decomposition.

As pointed out by Grayson and Wolf,[9] one must be very careful in distinguishing occluded and mechanochemical NE. Mechanochemical emission is intimately related to the energetics of fracture, although the emission of occluded gases is also useful for characterizing the material failure. In the case of a composite system, the presence of occluded gases/vapors at interfaces can strongly influence interfacial strength by either blocking bonding sites or chemically attacking existing bonds. Thus, NE can yield information about atomic scale processes such as plastic deformation, as well as microcracking, intergranular failure, and the role of voids and inclusions in the fracture process.

Accompanying discharges as well as rapidly migrating surface charge (surface breakdown) can yield intense bursts of phE, EE, and changing **B** fields which accompany the fracture of many materials. Indeed, the spectra of many triboluminescent materials show strong nitrogen lines, due to the ionization of atmospheric or occluded nitrogen. Recent results in our laboratory indicate that charge flow among locally charged regions in the absence of nitrogen can also excite surface states directly to yield intense luminescence (a form of electroluminescence).

phE from composite systems with transparent or semi-transparent materials can be exploited to determine the location and of failure at embedded interfaces. Multiple PMT measurements have been used to determine the direction and speed of debond cracks, for instance. Furthermore, comparison of photon emission (phE) signals with simultaneous electron emission (EE) measurements allow us to distinguish between interfacial failure at embedded interfaces and failure which communicates with the surface. As electrons cannot penetrate the matrix to any significant degree, EE is observed only from failure events at external surfaces.

Fracto-emission measurements during the failure of metal inorganic interfaces show that emission intensities can be very sensitive to the nature of the interface. Intense phE and EE is observed when Au is peeled from an oxidized Si substrate.[10] However, samples with very thin oxide layers showed orders of magnitude less emission, despite the fact that they were considerably stronger than the samples with thick oxides. The surfaces produced by interfacial failure are apparently charged due to the contact potential difference between the metal and the semiconductor. The thick oxide apparently serves as a barrier to the recombination of charge on either side of the interface, charge which subsequently is available for emission processes. The thin oxide allows rapid nonradiative recombination during crack growth, and thus yields little emission.

Our first work on electrical transient measurements accompanying fracture employed instrumented specimens of single crystal Si.[11] Electrical currents were observed during crack growth which we attributed to free carriers generated by the fracture event. These studies conclusively showed that bond breaking can yield excitations that bridge the bandgap.

EVIDENCE FOR CHAOS IN FRACTURE

In the case of the moving crack, we know from simple fractography that instabilities can arise at the crack front to cause deflection, branching, and even arrest. Dissipative, non-linear systems frequently display chaotic behavior, suggesting that these instabilities in crack growth are related to chaotic dynamics. Due to the difficulty of time-resolved imaging of the crack front at the high velocities where these instabilities occur, we have used the rapidly fluctuating photon emission signals to study the dynamics of this rapid crack growth. This approach rests upon the intuitive notion that photon emission signals indicate the rate of crack growth (area/time). Mathematically, this approach depends on the observation that the character of multidimensional chaotic systems can generally be inferred by analyzing measurements which reflect a single system variable or "degree of freedom."[12]

Typical phE data digitized at 10 ns time intervals during the fracture of TGDDM/DDS epoxy is shown in Fig. 1.[13] The autocorrelation function of this data, shown in Fig. 1(b), is consistent with a chaotic signal. The initial 30 ns spike in the autocorrelation corresponds to the finite pulse width of the photomultiplier tube's response to single photons. If the phE fluctuations were totally stochastic, this 30 ns spike should be the only significant feature in the autocorrelation, as purely random fluctuations are totally uncorrelated. However, the autocorrelation also shows a broad peak about 2 μs long, indicating that fluctuations tend to be correlated over a rather long time. The autocorrelation is also consistent with a lack of long term (greater than a few μs) periodicity or predictability, another hallmark of chaotic behavior. The autocorrelation of a photon emission signal accompanying fracture of single crystal MgO showed a similar broad peak about 200 ns long.

Additional analytic tests show that in fact the effective "phase space" dimension required to accommodate these fluctuations is quite small (correlation dimension ~ 3.3). Stochastic processes require an infinite number of dimensions, again indicating that the observed fluctuations are primarily non-stochastic in nature. Furthermore, this indicates the number of variables required in any "equation of motion" describing the phE fluctuations is small. Thus the underlying emission process, which must account for fluctuations in the rate of crack growth, is amendable to a description based on a relatively small number of variables. Although this analysis does not indicate which variables are important, the mechanics of crack growth suggest that these variable account for the local microstructure at the crack tip and the stress intensity along the crack front.

Perhaps the most decisive test for chaotic behavior is the existence of a positive Lyapunov exponent.[14] This exponent reflects our ability to predict future behavior on the basis of present knowledge of the system, viewed in terms of information theory. A positive exponent indicates that our ability to predict future behavior decreases exponentially with time. By definition such behavior is chaotic. Following a technique given by Wolf et al.[15] we have obtained a positive exponent for the epoxy phE data ($\lambda_{max} \sim +1.7$ bits/μs), again confirming the chaotic nature of this process.

The roughness of many fracture surfaces suggests that they are the result of chaotic crack deflection and branching. This motivated us to seek a direct connection between fracture surface roughness and fluctuations in the process of crack growth by comparing fractal dimension measurements of both the photon emission signals and the resulting fracture surfaces. As no one method for fractal analysis of time series data is completely reliable,[16] dimension measurements of the phE signals were carried out by both Fourier transform and box dimension methods. In particular, the Fourier transform of fractal time series data typically shows power law behavior ($1/f^{\beta}$) at high frequencies.[17,18] The Fourier spectrum, shown in Fig. 1(c) on a log-log plot, displays power law behavior at frequencies above about 20 Mhz. A least square fit of the high frequency coefficients on a log-log plot yielded $\beta = 2.1 \pm 0.1$. Mandelbrot has shown that the exponent in the inverse power law, β, is related to the fractal dimension, D_f, of the corresponding nominally one-dimensional data by

$$D_f = 2.5 - \beta/2. \tag{1}$$

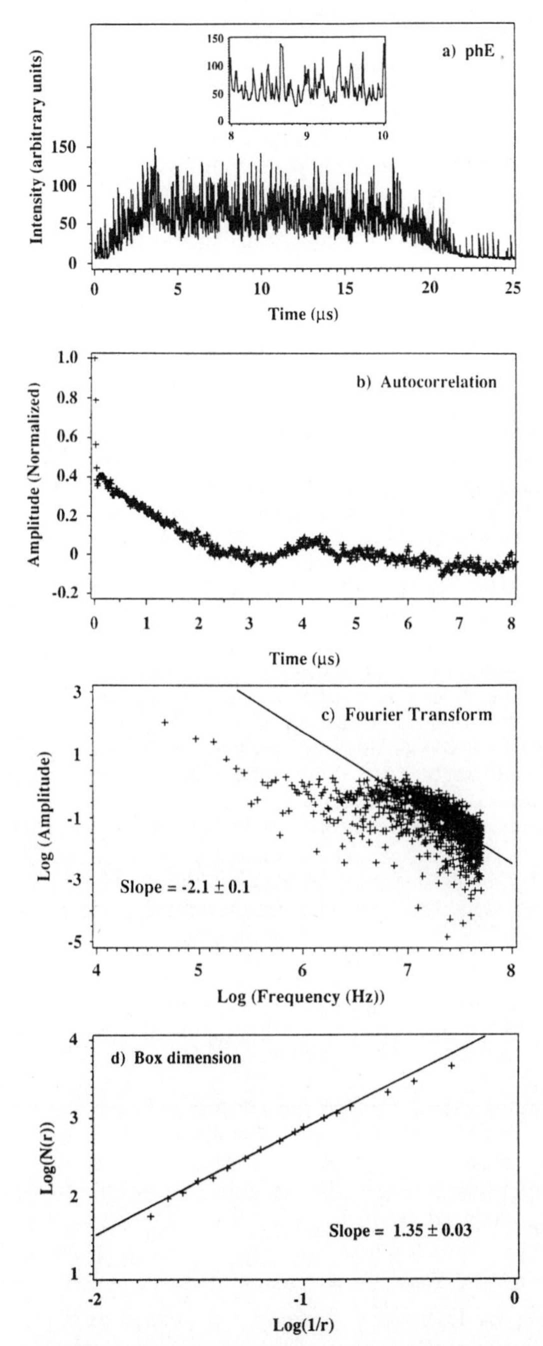

Fig. 1. Chaotic Behavior of the photon emission accompanying the fracture of an unfilled epoxy. (a) the phE during the time interval of crack growth, (b) the autocorrelation function of this data set, (c) the Fourier spectrum (log-log scale), and (d) the Box Dimensional analysis of this time series. The slope of 1.35 implies a fractal dimension of 1.45.

The subscript f is used to indicate that this is the fractal dimension computed from the Fourier transform. The value of β computed for these data thus corresponds to a fractal dimension of about 1.45.

A check on the reasonableness of this result was made using Mandelbrot's "box dimension," D_b.[19] The box dimension is computed by superimposing a square grid on the phE intensity curve and counting the number of grid squares intersected by the curve, N(r), as a function of grid square size, r. If the signal is fractal, N(r) proportional to $r^{-\beta}$, were $\beta = D_b$ is not an integer. The box dimension is thus equal to to slope of N(r) vs (1/r) on a log-log plot, shown in Fig. 1(d) for the epoxy phE data. A linear least squares fit of the linear portion yielded $D_b = 1.35 \pm 0.03$. Given the uncertainty in determining the linear portion of the Fourier spectrum and the problems inherent in box dimension computations, D_f and D_b are essentially in agreement.

The fractal dimension of a typical epoxy fracture surface was measured by the slit island method.[13] The fracture surface was coated with about 90 nm of gold and potted in a clear epoxy. The potted sample was then polished by hand to yield a flat surface roughly parallel to the fracture surface. Polishing was continued until the fracture surface was intersected by the polished surface, yielding 'islands' of the brown epoxy surrounded by as yet unbroken gold. Photographs of the island structure were analyzed to determine the fractal dimension of the surface, D_s. D_s is related to the slope of a log-log plot of the island area vs island perimeter, β, according to the relation

$$D_s = 2/\beta. \qquad (2)$$

An island area vs perimeter plot taken from photographs of a ~1 mm^2 area of the fracture surface yielded a slope of 1.51, implying that $D_s = 2.32$. D_s measurements made from different "cuts" of this same surface were within ±6% of this value. Thus the epoxy fracture surface is indeed fractal. Assuming the surface is isotropic, a one dimensional profile of such a fracture surface would have a fractal dimension of $(D_s - 1) = 1.32$. Since the phE signal is nominally one dimensional, we may compare $(D_s - 1)$ with the dimension of the phE signal, which was on the order of 1.35-1.45. The similarity of these numbers is suggestive, although we are not in a position to say exactly what the relationship should be. Thus, we are anxious to continue testing the use of phE as a probe of the fractal character of fracture processes.

Single crystal MgO also shows chaotic behavior in the phE during fracture, although the relatively short duration of fracture and the correspondingly fewer data points per fracture event prohibit a complete analysis. A box method analysis of the phE signal yielded $D_b = 1.11 \pm 0.1$. Macroscopic surface measurements using slit-island methods were not possible because of the relatively small relief of the surface features. We did succeed in obtaining fractal dimensions of MgO fracture surfaces using the scanning tunneling microscope (see below).

STM OBSERVATIONS OF FRACTAL SURFACES

STM and AFM are ideal techniques for fractal analyses. The numerical height vs position data are readily analyzed by either Fourier transform or 3-D box dimension methods. Insulating materials may be scanned by STM after Au coating, as shown by Jaklevic et al.[20,21] We have observed good stability and conductivity in films ~20-100 Å in thickness, with no evidence of Au clusters. Lateral resolution of 2 Å (in the plane of the surface) and vertical resolution of 5-10 Å (normal to the surface) are readily obtained. The major advantage of the STM relative to the atomic force microscope (AFM), which has potentially higher resolution, is the *reliability* of the STM. We have successfully obtained topographical information[22] on virtually all of the gold coated surfaces produced to date. Three STM scans of a gold-coated MgO fracture surface (single crystal broken in three point bend) are shown in Fig. 2. The scans shown in Figs. 2(a) and (b) were taken on the tensile side of the fracture surface, while that of Fig. 2(c) was taken on the compressive

STM Scans of an MgO Fracture Surface

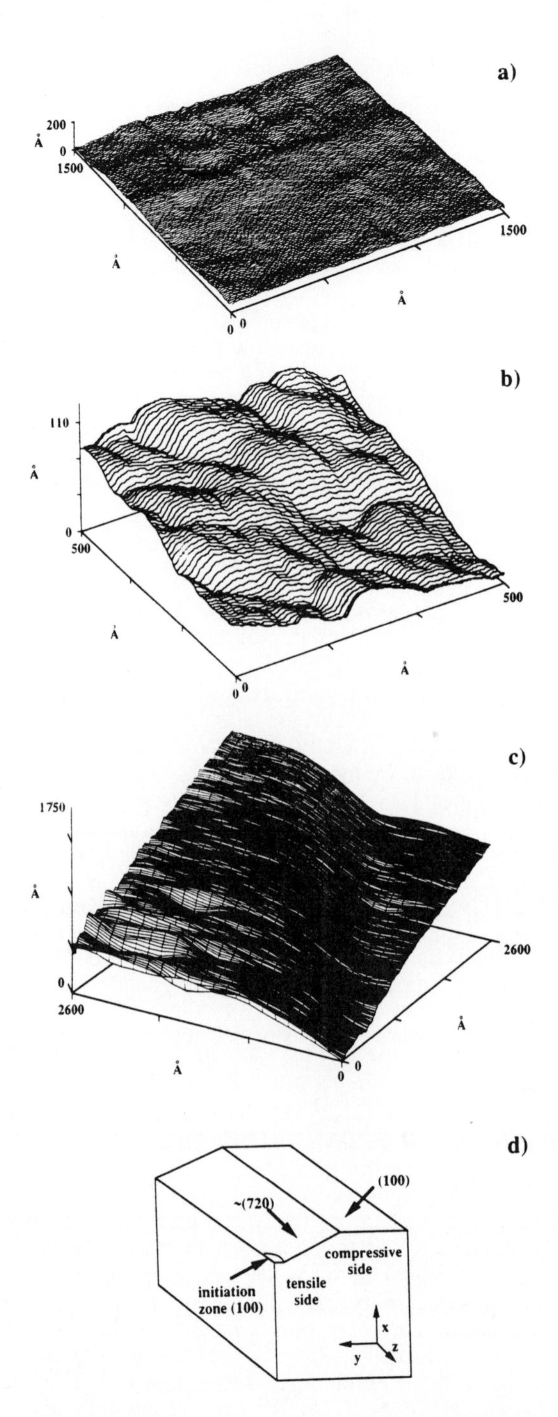

Fig. 2. Three STM scans of a gold-coated MgO fracture surface. (a) and (b) scans on the tensile side, (c) scans on the compressive side. (d) shows the orientation of the crystal after fracture; the initiation zone is at an edge flaw on the tensile side of the crystal.

side of the fracture surface. As indicated in Fig. 2(d), the tensile surface of samples broken in this geometry is inclined about 16° to the compressive surfaces. The compressive surfaces are macroscopically (100) in orientation, while the tensile surfaces are approximately (720). Fig. 2(a) shows a typical flat region of the tensile side. It is ~(720) in orientation on the scale of hundreds of nm, with deviations from the dominate plane of less than 36 Å. A number of similar planes were found over the entire tensile side and were typically of micron dimensions. Interspersed between the smooth regions were patches of rough topography, one of which is shown in Fig. 2(b). Again, the larger facets are of ~(720) orientation. At our present resolution (< 5 Å), we were unable to resolve the (001) atomic steps necessary to accommodate the surface inclination. However, atomic force microscope (AFM) measurements of similar MgO surfaces (*uncoated* with Au) did in fact show patches corresponding to steps of atomic height.[23]

Fractal analysis of scans from the tensile side of the fracture surface, outside the smooth fracture mirror, consistently yielded dimensions in the range of 2.3-2.4. However, smooth compressive side regions yielded dimensions of 2.0, consistent with the basically smooth, Euclidean surface created by fracture on {100} cleavage planes. Our significant departures form Euclidean dimensions determined by STM of Au coated surfaces were substantiated by similar values obtained by AFM scans on the uncoated MgO surfaces.[23] Tsai and Mecholsky have observed similar variations of fractal dimension along fracture surfaces of single crystal Si using optical photographs and the slit island technique.[24] Surfaces in the smooth fracture "mirror" and in the "post-branching" region were not fractal, while surfaces in the branching (hackle) region showed fractal behavior. The fractal dimensions of their Si surfaces (maximum $D_S = 1.10$) are also significantly lower than those observed in MgO. Tsai and Mecholsky showed that Si samples cut in different orientations to fracture along {110} planes showed different fractal dimensions, where surfaces with higher static fracture toughness (K_{IC}) showed higher fractal dimensions. Their results are consistent with the lower fractal dimensions of Si oriented to fracture along low energy {111} planes measured by Mitchell and Bonnell using STM techniques.[25] The questions of how these fractal dimensions relate to fracture energy, especially the energy required for *dynamic* crack growth (as opposed to the static fracture toughness) and the details of any interactions with plastic deformation at the crack tip are yet to be determined.

The potential correlation between fractal surface dimension and static fracture toughness suggest that the dynamics of crack growth can have substantial impact upon important engineering material properties. Correlations between toughness and fractal dimensions have been observed in both brittle materials[26] and metals.[27] In the main, toughness is enhanced in brittle materials by crack branching and in ductile materials by void growth. Both of these phenomena represent fluctuations in the process of crack growth which could easily take on a chaotic nature and affect surface roughness and fractal dimensions. Crack branching in brittle solids is poorly understood. Attempts to model crack branching in a predictive sense have met with limited success.[28,29] However, the nucleation, interaction, and coalescence of microcracks or voids in the process zone of an advancing crack have been observed in transparent materials.[28] Fractal models of crack growth promise to yield insight in rapid crack growth and its consequences.[30] For instance, fracture simulations involving triangular arrays of springs show fractal behavior.[31,32] The local strains in such models display interesting "multifractal" distributions immediately prior to ultimate failure.[33] Recently, analytic solutions for the crack tip stress field have also been found which show scaling behavior appropriate to fractal surface formation.[34] These approaches show promise in establishing the relation between static material properties and dynamic crack growth. Models which relate material toughness to fractal dimensions, for instance, can be checked on the basis of measurements in the literature. However, novel experimental methods will be required to verify the dynamic aspects of these models.

EMISSION OF NEUTRAL ATOMS AND MOLECULES

Fracture is often accompanied by the emission of neutral atoms and molecules. The energy required for rapid crack growth is well in excess of that required for surface formations, and some of the excess energy can result in the thermal or mechanical ejection of atoms from the fracture surface. Intense neutral emission can result in those cases when the stoichiometric fracture surface is not the lowest energy, equilibrium surface for this material. This appears to be the case for soda lime silicate glasses, for instance.[8] Of special interest are emissions which apparently reflect dislocation activity. Such emissions often appear as bursts delayed some milliseconds after fracture. This contrasts with most other fracto-emission components, which reach their peak *during* the fracture event and thereafter decay. Emissions which appear to be due to dislocation activity include alkali and alkali halide neutral emission accompanying the fracture of single crystal NaCl and LiF,[6] and Ge emission from single crystal Ge.[7]

As noted above, quadrupole mass spectrometry is generally used for time resolved neutral emission measurements. To unambiguously resolve the source of molecular emissions, we have recently employed a two quadrupole mass spectrometer system, to make simultaneous measurements at two masses. A single electron impact ionizer is symmetrically situated between two identical quadrupole mass filters as shown in Fig. 3(a), so that ions created in the ionizer will be attracted by the nearest mass filter. By comparing the emissions at various masses, we can often distinguish between ions created by direct ionization of molecules emitted from the fracture surface (parent molecules) and ions created by fragmentation of the parents (daughter molecules) by electron impact in the ionizer. The two-quadrupole system also allows direct comparison of the intensities of products emitted at different masses.

The fracture of single crystal NaCl yielded bursts at masses 23 (Na^+) and 58 ($NaCl^+$) are shown in Fig. 3. The origin of the time axis has been fixed at the time of fracture, as determined from load measurements made with a piezoelectric transducer. Note that the bursts occur over 100 ms *after* fracture. Intensity measurements indicate that these bursts are primarily composed of atomic Na and Cl, with small amounts of molecular NaCl. Similar bursts were also observed after the fracture of LiF. In contrast, the most intense signal from LiF was at mass 33 $Li(LiF)^+$. Alkali halide ions of the form $(LiF)_n$ are known to be unstable, dissociating to $Li(LiF)_{n-1}^+$.[35] Therefore, we attribute this signal to the emission of the alkali halide cluster $(LiF)_2$. Signals were also observed at masses 26 (LiF^+), 59 [$Li(LiF)_2^+$], and 85 [$Li(LiF)_3^+$] which we attribute to the emission of LiF, $(LiF)_3$ and $(LiF)_4$, respectively. Only small signals were observed at masses 7 (Li^+) and 10 (F^+), indicating relatively weak emission of the atomic species.

The long delay between fracture and these emission bursts can be accounted for if these emissions are due to the emergence of dislocation structures (e.g., loops, pileups, and slip bands) at the fracture surface.[6] Dislocations are produced in the fracture of alkali halides and are readily observed by etching the fracture surfaces.[36,37] Shallow dislocations and dislocation structures are attracted to the surface by a force which decreases as 1/distance from the surface. Dislocations are readily generated during the fracture of alkali halides, being driven into the bulk by the high stress field near the crack tip. After fracture, mobile dislocations close to the surface will be accelerated back to the surface.

To investigate the ability of dislocation motion to the surface to provide the energy required for emission bursts many ms after fracture, we constructed a numerical model of the process. The model dislocation was a half-loop lying in a plane normal to the surface of an isotropic material. The dislocation was allowed to contract under the influence of the image force from an initial radius R until it vanished, i.e. until "popout." Assuming the loop maintains its semicircular geometry as it contracts to the surface, the equation of motion for acceleration under the image force, normalized to unit dislocation length, is given by:

$$\frac{\alpha \ln[(4R/\rho) - 2]\cdot\ddot{R}}{c_t^2} + b\sigma_o\left(\frac{|\dot{R}|}{v_o}\right)^m + \beta|\dot{R}| + \frac{\alpha \ln[(4R/\rho) - 1]}{R} = 0. \quad (3)$$

where R is the loop radius, c_t is the shear wave velocity in the glide plane, and ρ is a core cutoff parameter. Here we take $\rho = b/2$, where b is the Burgers vector. The first term accounts for the "inertia" of the dislocation, and the fourth term gives the appropriate image force.[38] α is given by

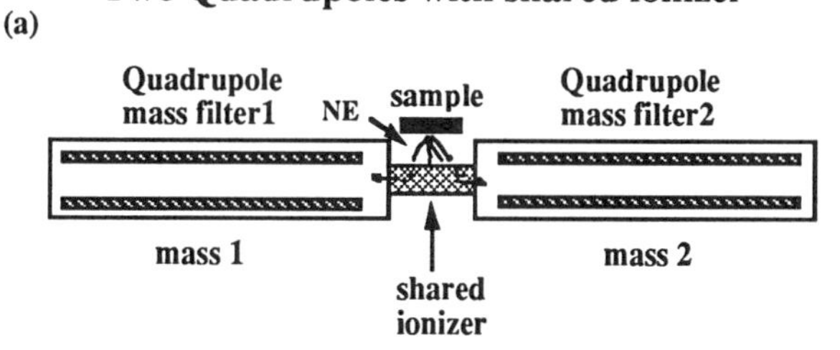

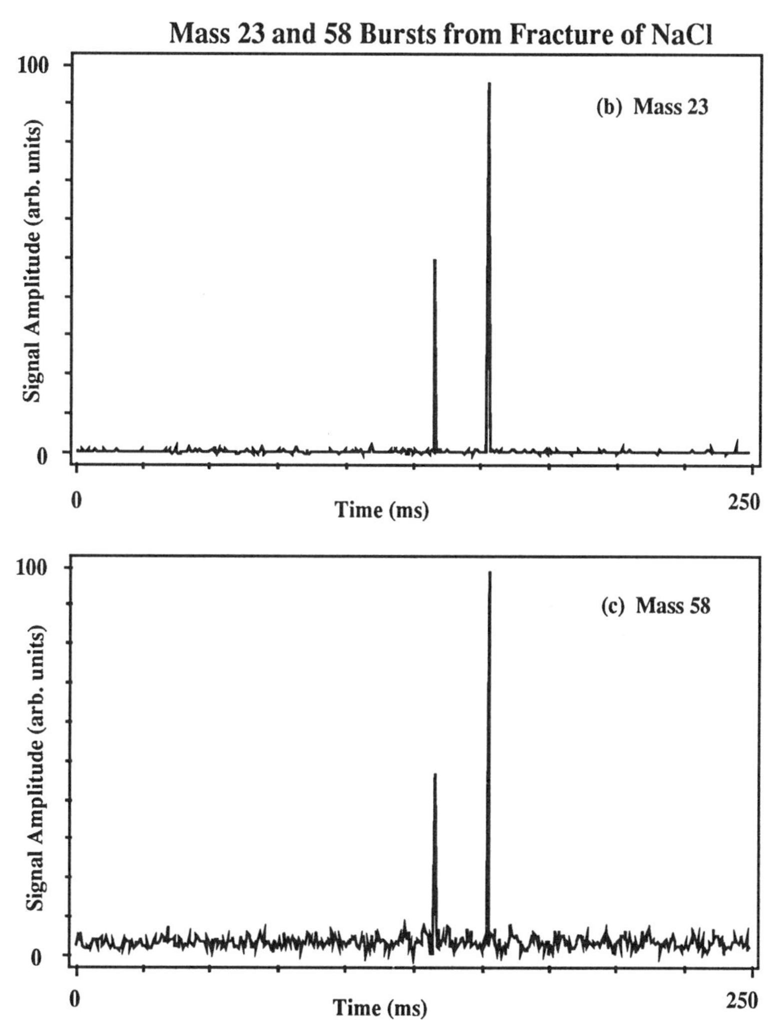

Fig. 3 (a) Two quadrupole experimental arrangement. Simultaneous, two-quadrupole emission measurements at (b) mass 23 (Na^+) and (c) mass 58 ($NaCl^+$) accompanying the fracture of single crystal NaCl.

$$\alpha = \frac{\mu b^2(2-\nu)}{8\pi(1-\nu)} , \qquad (4)$$

where μ is the shear modulus, b is the magnitude of the Burgers vector, and ν is Poisson's ratio. Using values appropriate to slip in LiF on the {110}<110> system, we obtain $\alpha = 1.19 \times 10^{-9}$ N. "Relativistic effects," which limit the dislocation velocity to speeds on the order of c_t, were neglected except as noted below.

In the absence of damping forces, such a loop disappears very rapidly. The middle two terms account for these damping forces on the basis of velocity vs stress measurements made by Johnston and Gilman.[39] Their data were modeled by a term linear in velocity ($v = \dot{R}$), which describes phonon damping effects, and a term in $(v/v_o)^m$, which empirically describes the effects of jog dragging and impurities. The parameters $\sigma_o = 6.5$ MPa, $v_o = 50$ μm/s, $\beta = 7\times10^{-5}$ Pa and m = (1/23) were chosen to fit Gilman's data for LiF. The parameter σ_o depends strongly upon the thermal history (and thus defect distribution) of the sample, while v_o is a function of dislocation character, differing a factor of 50 between screw and edge dislocations. No attempt was made to distinguish between screw and edge components of the dislocation half loop in this model; the value of v_o appropriate for screw dislocations was used.

This equation of motion is unstable near the surface, where the image force diverges, and at diameters greater than about 1 μm, where the damping forces are dominated by the term in v^m. To study the character of the solution at loop diameters greater than about 1 μm, we take advantage of the fact that the accelerations at these diameters are small enough that we can reasonably neglect the inertial term. The motion of the dislocation was modeled by patching the solution of Eq. 1 without the inertial term for diameters greater than 1 μm to the solution of the full equation at smaller diameters. Errors introduced by instabilities near the surface are small due to the high velocity of the dislocation in this region.

The solutions to these equations predict that shallow dislocation loops as small as 2 μm in diameter can require seconds to reach the surface. This is sufficient to explain the longest observed delays from fracture. Relativistic versions of this model indicate that the kinetic energy delivered by such a dislocation loop at "popout" can be on the order of several keV. Much of this energy would be dissipated by phonon interactions if it were not for the decoupling from the phonon field at very high dislocation velocities, where time required for dislocation "jumps" is shorter than that required for equilibration with the available phonon modes. The large number of alkali atoms involved in the emission bursts require significantly more energy than released by a single dislocation loop. Thus we propose that dislocation structures, such as dislocation pileups and slip bands, are involved.

Similar delayed bursts were observed to follow the fracture of single crystal Ge, as shown in Fig. 4. These bursts suggest that similar dislocation activity accompanies the fracture of single crystal Ge. This is significant in view of the extreme brittleness of this material. Normally, dislocations in Ge are quite immobile at room temperature.[40] However, dislocation production during fracture of very strong, chemically etched, samples has been verified by observation of the resulting etch pits.[41] (No special sample preparation beyond mechanical polishing was performed in our work.) Further, dislocations produced in fracture take the form of half-loops, which possess high kink densities to accommodate their curvature. Thus the "return" of these dislocation loops to the surface under the influence of the image force can be accomplished by kink motion alone. The high mobility of kinks in Ge at room temperature[42] would allow energetic dislocation emergence, which would yield the accompanying neutral Ge signal.

FAILURE AT INTERFACES

In ceramic and metal matrix composites the phenomena of adhesion and debonding, and dissipative processes such as frictional pullout, plastic deformation, microcracking, crack branching, and crack deflection can play critical roles in determining

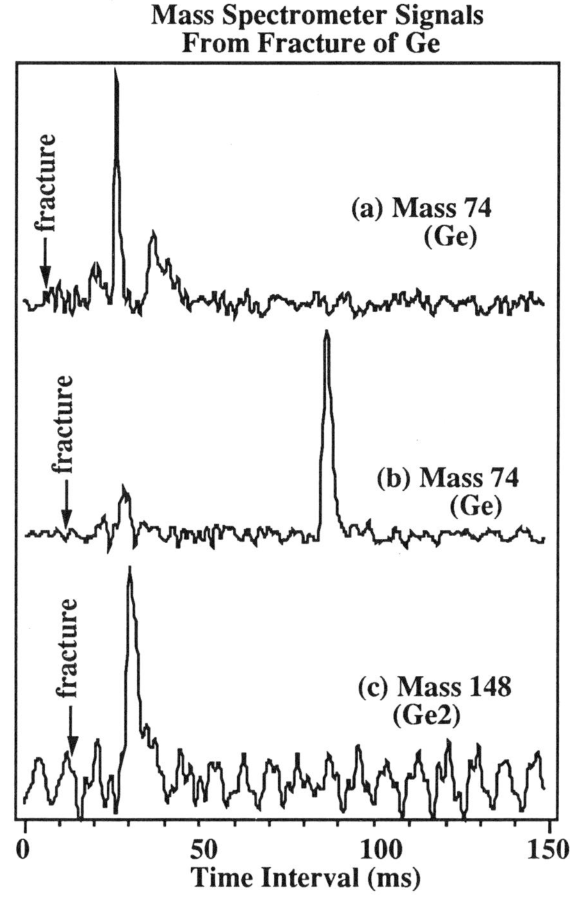

Fig. 4. Normalized quadrupole mass spectrometer signals accompanying the fracture of single crystal Ge. The signals of (a) and (b) are both for Mass 74 and were acquired during two representative fracture experiments. They are primarily from the emission of atomic Ge. (c) A similar signal for mass 148 (Ge_2) from the fracture of a third (larger cross section) sample. The peak electrometer currents were (a) 0.5 nA, (b) 0.6 nA, and (c) 4 nA, respectively.

strength and toughness. A classic example involves crack propagation in fiber-reinforced brittle-matrix composites, where loading of the crack leads to debonding of crack-bridging fibers, followed by frictional pullout which dissipates energy by generating heat at the sliding interface. The resulting load/deflection curve of such a material shows strong R-curve behavior, resulting in considerable enhancement of the total fracture energy (area under the curve) compared to simple matrix failure. The peak load and fracture energy are the parameters to be maximized. They depend on the inherent strength of the matrix and reinforcing fibers as well as the details of the interface, including residual stresses, and interfacial shear strength (dictating debonding). In cases where the debonded fiber experiences residual compression, the effective friction coefficient along the debonded interface, μ, and the length of intact, embedded fiber on each side of the crack are also important. The effective μ can be affected by the fiber microstructure, i.e., surface roughness and asperities.

For cracks oriented normal to the reinforcing fibers (if inclined, the fibers tend to fail in bending and display minimal pullout), the main contributions to toughening are:[43] the strain energy stored in the fiber on both sides of the crack (thought to be dissipated in the form of acoustic waves during debonding); the energy required for the formation of the debonded surface; the frictional energy dissipated in the pullout process (often the most important); and finally, increased matrix fracture due to crack deflection, branching, and microcracking. Obviously, the actual response of a particular material will be strongly influenced by loading rate and geometry, cyclic loading, temperature, and environment.

Although a number of workers have greatly contributed to the understanding a number of these effects and have developed methods to exploit them in the processing of actual materials, a better understanding of the contribution of each of these processes, particularly in a time resolved, dynamic fashion in a variety of environments would be highly beneficial. We are combining our fracto-emission techniques with the detection of *electrical transients* which accompany and follow the failure of interfaces. These techniques offer time and spatially resolved probes of a number of the events responsible for toughening in brittle matrix composites and can prove useful in determining failure mechanisms.

Metal/inorganic dielectric bonding is dominated by charge transfer which can be viewed from both contact charging or chemical bonding points of view.[44] Interfacial failure often results in intense phE and EE.[2,3,5] The phE, EE, and PIE measurements potentially offer information about these charge transfer processes. The link between FE and electronic interactions (which reflect bonding and thus micromechanical properties) is therefore of fundamental importance. To address this issue, we are currently measuring the energies of the charged particles emitted following fracture. These energies should be sensitive to the charge densities (small patches) on the newly formed fracture surfaces. These charge densities in turn depend on the occupation of surface and near surface states at the interface, which is in turn governed by the contact potential difference between the metal and dielectric phases.

After fracture has occurred along a metal/inorganic insulator interface, the insulator surface is still highly charged and very active in terms of emission. As a straight forward example, the electron emission from accompanying the peel of a Au metal film from an SiO_2 substrate[10] [100 nm of oxide on a single crystal Si(111) surface] is shown in Fig. 5. We have unambiguously determined that this emission is coming from the *oxide* surface. Probing the surface charge with low spatial resolution after the peel event, we find a high net positive charge on the oxide. This is consistent with previous studies of MOS structures, where charge separation is driven by the electron affinity of Si and the work function of Au. Despite the highly positive net charge, we observe electron emission from the oxide surface. This clearly requires the existence of negatively charged regions on the oxide. The local negative charge density must be sufficient to provide some electrons with enough kinetic energy to overcome the attraction of the surrounding positive charge. Thus, a fundamental question arises as to the homogeneity of a surface created by the fracture of an interface between two dissimilar materials, particularly when the bonding is not chemical, but involves van der Waals forces or forces due to transfer of free charge.
Previous studies of polymer surfaces indicate that the emission kinetics are consistent with thermally stimulated recombination. The binding energy of trapped carriers can be determined from the emission response to a linear temperature ramp.[45]

Higher resolution measurements of the charge distributions on the newly exposed insulator surfaces can also be probed for regions of net positive and negative charge. Contact charging literature, reviewed by Lowell and Rose-Innes,[46] indicate that patchiness in both charge density and sign is quite common and poorly understood. Scanning of the surfaces is done with miniature charge probes capacitively coupled to the input of a sensitive, high impedance, charge sensitive operational amplifier. Scanning force microscopy could be used to examine the charge distributions on much smaller scale,[47,48] where a lever arm with a conducting tip can be biased to detect charge densities on the order of 10^{-6} C/cm^2.

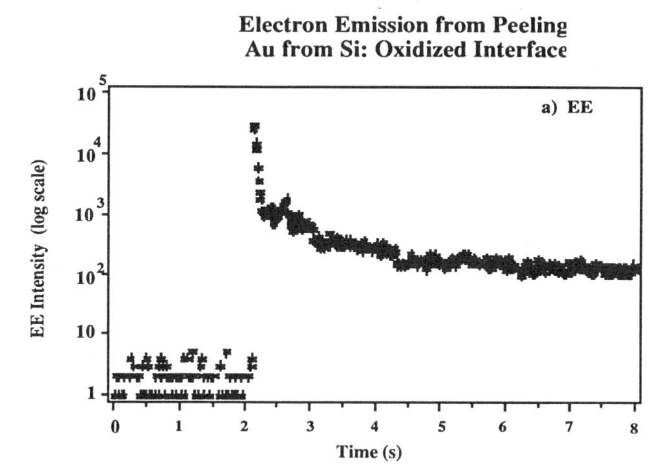

Fig. 5. Electron emission from peeling a gold film from a thermally oxidized Si (111) wafer. The peel area was 0.3 cm^2. The peak emission occurred during peeling, followed by a slow relaxation due to the excited oxide surface. Note that the overall surface charge on the oxide is *positive.*

Failure at Embedded Interfaces

We have recently obtained measurements of phE accompanying failure at internal interfaces in composites with transparent and semitransparent matrices, which allow for the transmission of light to the detector. Because photomultiplier tubes function in air as well as in vacuum, we are able to perform the phE experiments in both environments. By simultaneously measuring EE (electrons, which require vacuum), we can identify any failure event which exposes a surface to the outside world, i.e., a crack reaching the outer surface. Thus we can distinguish between internal and "external" failure. We have used these techniques to examine debonding events in glass fiber-epoxy composites.[49] These same methods are also proving helpful in studying the frictional pullout process.

As an example of these type of measurements, Fig. 6 shows data acquired for a simple system consisting of a small metal "pin" whose head was embedded in epoxy, as shown. During loading, we see prior to the pullout event, the phE was sustained for approximately 1s before the drop in force accompanying pullout. Only when the pin is pulled out of the matrix to we see the electron emission (which requires a pathway to the "outside world" and to the detector; i.e., pullout produces exposed surfaces capable of releasing electrons from the sample. The pre-failure emission of light is coming from *inside* of the epoxy. Tests stopped prior to pullout show that the end or tip of the pin has indeed detached (observed optically by means of interference fringes viewed from the right hand side of the diagram). Tests with larger, flat surfaces show that the first peak in the phE is due to a debonding crack on the head of the pin. It should be noted that in these experiments the metal pin was removed from the region of the detectors a few ms after pullout. Thus, the "tail" observed in both the phE and EE is coming from the epoxy. The hole left behind is glowing and shooting out electrons like a "rocket". Very similar results were obtained with short stainless steel rods embedded in epoxy ~ 1 rod diameter in distance. Thus, we see that combining the phE and EE measurements allows us to obtain sensitive tests of the onset of

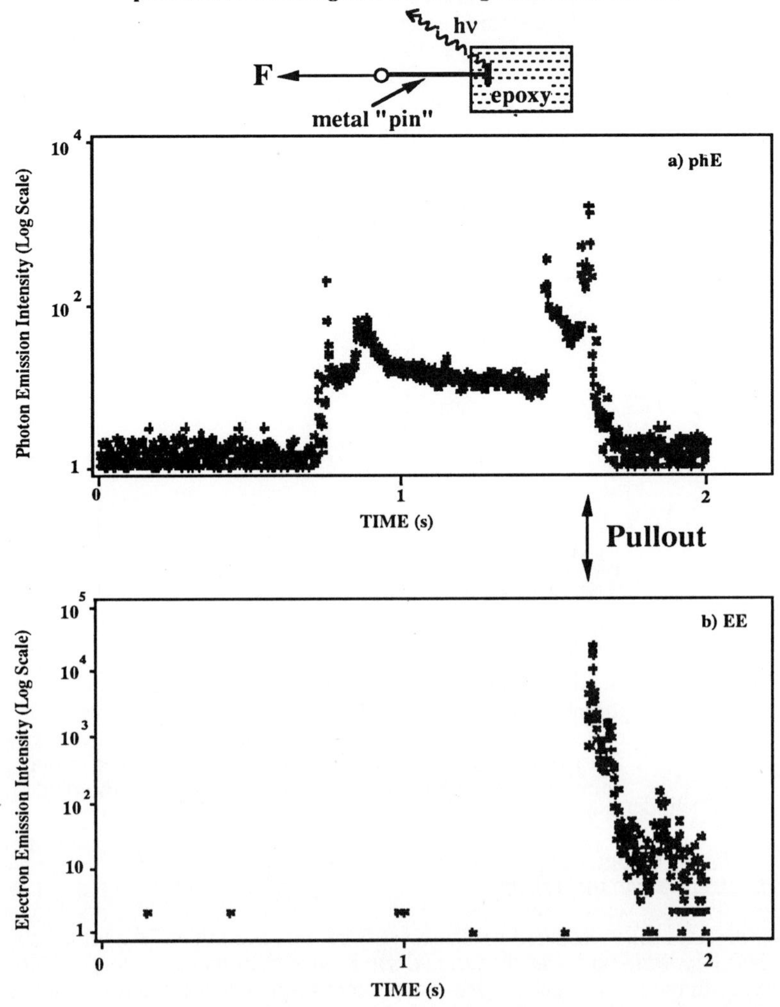

Fig. 6. phE and EE during loading and failure of a steel "pin" embedded in transparent epoxy. All of the photon emission preceding pullout is due to interfacial failure at the internal metal/epoxy interface.

internal failure and determination of when any fracture has occurred at or on the outside surface.

Metal Rod Pull Out from Transparent Matrix[50]

Debonding. Another geometry which results in internal interfacial failure is a fiber or cylindrical rod partially embedded into a material and pulled out. Depending on the ratio of the rod diameter to the length of the embedded portion of the rod, failure tends to occur either cohesively though the matrix or adhesively at the rod/matrix interface, the latter occurring in the case of short embedded lengths. Atkinson et al.[51] have discussed the fracture mechanics of this problem and have shown that debonding typically starts by detachment of the *end* or *tip* of the rod. This is generally followed by debonding from the sides of the rod, most frequently starting from the upper surface and moving down towards

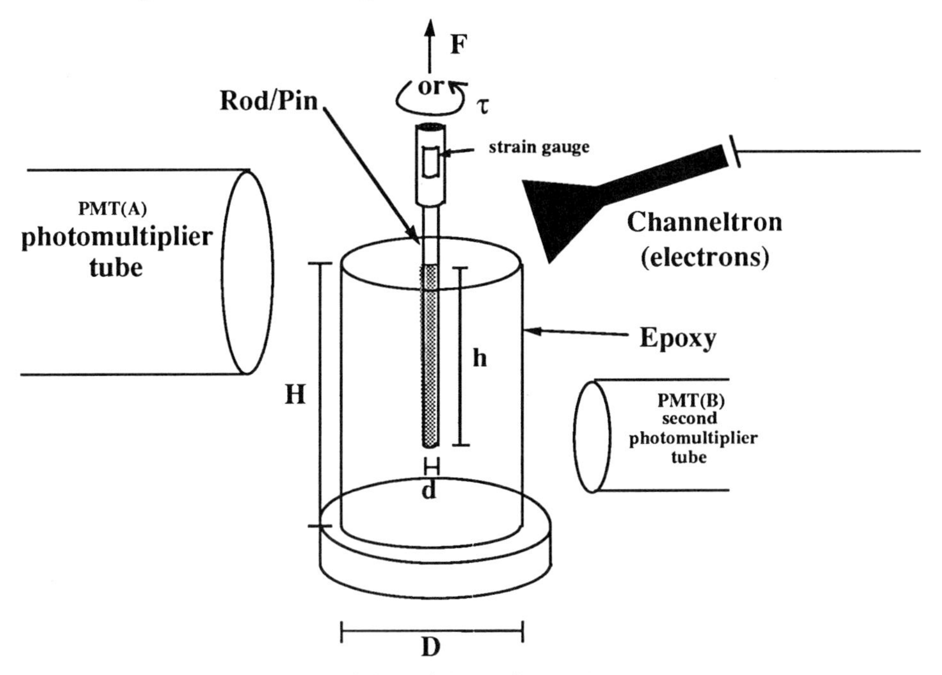

Fig. 7. Schematic of the rod debonding and pullout experiment.

the tip. Figure 7 shows the schematic of the experiment for testing the pullout of a metal (stainless steel) cylinder out of a transparent epoxy.

Two PMT Experiment.[50] To provide more detailed information on the sequence of failure events, we can acquire these signals at considerably faster rates and the use of two photomultiplier tubes with proper masking allows separate detection of light from the two ends of the rod in a time resolved fashion. In this experiment, we have used a stainless steel metal rod 50 mm in length and 2.5 mm in diameter which requires loads of ~5000 N to produce debonding. This leads to very fast debonding cracks and energetic, initial pullout events resulting in ~ 4 mm of pullout. Light shields were attached to the outer epoxy cylinder to limited the field of view of each PMT: PMT A was arranged to favor the detection of photons produced near the top (free surface) and PMT B (which has a much smaller photocathode) was positioned to favor the detection of photons produced near the tip (bottom) of the rod.

In Fig. 8, we see that the leading edge of the signal from PMT A precedes that from PMT B by 102 μs. This confirms that interfacial crack growth starts originates at or near the top (free surface) of the sample and propagates *towards* the embedded tip. The time interval between signals corresponds to an average crack speed of about 500 m/s, which is a reasonable interfacial crack velocity in this system.

In addition to the two PMT's, we added a CEM positioned near the epoxy-surface region as shown in Fig. 7. The resulting signals are shown in Fig. 9. The arrows indicate the onset of significant emission for each of the three detectors. The onset of the signal from PMT A precedes that of PMT B by ~68 μs, indicate crack motion from near the free end (top) of the rod towards the embedded end. This corresponds to an average crack velocity of ~700 m/s. The leading edge of the large burst in the signal from PMT A (actually off scale several times) coincides with the onset of emission from PMT B and with the drop in load accompanying rod pullout, as discussed above. The the onset of EE is delayed from the leading edge of the signal from PMT A by 150 μs. With the onset of EE, the debonding fracture event is nearly complete, and a surface crack finally forms which allows EE detection. Again, no evidence of a tip debond event prior to the fracture along the sides of the rod is observed.

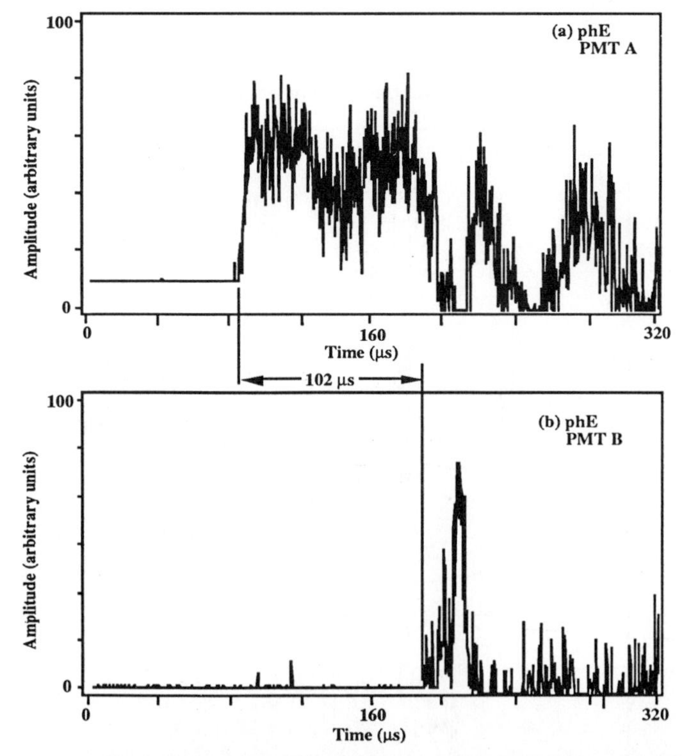

Fig. 8. Photon emission acquired on a fast time scale during the propagation of the debond interface crack between the stainless steel rod embedded in epoxy. (a) phE detected by PMT A located near the free surface; (b) phE detected by PMT B viewing the lower portion (tip region) of the metal rod.

We therefore conclude that the normal sequence of events for the tensile loading of the long stainless steel rod embedded in epoxy is as follows:

- debond crack formation at the metal/epoxy interface near the top but *below* the free surface; (For shorter rods and shallow embedding depths, failure initiated at the embedded tip).
- crack propagation along the metal/epoxy interface towards the embedded tip of the metal rod.
- arrival of the debonding crack at the free surface (and therefore electron emission), followed by rapid rod pullout. The duration of the initial pullout motion is typically ~50 μs.

Our conclusion that debond crack nucleation occurs below the surface is contrary to the fracture mechanics analyses of Atkinson et al.[51] and by Karbhari and Wilkins,[52] where the nature of the maximum in shear stress at the point where the fiber enters the matrix is discussed. Our results may be a result of sample geometry and/or inhomogeneous residual stresses within the epoxy due to thermal mismatches during the cure/cooling cycle.

Electrical Transients.[53] Recently, we have examined the electrical transients generated by debonding and pullout in the same model composite system described above (SS rod embedded in an epoxy matrix). Such transients are expected because of the charge

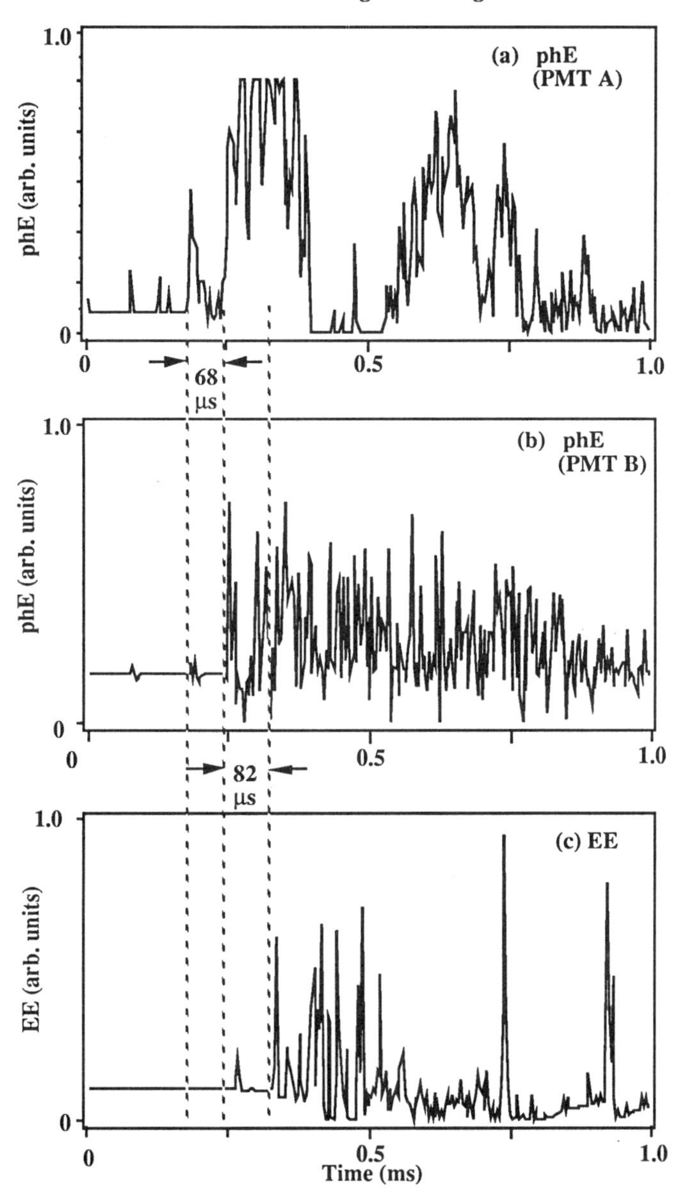

Fig. 9. phE *and* EE accompanying interfacial failure of stainless steel rod in epoxy on a fast time scale. (a) phE detected by PMT A located near the free surface; (b) phE detected by PMT B viewing the lower portion (tip region) of the metal rod; and (c) EE.

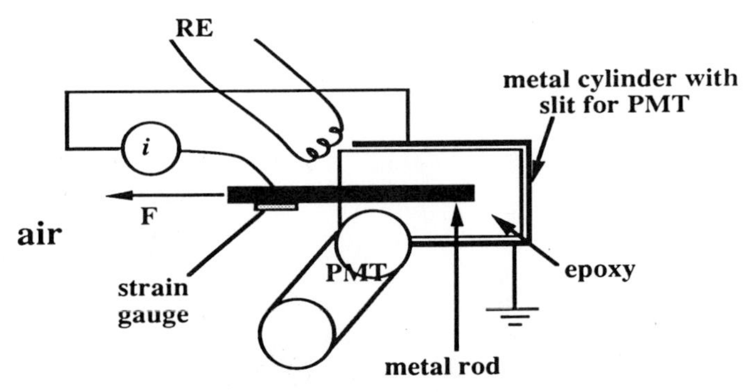

Fig. 10. Sample arrangement for transient electrical signal acquisition from interfacial failure between a metal rod and an epoxy matrix. In addition, a PMT, an RE antenna, and a strain gauge are shown.

separation that occurs whenever two dissimilar materials are pulled apart. Using charge probes following fracture, the stainless steel was found to have a net positive charge and the epoxy a net negative charge following separation. The arrangement of a typical experiment is shown in Fig. 10. The strain in the rod was monitored with a resistive strain gauge attached as shown. The time response of the strain gauge circuit was ~10 µs. A coil antenna was used to detect transient magnetic field bursts (denoted by RE -- Radiofrequency Emission) associated with charge separation activity. The tensile testing machine was enclosed in a light-tight, grounded aluminum box to minimize background levels in the current, phE, and RE measurements. The metal rod was insulated from the grips and a wire was attached for the transient current measurement. The epoxy sample was enclosed in a grounded metal cylindrical electrode which provided the "return" path for current from the rod. The cylindrical electrode was slotted along the side toward the photomultiplier tube to allow light detection.

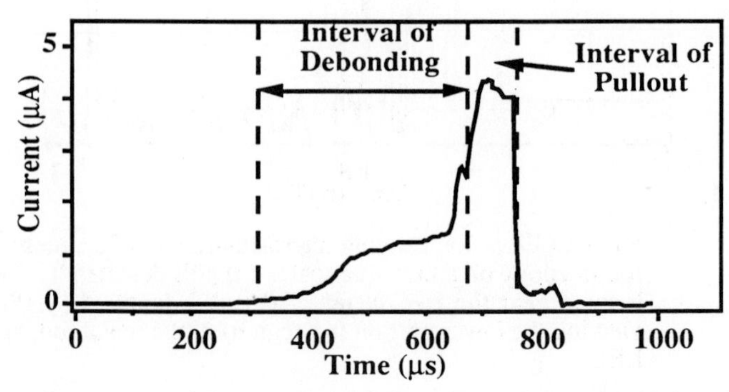

Fig. 11. Current to stainless steel rod during debonding and initial pullout of rod from epoxy.

Figure 11 shows a typical measurement of current flow from the metal rod to ground during interfacial failure. Simultaneous strain measurements indicate that the velocity of the debond crack was initially quite slow, then accelerated rapidly. The current is seen to follow this scenario. At completion of the debonding, the rod "lunges" out of the matrix, yielding additional current, generally higher than during debonding.

A display of simultaneous measurements of (a) rod strain, (b) phE, (c) current (made with the electrometer), and (d) RE are shown in Fig. 12. In Fig. 12(c), a positive-going signal corresponds to a positive current from the rod (a flow of electrons *to* the rod). The four arrows labeled A-D indicate the various stages in the progress of fracture. The first two arrows (A, B) mark the onset of phE and current, respectively. A very small drop in strain occurs with the onset of phE (see insert: Fig. 12(b)), and a somewhat larger drop in strain with the onset of the current signal. These two events are attributed to two stages of crack growth along the rod-epoxy interface. The delay (~35 μs) between the onset of phE and current is not typical. In most experiments, the first phE coincided with the onset of current. We suspect that the initial debonding event in this sample was not sufficiently strong to yield detectable current on the electrometer scale used.

Between the arrows marked B and C, interfacial crack growth initially appears to be continuous in the region where the strain drops linearly. The bump in the strain midway between the arrows is consistent with the passage of a reflected stress wave. From this point on, the strain is strongly influenced by reflected stress waves. During crack propagation, the current rises dramatically to about 0.55 μA. The phE in this region is weak but continuous, showing a modest burst during the steepest portion of the current rise. An RE signal is also observed in this region, but this was not the case for all samples.

At arrow C the load drops sharply to zero over about 20 μs: at this point, interfacial failure along the length of the rod is complete and rod pullout begins, associated with the first wave of stress relaxation along the rod/epoxy interface. A very large phE signal, an additional positive jump in the current, and a very large RE signal accompany this load drop. The coincident large phE and RE signals indicate electrical discharges accompanying relative motion of the rod and epoxy surfaces. The current does not begin to drop until arrow D, ~20 μs after the onset of pullout. Normally, the current would begin to drop within ~5 μs of the onset of pullout. The delay shown here may be due to a small portion of intact interface which finally detached; a sharp phE burst coincident with the leading edge of the current drop supports this interpretation.

Following arrow D, the strain oscillates at ~20 kHz and strong phE and RE signals occur in coincidence with each *drop* in strain. 20 kHz is approximately the frequency of a standing longitudinal wave in the 13 cm steel rod with both ends free. Due to the manner in which the loading apparatus is coupled to the specimen, both ends of the rod are effectively free following debonding so that stress waves can reflect from the rod ends and oscillate in a longitudinal direction. It should be noted that although the rod/epoxy surfaces are essentially separated, there is considerable contact between the two surfaces at asperities along the interface, most likely due to significant plastic deformation of the epoxy. The observed signals therefore result from small displacements of the rod in the direction of pullout accompanying successive stress waves as they pass through the rod. After several milliseconds these oscillations dampen to a background level of strain. Note that the small noise-like variations in the current after the main pullout event do not follow the 20 kHz oscillations in strain. This is consistent with a transient current source related to interfacial failure rather than rod displacement. Due to a discharging capacitance in the electrometer circuitry, the phE and current signals drop below zero after fracture; this particular behavior is thus an artifact of the detection circuitry.

To examine the origin of the current created during the debonding event, consider the geometry of the sample during fracture shown in Figs. 13(a) and (b), where we show the crack propagating from left to right. Assuming that interfacial failure occurs under conditions of constant charge, the opening of the gap between the rod and epoxy results in a change in the potential across this gap. In the region where no debonding has occurred, we assume the metal/epoxy interface has a separation of δ_0 (very small) which increases to a constant width $\delta = \delta_0 + \Delta\delta$, where a >> $\Delta\delta$ >> δ_0. If we integrate the electric field from the surface of the rod to the outer conductor, we obtain the potential difference between the rod and outer cylinder; because of the connection through the resistor R, shown in Fig. 13(b), this value is zero before and after charge flow. For cylindrical geometry, ignoring edge effects, the radial

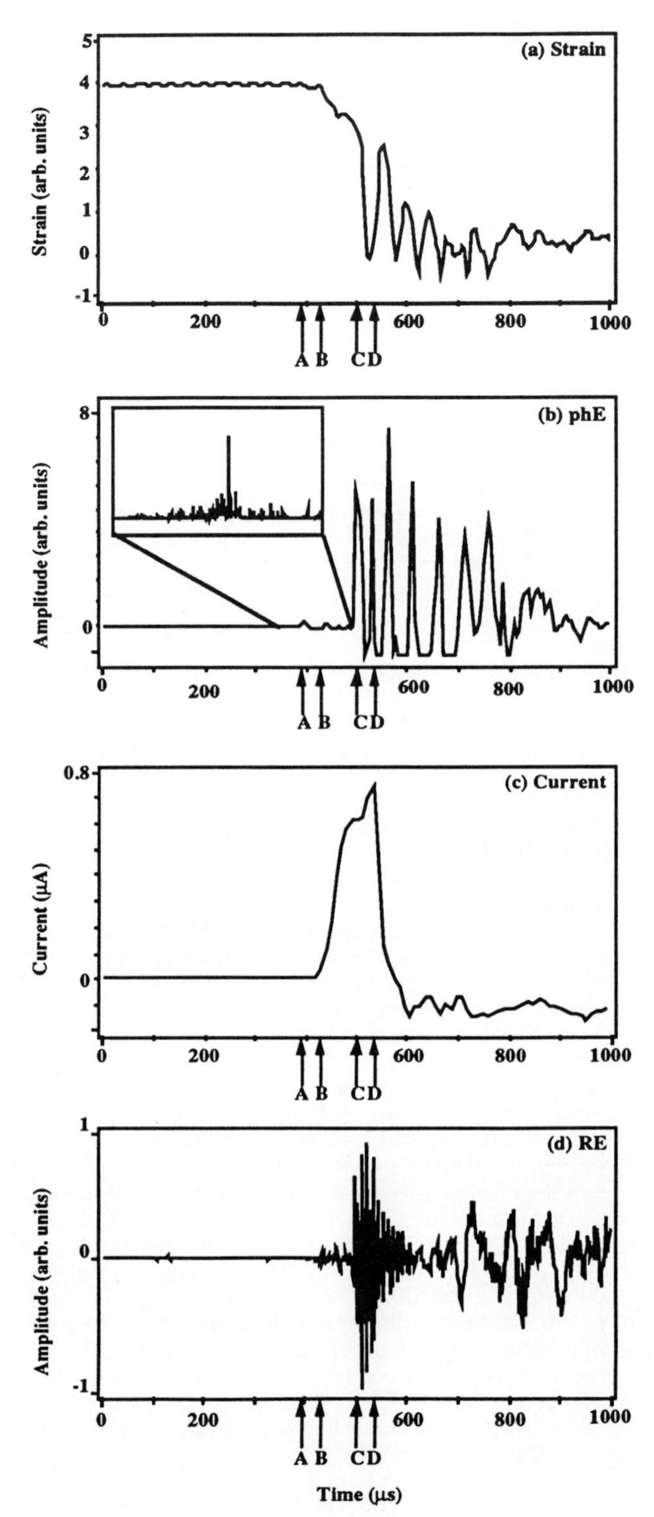

Fig.12. Signals for (a) strain, (b) phE, (c) rod current, and (d) RE. The current was measured with an electrometer. Insert in (b) shows phE in initial debond region.

(a) End View of Sample

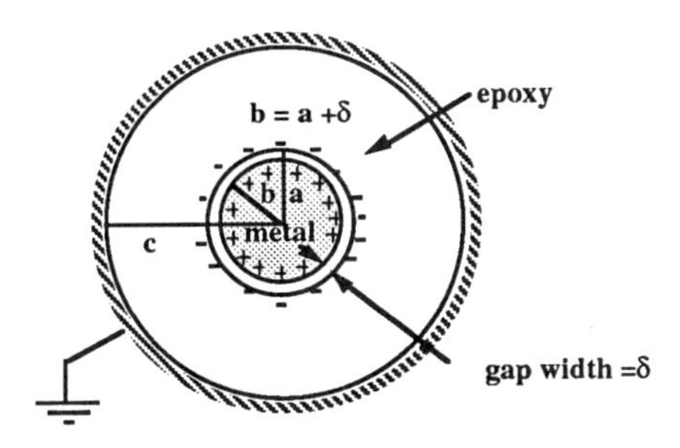

(b) Cross-section showing Debonded Region

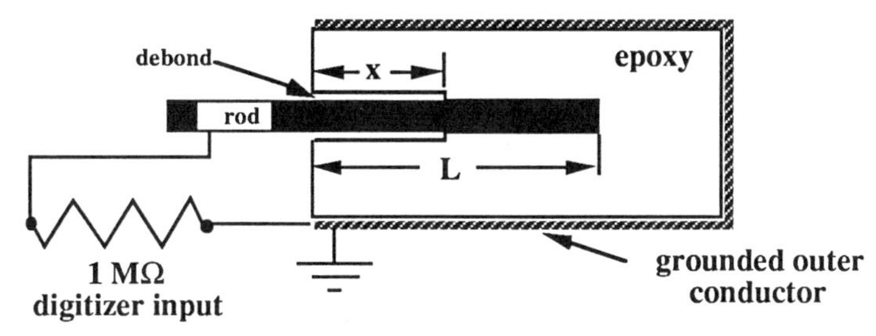

Fig. 13. Schematic of the (a) sample geometry and (b) electrical circuit for deriving the observed currents due to debonding and pullout.

components of the **E** fields in the separation region and in the epoxy at radial distance r are given in Equations 5 and 6:

$$E(r) = \frac{\rho^+}{2\pi\varepsilon_o r} \qquad a < r < a+\delta \tag{5}$$

$$E(r) = \frac{1}{2\pi\kappa\varepsilon_o}\left[\frac{\rho^+ + \rho^-}{r}\right] \qquad a+\delta < r < c \tag{6}$$

where ρ^+ and ρ^- are the charge/length on the metal rod and on the epoxy surfaces, respectively. ρ^- is the constant, free charge on the epoxy and κ is the relative dielectric constant of the epoxy. We assume that the gap is either vacuum or a low pressure gas with $\kappa = 1$. Before the crack is opened, the potential difference is given by

$$V = \int_a^{a+\delta} \frac{\rho^+}{2\pi\varepsilon_o r}\,dr + \frac{1}{2\pi\kappa\varepsilon_o}\int_{a+\delta}^{c}\left[\frac{\rho^+ + \rho^-}{r}\right]dr \tag{7}$$

or:

$$V = \frac{\rho^+}{2\pi\varepsilon_o}\ln\left(1+\frac{\delta}{a}\right) + \frac{(\rho^+ + \rho^-)}{2\pi\kappa\varepsilon_o}\ln\left[\frac{c}{a+\delta}\right] \tag{8}$$

Before any crack growth, $V = 0$, $\rho^+ = \rho^+_o$, and $\delta = \delta_o$, so that:

$$\rho_o^+ = \frac{\dfrac{\ln\left[\dfrac{c}{a+\delta_o}\right]}{\kappa}}{\left\{\ln(1+\dfrac{\delta_o}{a}) + \dfrac{\ln\left[\dfrac{c}{a+\delta_o}\right]}{\kappa}\right\}}\rho^- = K_o \cdot \rho^- \tag{9}$$

where K_o is the collection of constants shown. If a annular region debonds and opens to a width $\delta = \delta_o + \Delta\delta$, then to keep V=0, charge will flow, changing the local charge/length in the debonded region: i.e., $\rho^+_o \rightarrow \rho^+_f$, where ρ^+_f is given by:

$$\rho_f^+ = \frac{\dfrac{\ln\left[\dfrac{c}{a+\delta_o+\Delta\delta}\right]}{\kappa}}{\left\{\ln(1+\dfrac{\delta_o+\Delta\delta}{a}) + \dfrac{\ln\left[\dfrac{c}{a+\delta_o+\Delta\delta}\right]}{\kappa}\right\}}\rho^- = K_f \rho^- \tag{10}$$

where K_f represents the new collection of constants.

Thus, for an increment of length in the crack, Δx, the net charge that would flow (through R) would be:

$$\Delta q = (\rho_o^+ - \rho_f^+)\cdot\Delta x = (K_o - K_f)\cdot\rho^-\cdot\Delta x \tag{11}$$

For an increment of crack length Δx in time Δt ($v_c = \Delta x/\Delta t$), the current generated is:

$$i = (K_o - K_f)\cdot\rho^-\cdot v_c \tag{12}$$

We assume the time constant associated with R and the (sample + coax) capacitance is << time of fracture, which for R = 1 MΩ and C~5 pF is satisfied. Values of ρ^- for metal/polymer interfaces have been measured[14] and are typically $\sim 10^{-4}$ C/m^2 From the observed durations of the photon emission and strain decrease, we obtain average crack velocities of 200 m/s. We estimate $\Delta\delta$ to be ~1 µm. This spacing is consistent with Ohara's optical interferometer measurements[54] on polymer/glass structures undergoing shear deformation. Using known values of a, c, and κ (~4), Eq. 12 yields an estimated current of 0.2 µA and a total charge transfer of 60 pC. This is in good agreement (somewhat fortuitous, considering the approximations) with measured values for i (which ranged from 0.1 - 1 µA during debonding) and Δq (~50 pC). Furthermore, the predicted direction of charge transfer is electron flow *to* the rod, which is also in agreement with experiment.

As mentioned above, immediately after debonding, the constrained rod undergoes an initial, fast pullout due to the response of the rod and epoxy to the sudden stress release at the interface; this is followed by a damped longitudinal oscillation of the rod. Additional charge transfer occurs during this fast pullout because rod pullout shortens the length of the capacitor formed by the rod and shield. To maintain the charge density inside the capacitor at a constant value, ρ^+_f of equation 10 above, while maintaining zero potential difference between the rod and outer shield, additional charge must flow during pullout. For our initial conditions, electrons will again flow *to* the rod. Ignoring fringe fields and letting ΔL be the pullout distance and Δt be the duration of pullout, then the current generated during this event will be:

$$i = \frac{\Delta q}{\Delta t} = \rho_f^+ \cdot \frac{\Delta L}{\Delta t} \qquad (13)$$

For $\Delta L \sim 1.5$ mm, $\Delta t = 170$ μs, and $\rho^+{}_f = 1$ μC/m, this yields a current of 8 μA. We observe peak currents of 1-5 μA (corresponding to electron flow to the rod, as predicted) in good agreement with Eq. 13.

The time dependence of the current and phE during debonding may be due to changes in crack speed. phE intensities accompanying the fracture of neat epoxies depend strongly on crack velocity.[55] A similar dependence of EE intensities on crack velocity has been observed during the failure of filled polymers[56-58] and Au-SiO_2 interfaces.[33] This is consistent with the relatively intense phE and the rapidly rising current immediately prior to the completion of interfacial failure. In the final stages of crack growth, the stress intensity at the crack tip should be quite high and the crack velocity correspondingly high. Some of the phE in these final stages may be due to electrical discharges, but we note that phE bursts were also observed at small jumps in current long before the completion of debonding. The most natural interpretation of the initial phE bursts is that the initial debonding event involves localized, rapid crack growth at a stress concentrating defect. The new crack often relaxes the local stress, slowing subsequent crack growth and temporarily reducing the phE intensity.

The current and phE signals indicate that the epoxy surfaces are electrically charged during interfacial failure. The large, aligned phE and RE signals accompanying the completion of interfacial failure are attributed to electrical discharges between epoxy and rod surfaces as the surfaces are displaced with respect to one another. This displacement brings asperities on the two surfaces in close proximity, allowing the discharge of accumulated charge. phE and RE bursts after failure accompany additional displacements of the rod with respect to the epoxy as it oscillates with its natural frequency which depends on the clamping arrangements. The total charge involved in phE and RE events is a small fraction of the total charge on the two surfaces, as the large bursts following fracture do not affect the current signal significantly.

Thus, we see that interfacial failure in the metal rod/epoxy matrix system is accompanied by distinctive voltage and current signals which are correlated with interfacial failure events inferred from small changes in rod strain. These electrical transients are accompanied by phE signals which are sensitive indicators of the onset of interfacial crack growth. In the final stages of interfacial failure, intense phE and RE bursts reflect electrical discharges which are associated with the relative motion of rod and matrix accompanying pullout. Subsequent bursts are correlated with oscillations in rod strain which accompany further displacement.

The intense FE signals which generally accompany interfacial failure can provide important information on the fracture event, often with extremely good time resolution. By appropriately modifying the detection conditions one can readily extract information on the initial locus of debonding and its subsequent course. It might be useful, for example, to use these probes to examine the role of mixed mode and frictional stresses at the interface (discussed by Hutchinson and Jensen[59]) in crack propagation along that interface. Observations on macroscopic samples may eventually prove useful in understanding the fracture of fiber-based composites, where pullout contributes substantially to material toughness. We note that in principle, these measurements could also be carried out on fragile specimens (e.g., small, single fibers) and/or at high temperatures as long as electrical contact can be maintained.

Stick-Slip. If we continue to pull the rod at a constant strain rate following debonding, the rod experiences frictional pullout. For a variety of conditions, we found that during this pullout, *stick-slip* behavior is frequently observed.[50] A typical force vs elongation curve is shown in Fig. 14. The behavior is quite similar to the pullout of debonded Cu rods from epoxy reported by Cook et. al.[60] In the stick slip region, the rod "sticks" to the matrix until the load on the rod reaches a critical value, at which point the rod suddenly slips to a new position, where it is "grabbed", thereby sticking again. The load does not go to zero, and the oscillations tend to be quasi-periodic. The pattern of stick-slip cycles in Fig. 14 appears "chaotic". Like Cook et al., we observe that the frequency and amplitude of the load oscillations depend dramatically on the strain rate, which suggests that the frictional forces are v-dependent (as opposed to Coulombic friction). It should be noted that in our samples the areas under the stress-strain curves during pullout are actually equal to or larger than the areas prior to the debonding event, so that for at least this ideal system, pull-out can contribute significantly to the total energy of failure. Fig. 15 shows the applied force, phE, and EE vs time, acquired simultaneously on a slow time scale during stick-slip. The particle signals were acquired by pulse counting and the experiment was done in vacuum. The average pullout distance for a single stick-slip event was about 150 μm. On this time scale, the phE and EE bursts appear in coincidence with the drop in load, i.e. when the rod *slips*. The electrons were presumably generated right at the surface where the rod exits the epoxy.

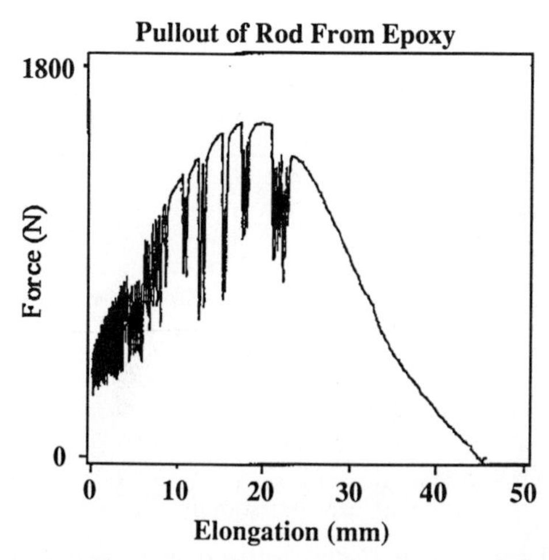

Fig. 14. Force vs elongation during pullout of a stainless steel rod vs displacement out of an epoxy matrix. The oscillations represent stick-slip behavior.

To establish the correlation between fracto-emission and the stick-slip motion on a much faster time scale, the phE was digitized at 5 ns intervals along with the strain in the rod which could be used to determine the "time-of-slip". Fig. 16 shows the resulting phE vs time, where the arrow indicates the onset of slip. Note that the phE actually starts before the onset of the load drop, indicating that there is interfacial activity prior to catastrophic release of the rod.

In this particular stick-slip event, a large (off-scale) photon burst was observed near the end of the trace at ~0.8 ms. An RE detector detected a burst in coincidence with this burst. Analyzing a large number of phE/RE records from stick-slip events, two emission patterns were observed. One pattern consisted of isolated single photons (no RE) with an onset slightly before the load drop and continuing a few milliseconds afterwards. The other pattern was that of Fig. 16, which includes 1-2 intense phE and RE bursts in coincidence (seen in about 15% of the slip-events). Every time we observed an RE signal, a large, intense phE burst was observed in coincidence. This coincidence indicates that these large bursts are produced by microdischarges between the metal and epoxy surfaces due to highly charged surfaces, generated during the slip process. The relative motion of the epoxy and metal surfaces during the slip portion of the cycle follows local adhesive failure where

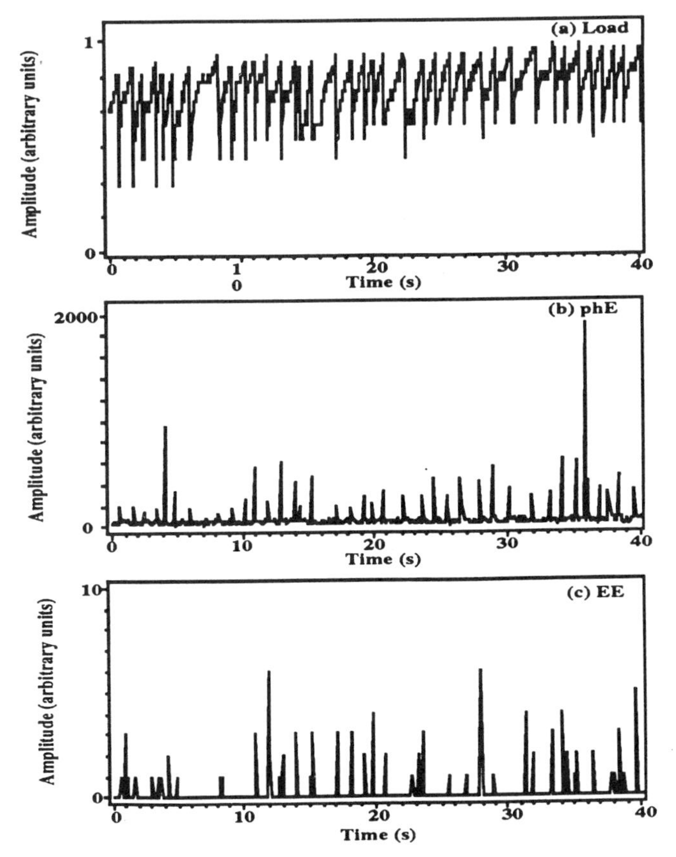

Fig. 15. (a) load, (b) phE and (c) EE during stick-slip motion of stainless steel rod pullout from epoxy matrix.

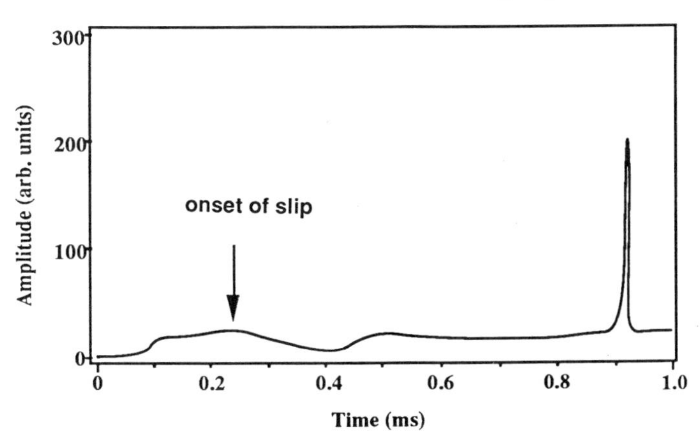

Fig. 16. Photon emission occurring ~100 μs prior and during the slip of the rod from the epoxy during one cycle of stick-slip motion. The arrow indicates the *onset* of slip.

interfacial contact has been established. The interplay between highly charged surfaces and the gases that are emitted subsequent to failure therefore frequently results in electrical breakdown.

Stick-slip motion involving solid-on-solid normally is associated with local instabilities in contact between surfaces, where "stick" is a consequence of static frictional forces being larger than during sliding.[61] Ohara[54] attributed stick-slip motion to surface deformation at points of frictional contact; using optical interferometry, he found that stick-slip motion along a glass-polymer interface was accompanied by the collective deformation of a multitude of semi-macroscopic contact regions. Schematically, one might envision these contact regions to involve asperities, as shown in Fig. 17(a). As thc rod is displaced, asperities on the epoxy surface deform, resisting the rod displacement and exerting this resistance via static friction. Locally, microscopic slip between the two surfaces eventually results, causing rapid variations (jittering) in the increasing shear stress as shown schematically in Fig. 17(b). Finally, a series of asperities slip, leading to a catastrophic "lunge" of the rod. Repositioning and new alignment of asperities on both the metal and epoxy surfaces results in the recapture of the moving rod before zero stress is reached and the cycle begins again. It is the during the lunge that rapid detachment and high velocity "rubbing" between regions of the surface occur, resulting in local charge separation and excitation of the polymer surface, which in turn yields phE. When the conditions for microdischarges are met (sufficient charge density and appropriate gaseous atmosphere), then discharges occur, producing intense, phE and RE bursts.

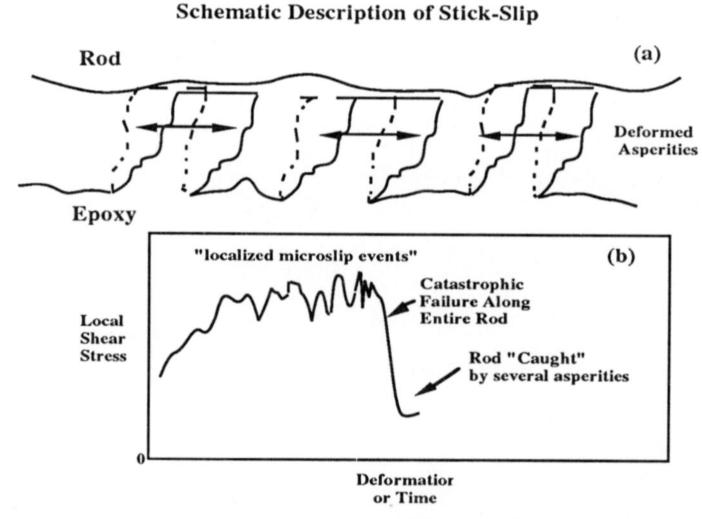

Fig. 17. Schematic of the interaction of asperities and events resulting in catastrophic "slip".

Scanning electron microscopy (SEM) was used to compare epoxy surfaces which had experienced debonding only (therefore, no pullout nor stick slip) with surfaces having experience extensive stick slip. The differences in appearance were significant. The debonded-only epoxy surfaces were very smooth with only small features which were presumably due to surface features of the rod surface molded into the epoxy. These surface features on the rod play an important role in the subsequent stick-slip activity. The epoxy surfaces having experienced extensive stick-slip, i.e., the rod has been pulled all of the way out of the epoxy showed numerous Herringbone-like fracture patterns aligned with the rod axis over a good portion of the fracture surface. The surface also shows occasional deep gouges, and some evidence of particulate matter gouged from the epoxy surface. The herringbone structure appears to have been formed as an asperity on the rod surface dragged across the epoxy, which was alternately gripped and released by the epoxy matrix, causing fracture in the form of the V-shaped markings which make up the herringbone pattern.

As mentioned previously, considerable energy is consumed during stick-slip motion (area under force-elongation plots of Fig. 14). The relatively high force required for pullout indicates considerable compression of the epoxy gripping the rod, such that the passage of asperities through the epoxy produces significant plastic deformation of the epoxy, as well as fracture, both dissipating significant energy. In many instances, enhanced matrix cracking during fiber pullout could be very beneficial in a composite. Additional amounts of energy are dissipated in the acoustic/thermal bursts which accompany each slip event, possibly associated with the deacceleration or "catching" part of the cycle when the rod motion is stopped. Our work on neutral particle emission (gases) released during cohesive failure of materials (including polymers) provides a useful tool for monitoring the growth of the microcracks during stick slip.

We have therefore shown that for a variety of geometries, we can detect photon emission generated at embedded metal/insulating matrix interfaces undergoing adhesive failure, transmitted through the matrix to external detectors. These signals along with other time resolved measurements (e.g., strain in the metal, radiowave emission, and electron emission) provide details on the sequence of failure events such as the direction and speed of debonding crack propagation, the onset of pullout, and the arrival of cracks to the matrix surface with sub-microsecond time resolution. In addition, we observed readily detectable fracto-emission signals accompanying the subsequent pullout process, which accompanied repetitive stick-slip activity. It is of some significance that the time-sequence and the force law dictating the rate of energy dissipation during pullout are both highly unpredictable. Thus, modeling the bridging of cracks with fibers or the latter stages of failure of embedded fixtures in glassy materials must obviously be done carefully. These types of measurements can provide details of the failure to help our understanding of both the debonding and pullout processes. We note that all of our emission experiments involve non-contact detection of events so that the testing of fragile specimens (e.g., single fibers) or testing materials at high temperatures can be done routinely.

Useful future work will be to use these probes to examine the role of mixed mode and frictional stresses at the interface (discussed by Hutchinson and Jensen[59]) in crack propagation along that interface. Also, it is possible that the rod pullout system might serve as a mechanical analog for a number of stick-slip systems, including fault motion in the earth; we point out that stick-slip plays an important role in fault dynamics and can lead to chaotic behavior in terms of spatial and frequency of slip in both nature and model systems.[62,63] Furthermore, during the time interval when the external force on the rod is increasing, the sequence of events leading up to the slip event appears to be an example of self-organized criticality,[64,65] wherein small failure events (localized slip) "mushroom" into catastrophic failure (massive slip). The photon emission prior to slip, shown in Fig. 16, may be a useful probe for examining on fast time scales these precursors to catastrophe in order to study in more detail the chain of events resulting in failure. As an example, it would be of interest to determine possible correlations between the duration and total intensity of these precursors and the magnitude of the drop in force at slip, analogous to the *magnitude* of the earthquake. It is obvious that as a "sentinel of failure" [Fig. 16], 0.1 ms is a very short time, although scaling to dimensions of the San Andres Fault would certainly result in significantly longer times. At a minimum, these experiments display failure and pullout processes typical of many fiber-matrix systems and other structures involving frictionally coupled materials, and offer new methods for studying the dynamic interactions between tribologically loaded surfaces.

CONCLUSION

We have shown above our recent studies involving microscopic behavior of cracks, often associated with dissipative phenomena such as crack branching, deflection, and plastic deformation, as well as similar phenomena at interfaces. The latter includes the fracture of embedded interfaces and frictional pullout which is of considerable importance in terms of toughening mechanisms in brittle matrix composites. The techniques described here can potentially provide details of failure mechanisms and assist in the interpretation of other methods such as acoustic emission and fractography probes, including detection of the micro-events occurring prior to failure and in composite materials, detecting instantaneous detection of the locus of fracture. Our goal is to continue to study the mechanisms and applications of FE and nano-scale topographical probes (STM and AFM) to the study of early stages of fracture and failure modes in a variety of materials, including composite systems and interfaces. In addition we continue to attempt to relate atomic level events during fracture (which result in fracto-emission) with the concepts of energy dissipation with both spatial and temporal resolution. These departures from *adiabatic* separation of atomic planes or molecular bonds are also reflected in the topographical, electrical, and chemical states of the resulting fracture surface. By correlating and analyzing these properties, we are thus striving to probe the dynamics of the most dangerous type of fracture, catastrophic crack growth, as cracks accelerate, arrest, and/or propagate through materials. The details provided by studies such as ours go hand-in-hand with the progress being made in the fracture mechanics community. As computer size increases, the mechanical description will naturally extend into the nanometer scale, perhaps down to the atomic level. Thus, the concepts of fractal geometry, chaotic behavior of crack tips, dynamic phenomena directly at the crack tip, and microscopic/dynamic descriptions of frictional interactions at interfaces may require new syntheses of mathematical formalisms which extend fracture mechanics into these other descriptive realms. Since traditionally, fracture is an interdisciplinary area of study, fruitful collaborations will form naturally among a wide range of engineering and scientific fields.

ACKNOWLEDGMENTS

This work was supported by the Ceramics and Electronics Materials Division of the National Sciences Foundation DMR 8912179, the Office of Naval Research Contract No. N00014-87-K-0514, the Dow Chemical Company, and the Washington Technology Center. We wish to thank our Washington State University colleagues Richard Hoagland and John Hirth for helpful discussions.

REFERENCES

1. J. F. Stebbins, Nature 351, 638 (1991).
2. J. T. Dickinson, "Fracto-Emission", in *Non-Destructive Testing of Fibre-Reinforced Plastic Composites - II,* J. Summerscales, ed. Elsevier Applied Science, London (1990), pp. 429-482.
3. J. T. Dickinson, "Fracto-Emission from Interfacial Failure", in Materials Research Society Symposium Proceedings 153, *Interfaces Between Polymers, Metals, and Ceramics*, edited by B. M. DeKoven, A. J. Gellman, and R. Rosenberg, pp. 331-344, (1989).
4. B. V. Deryagin, N. A. Krotova, and V. P. Smilga, *Adhesion of Solids*, (English Translation), (Consultants Bureau, New York, 1978).
5. J. T. Dickinson, "Fracto-emission accompanying adhesive failure," in *Adhesive Chemistry--Developments and Trends*, edited by L. H. Lee (Plenum Publishers, New York, 1984).
6. J. T. Dickinson, L. C. Jensen, S. C. Langford, and J. P. Hirth, J. Mater. Sci. 6, 112 (1991).
7. J. T. Dickinson, L. C. Jensen, and S. C. Langford, Phys. Rev. Lett. 66, 2120 (1991).
8. S. C. Langford, L. C. Jensen, J. T. Dickinson, and L. R. Pederson, J. Mater. Sci. 6, 1358-1368 (1991).
9. M. A. Grayson and C. J. Wolf, J. Polym. Sci., Polym. Phys. Ed. 23, 1087 (1985).
10. D. L. Doering, J. T. Dickinson, S. C. Langford, and P. Xiong-Skiba, J. Vac. Sci. Technol. 8, 2401 (1990).
11. S. C. Langford, D. L. Doering, and J. T. Dickinson, Phys. Rev. Lett. 59, 2795 (1987).
12. N. H. Packard, J. P. Crutchfield, J. D. Farmer, and R. S. Shaw, Phys. Rev. Lett. 45, 712-716 (1980).

13. S. C. Langford, Ma Zhenyi, and J. T. Dickinson, J. Mater. Research 4, 1276 (1989).
14. Robert Shaw, Z. Naturforsch. 36a, 80-112 (1981).
15. Alan Wolf, J. B. Swift, H. L. Swinney, and J. A. Vastano, Physica 16D, 285-317 (1985).
16. B. Dubuc, J. F. Quiniou, C. Roques-Carmes, C. Tricot, and S. W. Zucker, Phys. Rev. A 39, 1500 (1989).
17. Benoit B. Mandelbrot, *The Fractal Geometry of Nature*, (W. H. Freeman, New York, 1983).
18. B. B. Mandelbrot and R. F. Voss, "Why is nature fractal and when should noises be scaling," in *Noise in Physical Systems and 1/f Noise*, edited by M. Savelli, G. Lecoy, and J-P. Nougier, (Elsevier Science Publishers, Amsterdam, 1983), pp. 31-39.
19. B. B. Mandelbrot, Physica Scripta 32, 257-260 (1985).
20. R. C. Jaklevic, and L. Elie, J. Vac. Sci. Technol. A 6(2), 448 (1988).
21. R. C. Jaklevic, L. Elie, Weidian Shen, and J. T. Chen, Appl. Phys. Lett. 52(20), 1656 (1988).
22. S. C. Langford, Ma Zhenyi, L. C. Jensen, and J. T. Dickinson, J. Vac. Sci. Technol. A 8, 3470 (1990).
23. S. A. Anderson, "Development and Applications of a Compact Atomic Force Microscope," M. S. Thesis, Washington State University (1989).
24, Y. L. Tsai and J. J. Mecholsky, J. Mater. Res. 6, 1248 (1991).
25. M. W. Mitchell and D. A. Bonnell, J. Mater. Res. 5, 2244 (1990).
26. J. J. Mecholsky, T. J. Mackin, and D. E. Passoja, in *Fractography of Glasses and Ceramics*, edited by J. R. Varner and V. D. Frechette, (American Ceramic Society, Westerville, OH, USA, 1988), pp. 127-134. See also, J. J. Mecholsky, T. J. Mackin, and D. E. Passoja, J. Am. Ceramic Soc. 72, 60 (1989).
27. B. B. Mandelbrot, D. E. Passoja, and A. J. Paullay, Nature 308, 721 (1984).
28. K. Ravi-Chandar and W. G. Knauss, Int. J. Fracture 26, 141-154 (1984).
29. J. P. Dempsey and P. Burgers,Int. J. Fracture 27, 203 (1985).
30. H. Hara and S. Okayama, Phys. Rev. B 37, 9504 (1988).
31. E. Louis and F. Guinea, Europhys. Lett. 3, 871 (1987).
32. A. J. Markworth and J. K. McCoy, J. Mater. Res. 3, 675 (1988).
33. H. J. Herrmann, Physica A 163, 359 (1990).
34. R. C. Ball and R. Blumenfeld, Phys. Rev. Lett. 65(14), 1784 (1990).
35. J. Berkowitz, C. H. Batson, and G. L. Goodman, J. Chem. Phys. (Paris) 77, 631 (1980).
36. J. J. Gilman, C. Knudsen, and W. P. Walsh, J. Appl. Phys. 29, 601 (1958).
37. S. J. Burns and W. W. Webb, Trans. Metall. Soc. AIME 236, 1165 (1966).
38. J. P. Hirth and Jens Lothe, *Theory of Dislocations*, 2nd Ed., (John Wiley & Sons, New York, 1982), pp. 168-169.
39. W. G. Johnston and J. J. Gilman, J. Appl. Phys. 30, 129 (1959).
40. H. Alexander and P. Haasen, "Dislocations and plastic flow in the diamond structure," in *Solid State Physics, Vol. 22*, edited by F. Seitz and D. Turnbull, (Academic Press, New York, 1968), pp. 27-158.
41. O. W. Johnson and P. Gibbs, "Brittle fracture of germanium," in *Fracture of Solids*, edited by D. C. Drucker and J. J. Gilman, (Interscience, New York, 1963), pp. 315-338.
42. F. Louchet, D., Cochet Muchy, and Y. Brechet, Philos. Mag. A 57, 327 (1988).
43. G. H. Campbell, M. Rühle, B. J. Dalgleish, and A. G. Evans, J. Am. Ceram. Soc. 73, 521 (1990).
44. K. C. Russell, S.-Y. Oh, and A. Figueredo, Materials Research Society Bulletin 16, 46 (1991).
45. J. T. Dickinson and L. C. Jensen, J. Poly. Sci.: Poly. Phys. Ed. 23, 873 (1985).
46. J. Lowell and A. C. Rose-Innes, Adv. Phys. 29, 947 (1980).
47. J. E. Stern, B. D. Terris, H. J. Mamin, and D. Rugar, Appl. Phys. Lett. 53, 2717 (1988).
48. B. D. Terris, J. E. Stern, D. Rugar, and H. J. Mamin, Phys. Rev. Lett. 24, 2669 (1989).
49. A. S. Crasto, R. Corey, J. T. Dickinson, R. V. Subramanian, and Y. Eckstein, Composites Sci. & Technol., 30, 35 (1987).

50. Ma Zhen-Yi and J. T. Dickinson, "Fracto-emission from embedded interfaces," submitted to J. Appl. Phys.
51. C. Atkinson, J. Avila, F. Betz, and R. E. Smelser, J. Mech. Phys. Solids 30, 97 (1982).
52. V. M. Karbhari and D. J. Wilkins, Scripta Metallurgica and Materialia 24, 1197 (1990).
53. K. A. Zimmermann, S. C. Langford, and J. T. Dickinson, "Electrical transients during interfacial debonding and pullout of a metal rod from an epoxy matrix," submitted to J. Appl. Phys.
54. K. Ohara, Wear 50, 333 (1978).
55. J. T. Dickinson, L. C. Jensen, and S. K. Bhattacharya, Makromol. Chem., Macromol. Symp. 7, 129 (1987).
56. J. T. Dickinson and L. C. Jensen, J. Polymer Sci., Polymer Phys. Ed. 20, 1925 (1982).
57. J. T. Dickinson, L. C. Jensen, and A. Jahan-Latibari, Rubber Chem. and Tech. 56, 927 (1984).
58. J. T. Dickinson, L.C. Jensen, and S.K. Bhattacharya, J. Vac. Sci. Technol. A 3, 1398 (1985).
59. J. W. Hutchinson and H. M. Jensen, "Models of Fiber Debonding and Pullout in Brittle Composites with Friction," to be published.
60. R. F. Cook, M. D. Thouless, D. R. Clarke, and M. C. Kroll, Scripta Metall. 23,1725 (1989).
61. F. P. Bowdin and D. Tabor, *The Friction and Lubrication of Solids*, Clarendon Press, Oxford (1986), pp. 105-111.
62. J. M. Carlson and J. S. Langer, Phys. Rev. Lett. 62, 2632 (1989).
63. J. Huang and D. L. Turcotte, Nature 348 (234 (1990).
64. P. Bak and K. Chen, Sci. Am 264(1) 46 (1991).
65. P. Bak and Chao Tang, and Kurt Wiesenfeld, Phys. Rev.A 38, 364 (1988).

FRACTO-EMISSION FROM CERAMICS AT CRYOGENIC TEMPERATURES

Shigehiro Owaki and Toichi Okada
ISIR, Osaka University
8-1 Mihogaoka, Ibaraki, Osaka 567, Japan

Sumio Nakahara and Kiyoshi Sugihara
Dept. Mechanical Engineering, Kansai University
3-3-35 Yamatecho, Suita, Osaka 564, Japan

ABSTRACT

Electron emissions from fine ceramics of alumina during the fracture process (fracto-emission, FE) at cryogenic temperatures were observed and compared with those from brittle metals, glass and single crystals of SiO_2 etc. The features of FEs are intended to be classified in metals and insulators, and moreover, the latter are into single and poly-crystals, and non-crystalline (glassy state). However, the FEs from alumina ceramics of polycrystalline are very similar to those from sodium silicate glass and pure fused silica, but are different from those from synthetic quartz (single crystal) especially for fracture at room temperature. From experimental results, it is supposed that they depend on the fracture modes, fracture strength and electronic properties for generation of free electrons and its compensation on fracture surfaces of these materials. These problems are discussed here.

INTRODUCTION

Emissions of low energy electrons from material surfaces have been observed during and after various kinds of sample treatments such as deformation, abrasion and phase transformation.[1-3] Among them, the emission during and after the fracture of solid state materials, called fracto-emission (FE), in which emissions of ions and photons are involved, is supposed to depend on the fracture mode and the material characteristics.[4-8] Generation of free electrons and relaxation of the charge irregularity on the freshly created fracture surface are interesting subjects in not only fracture mechanism research but also surface physics. Although there are many techniques for the investigation of fracture mechanics of ceramics, FE measurement seems to become an available technique observing mechanical and electronic properties of materials at fracture.

In order to investigate fracture of alumina ceramics of polycrystalline, the fracture behavior should be compared with those of brittle metals (poly-crystalline) and inorganic materials of single crystal and glassy state. Thin plates of high-carbon steel and pure iron at low temperature were selected as brittle metals. Those of single crystals of quartz, fused quartz and sodium silicate glass were tested in the same experimental condition as that in case of alumina ceramics.

A three-point bending test provides electron detection during the fracture process and maintains the freshly created fracture surface if the test was performed in a high vacuum. The sample temperatures during the test affect the fracture mode depending on mechanical strength and electronic properties of materials as charge carrier mobility for relaxation of charge irregularity, especially at cryogenic temperatures. Moreover, the other thermally activated processes related with FE are expected to be suppressed at these temperatures. The test is performed in a cryostat chilled down to liquid He temperature with a clean vacuum lower than 10^{-5} Pa around the sample holder.

Macroscopic and microscopic observations of fracture surfaces were performed by an optical microscope and a scanning electron microscope (SEM) at RT after bending experiments. Correlation of fracto-emission and fracture mode were discussed based on the results.

The authors have performed experiments of electron emission from stainless steel during deformation (not fracture) at cryogenic temperatures and found some electron emission processes.[9-15] Investigation of FE described here were carried out using the same technique as in the above experiments.

EXPERIMENT

Deformation and fracture were performed by three point bending basically in the same manner and apparatus as the previous experiments.[9] Metal samples were notched and shaped, and those of insulating materials were rectangular plates shown in Fig. 1, where the directions of bend and electron emission, and sample chamber in a cryostat are illustrated. The cryostat was evacuated by a turbo-molecular pump to a pressure of 10^{-4} Pa at room temperature (RT). As the pressure decrease with the decrease of temperature (to 10^{-5} Pa), a considerably clean vacuum was secured around fracture surface of the sample. The temperature on sample surfaces decreased to 20 K for metals and 30 K for insulators with liquid He and to 90 K with liquid N_2 cooling.

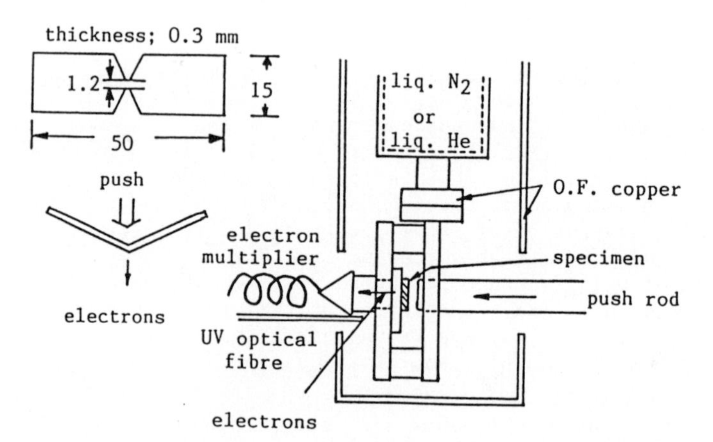

Fig. 1 Sample shape of metal and the direction of bend and electron emission, and sample chamber in the cryostat are illustrated.

Commercially available high-purity alumina plates (99.9 wt%, thickness; 15 mil) was cut into dimensions of 50 x 10 mm^2. The three-point bending test was performed with a span of 30 mm. The fracture initiated at edge flaws introduced by the cutting process.

Two kinds of glass were prepared as specimens of glassy state materials. One was sodium silicate glass, commercially available as deck glass plate for optical microscopes, and the other was pure fused quartz, in which no optical absorption band was observed in the UV and visible region of wavelength longer than 200 nm. The dimensions (mm) of the bending test pieces were 50 x 10 x (1.0 and 0.5).

As single crystal specimens, blocks were quarried from X- and Z-region of quartz crystals made by hydro-thermal synthesis. They were cut to plates of which planes are perpendicular to X or Z axis and of the same size as the above (we call them as X and Z specimens respectively). Generally Z-region crystals are known to be purer than those of the X-region. This was confirmed with the fact that an X-region plate was densely colored dark violet with high energy electron irradiation, while only small colored dots appeared in that of the Z-region.

Metal samples were heat-treated to remove residual stress. The shaped plates of pure iron (99.9 wt%) and high-carbon steel (SK-5; C, 0.84 wt%) were annealed at 923 K for 1 hr. in an Ar atmosphere. After that, they were electropolished in perchloric acid/acetic acid in order to normalize the surface state and to facilitate observations of surface change with a SEM, performed at RT after bending deformation experiments.

Some of specimens were irradiated by electrons of an energy of 21 MeV from a linear accelerator to dope lattice defects like color centers which may affect the features of FE. Among the specimens, only pure fused quartz were not visually colored except the optical absorption edge slightly shifted to longer wavelength. Sodium glass was colored densely brown but discolored gradually during storage at RT. After a few weeks storage for radiation cooling down, the flex test was performed using the specimens of slightly colored sodium glass.

Electrons emitted from the specimen surface opposite to the push rod controlled from outside the cryostat were detected by a ceramic electron multiplier and were counted as pulses after amplification. The counted pulses were stored as a time distribution before and after fracture in a personal computer. The branched signals were integrated and displayed on an analog recorder simultaneously with those of strain gage and an Au-Fe-chromel thermo-couple attached to the specimen.

A push rod for the three point flex test equipped with a micro-gage to monitor its stroke was operated slowly watching the electron emission every 0.25 mm. Different from FRP[10] and brittle metals, no electron emissions from glass and crystal specimens were observed before fracture. At the instant of fracture, electron emission showed a peak intensity and then decayed either quickly or slowly.

The load-displacement relation for the three-point bending test of alumina plates at RT and liquid N_2 temperatures was observed, and the results is shown in Fig. 2. At RT, the loads at breakage were 10 to 20 N. The same test was done by immersing the samples in liquid N_2 and the breakage load increased to 15 to 50 N as well as increase of the bending strength and Young's modulus.

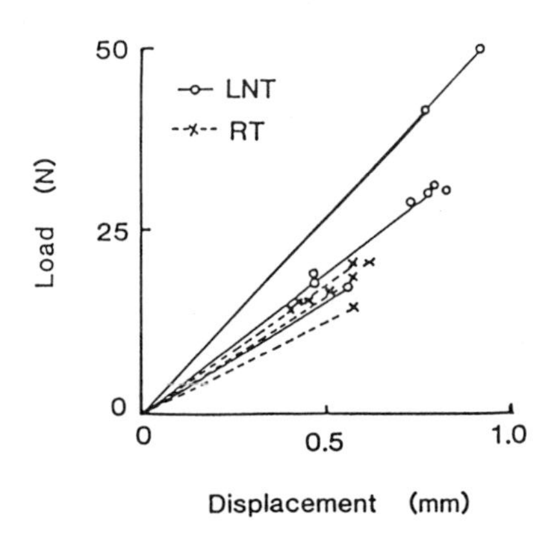

Fig. 2 Typical load-displacement profiles of alumina plates during three-point bending at RT and liquid N_2 temperatures.

As the uniformity and the reproducibility of fracto-emission data are known to be not so good, experiments in same condition were performed on about 3 to 5 specimens.

EXPERIMENTAL RESULTS

In general, a signal of electron emission shows a sharp pulse shape corresponding to local micro or macro fracture. Figures 3 (a) - (c) show typical time distribution of FE during the fracture processes of alumina plates taken with a acquisition rate of 1 ms/channel at the temperatures of 300 K, 90 K and 30 K respectively. Little FE was usually observed prior to the fracture except in the case of metals. Total intensity of FE integrated in time increased and the decay time of FE after fracture became longer with the decrease of temperature, and the emission continued for some tens minutes after the complete fracture of sample at 90 K. It should be noted that the fracture event itself occurs in several milliseconds as found from Fig. 4, which shows acoustic emission during the fracture.

Photographs of fracture surfaces of the samples by SEM are shown in Fig. 5 (a)fractured one at RT, (b) at 90 K and (c) at 30 K. The amount of micro-cracks increases and trans-granular facet dominates with the decrease of temperatures. Figures 6 (a) - (c) show macroscopic photographs of fracture surfaces taken by an optical microscope. From these photos, it is found that the fracture at RT produces smooth fracture surface, while rugged (or damaged, like a trace of tearing off) one appears at low temperature fracture. Therefore, the smooth fracture surface seems not to be responsible for the intense and prolonged FE, which are associated with the extent of rugged one and the trans-granular facets.

During RT deformation of pure iron plate which is ductile at that temperature, there are few peaks in the electron count rate. Only slip lines and no trace of fracture were observed in the SEM micrograph. However, during fracture at 20 K, many small peaks were observed in the initial stage and a large peak, at the final stage of the deformation, that is, fracture. Time distribution of the electron count rate is shown in Fig. 7, in which the emission coincides with push rod stroke (strain-gage signal delayed 5 sec on the chart). The decay of FE is apparently much shorter than those in the case of ceramics. In the SEM micrograph of

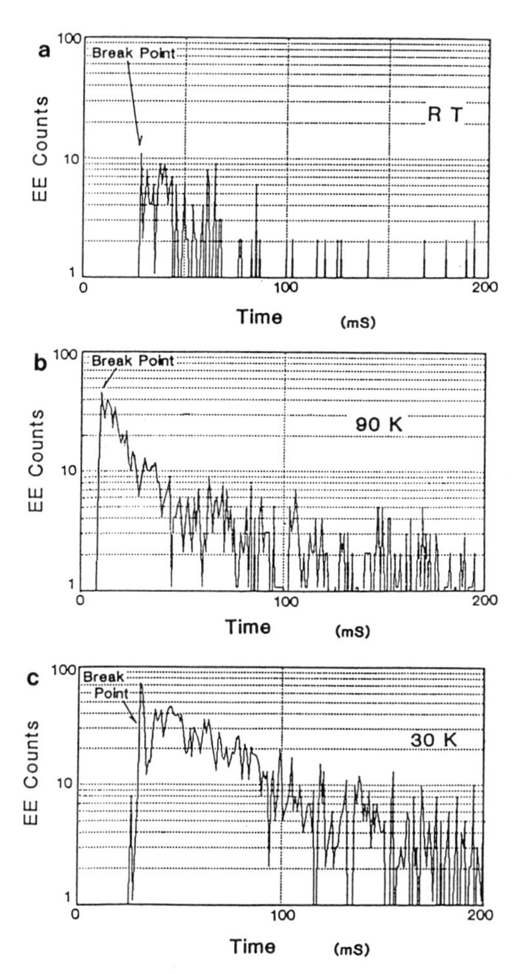

Fig. 3 FE from alumina plates fractured at (a) RT, (b) 90 K and (c) 30 K.

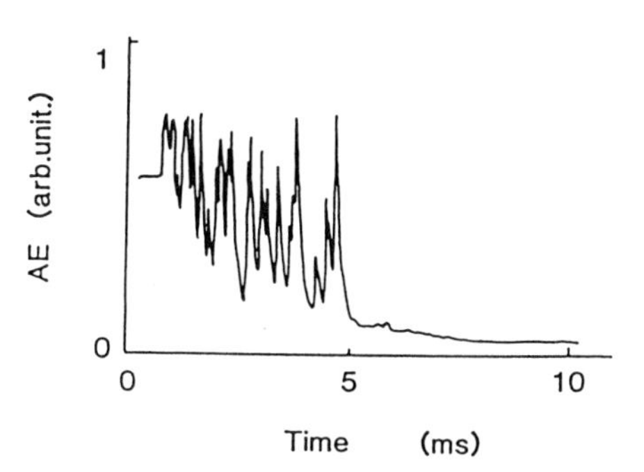

Fig. 4 Typical acoustic emission at complete fracture of alumina plate.

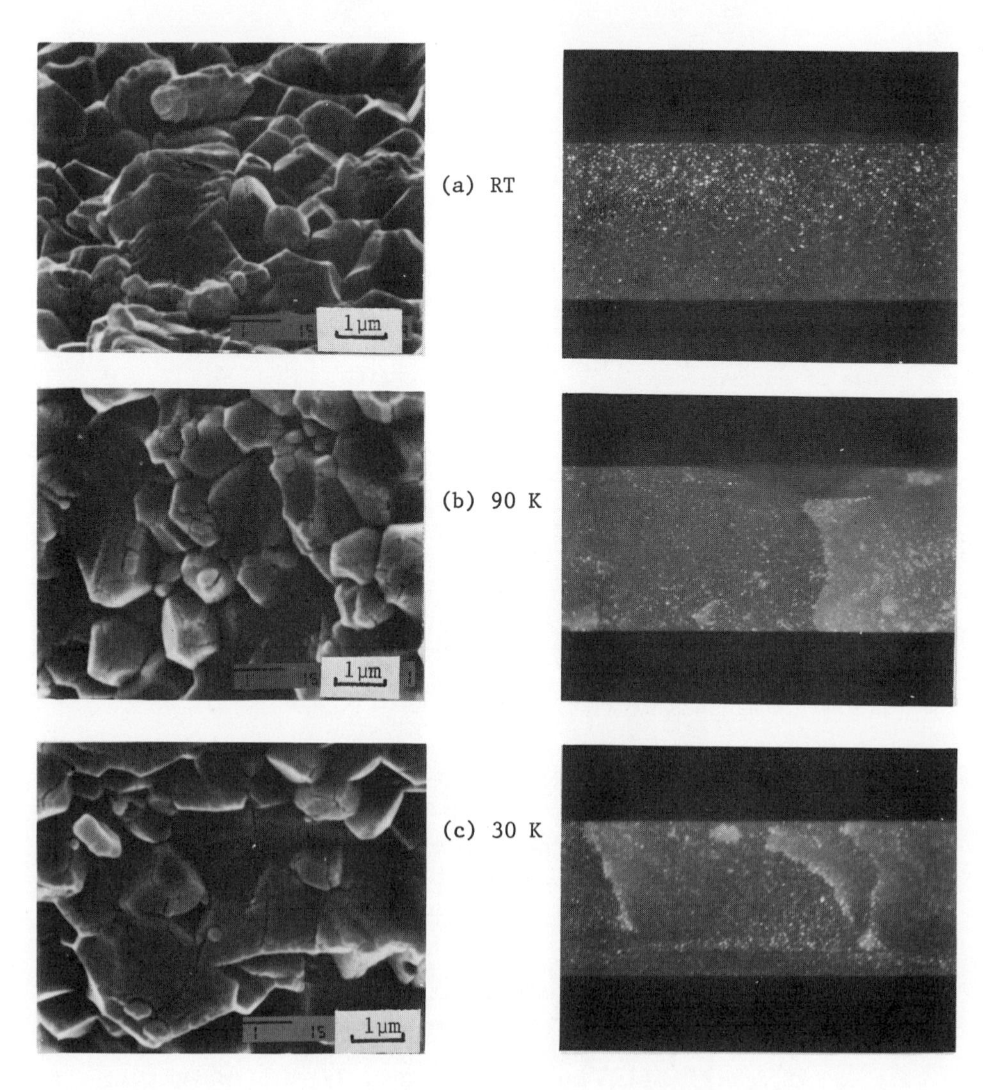

Fig. 5 SEM photos of fracture surfaces of alumina plates.

Fig. 6 Optical micrographs of fracture surfaces of alumina plates.

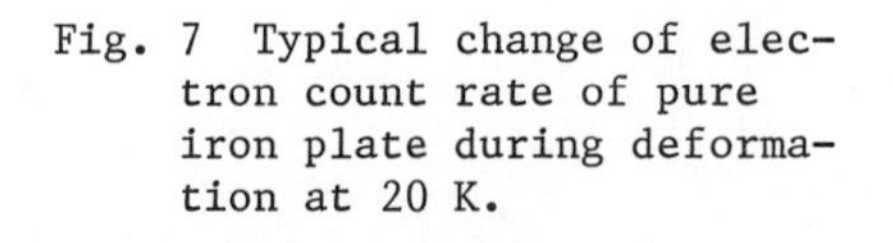

Fig. 7 Typical change of electron count rate of pure iron plate during deformation at 20 K.

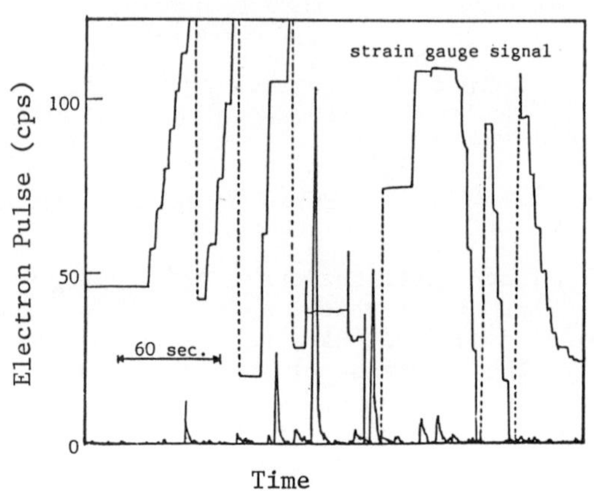

the pure iron fractured at 20 K, there are found some micro-cracks at the boundary of twins and many cleaved surfaces of crystal grains as shown in Fig. 8 (a) and (b). These shows a kind of phase transformation and typical brittle fracture.

In the case of high-carbon steel, many peaks were observed in initial stage of the deformation and a large peak appeared at macro-fracture in the final stage, that means the sample is perfectly divided at the notch portion as shown in Fig. 9. Both the intensity and frequency of the peaks increased with the sample temperature decrease, while the decay time remains unchanged within 1 minute. In the SEM micrograph of the sample fractured at 30 K as shown in Fig. 10 (a) and (b), there appeared such many kinds of surface change as cracks in ferrite, cracks within cementite (Fe_3C) and separation of the cementite- ferrite interface. These are supposed to be traces of origins of FE.

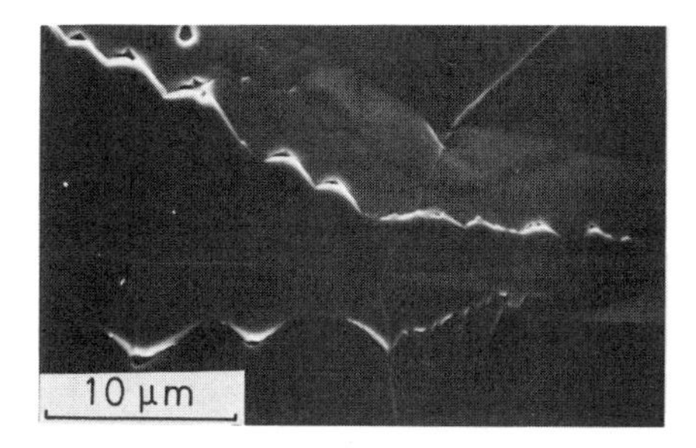

Fig. 8 (a) SEM micrograph of pure iron deformed at 20 K. There are found some micro cracks at the boundary of twins.

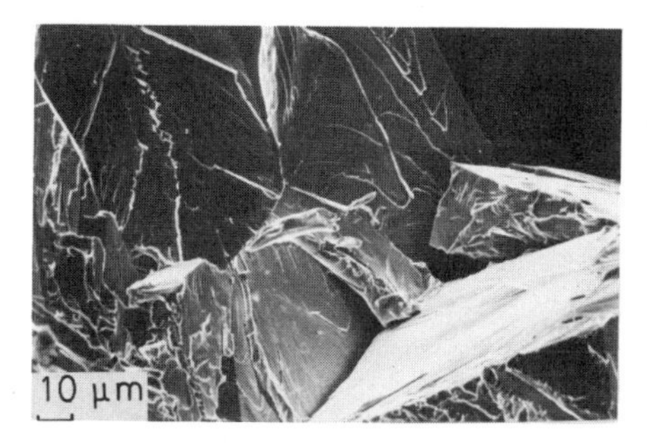

Fig. 8 (b) SEM micrograph of pure iron deformed at 20 K. There are found many cleaved surfaces of the crystal grains as the fracture.

Time distributions of fracto-emission from sodium glass plates are in Fig. 11, in which (a) was obtained during fracture process at RT. The emission does not have so high a peak intensity and ends in a short time. With the decrease of specimen temperature, however, the peak intensity increase and the decay time become longer as shown in Fig. 11 (b). This trend is more clear in the case of fused quartz as shown in Fig. 12, in which (a) shows the emission during fracture at RT, and (b) at 90 K and (c) at 30 K.

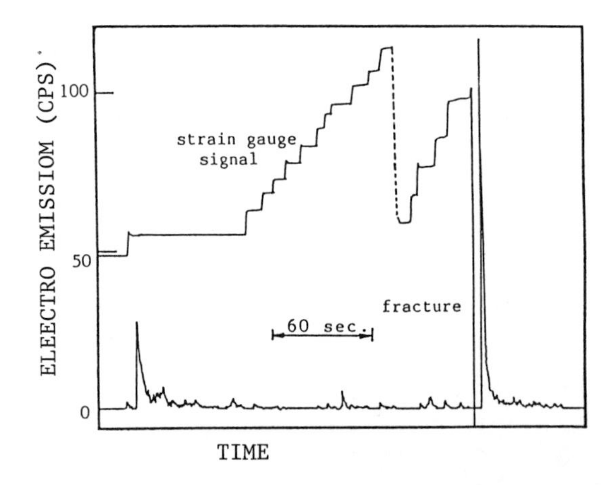

Fig. 9 Typical change in the electron count rate of high-carbon steel during deformation at 20 K.

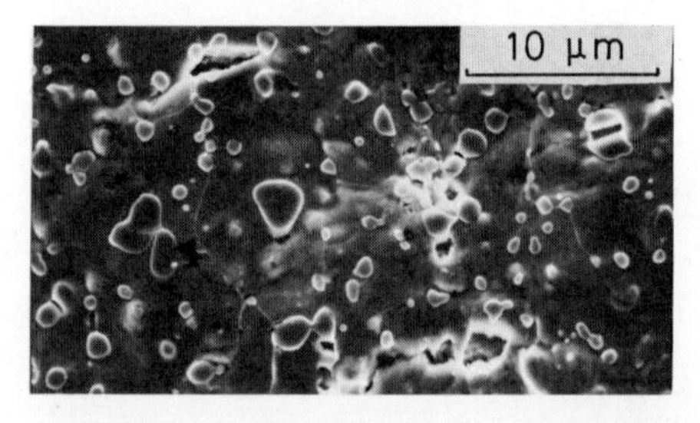

Fig. 10 (a) SEM micrograph of high-carbon steel deformed at 20 K : Cracks in ferrite, cracks within cementite and separation of the cementite-ferrite interface.

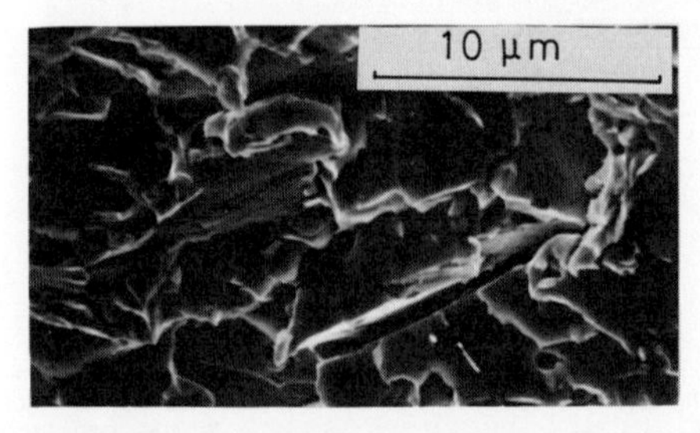

Fig. 10 (b) SEM micrograph of high-carbon steel deformed at 20 K : Cleaved surface of the fracture.

In the case of quartz single crystals, the FE features are considerably different from those of poly-crystalline ceramics and glasses. That is, the emission from X specimens even at RT features high peak intensity and long decay time as shown in Fig. 13 (a). Moreover, the decay time during fracture at 90 K became slightly shorter than that at RT as shown in Fig. 13 (b). The irradiation, however, induces longer decay times similar to those in poly-crystalline ceramics and fused quartz. The emission distributions from irradiated X specimens are shown in Fig. 13 (c). Figure 13 is displayed in log/log scale to clearly show the emission peak at the instant of fracture and the decay time depending on the relaxation process of the fracture surface.

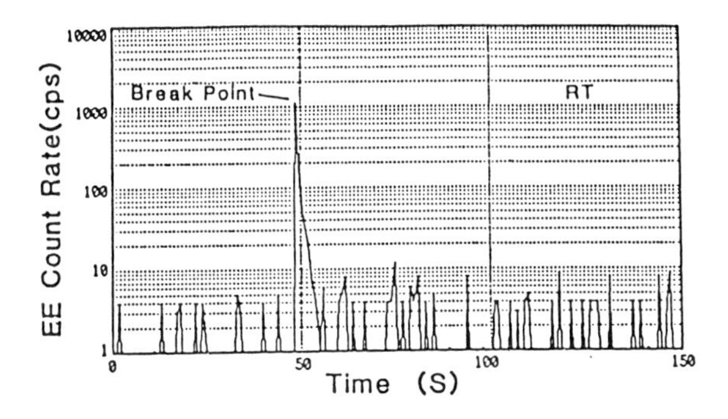

(a) FE was observed during fracture process at RT.

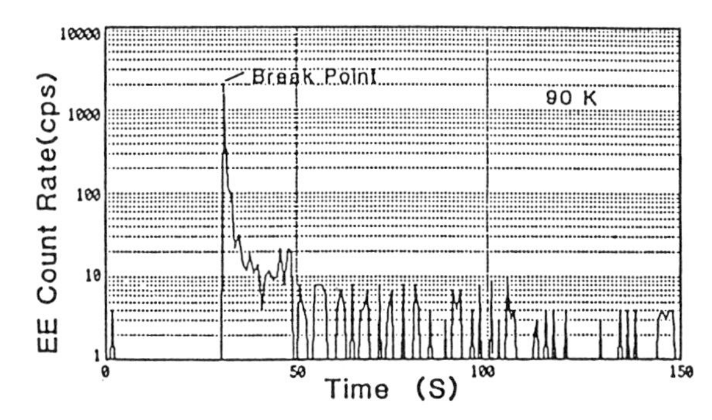

(b) FE at 90 K

Fig. 11 Time distribution of FE from sodium silicate glass plates.

DISCUSSION

One of reasons of the increase of FE intensity of crystalline alumina ceramics at low temperatures will be attributed to increase of the stiffness and the fracture strength with the decrease of temperature, because the increase of these mechanical properties induces much energy released at fracture. The released energy is dissipated in various objects and partially for phenomena related to FE origins as a localized rise in temperature, generation of charge separation and a few kinds of defects.[7,8,16] These situation is reasonably same in other materials like glass and single crystals.

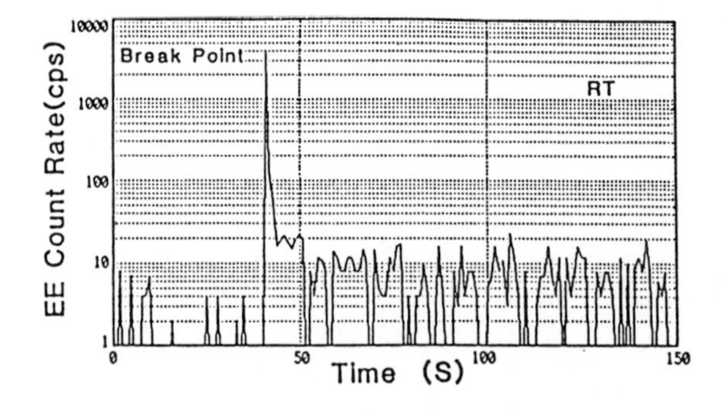

(a) FE was observed during fracture process at RT.

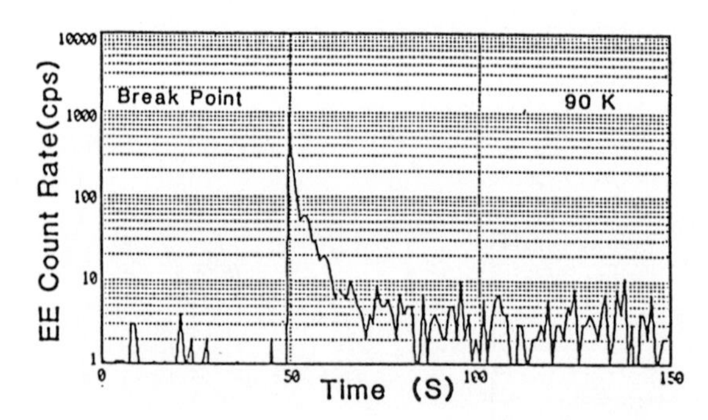

(b) FE at 90 K.

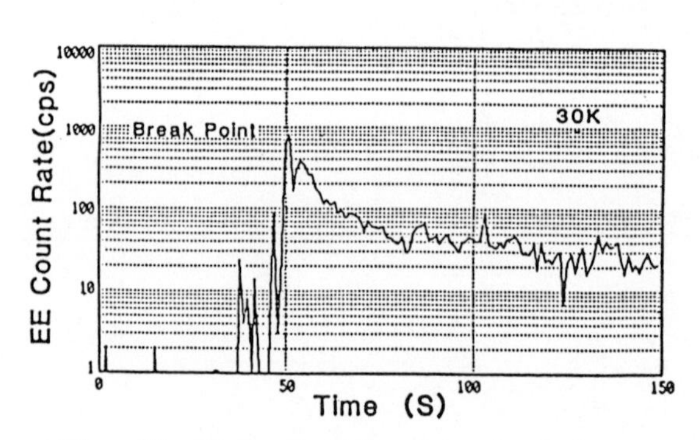

(c) FE at 30 K.

Fig. 12 Time distribution of FE from plates of pure fused quartz.

1E+05
10000
1000
100
10
1
10
100
1000
EE COUNT RATE (CPS)
TIME (S)

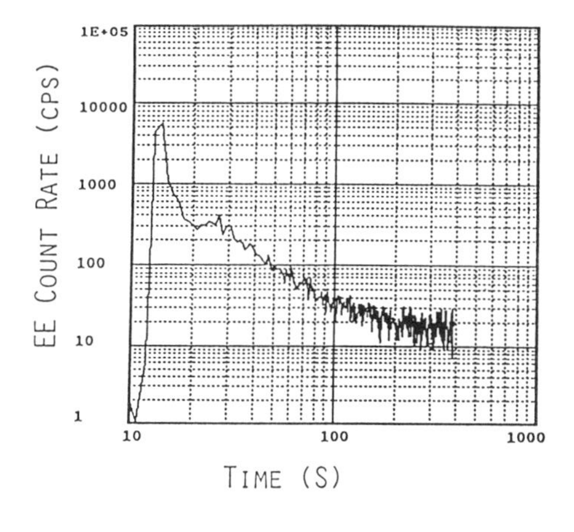

(a) FE was observed during fracture process at RT.

(b) FE at 90 K

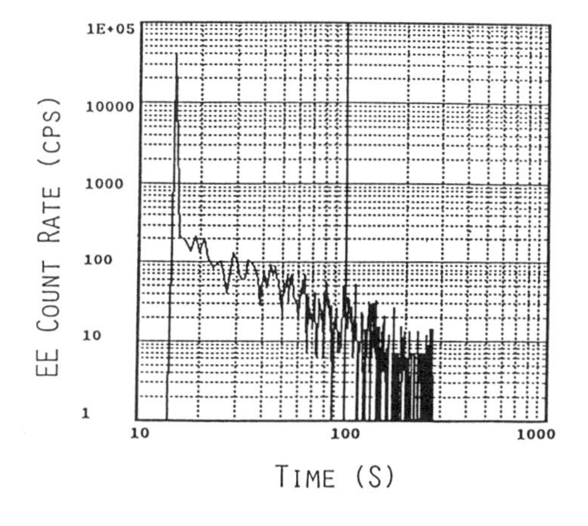

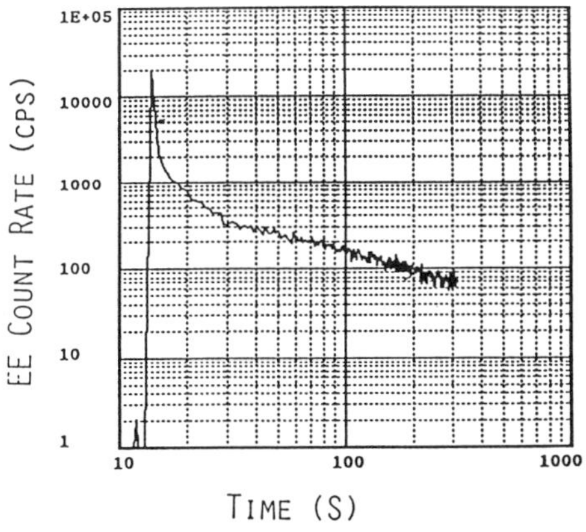

1E+05
10000
1000
100
10
1
10
100
1000
EE COUNT RATE (CPS)
TIME (S)

1E+05
10000
1000
100
10
1
10
100
1000
EE COUNT RATE (CPS)
TIME (S)

(c) FE from irradiated sample was observed at RT.

Fig. 13 Time distribution of FE from quartz single crystals of X region.

Another reason is macroscopic change of the fracture mode, in which the amount of micro-cracks increases and trans-granular facets dominate with the decrease of temperature.[12] In the crystal grains, many single crystals may be cracked and there may occur clear charge separation.

Moreover, it is easily supposed that decrease of charge carrier mobility prevents to compensate charge distribution on the fresh fracture surface and causes the longer decay time. Also the mechanical strength of glasses increases with decrease in temperature. Though the change of fracture mode has not yet been observed, mechanical strengths increase and they contribute to the FE change. Unique characteristics about fracture of glasses at low temperatures were not observed in this experiment.

Radiation effects on fracto-emission were examined for alumina ceramics, both glasses and quartz crystals. The effects are scarcely found in sodium glass as as shown in Fig. 14 (a), while in fused quartz, as shown in Fig. 14 (b), the effects provide a higher peak intensity and longer decay time of fracto-emission as same as in polycrystalline ceramics. Although the types of lattice defects like color centers induced by high energy radiation and fracture modes will be essentially different between fused quartz and poly-crystalline ceramics, the features of their FE are very similar. More experimental data are necessary for detailed discussion.

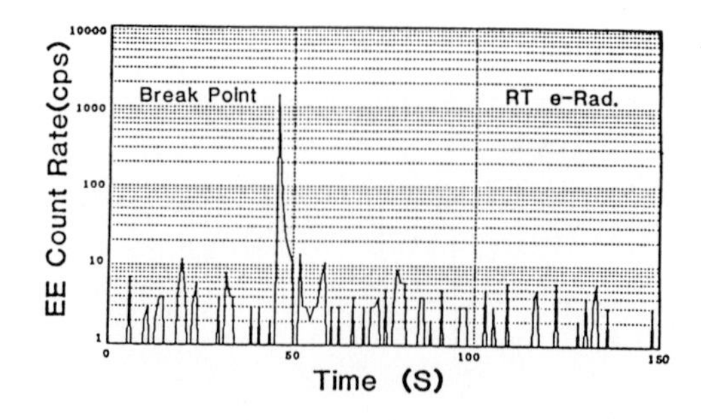

(a) FE from sodium silicate glass plate.

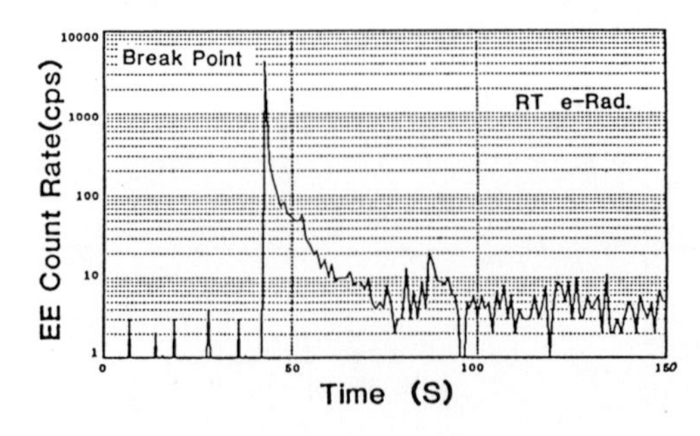

(b) FE from plate of pure fused quartz.

Fig. 14 Radiation effects on FE observed at RT.

The high intensity and the long decay time of the emission from quartz crystal during fracture at RT suggest that the charge separation are densely generated from fairly perfect crystalline lattice, and that there are only a few defects related with impurity filling role of the charge recombination on the fracture surface in the pure crystal. Though irradiation induces some defects which are different from the above defects, they may prevent the charge compensation rather than enhance it.

Although Dickinson et al. reported so much about the fracto-emission from single crystals like Si, MgO and LiF,[16,7,8] much more data are necessary for explanation of their phenomena. As for the emission from X specimens at 90 K, we have little information to explain the slightly shorter decay time which is quite different from those of ceramics and glasses. Of course, the stiffness and the fracture strength of quartz crystals are to increase with the decrease of temperature.

The electron emission from the quartz crystal Z specimen is basically not different from that of the X specimen, but often showed the anomalous behavior that the emission did not decay quasi-exponentially but continued at random for more than ten minutes after the fracture. The purity of the Z specimen is higher than that of the X specimen, and there is no cleavage facet in quartz single crystal. Crystal channeling may be related with these phenomena.

CONCLUSION

The electron emissions from poly-crystalline alumina and glasses of sodium silicate and fused quartz during fracture at RT have low peak intensity and short decay time. However, the intensity and the decay time increase with decrease of the sample temperature at fracture. These facts are attributed to macroscopic change of mechanical properties as increase of stiffness and fracture strength, and the fracture mode. As a microscopic view, clear charge separation and some kinds of defects are created in crystals and glassy materials consequently with the change of mechanical and electronic properties.

The decay time of FE from metals is much shorter than those from the insulators, because good electronic-conductivity on the freshly created fracture surface can cancel quickly the charge irregularity.

Radiation effects on the emission bring high intensity and long decay time in the ceramics, fused quartz and quartz of single crystal. These effects are not remarkable in sodium silicate glass because radiation induced defects may be annealed for period of the storage even at room temperature.

As for the emission from quartz single crystals, it is clear that the peak intensity is high and the decay time is long even in fracture at RT because of fracture of firmly binding lattice as other researchers have reported. However, the decay time scarcely changes at the 90 K fracture. Further experiments about this material have to be done, because there remain many problems to be solved.

ACKNOWLEDGMENT

The authors wish to thank Professor K. Katagiri, Iwate University for his help for the experiments related with metals. The irradiation on specimens was performed by an electron linear accelerator in the Radiation Laboratory, ISIR, Osaka University. The authors wish to thank the persons of the machine group there for their help. Also we wish to thank the graduate students in our group for their assistance in the completion of this paper.

REFERENCES

1) P. Braunlich and J. T. Dickinson ; "Proceedings of 6th International Symposium on Exoelectron Emission and Application" Rostock, East Germany (1979) 9
2) H. Glaefeke ; Exoemission in "Thermally Stimulated Relaxation in Solids" ed. P. Braunlich, Springer Verlag, Berlin, (1983)
3) J. T. Dickinson, M. K. Park, E. E. Donaldson and L. C. Jensen ; J. Vac. Sci. Technol. 20:436 (1982)
4) J. T. Dickinson, L.C. Jensen and A. Jahan-Latibari ; J. Vac. Technol. A 2:1112 (1984)
5) A. V. Poletaev and S. Z. Shmurak ; Sov. Phys. Solid State ; 26:2147 (1984)
6) J. T. Dickinson, A. Jahan-Latibari and L. C. Jensen ; J. Mater. Sci. 20:229 (1985)
7) S. C. Langford, J. T. Dickinson and L.C. Jensen ; J. Appl. Phys. 62: 1437 (1987)
8) J. P. Mathison, S. C. Langford and J. T. Dickinson ; J. Appl. Phys. 65:1923 (1989)
9) S. Owaki, K. Katagiri, T. Okada, S. Nakahara and K. Sugihara ; Adv. Cryog. Eng. Mater. 34:283 (1988)
10) S. Nakahara, T. Fujita, K. Sugihara, S. Owaki, K. Katagiri and T. Okada ; Adv. Cryog. Eng. Mater. 34:91 (1988)
11) S. Owaki, K. Katagiri, T. Okada, S. Nakahara and K. Sugihara ; Adv. Cryog. Eng. Mater. 36:1361 (1990)
12) S. Nakahara, T. Fujita, K. Sugihara, S. Owaki, K. Katagiri and T. Okada ; Adv. Cryog. Eng. Mater. 36:1201 (1990)
13) S. Nakahara, T. Fujita, K. Sugihara, S. Owaki, K. Katagiri and T. Okada ; Jpn. J. Appl. Phys. 24-4:198 (1985)
14) S. Owaki, K. Katagiri, T. Okada, S. Nakahara and K. Sugihara ; "Proceeding of 9th International Symposium on Exoelectron Emission and Application" Wroclaw, Poland (1988) 262
15) S. Nakahara, T. Fujita, K. Sugihara, S. Owaki, K. Katagiri and T. Okada ; "Proceeding of 9th International Symposium on Exoelectron Emission and Application" Wroclaw, Poland (1988) 254
16) S. C. Langford, D. L. Doering and J. T. Dickinson ; Phys. Rev. Lett. 59:2795 (1987)

THE INFLUENCE OF THE NETWORK OF MICROCRACKS UPON THE CRACK PROPAGATION BEHAVIOR INSIDE OF TRANSPARENT ZIRCONIA

K. Ahlborn

Carl Freudenberg
Central Research and Development
Weinheim, Germany

Y. Kagawa and A. Okura

The University of Tokyo
Institute of Industrial Science
Tokyo, Japan

INTRODUCTION

The fracture process in many ceramics is rather complicated and still subject of scientific research. The aim is to unveil fracture processes on micromechanic scale and to develop techniques, based on this knowledge, for the manufacture of ceramics with enhanced toughness.

Rice et al. (1980) demonstrated for cubic ceramics (CaF_2, ZnSe, $MgAl_2O_4$ and Y_2O_3) that a transition of the fracture surface energy occurs from the single crystal value (intragranular or grain boundary fracture) to the polycrystalline value. By the microscopical examination of the size of critical machining flaw on the fracture surface it was shown, that the fracture surface energy increases by a factor 3 to 4 at ratios of flaw size to grain size $a/\bar{D}$ between 1 and 6. After some transgranular propagation the young crack interacts with the grain boundaries and thus faces an increasing fracture resistance.

A more pronounced enhancement of the fracture surface energy occurs in noncubic ceramics of medium grain size due to the formation of microcracks (Al_2O_3, BeO, 10 to 100 μm, Rice, 1981) or phase transformations (PSZ, ZTA, Y-ZTP; Rühle et al., 1986). Fracture mechanic equations must be modified since these effects are not considered in LEFM. A crack emanating from a natural flow faces a fracture resistance growing with crack extention (R-curve) as measured on Al_2O_3 by Steinbrech et al. (1983), Swain (1986) and on PSZ by Rose

Fracture Mechanics of Ceramics, Vol. 10
Edited by R.C. Bradt *et al.*, Plenum Press, New York, 1992

et al. (1986). As a consequence the conventional fracture toughness K_{Ic} is not longer a material inherent parameter. It is dependent on test method, specimen type and notch shape.

Because of the absence of thermal expansion anisotropy and phase transformations neither microcracking nor R-curve behaviour is expected in cubic materials, and the fracture mechanical research has been concentrated on the above mentioned noncubic ceramics. The fracture surface energy is reported to be independent from grain size (Y_2O_3, 8 to 80 μm; Liang et al. 1991) or only a limited maximum is observed (Y_2O_3, 5 to 100 μm; Monroe and Smyth, 1978). Possible effects are in the same scale as the data scatter.

Techniques for the microscopical examination of the fracture were developed hand in hand with the fracture mechanical testing of ceramics. Usually a modified standard testing equipment is used and the fracture surfaces are examined after failure. DCB and double torsion specimen allow the in-situ examination with a travelling microscope attached to the test equipment. As an alternative several methods for the direct observation of the fracture processes have been developed. These methods allow the loading of small size specimen within the microscopical equipment as listed in Table 1.

This contribution discusses the fracture behaviour of a cubic zirconia ceramic, which contains thermally induced microcracks, as reported by Ahlborn et al. (1989). With a recently developed loading technique reported by Ahlborn et al. (1990) the interaction of the slowly growing crack with the microcrack network is observed in-situ inside of the transparent ceramic and recorded with video camera. The observed fracture mechanisms are correlated to the fracture toughness measured simultaneously.

MATERIALS AND METHODS

A transparent fully stabilized zirconia of the composition 83 mol % ZrO_2, 10 mol % TiO_2 and 7 mol % Y_2O_3 is prepared by sintering at 1673 K to 1973 K, followed by isostatically hot pressing in argon gas and final tempering for 4 h at 1273 K according to Nagasaki (1989). By changing the sintering temperature three average grainsizes are obtained: $\bar{D}$ = 25 ± 15 μm, 90 ± 40 μm and 130 ± 60 μm. The crystalline structure of the grains is cubic with dispersed tetragonal inclusions of 0,2 μm size.

Plates are hot pressed and polished to a nominal thickness of 1 mm. Then tapered cantilever beam specimen of three different specimen width W (small = 6.5 mm, medium = 13 mm, large = 20 mm) and proportional shape are cut. Finally an artificial notch of length a_o was cut to a depth ratio $a_o/W \geq 0.5$ with a diamond blade of 0.6 mm thickness leaving behind a triangular notch shape (Chevron notch type).

By a small size, compact and extremely stiff testing equipment the specimens are loaded at room temperature and relative humidity of about 40 % in mode I loading condition

Table 1. Overview of techniques for the direct observation of cracks in ceramic materials

	A	B	C	D	E	F
Reference	Wu (1981)	Wu (1981)	Swanson (1987)	Swanson (1987)	Frei (1989)	Ahlborn (1990)
Technique of Observation	X-Ray radiography	SEM	Microscope	Microscope	SEM	Microscope
Specimen type	Bendbar	TCB	TCB	Disc	Bendbar	TCB
Loading technique	Constant moment	Wedge	Wedge	Biaxial (3 point)	Bending outside	Mode I in-situ
Load Measurement	Yes	No	No	Possible	Possible	Yes
Crack stability in glass*	?	Unstable	Unstable	?	?	Stable
Resolution	10 μm	<1 μm	2 μm	2 μm	<1 μm	2 μm
Area of observation**	Surface + inside	Surface	Surface + inside	Surface + inside	Surface	Surface + inside
Atmosphere	Any	Vacuum	-------------Air---------------			
Consumption of material	Medium	Low	Low	High	Low	Low
Specimen preparation	Common	Difficult	---------Common--------			Difficult
Costs	High	High	Low	Low	High	Low

* Glass as representative of extremley brittle materials

** for transparent specimen

Not included: double torsion technique and DCB in standard testing equipment furnished with traveling microscope within an optical microscope as outlined by Ahlborn et al. (1990). A schematic drawing of the fixture is given in fig. 1. The tip of the artificial notch is continously observed, while the critical load is approached by small loading steps. As soon as the natural crack extends through the triangular section the specimen is unloaded. Then the natural crack is propagated slowly and stable by subsequent reloading/unloading steps through the specimen. The length of the natural crack Δ a and the load required for crack propagation are measured. Scattering of the measured K_I value due to deviations of the real specimen shape from the ideal symmetrical specimen shape and resulting deviations from pure mode I loading are less than 20 %.

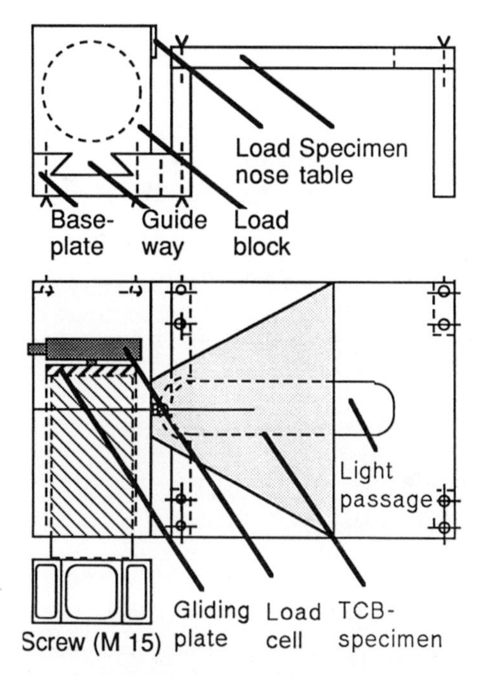

Fig. 1. Schematic drawing of the load fixture for mode I stable crack propagation and in-situ observation within an optical microscope, Ahlborn et al. 1990.

Experiments where the crack is propagated continously or instable are performed as well. During propagation all means of an optical microscope like different light sources, polarized light or Normanski technique are used in order to analyse in-situ the circumstances of fracture. Beside micrographs are video tapes an important tool of documentation, allowing the measurement of crack velocity. After failure the fracture surfaces are examined by SEM.

EXPERIMENTAL RESULTS

Microcracks

Microcracks located at the triplepoints of adjacent grains are detected by microscope for the zirconia with average grainsize of 90 and 130 μm as listed in Table 2. As can be concluded from the high scattering, the microcracks of the 130 μm ceramic are not equally distributed and exist preferentially near the surface. In the 25 μm-zirconia microcracks are not detectable by microscope, but get obvious by SEM-examination of the fracture surfaces.

The density and length of the microcracks in the 90 μm is increased by quenching in air from 1073 K.

For the 130 μm zirconia the density of microcracks near the surface is significantly increased by the applied load, whereas the interior network of microcracks is not affected. The microcrack network of the 90 μm zirconia is not affected at all.

Table 2. Density of cracked triplepoints (mm^{-3}) and scattering (%) as detected by microscope

$\bar{D}$ μm	As Received	After Loading	After Quenching	Increase %
25 ± 57 %	-	>1634*	-	-
90 ± 44 %	622 ± 44	622	-	0
90 ± 44 %	622 + 44	-	1126 ± 63	55
130 ± 46 %	47 ± 81	144 ± 75	-	306

* concluded from SEM examination of the fracture surfaces

Impurities of 2 to 20 μm size give rise to thermally induced stress fields as observed with polarized light. The grain boundaries appear to be stress and void free.

Crack propagation experiments

The fracture behaviour clearly depends on the grain size. Nevertheless some predominant fracture processes are in common for the different grain sizes examined, as described here. In the following it is strictly distinguished between the "artificial notch" of length a_o the "initial machining flaw" at the notch root (of minor length) and the "natural crack" emanating from this flaw over the extention Δ a. Altogether form the effective crack length

$$a = a_o + \Delta a,$$

which is required to calculate the stress intensity at crack propagation K_{Ir} via the common equation

$$K_{Ir} = \frac{P}{B\sqrt{W}} \cdot Y\ (a_o/W),$$

where P is the load at propagation, B is the specimen thickness, W the specimen width and Y the shape function as reported by Ahlborn et al. (1990).

For all grain sizes the natural crack emanates from a critical machining flaw located in the region of highest stress near the notch root. The size of the machining flaw in general is smaller than the grain size and not detectable by microscope. The natural crack propagates principally transgranular as observed by SEM.

As soon as the young crack reaches a precracked triplepoint located in its plane several interaction phenomena occur as shown in Fig. 2:

- The crack is twisted and forced into two different prefered fracture planes of the adjacent grains and
- as a result, the crack is splitted into two single fingers or lances.

Far behind the crack front

- a cleavage step is formed by the overlapping crack lances and
- the grain with the cracked triplepoint is pulled out by intergranular fracture, leaving a hollow of pyramidal shape.

This processes are repeated as soon as further precracked triplepoint located in the fracture plane are reached and repeated crack splitting occurs. Hence the portion of boundary fracture, the number of crack and the amount of cleavage steps increase.

After a certain extension of the natural crack the branches start to recombine and the number of cleavage steps decreases. The crack speeds up and final failure occurs at a relative effective crack length a/W between 0.7 and 0.8.

In the slow crack growth region the crack front is rugged, often forming a convex shape, whereas in the fast or instable region the crack front is more or less continous and concave shaped (Fig. 3).

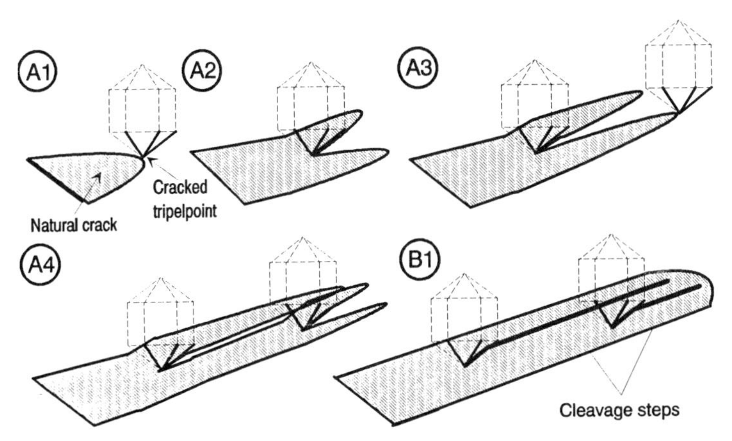

Fig. 2. Schematical drawing of the crack splitting process and the formation of lances (A1 to A4) followed by cleavage step formation and crack linking (B1).

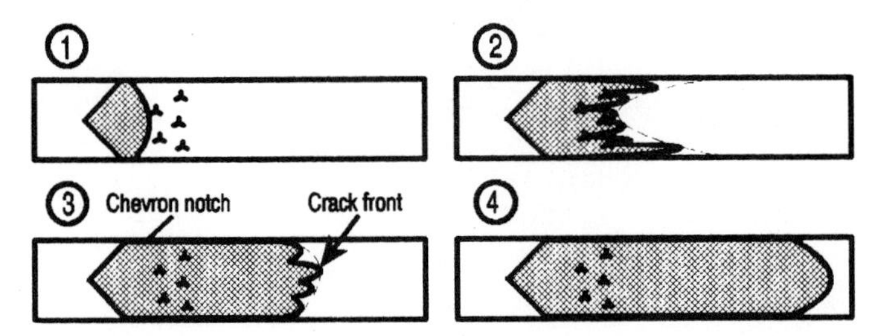

Fig. 3. Schematical drawing of the shape of the crack front at subsequent stages of fracture: 1. Young crack; 2. interaction with crack network and formation of lances; 3. recombination and crack linking; 4. final failure at enhanced crack speed.

25 μm-zirconia

The crack is splitted in more than 20 single crack lances propagating with nearly equal crack speed. Behind the crack front grain bridging and grain-pull-out effects are detectable by microscope. The portion of localized boundary fracture and the number of cleavage steps is high and proportional to the high amount of precracked triplepoints. The cleavage steps are comperatively short, because permanent crack splitting and recombination occurs. The fracture processes are due to an increase of the fracture toughness from about 2 to about 3.3 MPa$\sqrt{m}$ as shown in Fig. 4.

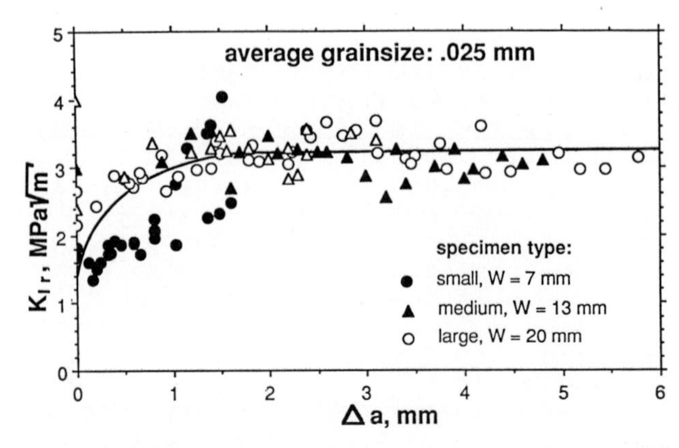

Fig. 4. Crack-growth-resistance versus crack extension for the 25 μm zirconia and three specimen width.

90 μm-zirconia

The crack front is severely rugged. The crack consists of 5 to 7 crack lances with smaller side branches. The crack lances near the surface propagate faster than the crack lances inside the specimen, thus forming a convex crack front. The crack speed differs by a factor of about 70 between the simultaneously propagating lances (from $6 \cdot 10^{-6}$ m/s to $4 \cdot 10^{-4}$ m/s).

130 μm-zirconia

Since most of the triplepoints near the surface are precracked grain boundary fracture is predominant in this region, whereas in the interior pure transgranular fracture occurs. The high microcrack density near the surfaces is probably originated by the polishing process; the microcrack network is sensitive to the stress field of the advancing crack (see Table 2).

INFLUENCE OF SPECIMEN SIZE AND LOADING TECHNIQUE

As can be seen from Fig. 4 the specimen size has no influence upon the measured fracture toughness. This result prooves that deviations from the ideal shape originated by manufacturing of the small size specimen do not superimpose the measured toughness.

Fracture toughness data measured by instable fracture or by steady propagation with subsequent forced instability fit in with the scattering of the measured R-curves, thus prooving that the experiments are performed near to the critical fracture toughness.

DISCUSSION

A network of microcracks is detected in the fully stabilized zirconia ceramic of cubic structure. As predicted by Clark (1980) the microcracks are located at the triplepoints of adjacent grains (four grain junction). The density of microcracks is reciprocal to the grain size, since the per volume amount of triplepoints increases with the per volume amount of grains. It is very likely that the microcracks are thermally induced, since the crack density is not affected by an applied external load, but by thermally induced internal stresses as generated by quenching of the 90 μm zirconia (see Table 2). This result is in disagreement with theory, which explains the existence of thermally induced microcracks with the thermal expansion anisotropy of the noncubic ceramics (note that the formation of microcracks by thermal cycling is observed also for aluminium alloys of cubic structure). The microcracks occur by boundary fracture only. The critical grain size for the formation of these microcracks is less than 25 μm.

Cracks are principally propagating transgranular, unless they interact with precracked grain boundaries. The interaction with the pre-existing microcracks generates crack front disturbances like crack twisting, crack splitting and crack pining. Behind the crack front in the wake of the process zone a cleavage step formation as reported by Swain et al. (1974) and grain-pull-out occurs. These latter processes most effectively increase the fracture toughness, since they lower the stress intensity at the crack front. The amount of crack lances, of grain-pull-out, of cleavage steps and the length of cleavage steps depend directly on the per volume amount of precracked triplepoints and hence are reciprocal to the grain size.

The observed fracture processes give rise to a R-curve behaviour. As shown in Fig. 5 an uniform R-curve is obtained for all measured grain and specimen sizes, when K_{Ir} is plotted against the ratio of crack extention to average grain size $\Delta a/\bar{D}$. The crack growth resistance increases from about 2 $MPa\sqrt{m}$ to 3,5 $MPa\sqrt{m}$.

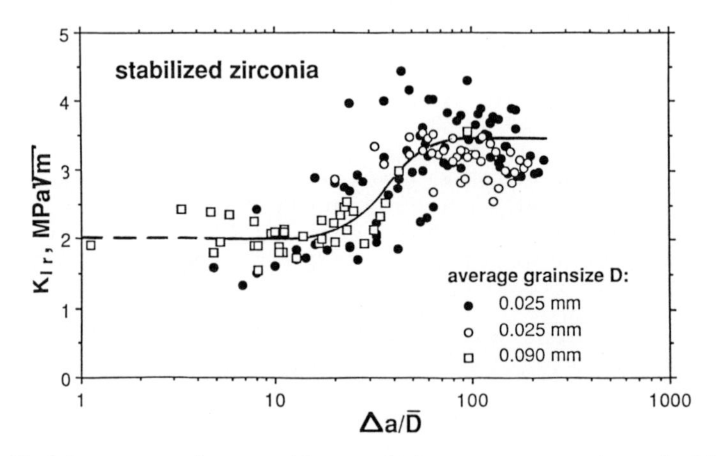

Fig. 5. Uniform crack-growth-resistance curve by plotting fracture toughness data versus the crack extension in relation to the average grain size

The natural crack has to extend over more than 20 grains until the toughening mechanisms come into action, since the most effective processes occur behind the crack tip. Or vice versa: Grain-pull-out and cleavage fracture occur about 20 grains behind the crack front. After a crack extension of about 200 grains instability occurs due to the recombination of cracklances. During fast crack propagation the length of cleavage steps decreases and the crack front gets uniform, whereas the amount of grains pulled out and the portion of boundary fracture keep constant.

CONCLUSION

By microscopical observation inside of a transparent zirconia it is shown that microcracks can exist in ceramics of cubic grain structure. The microcracks have substantial influence on the fracture behaviour as reported by Rice 1981 for noncubic ceramics. By the in-situ observation of the crack propagation the fracture processes are unveiled. Beside severe crack splitting processes especially cleavage fracture and grain-pull-out are due to the increase of the fracture resistance K_{Ir} from about 2 to about 3,5 MPa$\sqrt{m}$ with crack extension. This corresponds to an increase of the apparent fracture surface energy from about 10 J/m^2 to a high value of about 93 J/m^2, unusual for cubic ceramics.

The direct and in-situ observation of the fracture behaviour inside of the zirconia ceramics has prooved some fracture processes up to now deduced from the SEM examination of the fracture surfaces after failure (see list of references).

ACKNOWLEDGEMENT

The authors thankfully acknowledge the supply of the materials by Mr. Y. Itakura and Mr. Y. Nagasaki from Tosoh Company, Japan and the skilful but tedious specimen preparation by Mr. T. Horie, now with Akebono Brake, Japan. Dr. K. Ahlborn was supported with a post-doctoral fellowship in the frame of the "STP-program for young European scientists in Japan" by the European Communities (EC-DG XII).

REFERENCES

Ahlborn, K., Kagawa, Y., and Okura, A., Observation of the crack propagation of ZrO_2 at room temperature, 1989, in: "Proc. Annual Meet. Jap. Inst. of Metals".

Ahlborn, K., Kagawa, Y., and Okura, A., 1990, A technique for the in-situ observation of stable crack propagation in brittle solids, Scripta Metallurgica et Materialia, 24: 1063.

Clark, D.R., 1980, Microfracture in brittle solids resulting from anisotropic shape changes, Acta Metall., 28: 913.

Frei, H., Plappert, C. and Grathwohl, G., in: "Euro-Ceramics, 3", Elsvier Sci. Publishers, Barking.

Liang, K.M., Gu, K.F., Orange, G. and Fantozzi, G., Influence of the grain size on the crack resistance, 1991, Proc. 7th CIMTEC World Ceramic Congress, to be published, Elsevier Sci. Publishers, Barking.

Monroe, L.D., and Smyth, J.R., 1978, J. Am. Ceram. Soc., 61: 538.

Nagasaki, Y., Transparent zirconia, 1989, New Ceramics, 5: 85 (in Japanese).

Rice, R.W., Freiman, S.W., and Mecholsky, J.J., 1980, The dependence of strength-controlling fracture energy on the flaw-size to grain-size ratio, J. Am. Ceram. Soc., 63: 129.

Rice, R.W., Test - microstructural dependence of fracture energy measurements in ceramics, 1981, in: "ASTM STP 745", S.W. Freiman and E.R. Fuller, Eds., Am. Soc. Test. Mater., Philadelphia.

Rühle, M., Claussen, N., and Heuer, A.H., 1986, Transformation and microcrack toughening as complementary processes in ZrO_2-toughened Al_2O_3, J. Am. Ceram. Soc., 69: 195.

Steinbrech, R., Khehans, R., Schaarwächter, W., 1983, Increase of crack resistance during slow crack growth in Al_2O_3 bend specimens, J. Mater. Sci., 18: 265.

Swain, M.V., Lawn, B.R., Burns, S.J., 1974, Cleavage step deformation in brittle solids, J. Mater. Sci., 9: 175.

Swain, M.V., R-curve behaviour in a polycrystalline alumina material, 1986, J. Mater. Sci. Letters, 5: 1313.

Swanson, P.L., Fairbanks, C.J., Lawn, B.R., Mai, Y.-W., and Hockey, B.J., 1987, J. Am. Ceram. Soc., 70: 279.

Wu, C.C., Rice, R.W., and Becher, P.F., 1981, The character of cracks in fracture toughness measurements of ceramics, in: "ASTM STP 745", S.W. Freiman and E.R. Fuller, Eds., Am. Soc. Test. Mater., Philadelphia.

CRITICAL STRESS OF MICROCRACKING IN ALUMINA EVALUATED BY ACOUSTIC EMISSION

Shuichi Wakayama and Hisashi Nishimura

Department of Mechanical Engineering, Faculty of Technology
Tokyo Metropolitan University
Minami-Ohsawa 1-1, Hachioji-shi, Tokyo 192-03, Japan

ABSTRACT

Microfracture process during the bending tests of alumina ceramics were evaluated by acoustic emission technique. Different size specimens were used for the bending tests in order to investigate the dependence of microfracture process on the specimen size. A remarkable point in the AE generation pattern of each specimen, at which both AE events and energy increased rapidly, was observed before the final unstable fracture. It is important that the apparent stress at those points were independent of the AE threshold level and specimen size. Using the Fluorescent dye penetrant method, the fracture process on the surface was observed, then it has been understood that the stress corresponds to the critical stress for the maincrack formation due to the coalescence of microcracks and/or pores. Considering the microfracture process, statistical treatment for the strength of brittle materials has been discussed. Consequently, it was concluded that the critical stress can be the advanced evaluation parameter, which is equivalent to yield strength in metals, for ceramic materials.

INTRODUCTION

Toughening of the material is essential for the reliability of ceramics. Although there have been many approaches to the toughening mechanism, both theoretically [1-4] and experimentally [5-7], many important problems have still remained. The main reason is considered that nevertheless the parameters obtained experimentally are macroscopic, the parameters included in theoretical models are microscopic, such as the critical stress for microcrackings.

On the other hand, the statistical treatment of strength is one of the most important approaches to the reliability of ceramics. Weibull statistics [8] has been generally used for ceramic materials. Recently, Batdorf [9] and Lamon [10] developed the statistics which is available for multiaxial stress fields. In spite of these improvements, conventional statistical treatments neglect the fracture process, as pointed out by Freudenthal [11]. Therefore, it has become the most important subject to establish the experimental technique of the evaluation of microfracture process in ceramics.

Fracture Mechanics of Ceramics, Vol. 10
Edited by R.C. Bradt *et al.*, Plenum Press, New York, 1992

For the reasons mentioned above, non-destructive evaluation (NDE), which is one of the most powerful approaches for the reliability in itself, is also necessary for the development of the toughening mechanisms and the statistical treatment of ceramics. Especially, many acoustic emission studies on ceramics [12,13], including authors' [14,15], has been carried out because of its easiness in the detection of the nucleation times and locations of microcracks in the material.

In this study, microfracture process during bending tests of alumina ceramics, using the specimens with several dimensions, were evaluated. The nucleation time and locations of microcracks were determined using the arrival times and those differences between 2 ch transducers. The fluorescent dye penetrant method was also applied for the observation of fracture process on the surface. Bending tests are adopted since it is most usual in the measurement of the strength of ceramics and the bending strength is considered influenced by the fracture process on the surface. The purposes of this paper are to demonstrate the microfracture process, including the formation and growth of maincrack during the bending tests of alumina, to establish the evaluation technique of the critical stress of microcrackings and then to discuss the statistical treatment of the strength of brittle materials.

EXPERIMENTAL PROCEDURE

Materials

Materials used were two kinds of alumina ceramics with the different microstructures. Those mechanical properties and chemical compositions are tabulated in Table 1. Coarse grain alumina (ADS-10) has the mean grain size of 20 μm, Young's modulus of 385 GPa, porosity of 5 % and purity of 99.5 %. Those of fine grain alumina are 5 μm, 227 GPa, 9% and 92%, respectively. Specimens were cut from alumina blocks by diamond saw to the dimensions of 3 x 3 x 40 mm, 4 x 4 x 40 mm and 5 x 5 x 40 mm. Each specimens were chamfered to 0.1 mm and polished with diamond powder of 1 μm. In order to investigate the influence of corrosion by water on the microfracture process, "dry" specimens were also prepared by the dehydration in the vacuum of 10^{-4} Torr at 150 °C, 60 min.

Table 1. Mechanical properties (a) and chemical compositions (b) of materials.

(a) Mechanical properties

	Mean Grain Size / μm	Density / $kg \cdot m^{-3}$	Porosity / %	Young's Modulus / GPa	Poisson's Ratio
ADS-10	20	3.85×10^{-3}	5	377	0.231
ADS-80	5	3.57×10^{-3}	9	222	0.235

(b) Chemical compositions

	Al_2O_3	MgO	CaO	SiO_2	Na_2O	Fe_2O_3
ADS-10	99.5	0.25	-	0.03	0.01	0.021
ADS-80	92	2.5	1.5	1	0.2	0.02

Table 2. Specimen sizes.

(a) Coarse grain alumina (ADS-10)

Specimen Name	Width/mm	Height/mm	Lower Span/mm	Upper Span/mm
5x5 3-Point	5	5	24	-
3x3 4-Point	3	3	24	8
4x4 4-Point	4	4	24	8
5x5 4-Point	5	5	24	8

(b) Fine grain alumina (ADS-80) : Wet specimens

Specimen Name	Width /mm	Height /mm	Lower Span /mm	Upper Span /mm
3x3 3-Point	3	3	24	-
4x4 3-Point	4	4	24	-
3x3 4-Point	3	3	24	8
4x4 4-Point	4	4	24	8

(c) Fine grain alumina (ADS-80) : Dry specimens

Specimen Name	Width /mm	Height /mm	Lower Span /mm	Upper Span /mm
3x3 3-Point	3	3	24	-
3x3 S4-Point	3	3	9	3
3x3 4-Point	3	3	24	8
4x4 4-Point	4	4	24	8

Bending tests

Specimen sizes used for bending tests are tabulated in Table 2. In coarse grain alumina (a), 3, 4 and 5 mm squared specimens were used for 4-point bending tests and 5 mm squared specimens were also used for 3-point bending tests. On the other hand, in fine grain alumina (b,c), 3 and 4 mm squared specimens were used, considering that the strength of fine grain alumina are generally lower than that of coarse grain alumina, as mentioned in following sections. All of tests were carried out using the Instron-type tensile testing machine in air (temperature : 20 ^{0}C, relative humidity : 60 %) with a bending test equipment, where crosshead speed was controlled so carefully that the strain rate was constant as 4×10^{-6}/s.

AE Measurement

The AE measuring system used in this study is shown in Fig. 1, schematically. Two piezo-electric elements were used as AE sensors and attached directly on the both ends of the specimen. In this study, AE source locations were calculated from the difference of arrival times between two sensors. Therefore, sensitivity of two sensors, especially those equilibrium, was calibrated carefully enough using a pencil lead breaking as a simulated source, before each testing.

Since it is well known that the AE activity of ceramic materials is low, the minimization of noise level of the system is indispensable for AE measurement of such materials [14]. In this study, noise-filter-transformers were used and the connections between sensors and pre-

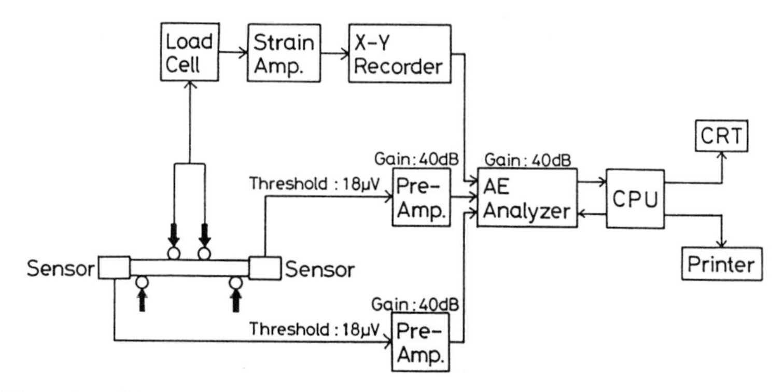

Fig. 1. Diagram of the AE measuring system. This system was developed especially for the low AE activity materials such as ceramics.

amplifiers were modified as same way as differential type transducers, in which the electromagnetic noise can be automatically canceled between the opposite phase signal cables. Consequently, the noise level at the input terminal of pre-amplifier was decreased to 14 μV, then threshold level was selected as 18 μV. AE signals were measured by the AE system (PAC:LOCAN) with load signals, sent to a personal computer (NEC:PC9801 LX4) through the RS-232C interface and analyzed with the computer.

RESULTS

AE Behavior

In this study, electromagnetic noise was minimized as mentioned above. On the other hand, mechanical noise was removed by teflon sheets between loading rods and a specimen, and then the negligibility was ascertained by the pre-loading using the specimen with twice height and same width. Furthermore, when AE generation pattern was analyzed, the AE source location was also used for the decrease of mechanical noise, especially for 4-point bending, i.e. the events located out of the upper span were neglected. Therefore it can be concluded that the sources of those AE events are microcrackings.

Figure 2 shows the result of 4-point bending test and typical AE generation pattern. In the figure, cumulative AE event counts are shown for the threshold level of 26 dB, 30 dB and 35 dB, respectively. It can be understood that the apparent bending stress, at which AE events increases rapidly, is independent of the threshold level. Therefore, those stress were determined from each specimen, as σ_C.

In conventional AE studies on the bending tests of ceramics [12,13], the beginning points of AE generation were focused. For an example, Gogotsi et al. [12] demonstrated that the stress at those point had the same distribution as bending strength. But it is clear in Fig. 2 that the stress at those points are influenced by threshold level in comparison with σ_C.

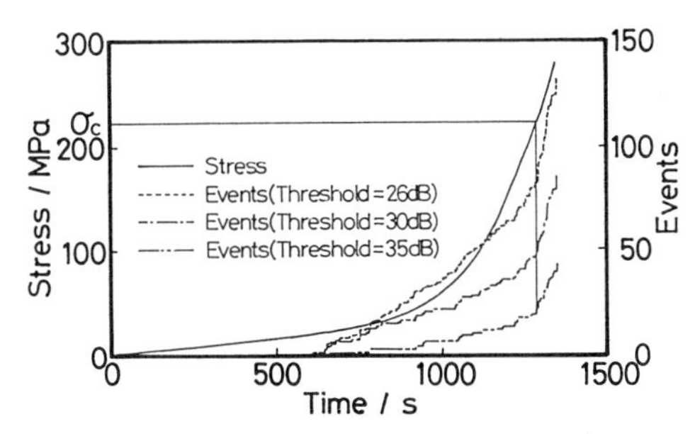

Fig. 2. AE generation pattern during 4-point bending test. A solid line is the stress-time curve. Cumulative AE events are drawn by 3 lines corresponding to respective AE threshold level as indicated in the figure.

AE Source Locations

In this figure, the AE source locations were determined by the product of the longitudinal wave velocity of the material (coarse grain alumina : 10,000 m/s, fine grain alumina : 9,000 m/s) and the differences of the arrival times between two transducers.

Figure 3 shows the relationship between the amplitude and the location obtained during the 4-point bending test of coarse grain alumina until the applied stress is lower than σ_C. It can be seen from the figure that several microcracks which generate the large amplitude AE (> 40 dB) are detected with in the range of upper span (± 4 mm). Therefore it is strongly suggested that several microcracks have been nucleate before the stress reaches σ_C.

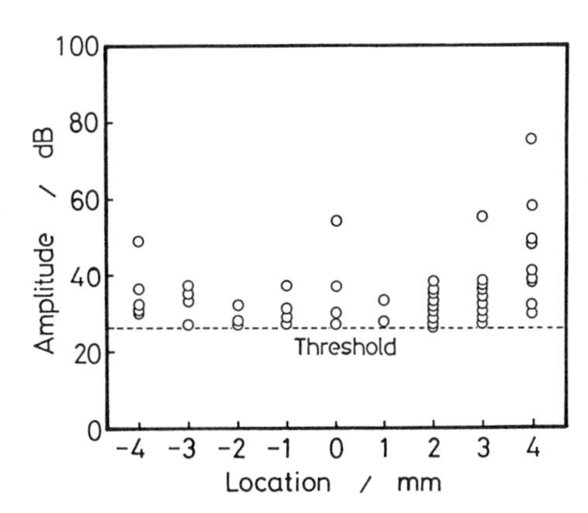

Fig. 3. Locations and amplitude of AE events detected before σ_C during the 4-point bending test of coarse grain alumina. The upper span of the specimen is 8 mm.

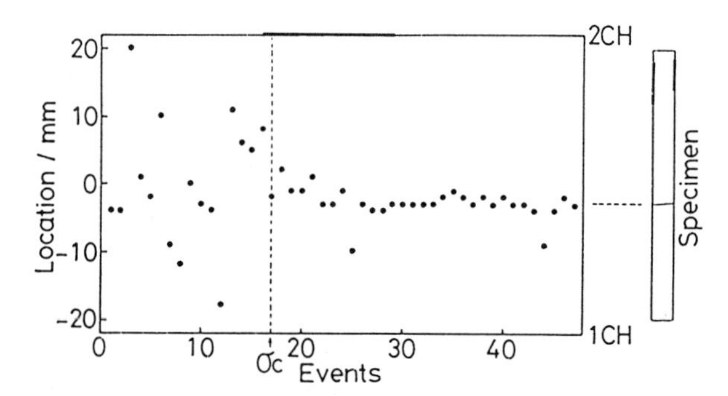

Fig. 4. Locations of microcracks detected during 4-point bending test of fine grain alumina. The horizontal axis indicate the order of the generation. 17 events were detected before σ_C.

The microcrack locations, which were nucleated during the 4-point bending test of fine grain alumina, are shown in Fig. 4. It can be seen from the figure that although microcracks distribute widely on the specimen surface before stress reached σ_C, microcrack locations concentrate within the narrow band after σ_C. It is important that the narrow band has excellent agreement with the origin of final fracture shown in the right part of the figure. It can be emphasized that the figure demonstrates the excellent ability of AE technique to monitor the time and location of the final failure.

Observation of Fracture Process

In order to understand the significance of σ_C, the fracture process on the surface under the tensile stress was observed using a fluorescent dye penetrant method according to the following sequences.

1. Specimens, covered with penetrant, were loaded to some predestinated value and unloaded immediately.

2. Penetrant was removed and the surface was observed using an optical microscope under an ultraviolet light.

3. Specimen was loaded to higher value than the former, then these sequences were repeated until the specimen fractured.

A similar technique was used for the eccentrically loaded column testing by Sines et al [16]. On the other hand, it was applied for the bending tests in this study, which is the most popular method to evaluate the strength of brittle materials.

More than 50 pictures were taken around the surface at each stage. After the failure, the maincrack, which caused the unstable fracture, was found and the fracture process was analyzed using the pictures of same positions. The results are shown in Fig. 5. Six photographs describe the same position as seen from the white spots which indicate the pores or microcracks. Within the stage from 198 MPa to 232 MPa, where the σ_C of this specimen is thought to exist, it is observed that the intergranular maincrack formation occurs due to the coalescence of pores and/or microcracks. Finally, the maincrack grows and becomes the origin of the final unstable fracture. Consequently, it can be concluded from the figure that σ_C corresponds to the macroscopic critical stress for maincrack formation.

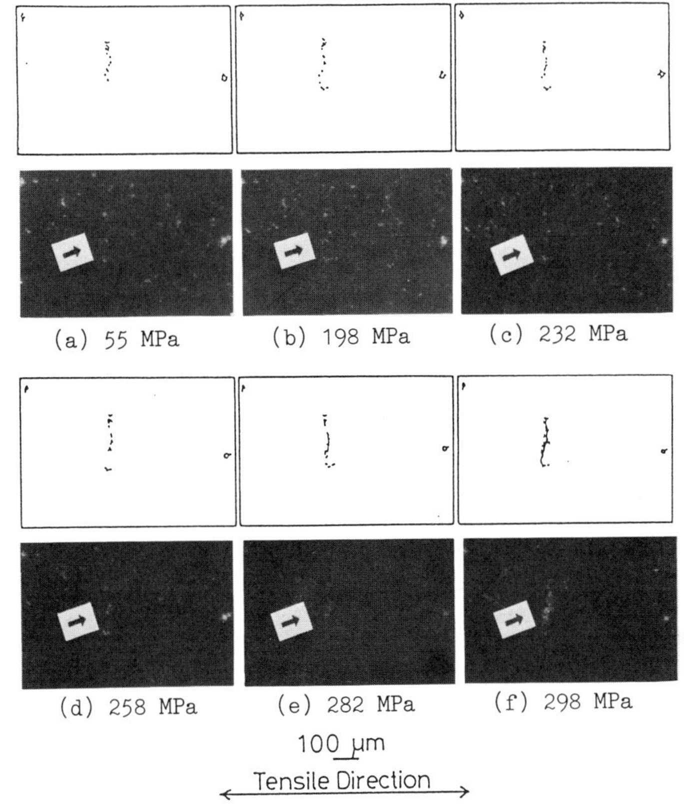

Fig. 5. Fluorescent dye penetrant observation of fracture process. Six pictures and corresponding sketches indicate the same position as seen from the white spots which are microcracks and pores. The main-crack formation is observed during the stage between (b) and (c) at the place indicated by arrows.

Bending Strength, σ_B

The strength of brittle materials are usually treated with Weibull statistics [8]. The Weibull plots of the bending strength obtained in this study is shown in Fig. 6. The strength of coarse grain alumina was higher than that of fine grain alumina. In coarse grain alumina (a), Weibull's coefficients are almost constant for 4-point bending, while it was twice for 3-point bending. In fine grain alumina (b,c), Weibull's coefficients are not changed by the loading configuration and the dehydration enhanced Weibull's coefficient but did not obviously vary the average strength.

In the case of bending tests, it is considered that the strength depends on the fracture process on the surface. Therefore, effective area S_E can be used as the measure of the specimen size, i.e.

$$\begin{aligned} S_E &= \int_S (\sigma/\sigma_R)^m \, dS \\ &= b[\, (L_1 - L_2) \,/\, (m+1) + L_2 \,] \end{aligned} \quad (1)$$

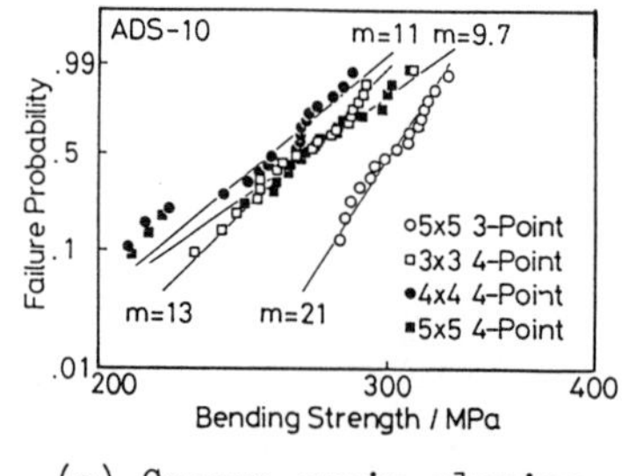

(a) Coarse grain alumina (ADS-10)

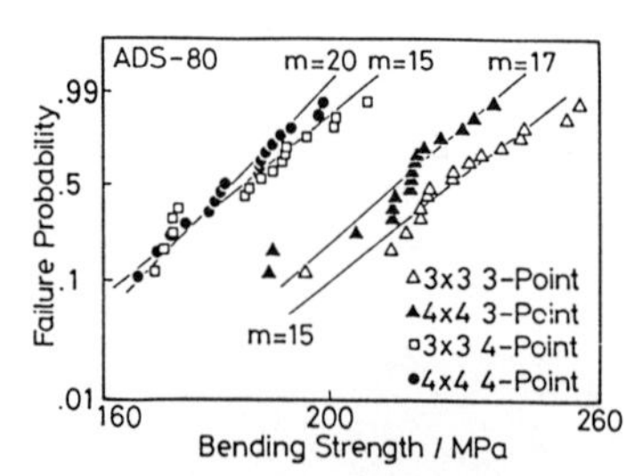

(b) Fine grain alumina (ADS-80) : Wet Specimens

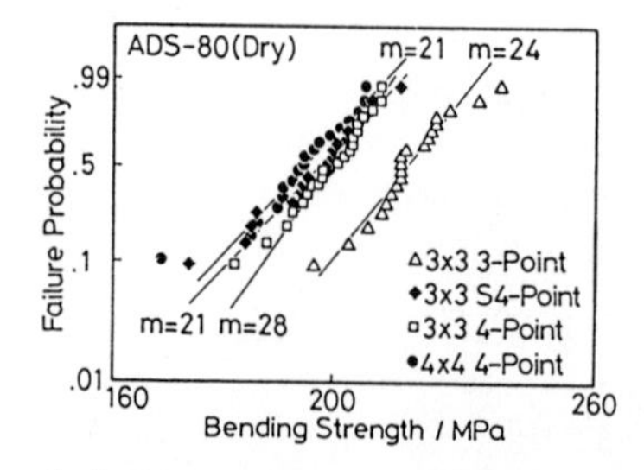

(c) Fine grain alumina (ADS-80) : Dry Specimens

Fig. 6. Weibull plots of the bending strength.

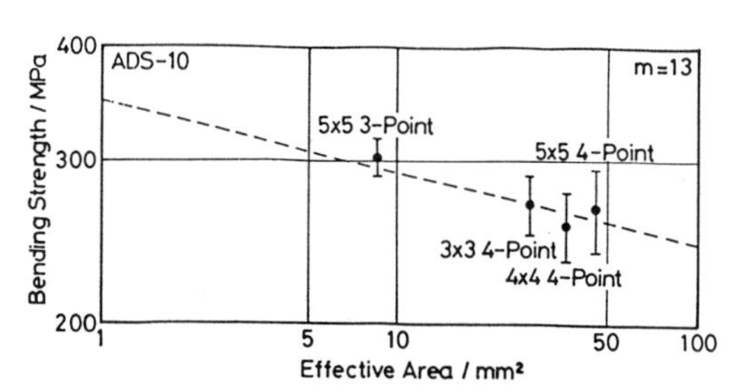

(a) Coarse grain alumina (ADS-10)

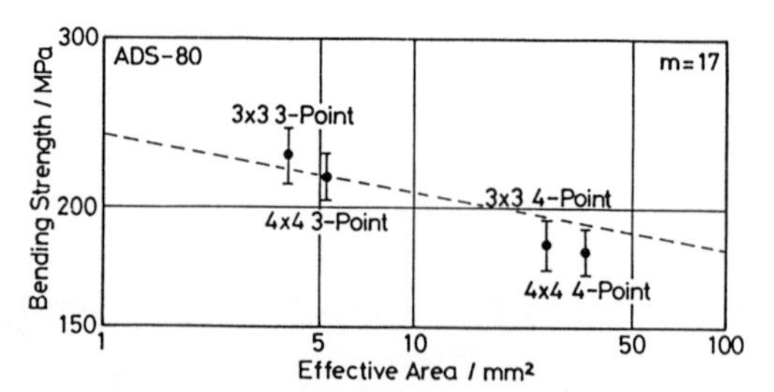

(b) Fine grain alumina (ADS-80) : Wet Specimens

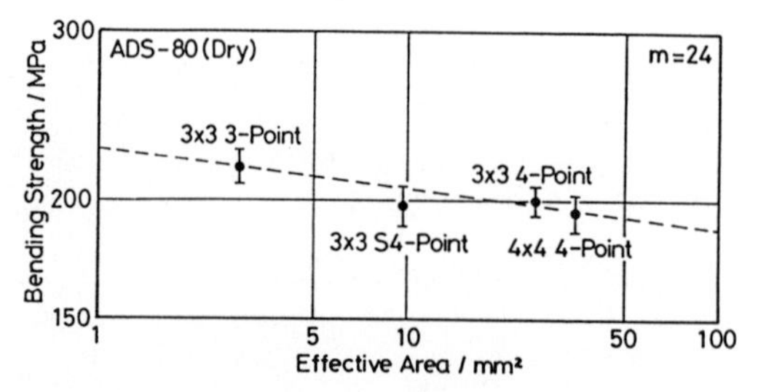

(c) Fine grain alumina (ADS-80) : Dry Specimens

Fig. 7. Size dependence of bending strength. The broken lines describe eq. (2).

where σ is the tensile stress on the surface area element dS, σ_R is the maximum stress, b is the width and L_1, L_2 are the lower, upper span, respectively. According to Weibull statistics [8], the mean strength, μ_1 and μ_2, obtained from specimens with two different sizes (S_{E1}, S_{E2}) are related as

$$\mu_2/\mu_1 = (S_{E1}/S_{E2})^{1/m} \tag{2}$$

The relationships between the bending strength and the specimen size are shown in Fig. 7. In the figure, the broken lines describe eq. (2) based on the data of (a) 3x3 4-point for coarse grain alumina, (b) 4x4 3-point for wet specimens of fine grain alumina and (c) 3x3 3-point for dry specimen of fine grain alumina.

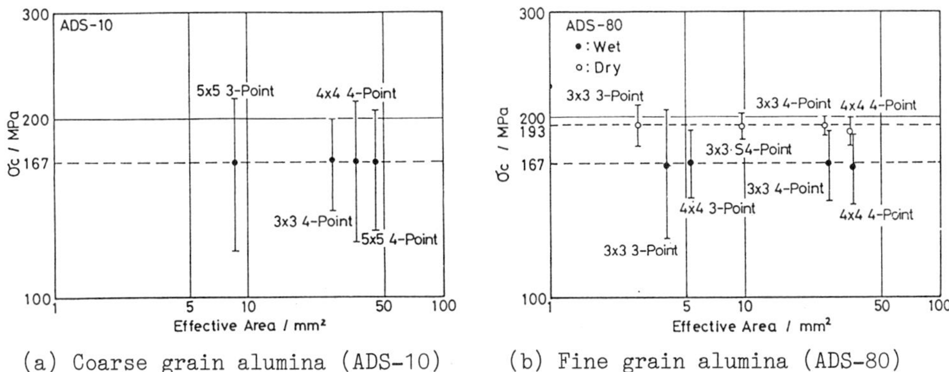

(a) Coarse grain alumina (ADS-10) (b) Fine grain alumina (ADS-80)

Fig. 8. Size dependence of σ_C. The averages, indicated by close circles, are independent of the specimen sizes in both materials.

Size and Strain Rate Dependence of σ_C

In this study, the value of σ_C was determined from the point at which both AE events and energy increase rapidly. In every case, the independence of AE threshold level was ascertained. When the AE increasing point was not clear, σ_C has not been determined. Consequently, the value of σ_C were obtained from 60 % of the whole specimens.

Figure 8 shows the size dependence of σ_C for (a) coarse and (b) fine grain alumina. Although it is well known that the bending strength, σ_B, of brittle materials depend on the specimen size strongly as shown in Fig. 7, it can be seen from the figures that σ_C distribute around the constant values independent of the specimen size for both materials, at least within the present study. It is interesting that the values of σ_C are same (167 MPa) for both coarse and fine grain materials. It can be understood from Fig. 8 (b) that the σ_C value is enhanced and the distributing range becomes small due to the dehydration of materials.

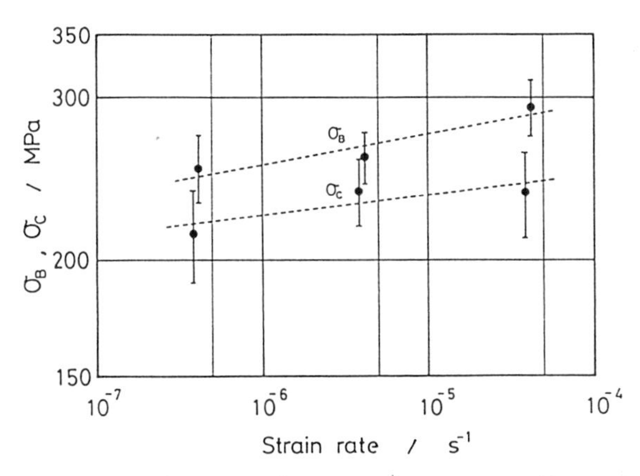

Fig. 9. Dependence of σ_B and σ_C on the strain rate. The specimens used here are coarse grain alumina with the dehydration.

On the other hand, the dependence of the bending strength, σ_B, and σ_C on the strain rate are investigated as shown in Fig. 9. The materials used were coarse grain alumina with dehydration (10^{-4} Torr, 150 ^{0}C, 60 min.). The temperature (20 ^{0}C) and relative humidity (60 %) during tests were controlled approximately constant. As the strain rate becomes larger, both σ_B and σ_C become higher as shown in the figure.

DISCUSSION

Microfracture Process During Bending Test

In this study, the nucleation times and locations of microcracks during bending tests of alumina were evaluated by acoustic emission technique. Then the microfracture process during bending test is discussed.

It has been made clear from Fig. 5 that the maincrack formation occurs when the stress is beyond σ_C. In this study, the locations of AE sources, i.e. microcracks were determined using the longitudinal wave velocity and the difference of the arrival times between 2 ch transducers, as shown in Figs. 3 and 4. Figure 3 describes that several microcracks has been nucleated before the applied stress reaches the σ_C. Furthermore, it can be seen from Fig. 4 that several AE events were detected before applied stress reached σ_C. Considering the shear and surface wave propagations, although some microcracks were detected out of the upper span apparently in Fig. 4, it is strongly suggested from the figures that microcracks have been nucleated before the stress reaches σ_C for both coarse and fine grain materials.

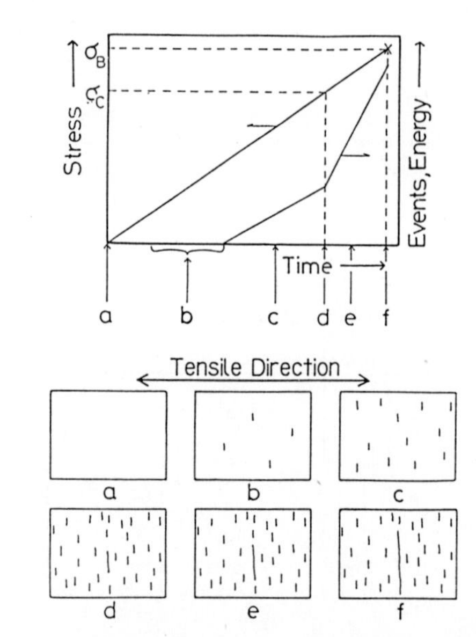

Fig. 10. Microfracture process during the bending test. Upper figure describes stress-time curve and AE generation pattern. Lower figures describe the microfracture process on the surface where tensile stress is applied.

From Figs. 3, 4 and 5, the microfracture process, during the bending test of alumina in this study, can be concluded as described schematically in Fig. 10.

1. The microcrack initiation has been occurred until the beginning point of AE detection at the latest (b).

2. Microcracks increase (c) and once the stress reaches σ_C, maincrack formation occurs due to the coalescence of microcracks and/or pores (d).

3. The maincrack grows (e) and finally causes the unstable fracture (f).

Statistical Treatment of Brittle Fracture

In Weibull statistics [8], only the uniaxial tensile stress field can be considered. Recently, Batdorf [9] and Lamon [10] developed the statistical treatments in which multiaxial stress field, including compressive and shear stress, can be considered. In such treatments, it is necessary to evaluate the critical stress for individual microcracking, at least their distribution. In this study, macroscopic stress at each microcracking were evaluated. Although further study is needed on the superposition of residual stress and the inclination of microcracks, it can be concluded that experimental technique indispensable for such approaches has been developed in this study.

Furthermore, the most of conventional statistical approaches neglect the fracture process such as the coalescence and growth of microcracks, as pointed out by Freudenthal. Actually, the maincrack formation was detected by acoustic emission technique in this study as shown in Fig. 10, then it can be understood that the fracture process before σ_C is statistical and that of after σ_C is fracture mechanical.

Since the criteria of crack coalescence has not been clear, it is difficult to establish the statistical treatment of the maincrack formation. However, σ_C can be treated on the way of the conventional statistical treatment as shown in Fig. 11, which describes the statistical distribution, i.e. Weibull plots. Figure 11 (a), in which the data are plotted for each specimen size individually, demonstrate that the data are included in an identified band. Therefore, all of the data were plotted together as show in Fig. 11 (b), in which the points are fitted by a solid line defined by the following equation.

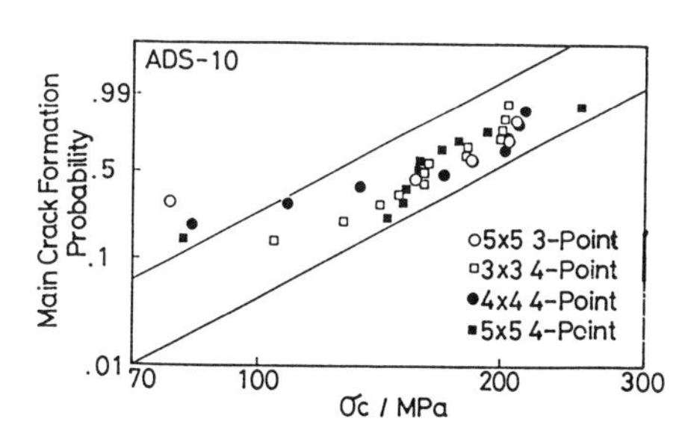

(a) Data from individual size specimens

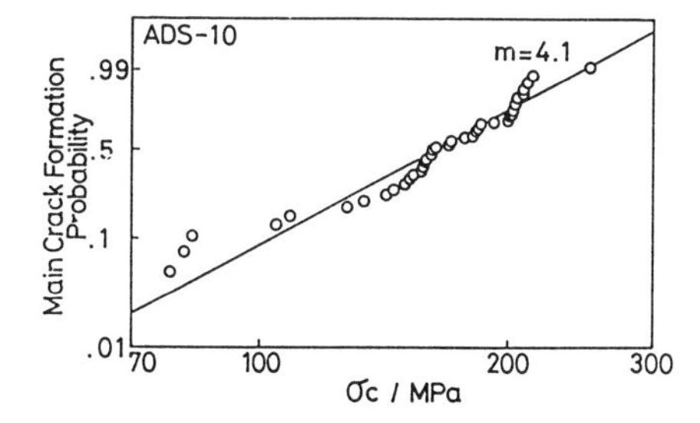

(b) Data from whole specimens

Fig. 11. Statistical distribution of σ_C. The symbols "m" indicate the shape parameter in eq. (3). The data are shown for (a) individual size specimens and (b) whole specimens.

$$P(\sigma_C) = 1 - \exp(-B\cdot\sigma_C{}^m) \quad (3)$$

where B (=0.84) and m (=4.1) are constants. Considering the similarity between Eq.(3) and Weibull's function, they can be concluded that B and m correspond to the scale and shape parameter, respectively, and that the σ_C, obtained from 4 different size specimens, distribute according to an identified distribution function.

On the other hand, it is considered that the final failure of the material corresponds to the catastrophically unstable growth of maincrack. A criterion of that can be represented as

$$K_I = Y\cdot\sigma\cdot\sqrt{a} > K_{IRf} \quad (4)$$

where σ, a and Y are apparent stress, crack length and the geometric parameter, respectively. K_{IRf} is the critical stress intensity at the unstable growth of a maincrack. From eq. (4), the strength, σ_B is written as

$$\sigma_B = K_{IRf}/(Y\cdot\sqrt{a_f}) \quad (5)$$

Consequently, it is understood that σ_B is determined from K_{IRf} and the crack length, a_f, at the final failure.

Knehans et al. [7] carried out the bending tests of alumina using the specimen with notches of various length and reported that those specimens fractured at the K_{IRf} and a_f of approximately constant values independent of the initial crack sizes. Therefore the size dependence of strength shown in Fig. 7 might derive from that of a_f. The fact that Weibull's coefficient of bending strength are 10 - 20 as shown in Fig. 6, and are much larger than that of σ_C (= 4.1), can be understood due to the difference of the scattering range between initial and final flows in the experiments of Knehans. The results by Knehans also explain that an obvious correlation between σ_C and σ_B, obtained from same specimen, has not been observed in this study.

Critical Stress for Maincrack Formation

Several theoretical models on the toughening mechanism have been proposed, e.g. a process zone [1,2], bridging [3] and others [4]. Most of those models include the critical stress concerning microcrackings. For an example, Evans et al. proposed the critical stress of microcrack saturation in a frontal process zone at the crack tip.

In this study, the macroscopic critical stress of microcracking, i.e. maincrack formation, σ_C, was evaluated. Since the maincrack formation in a bending specimen, i.e. microcrack coalescence, may be determined by same criteria as the crack growth in a process zone of the fracture toughness specimen, σ_C can be considered to control the size of process zone as the yield stress does the plastic zone size in metals. Therefore σ_C can be used as the measure of the pre-loading in a proof testing and also used for the design of ceramic structures. Consequently it can be concluded that σ_C is the advanced parameter for the materials evaluation in ceramic materials.

From the view point of the importance of σ_C in the engineering field, the investigation of the dependence on the strain rate and stress corrosion cracking is necessary. The strain rate dependency of σ_B and σ_C, shown in Fig. 9, are expected as the influence of the stress corrosion cracking due to the water. The Weibull plots of σ_C obtained from the wet

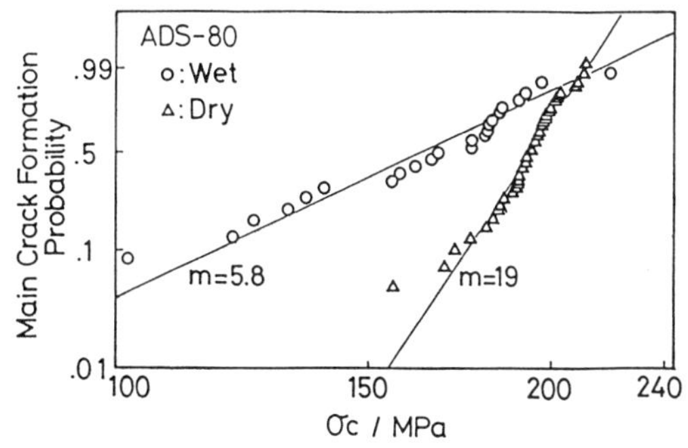

Fig. 12. Influence of the dehydration on the distribution of σ_C. The symbols "m" indicate the shape parameter in eq. (3). Only the lower values are changed by the dehydration.

and dry specimens of fine grain alumina are shown in Fig. 12. It has been made clear from Fig. 11 that the distribution function of σ_C is equivalent to Weibull function, therefore the slope of the fitting straight lines in Fig. 12 are considered as the shape parameters of Weibull function. The slope of σ_C of wet specimen is 5.8 and it becomes 19 due to the dehydration. On the other hand, the shape parameters of the bending strength, σ_B of wet specimen were 15 - 20 and those of dry specimen were 21 - 28, as shown in Fig. 6. Therefore it can be concluded that the influence of the stress corrosion cracking by water on σ_C are larger than that on σ_B. Furthermore, it has significant implications that only the lowest values of σ_C were influenced by the dehydration.

In this study, two kinds of alumina ceramics with different microstructures were used. It is interesting that the σ_C of both materials are the same value, as shown in Fig. 8. Although it is necessary to understand the influences of the microstructures, e.g. grain sizes, porosities and intergranular glass phase contents, on the σ_C, it can be concluded that the new characterization technique evaluating the advanced materials parameter, σ_C, has been established in this study.

CONCLUDING REMARKS

In this study, the nucleation times and location of microcracks during bending tests of two kinds of alumina were evaluated by acoustic emission technique. Specimens with various dimensions were used for bending tests, in order to investigate the dependence of microfracture process on the specimen size. Furthermore, the fracture process was observed using a fluorescent dye penetrant method. Consequently, following conclusions were obtained.

1. The remarkable point, at which both AE events and energy increase rapidly before the final unstable fracture, was observed. The apparent bending stress at the point, σ_C, was independent of AE threshold level and specimen size.

2. From the result of the fluorescent dye penetrant observation, it was understood that the formation of a maincrack due to the coalescence of microcracks and/or pores occurs at σ_C.

3. The microfracture process was made clear, i.e. the nucleation of microcracks, the formation and stable growth of a maincrack and the final unstable fracture occur respectively, although the weakest link assumption is included in conventional statistical treatments such as Weibull statistics.

4. The σ_C corresponds to the critical stress for maincrack formation due to the coalescence of microcracks and/or pores. The σ_C obtained from different size specimens distribute according to an identified function which is similar to Weibull function. Additionally, the influences of strain rate and dehydration on σ_C were investigated.

Finally, it is concluded that σ_C can be the new materials evaluation parameter in ceramics, which is significant equivalently to the yield stress in metals.

REFERENCES

1. R. G. Hoagland and J. D. Embury, J. Am. Ceram. Soc., 63:404(1980).
2. A. G. Evans and K. T. Faber, J. Am. Ceram. Soc., 67:255(1984).
3. Y. W. May and B. R. Lawn, J. Am. Ceram. Soc., 70:289(1987).
4. K. T. Faber and A. G. Evans, Acta Metall., 31:565(1983).
5. N. Claussen, J. Steeb and R. F. Pabst, Am. Ceram. Soc. Bul., 56:559 (1977).
6. H. Hubner and W. Jillek, J. Mat. Sci., 12:117(1977).
7. R. Knehans and R. Steinbrech, J. Mat. Sci. Lett., 1:327(1982).
8. W. A. Weibull, J. Appl. Mech., 18:293(1951).
9. S. B. Batdorf and H. L. Heinisch,JR., J. Am. Ceram. Soc., 61:355 (1978).
10. J. Lamon, J. Am. Ceram. Soc., 71:106(1988).
11. A. M. Freudenthal, Fracture, Vol. 2, H. Liebowits, ed. 592(1968).
12. G. A. Gogotsi, A. V. Drozdov and A. N. Negovskii, Proc. Ultrason. Int., 83:67(1983).
13. A. Katagiri, T. Nishiyama, T. Fukuhara and Y. Nozue, Proc. National Conf. on AE, JSNDI, Tokyo, 110(1983).
14. T. Kishi, S. Wakayama and S. Kohara, Fracture Mechanics of Ceramics, Vol. 8, R. C. Bradt, A. G. Evans, D. P. H. Hasselman and F. F. Lange ed., Plenum Press, New York, 85(1985).
15. S. Wakayama, T. Kishi and S. Kohara, Progress in AE III, K. Yamaguchi, K. Aoki and T. Kishi ed., JSNDI, Tokyo, 653(1986).
16. G. Sines and T. Okada, J. Am. Ceram. Soc., 66:228(1983).

ANALYSIS OF PRECRACKING PARAMETERS FOR CERAMIC

SINGLE-EDGE-PRECRACKED-BEAM SPECIMENS

Sung R. Choi* and Abhisak Chulya*
Cleveland State University, Cleveland, OH 44115, USA

Jonathan A. Salem
NASA Lewis Research Center, Cleveland, OH 44135, USA

ABSTRACT

The Single-Edge-Precracked-Beam (SEPB) method involves creation of a straight-through crack from an indentation crack. The straight-through crack is developed by applying a controlled bending load to a specimen via a precracking fixture. The fixture induces a sequence of stable growth of the initial indentation crack, pop-in, and then arrest, thereby forming the straight-through precrack. The effects of indentation load on precracking load as well as on precrack size were studied for experimental variables such as specimen width, precracker span and test material. The finite element method was used to obtain the stress distribution and stress intensity factor, thus providing a quantitative prediction of the experimental results obtained for silicon nitride, alumina, silicon carbide, and two SiC whisker-reinforced silicon nitrides.

I. INTRODUCTION

The Single-Edge-Precracked-Beam (SEPB) method, first applied to ceramics by Nose and Fujii,[1] involves creation of a straight-through crack in a ceramic beam specimen for evaluation of fracture toughness. In the SEPB method, originally termed the 'bridge'-indentation method,[2] a straight-through crack is developed from a Vickers indentation flaw placed in the center of the prospective tensile surface. The crack develops from a controlled bending load via a precracking fixture. The fixture induces a sequence of stable crack growth of the initial indent crack, pop-in, and then arrest, thereby producing a straight-through precrack.

The effects of precracking parameters such as indentation load, precracking load and precrack size have been studied experimentally and/or analytically by Warren and Johanesson,[2] and Bar-On et al.[3] However, their studies did not include in-depth analyses regarding those parameters on the basis of appropriate mechanics (eg. 'indentation' and conventional fracture mechanics principles).

In this work, we present the effects of indentation load on precracking load as well as on precrack size with respect to fixture and specimen

* NASA Resident Research Associates at Lewis Research Center, Cleveland, OH 44135, USA.

configurations. Strain gaging, finite element analysis (FEA), and indentation strength data were incorporated to obtain the stress distribution and stress intensity factor, thereby obtaining analytical solutions of the precracking parameters as a function of indentation load. The analytical results were compared to the experimental results obtained for silicon nitride, alumina, silicon carbide, and two SiC whisker-reinforced silicon nitrides.

II. EXPERIMENTAL PROCEDURES

The materials used in this study were Si_3N_4 with an equiaxed microstructure (Norton), Al_2O_3, siliconized SiC (Norton), and two SiC whisker-reinforced (30 Vol %) Si_3N_4's (Norton and Garrett). The physical properties of the test materials are summarized in Table 1. All the specimens were in the form of MOR bars with 4-6 mm width, 3 mm thickness, and 25-50 mm length. The bar specimens were finished with 320 grit diamond wheel, and the prospective tensile surface of each specimen where the indent was to be placed was hand-polished with 600 grit SiC paper.

Each specimen was indented with a Vickers microhardness indenter in the center of the polished surface with the indentation diagonals parallel and perpendicular to the prospective tensile stress direction. The indented specimen was then placed in a precracking fixture such that the indent site was located at the center of the lower fixture span (S) (see Fig. 1). The specimen was loaded gradually in the fixture to produce a controlled tensile stress induced by a compressive load via an Instron servohydraulic testing machine with a crosshead speed of 120 μm/min. Precracking was detected by an acoustic emission (AE) probe in conjunction with a CRT monitor, and the corresponding precracking load was subsequently monitored using the testing machine console. A wide range of indentation loads from P = 30 N to 440 N was used with experimental variables of span (S = 3-6 mm) and specimen width (W = 4-6 mm). The outer span as well as the contact length between the upper Si_3N_4 plate and the specimen were fixed to be L = 18 mm. The lengths of test specimens were typically 25 mm; however, it was found that specimen lengths between 18 mm and 50 mm did not have significantly different stress distributions. The precracked specimens were fractured using a four-point bend fixture to determine their precrack sizes as well as fracture toughness.

Table 1. Physical Properties of Test Materials

Material	Young Modulus* E (GPa)	Hardness# H(GPa)	$K_{IC}$$ (MPa$\sqrt{m}$)
Norton Si_3N_4	295	16.7	3.50(0.25)^
Norton SiC_w/Si_3N_4	305	19.4	4.62(0.42)
Garrett SiC_w/Si_3N_4	330	16.0	5.18(0.41)
96% Al_2O_3	324	10.0	3.09(0.17)
NC 433 SiC	311	23.5	-

* By strain gage.
\# By Vickers microhardness indenter.
$ By SEPB from this study.
^ The numbers in the parentheses indicate ±1.0 S.D.

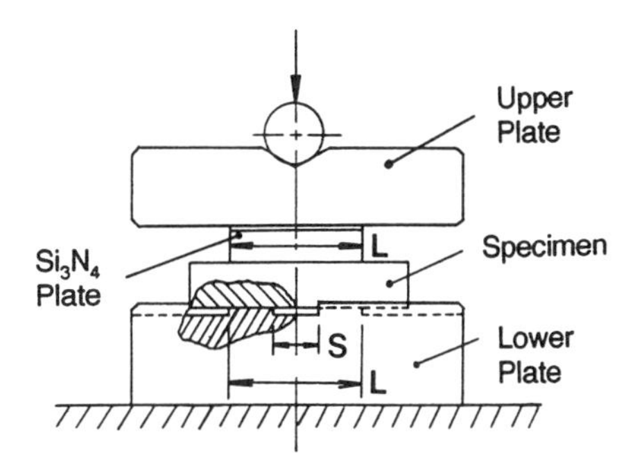

Figure 1. SEPB precracking fixture

The indentation strength data for each test material were determined using an indentation load range of P = 50 to 300 N to provide an additional information regarding stress distributions developed in the specimen with the aid of appropriate analysis. Three specimens were used for each indent load, which was considered sufficient in view of the very low standard deviation of indent strength (< 5%).

Finally, strain gages were utilized for some of the specimen and fixture configurations to estimate tensile stress as a function of applied compressive load and to compare with FEA and indent strength results.

III. ANALYSES

Indentation Fracture

For indentation cracks produced in ceramics and glasses by a Vickers indenter, subjected to an applied stress, the net stress intensity factor (K) representing a half-penny crack configuration consists of the two parts[5]

$$
\begin{aligned}
K &= K_r + K_a \\
&= \Sigma_r P/a^{3/2} + \Omega\sigma_a\sqrt{a} \qquad (1)
\end{aligned}
$$

where K_r is the Mode I residual stress intensity factor due to residual stresses resulting from the elastic/plastic mismatches of indentation. The residual stress is represented by a point force acting on the center of the half-penny crack plane. K_a is the Mode I applied stress intensity factor, Σ_r is the material/indenter geometry constant associated with residual contact stress, P the indentation load, 'a' the crack size, Ω the crack geometry factor, and σ_a the applied stress. The functional dependency of K to 'a' indicates that stable crack growth with conditions of $K \geq K_{IC}$ (K_{IC}, fracture toughness) and $dK/da < 0$ precedes during loading until the instability point where $dK/da = 0$ and $K = K_{IC}$ are fulfilled. Using this instability condition, one can obtain[5]

$$\sigma_f = 3K_{IC}/(4\Omega\sqrt{a_f}) \qquad (2a)$$

$$a_f = [4\Sigma_r P/K_{IC}]^{2/3} \qquad (2b)$$

where σ_f and a_f are, respectively, the fracture strength and the critical crack size at instability. The stable crack growth occurs from the as-indented initial crack size (a_i) to the final crack size (a_f). Also, from Eq. (2), the fracture strength as a function of indentation load is determined to be

$$\sigma_f = [3K_{IC}^{4/3}/(4^{4/3}\Omega\Sigma_r^{1/3})]\ P^{-1/3} \quad (3)$$

When the indented specimen is gradually loaded in the precracking fixture, the relationship between the compressive load (F_a) applied to the fixture via the testing machine and the maximum applied tensile stress induced in the specimen is elastic. Hence,

$$\sigma_a = \alpha\ F_a \quad (4)$$

with

$$\alpha = f(S, L, W, t, E, \nu)$$

where α is a proportionality constant that depends on inner span (S) and outer span (L) of the fixture, specimen width (W) and thickness (t), and Young's modulus (E) and Poisson's ratio (ν) of the specimen. With increasing applied load, the indent crack is subjected to continuous stable crack growth until the instability point (σ_f and a_f), where the stably grown half-penny shaped indent crack pops-in to form a straight-through crack. Therefore, precracking occurs when

$$F_a = F_p \quad \text{with} \quad \sigma_a = \sigma_f \quad (5)$$

where F_p is defined as the precracking load.

Now using Eqs. (3) through (5), the precracking load, F_p, is obtained

$$F_p = \frac{1}{\alpha}\sigma_f \quad (6a)$$

$$F_p = \frac{1}{\alpha(S, L, W, t, E, \nu)}\left[\frac{3K_{IC}^{4/3}}{4^{4/3}\Omega\Sigma_r^{1/3}}\right]P^{-1/3} \quad (6b)$$

The residual contact stress constant Σ_r is empirically expressed[6]

$$\Sigma_r = \phi\ (E/H)^{1/2} \quad (7)$$

where H is the material hardness, and ϕ is the calibration constant ($\phi \approx 0.016$ for the Vickers indenter). Substituting Eq. (7) to Eq. (6) yields

$$F_p = \frac{1}{\alpha(S, L, W, t, E, \nu)}\left[\frac{\Phi K_{IC}^{4/3}H^{1/6}}{\Omega E^{1/6}}\right]P^{-1/3} \quad (8)$$

where $\Phi = 3/(4^{4/3}\phi^{1/3})$. This equation shows that for a given indentation load and a given fixture configuration, the precracking load is a strong function of fracture toughness, but a weak function of hardness-to-Young's modulus ratio (H/E). Also, note that E and H do not differ significantly in advanced ceramic materials. For the given material and fixture configurations, a slope of -1/3 should yield in the plot of Log F_p versus Log P, based on Eq. (8).

The proportionality constant, α, can be evaluated by strain gaging, finite element analysis, or indentation strength measurements. In the indentation strength method, indentation strengths are determined as a function of indentation load, and then this indentation strength-indentation load data is combined with the precracking load-indentation load data, thereby yielding the constant α, based on Eqs. (3) and (6).

Finite Element Analysis

The finite element analysis was used to evaluate the proportionality constant (α) as a function of span (S), specimen width (W) and Young's modulus (E). The analysis was also used to estimate precrack sizes at precracking events based on stress intensity factors as a function of straight-through crack size with different levels of precracking load. The mesh used in this study is presented in Fig. 2. The following assumptions were made regarding boundary conditions : (1) two dimensional plane strain condition; (2) the upper fixture made of Si_3N_4 is rigid and provides the uniform compressive load at the upper surface of the specimen; (3) the lower fixture made of steel (hardened tool steel) is not rigid but deformable; (4) no friction exists in contact surfaces between the specimen and upper or lower fixture. The third assumption was found to be very important and realistic since a rigid body assumption resulted in erroneous stress distributions. The fixed values of L = 18 mm and t = 3 mm were used throughout the analysis.

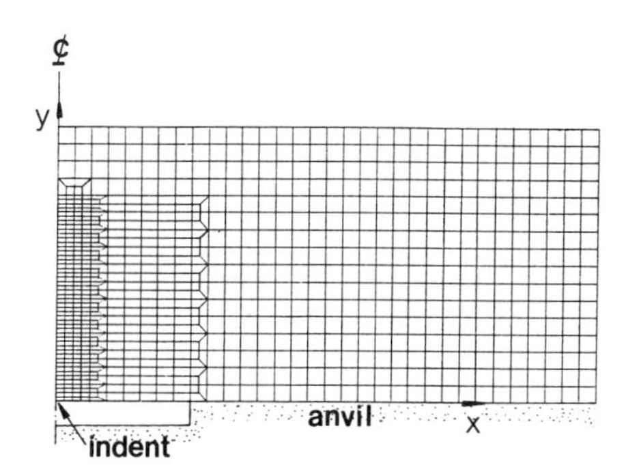

Figure 2. Finite element mesh (half-plane). Compressive load is applied from the top surface of a specimen.

Typical stress contours are shown in Fig. 3, where σ_{xx} is plotted for S = 4 mm and W = 4 mm for the Si_3N_4 material. This result shows that the stress distribution through the beam width (y axis) is not symmetric with respect to the central axis of the specimen, with the neutral axis shifted towards the y axis. Although not presented here, shear stress σ_{xy} along the y axis was found to be negligibly small compared to σ_{xx}. Note that the compressive stress, σ_{yy}, although present in the material body, was assumed to have no influence on the Mode I stress intensity factor since σ_{yy} acts in the direction parallel to the crack (indent or precrack) planes.

A summary of the evaluated α as a function of S and W is presented in Fig. 4. Here, α was obtained from Eq. (4) by evaluating σ_{xx}(max) as a function of F_a with a unit of [MPa/KN]. This figure shows that α increases with decreasing W and increasing S. Note that the simple beam theory is approximately applicable to σ_{xx}-S relation, but not applicable to σ_{xx}-W relation (note in simple beam theory: $\sigma_{xx} \alpha S/W^2$). The effect of E on α in a range of E = 290 to 330 GPa was found to be insignificant.

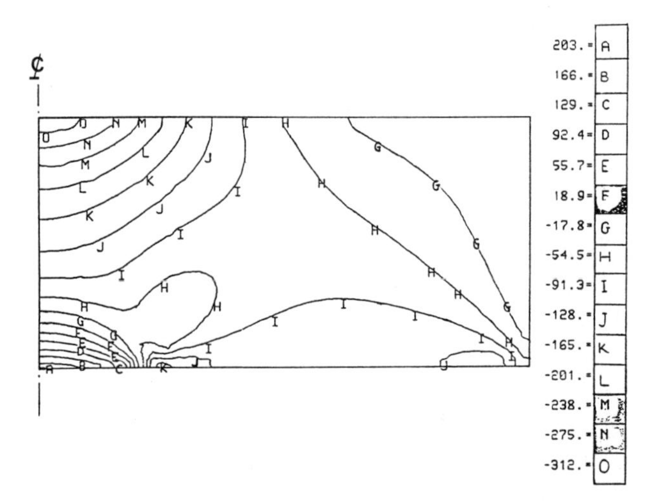

Figure 3. Contours of σ_{xx} (fiber stress) component (stress unit in MPa).

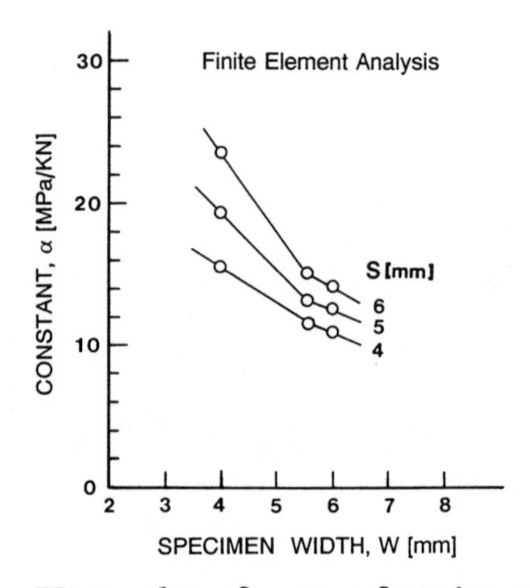

Figure 4. FEA results of α as a function of specimen width (W) for different fixture spans (S).

Figure 5 is a typical result of stress intensity factor versus straight-through crack size for different levels of precracking load (for S = W = 6 mm). The stress intensity factors were obtained by varying crack size for each level of precracking load. Also illustrated in the figure is the determination of a precrack size as a function of precracking load for a given K_{IC}. Here, the influence of residual contact stress due to indentation (that is, K_r field) was neglected, since precrack size is in general more than 10 times indentation crack size. This quasi-static stress intensity factor is utilized in order to predict a (popped-in) precrack size for a given indentation load. It should be noted here that the difference between K_{IC} and 'K_{Ia}' for dynamic crack arrest was assumed to be negligibly small. Figure 6, reconstructed from the results of Fig. 5, summarizes F_p versus precrack size with different levels of fracture toughness. This result indicates that precrack size increases with decreasing K_{IC} for a given precracking load. However, unlike the F_p relation in Eq. (8), there exists no closed form solution between precrack size and F_p.

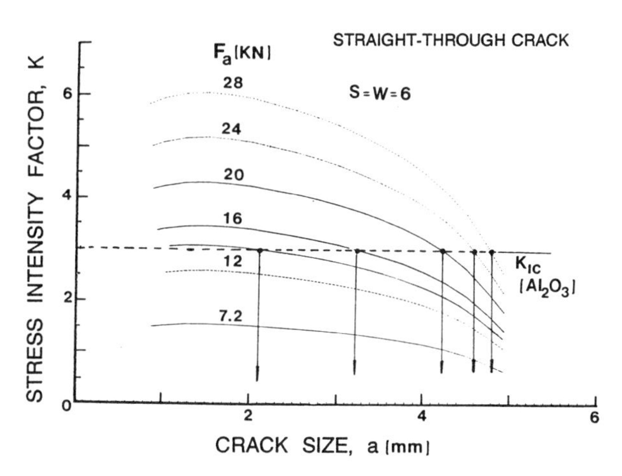

Figure 5. FEA results of stress intensity factor as a function of precrack size for different precracking loads (S = W = 6 mm).

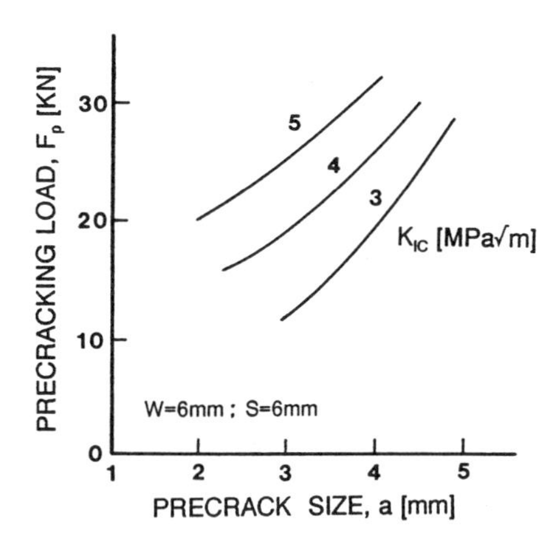

Figure 6. Precracking load versus precrack size for different K_{IC}'s, reconstructed from Figure 5.

Sequence of Precracking

The sequence of crack pop-in to form a straight-through crack now can be illustrated based on Eq. (1) and FEA results (Fig. 5), and is depicted in Fig. 7, where net stress intensity factor (K) is plotted as a function of crack size (a) with increasing applied stress. As the applied compressive load, or the applied tensile stress in the specimen increases, K of an indentation crack increases, resulting in stable crack growth, Δa_i, with conditions of $K = K_{Ic}$ and $dK/da < 0$ at $a = a_i + \Delta a_i$. With further increasing load, the crack continues to grow stably until the instability condition of $K = K_{IC}$ and $dK/da = 0$, where the indent crack pops-in to form a straight-through precrack with a size of a_{th}. If further subjected to increasing applied stress, the precrack grows stably to a crack size of $a = a_{th} + \Delta a_{th}$, since K decreases with increasing crack size, which eventually gives rise to a crack arrest condition of $K = K_{IC}$ and $dK/da < 0$.

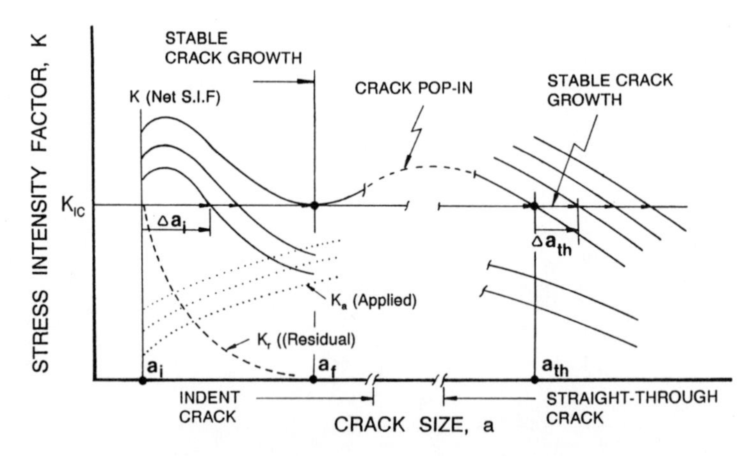

Figure 7. Net stress intensity factor as a function of crack size with different levels of applied stress.

IV. EXPERIMENTAL RESULTS AND DISCUSSION

Sequence of Precracking

Figure 8 is the results of crack size measurements as a function of applied compressive load (F_a) for the two types of indent flaws: as-indented, and indented and annealed in 1250°C N_2 gas to remove residual stress. For the as-indented flaws, the indent crack normal to the applied stress grows stably until F_p (= 11.9 KN), where the stably grown indent crack pops-in due to instability condition to form a straight-through crack with a size of a_{th} = 2 mm. With further increasing load, the straight-through crack precedes to grow stably. On the other hand, the annealed indent crack does not exhibit any stable crack growth due to the annealing out of residual stress; the annealed crack remains intact until instability at F_p (= 13.6 KN), where the intact indent crack pops-in to form a straight-through crack with a size of a_{th} = 2.5 mm. Note that both F_p and a_{th} are greater in the annealed indent flaw than in the as-indented flaw, indicative of the residual stress effect, as reflected in Eq. (1). This pre-cracking sequence for the indented flaw is in excellent agreement with the analytical results shown in Fig. 7.

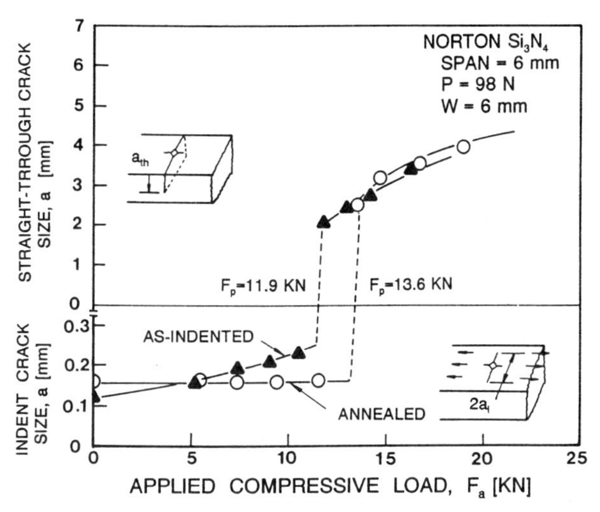

Figure 8. Experimental results of crack size measurements as a function of applied compressive load for two different types of indent flaws.

Configuration of Precrack

In order to ensure good precrack generation, it is essential that all loaded surfaces of specimen and fixtures are as plane parallel as possible. Likewise, the indentation, placed in the specimen center, should be aligned in the center of the fixture span if symmetrical and straight cracks are desired, as previously reported.[2] The acceptability of a precrack was based on the requirements specified in a standard test method.[4] It was found that good precracks could also be achieved by placing multiple indent cracks across the direction of specimen thickness. This is particularly useful for some ceramic materials such as SiC and Al_2O_3 with relatively high porosities that generally inhibit well-defined indent crack patterns due to chipping and crushing. The effects of number of indents on precrack size (a/W) and precracking load are shown in Fig. 9 for the SiC whisker-reinforced Si_3N_4 material. Note that a/W and F_p are almost independent of the number of indents (up to three) because interaction between the adjacent indent cracks is negligible; whereas, for the increased number of indents (up to seven), the effects were rapidly amplified due to pronounced crack interaction with decreased crack spacings.

Precracking Load versus Indentation Load

Figure 10 presents a summary of the experimental data on precracking load (Log F_p) as a function of indentation load (Log P) for all the test materials (S = W = 6 mm). This figure shows that as the indentation load increases the corresponding precracking load decreases. The solid lines in the figure represent the best-fit lines with slopes of -1/3 based on Eq. (8), indicating that data fit to the theoretical equation (Eq. (8)) is excellent. However, the SiC material does not seem to follow the theoretical curve due to an ill-defined indentation crack pattern produced on the specimen surface by chipping and crushing. It should be noted that the higher indent load produced the more ill-defined indent cracks, thereby enhancing the poor correlation between F_p and P.

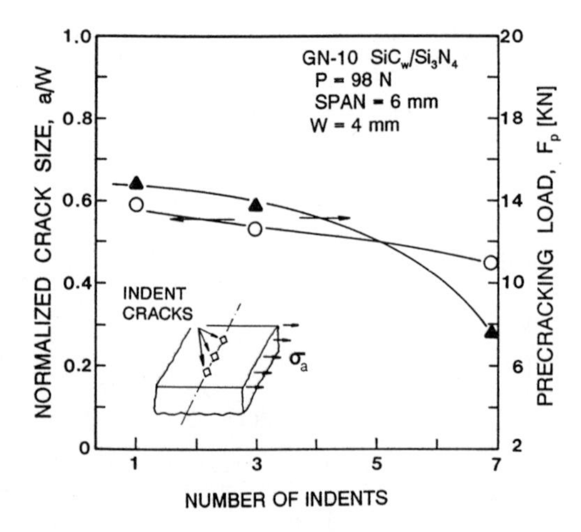

Figure 9. Effects of number of indents on precracking load and precrack size.

A typical result of α evaluated by strain gaging is presented in Fig. 11, where the maximum tensile stress (or strain) occurring in the specimen surface was plotted against applied compressive load (F_a), indicating that linear elasticity holds between F_a and σ_{max}. From this data, the slope was determined to be α = 13.2 ± 0.1 (MPa/KN) based on the linear regression analysis. This figure is for S = W = 6 mm in GN-10 SiC_w/Si_3N_4.

Results of indentation strength versus indentation load for each test material are summarized in Fig. 12. Similar to the F_p-P relation (Fig. 12), indentation strengths decrease with increasing indentation load with a slope of -1/3 in Log σ_f versus Log P plots (Eq. (3)). Now using this σ_f-P data in conjunction with the F_p-P data (Fig. 12), the constant α was estimated based on Eqs. (3) and (6), and is presented in Table 2 for all the test materials. Regardless of the test materials, α is almost constant with an average value of α = 13.3 ± 0.85 (MPa/KN). The reason for this is due to the fact that the major material property, E, does not differ significantly among the test materials, as seen in Table 1.

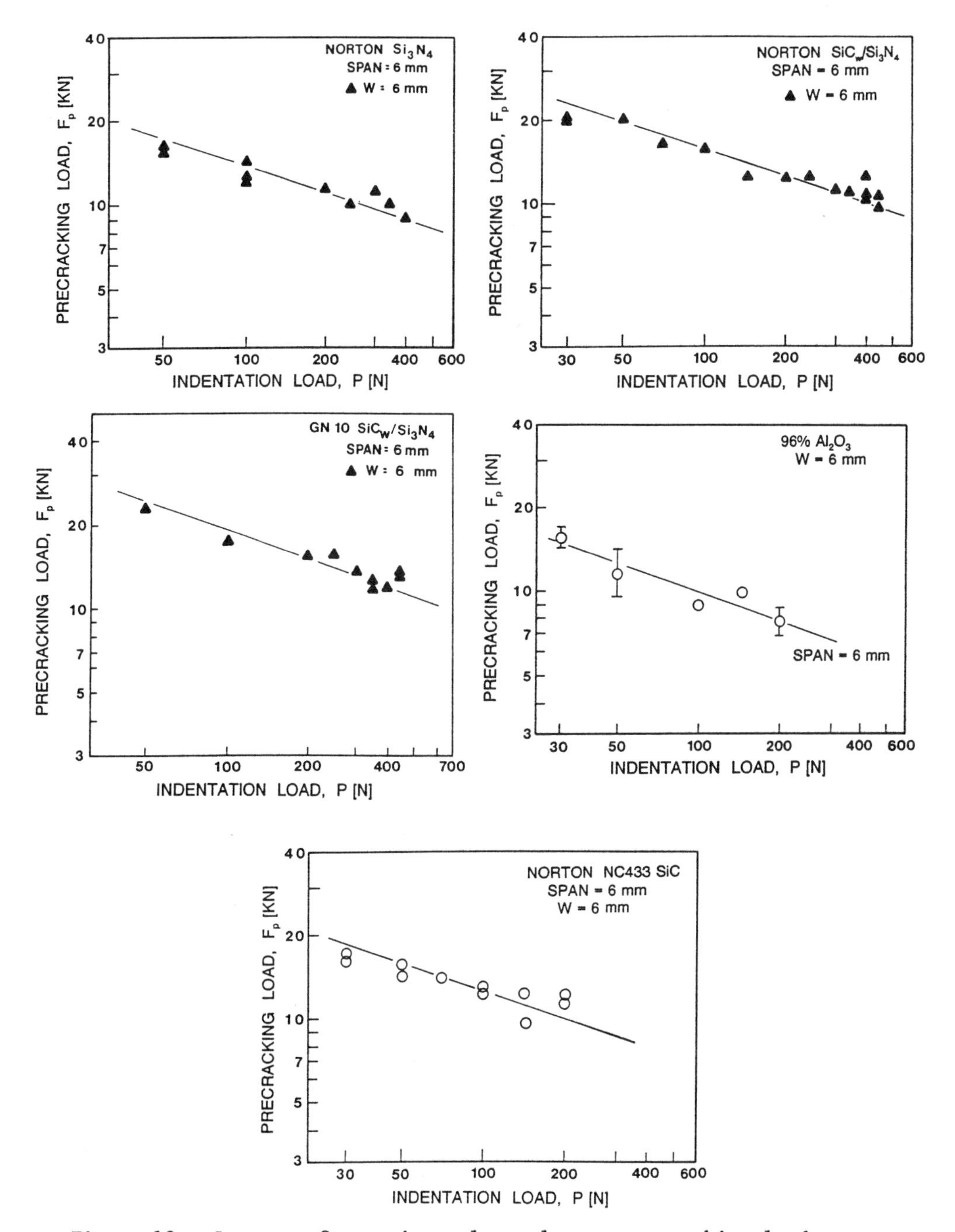

Figure 10. Summary of experimental results on precracking load versus indentation load for five test materials.

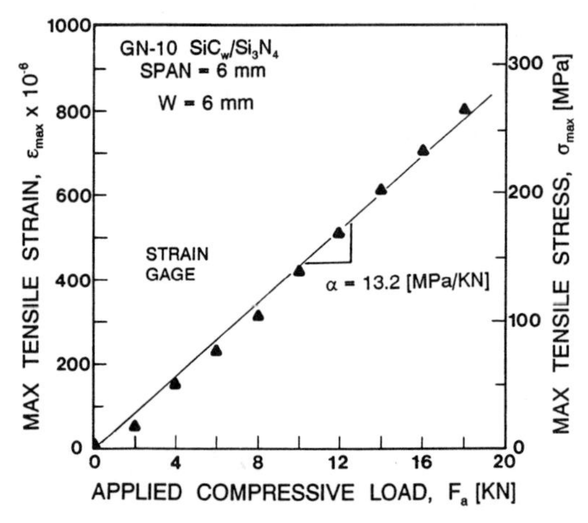

Figure 11. Maximum applied stress as a function of applied compressive load obtained with strain gaging. α represents a slope of σ_{max} vs. F_a curve.

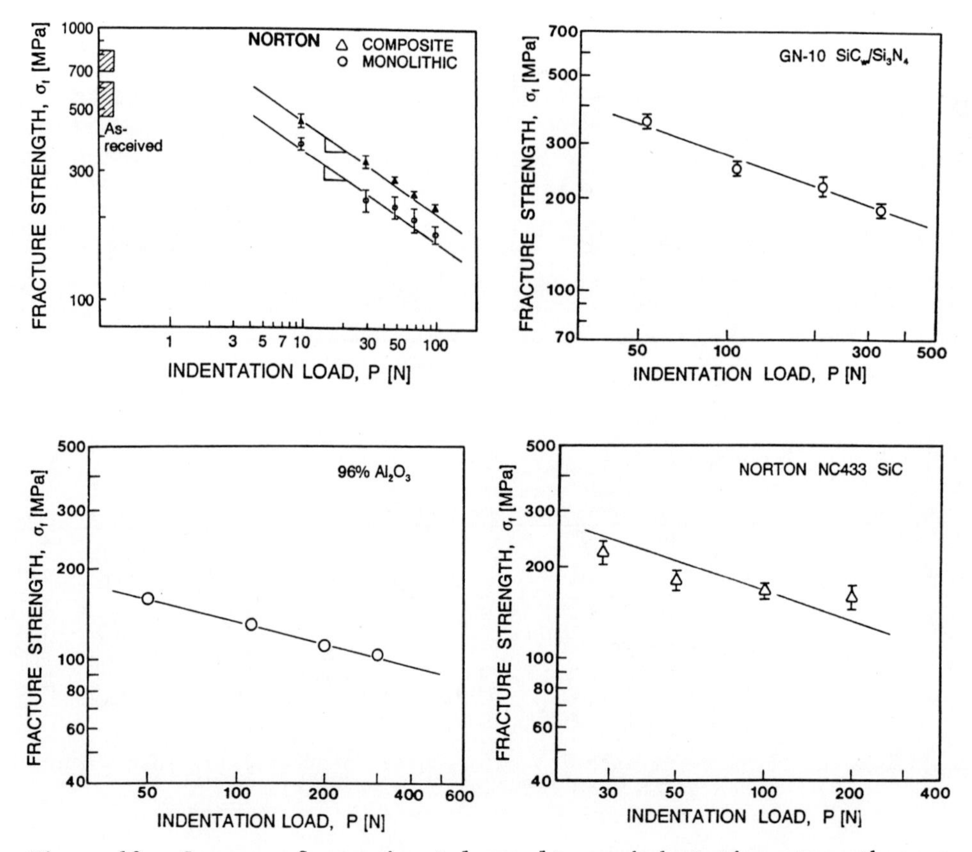

Figure 12. Summary of experimental results on indentation strength as a function of indent load for five test materials.

Table 2. Evaluation of α from Indentation Data (S = W = 6mm)

Material	(Precracking) $F_pP^{1/3}$ [KN.N$^{1/3}$]	(Strength) $\sigma_fP^{1/3}$ [MPa.N$^{1/3}$]	α [MPa/KN]
Norton Si_3N_4	65.7	787.7	12.01
Norton SiC_w/Si_3N_4	73.7	989.7	13.43
Garrett SiC_w/Si_3N_4	89.2	1277.3	14.32
96% Al_2O_3	46.4	632.3	13.62
NC 433 SiC	58.5	764.4	13.07
		Average α =	13.3±0.85

* $\alpha = (\sigma_fP^{1/3})/(F_pP^{1/3})$

Figure 13 summarizes the constant α evaluated by strain gaging, FEM and indent strength method for different S and W. Here, L = 18 mm and t = 3 mm. It can be seen here that the finite element solution agrees fairly well with experimentally evaluated data. Therefore, it can be concluded that Eq. (8) together with α in Fig. 13 can provide a closed form solution between F_p and P.

Precrack Size versus Indentation Load

A summary of the experimental results of precrack size versus indentation load is shown in Fig. 14. Here, the precrack size (a) is normalized with respect to the specimen width (W). This figure shows that precrack size decreases with increasing P, and that for a given indent load precrack size increases with decreasing W and increasing S. A poor correlation between precrack size and P was observed in the SiC material due to uncertainty in crack size measurements. For the given S and W, precrack size is shown to be a weak function of indentation load. The solid lines represent the best-fit lines to experimental data.

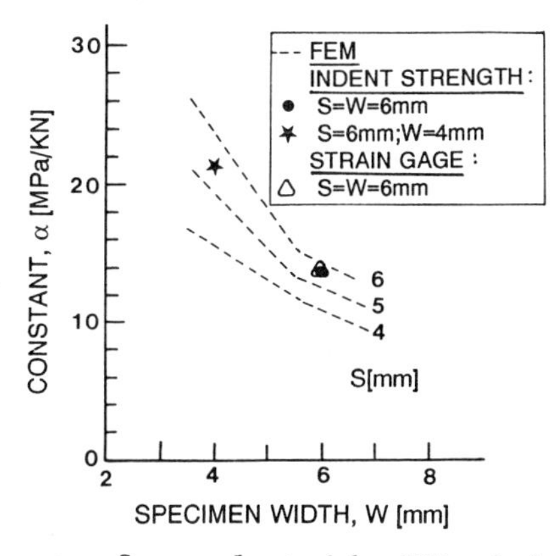

Figure 13. Summary of α evaluated by FEA, indentation strength method, and strain gaging.

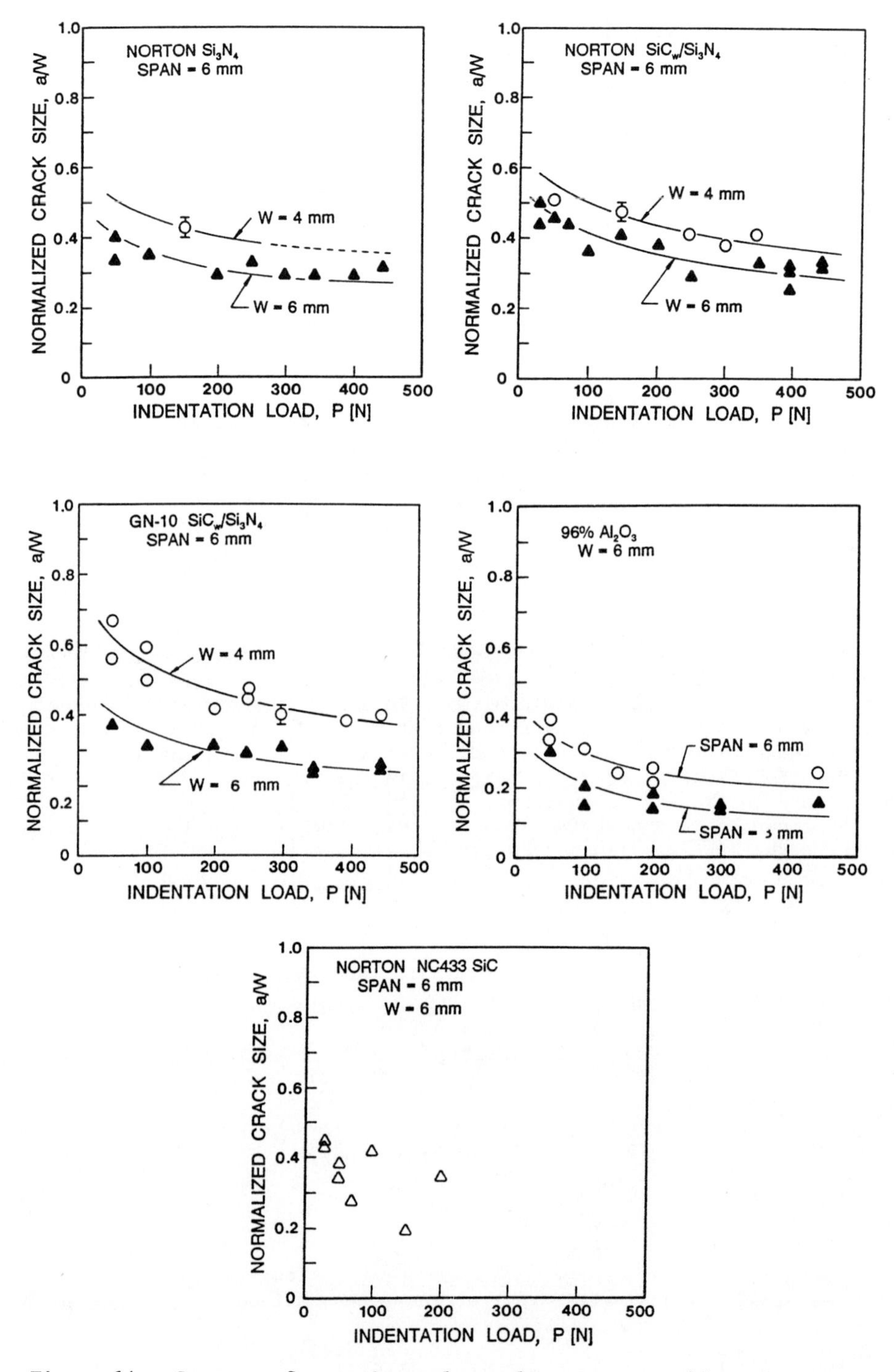

Figure 14. Summary of experimental results on precracking size versus indentation load for five test materials.

Figure 15 illustrates a typical example on how to estimate a precrack size for a given indentation load (and K_{IC}). The prediction here is made such that first F_p is determined from Eq. (8) with α for a given indentation load, and then with thus determined F_p, the corresponding precrack size is obtained from data as shown in Fig. 6. Unfortunately, no closed form solution is available in the a-P relation. An extensive work to obtain more data like in Fig. 6 with various combinations of S and W is in progress. It should be mentioned, however, that the experimental data shown in Figs. 10 and 14 can also be utilized as an engineering data base in SEPB precracking parameters.

V. SUMMARY

The precracking parameters for ceramic Single-Edge-Precracked-Beam specimens were analyzed using the FEM and indentation strength method.

(1). A relationship between precracking load and indentation load was derived as a function of specimen and fixture configurations, and material constants (E, H, and K_{IC}). An excellent agreement was found between theory and experiment.

(2). A prediction methodology regarding indentation load versus precrack size was presented based on the relation given in (1) and the numerically obtained precracking load-precrack size relation.

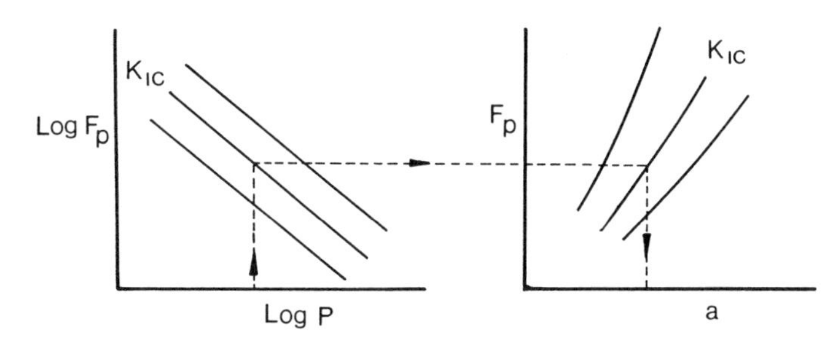

Figure 15. A schematic to illustrate a prediction of a required indent load for a given precrack size (and vice versa).

ACKNOWLEDGEMENT

The authors are very thankful to Mr. R. Pawlik at NASA Lewis for SEM and experimental work during the course of this study.

REFERENCES

1. T. Nose and T. Fujii, Evaluation of Fracture Toughness for Ceramic Materials by a Single-Edge-Precracked-Beam Method, J. Am. Ceram. Soc., 71: 328 (1988).
2. R. Warren and B. Johanneson, Creation of Stable Cracks in Hard Metals Using 'Bridge' Indentation, Powder Metall., 27: 25 (1984).
3. I. Bar-On, J. T. Beals, G. L. Leatherman, and C. M. Murray, Fracture Toughness of Ceramic Precracked Bend Bars, J. Am. Ceram. Soc., 73: 2519 (1990).

4. Standard Test Method for Plane-Strain Fracture Toughness of Metallic Materials, ASTM Designation: E 399, 1983.
5. D. B. Marshall, B. R. Lawn, and P. Chantikul, Residual Stress Effect in Sharp Contact Cracking: Part 2, J. Mater. Sci., 14: 2225 (1979).
6. G. R. Anstis, P. Chantikul, B. R. Lawn, and D. B. Marshall, A Critical Evaluation of Indentation Techniques for Measuring Fracture Toughness:I, J. Am. Ceram. Soc., 64: 533 (1981).

ELEVATED-TEMPERATURE FRACTURE RESISTANCES OF MONOLITHIC AND COMPOSITE CERAMICS USING CHEVRON-NOTCHED BEND TESTS

Asish Ghosh[1], Michael G. Jenkins[2], Mattison K. Ferber[2], Jouko Peussa[3] and Jonathan A. Salem[4]

[1]Philips Display Components Co., Ann Arbor Michigan, USA
[2]Oak Ridge National Laboratory, Oak Ridge Tennessee, USA
[3]FIMET, Helsinki, Finland
[4]NASA Lewis Research Center, Cleveland, Ohio USA

INTRODUCTION

Complete information regarding the elevated temperature fracture resistances of structural ceramics is essential for useful application of these materials in heat engine components[1,2]. Traditionally ceramics have been considered to fracture in a brittle catastrophic fashion and therefore have been characterized by the use of a single parameter such as plane strain fracture toughness, K_{IC}. However, innovative experimental work done on aluminum oxide[3] indicated the existence of increasing fracture resistance with increasing crack extension, that is R-curve behavior. Subsequent discoveries of R-curve behavior in other types of structural ceramics have precluded the adequate description of the fracture resistance of ceramics by fracture toughness alone.

The fracture history of a material can be generally divided into three regimes: i)crack initiation which is the onset of the stable propagation of the macro-crack; ii)extensive stable crack growth in which the material may exhibit increasing resistance to crack propagation; and iii)final fracture at which point the total energy dissipated during the fracture process causes the complete separation of the component. This description of these three distinct regimes makes it possible to characterize the fracture behavior of each material more completely by using three parameters, fracture toughness, R-curve and work-of-fracture. Figure 1 shows a schematic of the fracture process and the parameters required to effectively describe it.

This work presents the fracture resistances of a variety of brittle and quasi-brittle materials at temperatures from $20^{o}C$ to $1400^{o}C$. Results obtained by using chevron-notched specimens of monolithic and composite ceramics are detailed and linked to fractography and possible fracture micro-mechanisms.

EXPERIMENTAL PROCEDURE

Chevron-notched geometries have been used for fracture testing for more than two decades[4-11]. In this study, a primary reason for the use of the chevron-notched specimens is that a single chevron-notched specimen is adequate to

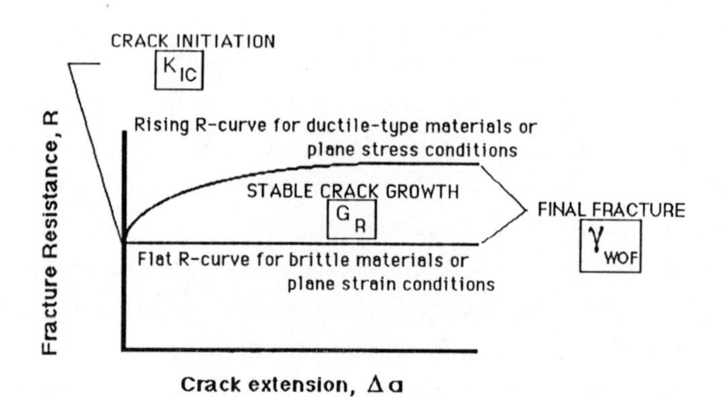

Figure 1. Schematic Illustration of Regimes of Fracture Resistance Characterization

obtain all three fracture resistance parameters viz. fracture toughness, R-curves, and work-of-fracture. In addition, the chevron-notched flexure bar is simple and efficient to fabricate. Finally, three-point flexure testing arrangement was chosen for ease of use at elevated temperatures.

A full chevron section was employed across the width of the flexure bar where the chevron-notch depth, $\alpha_1=a_1/W$, was 1.0 for all specimens. The initial notch depth, $\alpha_0=a_0/W$ ranged from 0.35 to 0.44 depending upon the test material. A schematic of the chevron geometry is shown in Figure 2. Specimens were generally square in cross section with nominal dimensions of 6 X 6 X 50 mm. Specimen material was usually tested in the as-received condition with the chevron notches fabricated with a diamond-grit circular saw blade.

Elevated-temperature testing was conducted in a resistance-heated furnace insulated with refractory brick. The upper part of the load fixture consisted of a single, solid, α-SiC push-rod machined to a single loading line at the free end and attached to a water-cooled load cell at the opposite end. The lower part of the fixture consisted of an α-SiC tube, 44 mm outer diameter, 6 mm wall thickness and 325 mm long, machined to produce two fixed knife edges of 40 mm span. This lower part of the fixture was attached to the moveable crosshead of the displacement-controlled, electro-mechanical test machine. The fracture tests were

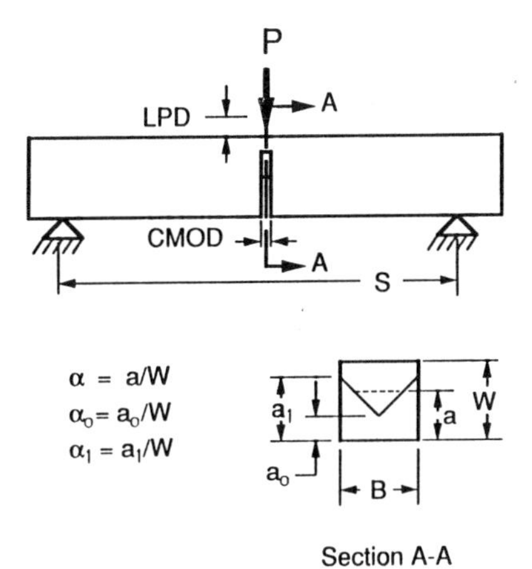

Figure 2. Chevron-notched Three Point Flexure Test Geometry

performed with the specimens first positioned on the lower fixture outside the furnace and then slowly raised (10 mm/min) into test position within the hot furnace where testing was conducted at a constant displacement rate of 0.01 mm/min.

A laser interferometric displacement gage (LIDG)[9,12-14] with an estimated resolution of 0.25μ was employed to determine specimen displacement at all temperatures. Details of the LIDG setup have been discussed extensively elsewhere[12-14].

The measured displacements were in the form of crack mouth opening displacements (CMODs), although some tests determined load point displacements (LPDs)[15] from the cross-head motion by subtracting the total machine/fixture compliance from total measured compliance. Fracture parameters were determined for each test from the digitized load-displacement plots[13]. All tests were conducted in ambient air with test temperatures ranging from 20°C to 1400°C.

FRACTURE PARAMETERS

Crack initiation can be characterized by the fracture toughness, K_{Ic} which has been shown[7] for the chevron-notch geometry and flat R-curve materials to be:

$$K_{Ic} = P_{max} Y_{min} / (BW^{1/2}) \quad (1)$$

where P_{max} is the maximum applied load, Y_{min} is the minimum geometry correction factor, and B and W are the specimen dimensions as shown in Figure 2.

Stable crack growth behavior can be described as crack growth resistance or R-curve. The R-curve is readily calculated as the strain energy release rate, G_R, plotted as a function of the incremental crack extension Δa. A global energy approach was used in which G_R^i for each Δa was calculated from the change in strain energy ΔU_i required to create an incremental fracture area, ΔA_i such that[13]

$$G_R^{\,i} = \Delta U_i / \Delta A_i \qquad (2)$$

At the point of complete fracture of the chevron section, the work-of-fracture can be determined from the total energy consumed during the entire fracture process divided by the total, projected fracture area, $2A_T$ such that[5,6]:

$$\gamma_{wof} = (1/2A_T)\int_a P \, d(LPD) \qquad (3)$$

where $\int_a$ is integrated over the total area under the load displacement curve.

TEST MATERIALS

Materials referenced in this study were commercially available at the time of fracture testing[15-24]. In one case[15] the material was given post-processing treatments after receipt from the manufacturer to illustrate certain material characteristics.

Types of materials tested included monolithic ceramics (α-SiC, Si_3N_4, $MgAl_2O_4$)[16-18] self-reinforced monoliths (acicular-grained Si_3N_4, acicular-grained mullite)[15,19-20] and ceramic matrix composites (SiC whisker/Al_2O_3 matrix, TiB_2 particulate/SiC matrix, SiC fiber/CVI SiC matrix, Al_2O_3 fibre/CVI SiC matrix)[21-24]. The materials are listed in Table 1.

RESULTS AND DISCUSSION

Apparent fracture toughnesses for the different materials are given in Table 2 and are plotted versus temperature in Figure 3. Figures 4 and 5 show the R-curves at two temperatures for linear elastic and non-linear elastic behaving materials respectively. Table 3 and Figure 6 show work-of-fracture values versus temperature.

TABLE 1--Summary of test materials

Type of Material	Material Designator	Material Description
Monoliths		
Mullite ($Al_2O_3 \bullet SiO_2$)	Mullite KM1[a] [15] Mullite KM3[a] [15] Mullite MM1[b] [15] Mullite MM3[b] [15]	As-processed Mullite KM1 + 1800°C heat treat [15] As-processed Mullite MM1 + 1800°C heat treat [15]
Spinel ($MgAl_2O_4$)	Spinel[c] [16]	Dense; bimodal grain size, optically transparent
α–Silicon Carbide (SiC)	Hexoloy SA[d][17]	Dense; fine, equiaxed grains
Silicon Nitride (Si_3N_4)	A2Y6[e] [18]	Dense; duplex grain size distribution, hot pressed
Silicon Nitride (Si_3N_4)	SN251[f] [19,20]	Acicular grain structure; cold isostatically pressed, sintered
Composites		
Silicon Carbide whiskers in Alumina Matrix (SiC_{ω}/Al_2O_3)	SA-25[g] [21,22]	25 wt% SiC whiskers in dense Al2O3 matrix
Titanium Diboride particles in Silicon Carbide Matrix (TiB_2/SiC)	Hexoloy ST[h] [23]	16 vol% TiB2 particles in dense SiC matrix
Silicon Carbide fibres in Silicon Carbide Matrix	Nicalon/CVI SiC III [i] [24]	30-34 vol% fibre/30-40 vol% CVI β-SiC
Alumina Fibres in Silicon Carbide Matrix	FP-Alumina/CVI SiC V [i] [24]	30-34 vol% fibre/30-40 vol% CVI β-SiC

[a] Mullite, Kyocera Corporation, Kyoto, Japan, 1989, [b] Mullite (7% free silica), McDanel Corporation, Beaver Falls, Pennsylvania, 1989,
[c] Spinel, Coors Ceramic Company, Golden, Colorado, 1986, [d] Hexoloy SA, Carborundum Co., Niagara Falls, New York, 1985,
[e] A2Y6 Silicon Nitride, GTE Laboratories, Inc.,Waltham Massachusetts, 1985, [f] SN251 Silicon Nitride, Kyocera Corporation, Kyoto, Japan, 1990.
[g] SA-25, Greenleaf Corporation, Saegertown, Pennsylvania, 1985, [h] Hexoloy ST, Carborundum Co., Niagara Falls, New York, 1986.
[i] CVI-β-Silcion Carbide, Refractory Composites, Inc., Whittier, California, 1986.

TABLE 2--Apparent fracture toughness (MPa $m^{1/2}$) at various temperatures

Material	Test Temperature (°C)							
	20	200	400	600	800	1000	1200	1400
Monoliths								
Mullite KM1 [15]	2.2± 0.3	. . .	1.9±0.2	. . .	1.9±0.5	1.7±0.2	2.0±0.3	2.3±0.3
Mullite KM3 [15]	2.3±0.3	. . .	1.8±0.2	. . .	1.7±0.2	2.1±0.1	2.1±0.2	1.9±0.2
Mullite MM1 [15]	1.9±0.3	. . .	2.0±0.2	. . .	2.1±0.2	1.9±0.2	2.1±0.2	2.2±0.2
Mullite MM3 [15]	2.3±0.2	. . .	1.7±0.2	. . .	2.3±0.2	2.1±0.2	1.9±0.2	2.0±0.2
Spinel [16]	1.8±0.1	1.8±0.1	1.7±0.1	1.6±0.2	1.3±0.2	1.3±0.4	1.2±0.1	1.3±0.1
Hexoloy SA [17]	2.9±0.3	2.6±0.3	2.9±0.4	3.2±0.5	2.7±0.7	2.8±0.7	3.4±0.1	3.0±0.9
A2Y6 [18]	6.9±0.1	7.0±0.3	6.6±0.2	6.8±0.2	6.0±0.2	6.6±0.1	5.2±0.1	2.8±0.1
SN251 [19,20]	7.9±0.4	. . .	. . .	. . .	7.1±0.7	6.9±0.4	6.0±0.3	10.4±0.5*
Composites								
SA-25 [21,22]	6.6±0.9	7.1±0.6	6.6±0.6	7.0±0.6	6.6±0.4	6.8±0.5	7.7±0.4	. . .
Hexoloy ST [23]	4.1±0.2	3.5±0.1	3.6±0.1	3.1±0.1	3.2±0.2	2.8±0.4	2.8±0.1	2.9±0.3
Nicalon/CVI SiC III [24]	2.9±0.3	2.6±0.3	2.5±0.5	2.5±0.4	2.4±0.6	3.0±0.7	2.8±0.4	2.8±0.4
FP-Alumina/CVI SiC V[24]	7.2±0.8	. . .	5.8±0.8	. . .	. . .	3.8±0.5	2.9±0.4	2.2±0.4

Note: Values shown are mean values ± one standard deviation.
* Temperature = 1371°C

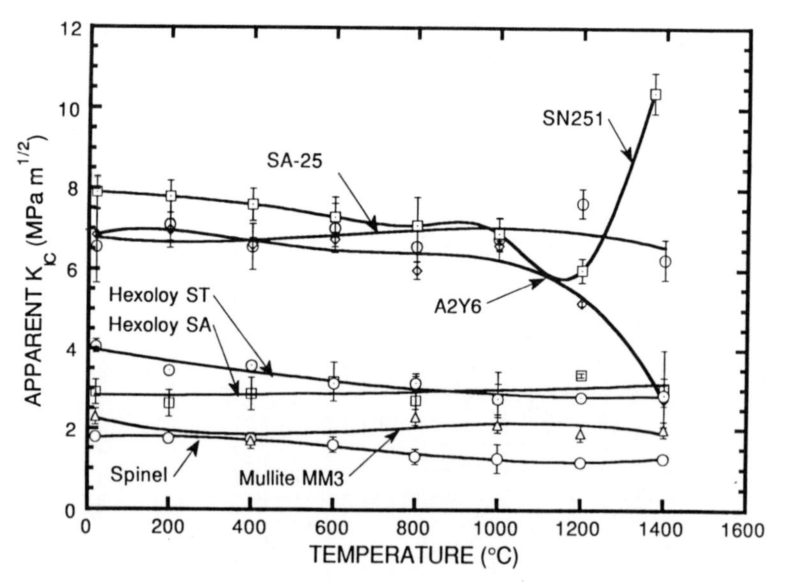

Figure 3. Apparent Fracture Toughness, K_{Ic}, versus Temperature for Various Materials

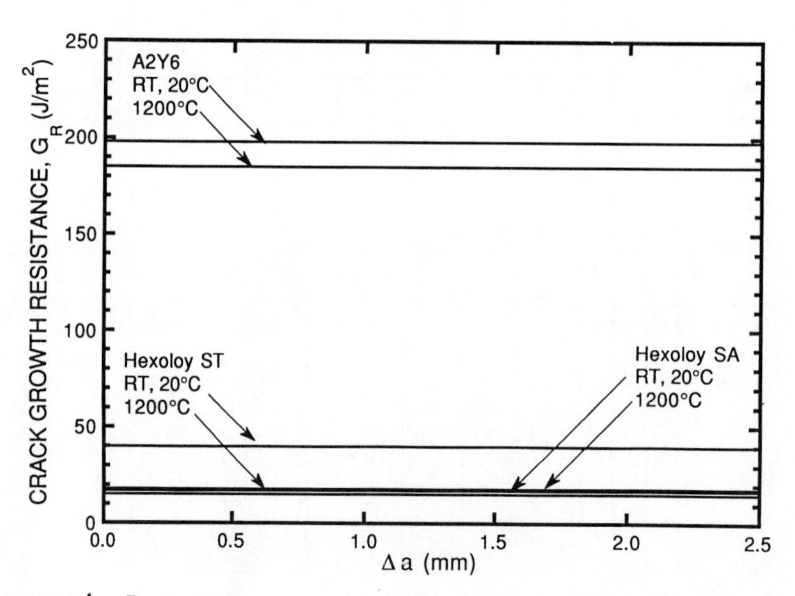

Figure 4. R-curves, G_R, at Two Temperatures for Various Materials Displaying Linear Elastic Behavior

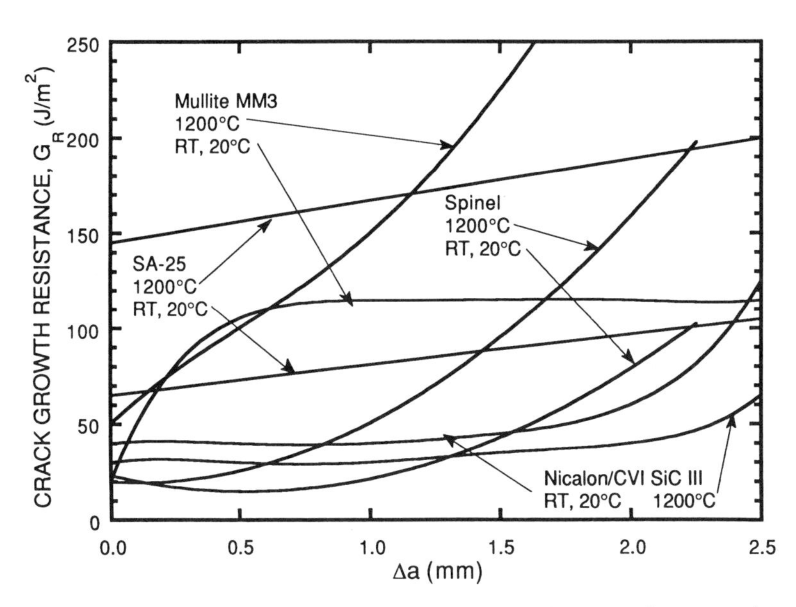

Figure 5. R-curves, G_R, at Two Temperatures for Various Materials Displaying Nonlinear Elastic Behavior

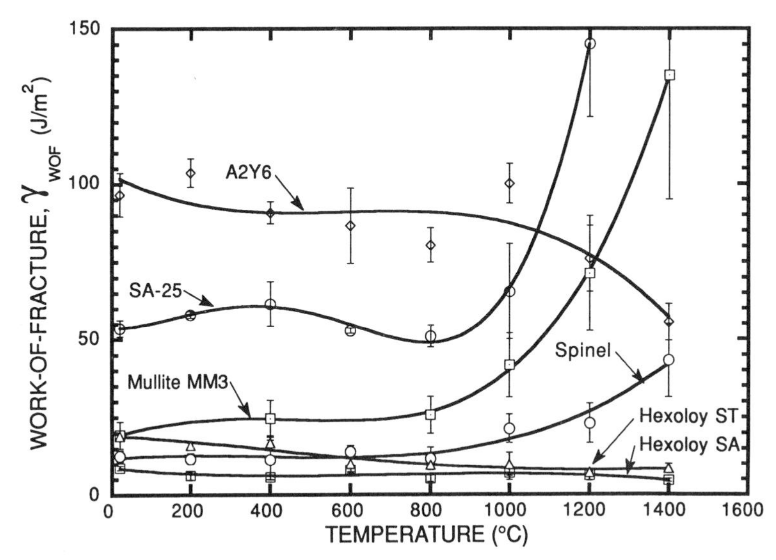

Figure 6. Work-of-fracture, γ_{wof}, versus Temperature for Various Materials

TABLE 3--Work-of-fracture (J/m^2) at various temperatures

Material	Test Temperature (°C)							
	20	200	400	600	800	1000	1200	1400
Monoliths								
Mullite KM1 [15]	10.5±3.8	...	9.6±2.8	...	10.0±3.6	11.7±3.8	17.0±5.5	26.2±9.1
Mullite KM3 [15]	18.5±4.1	...	20.9±5.2	...	18.2±4.1	22.9±4.6	40.1±12.3	90.5±32.6
Mullite MM1 [15]	17.4±4.1	...	20.4±5.5	...	20.2±4.1	23.6±5.2	26.4±7.3	29.8±8.5
Mullite MM3 [15]	19.5±4.3	...	24.7±5.9	...	25.8±5.9	41.7±10.3	71.3±18.5	135.2±40
Spinel [16]	12.7±1.8	11.8±1.8	11.5±6.4	14.1±1.8	11.8±3.6	21.4±4.6	23.2±6.4	43.2±11.8
Hexoloy SA [17]	8.8±0.6	6.4±1.4	5.8±0.8	8.0±0.5	5.3±1.0	7.1±2.2	6.3±1.7	4.6±0.5
A2Y6 [18]	96.7±6.9	103.2±5	90.9±3.5	86.7±12	80.5±5.5	100.8±6.4	76.1±10.7	55.5±6.0
SN251 [19,20]	84.5±8.3	...	...	...	...	...	...	...
Composites								
SA-25 [21,22]	53.7±2.7	58.1±1.1	61.6±7.2	52.8±0.8	51.1±3.6	65.5±15.4	145.3±24	...
Hexoloy ST [23]	19.2±1.5	16.2±0.8	16.9±2.3	10.1±3.0	9.8±1.4	9.9±3.8	7.3±0.1	8.5±1.6
Nicalon/CVI SiC III [24]	...	...	...	...	...	...	...	...
FP-Alumina/CVI SiC V[24]	...	...	...	...	...	...	...	...

Note: Values shown are mean values ± one standard deviation.

Monoliths

In comparisons[17] of various sharp crack techniques applied to the model, brittle, fine grained polycrystalline α-SiC, the apparent K_{Ic} at 20°C determined from the chevron-notched geometries(2.8-3.6 MPa $m^{1/2}$) compared well with other sharp crack techniques such as the controlled flaw (2.7-3.5 MPa $m^{1/2}$), precracked double cantilever beam specimens (2.5 MPa $m^{1/2}$) and precracked double torsion specimens (3.0-4.6 MPa $m^{1/2}$). Blunt notch techniques such as the single-edge-notched-beam(SENB) technique yielded apparent fracture toughnesses of 3.5-4.8 MPa $m^{1/2}$ while the Charpy impact specimen yielded a value of 3.9 MPa $m^{1/2}$. These comparisons indicated that for a fine-grained brittle material fracture toughness is a material property and the value obtained by any sharp crack technique is comparable to other sharp crack techniques. The R-curve for this fine-grained α-SiC is flat i.e. constant with crack extension as shown in Figure 4. It is also independent of temperature. Consequently the work-of-fracture, for fine grained α-SiC, is also constant with temperature as shown in Figure 6.

Large-grained monoliths such as $MgAl_2O_4$ spinel exhibit a different type of fracture behavior[16]. The apparent fracture toughness at 20°C is 1.8 MPa $m^{1/2}$ but decreases at 800°C to 1.3 MPa $m^{1/2}$. The R-curve, as shown in Figure 5, is rising with crack extension. The rising R-curves exhibited by this spinel are due to frictional effects of the large grains and are nearly identical from 20°C to 800°C, but increase with increasing temperature above 800°C. Similarly the work-of-fracture which is 12.7 J/m^2 at 20°C, remains approximately constant until 800°C, but increases consistently with temperature to 43.2 J/m^2 at 1400°C. The increased fracture resistance at higher temperatures is attributed to the energy dissipated in microcrack formation leading to grain boundary separation at those temperatures in addition to grain bridging and frictional effects which are the dominant "toughening" mechanisms at lower temperatures. The separation of grain boundary requires additiional energy which is reflected in the higher rising R-Curves and greater work-of-fracture values at elevated temperatures. Figure 7 shows fracture surfaces at 20°C and 1400°C for this spinel.

Self-Reinforced Monoliths

For self-reinforced monoliths such as mullite[15] the fracture toughness averaged 2.0 MPa $m^{1/2}$ remaining constant with temperature as summarized in Table 2 and shown in Figure 3. The R-curve increases with crack extension but reaches a plateau for temperatures from $20^{o}C$ until $1000^{o}C$, although, the R-curve increases monotonically after $1000^{o}C$ (Figure 5, MM3). The rising R-curve is due to the "wake effect" i.e. fracture processes occurring behind the advancing crack tip arising from mechanisms such as grain bridging and frictional interaction of grains. This behavior is observed for the as-received as well as the heat treated microstructures for both stoichiometric mullite(KM1,KM3) as well as the mullite with the glassy phase(MM1,MM3). The monotonic increase of R-curves at temperatures above $1000^{o}C$ is due to the sample size which was small compared to the zone size required for saturation of the fracture processes. The work-of-fracture yielded similar trends with temperature i.e. remained constant till $1000^{o}C$ but increased with temperature above $1000^{o}C$ as given in Table 3 and Figure 6.

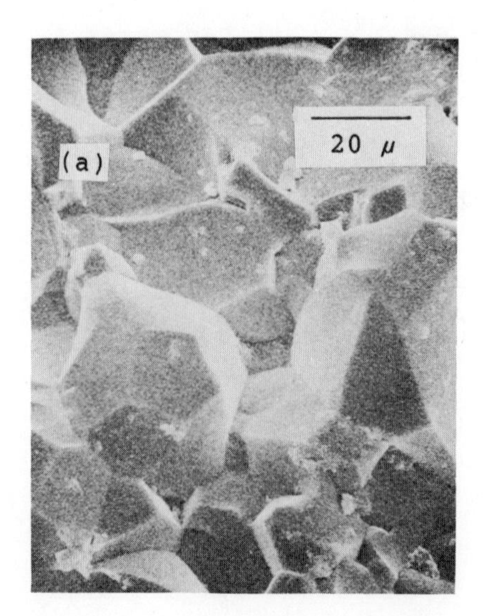

Figure 7. Fracture Surfaces of $MgAl_2O_4$ Spinel at (a)$20^{o}C$ and (b)$1400^{o}C$

Silicon nitride with acicular grains[19,20] exhibited a constant apparent fracture toughness from 20°C to 1000°C, averaging 7.3 MPa $m^{1/2}$, decreased at 1200°C to 6.0 MPa $m^{1/2}$, and finally increased to 10.4 MPa $m^{1/2}$ at 1371°C as given in Table 2 and shown in Figure 3. The artificial increase in fracture toughness at 1371°C is attributed to the softening of glassy secondary phase at higher temperatures preventing the formation of a sharp crack and the subsequent plastic hinging rather than fracture of the chevron section. Fracture surfaces at 20°C and 1371°C as shown in Figure 8 illustrate the flow of the glassy phase at high temperatures.

Composites

The TiB_2 particulate/SiC matrix composite[23] shows a consistent decrease in fracture toughness from 4.1 MPa $m^{1/2}$ at 20°C to 2.8 MPa $m^{1/2}$ at 1400°C (Table 2 and Figure 3). The R-curves are flat at all temperatures although the absolute values decrease with increasing temperature. The work-of-fracture also decreases consistently from 19.2 J/m^2 at 20°C to 8.5 J/m^2 at 1400°C. Residual stresses are responsible for crack deflection at lower temperatures resulting in "toughening" of the composite. Figure 9 shows the particle/matrix separation predominant at high temperatures. This particle/matrix separation relieves the residual stress thereby yielding lower fracture resistance at higher temperatures.

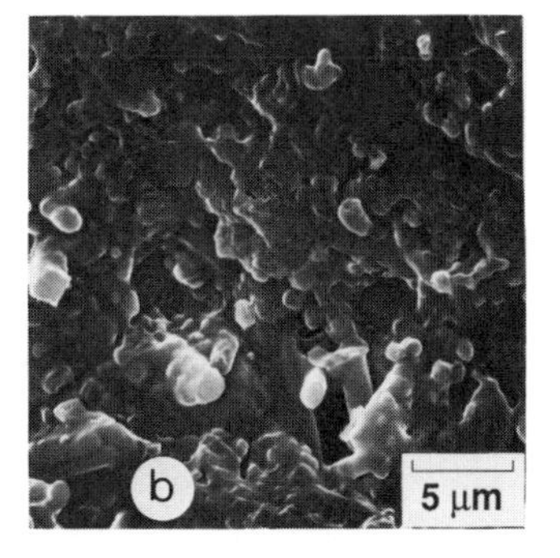

Figure 8. Fracture Surfaces of Silicon Nitride at (a)20°C and (b)1400°C

The whisker composite of SiC whisker/Al_2O_3 matrix[21,22] exhibit a constant apparent fracture toughness from 20°C to 1000°C averaging 6.78 MPa $m^{1/2}$ but increases to 7.7 MPa $m^{1/2}$ at 1200°C as summarized in Table 2 and shown in Figure 3. The R-curves rise monotonically with incremental crack extension for all temperatures, although, the R-curve at 1200°C is greater than the R-curve at 20°C. Similarly, the work-of-fracture is constant from 20°C to 1000°C averaging 57.13 J/m^2 although increasing to 145.3 J/m^2 at 1200°C as summarized in Table 3 and shown in Figure 6. The increased fracture resistance with crack extension is attributed to the whisker ligaments holding the opposing crack faces together behind the crack tip as shown in Figure 10. Higher fracture resistance at 1200°C is because of secondary cracking in addition to whisker bridging.

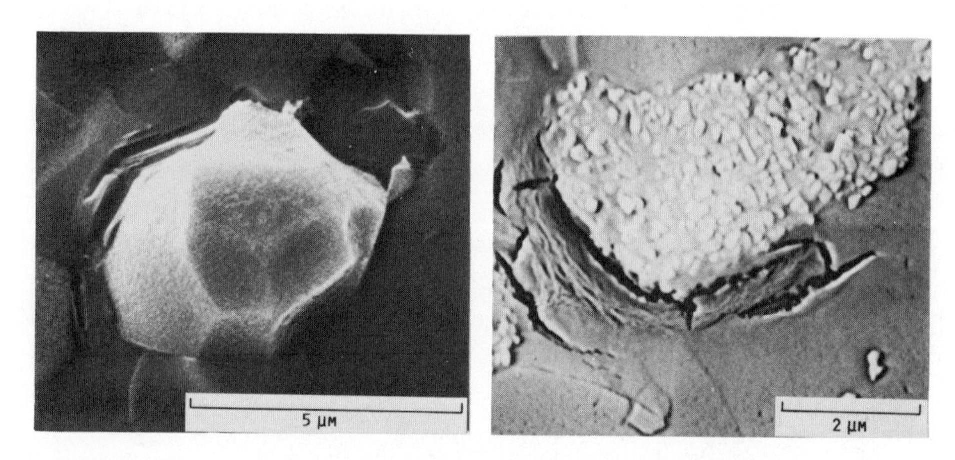

Figure 9. Fracture Surfaces of TiB_2 Particle/SiC Matrix Composite at 1400°C Showing Particle Matrix Separation

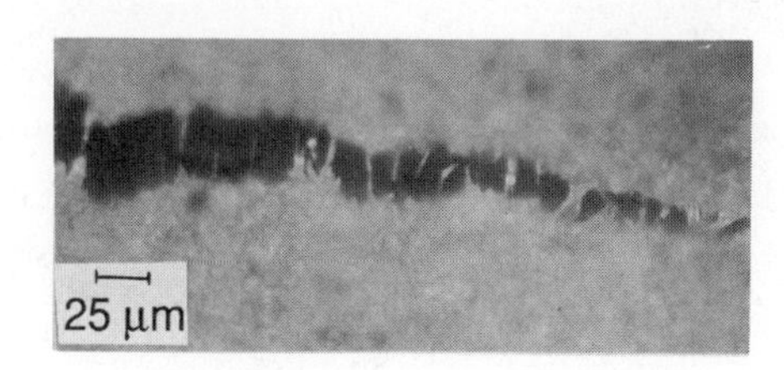

Figure 10. Crack Profiles of SiC Whisker/Al_2O_3 Matrix Composite Showing Crack Bridging Behind the Crack Tip

Fiber-reinforced composites exhibit fracture behavior which is dependent on the reinforcing materials[24]. While the Nicalon SiC fiber/CVI SiC matrix exhibited no dependence of apparent fracture toughness on temperature from 20°C to 1400°C averaged 2.7 MPa $m^{1/2}$, the apparent fracture toughness of FP Al_2O_3 fiber/CVI SiC matrix decreased from 7.2 MPa $m^{1/2}$ at 20°C to 2.2 MPa $m^{1/2}$ at 1400°C.

The R-curves for Nicalon SiC fiber/CVI SiC matrix monotonically increase with crack extension with the value at 1200°C being less than at 20°C. The rising R-curves in Nicalon SiC fiber/CVI SiC matrix are attributed to the fiber bridging of the opposite faces of the opposing macro-crack faces. The curves do not plateau because the "wake" process zone sizes are larger than the sample sizes. The decrease in fracture resistance of the FP Al_2O_3 fiber/CVI SiC matrix is due to environmental degradation of the alumina fibers which decrease the effect of fiber reinforcements.

CONCLUSIONS

The chevron-notched geometry yields apparent fracture toughness values which are consistent with the values measured by any other sharp crack techniques. This is specifically observed for fine grained α-SiC. A single, chevron-notched specimen can yield three different fracture parameters namely fracture toughness, R-curve and the work-of-fracture.

In this study, for brittle, fine-grained materials such as α-SiC, fracture toughness alone may be used to characterize the mechanical properties. However, for materials such as $MgAl_2O_4$ spinel, self-reinforced Si_3N_4, and self-reinforced mullite, which exhibit increasing fracture resistance with increasing crack extension, three parameters (fracture parameters are fracture toughness, R-curve and the work-of-fracture) are required to better characterize the fracture behavior. Fracture toughness in the case of mullite is not sensitive to microstructure but R-curves and work-of-fracture are dependent on microstructure and temperature. In addition, R-curves and work-of-fracture follow similar trends with microstructure and temperature irrespective of the material.

The fracture behavior of structural ceramics depends on the microstructure such as the grain size in monoliths (spinel) and the reinforcing material such as the fiber type as in composites(Nicalon SiC fiber/CVI SiC matrix as opposed to FP Al_2O_3 fiber/CVI SiC matrix). In addition temperature also affects the fracture behavior of the composites. Composites such as TiB_2 particulate/SiC matrix and Nicalon SiC fiber/CVI SiC matrix show lower R-curves at higher temperatures compared to at 20°C. However, SiC whisker/Al_2O_3 matrix exhibits a higher R-curve at 1200°C compared to at 20°C. In all cases however, the trends exhibited by R-curves and work-of-fracture are similar.

ACKNOWLEDGMENTS

The authors wish to thank the guidance and assistance of Professors R.C. Bradt and A.S. Kobayashi in the completion of much of the work described here which was carried out at the University of Washington. Research funded as part of the CTAHE of the Advanced Materials Development Program was partially performed in the HTML User Facility, both sponsored by the U.S.-DOE, Assistant Secretary for Conservation and Renewable Energy, Office of Transportation Technologies, under contract DE-AC05-84OR21400 managed by Martin Marietta Energy Systems, Inc.

REFERENCES

1. D.L. Vaccari, P.K. Khandelwal, "Life Prediction Methodology," Element 3.2.2.2, Ceramic Technology for Advances Heat Engines Project Semiannual Progress Report for October 1989 through March 1990, ORNL/TM-11586, Oak Ridge National Laboratory, Oak Ridge, TN pp 397-401, 1990
2. A. Comfort J. Cuccio, H. Fang, "Life Prediction

Methodology for Ceramic Components of Advances Engines," WBS Element 3.2.2.2, ibid, pp 402-411

3. R. Steinbrech, R. Knehans, W. Schaarwachter, "Increase of Crack Resistance During Slow Crack Growth in Al_2O_3 Bend Specimens," J. Mater. Sci., 18 [1] 265-270 (1983)
4. J.C. Newman Jr., "A Review of Chevron-Notched Fracture Specimens," Chevron-Notched Specimens: Testing and Stress Analysis, ASTM STP 855, Ed. J.H. Underwood, S.W. Freiman, F.I. Baratta, ASTM Philadelphia, pp 5-31, 1984
5. J. Nakayama, "Bending Method for Direct Measurement of Fracture Energy of Brittle Materials," Jap. J. of App. Physics, 3 [7] 422-423 (1964)
6. H.G. Tattersal, G. Tappin, "The Work of Fracture and its Measurements in Metals, Ceramics and other Materials," J. Mater. Sci., 1 296-301 (1966)
7. L.P. Pook, "An Approach to a Quality Control K_{Ic} Testpiece," Int. J. of Frac., 8 103-108 (1972)
8. D.G. Munz, J.L. Shannon Jr., R.T. Bubsey, "Fracture Toughness Calculation from Maximum Load in Four Point Bend Tests of Chevron Notch Specimens," Int. J. of Frac., 16 R137-R141 (1980)
9. A. Ghosh, M.G. Jenkins, K.W. White, A.S. Kobayashi and R.C. Bradt, "The Chevron Notched Bend Bar Technique for Fracture Resistance Measurements of Ceramics," Ceramic Materials and Components for Engines, Ed. V.J. Tennery, Amer. Cer. Soc., OH 592-603 (1989)
10. M. Sakai, K.K. Yamasaki, "Numerical Fracture Analysis of Chevron Notches: I and II," J. Amer. Cer. Soc. 66 371-375 (1989)
11. M.G. Jenkins, A.S. Kobayashi, K.W. White and R.C. Bradt, "A 3-D Finite Element Analysis of a Chevron-Notched, Three Point Bend Fracture Specimen for Ceramic Materials," Int. J. of Frac., 34 281-295 (1987)
12. M.G. Jenkins, A.S. Kobayashi, M. Sakai, K.W. White and R.C. Bradt, "Fracture Toughness Testing of Ceramics Using a Laser Interferometric Strain Gage," Amer. Cer. Soc. Bull., 12 [6] 1734-1738 (1987)
13. M.G. Jenkins, "Ceramic Crack Growth Resistance Determination Utilizing Laser Interferometry," Ph.D. Dissertation, Department of Mechanical Engineering, Univ. of Washington, Seattle, WA, USA 1987

14. M.G. Jenkins, A.S, Kobayashi, J.T. Peussa, J.A. Salem, A. Okura, "Laser Interferometry for Measuring Elevated Temperature Fracture Resistance of Ceramics," Proc. Int. Conf. on Adv. Cer. Matl., Tianjin Univ., Tianjin, People's Republic of China, May 16-20, C66-C71 (1988)
15. A. Ghosh, "Effect of Microstructure and Temperature on the Fracture Resistance of Duplex Microstructure Mullite," Ph.D. Dissertation, Department of Materials Science and Engineering, Univ. of Washington, Seattle, WA, USA 1989
16. A. Ghosh, K.W. White, M.G. Jenkins, A.S. Kobayashi and R.C. Bradt, "Fracture Resistance of a Transparent $MgAl_2O_4$," J. Amer. Cer. Soc., 74 [7] 1624-30 (1991)
17. A. Ghosh, M.G. Jenkins, K.W. White, A.S. Kobayashi and R.C. Bradt, "Elevated Temperature Fracture Resistance of a Sintered α-Silicon Carbide," J. Amer. Cer. Soc., 72 [2] 242-247 (1989)
18. A. Ghosh, M.G. Jenkins, A.S. Kobayashi and R.C. Bradt, "Elevated Temperature Fracture Resistance of a Hot Pressed Si_3N_4," Unpublished Work, Univ. of Washington, 1989
19. J.A. Salem, J.M. Manderscheid, M.R. Freedman and J.P. Gyekeneysi, "Reliability Analysis of a Structural Ceramic Combustion Chamber," NASA TM 103741 (1990)
20. J.A. Salem, S.R. Choi, M.R. Freedman and M.G. Jenkins, "Mechanical Behavior and Failure Phenomenon of an In-Situ-Toughened Silicon Nitride," NASA TM 103741, Feb 1991
21. M.G. Jenkins, K.W. White, A.S. Kobayashi and R.C. Bradt, "The R-Curve Behavior of SiC Whisker Polycrystalline Alumina Matrix Composite to 1400^oC," Whisker- and Fiber Toughened Ceramics, Proc. of an Int. Conf., Ed. R.A. Bradley, D.E. Clark, D.C. Larsen, J.O. Stiegler, ASM Int., pp 281-288, 1988
22. M.G. Jenkins, K.W. White, A.S. Kobayashi, R.C. Bradt, "Elevated Temperature Fracture Characteristics of a SiC Whisker/Al_2O_3 Matrix Composite," Eng. Frac. Mech., 30 505-510
23. M.G. Jenkins, J.A. Salem, S.G. Seshadri, "Fracture Resistance of a TiB_2 Particles/SiC Matrix Composite at

Elevated Temperatures," J. Comp. Mater., 23 [1] 77-91 (1989)

24. J.T. Peussa, "Elevated Temperature Fracture Properties of CVI Silicon Carbide Matrix Continuous Ceramic Fiber Composites," M.S.., Thesis, Department of Materials Science and Eng., Univ. of Wash., Seattle, WA USA 1987

RELIABILITY OF CERAMICS FRACTURE TOUGHNESS MEASUREMENTS BY INDENTATION

S.N. Dub and A.L. Maistrenko

Institute for Superhard Materials of the Ukrainian Academy of Sciences, Kiev 254153, USSR

ABSTRACT

Comparative fracture toughness tests of materials having different types of radial cracks have been carried out by conventional techniques as well as by indentation method. Some of the available equations for half-penny cracks are shown to hold good for $C/a > 2.1$ condition. At $C/a < 2.1$ none of the equations derived up to now is in good agreement with the results obtained by conventional fracture toughness measurement. An equation for K_{Ic} evaluation at $C/a < 2.1$ is proposed. It gives results which do not depend on indentation load and agree with conventional measurements.

INTRODUCTION

Indentation method has a wide use for fracture toughness testing of ceramics. K_{Ic} is calculated from the lengths of radial cracks initiating from the corners of Vickers impression in brittle materials. For this purpose about 20 semiempirical equations have been derived up till now[1-19]. The most common used equations can be classified into the following groups:

1. K_{Ic} is given by radial crack length C and indenter load P[5,15]:

$$K_{Ic} = (1/7)P/C^{3/2} \qquad (1)$$

$$K_{Ic} = 0.0725P/C^{3/2} \qquad (2)$$

These equations are based on the solution for the problem of stressed state in the tip of penny-like crack in infinite medium to the center of which a point force F is applied[20]:

$$K_I = F/(\pi C)^{3/2} \qquad (3)$$

Lawn and Fuller[1] substituted tangential component of the indenter load for F and received that

$$K_{Ic} = P/(\pi C)^{3/2}/tg\Psi \qquad (4)$$

As half-angle Ψ between the opposite facets of Vickers indenter is 74°, Eq.(4) results in the following equation

$$K_{Ic} = 0.0515P/C^{3/2} \qquad (5)$$

Fracture Mechanics of Ceramics, Vol. 10
Edited by R.C. Bradt *et al.*, Plenum Press, New York, 1992

which gives underestimated K_{Ic} values[21,22] If Ψ is taken equal to 68^0 (half-angle between the opposite edges of Vickers indenter), K_{Ic} is given by[22,23]:

$$K_{Ic} = 0.0726P/C^{3/2} \quad (6)$$

This equation holds quite good for half-penny cracks.

2. The second group comprises the equations which include also hardness H and impression radius a[4,12]

$$K_{Ic} = 0.15k(Ha^{1/2}/\phi)(C/a)^{-3/2} \quad (7)$$

where $k = 3.2$ and $\phi = 3$.

$$K_{Ic} = 0.203Ha^{1/2}(C/a)^{-3/2} \quad (8)$$

If hardness of a material being tested is load-independent then Eqs.(7) and (8) turn into those of the first group

$$K_{Ic} = 0.0742P/C^{3/2} \quad (9)$$

$$K_{Ic} = 0.0941P/C^{3/2} \quad (10)$$

3. The equations of the third group involve elastic modulus E[7,11]

$$K_{Ic} = 0.129(Ha^{1/2}/\phi)(E\phi/H)^{2/5}(C/a)^{-3/2} \quad (11)$$

$$K_{Ic} = 0.028Ha^{1/2}(E/H)^{1/2}(C/a)^{-3/2} \quad (12)$$

For the materials which hardness does not depend on indenter load Eqs.(11) and (12) are converted into equation of next type[8]

$$K_{Ic} = 0.016(E/H)^{1/2}P/C^{3/2} \quad (13)$$

4. The fourth group includes the equations for short (low C/a values) radial cracks derived by Niihara et at.[11,24]

$$K_{Ic} = 0.018Ha^{1/2}(E/H)^{2/5}(C/a-1)^{-1/2} \quad (14)$$

and by Shetty et al.[17]

$$K_{Ic} = \beta(HP/4l)^{1/2} \quad (15)$$

where $l = C - a$, and $\beta = 0.089$ for WC-Co alloys. Eq.(15) can be written as

$$K_{Ic} = 0.0303a^{-3/2}P(C/a-1)^{-1/2} \quad (16)$$

The problem is that all these equations give different results for the same materials. Therefore many works are aimed at testing these relationships and selection the most reliable ones. For this purpose K_{Ic} is determined by conventional techniques (double torsion, single edge notch, compact tension, etc.) as well as by indentation[25-32]. A failing of some of these works is that K_{Ic} tests by indentation are carried out at one indenter load, which is far from being sufficient for checking the above equations because K_{Ic} test results may vary strongly with indenter load[33-37]. Only those relationships will be valid for which K_{Ic} is independent on applied load and agrees with the results obtained by conventional fracture toughness testing. It can be achieved owing to a proper choice of A and B constants in a semiempirical equation relating K_{Ic} and radial crack length and indenter load:
for half-penny cracks

$$K_{Ic} = A_1 Ha^{1/2}(E/H)^{1/2}(C/a)^{-B_1} \quad (17)$$

and for Palmqvist cracks

$$K_{Ic} = A_2 Ha^{1/2}(E/H)^{1/2}(C/a-1)^{-B_2} \quad (18)$$

If exponent B is overestimated then K_{Ic} will decrease with increase in indenter load. Besides, the test results dependence on load can be attributed to :

- radial crack length being comparable to grain size at low indenter loads and much bigger than grain size at great loads
- radial crack shape changing with indenter load, and
- residual surface stresses existing in a specimen.

EXPERIMENTAL PROCEDURES

K_{Ic} for eight materials was measured by conventional techniques and by indentation (Table). These are: WC-Co alloys with different cobalt contents (3,6 and 8%), monocrystalline silicon, sapphire and lithium fluoride, glass (flint) and yttria-doped hot pressed silicon nitride. WC-Co alloys were tested by 3-point bending of prismatic specimens with fatigue crack at the tip of notch[38]. K_{Ic} measurements of Si_3N_4 ceramic were performed by compact tention of short bar specimens with shevron notches. The remaining materials were tested using Terratek fractomet tester. The specimens were chevron notched short rods. The chevron notch plane coincided with {111}, {11$\bar{2}$0}, and {011} planes for silicon, sapphire, and lithium fluoride, respectively. Then the testing of these materials was performed by indentation within the widest possible load range (TableI). At each load were made 5-9 impressions.

Table I. Parameters of Materials Studied

Material	Indenter load (N)	Hardness (GPa)	Young's Modulus (GPa)	Fracture Toughness ($MN/m^{1.5}$)	Crack shape	α	β
WC-8Co	245-1000	12.3	610	13.1	Palmq.*	.67	2.12
WC-6Co	147-1000	13.8	640	12.2	Palmq.	.80	2.25
WC-3Co	147-1000	15.7	660	9.0	Palmq.	.71	1.71
Al_2O_3	1-10	-	400	2.2	Palmq.	.81	2.25
LiF {011}	50-600	0.8	123	0.6	Palmq. +Median	.65	1.45
Si {111}	0.1-3	-	163	0.7	HalfP.**	.71	1.77
Si_3N_4(h.p.)	50-640	12.2	300	3.8	HalfP.	.71	1.59
Glass	0.4-20	-	54	0.5	HalfP.	.69	1.65

* - Palmqvist crack, ** - Half-penny crack

Silicon is known to undergo half-penny cracking[39]. This is typical of glass as well[2]. The results of 3-point bend tests of silicon nitride prismatic specimens with Vickers indentation made on one of lateral faces have shown that under 50 to 640 N loads half-elliptical cracks form in this materials (at 50 N the crack shape is close to half-disk while at 640 N it approaches half-ellipse, with axes ratio equal to 1.25).

In the case of monocrystalline LiF the impressions were produced on (001) cleavage plane. Palmqvist cracks at {011} planes and one penny crack at {001} plane were observed upon indentation. In LiF Palmqvist cracks show high C/a ratio (4÷6). In sapphire only Palmqvist cracks are reported to initiate from impression sides[40]. WC-Co alloys are also characterized by Palmqvist cracking[17,41].

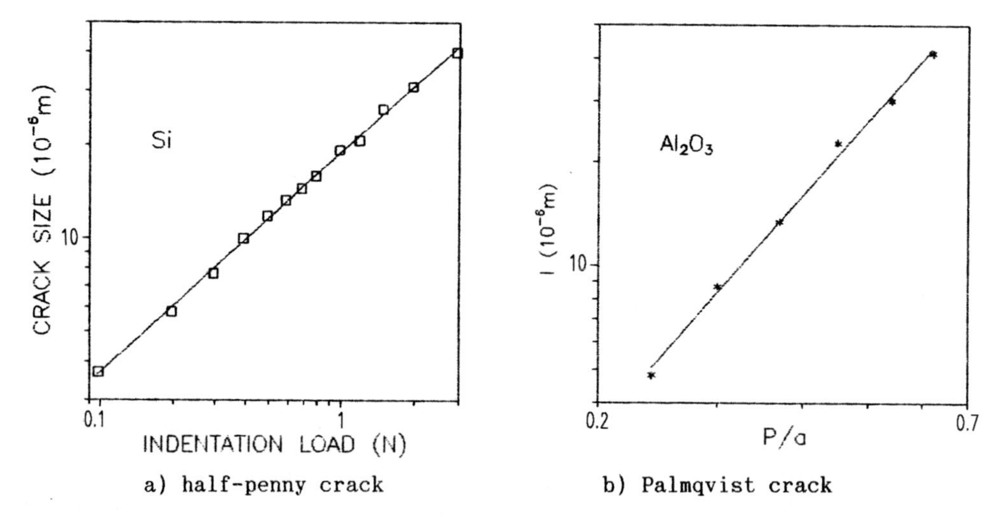

a) half-penny crack b) Palmqvist crack

Fig. 1 Radial crack length vs. Vickers indenter load.

RESULTS AND DISCUSSION

Fig. 1a shows the half-penny radial crack length as a function of indenter load. If it is true that K_{Ic} is proportional to $P/C^{3/2}$ for half-penny cracks then the following relation should hold

$$C = \mathrm{const}P^{\alpha} \qquad (19)$$

where $\alpha = 2/3$ (~ 0.67). It is evident from Table that actually for half-penny cracks α = 0.69-0.71. It is suggested[23] that α for Palmqvist cracks is higher than for half-penny ones. Table shows that this is not always the case. For WC-6Co and WC-3Co alloys α has the same value as for half-penny ones.

For Palmqvist cracks, if the equation (14) is true, then the following equality should be satisfied

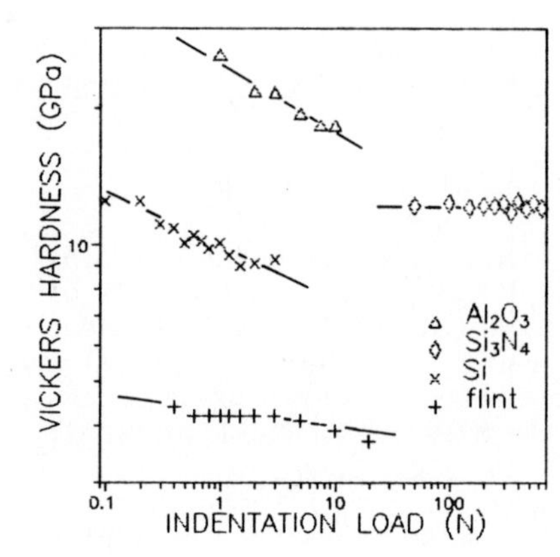

Fig. 2 Hardness as a function of indenter load.

$$l = C - a = \mathrm{const}(P/a)^{\beta} \qquad (20)$$

where $\beta = 2$. This equality is satisfied (Fig. 1b), but β value is far from 2 (Table). For WC-6Co, WC-8Co and Al_2O_3 β is markedly higher than 2 (2.12-2.25) while for WC-3Co it is the same as for materials with half-penny cracks (1.6-1.8). Thus, K_{1c} values calculated from Eqs.(14-16) should be expected to depend greatly on indenter load.

As to Vickers hardness dependence on indenter load the materials under study may be classfied into two groups. For hard alloys, silicon nitrides and LiF the hardness is load independent (Table), while the hardness of Al_2O_3, silicon and glass it dependents on indenter load (Fig. 2).

Knowing K_{1c}, H, E and P-dependence of a and C, the normalized fracture toughness $K_{1c}/Ha^{1/2}$ or $(K_{1c}/Ha^{1/2})(H/E)^{1/2}$ may be plotted versus normalized length of half-penny radial crack C/a and C/a-1 for Palmqvist cracks[7,24]. Fig.3 shows $K_{1c}/Ha^{1/2}$ as a function of C/a. The hardness values of sapphire, silicon and glass were taken equal to 22.0, 10 , and 4.1 GPa respectively. It is obvious that the points in the plot fall into 3 groups: the materials with half-penny cracks and sapphire; above - hard alloys, aside - lithium fluoride. Note, that straight line for WC-3Co points is of the same slope as for materials with half-penny cracks. The slope of the curves for WC-6Co and WC-8Co are lower than for other materials. It might be due to the fact that Palmqvist cracks may differ strongly from each other[41-43]. Tougher materials, WC-6Co and WC-8Co, are characterized by shallow Palmqvist cracks (their depths is less than their lengths). For more brittle WC-3Co the depths of Palmqvist cracks are comparable to their lengths and the cracks are closer to half-penny ones.

Fig. 4 shows $(K_{1c}/Ha^{1/2})(H/E)^{1/2}$ as a function of C/a for materials with half-penny cracks as well as for those with long Palmqvist cracks - WC-3Co, Al_2O_3 and LiF. It is evident that indentation fracture data for all these materials are described by relation

$$(K_{1c}/Ha^{1/2})(H/E)^{1/2} = 0.0298(C/a)^{-1.38} \qquad (21)$$

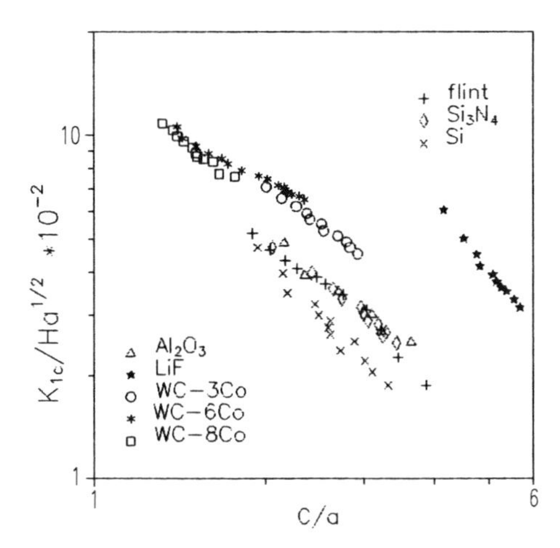

Fig. 3 The normalized crack length plotted as a function of normalized fracture toughness $K_{1c}/Ha^{1/2}$.

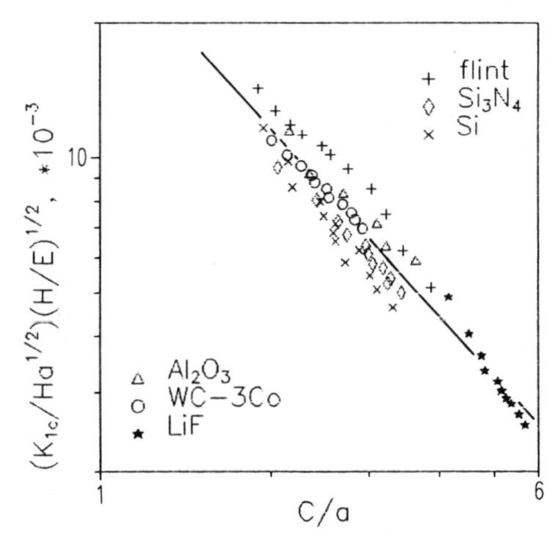

Fig. 4 Radial crack extension plotted using normalized coordinates $(K_{Ic}/Ha^{1/2})(H/E)^{1/2}$ and C/a.

hence

$$K_{Ic} = 0.0298Ha^{1/2}(E/H)^{1/2}(C/a)^{-1.38}, \quad C/a > 2.1 \qquad (22)$$

The exponent was obtained empirically to be 1.38 which is lower than that (1.50) resulting from the solution of the problem of penny crack in an infinite medium to the center of which the point force F is applied. This discrepancy can be explained by the fact that indentation results in two intersecting surface half-penny cracks. It is this difference that is taken by empirical exponent. However, the difference in exponents between Eq.(22) and Eqs(2),(7),(9),(12), and (13) is not great and K_{Ic} values calculated using Eqs(2)-(13) at C/a > 2.1 can be expected not to depend significantly on indenter load. This is confirmed by the data given in Fig. 5. From Eqs.(1)-(13), Eq.(13), derived by Antis et al., gives the best coincidence with conventional fracture toughness test results (Fig. 6a). Variation of calculated K_{Ic} as a function of indentation load for Eq.(22) is shown in Fig. 6b. It is seen that K_{Ic}, calculated from Eq.(22) depends only slightly on load and also are in agreement with conventional fracture toughness test results.

For short (C/a < 2.1) Palmqvist cracks initiating in WC-6Co and WC-8Co hard alloys the following equation (Fig. 7) has been derived

$$K_{Ic}/Ha^{1/2} = 0.0727(C/a-1)^{-0.34} \qquad (23)$$

hence

$$K_{Ic} = 0.0727Ha^{1/2}(C/a-1)^{-0.34}, \quad C/a < 2.1 \qquad (24)$$

Empirically determined exponent (0.34) is much lower than that (0.50) used in Eqs.(14) and (16). Consequently, for materials with C/a < 2.1 the values of K_{Ic} evaluated from Eq.(14) and (16) must decrease with increase in Vickers indenter load. The data shown in Fig. 8 indicate that this is actually the case. K_{Ic} values estimated from Eq.(24) for WC-6Co do not depend on indenter load and agree with those obtained by conventional fracture toughness testing (Fig. 8).

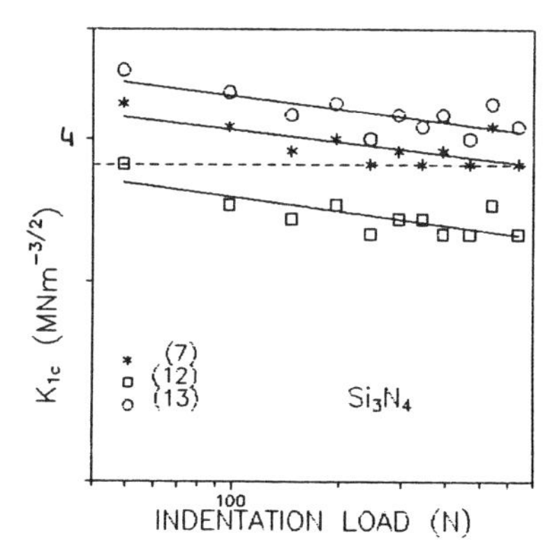

Fig. 5 Variation of calculated K_{1C} as a function of indentation load for Si_3N_4 ceramic.

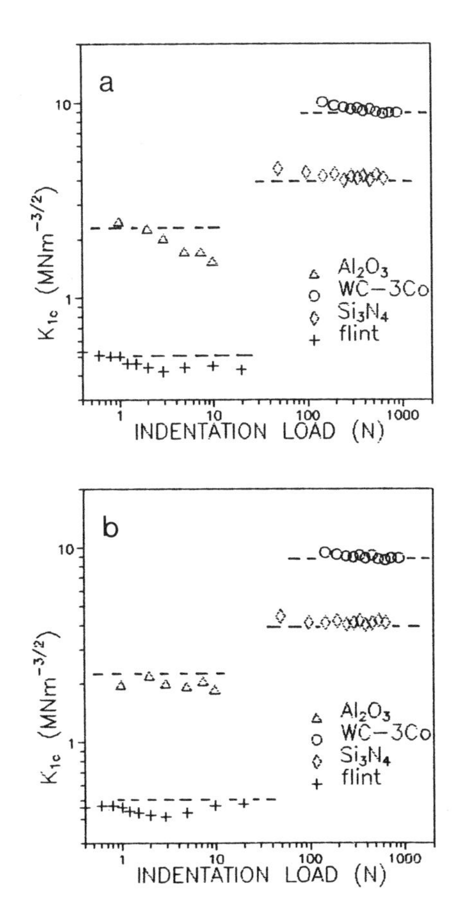

Fig. 6 Variation of calculated K_{1C} as a function of indentation load for a) Eq. (13) and b) Eq. (22).

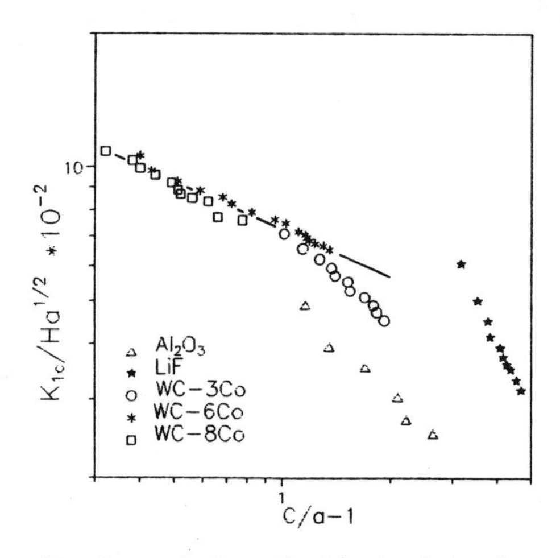

Fig. 7 The normalized crack length C/a-1 plotted as a function of normalized fracture toughness $K_{Ic}/Ha^{1/2}$ for Palmqvist crack

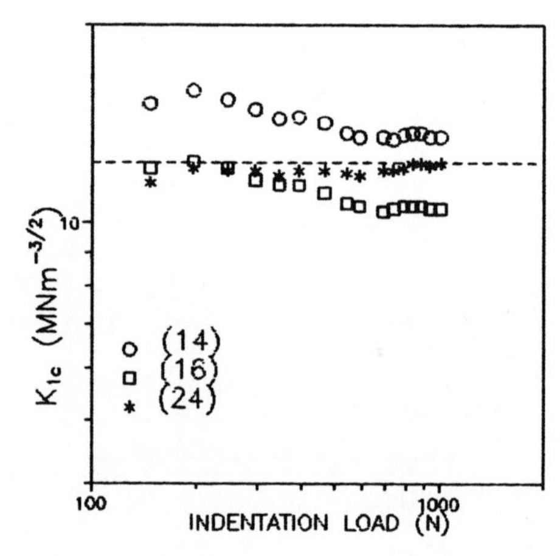

Fig. 8 Variation of calculated K_{Ic} as a function of indentation load for WC-6Co hard alloy.

CONCLUSIONS

For brittle materials, when $C/a > 2.1$, fracture toughness determination by indentation method can be carried out using Eqs. (7),(13), and (22) for both half-penny and Palmqvist cracks. Eq.(12) yields 10 - 20% underestimation of K_{Ic}. As for Eqs.(2) and (9), they can be used for half-penny cracks in the case when hardness is load independent. When $C/a < 2.1$ Eq.(24) should be used.

ACKNOWLEDGEMENTS

The authors would like to thank A. Isakov for providing the specimens of hot pressed silicon nitride, and Ye.R.Dobrovinskaya for preparing the sapphire specimens.

REFERENCES

1. B. R. Lawn and E. R. Fuller, Equilibrium Penny-like Cracks in Indentation Fracture, J. Mater. Sci., 10:2016 (1975).
2. B. R. Lawn and M. V. Swain, Microfracture Beneath Point Indentations in Brittle Solids, J. Mater. Sci., 10:113 (1975).
3. A. G. Evans and T. R. Wilshaw, Quasi-static Solid Particle Damage in Brittle Solids - 1. Observation, Analysis and Implications, Acta Met., 24:939 (1976).
4. A. G. Evans and E. A. Charles, Fracture Toughness Determination by Indentation, J. Am. Ceram. Soc., 59:371 (1976).
5. B. R. Lawn and D. B. Marshall, Hardness, Toughness and Brittleness: an Indentation Analysis, J. Am. Ceram. Soc., 62:347 (1979).
6. A. G. Evans, Brittle Fracture: Micromechanics, in: "Encyclopedia of Materials Science and Engineering" vol. 1 , M. B. Bever, ed., Pergamon Press, Oxford (1986).
7. B. R. Lawn, A. G. Evans and D. B. Marshall, Elastic/Plastic Indentation Damage in Ceramics: the Median/Radial Crack System, J. Am. Ceram. Soc. 63:574 (1980).
8. G. R. Antis, P. Chantikul, B. R. Lawn and D. B. Marshall, A Critical Evaluation of Indentation Techniques for Measuring Fracture Toughness: 1, Direct Crack Measurement, J. Am. Ceram. Soc., 64:533 (1981).
9. D. G. Bhat, Comment on "Elastic/Plastic Indentation Damage in Ceramics: the Median/Radial Crack System", J. Am. Ceram. Soc., 64:C-165 (1981).
10. D. B. Marshall and A. G. Evans, Reply to "Comment on 'Elastic/Plastic Indentation Damage in Ceramics: the Median/Radial Crack Systems'", J. Am. Ceram. Soc., 64:C-182 (1981).
11. K. Niihara, R. Morena and D. P. H. Hasselman, Evaluation of K_{IC} of Brittle Solids by the Indentation Method with Low Crack - to- Indent Ratios, J. Mater. Sci. Lett. 1:13 (1982).
12. K. Niihara, A. Nakahira and T. Hirai, The Effect of Stoichiometry on Mechanical Properties of Boron Carbide, J. Am. Ceram. Soc., 67:C-13 (1984).
13. P. Miranzo and J. S. Moya, Elastic/Plastic Indentation in Ceramics: a Fracture Toughness Determination Method, Ceram. Int., 10:147 (1984).
14. A. G. Evans, Fracture Toughness: the Role of Indentation Techniques, in: "Fracture Mechanics Applied to Brittle Materials, ASTM 678", S. W. Freiman ed., American Society for Testing and Materials, Philadelphia (1979).
15. K. Tanaka, Elastic/Plastic Indentation Hardness and Indentation Fracture Toughness: the Inclusion Core Model, J. Mater. Sci., 22:1501 (1987).
16. J. Lankford, Indentation Microfracture in the Palmqvist Crack Regime: Implications for Fracture Toughness Evaluation by the Indentation Method, J. Mater. Sci. Lett., 1:493 (1982).
17. D. K. Shetty, I. G. Wright, P. N. Mincer and A. H. Claner, Indentation Fracture of WC - Co Cermets, J. Mater. Sci., 20:1873 (1985).
18. M. T. Laugier, New Formula for Indentation Toughness in Ceramics, J. Mater. Sci. Lett., 6:355 (1987).
19. G. Orange, K. M. Liang and G. Fantozzy, Crack Resistance and Fracture Toughness of Alumina and Zirconia Ceramics: Comparison of Notched-Beam and Indentation Techniques, in :"Sci. Ceram. 14: Proc. 14th Int. Conf., Canterburg, Sept. 7-9th 1987". Stoke on Trent (1988).
20. G. C. Sih, "Handbook of Stress Intensity Factors", Institute of Fracture and Solid Mechanics, Lehigh Univ. (1973).
21. N. Miyata and H. Jinno, Fracture Toughness and Fracture Surface Energy of Lead Borate Glasses, J. Mater. Sci. Lett., 1:156 (1982).
22. M. H. Lewis, R. Fung and D. M. Taplin, Indentation Plasticity and fracture of Si_3N_4 Ceramics Alloys, J. Mater. Sci., 16:3437 (1981).

23. Z. Li, A. Ghosh, A. S. Kobayashi and R. C. Bradt, Indentation Fracture Toughness of Sintered Silicon Carbide in the Palmqvist Crack Regime, J. Am. Ceram. Soc., 72:904 (1989).
24. K. Niihara, Indentation Fracture of Ceramics, Ceramics Japan, 20:12 (1985).
25. M. Majdic and G. Ziegler, Correlation of Crack Lengths from Vickers Hardness Indentation and Fracture Toughness of Ceramics Materials, Ber. Dtsch. Keram. Ges., 57:140 (1980).
26. R. P. Ingel, R. W. Rice and D. Lewis, Room-temperature Strength and Fracture of ZrO_2-Y_2O_3 single Crystals, J. Am. Ceram. Soc., 65:C-108 (1982).
27. M. V. Swain, Grain-size Dependence of Toughness and Transformability of 2 mol% - TZP Ceramics, J. Mater. Sci. Lett., 5:1159 (1986).
28. R. L. K. Matsumoto, Evaluation of Fracture Toughness Determination Methods as Applied to Ceria-stabilized Tetragonal Zirconia Polycrystal J. Am. Ceram. Soc., 70:C-366 (1987).
29. A. A. Morrone, S. R. Nutt and S. Suresh, Fracture Toughness and Fatigue Crack Growth Behaviour of an Al_2O_3-SiC Composite, J. Mater. Sci. 23: 3206 (1988).
30. M. Lee and M. K. Brun, Fracture Toughness Measurement by Microindentation and Three-point Bend Methods, Mater. Sci. and Eng., A105/106:369 (1988).
31. P. Lemaitre and R. Piller, Comparison of the Fracture Toughness of Alumina Measured by Three Different Methods, J. Mater. Sci. Lett., 7: 772 (1988).
32. A. Pajares, F. Guiberteau, A. Dominguez-Rodriguez and A. H. Heuer, Microhardness and Fracture Toughness Anisotropy in Cubic Zirconium Oxide Single Crystals, J. Am. Ceram. Soc., 71:C-332 (1988).
33. R. Moussa, I. Coppolani and F. Osterstock, Indentation Techniques Applied to Silicon Carbides, Proc. Brit. Ceram. Soc., 32:237 (1982).
34. K. Homma and T. Torigoe, An Evaluation of Crack Growth Behavior for Silicon Nitride Subjected to Double Torsion and Vickers Indentation by Means of Acoustic Emission Monitoring, Trans. Jap. Soc. Mech. Eng. A51:499 (1985).
35. H. Beall, K. Chyung, R. L. Steward, K. Y. Donaldson, H. L. Lee, S. Baskaran and D. P. H. Hasselman, Effect of Test Method and Crack Size on the Fracture Toughness of a Chain-Silicate Glass-Ceramics, J. Mater. Sci., 21:2365 (1986).
36. K. Hayashi, M. Nomura and T. Nishikawa, Fracture Toughness of LiF Single Crystal, J. Soc. Mater. Sci. Jap., 35:66 (1986).
37. Y. Ochi, A. Ishii, S. K. Sasaki, S. Kurakazu and M. Kawai, Effects of Vickers Indenter Load and Microstructure on Bending Strength and Fracture Toughness in Normal-Sintered Silicon Carbide, Trans. Jap. Soc. Mech. Eng. A, 56:488 (1990).
38. S. N. Dub and G. I. Chepovetskiy, Fracture Toughness Measurement of Brittle Solids from the Lengths of Palmqvist Cracks, in: "Fracture Mechanics and Physics of Brittle Solids", Institute of Material Science Problem, Kiev (1990).
39. B. R. Lawn, D. B. Marshall and P. Chantikul, Mechanics of Strength-Degrading Contact Flaws in Silicon, J. Mater. Sci., 16:1769 (1981).
40. R. F. Cook and G. M. Pharr, Direct Observation and Analysis of Indentation Cracking in Glasses and Ceramics, J. Am. Ceram. Soc., 73:787 (1990).
41. M. T. Laugier, Indentation Cracking in Ceramics and Cermets, Inst. Phys. Conf. Ser., 75:449 (1986).
42. J. D. Sullivan and P.H. Lauzon, Shape Profiles of Cracks Formed Under a Vickers Pyramid Indenter, J. Mater. Sci. Lett., 5:247 (1986).
43. S. L. Jones, C. J. Norman and R. Shahani, Crack - Profile Shapes Formed Under a Vickers Indent Pyramid, J. Mater. Sci. Lett. 6:721 (1987).

THE INDENTATION FRACTURE RESISTANCE OF SELF-REINFORCED MULLITES

Tatsuo Sakai*, Asish Ghosh** and Richard C. Bradt***

* Faculty of Science and Engineering
Ritsumeikan University
Kita-ku, Kyoto, 603 JAPAN

** Philips Display Components Co.
Ann Arbor, MI 48106, USA

*** Mackay School of Mines, University of Nevada-Reno
Reno, NV 89557, USA

INTRODUCTION

A materials science goal for structural ceramics researches is to develop materials with ever-increasing fracture toughnesses. Mullite($3Al_2O_3 \cdot 2SiO_2$) is a promising structural ceramic that develops a duplex microstructure of large acicular, or needle-like crystals within an equiaxed finer grain size matrix during grain growth. By selecting the appropriate thermal treatments, one can develop this duplex microstructure consisting of quite large "fiber-like"(acicular or needle-like) crystals in an equiaxed fine grain size matrix. The sizes of the acicular grains and their volume fraction maybe controlled by the annealing process that promotes grain growth. This microstructure is a special kind of self-reinforced fiber composite in which the material is reinforced by its own fibrous crystals. Thus it is possible to prepare a monolithic, yet fiber reinforced mullite structure having different mechanical properties.

When brittle materials such as mullite are used in structural applications, a high fracture toughness is one of the most important properties to ensure reliability. Consequently, a number of experimental techniques have been proposed and applied to evaluate the fracture toughnesses of structural ceramic materials[1-4]. These include the double cantilever beam(DCB), the single edge notch beam(SENB) and the chevron-notched specimen(CNS). The chevron-notched technique has an interesting feature in that stable crack growth can be easily achieved during the fracture test and thus the R-curve can be readily obtained. In all of these techniques, specific specimen geometries must be prepared for the fracture mechanics tests. However, another thechnique, that of the micro-Vickers indentation method does not require any specific geometry for the test specimen[5-9].

The former specific geometry methods utilizing large fracture mechanics specimens provide a macro-fracture toughness for the material since the crack is a large one and the fracture path proceeds directly through the entire body of the specimen. However, the micro-Vickers indentation technique is limited to a small area at the specimen surface in the vicinity of the indentation. It provides a sort of local fracture toughness value for only a small portion of the material. But, due to the simplicity of the micro-Vickers indentation test, there has been a steadily increasing interest in its application. It has been established that two different types(geometries) of cracks may form in the vicinity of Vickers indentations. One type is the Palmqvist cracks which occur at each of the four indent corners at low loads, while the other is the median type crack which forms in a half-penny shape[10-12]. Median cracks are created only when the applied indentation load exceeds a certain critical or threshold level. Numerous equations have been proposed to calculate the fracture toughnesses for the two respective crack systems[5,9,13-15].

This paper reports micro-Vickers indentation measurements of the fracture toughnesses of stoichiometric mullite processed to have different duplex microstructures containing different volume fractions of large acicular, needle-like grains. Fracture toughness values were calculated by applying equations proposed for the Vickers indentation crack system and the results were compared with values previously determined by the chevron-notched beam technique. Thus, the applicability of the various equations proposed for indentation fracture mechanics can be compared. Furthermore, the most preferable equations are indicated along with some requirements to obtain reasonable fracture toughness through this very convenient and simple technique.

MATERIALS AND EXPERIMENTAL PROCEDURES

Materials

A stoichiometric translucent mullite($3Al_2O_3 \cdot 2SiO_2$) was obtained from a commercial supplier$^+$ in the form of rectangular bars with nominal dimensions of 7mmx7mmx50mm. This mullite was processed from powders prepared by the sol-gel technique followed by spray drying, and subsequent pressing and sintering. The as-received density was 3.14g/cm^3, while the room temperature Young's modulus is 210GPa and the Poisson ratio is 0.29. As-received samples were divided into three sets. One third of the specimens were tested as-received(M-AR), while the remainder were processed to yield self-reinforcing microstructures. A second third was annealed at 1750°C for five hours(M-75), while the remainder were annealed at 1800°C for five hours(M-80) to obtain the acicular grain growth. Thus, three similar microstructures, but with different sizes and volume fractions of acicular needle-like grains were obtained. Further details of the annealing process are described elsewhere[16].

Indentation Toughness Testing

Samples for the indentation toughness measurements were mounted in a transoptic powder mount and polished with successively finer SiC abrassives and finally with 0.3μm alumina in an automatic vibratory polisher until a mirror-like surface was achieved. Vickers indentations were made at test loads of 50, 100, 200, 300, 400, 500, 600, 800 and 1000g, using a commercial microhardness testing machine. The loading time was fixed at 10s. The indentation impression sizes and the crack lengths, as illustrated in Figure 1, were measured immediately after unloading. Only perfect indentations with clearly symmetrical indentations and similarly symmetrical crack patterns were utilized in the final calculations for the fracture toughnesses. The Vickers micro-hardnesses were calculated by the following equation;

$$H_V = 463.6\ P/a^2 \quad \text{(GPa)} \qquad (1)$$

Knoop indentations were similarly made at loads of 50, 100, 200, 300, 400, 500 and 600g. The Knoop micro-hardnesses were calculated from:

$$H_K = 14229\ P/d^2 \quad \text{(GPa)} \qquad (2)$$

$^+$ Kyocera Corp., Kagoshima, Japan

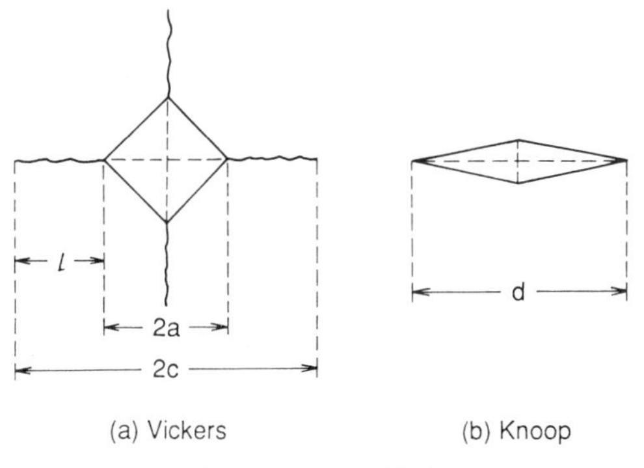

Figure 1 Characteristic lengths of Vickers and Knoop indentations

The true hardness, Ho, was determined based on the load dependence of the Knoop hardness measurements.

Figure 2(a) illustrates Vickers indentations having symmetrical crack patterns for the test loads of 100g, 300g and 500g, whereas Figure 2(b) illustrates Knoop indentations at the same test loads. The Knoop indentations do not exhibit any cracks at these test loads. Examples of Vickers indentations with unacceptable asymmetrical crack patterns are shown in Figure 3, where cracks originate at the sides of the indentation instead of the corners and some cracks exhibit branching. Asymmetrical crack patterns such as those in Figure 3 are not suitable to evaluate the fracture toughness. Therefore, Vickers indentations were repeatedly made until twenty-five acceptable symmetrical crack patterns were obtained. The fracture toughnesses, their averages and confidence intervals were calculated from those 25 patterns.

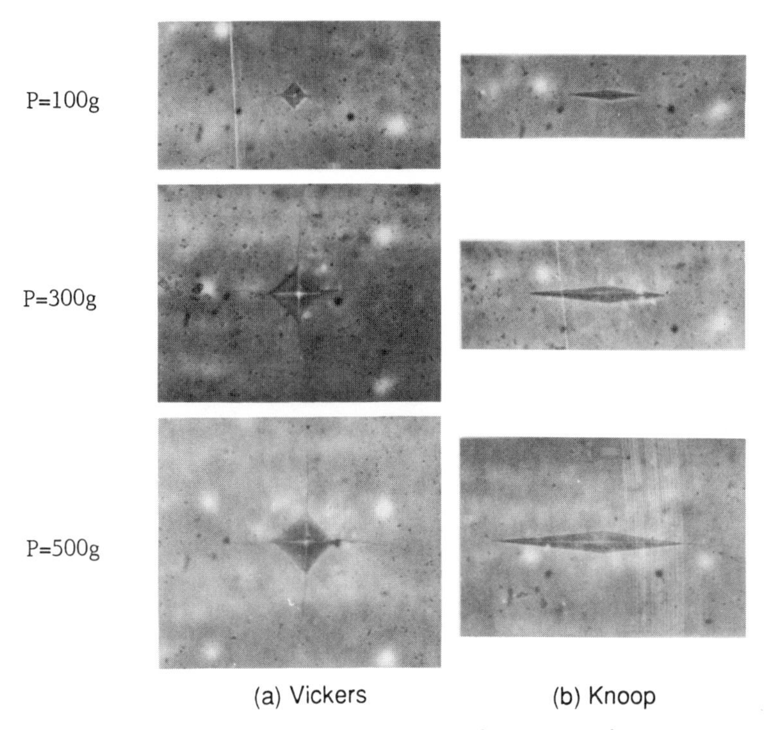

Figure 2 Examples of Vickers and Knoop indentations

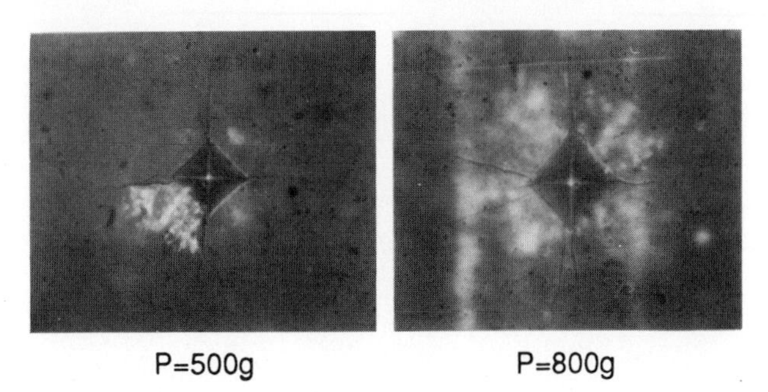

Figure 3 Examples of unacceptable crack patterns around Vickers indentations

Vickers indentation cracks can be of either the Palmqvist or median crack variety. The identification of the type is essential to the evaluation of the fracture toughness by the indentation microfracture technique. A number of similar equations have been proposed to calculate the fracture toughness from the indentation crack lengths. These have been summarized by Ponton et al.[5] and by Li et al.[9]. They can be categorized by three different approaches. Palmqvist cracks are addressed as either a semi-elliptical crack[10,17] or a 2-dimensional through crack[7]. One approach considers median cracks as half-penny shapes[8,18-22]. A third group of equations is based on empirical curve fitting methods[13-15]. The equations can be summarized along with their proposers as follows;

(1) Palmqvist Cracks

Niihara et al.[11]:

$$(K_{IC}\phi/Ha^{1/2})(H/E\phi)^{0.4} = 0.035\ (l/a)^{-1/2} \tag{3}$$

and Shetty et al.[7]:

$$K_{IC} = (HP/4l)^{1/2}/[3\pi(1-\nu^2)(2^{1/2}\tan\Psi)^{1/2}] \tag{4}$$

(2) Median Cracks

Lawn and Swain[8]:

$$K_{IC} = \frac{(1-2\nu)(\eta H)^{1/2}}{2^{1/2}\pi^2\ \xi}\ (\frac{P}{c})^{1/2} \tag{5}$$

Lawn and Fuller[18]:

$$K_{IC} = (1/\pi^{3/2}\ \tan\Psi)P/c^{3/2} \tag{6}$$

Evans and Charles[6]:

$$(K_{IC}\phi/Ha^{1/2})(H/E\phi)^{0.4} = k(c/a)^{-3/2} \tag{7}$$

Niihara et al.[11]:

$$(K_{IC}\phi/Ha^{1/2})(H/E\phi)^{0.4} = 0.129\ (c/a)^{-3/2} \tag{8}$$

Lawn, Evans and Marshall[19]:

$$K_{IC} = 0.028\ Ha^{1/2}\ (E/H)^{0.5}(c/a)^{-3/2} \tag{9}$$

Anstis et al.[20]:

$$K_{IC} = 0.016\ (E/H)^{0.5} P/c^{3/2} \quad (10)$$

Japanese Industrial Standard(JIS)[21]:

$$K_{IC} = 0.018\ (E/H)^{0.5} P/c^{3/2} \quad (11)$$

and Tanaka[22]:

$$K_{IC} = 0.0725\ P/c^{3/2} \quad (12)$$

(3) Curve-Fitting Approaches

Blendell[13]:

$$(K_{IC}\phi/Ha^{1/2})(H/E\phi)^{0.4} = 0.055\ \mathrm{Log}(8.4a/c) \quad (13)$$

Evans[14]:

$$(K_{IC}/Ha^{1/2})(H/E)^{0.4} = 10^{y} \quad (14)$$

where

$$y = -1.59+0.34x-2.02x^2+11.23x^3-24.97x^4+15.32x^5 \quad (15)$$

and Lankford[15]:

$$(K_{IC}\phi/Ha^{1/2})(H/E\phi)^{0.4} = 0.142\ (c/a)^{-1.56} \quad (16)$$

In the above equations: ϕ is the ratio of the hardness to the yield stress, (H/σ_y), which is about 3; Ψ is the half angle of the Vickers indenter equal to 68°; η and ξ are dimensionless constants which are equal to $2/\pi$ and 2, respectively, and $x=\mathrm{Log}(c/a)$. The above equations for the indentation fracture toughness were reduced to similar forms by substituting the required parameters and are summarized in Table 1.

RESULTS AND DISCUSSIONS

Vickers and Knoop Microhardnesses

The Vickers indentations and their crack lengths were measured for the nine levels of the test load. A summary of the percentages of perfect and unacceptable patterns are shown in Table 2. In the case of the P=50g for the as received mullite(M-AR), no cracks were observed at the corners of the indentations, so no results are presented in the corresponding column. It is observed that the percentage of unacceptable crack patterns increases with an increase in the indentation testing load. For the same test load, the number of unacceptable patterns increases in the microstructural order M-AR, M-75 and M-80. For higher testing loads, only about two thirds of the indentations are acceptable for toughness calculations.

Characteristic dimensions(a, c and l) of the Vickers indentations are summarized in Table 3. The Vickers microhardness calculated by Equation(4) are depicted in Figure 4 as the function of the testing load. The Vickers microhardnesses appear to be independent of the test load for each microstructure. However, a slight decrease of the Vickers microhardness occurs following the high temperature grain growth annealing heat treatments. When cracks accompany the indentations, those cracks appear to have an effect on the size of the indentation. This is not surprising as most researchers believe that cracked indentations do not yield the correct hardness values.

Knoop microhardness measurements were also completed for several levels of test loads: 50, 100, 200, 300, 400, 500 and 600g. For loads of P>600g, occasional cracks were observed to originate from both tips of the

Table 1 Equations for calculation of K_{IC} values from Vickers indentation crack systems

Palmqvist crack system equations		
P-1	Niihara, et al. [11]	* $K_{IC}=0.0433(HP/l)^{1/2}$ $K_{IC}=0.0436(HP/l)^{1/2}$ $K_{IC}=0.0441(HP/l)^{1/2}$
P-2	Shetty, et al. [7]	$K_{IC}=0.0310(HP/l)^{1/2}$
Median crack system equations		
M-1	Lawn & Swain [8]	$K_{IC}=0.0120(HP/c)^{1/2}$
M-2	Lawn & Fuller [18]	$K_{IC}=0.0726P/c^{3/2}$
M-3	Evans & Charles [6]	$K_{IC}=0.0742P/c^{3/2}$
M-4	Niihara, et al. [11]	* $K_{IC}=0.109P/c^{3/2}$ $K_{IC}=0.109P/c^{3/2}$ $K_{IC}=0.111P/c^{3/2}$
M-5	Lawn, et al. [19]	* $K_{IC}=0.0674P/c^{3/2}$ $K_{IC}=0.0679P/c^{3/2}$ $K_{IC}=0.0689P/c^{3/2}$
M-6	Anstis et al. [20]	* $K_{IC}=0.0770P/c^{3/2}$ $K_{IC}=0.0776P/c^{3/2}$ $K_{IC}=0.0788P/c^{3/2}$
M-7	JIS [21]	* $K_{IC}=0.0866P/c^{3/2}$ $K_{IC}=0.0873P/c^{3/2}$ $K_{IC}=0.0887P/c^{3/2}$
M-8	Tanaka [22]	$K_{IC}=0.0725P/c^{3/2}$
Curve fitting equations		
C-1	Blendell [13]	$K_{IC}=0.0285H^{.6}E^{.4}a^{.5}Log(8.4a/c)$
C-2	Evans [14]	$K_{IC}=H^{.6}E^{.4}a^{.5}10^{Y}$
C-3	Lankford [15]	$K_{IC}=0.0735H^{.6}E^{.4}a^{.5}(a/c)^{1.56}$

* Since coefficient depends on hardness, equation is provided on each of M-AR, M-75 and M-80.

Table 2 Percentage of perfect crack patterns of total measurements at each load

Load (g) (N)	50 0.49	100 0.98	200 1.96	300 2.94	400 3.92	500 4.90	600 5.88	800 7.84	1000 9.80
M-AR mullite									
No. perfect	*	25	25	25	25	25	25	25	25
No. unaccept.	*	0	5	6	7	6	10	10	13
% perfect	*	100	83	81	78	81	71	71	66
M-75 mullite									
No. perfect	25	25	25	25	25	25	25	25	25
No. unaccept.	1	3	9	9	10	11	11	12	13
% perfect	96	89	74	74	71	69	69	68	66
M-80 mullite									
No. perfect	25	25	25	25	25	25	25	25	25
No. unaccept.	2	2	9	11	11	12	12	13	14
% perfect	93	93	74	69	69	68	68	66	64

* No cracks were observed for this case.

Table 3 Characteristic lengths a, c, l and Vickers hardnesses for different test loads

Load	(g)	50	100	200	300	400	500	600	800	1000
	(N)	0.49	0.98	1.96	2.94	3.92	4.90	5.88	7.84	9.80
M-AR mullite										
a	(μm)	4.17	5.90	8.60	10.65	12.05	13.32	14.96	17.30	19.55
σa	(μm)	0.14	0.17	0.19	0.18	0.19	0.17	0.15	0.19	0.11
c	(μm)	---	10.23	17.66	24.29	30.53	34.35	38.17	46.07	53.52
σc	(μm)	---	3.90	1.15	0.78	1.82	1.02	1.09	1.33	1.28
l=c-a	(μm)	---	4.34	9.06	12.89	18.48	21.03	23.21	28.77	33.97
Hv	(GPa)	13.1	13.1	12.3	12.0	12.5	12.8	12.2	12.2	11.9
M-75 mullite										
a	(μm)	4.28	6.15	8.77	10.77	12.28	13.93	15.48	17.92	20.47
σa	(μm)	0.17	0.28	0.22	0.23	0.20	0.30	0.33	0.39	0.53
c	(μm)	6.92	11.48	19.58	26.66	30.59	35.69	40.97	49.15	56.46
σc	(μm)	3.14	3.59	0.87	1.50	1.46	1.57	1.41	2.32	2.20
l=c-a	(μm)	2.64	5.32	10.81	15.89	18.31	21.76	25.49	31.24	35.99
Hv	(GPa)	12.4	12.1	11.8	11.8	12.1	11.7	11.4	11.3	10.9
M-80 mullite										
a	(μm)	4.58	6.25	8.89	11.19	12.41	14.23	16.08	18.60	20.90
σa	(μm)	0.29	0.36	0.36	0.39	0.35	0.47	0.96	0.92	0.80
c	(μm)	7.41	11.87	20.28	28.03	32.15	38.12	42.89	51.46	58.92
σc	(μm)	1.69	3.64	1.29	1.92	1.99	1.98	2.27	1.95	2.85
l=c-a	(μm)	2.83	5.62	11.40	16.84	19.74	23.88	26.81	32.86	38.01
Hv	(GPa)	10.9	11.7	11.6	10.9	11.8	11.3	10.7	10.6	10.4

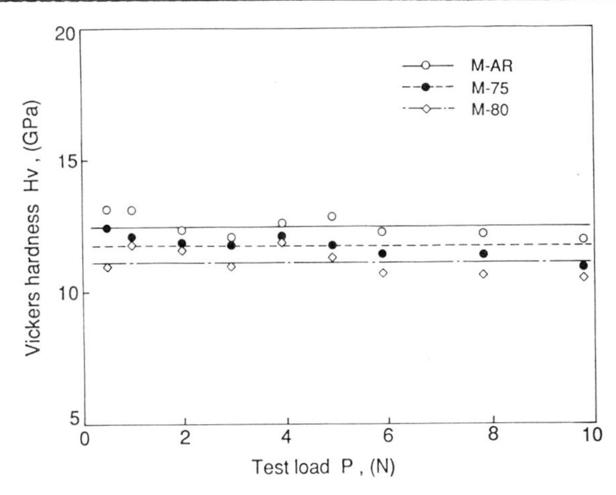

Figure 4 Vickers hardness as function of test load

long diagonal. Since the objective of this test is to determine the true hardness as explained later, the Knoop indentations with cracks were not used for the hardness measurements. By repeating twenty five measurements, the mean value and standard deviations of the Knoop hardness were obtained for each of the three mullite microstructures. Table 4 summarizes the experimental results. The Knoop hardnesses are plotted in Figure 5 as a function of the test load. A rapid decrease with an increase of the test load is found in the low test load regime, but tends to be test load independent as the load increases. This trend is similar for all three of the mullite microstructures. Again, similar to the Vickers microhardness, the Knoop microhardness is decreased by the grain growth anneals.

From the Knoop microhardness, which does not have cracks associated with the indentations, the true hardness, Ho, can be determined for each mullite as Li et al.[9] have done for SiC. True hardness can be successfully obtained from the above results by the following equation;

$$d = (14229/H_o)^{1/2} P^{1/2} - d_e \quad (17)$$

Table 4 Knoop hardnesses at different indentation test loads

Load		50	100	200	300	400	500	600
	(g)	50	100	200	300	400	500	600
	(N)	0.49	0.98	1.96	2.94	3.92	4.90	5.88
M-AR mullite								
d	(μm)	19.95	30.98	47.58	60.20	70.16	80.16	87.87
σd	(μm)	0.69	0.63	0.83	0.54	0.86	0.48	0.99
Hk	(GPa)	17.6	14.6	12.3	11.6	11.3	10.9	10.8
M-75 mullite								
d	(μm)	21.60	33.09	50.15	62.14	72.21	82.30	90.63
σd	(μm)	0.86	1.12	1.52	1.17	1.07	1.37	2.03
Hk	(GPa)	15.0	12.8	11.1	10.9	10.7	10.3	10.2
M-80 mullite								
d	(μm)	22.14	33.91	51.08	63.44	74.02	83.26	92.45
σd	(μm)	0.80	0.82	0.75	1.57	4.15	1.53	1.89
Hk	(GPa)	14.3	12.2	10.7	10.4	10.3	10.1	9.8

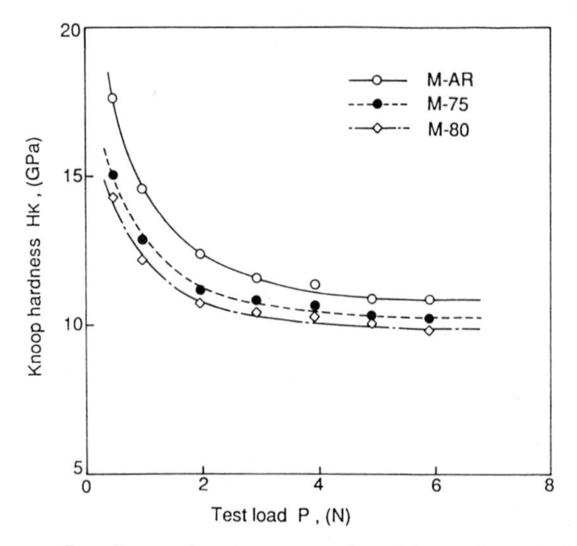

Figure 5 Knoop hardness as function of test load

Therefore, a plot of the test load, $P^{1/2}$, versus the indentation length, d, yields a slope which is equal to $(14229/Ho)^{1/2}$. Figure 6 illustrates the relationships between $P^{1/2}$ and d for these three mullites, in which excellent linearity is evident. From the slopes of each regression line, the true hardnesses of the respective microstructures were calculated by Equation(17). These values are indicated in Table 5 along with the apparent values of the Vickers microhardness H_V(P=1000g) and the Knoop microhardness H_K(P=600g). The true hardnesses for these three mullite microstructures are 9.1, 8.9 and 8.7GPa, decreasing as the grain size increases from annealing.

The three different hardness values which have been determined for these three mullite microstructures, namely the Vickers microhardness, the Knoop microhardness and the calculated true hardness have some characteristics which merit further discussion and consideration. This is the situation both for a comparison of the three mullite microstructures and for a comparison of the hardness measurement techniques. It is evident that the finer grain size mullites are harder by each of the three measures of hardness. This is to be expected, for it is generally accepted that the fine grain sizes of crystalline solids are harder than coarser grain sizes. Unfortunately, because of the duplex nature of the microstructures of these

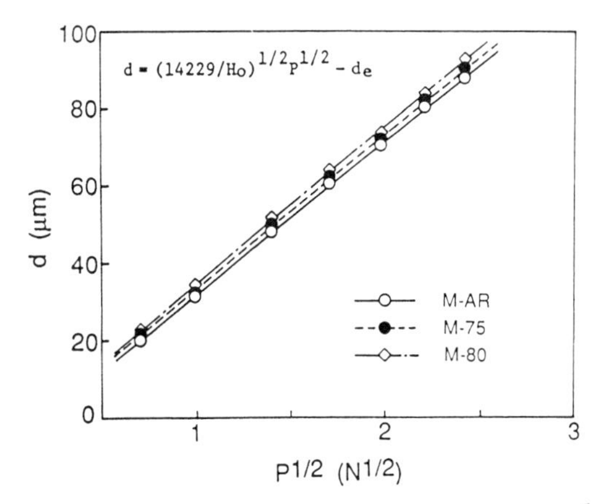

Figure 6 Relationships between d and $P^{1/2}$

Table 5 Vickers, Knoop and true hardnesses (GPa)

Material	Vickers H_V(1000g)	Knoop H_K(600g)	True Ho
M-AR	11.9	10.8	9.1
M-75	10.9	10.2	8.9
M-80	10.4	9.8	8.7

mullites, it is not convenient to address the hardness/microstructure relationship on a quantitative basis such as the usual $d^{-1/2}$ analysis. It will suffice to note that these mullites ascribe to the expected hardness/grain size trend.

The Vickers(Figure 4) and the Knoop(Figure 5) microhardnesses reveal distinctly different trends of hardness versus testing load for the same range of test loads. The Vickers microhardness values always exceed the Knoop microhardness. The Vickers microhardness is essentially independent of the test load, whereas the Knoop microhardness exhibits a strong indentation size(test load) effect on the microhardness. The reason for this is that the Vickers indentations are much deeper and all are associated with crack patterns at loads of 1N and above. The cracked-pattern Vickers hardnesses are not valid hardnesses, other than perhaps for a very cursory comparison with one another. However, in spite of the extensive cracking it is appropriate to note that all three of these mullite structures do retain their same order of hardness for each of the testing loads. The much shallower-indentation Knoop microhardnesses reveal a strong load dependence and still maintain the order expected on the basis of grain size for all of the test loads. The Knoop microhardness values also approach the true hardnesses at the higher test loads.

Fracture Resistance

The crack growth resistance curves and the fracture toughnesses have been determined for the present mullites by Ghosh[16] using the chevron-notched specimen technique. The fracture toughnesses of these three mullites are the same regardless of the microstructural differences, yielding 2.24±0.39 MPa·$m^{1/2}$. Measurements of the fracture toughness by fracture mechanics specimens have been made on mullites by several other researchers as shown in Table 6. Kanzaki et al.[23] examined the effect of Al_2O_3 content on the fracture toughness by means of the single edge notch

Table 6 Fracture toughnesses of mullite reported in references

Reference	Method	Fracture toughness
Kanzaki et al. [23]	SENB	2.4 ∿ 2.8 $MPa \cdot m^{1/2}$
Mah and Mazdiyasni[24]	CSF	1.8 $MPa \cdot m^{1/2}$
Ismail et al.[25]	VIM	2.73 $MPa \cdot m^{1/2}$
Mizuno and Saito[26]	VIM	2.02 $MPa \cdot m^{1/2}$
Sato et al.[27]	VIM	2.2 $MPa \cdot m^{1/2}$

SENB ; Single edge notch beam method
CSF ; Controlled surface microflaw method
VIM ; Vickers microhardness indentation method

beam(SENB) technique, and reported that the fracture toughness at room temperature varied from 2.4 to 2.8$MPa \cdot m^{1/2}$ depending on the stoichiometry of the mullite. Mah and Mazdiyasni[24] investigated the fracture toughness of mullite by using a controlled surface microflaw(CSF) technique, where a semicircular microcrack was provided by a Knoop microhardness indentation. They reported the fracture toughness to be 1.8$MPa \cdot m^{1/2}$ at room temperature. Ismail, et al.[25], Mizuno, et al.[26] and Sato, et al.[27] have reported some other experimental results for the fracture toughness obtained by the Vickers microhardness indentation method. These are also summarized in Table 6.

Indentation Fracture Toughness

To evaluate the indentation fracture toughnesses, it is essential to distinguish the Palmqvist and the median crack types. It is often difficult from only a direct observation of the cracks, because both types of cracks appear quite similar on the specimen surface. However, as first noted by Niihara et al.[11,17], a significant difference exists in the relationship of the crack length, l, versus the test load, P. For low test loads, the relationship is proportional with a steep slope, but when the load exceeds a certain critical value, the slope changes abruptly. The initial, low-load, portion originates from Palmqvist cracks, whereas the latter is indicative of median cracks. It is, therefore, possible to distinguish the two crack types by plotting the crack length, l, as a function of the test load, P.

The relationships for the three mullite microstructures are depicted in Figure 7, where each is presented by shifting the abscissa. It is apparent that the relationship of l versus P possesses a distinct knee of transition at a critical test load. Distinct linearity exists in the region where P is less than that of the knee for which the solid regression lines are presented. In the regions higher than the critical load, the linearity of the data is not as good, so that dashed regression lines are similarly applied to distinguish those regions. The test loads at the knee points are approximately the same for all three of the mullite microstructures.

According to the experimental results by Breval et al.[28] and the summary of equations[9], the following relationship has been confirmed for median cracks;

$$c = A P^{2/3} \qquad (18)$$

whereas Palmqvist cracks are described by

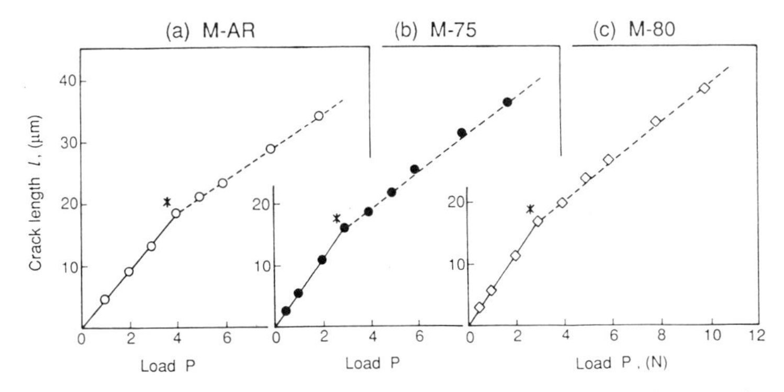

Figure 7 Relationships between crack length l and test load P

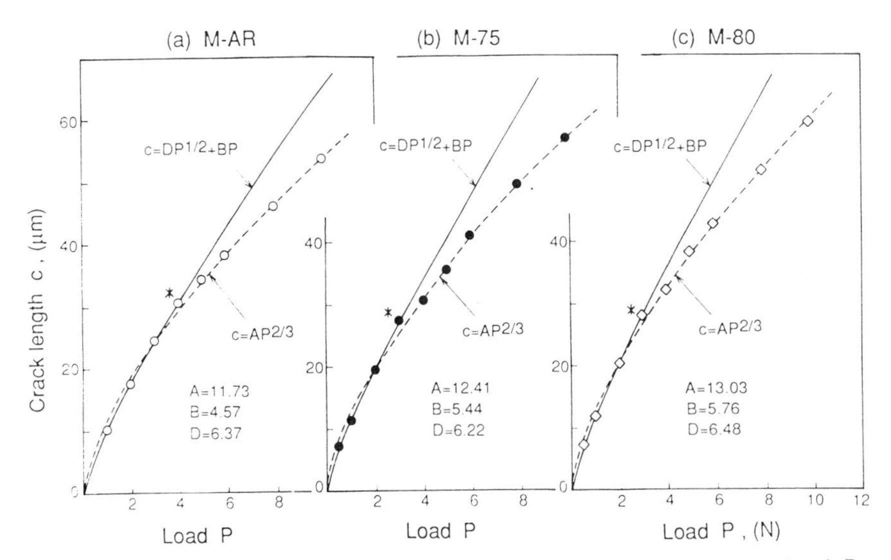

Figure 8 Relationships between crack length c and test load P

$$l = B P \qquad (19)$$

The solid lines in Figure 7 correspond to the latter equation with the material constant B. Since c is equal to $(a+l)$, the relationship between P and c for the Palmqvist crack system can be provided by

$$c = a + l = D P^{1/2} + B P \qquad (20)$$

Thus the relationships between c and P for both types of cracks are known. Figure 8 depicts these two relationships. The constant B is readily determined from the slope of the solid line in Figure 7, whereas A is obtained from the results for the median cracks in Figure 8 by regression analysis. The other constant D can be also determined such that the two regression lines pass through the respective knee points. The values of A, B and D are presented in each diagram.

Once the crack types have been identified, the fracture toughnesses can be evaluated. The toughnesses calculated from the equations in Table 1 are depicted as the function of the test load P in Figures 9, 10 and 11. These results exhibit a wide range of values from 0.5 to $4.0 MPa{\cdot}m^{1/2}$. The

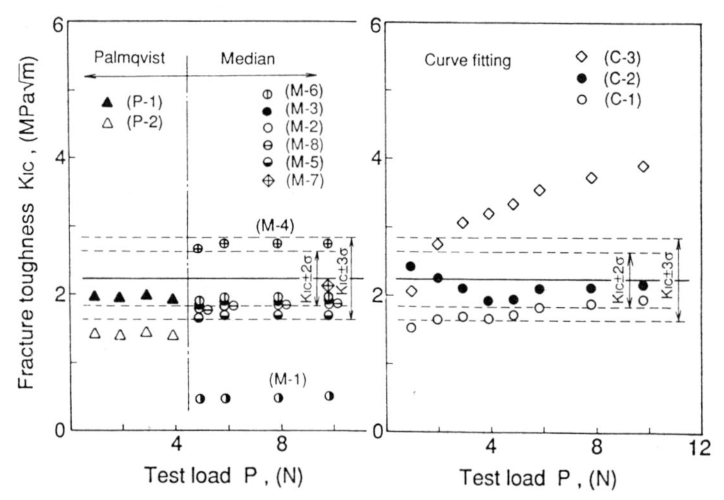

Figure 9 Fracture toughness as function of test load(M-AR)

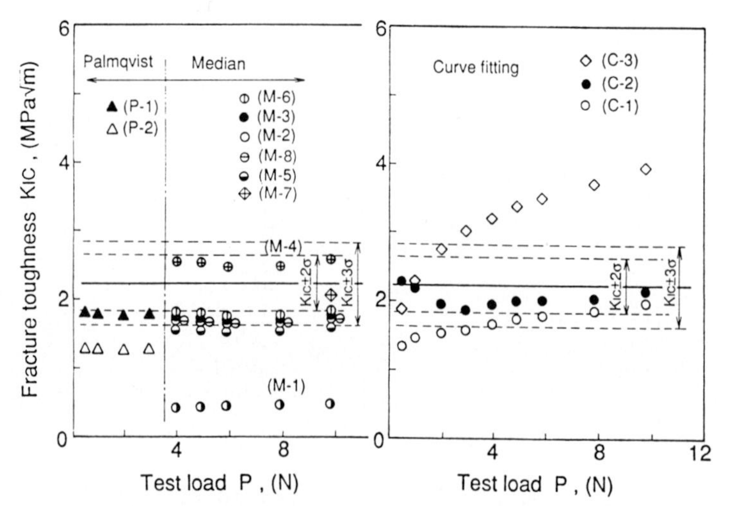

Figure 10 Fracture toughness as function of test load(M-75)

fracture toughness by JIS[Equation(11)] was calculated only for the result at the highest test load due to the recommendation of the standard. This value was indicated by the mark of ⊕ in each diagram. The macroscopic fracture toughnesses after Ghosh[16] which were determined by the chevron-notched beam technique are presented as the averages and bounds of $K_{IC}+2\sigma$ and $K_{IC}+3\sigma$ by the solid and dashed lines. Although only the Niihara equation(P-1) provides a similar result for Palmqvist cracks, six different equations(M-2, -3, -5, -6, -7, -8) yield comparable values for the median cracks. Among them, the fracture toughness based on JIS indicates the remarkable agreement with the average value obtained by the chevron-notched beam technique. However, for median cracks, Equation(M-4) gives too high a value, while Equation(M-1) yields much too low a value.

For the indentation fracture toughnesses calculated from the curve-fitting equations(C-1 and C-3), a test load dependence monotonically increasing with an increase of the load is observed. For the case of Equation(C-2), the fracture toughness result also indicates a test load

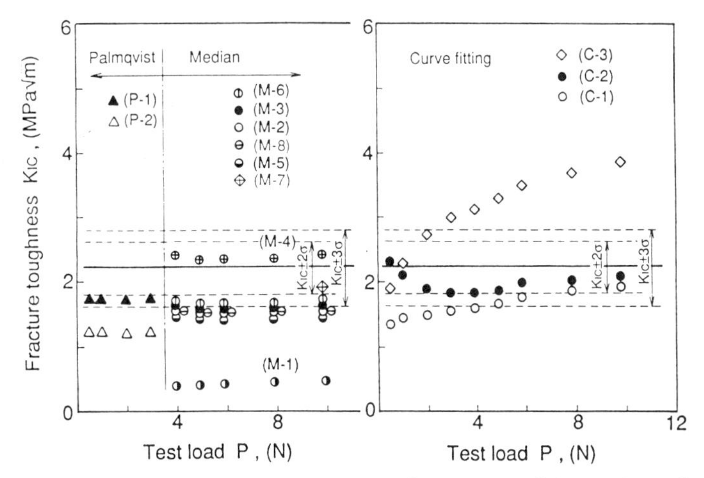

Figure 11 Fracture toughness as function of test load(M-80)

dependence, but one such that a minimum occurs around the point of the crack-type transition from Palmqvist to median cracks. This load dependence occurs because of the estimate of the fracture toughness without regard for the crack type. This emphasizes that, for the evaluation of fracture toughness by the indentation technique, it is essential to identify the crack type in advance. Examination of the relationship between the crack length and the test load is useful to distinguish between the respective types of cracks.

SUMMARY AND CONCLUSIONS

The fracture resistances of three stoichiometric($3Al_2O_3 \cdot 2SiO_2$) mullites having different microstructures were determined by the micro-Vickers indentation fracture method, and the microhardnesses of the structures were determined. The results were compared with one another and previously published results for mullite. The main conclusions of this study are as follows;

Vickers hardnesses calculated from indentations with cracks are almost constant regardless of the test load, while the Knoop hardness decreases with increasing test load. The true hardness is successfully obtained by compensating for the testing load dependence, and it decreases slightly with an increase of the size of the microstructure.

The type of indentation crack can be distinguished, depending on the test load P. When P is lower than a critical value, Palmqvist cracks occur, but median cracks develop when P exceeds the critical level. The ratio of c/a at the critical load is about 2.5. It is necessary to identify the crack type, when the fracture toughness is estimated by the indentation technique. By examining the applicability of the numerous equations proposed to calculate the indentation fracture toughness, the Niihara equation(P-1) provides the most reasonable result for Palmqvist cracks. For median cracks, several equations(M-2, -3, -5, -6, -7, -8) provide satisfactory results, in agreement with those obtained by the chevron-notched conventional fracture mechanics specimen technique. Among them, the remarkable agreement was confirmed for the equation(M-7) proposed in the Japanese Industrial Standard.

ACKNOWLEDGEMENT

The authors express their acknowledgements to the Kyocera Corporation for supplying the experimental materials and to Ritsumeikan University and NASA for their financial support of this study.

REFERENCES

(1) R.W.Davidge, "Mechanical Behaviour of Ceramics", Cambridge University Press, 1979

(2) M.Sakai, "Evaluation Methods for Ceramics : Fracture Toughness of Polycrystalline Ceramics", Bull. Ceram. Soc. Japan, 23[5], 480-488 (1988)

(3) M.G.Jenkins, A.S.Kobayashi, M.Sakai, K.W.White and R.C.Bradt, "Fracture Toughness Testing of Ceramics Using a Laser Interferometric Strain Gage", Amer. Ceram. Soc. Bull., 66[12], 1734-1738 (1987)

(4) M.G.Jenkins, A.Ghosh, K.W.White, A.S.Kobayashi and R.C.Bradt, "The Fracture Resistance of A Sintered Silicon Carbide Using The Chevron-Notch Bend Specimens", Proceedings of the Symposium on Silicon Carbide held at Columbus, OH, Aug.1-3, (1987)

(5) C.B.Ponton and R.D.Rawlings, "Dependence of the Vickers Indentation Fracture Toughness on the Surface Crack Length", Trans. Brit. Ceram. Soc., 88[3], 83-90 (1989)

(6) A.G.Evans and E.A.Charles, "Fracture Toughness Determinations by Indentation", J. Amer. Ceram. Soc., 59[7-8], 371-372 (1976)

(7) D.K.Shetty, I.G.Wright, P.N.Mincer and A.H.Clauer, "Indentation Fracture of WC-Co Cermets", J. Mat. Sci., 20, 1873-1882 (1985)

(8) B.R.Lawn and M.V.Swain, "Microstructure Beneath Point Indentations in Brittle Solids", J. Mat. Sci., 10, 113-122 (1975)

(9) Z.Li, A.Ghosh, A.S.Kobayashi and R.C.Bradt, "Indentation Fracture Toughness of Sintered SiC in the Palmqvist Crack Regime", J. Amer. Ceram. Soc., 72[6], 904-911 (1989)

(10) S.Palmqvist, "Method att bestamma segheten hos sproda material, sarskilt hardmetaller", Jernkontorets Ann., 141, 300-307 (1957)

(11) K.Niihara, R.Morena and D.P.H.Hasselman, "Evaluation of K_{IC} of Brittle Solids by the Indentation Method with Low Crack-to-Indent Ratios", J. Mat. Sci. Lett., 1, 13-16 (1982)

(12) J.G.P.Binner and R.Stevens, "The Measurement of Toughness by Indentation", Trans. Brit. Ceram. Soc., 83[6], 167-172 (1985)

(13) J.E.Blendell, "The Origins of Internal Stresses in Polycrystalline Alumina and Their Effects on Mechanical Properties", PhD Thesis, MIT (1979)

(14) A.G.Evans, "Fracture Toughness : The Role of Indentation Techniques", 112-135, in Fracture Mechanics Applied to Brittle Materials, ASTM STP 678, S.W.Freiman, Ed., (1979)

(15) J.Lankford, "Indentation Microstructure in the Palmqvist Crack Regime : Implications for Fracture Toughness Evaluation by the Indentation Method", J. Mat. Sci. Lett., 1, 493-496 (1982)

(16) A.Ghosh, "Effect of Microstructure and Temperature on the Fracture Resistance of Duplex Microstructure Mullite", Ph.D. Thesis, Univ. of Washington, Seattle, WA, U.S.A. (1989)

(17) K.Niihara, "A Fracture Mechanics Analysis of Indentation-Induced Palmqvist Crack in Ceramics", J. Mat. Sci. Lett., 2, 221-223 (1983)

(18) B.R.Lawn and E.R.Fuller, "Equilibrium Penny-like Cracks in Indentation Fracture", J. Mat. Sci., 10, 2016-2024 (1975)

(19) B.R.Lawn, A.G.Evans and D.B.Marshall, "Elastic/Plastic Indentation Damage in Ceramics : The Median/Radial Crack System", J. Amer. Ceram. Soc., 63[9-10], 574-581 (1980)

(20) G.R.Anstis, P.Chantikul, B.R.Lawn and D.B.Marshall, "A Critical Evaluation of Indentation Techniques for Measuring Fracture Toughness : I , Direct Crack Measurements", J. Amer. Ceram. Soc., 64[9], 533-538 (1981)

(21) Japanese Industrial Standard, JIS R 1607, "Testing Methods for Fracture Tiughness of High Performance Ceramics", Japanese Standard Association, (1990)

(22) K.Tanaka, "Elastic/Plastic Indentation Hardness and Indentation Fracture Toughness : The Inclusion Core Model", J. Mat. Sci., 22, 1501-1508 (1987)
(23) S.Kanzaki, H.Tabata, T.Kumazawa and S.Ohta, "Sintering and Mechanical Properties of Stoichiometric Mullite", J. Amer. Ceram. Soc., 68[1], C6-C7 (1985)
(24) T.I.Mah and K.S.Mazdiyasni, "Mechanical Properties of Mullite", J. Amer. Ceram. Soc., 66[10], 699-703 (1983)
(25) M.G.M.U.Ismail, Z.Nakai and S.Somiya, "Microstructure and Mechanical Properties of Mullite Prepared by The Sol-Gel Method", J. Amer. Ceram. Soc., 70[1], C7-C8 (1987)
(26) M.Mizuno and H.Saito, "Preparation of Highly Pure Fine Mullite Powder", J. Amer. Ceram. Soc., 72[3], 377-382 (1989)
(27) T.Sato, M.Ishizuka and M.Shimada, "Sintering and Characterization of Mullite-Alumina Composites", Ceram. Int., 12, 61-65 (1986)
(28) E.Breval, G.C.Dodds and N.H.MacMillan, "The Hardness, Stiffness and Toughness of Diphasic Abrasive Materials Prepared by Sol-Gel Techniques", Mat. Res. Bull., 20[4], 413-429 (1985)

CLEAVAGE OF CERAMIC AND MINERAL SINGLE CRYSTALS

Richard A. Schultz and Richard C. Bradt

Mackay School of Mines
University of Nevada, Reno
Reno, Nevada 89557 U.S.A.

ABSTRACT

The general aspects of single crystal cleavage are reviewed and the various criteria which have been proposed for the phenomenon are considered. It is suggested that a K_{IC} (fracture toughness) criterion is the most appropriate one for the description of cleavage. The traditional mineralogical structural classification of cleavage types which is based on fracture directionality (dimensionality) is then applied to several crystal structures in each catagory, from which it is concluded that the cleavage toughnesses of single crystals are often less than 1 MPa $m^{1/2}$.

In several cases, reliable measurements of polycrystalline fracture toughnesses also exist which can be compared to the single crystal cleavage toughness values. Polycrystalline toughnesses appear to be consistently greater than the single crystal cleavage toughnesses. The role of texture or preferred orientation is also briefly addressed. It is concluded that toughnesses may be highly anisotropic and that the toughness values for intensely oriented (textured) microstructures may approach those for single crystal cleavage.

INTRODUCTION

The cleavage of single crystals is a fascinating phenomenon that readily arouses the curiosity of every scientist who considers the fracture process. Practically every mineralogy textbook and guidebook for rock and mineral collectors has a section on this interesting phenomenon[1-5]. In its purest form, cleavage occurs when a crystalline material preferentially fractures along a planar surface which is determined by characteristics of the crystal structure. What appear to be very smooth planar cleavage surfaces may extend over substantial areas. These areas often are accompanied by finer scale very sharp and distinct cleavage steplike patterns when a multiplicity of cleavage planes exists. Figure 1 illustrates cleavage surfaces of calcite ($CaCO_3$) viewed at several different levels of magnification. The distinctive features of cleavage are easily recognized during the examination of those fracture surfaces (fractography). It is evident that fundamental elements of the crystal structure must dominate the fracture process whenever well defined cleavage occurs.

Fracture Mechanics of Ceramics, Vol. 10
Edited by R.C. Bradt *et al.*, Plenum Press, New York, 1992

Although this brief description of cleavage along with the definitive illustration in Figure 1 may imply that cleavage is a rather perfect phenomenon, it in fact may vary widely in quality from one structure to another. Even though the individual members of isostructural groups of crystals usually exhibit cleavage on the same crystal planes (rock salt or halite, NaCl, $\{100\}$; zinc blende, ZnS, $\{110\}$; fluorite, CaF_2, $\{111\}$ etc.), the cleavage is not

Figure 1. The ever-decreasing scale of the geometry of the cleavage of calcite ($CaCO_3$) on the $\{10\bar{1}1\}$ rhombohedral cleavage plane.

always uniform in its characteristics. The quality of cleavage is frequently catagorized as *perfect*, *good*, *distinct*, and *indistinct*. The term *parting* is also sometimes used to refer to less well defined cleavage, yet describing a fracture which may occur in a crystallographically dominated sense. Frequently, an isostructural group of crystals may exhibit a transition from perfect to indistinct cleavage with a trend of some characteristic of the structural components such as ionic size, charge, or polarizability.

It is perhaps appropriate to briefly consider the non-cleavage fracture process. As cleavage is a process dominated by the crystal structure, fracture in a non-crystalline or amorphous material such as glass may be expected to exhibit a fracture character that is free of the geometric restraints of crystal structure. In fact, glass does just that, yielding a characteristic type of fracture surface that is known as *conchoidal.* Very fine grain size polycrystalline ceramics and single crystals that do not have strongly preferred cleavage planes such as quartz (SiO_2) may also exhibit conchoidal fracture under certain conditions of failure. The features of conchoidal fracture surfaces are usually dominated by the macroscopic externally applied stress state rather than a specific feature of the structure of the material. It is also appropriate to note that some crystal structures simply do not possess or exhibit distinctive cleavage planes. One such very common crystal structure is that of garnet ($Fe_3Al_2Si_3O_{12}$), a cubic crystal structure that finds commercial application as an abrasive. Garnet is hard and reportedly tough, but it does not possess a distinctive cleavage plane, although it is sometimes described as exhibiting {110} parting with uneven fracture that is conchoidal in nature[5].

In addition to the obvious asthetic and scientific aspects of cleavage, the phenomenon also has features that are of practical interest. The cleavage of diamond on the {111} plane is of course paramount to its use as a gemstone, as is the cleavage of other gemstones for their beauty. In the crushing and grinding of minerals during their beneficiation, the process of cleavage is every bit as critical as it is to the gemstone industry. Cleavage is also important to the abrasives field, as most commercial abrasives owe their performance to their cleavage characteristics; frequently this is because a cleavage type of fracture continually renews the sharp cutting edges of the grain during industrial processes. In contrast, as previously noted, the garnet structures, both the pyralspite and ugrandite series, reportedly owe their abrasive qualities to their lack of distinct renewable cleavage planes. Thus, one can conveniently support and promote the study of cleavage not only for its occurrence, but also for its lack of presence in some crystal structures or in specific instances.

THE CRITERION FOR CLEAVAGE

For most researchers, observing a single example of cleavage generates the curiosity to ask what determines the crystallographic planes on which cleavage occurs and many related questions as well. The answers have not always been very gratifying, but the proposed criteria certainly merit review if for no other reason than to put the entire crystal cleavage process into proper perspective[6,7]. As might be imagined, the distinctive geometry of the cleavage process and the *external geometry* of naturally occurring crystals has lead to some associations of these two features. However, the crystal planes which predominate during the natural growth of crystals are not consistently the same planes that are prominent during cleavage. Nevertheless, a few instances of the correspondence of the two, namely cleavage planes and crystal growth planes, are known to most crystallographers. The {100} plane for the halite (NaCl) structure and the {111} for the fluorite (CaF_2) structure are two familiar examples, as halite cubes and fluorite octahedra can be found in abundance. As tempting as this correlation may seem, any critical examination of the two processes quickly informs the scientist that the correspondence of cleavage planes and growth planes must be a fortuitous one for those few instances where in fact they are the same.

Once it is concluded that the growth planes aren't the criteria for the cleavage planes of crystals, then it is only logical to consider the *planes bounding the unit cell* as a possibility, for they also exhibit the distinctive geometry that is shared by cleavage planes. One has only to consider the case of the halite (NaCl) structure to find a familiar example of the correspondence between the unit cell planes and the cleavage planes; i.e., the {100}. However, applying the bounding planes of the unit cell as a criterion for cleavage fails once a wider examination of structures is considered. Obviously, the fluorite (CaF_2) structure, which is cubic yet cleaves on the {111}, and also the zinc blende (ZnS) structure, which is also cubic but cleaves on the {110}, provide clear contradictions to this unit cell bounding plane criterion. It must be rejected as the determinant criterion for the cleavage planes of crystals.

Continuing with the structural approach to cleavage, it is appropriate to consider the *most closely packed crystal planes* as a possibility. As the slip planes for dislocation motion are usually the closely packed planes and thus have clearly demonstrated their ability to experience the breaking of bonds during plastic flow, there is a natural tendency to suspect that they may also be susceptible to cleavage. Similar to the previous suggestions, the most closely packed plane criterion does not yield a consistent result for many crystal structures. The halite (NaCl) structure cubes and the classical cleavage rhombs of the calcite ($CaCO_3$) structure are just a couple of the numerous contradictions. Other simple structural criteria which have been considered and which are familiar to all who study crystallography can also be proposed and, in fact, many do apply in specific instances. However, continuing pursuit of a critical structural feature that defines the cleavage planes for all crystal structures leads to repeated discouragement and finally to the conclusion that it must be the *properties* of the crystal structure which should be examined rather than the crystal structure itself. Of course the two are related.

Once it is realized that consideration of the crystal properties is perhaps the correct approach to understanding cleavage, then various crystal properties can be examined as possible criteria. However, since cleavage involves the breaking of bonds, it naturally follows that a cleavage criterion must necessarily physically relate to the bonds and the ions, groups of ions, or atoms which constitute those bonds. As simple as it may seem, it is essential to understand that cleavage constitutes the rupture or breaking of the bonds, not the "cracking" of the ions or the atoms. Once cleavage occurs, then the two resulting fracture surfaces consist of matching pairs of ions or atoms across the newly created surfaces and which previously constituted the structure. This paired matching of surface features is a basic factor which prevents certain crystal planes from experiencing cleavage in simple ionic structures. Ionic crystals will not cleave on those crystal planes which result in all positive charges on one fracture face and all negative charges on the other. From the charge perspective, these planes are known as non-neutral planes. Examples of non-neutral planes in several common ionic structures are the {111} in the halite (NaCl) structure, both the {110} and {100} in the fluorite (CaF_2) structure, and the {100} and {111} in the zinc blende (ZnS) structure. The converse of this argument must be that only neutral planes, in the sense of their ionic charge distribution, can be cleavage planes. This is easily recognized on the {100} planes of the halite (NaCl) structure where cations and anions alternate. It is equally obvious that if the bonds perpendicular to the {111} in the halite (NaCl) structure were to cleave, then planes of all cations would separate from planes of all anions, yielding two oppositely charged cleavage surfaces. The *charge effect* just described is not a positive criterion for cleavage; rather, it is an exclusionary one which specifies that non-neutral planes simply will not experience cleavage in any highly ionic crystal structure.

Having focused on the bonds within the crystal structures, it is appropriate to address several of the various cleavage criteria that have been suggested on the basis of the bonds. These criteria all suffer by the rather imprecise nature of the bonds in ceramics and minerals, their varying strengths, a balance of ionicity versus covalency, and the anisotropy which they impart to the various crystal structures. Nonetheless, this approach does provide some additional structural insight and merits review and consideration. The *bond density concept* is perhaps the first of several ideas which have been advanced. It simply proposes that the minimum number of bonds per unit area determines the cleavage planes. Lower bond densities are associated with weaker bonding crystal planes that are more readily cleaved. This bond density approach is not very satisfactory, nor has it been very successful, for it cannot readily incorporate the different bond strengths and the anisotropy of those bond strengths into the criterion.

Of course, the natural extension of applying the bond density concept is to consider that property which is directly related to the bond strengths, the *elastic modulus*. Stronger bonds should result in a higher Young's modulus perpendicular to the plane. In contrast, weaker bonds should have lower elastic moduli and thus identify the cleavage planes as those with the minimum Young's modulus. As appealing as this may seem, it doesn't consistently apply even to the most simple structures. For example, sylvite (KCl) and galena (PbS), which both have the halite (NaCl) structure and a {100} cleavage plane, also have their Young's modulus maximum in the <100>. The diamond cubic structure, which has the Young's modulus maximum in the <111>, also cleaves on the {111}. Cleavage in these examples is exactly the opposite to the behavior which is expected from the application of an elastic modulus criterion. It is evident that the elastic modulus is not, by itself, capable of consistently predicting the cleavage planes of all crystal structures. The same type of summary statement could perhaps also be applied to a criterion based on the *surface energy* of the crystal planes. It can be argued that an energy criterion, in which the cleavage plane is the plane of the minimum surface energy, should apply. However, like the previously discussed criteria, Gilman[6] applies the minimum surface energy concept and notes that it too has some exceptions, noticeably the potassium halides. The minimum surface energy concept is not a sufficient one to consistently predict single crystal cleavage planes.

In summarizing this examination of the various cleavage criteria which are based on structure and properties, it may be concluded that each of the several aforementioned criteria can correctly predict the cleavage plane in certain crystal structures, but fails when applied in a universal sense. In retrospect, this should not be surprising because cleavage is a dynamic fracture process and none of the aforementioned structural features or physical properties actually describe a fracture process. If a cleavage criterion is to be specified or considered, then a *fracture criterion* must be applied. Because cleavage occurs rapidly and catastrophically, application of the critical crack growth resistance parameters for the opening mode such as K_{IC}, G_{IC}, or J_{IC} seems to be appropriate. As the focus of this paper is primarily on brittle engineering ceramics and because most ceramic scientists and engineers are currently applying the stress intensity concept in design, materials evaluation, and materials testing, the remainder of this approach to cleavage will utilize K_{IC}, the *fracture toughness*, as the cleavage criterion of choice.

To justify the choice of K_{IC} as a cleavage criterion, it is appropriate to directly refer to several experimental studies where the researchers have actually made experimental measurements on several different planes of a crystal structure. Usually these measurements are made by some form of indentation precracking or by utilizing guide notches to force the incipient crack to initiate on the crystallographic plane of interest. Of course, after only a very short distance of propagation, the crack is usually redirected by the crystal structure onto the cleavage plane as several of the following authors illustrate in their publications. In Table I, the experimental cleavage toughnesses are listed for several low indice crystal planes each for lithium fluoride (LiF), gallium phosphide (GaP), silicon (Si), and spinel ($MgAl_2O_4$). In every instance, the well established cleavage plane has the *lowest* fracture toughness value. Very interestingly, for the primarily ionic halite structure of lithium fluoride (LiF), where charge effects are dominant, the cleavage toughness differences between the crystal planes are much more pronounced than for the more covalently bonded gallium phosphide (GaP) and silicon (Si).

Table I. Experimental Cleavage Toughnesses for Several Single Crystals

Crystal	*Plane*	K_{IC} *(MPa m$^{1/2}$)*	*Authors*
LiF	{100}*	0.50	Hayashi et al. [8]
	{110}	0.70	
	{111}	1.50	
GaP	{100}	0.73	Hayashi et al. [9]
	{110}*	0.65	
	{111}	0.81	
Si	{100}	0.95	Chen and Leipold [10]
	{110}	0.90	
	{111}*	0.82	
$MgAl_2O_4$	{100}*	1.18	Stewart and Bradt [11]
	{110}	1.54	
	{111}	1.90	

*Indicates cleavage plane.

As the literature is not always specific, nor very clear as to the technique by which the K_{IC} value is actually measured or calculated, the original references are included in this paper for the reader to consult for further details. For those cases from the literature where some form of a surface energy value has been converted to a K_{IC} value, the fracture toughnesses have been obtained for the various tables included in this paper by applying the formula:

$$K_{IC} = (2\,E\,\gamma)^{1/2} \qquad (1)$$

where γ is the fracture surface energy and E is the Young's modulus perpendicular to the cleavage plane. Equation (1) does not include the $(1-\nu^2)$ term in the denominator of the square root term for the plane strain condition, which is assumed in many instances[12]. As most brittle ceramics have a Poisson's ratio, ν, of about 0.2, that term is 0.96 or a difference of only 1.02 once the square root is considered. That minor factor is not a significant difference in terms of the concepts which are advanced in this paper. As the reader will appreciate, neither is the fact that some of the measured toughnesses which are referenced in the summaries may not fully meet some of the specimen size restrictions which are imposed by standards for obtaining valid measurements of K_{IC}.

STRUCTURAL EFFECTS ON SINGLE CRYSTAL CLEAVAGE

To address the single crystal *cleavage toughnesses* of ceramics in terms of their structure, it is not only convenient, but also pedagogically sound, to initially consider the cleavage of various crystal structures by using the mineralogical approach which has been presented by Sorrell[1]. Figure 2 illustrates a slight modification of that approach to cleavage which is based on the geometric aspect of the directionality (dimensionality) of cleavage. The structural catagories begin with the cleavage of layer structures, including mica and graphite, two materials which are well known for their ease of cleavage. The classification then proceeds with two-directional cleavage and then the situation for cleavage planes which constitute three-directional (dimensional) geometries that are familiar to all crystallographers.

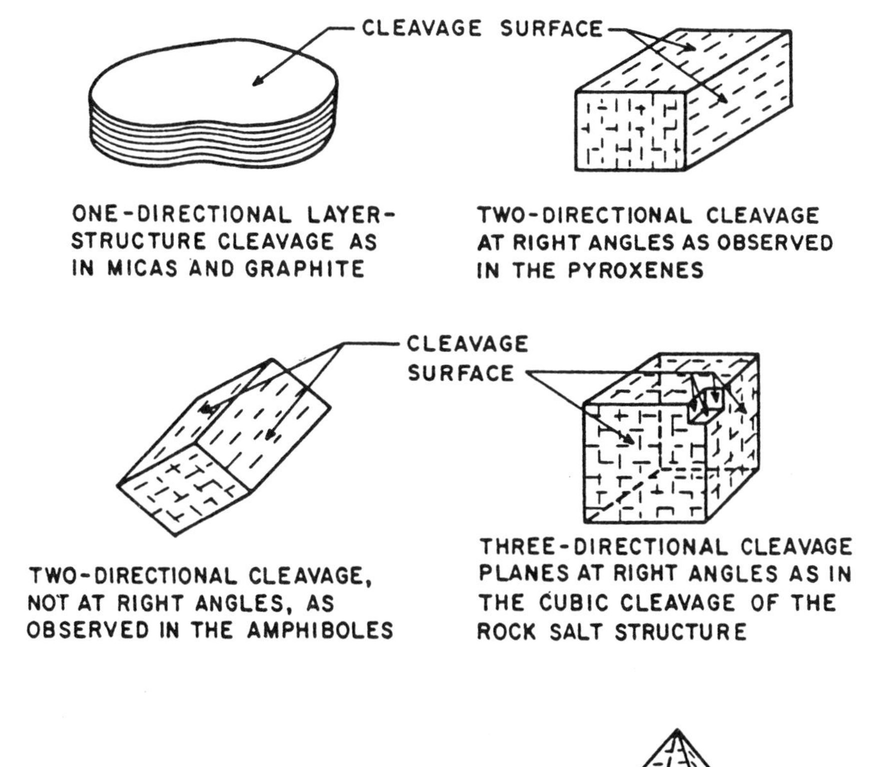

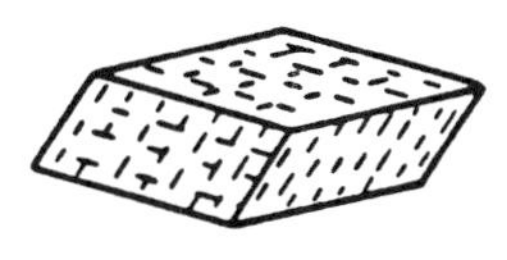

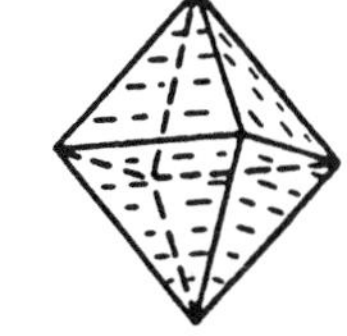

Figure 2. A mineralogical structural approach to the classification of single crystal cleavage, after Sorrell[1].

However, not all ceramics and minerals of interest, nor those for which toughnesses have been measured, can be conveniently catagorized by the three three-directional (dimensional) groups presented in Figure 2, namely those with multiple cubic, rhombohedral, and octahedral cleavage planes. Consequently, several crystal structures must be discussed in an individual sense beyond the initial mineralogical structural classification which Sorrell[1] has presented.

Cleavage in the *layer structures* such as the numerous micas and graphite may be expected to be similar for many other layer structures such as those of various clays and molybdenite (MoS_2). Unfortunately, only a few of these structures have been the subject of very thorough studies, yet the results to date for layer structures do provide a reliable baseline for the discussion of other single crystal cleavage toughness measurements. These structures also provide a physical basis to understand just how easy the single crystal cleavage process may be under nearly ideal conditions.

Table II summarizes the cleavage toughness values for several layer structures. The first graphite toughness is estimated from theoretical surface energies and elastic moduli in Gilman's review article[6] and seems to be extraordinarily low. However, an explanation can be offered which suggests that actual experimental results may be preferable to theoretical calculations. For example, if the graphite toughness calculated from theory (K_{IC} = 0.03 MPa $m^{1/2}$) is compared with that which Sakai et al.[13] have measured for pyrolytic carbon (K_{IC} = 0.53 MPa $m^{1/2}$), it is evident that about an order of magnitude difference exists. As most theoretical approaches to the calculation of surface energies are based on equilibrium models and the fracture process which leads to the formation of cleavage surfaces is a dynamic, non-equilibrium one that is not purely planar on a local microscopic basis, it should not be surprising that a substantial difference is observed. In retrospect it would be remarkable if theory and experiment yielded the same cleavage toughnesses. The order of magnitude difference in Table II for the two graphite cleavage toughnesses is not at all unreasonable. However, the most important message from the summary of layer structure cleavages is that their toughnesses are probably less than 1 MPa $m^{1/2}$.

Table II. Cleavage Toughnesses of Several Layer Structures

Material	*Plane*	*K_{IC} (MPa $m^{1/2}$)*	*Authors*
graphite	{0001}	0.03	Gilman [6]
pyrolytic carbon	{0001}	0.53	Sakai et al. [13]
muscovite mica	{0001}	0.19	Gilman [6]; Wan et al. [14]
$YBa_2Cu_3O_x$	{0001}	1.10	Cook et al. [15]
ice		0.11	Liu and Miller [16]

Also listed in Table II is the reported cleavage toughness for the ceramic superconductor, $YBa_2Cu_3O_x$ [15]. Similar to the other layer structure toughnesses, it is also low, although it appears to be somewhat higher than the graphite and the mica. It might be argued that the measurement is perhaps too high. However, the cleavage toughness of this structure is not too different than the other layer structures and is still relatively low in comparison to many other toughnesses[12]. The $YBa_2Cu_3O_x$ result supports the conclusion that the cleavage of layer structures is not very high in fracture toughness. Most cleavage toughnesses of layer structures appear to be less than 1 MPa $m^{1/2}$.

Recently, Lawn and coworkers[14] have been studying the cleavage of muscovite mica. They have observed that moisture has a significant effect on the crack growth resistance, often by as much as a factor of five. That finding and its relationship to charging effects on the mica fracture surfaces is highly significant from a fundamental perspective. However, a factor of five times the reported fracture toughness given in Table II for mica does not alter the conclusion that the cleavage toughnessses of layer structures are rather low.

The cleavage of *chain structures* is usually parallel to the chains and is often called prismatic for prism-like geometric forms result. Cleavage toughness values for single or double chain silicate structures as specified in the classification by Sorrell[1] apparently have not been measured, as none seem to be reported in the literature. Nevertheless, some facts have been established regarding the cleavage of these structures. The cleavage planes are parallel to the chain lengths as the chains contain the strongest bonds in the structures. Crystals having the *single chain structure* are typified by the pyroxene group of rock forming minerals, including enstatite ($MgSiO_3$), diopside ($CaMgSi_2O_6$), wollastonite ($CaSiO_3$), hedenbergite ($CaFeSi_2O_6$), ferrosilite ($FeSiO_3$), hypersthene ($(Mg,Fe)SiO_3$), and jadeite ($NaAl(SiO_3)_2$). As a practical note, several of the above pyroxenes are also known to form in slags during steel production. Cleavage in the single chain silicate structures is two-directional and nearly at right angles, as shown in Figure 1, although one cleavage plane usually predominates. Crystals having the *double chain structure* are typified by the amphibole group, including tremolite ($Ca_2Mg_5(Si_8O_{22})(OH,F)_2$), actinolite ($(CaFe_5Si_8O_{22})(OH,F)_2$), hornblende ($(Ca,Na,K)_{2\text{-}3}(Mg,Fe,Al)_5(Si,Al)_8O_{22}(OH,F)_2$), and glaucophane ($Na_2Mg_3Al_2Si_8O_{22}(OH,F)_2$).

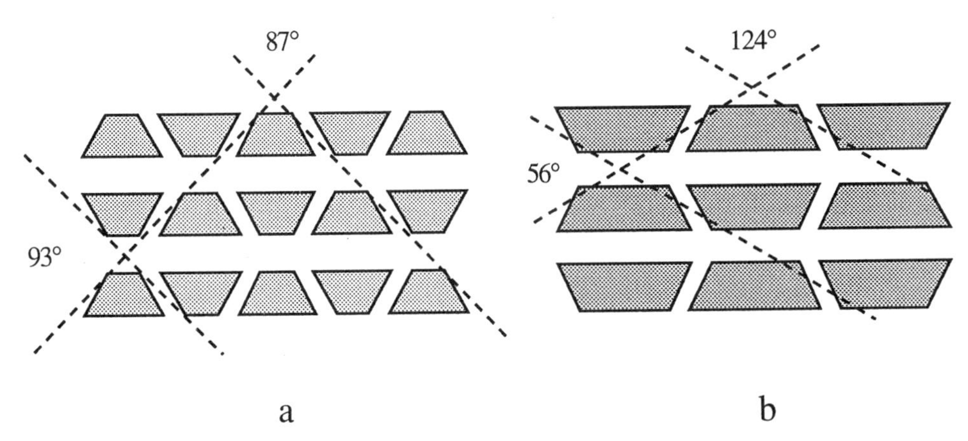

Figure 3. Relationship of cleavage planes to the silica tetrahedra linkages in the chain silicates. Chains are depicted on end, showing rough trapezoidal cross sections. (a) Single chain cleavage. (b) Double chain cleavage. Note the angles.

Cleavage in double chain crystals is also two-directional, but at angles different from 90°. It is well established that the chain silicates cleave in a manner to yield surfaces with characteristic angles. As noted above, the pyroxenes appear to exhibit distinctive right angles, but actually cleave at angles of 93° and 87°. These angles are very different from the amphiboles which appear to yield angles of about 60° and 120°, although actually 56° and 124°. Cleavage toughnesses of the single and double chain silicate structures are likely to be less than 1 MPa $m^{1/2}$ once they are measured and reported.

Following the cleavage of the chain structures, Sorrell[1] addresses the three-directional (dimensional) cleavage situations. The simplest of these geometric cleavages is the *cubic* variety as commonly observed for the halite (NaCl) structure, a cubic crystal structure which also cleaves on the {100}. Table III lists a number of crystals for which cubic cleavage is observed and fracture toughnesses are available. Structures other than the halite (NaCl) structure can also exhibit cleavage that displays the cubic geometry. As for the previous layer cleavages, these fracture toughnesses are not very large either. In fact, most are less than 1 MPa $m^{1/2}$, especially for those crystals with little or no covalent bonding.

Table III. Cleavage Toughnesses for Some Cubic Cleavages

Crystal	*Plane*	K_{IC} *(MPa* $m^{1/2}$*)*	*Authors*
KBr	{100}	0.12	Freiman/Becher/Klein [17]
KCl	{100}	0.15	Freiman/Becher/Klein [17]
NaCl	{100}	0.17	Freiman/Becher/Klein [17]
LiF	{100}	0.31	Ahlquist [18]
	{100}	0.50	Hayashi et al. [8]
MgO	{100}	0.81	Gilman [6]
PbS	{100}	0.18	Atkinson and Avdis [19]
	{100}	0.32	Gilman [6]
FeS_2	{100}	0.96	Atkinson and Avdis [19]
$MgAl_2O_4$	{100}	1.18	Stewart and Bradt [11]

The values in Table III suggest a superimposed ionic size effect on these cleavage toughness results. That effect can be anticipated if the toughnesses of lithium fluoride (LiF) and magnesia (MgO) are compared with those of potassium bromide (KBr) and galena (PbS). The former two crystals consist of higher elastic modulus structures with smaller ions, whereas the latter two consist of larger, more polarizable ions. The cleavage toughnesses of the former (LiF and MgO) are about three times those of the latter (KBr and PbS). The more tightly bonded, higher elastic modulus crystal structures have substantially larger cleavage toughnesses.

Pyrite (FeS_2) and spinel ($MgAl_2O_4$) single crystals do not have the halite (NaCl) structure, but do exhibit cubic cleavage on the {100} planes. It is of interest that these two crystal structures appear to have higher cleavage toughnesses, both about 1 MPa $m^{1/2}$. Pyrite (FeS_2) is a metallic-like crystal, where in the case of the spinel ($MgAl_2O_4$) the increased toughness may arise from the stronger bonds with the trivalent cations. This is partially speculation, but the sequence of increasing cleavage fracture toughness in the series LiF : MgO : $MgAl_2O_4$ could be related to the ionic charge contributions. These subtleties notwithstanding, the cleavage toughnesses for the cube geometry or mode, i.e., the {100} of cubic crystal structures is rather similar to that of the layer structure cleavage toughnesses, but perhaps slightly larger. Some cubic cleavages appear to have toughnesses of about 1 MPa $m^{1/2}$, but in general, cubic cleavages are not very tough.

Following the cubic cleavage geometry, Sorrell[1] lists *rhombohedral* and *octahedral* cleavage geometries. However, there are several published results of cleavage toughnesses for the zinc blende (ZnS) structure, a cubic crystal structure which cleaves on the {110}. If the mineralogical structural classification were maintained, then this would be *dodecahedral* cleavage, which is sort of a three-dimensional type of cleavage with some of the angles being right angles and some angles which are not. The values of several of these cleavage toughness are listed in Table IV.

Table IV. Cleavage Toughnesses for the Dodecahedral Cleavage Geometry

Crystal	*Plane*	K_{IC} *(MPa* $m^{1/2}$*)*	*Authors*
GaP	{110}	0.65	Hayashi et al. [9]
ZnS	{110}	0.23	Gilman [6]
InS	{110}	0.21	Gilman [6]

The cleavage toughnesses of the *dodecahedral* geometry listed in Table IV are all less than 1 MPa $m^{1/2}$, similar to many of the previous cleavage toughness values. This level of toughness is not very impressive, yet having previously summarized other cleavage toughnesses, it is perhaps the magnitude of toughness which should be expected. From the values in Table IV, one can imagine a Coulomb-like charge effect as GaP ("3x5") slightly exceeds ZnS ("2x6") and thus might be expected to be slightly tougher (15/12). Of course, additional experimental results are needed to substantiate that hypothesis.

Rhombohedral cleavage as exhibited by the calcite ($CaCO_3$) structure on the $\{10\bar{1}1\}$ is Sorrell's[1] next catagory. That geometry is illustrated on several different levels in Figure 1. It is quite familiar to every mineralogist and mineral collector as the classical cleavage rhombs of $CaCO_3$ which are often displayed for their doubly refracting optical characteristics. The entire structural series from calcite ($CaCO_3$) to magnesite ($MgCO_3$), including rhodochrosite ($MnCO_3$), siderite ($FeCO_3$), and smithsonite ($ZnCO_3$), exhibit perfect $\{10\bar{1}1\}$ rhombohedral cleavage, as does dolomite, the double carbonate of calcium and magnesium ($Ca,Mg(CO_3)_2$). However, only calcite ($CaCO_3$) has been measured for its cleavage fracture toughness. Three different authors report experimental results, including

Gilman[6] (0.18 MPa $m^{1/2}$), Santhanam and Gupta[20] (0.22 MPa $m^{1/2}$), and Atkinson and Avdis[19] (0.19 MPa $m^{1/2}$). Obviously, calcite ($CaCO_3$) cleaves very easily as it has a quite low cleavage toughness. No doubt that is why such perfect cleavage rhombs abound in nature and the crystallographic dominance of the cleavage persists to the submicron levels illustrated in Figure 1. As calcite ($CaCO_3$) has the lowest elastic modulus of the structural series, simply on the basis of Equation (1) it might be surmised that the other crystals in that structural group are tougher. They are probably only marginally so, for all have a tendency to exhibit perfect cleavage.

The *octahedral* cleavage geometry as evidenced by fluorite (CaF_2) on its {111} is the last element of Sorrell's[1] mineralogical structural classification for cleavage. Several different crystals which cleave in that geometrical manner have had their toughnesses measured. Table V summarizes those results. The three alkaline earth fluorides seem to have toughnesses of magnitudes expected for cleavages, less than 1 MPa $m^{1/2}$ and decreasing from Ca through Ba as do the elastic moduli and the fracture surface energies. The covalent bonding which is present in both silicon (Si) and diamond (C) yields significantly higher cleavage toughnesses, as may be expected from some of the previous cleavage toughness summaries. The diamond (C) cleavage toughnesses which are in the 3-4 MPa $m^{1/2}$ range are some of the highest single crystal cleavage toughnesses reported in the literature. As the hardness of diamond is well known, it is evident that coupling this high cleavage toughness with the extraordinary hardness makes it almost ideal for the abrasive applications for which it is well known to excel.

Table V. Octahedral Cleavage Toughnesses

Crystal	*Plane*	K_{IC} *(MPa* $m^{1/2}$*)*	*Authors*
CaF_2	{111}	0.45	Gilman [6]
	{111}	0.33	Becher and Freiman [21]
SrF_2	{111}	0.27	Becher and Freiman [21]
BaF_2	{111}	0.22	Becher and Freiman [21]
Si	{111}	0.82	Chen and Leopold [10]
C (diamond)	{111}	3.60	Field and Freeman [22]
	{111}	2.90	Gilman [6]
	{111}	4.10	Field and Freeman [22]

Another structure, one which might be expected to exhibit the octahedral cleavage geometry, is that of stabilized cubic zirconia (ZrO_2), which has a defect fluorite structure. Guillou et al.[23] have measured calcia (CaO) stabilized cubic zirconia and Pajares et al.[24] have considered the yttria (Y_2O_3) stabilized form. The measured toughnesses of those single crystals varied from about 0.9 to 1.9 MPa $m^{1/2}$. In another study[25] it was reported as 1.5 MPa $m^{1/2}$. Very interestingly, those crystals did not readily cleave on the {111} as might be expected for the fluorite (CaF_2) structure. It might be imagined that the introduction of the numerous anion vacancies has a profound effect on the role which the crystal structure assumes in defining the cleavage plane. That effect is not understood, but for it to create a non-{111} cleavage tendency in the fluorite (CaF_2) structure it must be significant.

There are, of course, numerous crystals which exhibit cleavage characteristics that can not be catagorized within the original Sorrell mineralogical structural classification scheme. For example, Atkinson and Avdis[19] report the cleavage toughnesses of single crystal microcline feldspar ($KAlSi_3O_8$), a triclinic structure, to be 0.39 MPa $m^{1/2}$ on the {001}, also 0.39 MPa $m^{1/2}$ on the {010}, and 0.31 MPa $m^{1/2}$ on the {110}. Interestingly, some mineralogy texts suggest that its {001} cleavage is perfect and the {010} is good. The {110} is usually not even mentioned, but occasionally is given credit for parting. No further attempt will be made to summarize other mineral cleavage toughnesses here; rather, two more familiar ceramic crystal structures which do not fit Sorrell's classification will be discussed: quartz (SiO_2) and alumina or sapphire (Al_2O_3). Each will be addressed separately.

Single crystal quartz (SiO_2) is synthetically produced by hydrothermal methods for piezoelectric devices. It also occurs naturally at many locations throughout the world. The fracture of single crystal quartz often takes on the appearance of being totally conchoidal in nature. However, Bloss and Gibbs[26] have suggested that those apparent conchoidal fractures are really a submicroscopic combination of cleavage planes not unlike those depicted in Figure 1 for calcite ($CaCO_3$). Table VI summarizes some of the measured cleavage fracture toughnesses for the fracture of quartz on specific planes at room temperature. Atkinson[7] lists another dozen or so toughness values for quartz, but many are measured by questionable techniques and are not included in Table VI. Most of those values, however, are similar to the results given in Table VI.

Table VI. Measured Cleavage Toughnesses of Single Crystal Quartz

Type	*Plane*	K_{IC} *(MPa* $m^{1/2}$*)*	*Authors*
Brazilian	{0001}	1.15	Iwasa and Bradt [27]
	$\{01\bar{1}0\}$	0.97	
	$\{01\bar{1}1\}$	0.86	
	$\{11\bar{2}0\}$	0.85	
	$\{11\bar{2}1\}$	0.94	
synthetic	{0001}	1.17	Iwasa and Bradt [27]
	$\{11\bar{2}0\}$	0.96	
	{0001}	1.24	
	$\{11\bar{2}0\}$	0.95	
synthetic	$\{01\bar{1}1\}$	0.85	Atkinson[28]
	$\{0\bar{1}11\}$	1.00	Atkinson[28]
	$\{0\bar{1}11\}$	1.01	Meredith and Atkinson[29]

The summary of toughnesses in Table VI is an excellent point from which to address the occurrence of macroscopic conchoidal fracture as opposed to distinct planar cleavage for single crystal quartz. From the cleavage perspective, Brace and Walsh[30], Bloss and Gibbs[26], and Martin and Durham[31] all report cleavage on the rhombohedral faces $\{10\bar{1}1\}$. Crack zig-zag behavior on these planes prior to macroscopic crack bifurcation has also been reported by Ball and Payne[32]. However, equally or more significant is the additional observation by Ball and Payne that for single crystals oriented to fracture on the $\{11\bar{2}0\}$, a completely smooth planar fracture surface occurs and no bifurcation results. They reported that the $\{11\bar{2}0\}$ fracture surfaces of single crystal quartz are very much like those of glass. For natural quartz, the results in Table VI just barely support the $\{11\bar{2}0\}$ as the preferred cleavage plane over the $\{01\bar{1}1\}$, but admittedly, the toughness differences between the different planes are not very great. It must be concluded that it is the near equality of the cleavage fracture toughnesses of quartz on many different planes that causes the frequent conchoidal macroscopic fractures, but yet results in distinct cleavage for some specific conditions.

The cleavage of single crystal alumina or sapphire (Al_2O_3) has been addressed from several different perspectives by Iwasa and Bradt[33] in their review of its fracture characteristics. All viewpoints agree, however, that the rhombohedral plane $\{\bar{1}012\}$ is the cleavage plane at room temperature and that the basal plane {0001} is considerably tougher. The *r*-plane has a cleavage fracture toughness of about 2.38 MPa $m^{1/2}$ while the *c*-plane is reported to be 4.54 MPa $m^{1/2}$. The *m*-plane and the *a*-plane are intermediate at 3.14 MPa $m^{1/2}$ and 2.43 MPa $m^{1/2}$, respectively. These cleavage fracture toughnesses are much larger than any of the previously reported values in this paper, except for those of diamond (C) on the {111}. Part of the reason for this is undoubtedly the strength of the bonding in aluminum oxide, which leads to high Young's modulus values for the structure. Another interesting aspect of the cleavage of sapphire is that temperature affects the preferred cleavage plane. As Bradt and Scott[34] point out in their review, between about 800° C and 1500° C, the basal *c*-plane becomes the lowest toughness plane and exhibits distinct cleavage. Perhaps at different temperatures or pressures some of the previously discussed crystal structures will also exhibit cleavage plane transitions.

Having addressed the single crystal cleavage toughnesses for a number of ceramics and minerals, it is appropriate to reach some general conclusions. In a specific sense, it is not an easy task; however, one point is an obvious one and that is that the fracture toughnesses for cleavage of single crystals is consistently less than about 1 MPa $m^{1/2}$. Cleavage planes are simply not very tough by any standards, which is why they are easily created by the fracture process and why they are so prevalent in nature. There seems to be a general tendency for the more complex crystal structures with multiple ions and a higher level of covalent bonding to exhibit slightly higher cleavage toughnesses than the simple crystal structures. As not every crystal structure has been measured, it is not possible to assign an upper bound to the cleavage toughnesses of all single crystals. However, on the basis of the results for diamond (C) and sapphire (Al_2O_3), an argument for a conservative upper estimate (limit) of about 5 MPa $m^{1/2}$ could be advanced. One might speculate that it would be some of the higher elastic modulus compounds such as the borides, carbides, nitrides, beryllides, and aluminides that would reach or perhaps exceed that estimated level.

POLYCRYSTALLINE TOUGHNESSES

The fracture of polycrystalline materials has often also been described as cleavage whenever failure occurs in an opening mode and in a brittle fashion. While the authors do not advocate the utilization of that terminology, there are a number of reliable fracture toughness measurements in the literature for polycrystalline specimens of the same materials which have just been considered from the single crystal cleavage point of view. It seems natural to consider those polycrystalline fracture toughnesses, if for no other reason than to compare them with the single crystal cleavage toughnesses for the same crystal structures. The following comparison is an interesting one although it is not all inclusive of the toughness measurements of brittle polycrystalline ceramics and minerals. Neither is it able to address the effects of the polycrystalline microstructure in a systematic fashion, for that has not been the subject of very many studies to date.

A good starting point is the comparison of the fracture toughnesses of polycrystalline graphites with the cleavage toughness of graphite in Table II. Wood et al.[35] have measured the fracture toughnesses of a number of commercial graphites and summarized other researchers' published values, too. There is a strong textural dependence relative to the molding and extrusion directions such that the toughness transverse to the forming direction is as much as 50% greater than the toughness parallel to it. This is because fracture in the latter case tends to follow the cleavage planes of the individual graphite flakes or crystals. However, even in their toughest states, the fracture toughnessses of polycrystalline graphites are only about 1.4-1.5 MPa $m^{1/2}$. This is an increase over the cleavage toughness to be sure, but polycrystalline graphite is not a very tough material by any standard.

Approximately following the mineralogical structural classification of Sorrell, there have been several reports of the fracture toughnesses of polycrystalline spinel ($MgAl_2O_4$)[36-38]. At room temperature the fracture toughness of polycrystalline $MgAl_2O_4$ appears to be between about 1.8 and 2.0 MPa $m^{1/2}$, independent of the grain size of that material. Again, similar to the carbons, this is not a very significant increase beyond the cleavage toughness of single crystal spinel as has been previously noted by Stewart et al.[39] In fact, the polycrystalline to single crystal toughness ratio for graphite seems to be much greater than that for spinel. Perhaps this is because of the multiplicity of cleavage planes and their various orientations for the cubic cleavage geometry of spinel as opposed to the single layer plane cleavage for graphite. Additional data are needed to test this hypothesis in a critical manner.

In their article on the cleavage of the alkaline earth fluorides, Becher and Freiman[21] also report several fracture toughness values for dense, transparent polycrystalline fluorite (CaF_2). Values from 1.6 to 3.6 MPa $m^{1/2}$ were reported which are significantly greater than the 0.33 to 0.45 MPa $m^{1/2}$ obtained for the single crystal cleavage of CaF_2 on its {111}. As the octahedral cleavage geometry of CaF_2 has an even greater multiplicity of cleavage planes than does the cubic cleavage of the spinel, this polycrystalline to single crystal ratio appears inordinately large. However, a tortuous crack path in spinel can perhaps also readily orient to cleave on the {110} and {111} planes in that structure, albeit in a more difficult fashion. In fluorite (CaF_2), however, the {100} and {110} planes are both non-neutral and cannot cleave, thus the cracks must experience a much more severe reorientation process in polycrystalline fluorite than in the single crystal to propagate catastrophically.

Fracture of natural polycrystalline quartz has been the subject of study by geologists. The fracture toughness of Arkansas novaculite[40,41] has been reported in several studies and seems to range from 1.3 to 1.8 MPa $m^{1/2}$, while a Mojave quartzite[42] has been measured at 2.10 MPa $m^{1/2}$. The approximate two to one toughness ratio for polycrystalline quartz fracture versus the single crystal cleavage of quartz lends additional insight to the effect of microstructure on the crack propagation resistance. As the cleavage of single crystal quartz is practically isotropic and the toughness of the polycrystalline quartz is only about twice that of the single crystals, it may be reasonable to assume that polycrystalline microstructures are capable of doubling the toughness of single crystals, or at least increasing the toughness about 1 MPa $m^{1/2}$. Neither level of increase is a very remarkable toughening effect.

Many and varied measurements of fracture toughness have been made for polycrystalline alumina. It is probably the model material for most ceramists. Bradt and Scott[34] have reviewed all of those test results. Reliable toughness measurements for dense polycrystalline alumina bodies appear to vary from 3 to 5 MPa $m^{1/2}$. This is perhaps not the level of increase expected beyond the single crystal cleavage toughnesses. In fact it is only about a factor of two tougher than cleavage on the r-plane, not unlike the level of toughness increment observed for quartz (SiO_2). It must be concluded that the microstructural features which have been studied or developed at the present do not yield very significant toughness increases for the polycrystalline forms of brittle crystal structures which readily experience cleavage fracture in single crystal form.

In the discussion of the toughness of polycrystalline graphite the effect of *texture* was briefly noted. Texture is a significant factor in the toughness of brittle materials and merits discussion in terms of single crystal cleavage toughnesses. Unfortunately, there is not very much reliable data in the literature, particularly for examples of crystal structures where single crystal cleavage toughnesses have also been measured. Three excellent studies which address the toughness anisotropy of textured (oriented) ceramics are those of Virkar and Gordon[43] on beta-alumina ($Na_2O \cdot 11Al_2O_3$), that of Lange[44] on a hot pressed silicon nitride (Si_3N_4), and the study by Veldkamp and Hattu[45] for a strontium hexaferrite ($SrO \cdot 6Fe_2O_3$). These studies all suggest that the toughness anisotropy can be as great as 50%.

Iwasa et al.[46] have made extensive fracture toughness measurements of a highly textured barium hexaferrite ($BaO \cdot 6Fe_2O_3$) and for the same material in an isotropic state. This material has the magnetoplumbite structure, which is a layered hexagonal structure, thus one might compare its fracture toughness to those of the layer structures in Table II. In the isotropic state, that ferrite has a fracture toughness of about 1.6 MPa $m^{1/2}$. However, when the microstructure is a highly textured one that has been processed in a magnetic field, the toughness parallel to the basal plane alignment was only 0.9 MPa $m^{1/2}$, but the toughness at right angles was 2.8 MPa $m^{1/2}$. Unfortunately, single crystal cleavage toughnesses are not available for a more focused comparison or contrast; however, the potential toughening effect of texture development is nonetheless evident. Of course, the increase in toughness may occur at the expense of the toughness in the complementary orientation.

Salem et al.[47] addressed the effect of texture of an extruded alumina (Al_2O_3) on the *R*-curve characteristics, but at the same time measured fracture toughnesses for the textured specimens. For the orthogonal directions, they report fracture toughnesss of 3.6, 3.9, and 4.9 MPa $m^{1/2}$, which compare favorably with the cleavage toughnesses previously discussed for sapphire and also exhibit differences comparable to those for the textured hexaferrite. The largest toughness exceeds the basal plane cleavage toughness of sapphire, which is not surprising as both grain size and textural effects may be expected to provide toughening.

There are not a large number of well documented studies of textural effects on the fracture toughness of minerals, although Atkinson[7] does list a few in his text. Several are worth noting and one is meritorious for comparison with single crystal cleavage toughnesses. In the former group are a coal which is reported to have fracture toughnesses of 0.063 and 0.047 MPa $m^{1/2}$, normal and parallel to the bedding plane; a marble with fracture toughnesses of 0.7 and 1.4 MPa $m^{1/2}$, and a shale with fracture toughnesses which varied from 1.1 to 1.9 MPa $m^{1/2}$. These are all orientation dependent values which are comparable to the hexaferrite and alumina which were previously discussed.

Atkinson[7] also lists fracture toughness values for natural salt (NaCl) normal and parallel to the layering after Rummel and Müller[48]. The toughness normal to the layering is 0.57 MPa $m^{1/2}$, while parallel values of 0.23 and 0.40 MPa $m^{1/2}$ are reported. These are only slightly greater than the values of 0.17 MPa $m^{1/2}$ for the cleavage toughness of NaCl on its {100}. Again this confirms that the microstructural effects are capable of only modest increases in the fracture toughnesses above the single crystal cleavage toughness values.

CONCLUSIONS

The fracture toughnesses of ceramic and mineral crystal structures for specific crystallographic planes are a good criterion for predicting cleavage. Addressing cleavage from the mineralogical structural classification that has been proposed by Sorrell is an excellent approach to understanding the phenomenon. It is evident that the fracture toughness for cleavage planes is not very large for most simple ionic crystal structures, usually less than 1 MPa $m^{1/2}$. Even the more strongly bonded covalent structures such as diamond and sapphire have cleavage toughnesses less than 5 MPa $m^{1/2}$. Single crystals simply do not appear to be very tough on their cleavage planes, which is why they experience cleavage in natural conditions and utilitarian applications.

When the fracture toughnesses of polycrystalline ceramics and minerals are compared with single crystal cleavage toughnesses, it is obvious that microstructural factors have a toughening effect. That effect can be considered to be a two-fold one, at least. The first aspect is simply one of the microstructure, perhaps related to the point that a crack in a polycrystalline material does not lie fully on one cleavage plane of a single grain. This toughening mechanism has only a modest effect of increasing the toughness, perhaps doubling its value or increasing its level by 1 or 2 MPa $m^{1/2}$. A related effect is that of preferred crystallographic orientation or texture of the grains or crystals. This factor appears to provide an additional increment of toughening for the non-cleavage plane orientations, but reduces the toughness parallel to the cleavage planes. Texture creates a significant toughness anisotropy which can reach levels of several MPa $m^{1/2}$.

ACKNOWLEDGMENT

The authors gratefully acknowledge the inspiration provided by C.A. Sorrell from his mineralogical structural classification of cleavage. Assistance of M. Jensen with the fractography is also appreciated as is the supply of mica results by B.R. Lawn.

REFERENCES

1. C.A. Sorrell, *Minerals of the World*, Golden Press, New York (1973).
2. W.C. Roberts, G.R. Rapp, Jr., and J. Weber, *Encyclopedia of Minerals*, Van Nostrand, New York (1974).
3. L.B. Berry and B. Mason, *Mineralogy*, Freeman, San Francisco (1959).
4. K. Frye, *Encyclopedia of Mineralogy*, Hutchinson Ross, Stroudsburg, Penn. (1981).
5. W.A. Deer, R.A. Howie, and J. Zussman, *Rock Forming Minerals*, vol. 1, Wiley, New York (1962).

6. J.J. Gilman, Cleavage, ductility, and tenacity in crystals, in *Fracture*, B.L. Averbach, D.K. Felbeck, G.T. Hahn, and D.A. Thomas, eds., Wiley, New York, pp. 193-222 (1959).
7. B.K. Atkinson, ed., *Fracture Mechanics of Rock*, Academic, New York (1987).
8. K. Hayashi, M. Namura, and T. Nishikawa, Fracture toughness of LiF single crystals, *J. Mat. Sci. Japan* **35** (388): 66-72 (1986).
9. K. Hayashi, M. Ashizuka, R.C. Bradt, and H. Hirano, Cleavage of gallium phosphide, *Mat. Lett.* **1** (3-4): 116-118 (1982).
10. C.P. Chen and M.H. Leipold, Fracture toughness of silicon, *Bull. Am. Ceram. Soc.* **59** (4): 469-472 (1980).
11. R.L. Stewart and R.C. Bradt, Fracture of single crystal $MgAl_2O_4$, *J. Mat. Sci.* **15** (1): 67-72 (1980).
12. R.W. Hertzberg, *Deformation and Fracture Mechanics of Engineering Materials*, Wiley, New York (1976).
13. M. Sakai, R.C. Bradt, and D.B. Fischbach, Fracture toughness anisotropy of a pyrolytic carbon, *J. Mat. Sci.* **21** (7): 1491-1501 (1986).
14. K.T. Wan, N. Aimard, S. Lathabac, R.G. Horn, and B.R. Lawn, Interfacial energy states of moisture exposed cracks in mica, *J. Mat. Res.* **5** (1): 172-182 (1990).
15. R.F. Cook, T.R. Dinger, and D.R. Clarke, Fracture toughness measurements of $YBa_2Cu_3O_x$ single crystals, *App. Phys. Lett.* **51** (6): 454-456 (1987).
16. H.W. Liu and K.J. Miller, Fracture toughness of fresh-water ice, *J. Glaciol.* **22** (86): 135-143 (1979).
17. S.W. Freiman, P.F. Becher, and P.H. Klein, Initiation of crack propagation in KCl, *Phil. Mag.* **31** (4): 829837 (1975).
18. C.N. Ahlquist, The influence of yield strength on fracture of semi-brittle ceramic crystals, *Acta Met.* **22** (9): 1133-1137 (1974).
19. B.K. Atkinson and V. Avdis, Fracture mechanics parameters of some rock-forming minerals determined using an indentation technique, *Int. J. Rock Mech. Min. Sci. Geomech. Abs.* **17** (6): 383-386 (1980).
20. A.T. Santhanam and Y.P. Gupta, Cleavage surface energy of calcite, *Int. J. Rock Mech. Min. Sci.* **5** (3): 253-258 (1968).
21. P.F. Becher and S.W. Freiman, Crack propagation in alkaline-earth fluorides, *J. Appl. Phys.* **49** (7): 3779-3783 (1978).
22. J.E. Field and C.J. Freeman, Strength and fracture properties of diamond, *Phil. Mag.* **43** (3): 595-618 (1981).
23. M.O. Guillou, G.M. Carter, R.M. Hooper, and J.L. Henshel, Hardness and fracture anisotropy in single crystal zirconia, *J. Hard Mat.* **1** (1): 65-78 (1990).
24. A. Pajares, F. Guibearteau, A. Dominguez-Rodriguez, and A.H. Hewer, Microhardness and fracture toughness anisotropy in cubic zirconia single crystals, *J. Am. Ceram. Soc.* **71** (7): C332-C333 (1988).
25. R.P. Ingal, D. Lewis, B.A. Bender, and R.W. Rice, Temperature dependence of the strength and toughness of ZrO_2 single crystals, *J. Am. Ceram. Soc.* **65** (9): C150-C152 (1982).
26. F.O. Bloss and G.V. Gibbs, Cleavage in quartz, *Am. Min.* **48** (7,8): 821-838 (1963).
27. M. Iwasa and R.C. Bradt, Cleavage of natural and synthetic single crystal quartz, *Mat. Res. Bull.* **22** (9): 1241-1248 (1987).
28. B.K. Atkinson, A fracture mechanics study of subcritical tensile cracking of quartz in wet environments, *Pure Appl. Geophys.* **117**: 1011-1024 (1979).
29. P.G. Meredith and B.K. Atkinson, High-temperature tensile crack propagation in quartz: Experimental results and application to time-dependent earthquake rupture, *Earthquake Predic. Res.* **1**: 377-391 (1982).

30. W.F. Brace and J.B. Walsh, Some direct measurements of the surface energy of quartz and orthoclase, *Am. Min.* **47** (9,10): 1111-1122 (1962).
31. R.J. Martin, III and W.B. Durham, Mechanisms of crack growth in quartz, *J. Geophys. Res.* **80** (35): 4837-4844 (1975).
32. A. Ball and B.W. Payne, The tensile fracture of quartz crystals, *J. Mat. Sci.* **11** (4): 731-740 (1976).
33. M. Iwasa and R.C. Bradt, Fracture toughness of single-crystal alumina, in *Advances in Ceramics*, **10**, Am. Ceram. Soc., pp. 767-779 (1983).
34. R.C. Bradt and W.D. Scott, Mechanical properties of alumina, in *Alumina*, Am. Ceram. Soc., pp. 23-39 (1990).
35. J.L. Wood, R.C. Bradt, and P.L. Walker, Jr., Subcritical crack growth in fine grained polycrystalline graphite, *Carbon* **18** (3): 169-178 (1980).
36. R.L. Stewart and R.C. Bradt, Fracture of polycrystalline $MgAl_2O_4$, *J. Am. Ceram. Soc.* **63** (11-12): 619-623 (1980).
37. M. Sakai, R.C. Bradt, and A.S. Kobayashi, The toughness of polycrystalline $MgAl_2O_4$, *Nip. Ser. Kyo. Ron.* **96** (5): 525-531 (1988).
38. A. Ghosh, K.W. White, M.J. Jenkins, A.S. Kobayashi, and R.C. Bradt, The fracture resistance of transparent $MgAl_2O_4$, to be published in *J. Am. Ceram. Soc.*
39. R.L. Stewart, M. Iwasa, and R.C. Bradt, Room temperature K_{IC} values for single crystal and polycrystalline $MgAl_2O_4$, *J. Am. Ceram. Soc.* **64** (2): C22 (1981).
40. B.K. Atkinson, Stress corrosion and the rate-dependent tensile failure of a fine-grained quartz rock, *Tectonophysics* **65** (3/4): 281-290 (1980).
41. P.G. Meredith, B.K. Atkinson, and N.B. Hillman, K_{IC} measurements on rock using a short-rod apparatus, in Experimental Petrology, 5th report, Nat. Environ. Res. Council, Swindon, Britain, *Publ. D.18.1981*, pp. 297-298 (1981).
42. B.K. Atkinson, Subcritical crack growth in geological materials, *J. Geophys. Res.* **89** (B6): 4077-4114 (1984).
43. A.V. Virkar and R.S. Gordon, Fracture properties of polycrystalline lithia-stabilized β'' alumina, *J. Am. Ceram. Soc.* **60** (1-2): 58-61 (1977).
44. F.F. Lange, Strong, high-temperature ceramics, *Annu. Rev. Mater. Sci.* **4**: 365-390 (1974).
45. J.D.P. Veldkamp and N. Hattu, On the fracture toughness of brittle materials, *Phillips J. Res.* **34** (1-2): 1-25 (1979).
46. M. Iwasa, E.C. Liang, R.C. Bradt, and Y. Nakamura, Fracture of isotropic and textured Ba hexaferrite, *J. Am. Ceram. Soc.* **64** (7): 390-393 (1981).
47. J.A. Salem, J.L. Shannon, Jr., and R.C. Bradt, Crack growth resistance of textured alumina, *J. Am. Ceram. Soc.* **72** (1): 20-27 (1989).
48. F. Rummel and W. Müller, Bruchmechanische Untersuchungen fur Salzgestein in der Schachtanlage Asse II. 1. Ber. zum GSF-Auftrag, Best. Nr. 31/139081/83 (1984).

INDENTATION FRACTURE OF PURE AND MeV ENERGY ION IMPLANTED SAPPHIRE

R. Nowak*, K. Ueno and M. Kinoshita

Government Industrial Research Institute, Osaka
1-8-31 Midorigaoka, Ikeda-shi, Osaka 563, Japan

INTRODUCTION

The indentation test is considered to be one of the simplest and fastest techniques for gathering a variety of experimental data on the mechanical properties of solids. This method has been used for years to estimate the plastic properties of materials. However, when testing brittle materials (including most ceramics at room temperature) the fracture usually occurs around and beneath the indenter. The generation of cracks during the indentation cycle may influence the deformation process, and in consequence affect the hardness value [1]. This is frequently overlooked in most reports since the standard procedure is developed for metallic materials which do no consider the effect of cracks, a criticism pointed out by Clinton and Morrell [2].

On the other hand, the fracture process during indentation is a field of growing interest with an increasing number of applications. The most common is toughness measurements of brittle materials.
Toughness evaluation of materials was the catalyst for a large number of advanced studies on the indentation process and associated phenomena from the perspective of both mechanics [3]-[16] and materials science [17]-[21]. Within the last decade, the increasing use of the indentation test for brittle materials toughness measurement has created a need for a detailed description of the processes associated with the penetration of the indenter of a given shape into the solid.

Although major increases in the knowledge of indentation processes have been achieved, a general theory of indentation deformation and fracture has yet to be proposed, and there are many effects whose origins are still not clearly understood. One example of a not well understood phenomenon is the hardness anisotropy and indentation fracture in hexagonal single crystals.

Great progress in the indentation field has recently taken place. The rapid development of so called "new materials" has added new requirements for the theoretical description of the materials structure and mechanical behaviour as well as new restrictions on the testing methods.

---

* on leave from the Institute of Nuclear Physics & Techniques,
University of Mining & Metallurgy, Cracow, Poland

Fracture Mechanics of Ceramics, Vol. 10
Edited by R.C. Bradt *et al*., Plenum Press, New York, 1992

In "new materials", quite often the mass and volume available for experimental investigations is very small (eg. materials with surfaces modified within the micron range or small volume crystals for microelectronics). Therefore the micro or even the nano indentation technique is often the only method available for studying the mechanical properties within the region of interest.

Recent developments in hardness testing have enabled continuous monitoring of the load and depth of penetration by the indenter during the indentation process. This technique seems quite revolutionary in indentation fracture research and makes it possible to derive a new set of information about the mechanical behavior of materials.

Studies on dynamic and ultra-micro indentation processes and new techniques for surface examination constitute a new and expanding area of interest. The sophisticated reports published recently by Cook, Pharr, Clarke, Oliver and co-workers give a completely original view of the subject of indentation. In-situ observations of the propagation of indentation cracks in ceramics, together with the ability to measure load versus depth curves [22]-[23], the measurement of the stresses generated in sapphire beneath the indenter by the Raman microprobe using ruby fluorescence or cathodoluminescence [24] and through spectroscopic investigations by scanning electron microscopy [25] serve as examples of the recent trends.

Although the works of Evans, Lawn, Marshall and Swain [3]-[16] have been widely accepted as the basis for a theory of the fracture mechanics of ceramics, the development of new, sophisticated materials have led to invaluable progress in the theory during the past two years. Recent theoretical considerations [26] have led for the first time to a solution of the three dimensional contact fracture problem using the body force method.

The aim of the present study is to report the results of the above mentioned methods (classic and the new dynamic indentation tests) for systematic studies of fracture in pure sapphire and in a new composite material - sapphire implanted with metallic ions of MeV energy.

Since fracture of ceramics almost always starts at the surface, a number of efforts have been made in order to find suitable methods for surface modification to reduce the brittleness in these solids.
It has been generally observed that good methods for improving the mechanical properties of solids by surface modification fall into three categories: 1) methods which reduce the number of preexisting surface flaws, 2) methods which generate a layer of material at or near the surface with better plastic properties than the matrix, and 3) methods which create a large compressive surface stress. Ion implantation provides a powerful tool for inducing compressive residual stresses near the surfaces and also for improving the plastic properties in this region by producing near surface amorphization.

The mechanical properties of pure and ion implanted sapphire have been extensively investigated [27]-[44], but studies in the MeV energy range are relatively scarce [40][43] (including previous papers from this laboratory). This is not surprising since such energies for heavy ions are currently available at only a few sites.

In this work as well as in our previous study [43] the indentation test was used not only to measure the plastic deformation of the sapphire, but also as a tool for measuring the residual stress generated as a consequence of implantation, using the method of Lawn and Fuller [44].

Pure Sapphire

Single crystals of aluminum oxide have been chosen as a model material for our studies for the following reasons:

- the structure and mechanical behavior of sapphire has been extensively studied during the last three decades [17][19][20][22][46-58]
- the crystals have a hexagonal structure (see Fig. 1a) with a distinct anisotropy (the changes of anisotropic behaviour were used in the present study as a tool for studying the effect of ion implantation on mechanical behaviour of crystal)
- implanted sapphire (in the keV energy range) has recently been studied in detail [27]-[44]
- although the principles of deformation and fracture in sapphire are well understood (Fig 1b), there are many unsolved problems associated with the mechanical behavior of indented sapphire, one example of which is the anomalous, not predicted by crystallography, direction of crack propagation when basal plane is indented by Vickers pyramid [56]
- the transparency of alumina crystals allow optical methods to be used for crack observation
- high quality sapphire single crystals are commercially available
- alumina is a component in a large number of ceramics (applications).

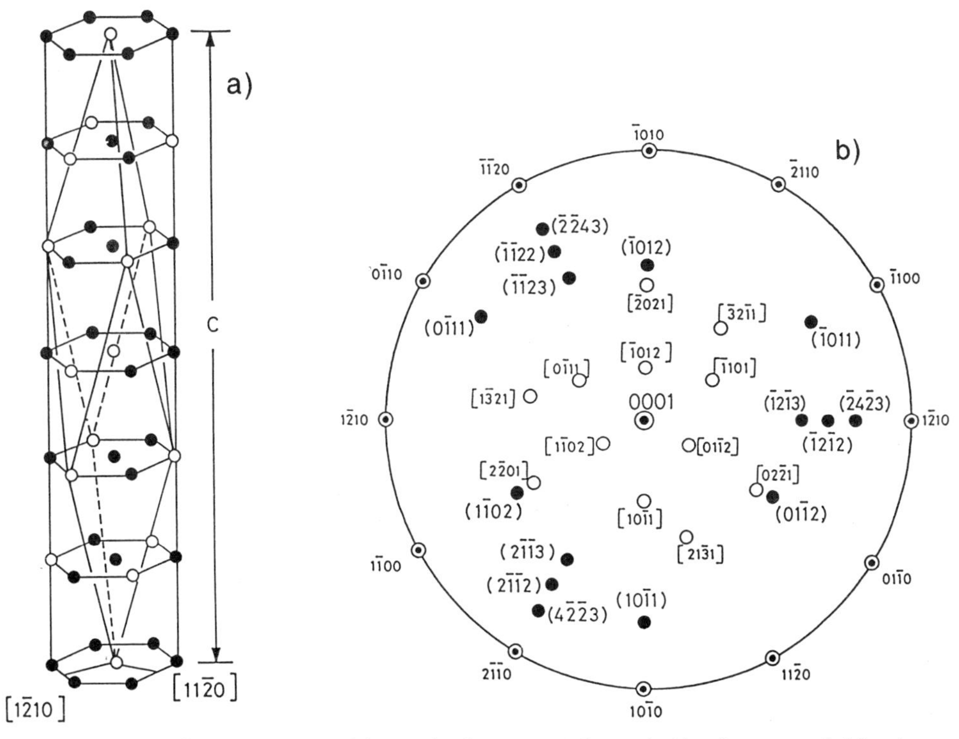

Figure 1. Structure of sapphire single crystals. a) the hexagonal/rhombohedral unit cell (c/a=2.73) showing the empty and filled cation sites, (after Kronberg [47]); b) the orientation of the prominent slip systems in sapphire shown on a standard projection (Snow and Heuer [48]) - open and filled circles denote the position of the slip direction and slip plane respectively.

Large, pure, transparent sapphire single crystals were grown by the Edge-Defined, Film-Fed Growth method by the KYOCERA CORPORATION and high quality specimens were provided as two identical parallelepipeds with edge length 15mm.

A special sequence of thermal treatments and polishing processes (including effective mechano-chemical polishing) were applied to the crystals to obtain a smooth, pair parallel, defect free surface (information courtesy of KYOCERA). The condition of the surface was examined by means of electron diffraction which proved that residual stresses were not present in the surface region. The orientation of the crystal surfaces was checked by the Laue X-ray diffraction technique. The deviation from the exact orientation for the basal (0001) and prismatic $(10\bar{1}0)$ planes was less than 1 degree.

Ion Implanted Crystals

One of the two crystals of sapphire was exposed to high energy ion bombardment in vacuum in a beam generated in the tandem-type accelerator at the Government Industrial Research Institute, Osaka. The implantation was performed in high vacuum (pressure less that $3x10^{-6}$ Torr) at room temperature. The sample temperature did not significantly increase due to a special cooling system in the specimen chamber and the low ion current density (around 0.35 $\mu A/cm^2$. For more details please refer to our previous work [43].

The implanted areas were restricted to dimensions of 5 mm x 5 mm through the use of a stainless steel mask (see Fig. 2). Both the basal and pyramidal planes were implanted with Au(2+) and Ni(2+) ions. The implantation conditions for the areas marked schematically in Fig. 2. are listed in the Table 1.

Table 1. Implantation conditions of the sapphire crystal

Symbol	Fluence [ions/cm^2]	Species	Ion Energy	Temperature
Ni LD_1	$1x10^{15}$	Ni^{2+}	3 MeV	273K
Au LD_1	$1x10^{15}$	Au^{2+}	3 MeV	273K
Ni LD_2	$2x10^{15}$	Ni^{2+}	3 MeV	273K
Au LD_2	$2x10^{15}$	Au^{2+}	3 MeV	273K
Ni HD	$2x10^{16}$	Ni^{2+}	3 MeV	273K
Au HD	$2x10^{16}$	Au^{2+}	3 MeV	273K

The depth profile of the implanted ions in alumina was estimated from computer simulation. The TRIM simulation code [59] was applied. The calculations used an analytic formula for determining nuclear scattering angles. They based on the Molière potential V [59]:

$$V(R)= (Z_1Z_2\ e\ /\ a\ R)\ \Phi(R)$$

where R is the reduced atomic separation, $\phi(R)$ is the Mollière screening function, Z_1 and Z_2 are incident particle and target atomic numbers, respectively, a is the screening length and e is the electronic charge. The projected range of implantation with Ni ions was estimated to be 1.14 µm with a standard deviation of 1460 Å, while for Au ions it was 0.68 µm with a standard deviation 1090 Å. The results of calculations showing the expected depth distribution of the implanted atoms in the polycrystalline target has been previously reported by us [43] and are repeated on Fig. 3.

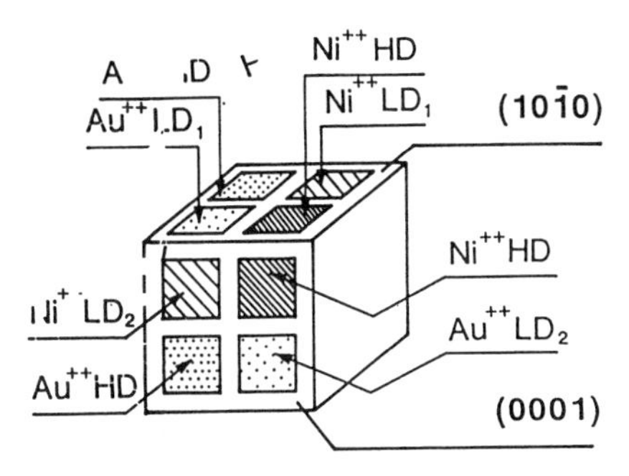

Figure 2. Location of the implanted areas on the low index surfaces of the sapphire specimen. The implantation conditions and their relation to the symbols used are explained in Table 1. Symbols LD and HD refer to low dose and high dose respectively

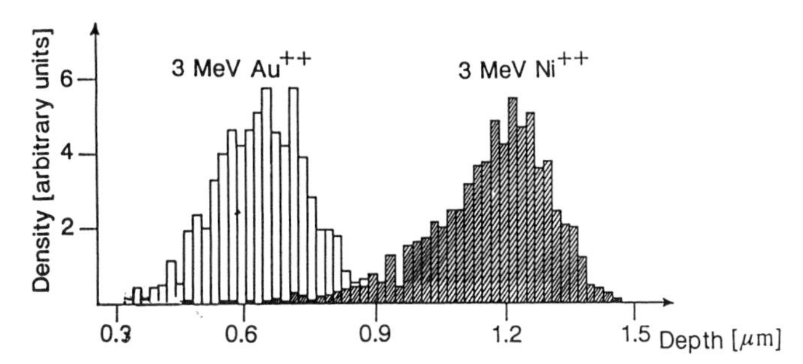

Figure 3. The distribution profile of 3 MeV Ni and Au ions implanted into polycrystalline alumina calculated by the TRIM simulation code (after Nowak et al. [43])

EXPERIMENTAL PROCEDURE

Conventional and Dynamic Indentation

Two types of hardness testers from the SHIMADZU CORPORATION were used in the investigation of the indentation fracture of sapphire.
The conventional equipment enabled us to make indents under higher loads (up to 1 kgf) and served as a reference point for the new data obtained by the dynamic method. Both Vickers and Knoop indenters were used. The indentations were made under various static loads (for Vickers: 50, 100, 200, 500 and 1000 gf, for Knoop: 50, 100, 200 gf). The size of indentation marks as well as the length of the generated cracks was measured using the optical system attached to the tester. In order to have more confidence in the data, about 10 indentation have been made for each data point.

Subsequent indentation experiments were carried out on the newest SHIMADZU computerized dynamic ultra-microhardness tester (model DUH-200) which enabled continuous monitoring of the load and depth of penetration during the indentation process. The ultra-microhardness tester had an indentation depth resolution of 0.1 μm which enabled it to register crack nucleation depth and load. The indentations were made to upper load limits of 50, 100 and 200 gf. However, the number of tests for each data point was limited to 5 or 3 repetitions due to time limitations. The dynamic measurements also used both types of penetrating pyramids (Knoop and Vickers) as each generates quite different effects in sapphire because of differences in the generated stress state and depth of penetration. In order to see completely plastic, crack free indentation into sapphire several low load indentation were made. However, the relatively deep location of the implanted layer forced us to work at a higher range of loads (up to 200gf). The load-depth curves for both the loading and unloading cycles were registered.

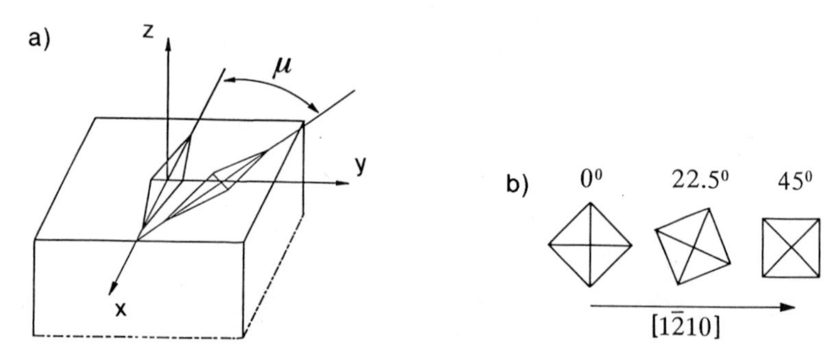

Figure 4. The method of measuring the anisotropy of indentation properties on the given crystal surface using a) Knoop pyramid b) Vickers indenter. Axis x is parallel to the direction [1$\bar{2}$10] in sapphire and angle μ denotes current position of the indenter

Anisotropy of the Indentation

When investigating single crystals, one must always deal with problems associated with anisotropy. In particular in the hexagonal structure which is only transversally isotropic this effect will be very pronounced in the mechanical behaviour of the crystal. Within this work the anisotropic response of a material during the indentation process was measured the opposite of other authors (eg. [40]) who tried to avoid the effects of anisotropy by keeping the orientation of the indenter unchanged with respect to the crystal. The present report is the first concerning the effect of MeV energy ion implantation on indentation anisotropy.

In order to measure the anisotropic behaviour of indented sapphire, the specimens were mounted on a jig which permitted rotation of the sample about the vertical axis to any desired position. The Knoop pyramid indentations were performed on the (10$\bar{1}$0) surface in increments of 5° and 10° while for the the Vickers indenter only three positions were chosen (Fig. 4). For dynamic measurements which are much more time consuming only four (0°, 30°, 60°, 90°) or three (0°, 45°, 90°) orientations were considered.

Fracture Toughness

The fracture toughness of pure and implanted crystals was calculated from the crack length generated during indentation using the equation proposed by Lawn and co-workers [8][10].

$$K_c = 0.0139\ (E/H)^{1/2} PC^{-3/2}$$

where C is the length of radial crack, H is hardness, P is the applied load and E is Young's modulus.

Residual Stresses

The relative values of the residual stresses induced into sapphire by ion implantation have been estimated by two experiments:

1) measurements of the material expansion due to implantation,
2) measurements of the surface traces of radial/median cracks produced by the indentation of a Vickers pyramid on implanted surfaces.

The steps due to the irradiation growth of sapphire, produced on both the basal and prismatic planes were measured by a surface profilometer Talystep and served as a first rough estimate of the damage and stresses induced by implantation.

Indentation tests (Vickers pyramid) were performed in order to estimate the level of residual stresses (s_r) induced by implantation. According to the method developed by Lawn and Fuller [44] s_r values satisfy the equation:

$$2\ \phi\ s_r\ d^{1/2}/K_c = 1 - (c_o/c)^{3/2}$$

where ϕ is a crack geometry term of about unity, d is a layer depth, K_c is toughness, and c and c_o are the indentation crack lengths in implanted and unimplanted material, respectively.

Indentation Crack Observations

The indentation marks were observed in an optical microscope with Nomarski contrast by using both reflection and transmission techniques. The transparency of sapphire also allows a side illumination study in order to enable the observation of sub-surface lateral cracks.

RESULTS AND DISCUSSION

Pure Sapphire

Conventional Knoop pyramid indentation. Systematic studies of the anisotropy of Knoop indentation were performed on the prismatic plane (1010). The experimental results obtained by the conventional tester with applied static loads of 50 and 200 gf are shown in Fig. 5. The results for 200 gf indentation into sapphire are consistent with the Knoop hardness variation registered for 300 gf indentation by Brookes [55]. However, Brookes suggestion that the above mentioned anisotropy is governed by deformation by basal slip and can be easy predicted by CRSS-model is in contradiction with Kollenberg [57] and our results [1].

Although the Knoop indenter penetration is quite shallow and generates large compressive stresses, fracture occurs even in low load experiments. Application of a side illumination system enabled us to also detect sub-surface cracks which are not visible when viewed normal to the surface. The scale of fracture is strongly dependent on the indenter orientation and on the applied indentation load.

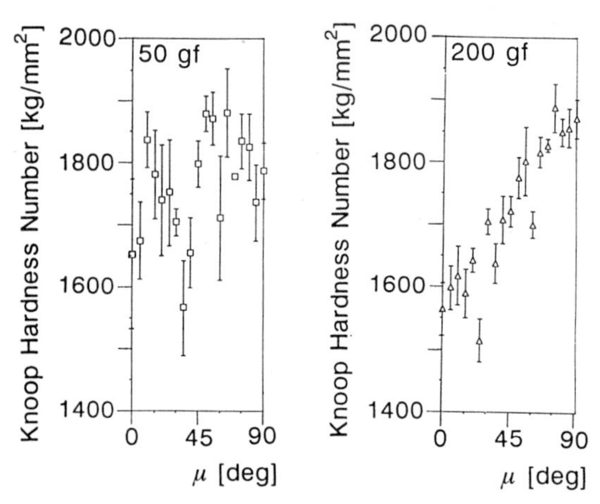

Figure 5. Knoop hardness variation measured for $(10\bar{1}0)$ plane in sapphire as a function of the orientation of the indenter denoted by the angle between the $[1\bar{2}10]$ direction and indenter's long diagonal

When the long diagonal is parallel to the $[1\bar{2}10]$ direction extensive fracture occurs particularly in form of large lateral cracks (Fig. 6). For the indentation perpendicular to the $[1\bar{2}10]$ small scale fracture was detected only in the side illumination system. The middle orientation (μ=35°) indent shows more developed cracks together with large basal twins. The activation of basal twinning system for this orientation was predicted in our previous theoretical treatment [1] and is in full agreement with the results of Kollenberg [57].

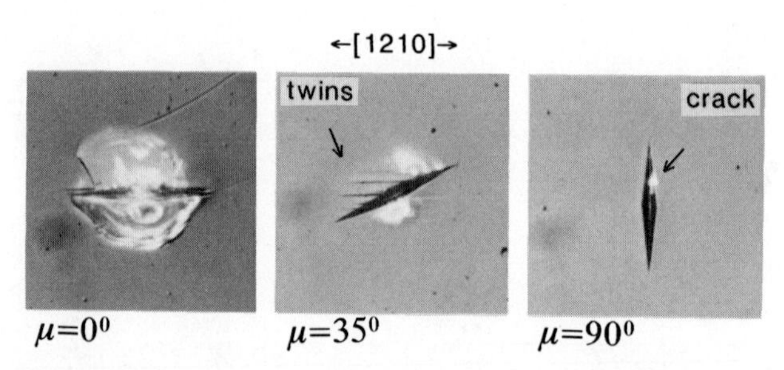

Figure 6. Micrographs of indentation fracture patterns (load 200 gf) as function of orientation of Knoop indenter (μ).

Although the deformation and fracture patterns of both the 200 gf and 50 gf indentation marks are quite similar, in the low load measurements no hardness anisotropy is observed (Fig. 5). This is due to the low precision of the optical system associated with the conventional tester which leads to large errors in measuring such small indents and also due to occuring fracture which influences results.

This situation is a good example of the need for a more sophisticated tester with a non optical measurement of the indent parameter. The SHIMADZU DUH-200 used in the studies described below met these requirements.

Knoop indentation anisotropy on the basal plane of sapphire has been reported by many researchers [54][55], nevertheless their results are in contradiction. Recently, Kaji and Bradt [58] have taken time to clarify the problem of various reports of the Knoop hardness anisotropy on the basal plane of sapphire and gave the most consistent analysis of the indentation induced deformation through the potential activation of the various slip systems. Within the literature, however, no studies have as yet focused on fracture problems (which are in our opinion very essential particularly for indentation on the basal plane of sapphire) but considered only slip systems by applying advanced CRSS models.

Dynamic Knoop Pyramid Indentation. Typical examples of the registered load vs. depth curves for indentation into sapphire are shown in Fig. 7.

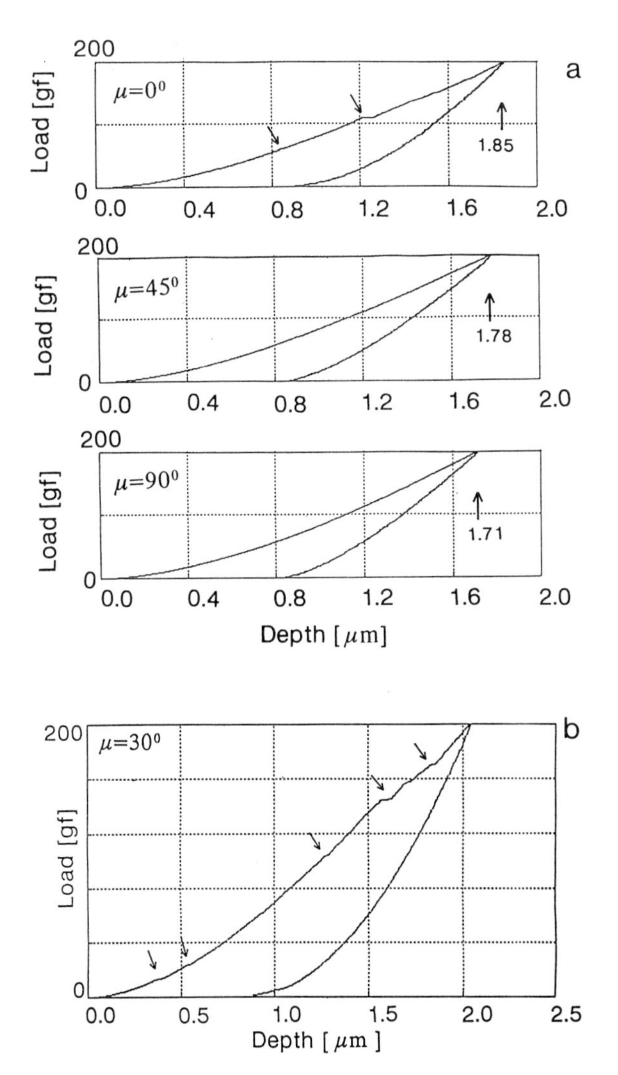

Figure 7. Typical load-displacement curves for the indentation in sapphire registered for: a) different orientations (μ) of the Knoop indenter on the $(10\bar{1}0)$ plane b) indentation on (0001) plane ($\mu=30^{\circ}$) The kinks on the curves are due to onset of a crack propagation.

The hardness values calculated for the dynamic indentations based on the depth of penetration in general differ from those obtained by the classic method [60] including the present results. The size of the indentation traces and the crack patterns around them are identical in the two sets of data by conventional and dynamic techniques. The difference is due to the presence of different phenomena: dynamic hardness is expressed only by maximum depth of indenter penetration into the solid while the conventional measurements are also influenced by such processes as post-indentation recovery by elastic, plastic and fracture response.
The hardness anisotropy trends observed by both techniques remains. Nevertheless anisotropy is much less pronounced in the case of dynamic measurements.

It was also concluded [61] that indentation anisotropy manifests itself in the dynamic test through the variations of the areas bounded by loading and unloading curves which are believed [61] to be a function of the amount of energy consumed for plastic deformation of indented sapphire. If there are kinks on the loading curve (Fig. 7), the bounded area is related to the energy expended in plastic deformation plus energy dissipated in fracture.

The great advantage of using the Shimadzu dynamic hardness tester was the ability to register the onset of crack propagation during the loading cycle (Fig. 7). It is believed that the kinks on the curves are associated with the development of cracks. The indentation fracture in sapphire starts at surprisingly low indentations loads (Fig. 7b). These results are in excellent agreement with a recent observation of Cook and Pharr who stated that "the most significant observation was that radial cracks initiated almost immediately in the loading cycle below the optical resolution of the instrument" [22]. However, indentation by Cook and Pharr [22] was done only on the basal plane of sapphire under very high maximum load (around 4 kgf). We observed that for crack initiation, higher loads are necessary when the prismatic plane is indented. For some of the indents the acoustic emission detector was used in order to register the onset of crack propagation.

Conventional Vickers pyramid indentation. When the same load is applied, the Vickers indenter penetrates much deeper into the material and produces the better developed fracture pattern around and beneath the indenter compared to the Knoop pyramid. The sets of Vickers indentations made on both the basal and pyramidal planes are shown in Fig. 8 a) and b) respectively. The micrograph f) is presented twice (f - reflection, g - transmission micrographs) in order to reveal more details.
The above mentioned results are comparable with those reported previously by Smith and Pletka [56]. The differences between these two observations of crack patterns are due to following: the former authors have used pure and variously doped sapphire samples, the orientation of their indents was only with a diagonal parallel to the [1$\bar{2}$10] direction, the surface condition of their crystals made it impossible to observe sub-surface cracks, and in addition the surroundings of the indents were not entirely visible.

Although the indentation crack patterns observed in our pure sapphire crystals are more complicated that those shown by previous authors, the essential features remain eg. the radial crack orientation is less dependent of the Vickers indenter position and associated only with the crystal structure [22] [56]. The anisotropic properties of sapphire manifest themselves in every indent on the prismatic plane which is quite asymmetric (Fig. 8). This fact made difficult to measure the hardness (which is in addition largely influenced by fracture). The results of such an "improper measurement" are shown in Figure 9a. Figure 9b presents similar data measured on the basal plane of sapphire. Both curves show anoma-

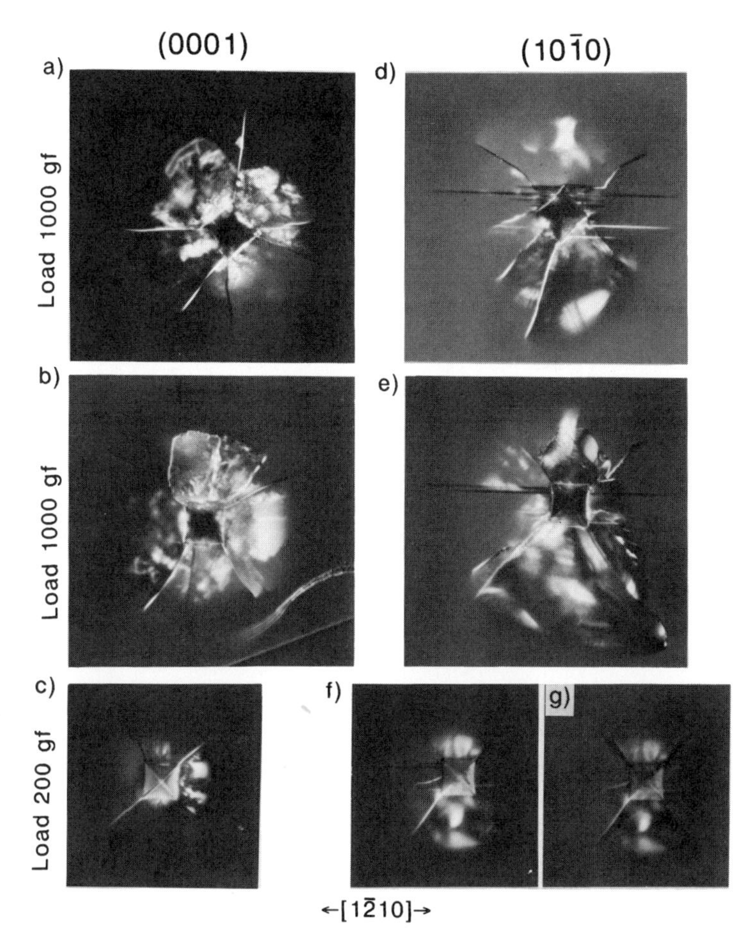

Figure 8. Micrographs of crack patterns around Vickers indents made on (0001) plane { a) P=1000 gf, μ=0°; b) P=1000 gf, μ=90°; c) P=200gf, μ=90°} and on (10$\bar{1}$0) plane {d) P=1000 gf, μ=0°; e) P=1000 gf, μ=90°; f) P=200 gf, μ=90°; g) view of trace f) in transition light} differing in orientation with respect to [1$\bar{2}$10] direction and maximum load applied to indenter.

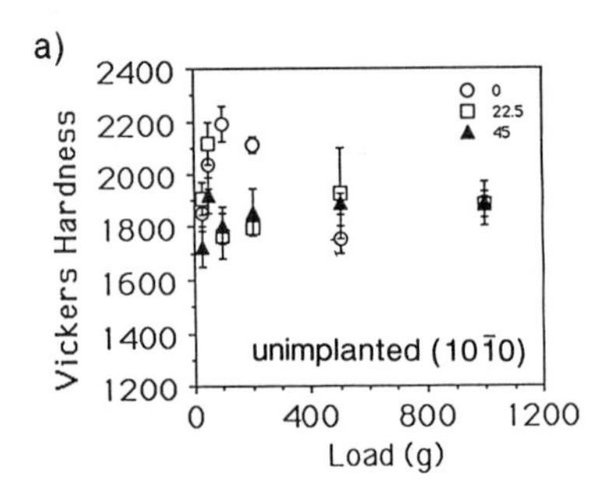

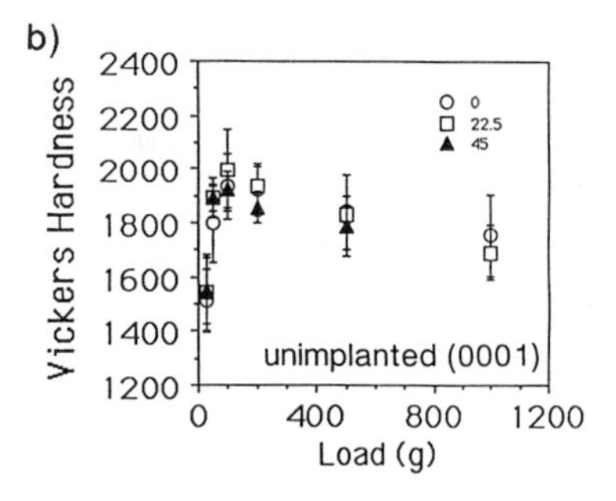

Figure 9. The results of Vickers hardness measurements on basal and pyramidal plane of sapphire as a function of indentation load.

lous shapes at low loads, possibly associated with the low accuracy of hardness measurements for very small indents (Fig. 5a). The distinct anisotropy of Vickers hardness on the prismatic plane is registered.

Implanted Sapphire

Residual stresses The first rough estimation of the residual stresses generated in the subsurface region due to incorporation a large number of foreign atoms into the sapphire lattice was done on the basis of measurements of the vertical expansion of material due to irradiation. Following the idea of Burnett and Page [36] and assuming after them homogeneous expansion of the polyhedral about each site while implanted atoms are assumed to be located in interstitial positions, the stresses in the implanted layers were calculated. The highest estimated maximum stress on the prismatic plane was found for gold implanted region (Au HD) while on the basal plane the nickel implanted area (Ni HD) had the highest stress. The results are collected in Table 2.

Table 2. Maximal residual stresses estimated according to Burnett and Page method [36]

maximal estimated residual stress [GPa]				
implanted area	Ni HD	Ni LD	Au HD	Au LD
(0001) plane	7.4	2.9	5.6	2.1
(10$\bar{1}$0) plane	6.6	3.5	10.1	2.0

The measured height of the surface steps which is the basis for a very approximate evaluation of the maximum stresses generated in the implanted layer is shown in Fig. 10 as two separate diagrams for the two crystal planes.

It is not surprising that the height of surface steps correspond with previously calculated values of the residual stresses as it was considered during calculation. However, very striking is the fact that the above reported relations are in excellent agreement with the dependence of average stresses (Fig. 11a) measured by indentation according to the method of Lawn and Fuller [44] on implantation parameters.

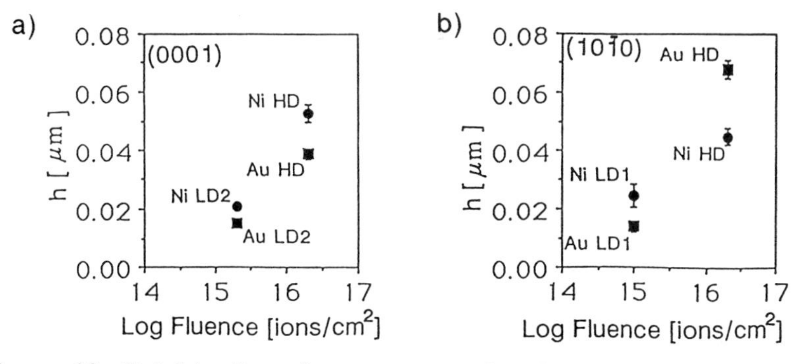

Figure 10. Height of surface steps produced by ion implantation

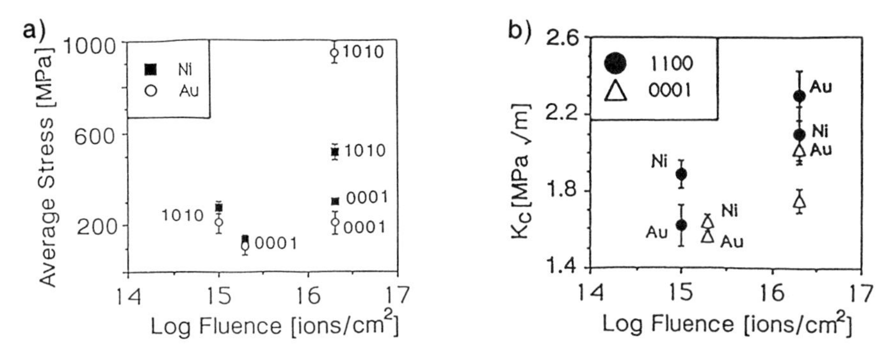

Figure 11. Average residual stresses (a) and fracture toughness (K_c) measured by the indentation fracture method, as a function of ion fluence, implanted species and crystallographic indices of the investigated surfaces

The measured values of residual stresses correspond well with the results obtained by other authors for similar type species [36][40]. Although the variety of applied fluences and species is not very wide, significant information is provided. The compressive nature of these stresses and their magnitude has been estimated.

The measured values of average residual stresses are lower than those quoted by McHargue [40] for sapphire implanted by Fe ions with 1 MeV energy. This is due to the application of the indentation method which usually gives [36] a 10 times lower estimate of the stress level than the cantilever beam technique. Another reason for the lower estimation of the residual stresses is our assumption that the implanted region reaches from the surface to a depth of R_p+2*s i.e. deeper than in the calculations of the other authors. Disregarding the above mentioned discrepancies, the results correlate quite well with these obtained by McHargue [40]. However, the applied doses, species and exact values of ion energies are different in these two works.

Summarizing, one can conclude that the implantation growth of sapphire and residual stress induced by ion implantation is a function of ion fluence, crystallographic orientation of the implanted surface and type species applied. Therefore the ion implantation technique seems to be useful for generating residual stresses in the surface region in a controlled, reproducible manner (by changing the implantation parameters) with a substantial effect on mechanical (especially fracture) properties.

Fracture Toughness The variation of the fracture toughness values (K_C) measured by the indentation method in different implanted areas is shown in Fig. 11b. The Vickers indenter orientation was chosen as 90° (see Fig. 4) in order to come closest to an ideal crack pattern - well developed radial/median cracks and minimizing the lateral crack number and area. Improper choice of the crack patterns for the evaluation of the fracture toughness value has been frequently criticized (eg. in this volume by Sakai, Ghosh and Bradt [62]).

The observed variations in fracture toughness have been expected since measurements of the residual stresses by indentation reveals an influence of implantation on the length of surface traces of the radial cracks. Implantation improves fracture toughness. The large compressive stresses generated in the implanted layer are responsible for the limitations in the development of cracks. Post-fracture phenomena such as crack healing are also expected, but have not yet been confirmed due to experimental difficulties in making this observation.

Conventional indentation. In order to estimate the influence of implantation layer with high compressive residual stresses on the mechanical behaviour of sapphire conventional Knoop and Vickers indentation tests were performed.

The Knoop indentation anisotropy was assumed to be a sensitive method to estimate the influence of implantation on mechanical behaviour of sapphire. The results of Knoop hardness anisotropy measurements in variously implanted areas are shown in Fig. 12. The result is quite surprising as after low dose implantation (both species) the material became harder and the hardness anisotropy was less pronounced than in original pure sapphire. After high dose implantation the effect is different as the material retains its anisotropic properties. The type of species also has a distinct influence only at high fluences. Nickel ions (which penetrate quite deep) make sapphire harder (Ni HD) while softening is observed for implantation with gold (Au HD). The Au ions are much bigger than Ni, therefore the damage produced by them is much larger. This is the reason for the different effects on the mechanical properties of sapphire. It seems that under Au implantation the onset of amorphisation occurs and softer phase is formed. However, according to the literature, the dose $2x10^{16}$ ions/c^2 seems to be

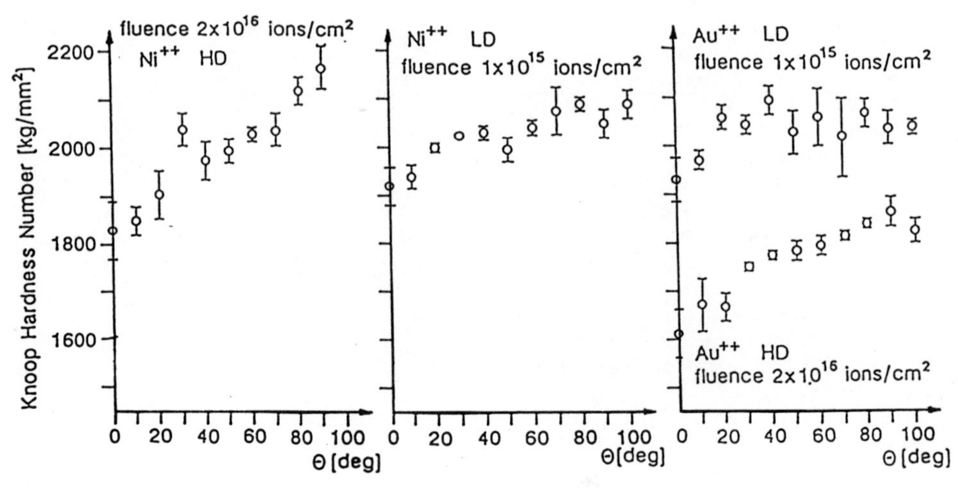

Figure 12. Variation of Knoop hardness anisotropy on the $(10\bar{1}0)$ plane of sapphire due to implantation parameters (load 200 gf)

too low to initiate amorphisation of the sapphire. On the other hand, there is not much information about ion implantation in the MeV energy range and to our best knowledge there is the first report about implantation of Au into sapphire (E=3 MeV).

Also surprising is the fact that the behavior of the two materials after low dose implantation is quite similar although the implanted species are different. The contributions of two independent phenomena is suspected.

Careful observation of these indentations together with those made by the Vickers pyramid have supported our confidence to the results (Fig. 13). There is no doubt that implantation affects indentation fracture very much. This influence is a function of the applied fluence. The crack pattern in implanted specimens is very much reduced (both radial and lateral cracks). When implanted with high doses we see almost completely plastic indentation even in the "brittle direction" ($\mu=0^{\circ}$); for low doses these effects are moderate. Significant shortening of the radial cracks generated by the Vickers indenter is also visible (Fig. 13). In addition, the Vickers hardness load dependence has been measured. This experiment was possible only on the basal plane since the strong deformation of the Vickers traces due to anisotropy made the tests on $(10\bar{1}0)$ plane invalid. In Figure 14 two diagrams of the load dependence of basal plane Vickers hardness for high dose gold and nickel implanted sapphire are shown. In this case anisotropy is not distinct (please compare with Fig. 9) as the (0001) plane is transversally isotropic. However differences between the areas implanted with different species are clearly visible. The results begin to converge at higher loads, when the Vickers pyramid penetrates very deep and the influence of the thin subsurface layer is negligibly small. The softening of sapphire implanted with Au at high dose and hardening when implanted with nickel is also noticeable (Fig. 14) within the proper range of loads.

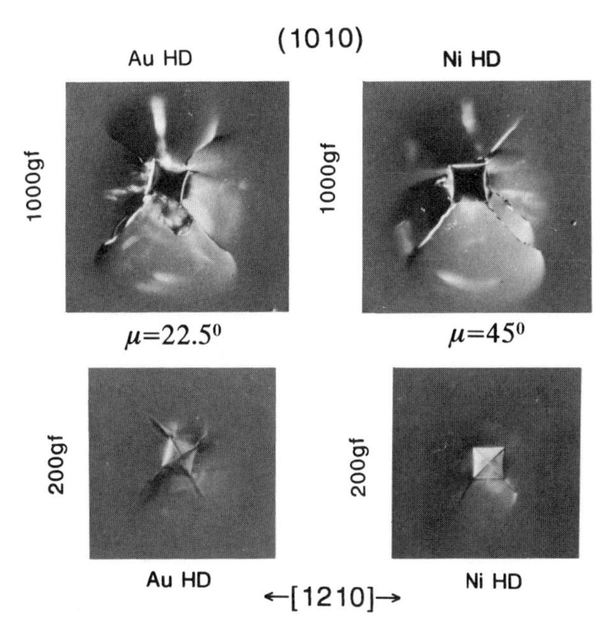

Figure 13. Micrographs of indentation fracture patterns as a function of implantation parameters, orientation of indenter and crystallographic orientation of the investigated plane.

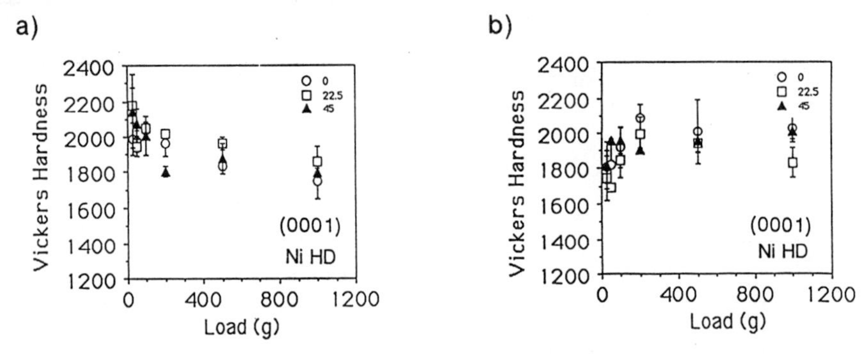

igure 14. Vickers hardness as a function of test load for: a) gold implanted (Au HD) and b) nickel implanted (Ni HD) sapphire

Dynamic Knoop Pyramid Indentation. Typical load depth curves for sapphire implanted with nickel (Ni HD) serve as an example of the dynamic indentation test results in the implanted areas (Fig. 15). The maximum indentation depth reached by the indenter differs very little from the one reported for unimplanted crystal (please compare with Fig. 7). However, a small relative hardening in the Ni implanted region and a small softening in the gold implanted area was observed.

In order to explain these original results it is indispensable to have detailed information about the location of the implanted layer under the surface of sapphire from Rutherford back scattering and about the location of the incorporated ions in the sapphire lattice (eg. channeling method). Information on the state of the damaged surface layer (amorphous or crystalline) should be also obtained (electron channeling patterns in a scanning electron microscope [63]).

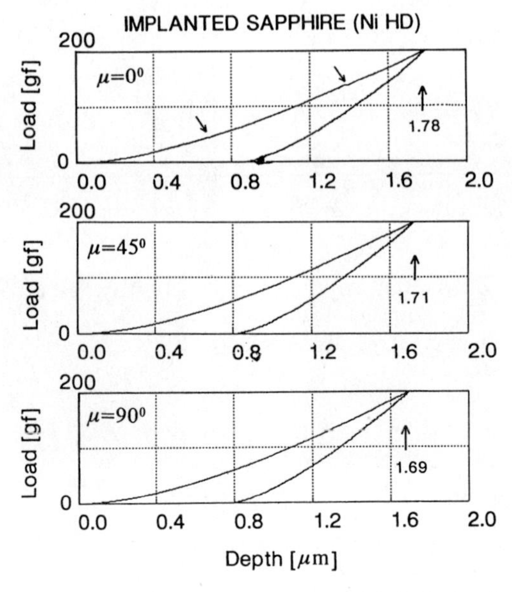

Figure 15. Examples of load-depth curves for nickel implanted (NiHD) sample

SUMMARY AND CONCLUSIONS

The mechanical behaviour of pure and and ion implanted sapphire crystals during and after indentation process have been studied. The investigations were performed on two crystallographic planes : basal (0001) and pyramidal ($10\bar{1}0$) which were specially prepared (smooth, transparent, stress free).

One advantage in these studies was the use of the newest dynamic ultra-micro hardness tester (SHIMADZU DUH-200) which enabled registration of the load-indentation depth history within the entire indentation cycle.

The advantages offered by the new equipment together with careful interpretation of the data from conventional techniques (Knoop, Vickers hardness, Nomarski contrast-optical microscopy) resulted in these observations on the deformation and fracture in indented pure sapphire:

- the generation of microcracks in the initial stages of indentation was observed and confirmed by acoustic emission (AE) studies
- a strong influence of crystal anisotropy (index of surface, position of the indenter) on all the observed phenomena (initiation of cracks, activation of deformation systems, final crack pattern, hardness, indentation history, toughness) was found.

The most interesting result coming from this part is a conclusion that indentation deformation of sapphire (both Knoop and Vickers) is almost always (starting from low loads of 10-20 gf) accompanied by fracture which influence the deformation (eg. hardness data).

In order to modify the mechanical properties of sapphire which was found to be even more brittle than it is commonly considered (see above) the crystal was implanted with various doses by Au and Ni ions of 3 MeV energy.

Since the experimental data from this energy range does not exist, the application of the new dynamic hardness tester the experimental approach yielded novel results.

The level of the residual stresses induced by ion implantation was found to be dependent on the crystallographic orientation of the surface and on the implantation parameters.

The clear effect of 3 MeV gold as well as nickel implantation into sapphire on vertical expansion, measured average and integrated residual stresses, fracture toughness, hardness, indentation depth, length of the radial cracks as well as median and lateral crack patterns has been registered.

The present experimental work shows broad spectrum of problems which need an interdisciplinary approach and further careful examination. However, the large amount of experimental data and their complex interpretation should not undermine the clearness of considerations and validity of the collected results.

ACKNOWLEDGEMENTS

We are indebted to Dr. S. N. Crichton for many valuable suggestions and also for reading the manuscript. The specimen implantation by Dr. Y. Horino is gratefully acknowledged. This work could not have been completed without the kind support of the SHIMADZU CORPORATION who permitted the use of their newest ultra-microhardness tester. The suggestions and assistance with hardness measurements by Mr. N. Abeno and M. Souma (SHIMADZU) is appreciated. Appreciation is also extended to Prof. R.C. Bradt and Prof. M.V. Swain for helpful discussions. One of us (R. Nowak) is grateful to Japanese Government for support through on STA Fellowship.

REFERENCES

1. R. Nowak, K. Ueno, and M. Kinoshita, On the anisotropy of indentation deformation and fracture of sapphire and silicon carbide single crystals, in: "Mechanical Behaviour of Materials", M. Jono and T. Inoue, ed., Pergamon Press, 551-556, (1991).
2. D. J. Clinton and R. Morrell, The hardness of alumina ceramics, Proc. Brit. Cer. Soc., 34:113 (1984).
3. B. R. Lawn and R. Wilshaw, Indentation fracture: principles and applications, J. Mat. Sci., 10:1049-1081 (1975).
4. B. R. Lawn and M. V. Swain, Microfracture beneath point indentations in brittle solids, J. Mat. Sci., 10:113-122 (1975).
5. M. V. Swain, A note on the residual stress about a pointed indentation impression in a brittle solid, J. Mat. Sci., 11:2345-2348 (1976).
6. B. R. Lawn and A. G. Evans, A model for crack initiation in elastic/plastic indentation field, J. Mat. Sci., 12:2195-2199 (1977).
7. M. V. Swain and J. T. Hagan, The origin of median and lateral cracks around indents in brittle materials, J. Phys. D., 11:2091 (1978).
8. B. R. Lawn and D. B. Marshall, Hardness, toughness, and brittleness: An indentation analysis, J. Am. Ceram. Soc., 62:347-350 (1979).
9. D. B. Marshall, B. R. Lawn and P. Chantikul, Residual stress effect in sharp contact cracking, J. Mat. Sci., 14:2225-2235 (1979).
10. B. R. Lawn, A. G. Evans and D. B. Marshall, Elastic/plastic indentation damage in ceramics: The median/radial crack system, J. Am. Ceram. Soc., 63:574-581 (1980).
11. G. R. Anstis, P. Chantikul, B. R. Lawn and D. B. Marshall, A critical evaluation of indentation techniques for measuring fracture toughness: I, Direct crack measurements, J. Am. Cer. Soc., 64:533 (1981).
12. P. Chantikul, G. R. Anstis, B. R. Lawn and D. B. Marshall, A critical evaluation of indentation techniques for measuring fracture toughness: II, Strength method, J. Am. Ceram. Soc., 64:539-543 (1981).
13. S. S. Chiang, D. B. Marshall and A. G. Evans, The response of solids to elastic/plastic indentation. I. Stresses and residual stresses, J. Appl. Phys., 53:298-311 (1982).
14. S. S. Chiang, D. B. Marshall and A. G. Evans, The response of solids to elastic/plastic indentation. II. Fracture initiation, J. Appl. Phys., 53:312-317 (1982).
15. D. B. Marshall, B. R. Lawn and A. G. Evans, Elastic/plastic indentation damage in ceramics: The lateral crack system, J. Am. Ceram. Soc., 65:561-566 (1982).
16. D. B. Marshall, T. Noma and A. G. Evans, A simple method for determining elastic-modulus-to-hardnes ratios using Knoop indentation measurements, Comm. Am. Ceram. Soc., C-175-C-176 (1982).
17. B. J. Hockey and B. R. Lawn, Electron microscopy of microcracking about indentations in aluminum oxide and silicon carbide. J. Mat. Sci., 10:1275-1284 (1975).
18. K. Tanaka, Y. Kitahara, Y. Ichinosc and T. Iimura, Fracture analysis of Manganese Zinc Ferrite using indentation, Acta Met., 32:1719 (1984).
19. M. Iwasa and R. C. Bradt, Fracture toughness of single-crystal alumina. Adv. Ceram., 10:767-779 (1985).
20. R. C. Bradt and W.D. Scott, Mechanical properties of alumina. in: Alumina Chemicals: Science and Technology Handbook, (ed. LeRoy D. Hart), Am. Ceram. Soc. Inc., Westerville, Ohio, 1990, 23-39.

21. A. Pajares, F. Guibertau, A. D. Rodriguez and A. H. Heuer, Indentation induced cracks and the toughness anisotropy of 9.4-mol% Yttria stabilized cubic zirconia crystals, J. Am. Cer. Soc., 74:859 (1991).
22. R. F. Cook and G. M. Pharr, Direct observation and analysis of indentation cracking in glasses and ceramics. J. Am. Ceram. Soc., 73:787-817 (1990).
23. R. Tandon, D. J. Green and R. F. Cook, Surface stress effects on indentation fracture sequences. J. Am. Ceram. Soc.,73:2619-2627 (1990).
24. S. E. Molis and D. R. Clarke, Measurement of stresses using fluorescence in a optical microprobe: Stresses around indentations in a chromium doped sapphire. J. Am. Ceram. Soc., 73:3189-3194 (1990).
25. C. P. Ostertag, L. H. Robins and L. P. Cook, Cathodoluminescence measurement of strained Alumina single Crystals, J. Eur. Ceram. Soc., 7:109-116 (1991).
26. L. Yingzhi and D.A. Hills, The analysis of three-dimentional cracks generated by sharp indentation. J. Mech. Phys. Solids, 38, 255-272 (1990).
27. G. B. Krefft and E. P. EerNisse, Volume expansion and annealing compaction of ion-bombarded single-crystal and polycrystalline Al_2O_3, J. Appl. Phys., 49:2725-2730 (1978).
28. C. J. McHargue, H. Naramoto, B.R. Appleton, C.W. White and J. Williams, Structure and properties of Single Crystal Al_2O_3 implanted with chromium and zirconium, Proc. Mat. Res. Soc., 7:147-153 (1982).
29. H. Naramoto, C. W. White, J. M. Williams, C. J. McHargue, O. Holland, M. M. Abraham, and B. R. Appleton, Ion implantation and thermal annealing of Al_2O_3 crystals. J. Appl. Phys., 54:683-696 (1983).
30. P. J. Burnett and T. F. Page, Modifying the tribological properties properties of ceramics by ion implantation. Proc. Br. Ceram. Soc., 34:65-76 (1984).
31. G. C. Farlow, C. W. White, C. J. McHargue and B. R. Appleton, Behaviour of Implanted Al_2O_3 in an oxidizing annealing environment, Proc. Mat. Res. Soc., 27:395-400 (1984).
32. T. Hioki, A. Itoh, S. Noda, J. Kawamoto and O. Kamigato, Strengthening of Al_2O_3 by ion implantation, J.Mat.Sci.Lett., 3:1099 (1984).
33. P. J. Burnett and T. F. Page, Surface softening in silicon by ion implantation. J. Mat. Sci., 19:845-860 (1984).
34. P. J. Burnett and T. F. Page, Changing the surface mechanical properties of silicon and Al_2O_3 by ion implantation. J. Mat. Sci., 19:3524-3545 (1984).
35. P. J. Burnett and T. F. Page, Surface softening in silicon by ion implantation. J. Mat. Sci., 19:845-860 (1984).
36. P. J. Burnett and T. F. Page, An investigation of ion implantation induced near-surface stresses and their effects in sapphire and glass. J. Mat. Sci., 20:4624-4646 (1985).
37. T. Hioki, A. Itoh, M. Okubo, S. Noda, H. Doi, J. Kawamoto and O. Kamigato, Mechanical property changes in sapphire by nickel ion implantation and their dependence on implantation temperature. J. Mat. Sci., 21:1321-1328 (1985).
38. T. Hioki, A. Itoh, S. Noda, H. Doi, J. Kawamoto and O. Kamigato, Effect of ion implantation on fracture stress of Al_2O_3. Nucl. Inst. Meth. Phys. Res., B7/8:521-525 (1985).
39. C. J. McHargue, Structure and mechanical properties of ion implanted ceramics. Nucl. Instrum. & Methods , 19/20:797-804 (1987).
40. C. J. McHargue, M. E. O'Hern, C. W. White and M. B. Lewis, Ion implantation in ceramics-residual stress and properties. Mat. Sci. Eng., A115:361-367 (1989).
41. T. Hioki, A. Itoh, S. Noda, H. Doi, J. Kawamoto and O. Kamigato, Modification of the mechanical properties of ceramics by ion implantation. Nucl. Instrum. & Methods , 39:657-64 (1989).

42. S. J. Bull and T. F. Page, Thermal effects on the microstructure and mechanical properties of ion implanted ceramics. J. Mat. Sci., 26:3086-3106 (1991)
43. R. Nowak, Y. Horino, K. Ueno, M. Kinoshita, W. Ensinger and M. Satou, Indentation fracture of high-energy ion implanted alpha-alumina single crystals, in: "Fracture of Engineering Materials and Structures", S. H. Teoh and K. H. Lee, ed., Elsevier Applied Sci., 361-367 London and New York, (1991)
44. B. R. Lawn and E. R. Fuller, Measurement of thin-layer surface stresses by indentation fracture. J. Mat. Sci.,19:4061-4067 (1984)
46. M. L. Kronberg, Plastic deformation of single crystals of sapphire: basal slip and twinning, Acta Met., 5:507-524 (1957)
47. J. D. Snow and A. H. Heuer, Slip systems in $Al_{20}O_3$, J. Am. Ceram. Soc., 56:153-157 (1973)
48. R. Schuplein and P. Gibbs, Surface structure in corundum: II, Dislocation structure and fracture of deformed single crystals, J. Am. Ceram. Soc., 45:439-452 (1962)
49. E. Stofel and H. Conrad, Fracture and Twinning in sapphire, Trans. AIME, 227:1053-1060 (1963)
48. H. Conrad, Mechanical behaviour of sapphire, J. Am. Ceram. Soc., 48:195-201 (1965)
49. D. J. Barber and N.J. Tighe, Electron microscopy of synthetic corundum crystals, Phil. Mag., 11:495-512 (1965)
50. A. H. Heuer, Deformation Twinning in corundum, Phil. Mag., 13:379-393 (1966)
51. B. J. Pletka, T. E. Mitchell and A. H. Heuer, Dislocation structures in sapphire deformed by basal slip, J. Am. Ceram. Soc., 57:388 (1974)
52. D. J. Gooch and G. W. Groves, Non-basal slip in sapphire, Phil. Mag., 28:623-637 (1973)
53. S. M. Wiederhorn, B.J. Hockey and D. E. Roberts, Effect of temperature on the fracture of sapphire, Phil. Mag., 29:783-796 (1974)
54. R. J. Bratton, Precipitation and hardening behavior of Czochlarski star sapphire, J. Appl. Phys., 42:211-216 (1971)
55. C. A. Brookes, J. B. O'Neill and A.W. Redfern, Anisotropy in the hardness of single crystals. Proc. Roy. Soc., A322:73-88 (1971).
56. S. S. Smith and B. J. Pletka, Indentation fracture of single crystal and polycrystalline aluminum oxide, in: "Fracture Mechanics of Ceramics", R. C. Bradt, A. G. Evans and D. P. H. Hasselman, ed., Plenum, (1983)
57. W. Kollenberg, Plastic deformation of Al_2O_3 crystals by indentation at temperatures up to 750°C, J. Mat. Sci., 23:3321-3325 (1988)
58. M. Kaji and R. C. Bradt, Microhardness of single crystal Al_2O_3 on the basal plane, in press
59. J. P. Biersack and L. G. Haggmark, A Monte-Carlo computer program for the transport of energetic ions in amorphous targets. Nucl. Instrum. & Methods ,174:257-69 (1980)
60. D. Stone, W. R. LaFontaine, P. Alexopoulos, T. W. Wu and Che-Yu Li, An investigation of hardness and adhesion of sputter-deposited aluminum on silicon by utilizing a continuous indentation test, J. Mater. Res., 3:141-147 (1988)
61. M. Sakai - private information
62. T. Sakai, A. Ghosh and R. C. Bradt, The indentation fracture resistance of self-reinforced mullites, in this volume
63. D. C. Joy, D. E. Newbury and D. L. Davison, Electron channeling patterns in the scanning electron microscope, J. Appl. Phys., 53:R81-R122 (1982)

EFFECTS OF PROCESS ZONE AND SPECIMEN GEOMETRY ON

FRACTURE TOUGHNESS OF SILICON NITRIDE CERAMIC

Shinji Yamauchi and Toshiro Kobayashi

Department of Production Systems Engineering
Toyohashi University of Technology
Tempaku-cho, Toyohashi, 441 JAPAN

ABSTRACT

A precise fracture toughness testing method in ceramics should be established in an early period.

In the present study, requirements to obtain the valid fracture toughness values of sintered silicon nitride ceramic are examined in the static three point bending test and instrumented impact test. Moreover, it is generally considered that the toughness of ceramics is influenced by microcracking or phase transformation at a crack tip process zone. It is important, therefore, to understand and clarify the feature of process zone. Measurement of residual stress by X-ray method and transmission electron microscope (TEM) observation are carried out for this purpose.

It was shown that fracture toughness was not affected by crack length (a) to specimen width (W) ratio a/W and span length (S) to specimen width ratio S/W. However, the fracture toughness was affected by specimen thickness (B) and notch root radius (ρ). Static and dynamic fracture toughnesses increased with increasing the process zone size.

The valid fracture toughness value was obtained by precracked type specimen thicker than 4mm. This condition was represented by $B \geqq 70(K_{IC}/\sigma_{mc})^2$, where σ_{mc} is a local critical fracture stress at the process zone.

INTRODUCTION

High strength ceramic materials which are superior in thermal resistance and corrosion resistance have been developed and to be utilized in various fields, for examples in gas turbine, aerospace and biomaterials.

On the other hand, fracture of ceramics is generally brittle; moreover when it is used for structural components, fracture toughness value is needed for design, but its evaluation method has not been established yet. Therefore, toughening and establishment of fracture toughness testing method in ceramics are key problems to be overcome(1).

Specimen size criterion in fracture toughness test of ceramics has not been established yet. Reports of effect of specimen geometry on fracture toughness is limited. In fact each investigator has used each specimen geometry from experience.

The one purpose of this paper is to make clear the effect of the specimen geometry on the fracture toughness value. That is, effects of the ratio of specimen width to span length, pre-crack length to specimen width, specimen thickness and notch root radius are studied under static and dynamic loading conditions.

Moreover, it has been recognized that brittle fracture of ceramics is affected by occurrence of microcracking or phase transformation[2]. It is very important to understand the crack tip process zone not only to develop strengthened ceramics but also to evaluate meaningful fracture toughness value.Therefore, studies by X-ray method and transmission electron microscope(TEM) observation in the process zone at a crack tip are made.

EXPERIMENTAL PROCEDURE

Specimen

Sintered silicon nitride ceramic is used in the present study. Mechanical properties are shown in Table 1. Geometries of test specimens for static three-point bending and instrumented Charpy impact tests are shown in Fig.1. The pre-crack(ρ=0)is made by the pop-in method(SEPB)[3] applying a load to the surface indentation of the specimen placed onto the grooved anvil.The ratios of specimen width(W) to span length(S){S/W=4,5,6,7and8}, pre-crack length(a) to specimen width { a/W = 0.3,0.4,0.5,0.6 and 0.7 } , specimen thickness (B){B=2,4,6 and 8mm} and notch root radius(ρ){ρ=0,40,50,100 and 150μm} are widely varied. Then,the effect of specimen geometry on the fracture toughness value is examined.

Table 1 Mechanical properties of test material.

Density (g/cm^3)	Bending stress (MPa)	Young's modulus (MPa)	Poisson's ratio
3.22	880	3.1 x 10^5	0.24

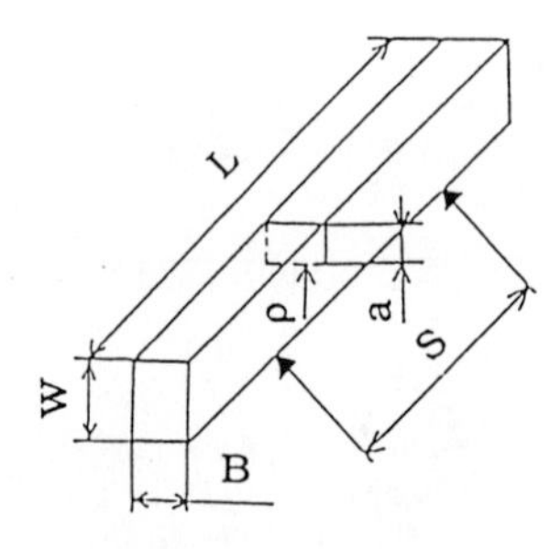

B:specimen thickness(mm)
W:specimen width(mm)
L:specimen length(mm)
S:span length(mm)
a:crack length(mm)
ρ:notch root radius(μm)

I	S/W	W	S	L	a	B	ρ
	4	7.5	30	40	3.8	4	0
	5		37.5	50			
	6		45	60			
	7		52.5	70			
	8		60	80			

II	a/W	W	S	L	a	B	ρ
	0.3	8	30	40	2.4	4	0
	0.4				3.2		
	0.5				4.0		
	0.6				4.8		
	0.7				5.6		

III	B	W	S	L	a	ρ
	2	8	30	40	4	0
	4					
	6					
	8					

IV	ρ	W	S	L	a	B
	0	8	30	40	4	4
	40					
	50					
	100					
	150					

Fig.1 Geometry of test specimen.

Fracture toughness test

The Stress intensity factor of a pre-cracked specimen is obtained by the following equations(1) and (2).

$$K_{IC} = 3\ P_f\ S\ a^{1/2}/2\ B\ W^2 \cdot Y \quad (1)$$
$$Y = A_0 - A_1\lambda + A_2\lambda^2 - A_3\lambda^3 + A_4\lambda^4 \quad (2)$$
$$\lambda = (a/W)$$

where,

P_f : fracture load

A_n : coefficient for the approximate function.

On the other hand, coefficient A_n of the three-point bending test has been generally given only for the ratios of specimen width to span length are 4 and 8 ; therefore, in other ratios, the correction equation which explicated by Wakai *et al.*(4) is used for calculating of the stress intensity factor K_{IC}.

The static three-point bending test is carried out at room temperature by an Instron type testing machine. The fracture load is measured with cross head speed of 0.1mm/min. Load-deflection curves are recorded by an X-Y recorder.

Moreover, AE signal is detected in the specimens with various notch root radii. The AE signal is measured by the SHIMADZU SAE-1000A; the AE sensor is set onto both ends of a specimen and AE energy and AE count are detected.

The dynamic fracture toughness test is carried out with 3.92J instrumented Charpy impact testing machine at room temperature. The impact velocity (V_0:0.2 m/s ~ 0.57 m/s) satisfied the recommended condition of $E_0 \geq 3E_t$ (where, E_0 : applied energy, E_t : total absorbed energy)(5).The dynamic fracture toughness is evaluated by the impact response curve method which was presented by J.F.Kalthoff applying to instrumented Charpy impact test(6). As a variation of this method, the small strain gage method(6) developed by Kobayashi *et al.* is applied, where a small semi-conductor gage is put directly just beside the crack tip on the specimen surface and calibration test is performed in the static loading condition. The strain gage signal is directly transferred to the digital storage oscilloscope via bridge circuit. The dynamic fracture toughness is evaluated from the impact response curve and the time-to-fracture(t_f).

X-ray residual stress measurement

In general, the X-ray residual stress measurement method which has been used in metals is applied to ceramics(7). The residual stress on fracture surface after the fracture toughness test is measured for various specimen thickness and notch root radius specimens. The residual stress is measured on the diffraction plane(321) with Cr-Kα radiation. The diffraction angle is determined by the middle point method and the residual stress is calculated by the iso-inclination technique of $\sin^2\psi$ method by the following equation(3)(8). An incident beam of X-ray (ψ) is 0° and changed at every 5° interval from 20° to45°(9).

$$\sigma_R = -\ [E/2(1+\nu)] \cdot \cot\theta_0 \cdot (\pi\ /180\ ) \cdot [d\ (2\theta)/d(\sin^2\psi)] \quad (3)$$

where,

σ_R : residual stress

E : Young's modulus

ν: Poisson's ratio

θ: Bragg angle.

In this study, in order to measure the residual stress distribution below the fracture surface, the surface layer is repeatedly removed by hydrofluoric acid etching of about ten minutes at 100°C. In this case, etched amount at the notched portion is measured with a micrometer and a distance below the fracture surface is estimated on the assumption that also the fracture surface portion is similarly removed by etching.

Microstructural observation

The fracture surfaces after fracture toughness test is observed by SEM. Moreover, measurement of the ratio of the real fracture surface area(A_e) to original cross sectional area(A_0) which relates to the fracture surface roughness is carried out by the three-dimensional image analyzer. A small portion of the specimen cut off vertical direction against fracture surface is used for a thin film preparation by the ion thinning device. The microstructure near the fracture surface is observed by transmission electron microscope(TEM).

RESULTS

Results of the fracture toughness testing

Figures 2 and 3 show that the fracture toughness values are almost constant independent of the ratios of span length to specimen width (S/W) and pre-crack length to specimen width(a/W). Fracture load value was so low that dynamic fracture toughness value was not evaluated in the case of impact testing at S/W=8.

On the other hand, the fracture toughness value depended on the specimen thickness and the notch root radius. The variation of the fracture toughness value by changing the specimen thickness is shown in Fig.4. The static and dynamic fracture toughness values are constant at B≧4mm. The variation of the fracture toughness values by changing the notch root radius is shown in Fig.5. The static and dynamic fracture toughness values increased with increasing of notch root radius. The similar result in the previous report[6] has been ascertained also in this study.

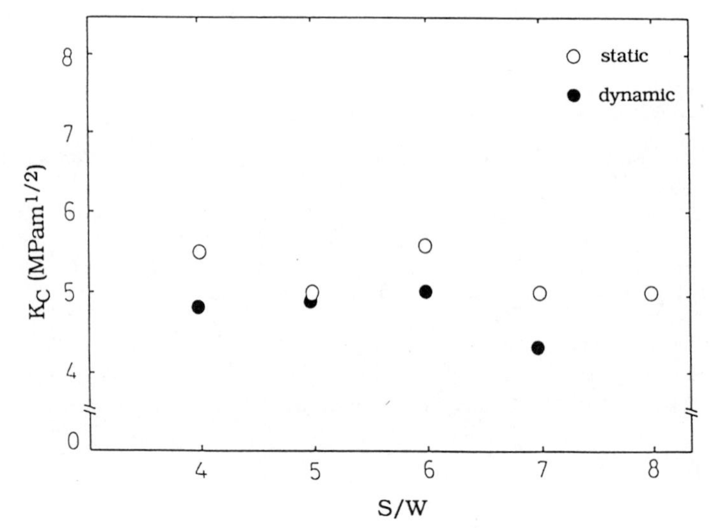

Fig.2 Relationship between S/W and fracture toughness K_C.

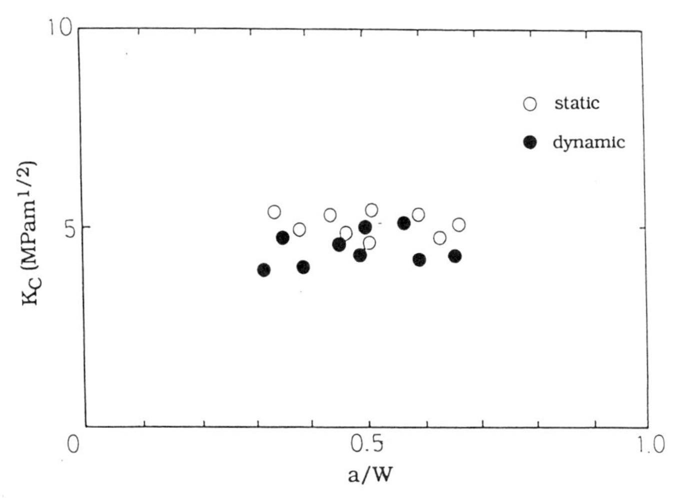

Fig.3 Relationship between a/W and fracture toughness K_C.

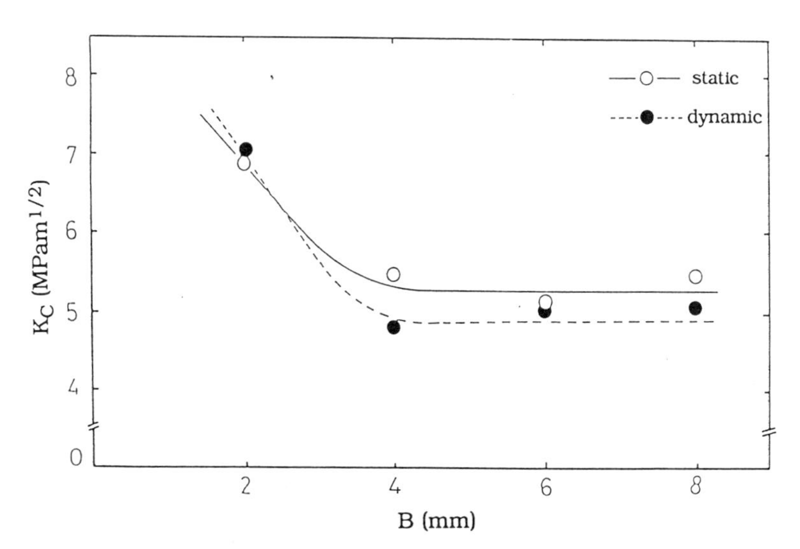

Fig.4 Relationship between specimen thickness and fracture toughness K_C.

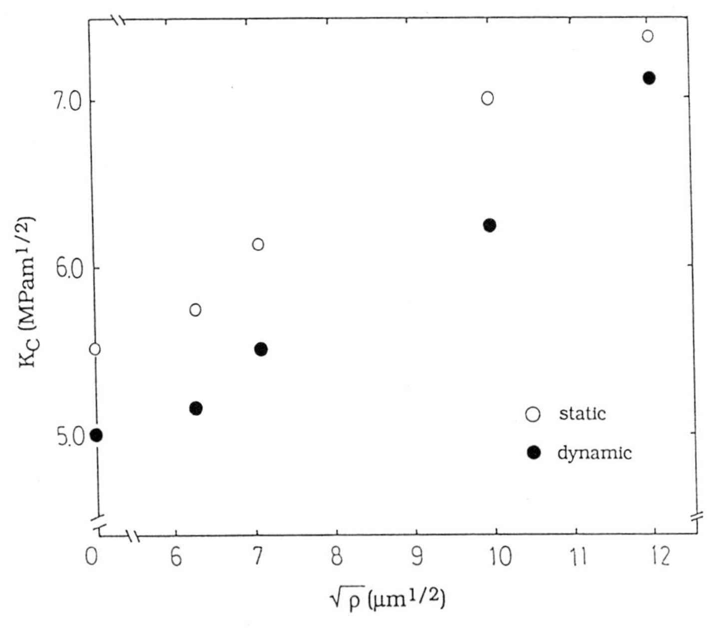

Fig.5 Relationship between notch root radius and fracture toughness K_C.

Results of X-ray residual stress measurement

The slope $d(2\theta)/d(\sin^2\psi)$ of straight line was measured by the method of least squares and the residual stress was determined by equation(3). The residual stress at specimen thickness from 2 to 8mm is shown in Fig.6. In specimen thickness =2mm, the residual stress at fracture surface was tensile, then changed into compressive and decreased to zero. On the other hand, in specimen thickness ≧4mm , the residual stress at fracture surface was almost zero, then compressive and decreased to zero. The residual stress regions are observed to extend under the static and dynamic loading conditions at B=2mm.

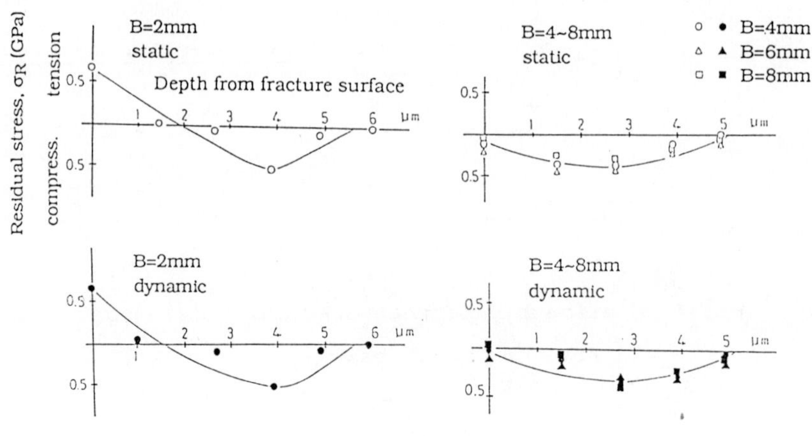

Fig.6 Variation of residual stress distribution with specimen thickness B.

The change of residual stress with notch root radius from 0 to 150μm at B=4mm specimen is shown in Fig.7 (The residual stress in Fig.7 is another data from Fig.6 in the case of ρ=0). The compressive residual stress at the fracture surface has a tendency to increase with increasing ρ; moreover, the residual stress region also increased. Significant difference by loading condition has not been recognized in the X-ray residual stress measurement.

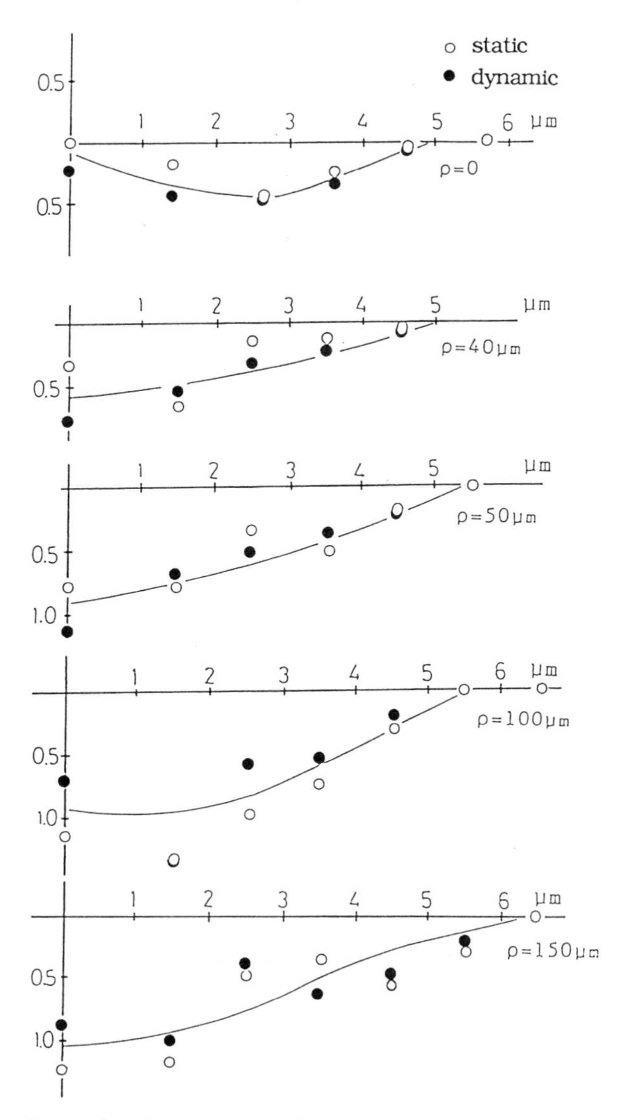

Fig.7 Variation of residual stress distribution with notch root radius ρ.

Results of microstructural observation

The microstructure of test material by TEM is shown in Fig.8. In the test material, the columnar α phase (arrow) was coexisted with the equiaxed β phase. The columnar α phase is produced at relatively low temperature under 1300°C and the equiaxed β phase is produced under high temperature. However, their formations depend on sintered condition, used powder and size of sintered body; therefore, the control of phase distribution has been difficult(10). It is known that the columnar α phase participates for crack bridging,debonding and crack deflection in fracture process; it has been reported therefore that bending stress and fracture toughness value increase with increasing these phenomena(11)(12).

Fig.8 Typical microstructure of test material.

DISCUSSION

Measuring condition of valid fracture toughness value

The effect of inertia load by impact is a problem on the valid fracture toughness measurement under the dynamic condition(13). Therefore,one of the authors has applied the impact response curve method, which uses a small semi-conductor strain gage placed just beside the crack tip on the specimen surface. Its details have been reported elsewhere(6). Under the dynamic condition in this study, effect of the inertia load on a hammer load signal was recognized larger with increasing specimen thickness. As increasing of S/W ratio, the fracture load was reduced remarkably and the hammer signal was easily affected by an electric noise. Therefore, it was not desirable to increase the S/W ratio.

The fracture toughness value was almost constant for various ratios of a/W and S/W; however, ASTM E399 gives encouragement to standardize for S/W=4 or 8 and a/W=0.5 in metallic materials. It is generally considered that the fracture toughness test has a difficulty in increasing the specimen thickness in many cases. In this study, therefore, it has been recommended that the condition of B=4 or 6mm, S/W=4 and a/W=0.5 is suitable to measure the valid fracture toughness.

By the way, the variation of the impact response curve with notch root radius (ρ) = 0~150mm is shown in Fig.9. The time history was almost similar. The impact response curve depends only on the elastic reaction of the specimen-striker system. Therefore, this method is based on the concept that time variation until fracture is same under the same impact condition(14). The fracture toughness value increased with increasing the notch root radius until ρ=100mm and the time history was similar. Therefore, once the stress intensity-time curve is determined for a dull

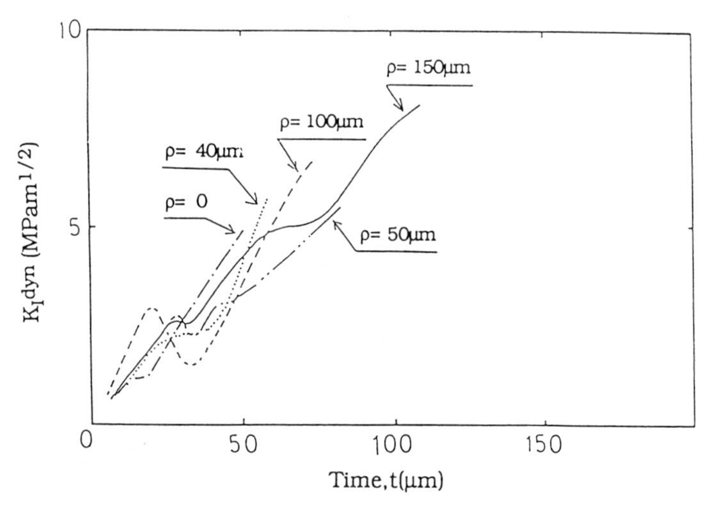

Fig.9 Variation of impact response curve with notch root radius ρ.

slit specimen of ρ≦100mm,then the fracture toughness value can be estimated only from a measurement of time-to-fracture in the pre-cracked specimen.

Characterization of the process zone

The fracture behavior of the test specimen was linear elastic regardless of the testing condition. The linear fracture mechanics which was presented by Irwin can be applied to the brittle material[15]. The local stress at a crack tip σ_{yy} is evaluated by the following equation(4).

$$\sigma_{yy} = K_I / (2 \pi r)^{1/2} \qquad (4)$$

where,

$$\sigma_{yy} \longrightarrow \infty \text{ at a crack tip } r \longrightarrow 0.$$

However,microcracking or phase transformation occurs at the main crack tip process zone. Therefore, a finite local critical fracture stress σ_{mc} must exist[16]. TEM micrograph near the fracture surface after the fracture toughness test is shown in Fig.10. Microcracking with the phase transformation in TZP has been already reported by T.Kobayashi[1] or A.G.Evans[17]. It is recognized from Fig.10 that the process zone is formed with a secondary cracking at a main crack tip[18] also in this material. The secondary microcracking at a crack tip in Fig.10 is closed and the measured residual stress distribution corresponded to the result of TEM observation. It has been assumed therefore that the measurement of the residual stress at the fracture surface is an effective method for measuring of the process zone size.

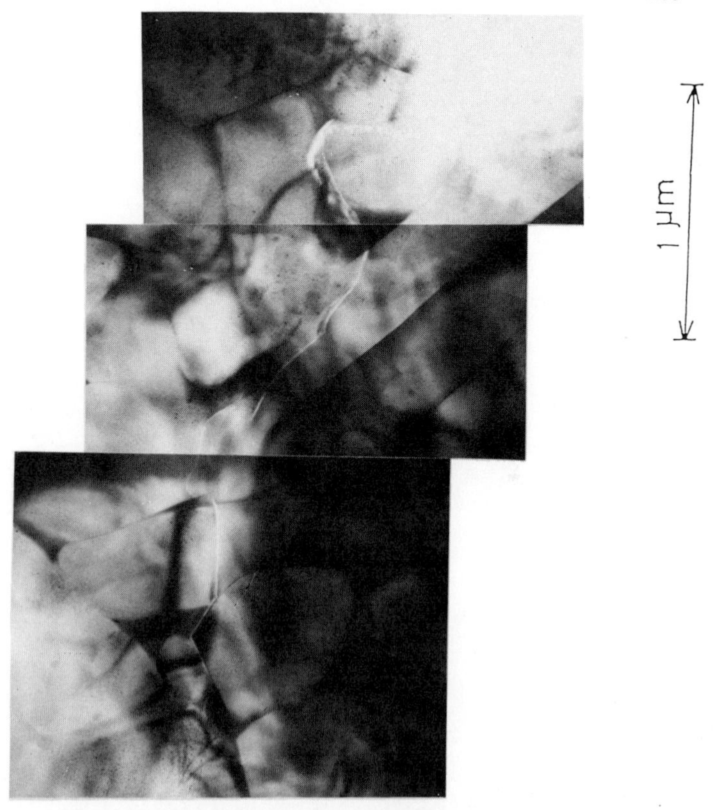

Fig.10 Transmission electron micrograph showing a secondary microcracking near the fracture surface.

Effect of the specimen geometry on the fracture toughness value

By the way, ASTM E399 gives generally the valid condition for the specimen size by the equation(5).

$$B \geq 2.5 \, (K_{IC} / \sigma_Y)^2 \qquad (5)$$

where,

σ_Y = tensile yield stress.

However, equation(5) is not applicable to the brittle material as ceramics. The relationship between specimen thickness(B) and $(K_C/\sigma_{mc})^2$ is shown in Fig.11. Where, the local critical fracture stress σ_{mc} is estimated from substituting the process zone size r_p in the equation(4), where r_p is a distance from a main crack tip to the point where the residual compressive stress becomes zero. As the result, the coefficient α becomes about 70 and the valid condition for the specimen size in the test material is given as follow.

$$B \geqq 70 \; (K_{IC} \; / \; \sigma_{mc})^2 \qquad (6)$$

If $\sigma_Y = \sigma_{mc}$, the coefficient α is about 30 times larger than the one in the ASTM E399 standard. However, σ_{mc} is not considered to be same with σ_y and the process zone size is very small in comparison with the specimen thickness in the present study. Moreover, according to the result which examined the relationship between K_{IC} and specimen thickness B for Al_2O_3 by F.E.Buresh[19], the coefficient α=50~60 has been reported for a valid value. This result was almost coincident with the one in the present study.

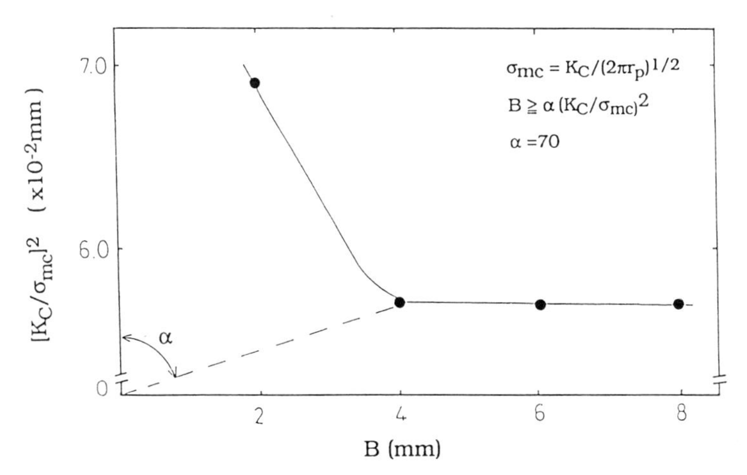

Fig.11 Relationship between specimen thickness B and $(K_C/\sigma_{mc})^2$.

Effect of the process zone on the fracture toughness value

In general, γ_s, which is the fracture surface energy, γ_p, which is the forming energy of a process zone, γ_D, emanating energy which is dissipated as tossing of broken halves, light and heat, and $\gamma_{etc.}$, which is other energy correlated with the fracture process of a ceramics. The total needed energy Γ is therefore given by the following equation(7).

$$\Gamma = \gamma_s + \gamma_p + \gamma_D + \gamma_{etc.} \qquad (7)$$

In the case of ceramics, γ_s and γ_p are very large comparing with other terms. Therefore, the total needed energy Γ may be evaluated as the equation(8),

$$\Gamma \approx \gamma_s + \gamma_p . \quad (8)$$

That is, it is considered that the fracture surface morphology and the process zone affect intensely to the fracture toughness value. The energy consumed by forming the process zone is known to be the source of the AE, the energy is usually required to cause the microcracking or the phase transformation. The result of the AE measurement at the static fracture toughness test under each different condition of the notch root radius is shown in Fig.12. The load, total AE energy and AE cumulative event count in the upper part and AE event rate(AE event count per time) in the lower part are shown. As total AE energy and AE cumulative event count increased with increasing the load, it seems that they correspond with forming of the process zone at a crack tip. Moreover, occurrence of AE signal increased with increasing the notch root radius and coincided with the result of measurement of the residual stress.

The specimens of ρ=150mm (4x8x40:slit specimen) and B=2mm(2x4x40: pre-cracked specimen) which showed high fracture toughness values were compared with the one of B=4mm(4x8x40:pre-cracked specimen) which showed low fracture toughness value in SEM micrograph on the fracture surface after the static and dynamic fracture toughness tests (Fig.13). The specimens of B=2 and 4mm show fracture surface with interface decohesion at columnar α phase, on the other hand, the specimen of ρ=150mm shows transgranular fracture surface with a less of decohesion. The notch root radius is so blunt in the latter specimen that the applied load increased and the process zone extended and resulted in the increasing of the residual compressive stress and changing of the AE measuring result. However, according to the above fracture surface morphology, fracture initiation mechanism of each specimen can not be identified completely. It will be necessary to investigate more details on this point in the future.

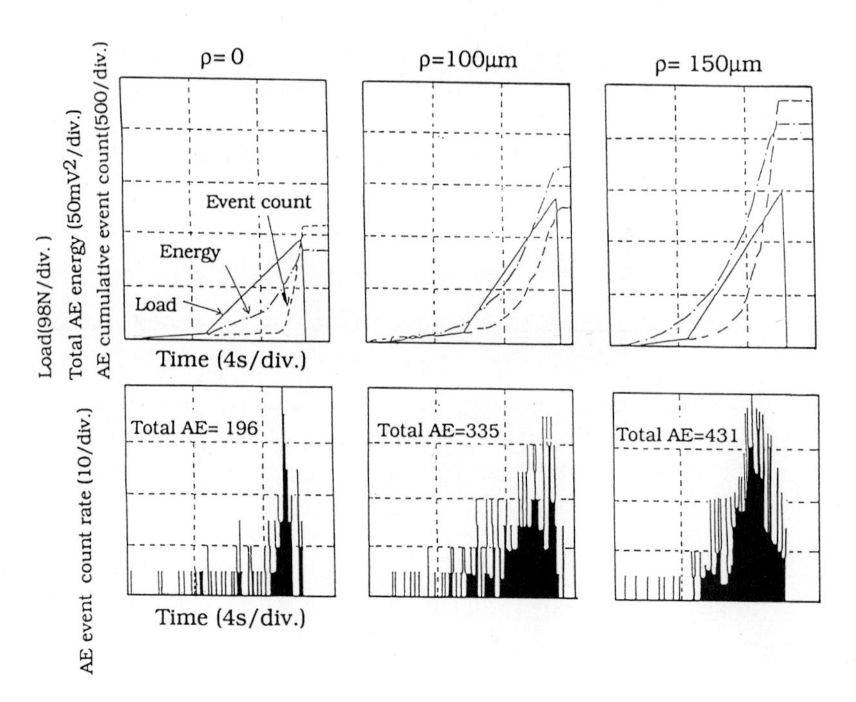

Fig.12 AE behavior in static fracture toughness test.

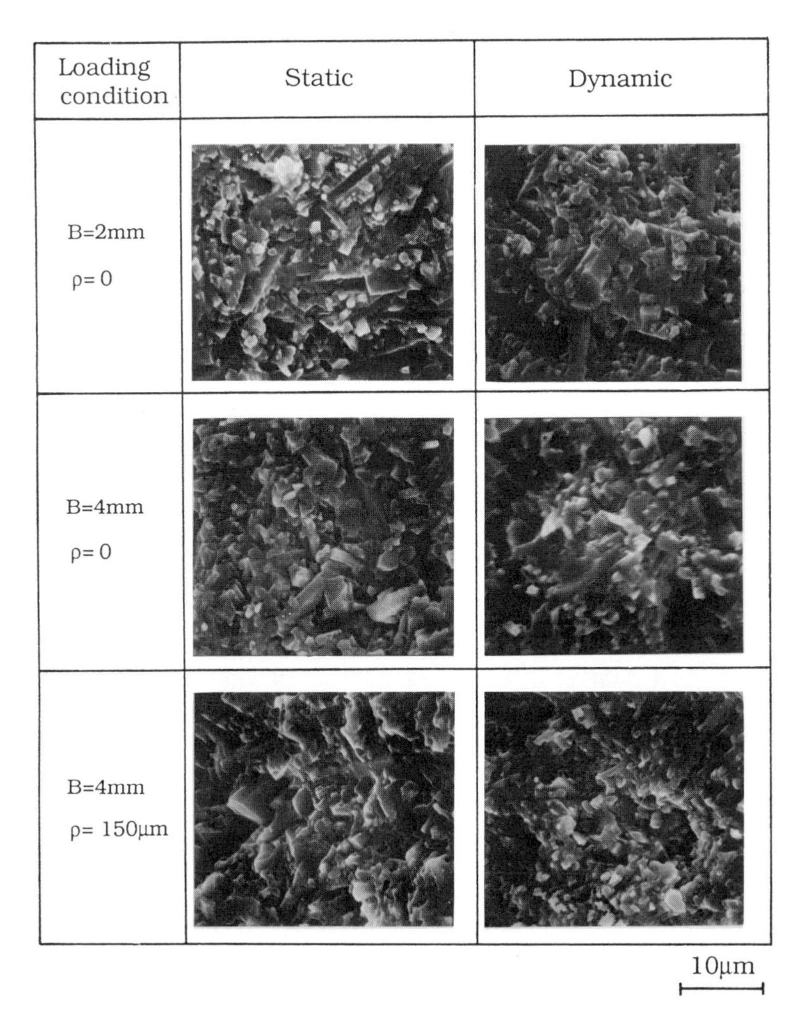

Fig.13 Typical SEM micrographs of fracture surface.

Table 2 Variation of specific area A_e/A_o and critical strain energy release rate G_c with specimen geometry.

Loading condition	B=2mm ,ρ= 0		B=4mm ,ρ= 0		B=4mm, ρ= 150μm	
	A_e/A_o	$G_c(J/m^2)$	A_e/A_o	$G_c(J/m^2)$	A_e/A_o	$G_c(J/m^2)$
Static	10.0	162	9.7	98	9.6	176
Dynamic	11.1	149	9.9	83	9.6	167

The specimen of B=2mm will be held in a plane stress condition so that microcracking will occur more easily and the process zone will extend. This will result in $K_{B=2} > K_{B\geqq 4}$. This assumption is supported by the fact that a residual stress is a tensile side near the fracture surface, where a plane stress condition prevails.

The typical measuring results of the critical strain energy release rate G_c and the specific area A_e/A_0 by the image analyzer are shown in Table 2. The specific area in the specimen of B=2mm is increased and the result also supports the above explanation.

CONCLUSION

The fracture toughness test of sillicon nitride ceramic has been conducted and the following results have been obtained.

(1) It has been shown that the fracture toughness value independs on the ratios of pre-crack length to specimen width (a/W) and span length to specimen width(S/W).

(2) It has been shown that the residual stress distribution measured by the X-ray method is identified with the result of the fracture surface observation by the transmission electron microscope(TEM).

(3) It has been shown that a pre-cracked specimen thicker than 4mm is recommended to obtain the valid fracture toughness value for the test material. Moreover, a specimen thickness and the ratio of span length to width are limited by the effect of the inertia load under the dynamic condition and the accuracy of the measuring system.

(4) The local critical fracture stress (σ_{mc}) has been evaluated from the measurement of process zone size. The valid condition for the specimen size in the test material was given as $B \geqq 70(K_{IC}/\sigma_{mc})^2$. The difference by the loading rate was not recognized clearly.

REFERENCES

(1) T.Kobayashi : Tetsu-to-Hagane,76(1990),149.
(2) K.T.Faber and A.G.Evans : Acta Met.,31(1983),57.
(3) T.Nose and T.Fujii : J.Am.Ceram.Soc.,71(1988),328.

(4) F.Wakai, S.Sakaguchi and Y.Matsuno : Yogyo-Kyokai-Shi, 93(1985),479.
(5) T.Kobayashi and M.Niinomi : Nuclear Eng.and Design, 111(1989),27.
(6) T.Kobayashi, K.Matsunuma, H.Ikawa and K.Motoyoshi : Eng.Frac.Mech.,31(1988),873.
(7) T.Mishima, Y.Nanayama, Y.Hirose and K.Tanaka : Zairyo (J.Mat.Sci.Japan)36(1987),805.
(8) Y.Miyoshi : Zairyo(J.Mat.Sci.Japan)37(1988),75.
(9) H.Kishimoto, A.Ueno, H.Kawanoto and S.kondo : Zairyo (J.Mat.Sci.Japan)36(1987), 810.
(10) M.Koizumi, H.Yanagida : "The fundamental of ceramics" (in Japanese) vol.1,OHM,(1987).
(11) A.G.Evans and R.W.Davidge : J.Am.Ceram.Soc.,5(1970),314.
(12) F.E.Lange : J.Am.Ceram.Soc.,62 (1979),428.
(13) T.Kobayashi, Y.Koide,Y.Daicho and R.Ikeda : Eng.Frac. Mech.,28(1987),21.
(14) J.L.Kalthoff : Metals Handbook,8(1985,ASM),269.
(15) G.R.Irwin : Appl.Mat.Res.,3(1964),65.
(16) M.Sakai : Ceramics(in Japanese),20(1985),33.
(17) A.G.Evans : Mater.Sci.Res.,21(1986),775.
(18) N.Miyata : Zairyo (J.Mat.Sci.Japan),37(1987),361.
(19) F.E.Buresch : ASTM STP 678(1978),151.

BIAXIAL COMPRESSIVE STRENGTH OF SILICON NITRIDE:

NEW DATA - MICROMECHANICS MODELS

Yong-Nian Yan

Tsinghua University
Beijing, P.R. China

George Sines

University of California
Los Angeles, USA

ABSTRACT

Studies on the biaxial compressive strength of alumina showed that the compressive failure was insensitive to the presence of the secondary compressive stress except possibly close to the equi-biaxial compressive state. It was found in another study that, in contrast to alumina and siliconized silicon carbide, the compressive failure of silicon nitride is not induced by the growth of cracks from Griffith flaws. Tests were conducted to discover whether the compressive biaxial strength of silicon nitride had a behavior in biaxial compression similar to that of alumina even though the micromechanisms of failure were different. The behavior was similar to that of alumina, but the behavior in the critical region near to the equibiaxial state had not been determined with sufficient confidence.

Recently eight specimens of the original batch became available and they were tested near to the equibiaxial state; this data is now presented and it shows some, but not a strong, interaction of the compressive stresses. Possible fracture mechanics micromechanisms, other than Griffith flaws, for the compressive failure of silicon nitride are presented. Test devices for testing brittle materials in various states of biaxial compression are described.

INTRODUCTION

In order to attempt to distinguish between micromechanical models, strength tests must be conducted over a range of stress states with sufficient replication of tests to establish statistical significance. We have conducted tests on alumina under biaxial tension,[1,2] on porous zirconia under biaxial tension-compression,[3] and on alumina[4] and silicon nitride[5] under biaxial compression. The tests on the silicon nitride in compression were of special interest because our study showed that a hot-pressed silicon nitride did not fail in compression by the classical Griffith flaw theory in which cracks extend stably from the flaws until their interaction causes failure, even though a dense alumina and a siliconized silicon carbide did.[6,7] The tests on the biaxial compressive strength of the silicon nitride were somewhat flawed

Fracture Mechanics of Ceramics, Vol. 10
Edited by R.C. Bradt *et al.*, Plenum Press, New York, 1992

by lack of sufficient replication of tests in the region around the equi-biaxial state. Recently eight specimens from the original batch became available to further explore this critical region. The test apparatus and procedures are presented in detail in Reference 8 and summarized in Reference 5. They will be briefly reviewed here.

EXPERIMENT

Material

The material investigated was isostatically pressed, sintered silicon nitride, manufactured by the Kyocera Corporation of Kyoto, Japan. This material is designated by them (SN-220) and is recommended for applications requiring high mechanical strength, high wear resistance and high heat shock resistance. The porosity is about 0.1 volume percent. The bulk density is typically 3.2 g/cm^3.

Specimen

The machine drawing for the test specimen is shown is Figure 1. Notice that the outer diameter of the specimen is only 1 cm and that the inner diameter is 0.629 cm. It is important that the specimen should break in the middle region, i.e, the test section; to ensure that this happens, the cross section of the ends of the specimen is made 2.57 times that of the test section. The thickness increases from test section to the end smoothly with a radius of 1.5 cm which is about 20 times the thickness, thereby giving a negligible stress concentration.

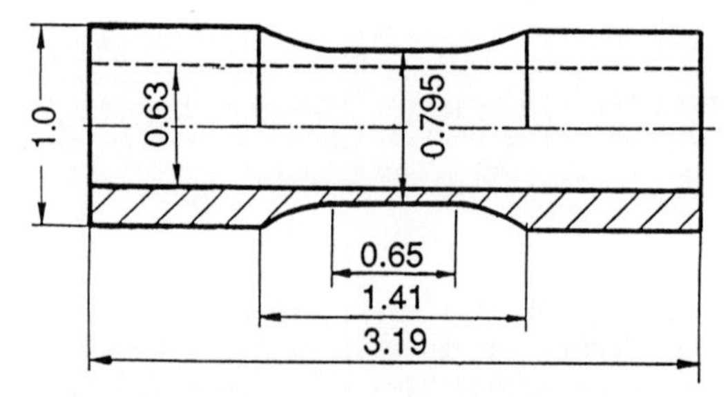

Fig. 1. Ceramic specimen for biaxial stress test. (centimeters)

In the test, the thin-walled test section is subjected to high external pressure; therefore, a check was made to ensure that it did not fail by buckling.[8]

Pressure Vessel

Central to the various pieces of apparatus used to load the specimen under biaxial compression was the application of high external hydrostatic pressures. The pressure anticipated necessary to break this strong ceramic by external pressure on the hollow specimen approached 1.2 GPa (170 ksi). Achieving this high pressure required development. The pressure vessel is a thick-walled cylinder is shown in Figure 2. It is enclosed in two cylinders to protect the operators from flying fragments in case it bursts. The desired design pressure causes stresses in the vessel that exceed the yield strength. This does not constitute failure, but induces favorable residual stresses. This process is called "autofrettage". The thick-walled cylinder is made of 4340 steel hardened to Rockwell C50. The protective cylinders are made of soft steel and are fastened to the cylinder with epoxy.

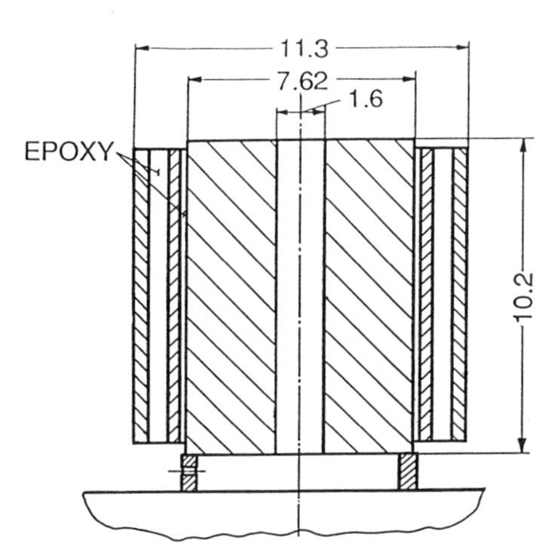

Fig. 2. The high pressure testing cylinder with external safety jackets. (centimeters)

Pressurizing Medium

The sealing problem for a liquid at such high pressure in a piston-cylinder pressurized chamber is formidable; however, there is an alternative--rubber is a good pressure transmission medium. It behaves very much like a fluid because of its low shear modulus. It also does not leak like a fluid, but extrudes into the small clearance between piston and cylinder until a stable condition is reached in which the force on the extruded rubber is balanced by the friction of the rubber against the surface of the cylinder and piston. Tests were conducted to see the degree to which rubber behaved as a hydrostatic medium. It was satisfactory, but when the rubber was compounded with oil, it was even better.[8]

Using a piston acting directly onto the rubber, also simplifies the system by eliminating the external high pressure hydraulic system and its control system that would be needed using a liquid medium. Force was applied to the piston by a 100 klb Riehle beam machine.

In the study of Reference 8, circumferential strain gages in the middle on the outside surface of the pressure vessel were used to measure the internal pressure; the pressure was related to the strain by an elasticity, finite element program. In the device used to obtain the new data to be presented here, strain was measured directly by strain gages in the bore of the specimen.

TEST APPARATUS FOR EQUIBIAXIAL STRESS

State $\sigma_{\theta\theta} : \sigma_{zz} = 1 : 1$

In the course of the study, a number of test devices were developed. Each of the devices that were developed first can give only one stress state or only a very narrow range of stress states. These devices are much simpler than the one developed for the stress states near to the equi-biaxial one. Because of their simplicity, they may have value for studies that are limited to those stress states. They are described in Appendix A.

For biaxial compressive stress states greater than 1:0.5, we were unable to design a device having only a single pressure chamber; all of the feasible designs required two pressure chambers containing the rubber pressurizing medium at two different pressures. Measurement of strain on the outer wall of the vessel could no longer be used to determine the two pressures; therefore, it was imperative to devise methods to measure strain directly on the specimen.

Techniques to insert and apply strain gages in the small bore of 0.63 cm had to be developed and also provisions had to be made to have the strain gage leads pass out of the cylinder. Appendix B presents the procedure for mounting the strain gages. The description of the apparatus follows.

First notice in Figure 3 the bottom piston has an axial hole and then a radial groove that permits the strain gage leads to pass out of the apparatus. The leads, by following this path, are not subjected to any high pressures so there is no need to compensate the strain readings for the effect of pressure on their resistivity.

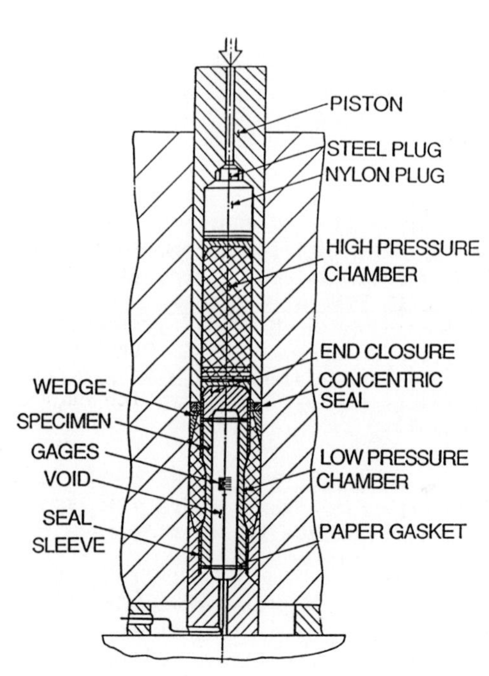

Fig. 3. Apparatus for a wide range of biaxial compressive stress states.

Notice that the load applied to the upper piston applies pressure to two pressure chambers: The high pressure chamber applies an axial load to the specimen and the low pressure chamber applies an external radial pressure to the specimen. (The "high pressure chamber" is so named because the pressure in this chamber must be higher than that in the other to attain equibiaxial compressive stress, State 1:1, in the test section.)

In one extreme case, the high pressure chamber could be empty, thus giving no axial load to the specimen. The hollow piston in this case only pressurizes the low pressure chamber resulting in a uniaxial compressive circumferential stress, State 1:0. The other extreme case is no rubber in the low pressure chamber and rubber in only the high pressure chamber resulting in a uniaxial, axial, compressive stress, State 0:1. In the stress states for which this device was employed, that is, in the vicinity of the State 1:1, the relative pressures have to be adjusted by reducing the compressibility of the high pressure chamber by inserting disks of nylon, which is less compressible than the rubber that it replaces. This adjustment to obtain the desired stress state was made by trial and error at stresses less than half those anticipated to cause fracture.

This latest developed testing apparatus, intended for stress states in the vicinity of 1:1, is capable of obtaining all the other stress states at which specimens were tested in the study.[8]

In the test, the amount of rubber in the high pressure chamber had to be adjusted several times, thus, it was necessary to unload several times to adjust the volume. During unloading the pressure in the low pressure chamber must decrease faster than that in high pressure chamber, otherwise, the specimen will fracture owing to axial tensile stress. For this reason, a concentric nylon seal, made up of an inner and outer nylon ring, was developed.(See Figure 3) When the high pressure piston moved up, because of the difference friction between the outer and inner of the concentric seal, these two rings easily separate and the high pressure rubber extrudes out of the clearance between the inner and outer rings. Thus, the pressure in the low pressure chamber can be reduced when the piston moves up.

The mating surfaces of the specimen, end closure and lower piston were lapped smooth. The checking method and assembling procedure are described in Reference 8. There are two seal sleeves, the upper connective seal sleeve and lower connective seal sleeve. Their functions are as follows:

(1) The end closure, specimen and lower piston are formed into an entity with connective seal sleeves and held together by epoxy. When the piston moves up upon unloading, they do not move separately from their original correct position and thereby damage the leads of strain gages.

(2) Sometimes it is possible that the pressure in the low pressure chamber is larger than that in the high pressure chamber. So, the rubber in the low pressure chamber could be extruded into the gap between the end surface of the specimen and the end closure. This extrusion can be prevented by the use of the sleeves. The sealing is improved because the sleeves are under compression by the pressure of the rubber.

(3). The sleeve is made of brass with thickness of 0.025 cm (0.010 in.). It serves to align the piston, the end closures, and the specimen together.

An axial hole runs through the upper piston. By use of a rod through this hole, the rubber in the high pressure chamber can be pushed out. A soft steel plug is inserted into this hole to prevent the nylon plug from extruding out this hole. The nylon plug and several nylon disks are used to adjust the volume of rubber in the high pressure chamber.

A lubricant, molybdenum disulfide powder mixed with extreme pressure grease, was applied to the surface of the piston, the inside and outside surfaces of the specimen, the pressurization wedge, the rubber, and the nylon disk and plug. The air inside the device must be removed by pushing down the piston and holding it several seconds after the device is assembled. During testing, the strain values, $\epsilon_{\theta\theta}$ and ϵ_{zz}, of the specimen must be measured simultaneously. According to the ratio of these values, the rigidity of the two chambers can be adjusted to reach the desired ratio. Because the low pressure chamber is difficult to adjust, only the high pressure chamber was adjusted.

In this study, seven additional tests were performed. (One of the specimens of the eight was inadvertently broken.) These tests have been added to the previously obtained data in the Table. They are shown as solid points on the biaxial fracture stress map.

TEST RESULTS

All of the specimens failed in the test section, and in no case did the fracture extend into the thickened ends. There was no difficulty recognizing failure; it was announced by a noise that ranged from a bang to a loud click depending on the stress state. Typically, the fractured region extended transversely across the specimen with a region finely pulverized about equal to the wall thickness.

The test results are presented in Table 1 and plotted on Figure 4. The uniaxial tensile strength in the circumferential direction had been determined in the previous study;[6] it is assumed that the tensile strength in the axial direction is approximately the same. Note that the average of the tests for State 1:-0.3 failed at a tensile stress near to that of the uniaxial test. Also the uniaxial compressive strength in the axial direction had been determined in the previous study and is presented here. The tests in the previous study were done on the same batch of specimens as the present study; therefore, it is meaningful to include them with the data for the present study.

The seven new data points, indicated by the solid points on Figure 4, along with the three previous data points, were taken in the region near to the 1:1 state. It is in this region that the most interaction between the compressive stresses might be expected, because they both have high values here.

The Failure Envelope for Biaxial Stress States

The failure envelope is plotted on Figure 4 for the experimental results on silicon nitride given in Table 1. The vertical axis is the hoop stress $\sigma_{\theta\theta}$; the horizontal axis is the axial stress σ_{zz}. The uniaxial tensile strength and the uniaxial compressive strength of the silicon nitride are 282 MPa (41 ksi) and 5.51 GPa (800 ksi), respectively, and marked by points T and C. The test device shown in Figure 3 maintained approximately the same stress ratio during the loading.

The dotted line OA is for the stress state $\sigma_{\theta\theta}:\sigma_{zz} = 1:1$. In order to determine the influence of the direction of the principal stress on the compressive strength under biaxial stress, tests were made not only for the stress state having $\sigma_{zz}<\sigma_{\theta\theta}$, which is the region of the failure envelope

Table 1. Test Results for Silicon Nitride.

Stree Ratio $\sigma_{\theta\theta}$	: σ_{zz}	The Larger Compressive Stress (GPa) Individual	Average
1	0.07	5.49 6.05	5.79
1	0.186	5.68 5.88	5.78
1	0.5	5.73 5.60 5.60	5.64
1	0.92	4.82	
0.79	1	4.90	
0.71	1	5.23	
0	1		5.12[1]
0.718	1	5.56*	
0.762	1	5.36*	
0.882	1	5.69*	
0.884	1	4.82*	
0.889	1	5.40*	
0.927	1	4.85*	
1	0.989	4.32*	
1	-0.3 (tension)	1.086 0.843 1.259	1.063 σ_{zz}=0.311 tension
Uniaxial Tension [1] -1	0	Six Specimens	0.282 (tension) Std. Dev. 0.028

* New Data
[1] Reference 6 and 7

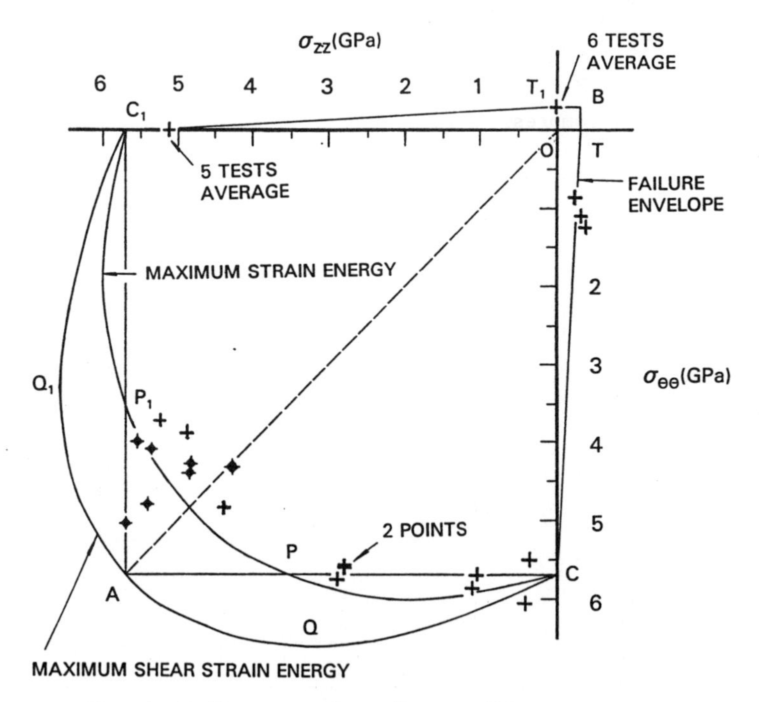

Fig. 4. Failure envelope for a silicon nitride.

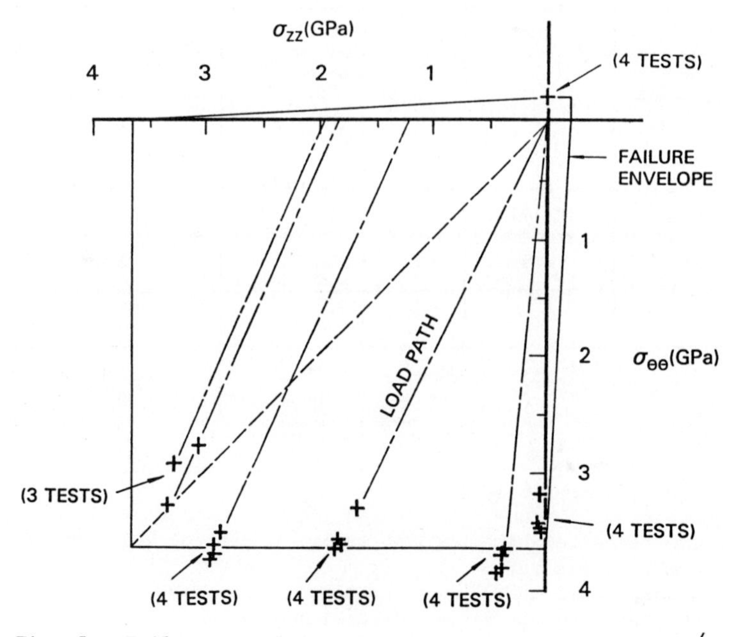

Fig. 5. Failure envelope for a dense sintered alumina.[4]

below the dotted line OA, but also for the stress state having $\sigma_{zz} > \sigma_{\Theta\Theta}$, the area above OA.

Under biaxial compression, the test data appear in the plot presented in Figure 4 to be almost independent of the secondary compressive stress for $\sigma_{\Theta\Theta}:\sigma_{zz}$ = 1:0.07, 1:0.186, and 1:0.5,; the averages of the test points for each state are almost along a straight line. The line CA is drawn at the average value of $\sigma_{\Theta\Theta}$ for these three stress states; line AC_1 for σ_{zz} is drawn at the same value. However, in the region near to equibiaxial compression, there does appear to be some interaction. This is a region that is critical to the determination of either a micromechanic theory of failure or an energetic criterion.

The strength of "WESGO" AL-995 alumina was measured by Adams and Sines under biaxial compression, for the stress ratios 1:0.01, 1:0.1, 1:0:5, 1:0:8 and 1:1.1, and their results are shown in Figure 5.[4] The specimen in that study was similar in shape to that of the current study, but about five times larger. The test apparatus was of a different design. In the tests having $\sigma_{zz}/\sigma_{\Theta\Theta} > 0.5$, an axial load was applied first and then additional axial loading and the circumferential loading was applied by external fluid pressure. Here it also can be seen that there is little influence of the intermediate stress σ_{zz} on fracture, when the ratio $\sigma_{zz}/\sigma_{\Theta\Theta}$ is less than 0.8.

The biaxial tensile quadrant T_1 B T in Figure 4 is shown as a square. It has been shown that usually there is some rounding in the region near to B and this would be consistent with Weibull statistics applied to the Griffith flaw theory,[1] but on the scale of the envelope as shown here, the rounding would not be visible. We have no data on this ceramic in this region.

There are only three test points in the compression-tension quadrant, all of which were tested at the stress state $\sigma_{\Theta\Theta}:\sigma_{zz}$=1:-0.3, $\sigma_{\Theta\Theta}<0$, $\sigma_{zz}>0$. In the biaxial compression-tension regions, point T is connected to point C, and T_1 to C_1, consistent with the Coulomb-Mohr construction. Here also the scale of the drawing can not show the details of the envelope based on the Griffith criterion as was done in Figure 6 of Reference 3. Also the more detailed construction could not be justified because of the lack of sufficient data for this region.

One result of the present study is that the ratio of the uniaxial compressive strength to the uniaxial tensile strength was found to be about 20 in the circumferential direction and about 17 in the axial direction. The ratio was not 8 as it should be according to the original Griffith theory. (The tensile strength in the circumferential direction was 282 MPa, while that measured in the axial direction in the biaxial test, State 1:-0.3, was 311 MPa; the uniaxial compressive strength in the axial direction was 5.2 GPa and the compressive strength in the circumferential direction for most of the biaxial region was 5.7 GPa.)

MECHANICS OF COMPRESSIVE FAILURE

Griffith Flaw Theory

A result of this study is that the uniaxial compressive-uniaxial tensile strength ratio is not eight according to the Griffith theory but a number closer to twenty. This ratio is an important feature of ceramic behavior and is about the same ratio that has been observed for dense alumina.[4] In 1924, when Griffith explained the marked discrepancy between the theoretical and actual strength of brittle material, he showed that the tip of the flaw in the material acts as a tensile stress concentration. When the concentrated tensile stress equals the intrinsic or theoretical bond strength of the flaw-free material, a crack initiates and finally leads to fracture.[9] Inglis first obtained the expression for the tensile stress on the surface of an elliptical hole subjected to stresses applied transverse to the axes of the hole; the stresses are a function of elliptical shape, orientation to the applied stress and the applied stress field.[10] Using this analysis, Babel and Sines derived a failure criterion for porous ceramic under biaxial stress states.[3] Depending on the openness of the crack, one can predict that the uniaxial compressive strength should be from three to eight times the uniaxial tensile strength. If the elliptical flaw is circular the ratio is 3 and if it is a closed sharp crack, the ratio is 8.

The ratio for the silicon nitride, SN-220, siliconized silicon carbide and alumina was approximately 20; the Griffith theory and Babel-Sines failure criterion do not explain this value. In compression, these theories are damage criteria, not fracture criteria.[7] Failure does not occur until so many flaws extend cracks that the material behaves more as a granular material from the interaction of the multitude of cracks. In contrast to this, when under an applied tensile stress, the first crack initiated from the worst flaw results in failure.

There may be a relationship between the strength ratio and the grain size and microplastic behavior. R.W. Rice analyzed the strength ratio of eight kinds of brittle materials with different grain sizes. He pointed out that the larger the grain size of the material, the larger the strength ratio.[11]

McClintock and Walsh propose that friction between the contacting surfaces of the flaw strongly influence the crack initiation and propagation near the tip of the closed flaw. The concentrated tensile stress is caused primarily by the resolved shear stress and the zone of compressive stress by the normal stress. If the normal stress is large enough to close the flaw and if the friction is large enough to stop the slip, there will be no tensile stress concentration at the tip of the closed flaw. This criterion can predict compressive strengths greater than 8 times the tensile strength.[12]

Sines and Taira have discovered degradation of the tensile strength of NC-430 siliconized silicon carbide and a dense alumina from prior compression, which implies crack initiation and growth under compressive loading; but there was no degradation of the tensile strength of the silicon nitride, SN-220, which implies no crack initiation or no crack growth from compressive loading.[6,7] It is possible that this silicon nitride has a much greater coefficient of friction which prevents crack growth under compressive loading. For this silicon nitride, compressive failure must occur from some yet undiscovered mechanism.

Non-Griffith Compressive Failure Mechanisms

Under the very high compressive stress of almost 6 GPa, two modes of failure might occur for silicon nitride: phase transformation and twinning.

(1) Transformation: Silicon nitride, Si_3N_4, exists in two polymorphic forms, alpha and beta; both have hexagonal structure, but they differ slightly in lattice dimensions. The alpha, the low temperature phase, transforms to beta at about 1500°C (2730°F); the phase change is not considered reversible. Both phases are usually found mixed in different proportions, depending on the temperature of preparation of the nitride. The main parameters of the two phases are given in Table 2.[13]

Table 2. Structure of Silicon Nitride.[13]

	Phase	a (Å)	c (Å)	c/a	Density (g/cm)	Stack Sequence
	α	7.753	5.618	0.725	3.18	ABCDA
BCD						
	ß	7.606	2.909 (5.818)[1]	0.382	(3.19)[2]	ABAB

[1] Note that the ß unit cell contains 1/2 the atoms as does α; therefore, 2c of the ß should be used for comparison.
[2] Calculated from the lattice measurements, ß is 0.33% more dense than α.

Under a very high compressive stress, phase transformation might occur at room temperature in silicon nitride, which usually is composed of grains of both phases. Phase transformations induced by pressure tend to the denser phase. Transformation of silicon nitride from hydrostatic pressure would be from the α to the ß phase, which Table 2 shows is about 0.33% more dense. However, there is a stronger possibility that it is not the hydrostatic pressure component acting on the volume change that causes the transformation, but that it is the compressive stress acting along the crystal axes. Notice in Table 2, that the a-axis of the ß phase is 1.9% smaller that of the α phase. A high compressive stress in the a-direction might cause an α grain to transform to the ß phase. Also notice that the c-axis of the α phase is 3.4% smaller than that of the ß phase. A compressive stress in the c-direction on a ß grain might cause it to transform to an α grain. Such sudden and inhomogeneous transformations in such a brittle material might well lead to catastrophic failure.

2) Twinning At the very high stresses attained in these compression tests, it may be possible for twinning to be initiated. The sudden occurrence of twinning might cause the boundaries of the twinned regions to initiate cracks that could lead to immediate failure. Twinning in hexagonal structures, such as that for silicon nitride, usually occurs on the {1012} planes in the <1011> directions.

We can find no studies on the occurrence of phase transformations of silicon nitride induced by applied compression, nor can we find studies on its twinning. These would be two very interesting subjects to study with the diamond anvil apparatus.

Statistical Effects

Near the 1:1 ratio, our tests had been few, but now we have 10 test points in this region that show a reduction in compressive strength. Strength reduction might occur from statistical effects. Assume that there is some kind of flaw that initiates fracture from applied shear and has a statistical distribution in severity and random orientation. Under equibiaxial compression, all flaws, whose normals form a cone about 45 degrees from the normal to the free surface, are under the same high shear stress. It is more probable that fracture will be initiated under this stress state, than at lower ratios where only the category of flaws that have their normals at 45 degrees to the direction of the greatest compressive stress and perpendicular to that of the secondary compressive stress have the maximum shear stress. Because more of them are being highly stressed under the 1:1 ratio than at other ratios, the compressive strength under the equibiaxial compressive stress state should be less from the statistical point of view.

If the micromechanical failure mechanism is triggered by compressive stress acting on flaws, then here too the number of planes subjected to high compressive stress increases as the 1:1 state is approached. At that state, all planes whose normals are perpendicular to the radial direction have the same high stress. In contrast, away from the stress region of the 1:1 state, only the few planes whose normals are parallel to the greatest compressive stress might have their failure mechanism triggered. Thus by statistics applied to the sensitivity of the flaws, the equi-biaxial compressive strength should be less than that for the uniaxial state and the other states away from the 1:1 region.

We have studied in some detail the shape of the biaxial compressive envelope if the failure mechanism is by crack extension from Griffith flaws[14], which we have shown are the micromechanism of failure for alumina and siliconized silicon carbide.[6,7] Consideration of biaxial tension and tension-compression is much simpler[1], because here failure occurs from the first flaw to initiate a crack instead of the complicated interaction of many cracks that must be considered in compression.[15]

It must be noted that under compression, crack initiation from inherent Griffith flaws, their growth and interaction have never been observed; there exists only indirect evidence for some ceramics consistent with these models. However, in dense alumina under tension, crack initiation from inherent flaws, and their growth and interaction have been observed.[16,17]

Criterion Shape, Micromechanical, Non-statistical

The two straight lines, CA and C_1A in Figure 4, in the biaxial compression quadrant correspond to several criteria that are based upon micromechanical theories. They are the original Griffith (local micro-tensile stress) flaw theory[9], the McClintock-Walsh theory[12], and the Babel-Sines open flaw theory[3]. In this quadrant, these three theories are crack initiation damage theories and not necessarily fracture theories leading to catastrophic failure based upon crack growth and interaction; however, crack growth and crack interaction leading to fracture should be related to conditions for crack initiation.

The two lines might also be considered to be the statement that the criterion for failure in this quadrant is the maximum compressive stress, the maximum shear stress or the maximum shear strain.

Criterion Shape, Energetic

The maximum distortion energy criterion, line $CQAQ_1C_1$ in Figure 4 is superimposed on the data for the compression-compression quadrant. This is, of course, the well-known von Mises criterion for plastic yielding of a ductile material.

$$\sigma_2{}^2 - \sigma_2\sigma_3 + \sigma_3{}^2 = \sigma_c{}^2$$

where σ_c is the uniaxial compressive strength.
Another significance can be given to this mathematical expression: it is the root-mean-square of the shear stress averaged over all planes in all directions.[18,19] It does not fit the data in Figure 4 or that in Figure 5 very well. Another energetic criterion is compared to the data: the total elastic strain energy criterion, which is expressed by

$$\sigma_2{}^2 - 2\mu\sigma_2\sigma_3 + \sigma_3{}^2 = \sigma_c{}^2$$

where μ is Poisson's ratio.

This is shown by Line CPP_1C_1 on Figure 4. Notice that this criterion does fit the data fairly well. At the high stresses at which failure occurs in these advanced ceramics, the energy released by a local event may be enough to initiate pulverization. There is no obvious reason to limit the energy release to only the elastic shear strain energy.

<u>Summary of Fracture Mechanisms in Compression</u>

Two different types of criteria can be used to fit the data for the fracture of high strength, dense ceramics in biaxial compression. There are several competing micromechanical theories, but not mutually exclusive theories, modified by statistical analysis that fit the data reasonably well. The total elastic strain energy criterion also fits the data.

The shape of the failure envelope can not be used to distinguish between the micromechanical mechanisms. One method of distinguishing, or at least eliminating the Griffith flaw mechanism, is to see whether a prior applied compressive stress degrades the transverse tensile strength. The possibility of phase transformations or twinning being mechanisms should be studied by diamond anvil apparatus.

The above consideration pertain only to failure under biaxial compression. In biaxial tension-compression and in biaxial tension, the Griffith flaw theory or a modification of it, appears to the controlling micromechanism.

SUMMARY

Biaxial compression testing of ceramics by miniature specimens is feasible. The very high hydrostatic pressures required in these tests can be attained by using rubber as the pressurizing medium.

An apparatus has been developed that can be used to apply the complete range of biaxiality in compression. Simpler apparatus has been developed for particular biaxial compression stress states.

There is indirect evidence that some ceramics fail in compression from crack growth from inherent flaws; however, there is indirect evidence that hot-pressed silicon nitride does not fail by this micromechanism. It is suggested that the extremely high stresses, greater that 5 GPa, may induce phase transformations or twinning that would lead to fracture.

Alumina, which appears to fail by crack extension from Griffith flaws, and isostatically pressed, sintered silicon nitride, which appears to fail by some other micromechanism, have similar behavior in strength under biaxial compression: there is very little effect of the secondary compression until the region near to equi-biaxiality it reached. The interaction decreases the compressive strength. This can be explained either by statistical consideration of flaws, or by the elastic strain energy criterion.

ACKNOWLEDGEMENT

This study was partially supported by Kyocera Award No. 980822, and by funds from Tsinghua University.

REFERENCES

1. M. N. Giovan and G. Sines, Biaxial and Uniaxial Data for Statistical Comparisons of a Ceramic's Strength, J. Am. Ceram. Soc., 62:510-515 (1979).
2 M. N. Giovan and G. Sines, Strength of a Ceramic at High Temperatures Under Biaxial and Uniaxial Tension, J. Am. Ceram. Soc., 64:68-73 (1981).
3. H. W. Babel and G. Sines, Biaxial Fracture Criterion for Porous Brittle Materials, J. Basic Eng. ASME, 90:285-292 (1968).
4. M. Adams and G. Sines, Determination of Biaxial Compressive Strength of a Sintered Alumina Ceramic, J. Am Ceram. Soc. 59:300-304. (1976).
5. G. Sines and Y. N. Yan, The Strength and Fracture of Hot Isostatically Pressed Silicon Nitride in Biaxial Compression in "Fractography of Glasses and and Ceramics II," V. D. Frechette and J. R. Varner, ed., to be published by the Am. Ceram. Soc.
6. G. Sines and T. Taira, The Degradation of Tensile Strength of a Silicon Nitride Ceramic from Prior Compression, UCLA Engineering Report 86-35, August 1986, 31 pp.
7. G. Sines and T. Taira, The Degradation of Tensile Strength of Ceramics from Prior Compression in Materials Science Monograph 38B (High Tech. Ceramics, Pt. B), Elsevier, Amsterdam (1987).
8. G. Sines and Y. N. Yan, The Strength of Hot-Isostatically Pressed Silicon Nitride SN-220 in Biaxial Compressive Stress States, UCLA Engineering Report 90-07, August 1989, 85 pp.
9. A. Griffith, The Theory of Rupture, Proc. of the First Int. Congress for Applied Mechanics, Delft, 55-63 (1924).
10. C.E. Inglis, Stresses in a Plate Due to the Presence of Cracks and Sharp Corners, Proc. of the Institute for Naval Architects, 55:219-230, (1915).
11. R.W. Rice, The Compressive Strength of Ceramics in "Materials Science Research, Proc. of the Sixth University Conference on Ceramic Science," W.W. Kriegel and H. PalmourIII, ed. Vol. 5, Plenum Press, New York, 195-229 (1971).
12. F.A. McClintock, and J.B. Walsh, Friction on Griffith Cracks in Rocks Under Pressure, Proc. of the Fourth U.S. National Congress of Applied Mechanics, 1015-1021, (1962).
13. S.N. Ruddlesden and P. Popper, On the Crystal Structure of the Nitrides of Silicon and Germanium, Acta cryst. 11:465-468, (1958).
14. M. Adams and G. Sines, A Statistical, Micromechanical Theory of the Compressive Strength of Brittle Materials, J. Am. Ceram. Soc., 61:126-131, (1978).
15. H. Horii and S. Nemat-Nasser, Brittle Failure in Compression: Splitting, Faulting and Brittle-Ductile Transition, Phil. Trans. R. Soc. Lond. A319:337-374, (1986).
16. G. Sines and T. Okada, Flaws Responsible for Slow Cracking in the Delayed Fracture of Alumina, J. Am. Ceram. Soc. 66:228-232 (1983).
17. T. Okada and G. Sines, Crack Coalescence and Microscopic Crack Growth in the Delayed Fracture of Alumina, J. Am. Ceram. Soc. 66:719-723 (1983).
18. G. Sines, Yielding and Plastic Instability Under Biaxial Stress in Design of Metal Pressure Vessels, Appendix 1, Journal of Materials, 4:377-392 (1969).
19. U. Dehlinger, Die Fliessbedingung bei mehachsigem Spannunszustand viel kristalliner Metalle, Zeitschrift für Metallkunde, 35:182-184 (1943).

APPENDIX A

TEST DEVICES FOR SPECIFIC STRESS STATES

Although the device described in the text and illustrated in Figure 3 can be used to attain all the biaxial compressive stress states, it is more complicated to manufacture and to use than some of the special ones described below. Along with that presented in Figure 3, they were used to obtain the data presented in the Table. They are described here because one of them may fill a special testing need.

State 1:0.5.

For the biaxial compressive stress state 1:0.5, a closed-end cylinder loaded by external fluid pressure is used. The specimen is shown assembled with the hardened steel end closures in Figure A-1. Paper gaskets are between the closures and the specimen's ends to minimize local stresses from disparities on the mating surfaces. The assembly of specimen, end closures, and gaskets are held together by a central tie rod. Because the specimen with its end closures has to be surrounded by rubber, the rubber is cast into two parts and then put around the specimen and then the rubber and specimen assembly is inserted into the chamber.

Call the external diameter at the test section D_m and the internal one D_i. The net axial force acting on the test section is

$$F = p \pi D_m^2/4,$$

which when divided by the net cross-sectional area of the test section gives the axial stress

$$\sigma_{zz} = p D_m^2/(D_m^2 - D_i^2),$$

which is constant through the thickness. The analysis by Lame' gives the

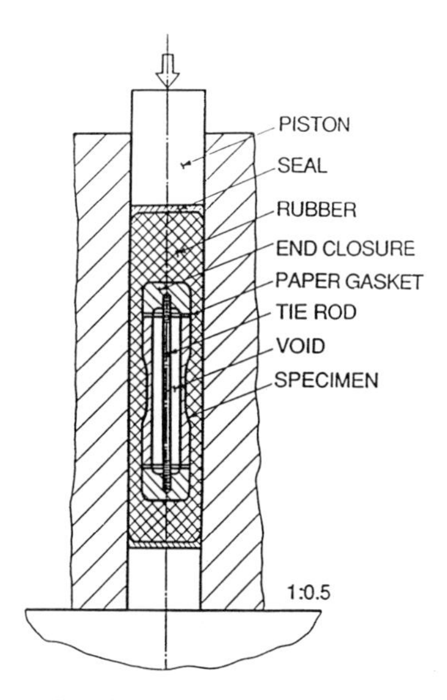

Fig. A-1. Apparatus for biaxial compressive stress state 1:0.5.

circumferential stress, which has it maximum value at the internal wall.

$$\sigma_{\theta\theta} = 2\ \sigma_{zz} = 2p\ D_m^2/(D_m^2 - D_i^2)$$

The radial stress is maximum at the external wall and equal to p, while at the inner wall, where the circumferential stress is maximum, it is zero. For the specimen used, which has a relatively thin wall, the radial stress is small compared to the circumferential stress.

1:0.186 Biaxial Compressive Stress Test

The device for the stress state 1:0.186 is shown in Figure A-2. Part of the axial load is carried by a center column. A calculation must be made to be sure that it will not buckle when the axial load from the pressure is applied to it. Here the pressure applied to the end of the bore area of the specimen is carried by a double-ended piston and not by the test section. Each end is sealed by a thin soft copper disk.

The pressure acts only on the net area of the butt end of the specimen to give an axial compressive force of

$$F_b = p\ \pi(D_o^2 - D_i^2)/4,$$

where D_o is the outer diameter at the end. This is partially counter-acted by the force on the shoulder

$$F_s = p\ \pi(D_o^2 - D_m^2)/4,$$

The sum of the forces acts on the net area of the test section to give an axial stress

$$\sigma_{zz} = (F_b - F_s)\ 4/\pi(D_m^2 - D_i^2) = p\ (D_o^2 - D_i^2)/(D_m^2 - D_i^2)$$

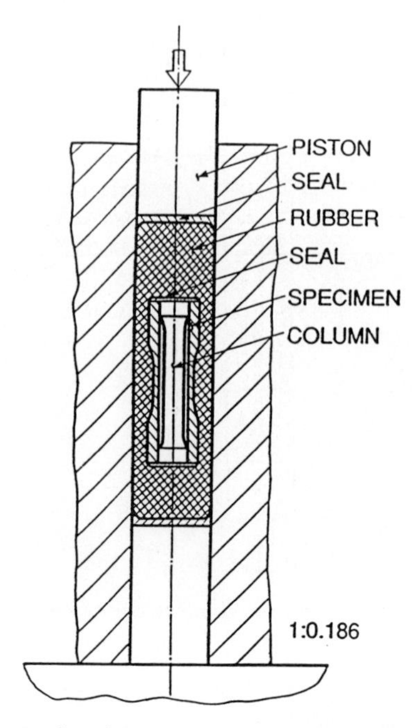

Fig. A-2. Apparatus for biaxial compressive stress state 1:0.186.

State 1:0.07

In Figure A-3, another method is presented to reduce the axial compression, so that stress states even closer to uniaxial circumferential compression can be achieved.

The diameter of the internal piston and loading ring are selected to give the desired reduced axial force. The arrangement keeps the top surface of this internal piston free of pressure; only the loading ring applies pressure to the rubber. The diameter is chosen to give the very small axial compressive stress that is 0.07 of the circumferential stress.

This design presents another sealing problem. Because the upper annular area of the bottom part of the loading piston is smaller than its lower annular area, the reliability of seal is not increased as the pressure increases. The seal problem can be solved by having two thin copper seal sleeves, whose inner diameter fit closely with that of the loading piston and specimen.

State 1:-0.03

The extreme case for the configuration used for the 1:0.07 is shown in Figure A-4; it gives the compressive-tensile state of 1:-0.03. Here a hollow piston extends around the butt end of the specimen so that the end of the specimen has no pressure acting on it.

The axial force is caused only by the pressure acting on the shoulder between the butt diameter and the net section. This gives an axial tensile force

$$F = p\pi(D_o^2 - D_m^2)/4,$$

which acts on the net cross-sectional area of the test section, to give an axial tensile stress

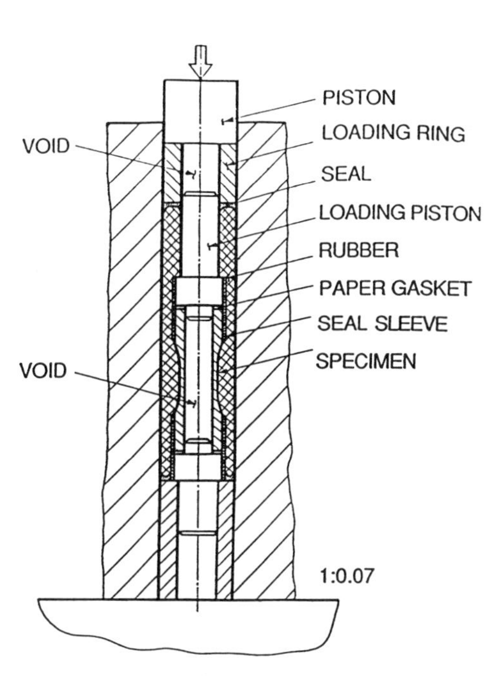

Fig. A-3. For State 1:0.07
Comp.-Comp.

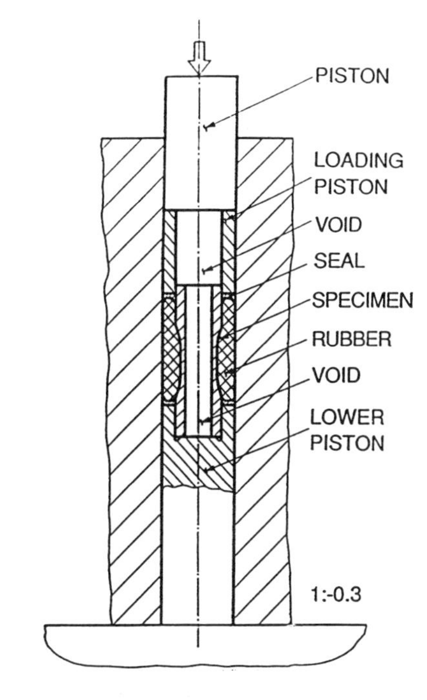

Fig. A-4. For State 1:-0.3
Comp.-Tens.

$$\sigma_{zz} = F\ 4/(D_m^2 - D_i^2)/\pi = p\ (D_o^2 - D_m^2)/(D_m^2 - D_i^2).$$

The axial tensile stress is 0.03 of the circumferential compressive stress for this particular loading.

States 0:1 and 1:0.

In a previous study on the same batch of specimens, uniaxial axial compression tests were conducted as well as uniaxial circumferential tensile tests.[6] The compression tests were conducted on a conventional testing machine using special alignment fixtures. The tensile tests were performed by casting a rubber plug into the bore of specimen and compressing the rubber by a piston fitted into the bore. The hydrostatic pressure in the rubber created a circumferential tensile stress in the specimen.

In Figure A-5, is a device used to determine the Poisson's ratio. It also can be used to measure the uniaxial compressive strength in the axial direction. All of the four devices presented here in the Appendix can be modified to have strain gages applied directly to the bore of the specimen by the same method used in this device. The bottom piston has its end shaped to exactly match the dimensions of the end of the specimen and to approximately match the transverse compliance of the specimen. This is similar to the bottom piston shown in Figure 3, which is longer, because of the space needed for the several chambers.

The stress state is fully determined by the loading on the upper end of the specimen; therefore, this bottom piston can be the same for all the devices, except a seal sleeve may be necessary for 1:0.07 apparatus shown in Figure A-3.

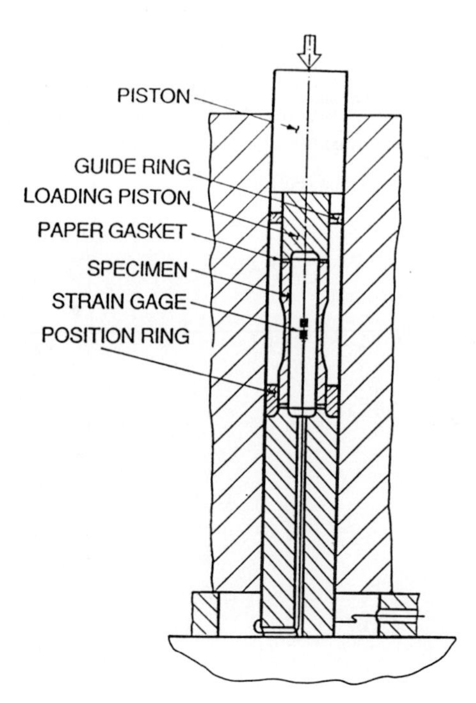

Fig. A-5. Device for determining Poisson's ratio, uniaxial compressive strength in the axial direction, and an illustration of method for using strain gages in the bore of the specimen.

Some general considerations

In the devices in which hardened steel contacts the end of the specimen, it is necessary to have a compliant layer to minimize local stresses from disparities on the contacting surfaces. The advantages of paper over soft metal are discussed in Chapter 1 of Reference 8.

In the biaxial compression tests, plasticine was inserted into the bore of the specimen with a controlled amount of empty space so that it would not be put under pressure. When the specimen breaks, the plasticine will absorb energy that is released from the highly compressed rubber and from the testing machine. This is a good way to minimize fragmentation of the specimen and reduce shock to the testing machine.

All specimens in all of the tests failed in the test section; failure was not induced at the ends because the butt ends remained intact. The reduction in shock by the plasticine prevented fragmentation of the butt ends.

APPENDIX B

To mount the gages in the bore of the specimen, the following method was developed: A rubber cylinder was cast with a diameter of 0.58 cm, thus giving a clearance between the bore of the specimen and the rubber cylinder of 0.025 cm. The gages with the connecting leads were attached to the surface of the rubber cylinder with the mounting side of the gages to the outside. Strain gage adhesive was applied to the mounting side of the gages, and the assembly of the rubber cylinder, gages, and leads inserted into the bore of the specimen. Opposing pistons were inserted into the ends of the specimen bore. Force on the pistons axially compressed the rubber and expanded it against the bore to place the strain gages in contact with the surface of the bore. The adhesive was cured while maintaining the pressure. The rubber cylinder was left in the bore during testing. It was short enough so that a space was left in the bore to avoid developing internal pressure in the specimen. See Appendix F of Reference 8.

FRACTURE BEHAVIOR OF NON-OXIDE CERAMICS UNDER BIAXIAL STRESSES

Y. Nakasuji, N. Yamada, H. Tsuruta, M. Masuda and M. Matsui

NGK Insulators, Ltd.
2-56 Suda-cho, Mizuho-ku, Nagoya, 467 Japan

ABSTRACT

In order to apply non-oxide ceramics to structural uses, such as gas turbine components and diesel engine components, greater understanding of their fracture behavior is needed. The fracture behavior under multiaxial stresses is important to component design because complicate stresses are established in real parts. Torsional, combined tensile/torsional and combined compressive/ torsional tests were performed for the strength tests under tension/compression biaxial stresses. Ball-on-ring and ring-on-ring tests were performed for the strength tests under tension/tension biaxial stresses. Tested materials were two kinds of sintered silicon nitride and one kind of HIPed silicon carbide. The fracture strength was predicted using the multiaxial distribution function based on various proposed fracture criteria from the data of uniaxial strength. Moreover, the authors propose modified G-Criterion which is expressed as the equivalent stress $Z = (\sigma_n^2 + (\beta \tau_n)^2)^{1/2}$. β indicates the degree of the influence of shear stress on a fracture. β value of each ceramic was determined by the strength data tested under tension/compression biaxial stresses. The predicted fracture strength was compared with experimental results under both tension/compression and tension/tension biaxial stresses. The fracture behavior of non-oxide ceramics under biaxial stresses was discussed.

INTRODUCTION

Silicon nitride and silicon carbide among structural non-oxide ceramics have high strength at high temperature, so the gas turbine components and diesel engine components using these ceramics are been developing actively. In order to apply silicon nitride and silicon carbide to structural uses, greater understanding of the mechanical strength is needed. In mechanical parts, the stress state is not simple uniaxial, but complicated multiaxial. For example, in a gas turbine rotor, the complicated multiaxial stresses occur because of the conbination of centrifugal stresses by rotating, the complex shape of rotor and the thermal stresses. However, the fracture behavior of silicon nitride and silicon carbide under multiaxial stresses is not made clear, the standard of design methodology for ceramic components is not developed.

A variety of fracture criteria have been proposed, and many researchers have discussed fracture behavior under mixed modes.[1-13] In addition, several methods have been proposed to predict the reliability of ceramic components

Fracture Mechanics of Ceramics, Vol. 10
Edited by R.C. Bradt *et al.*, Plenum Press, New York, 1992

under multiaxial stresses from the strength determined by tensile and bending test under uniaxial stresses, but they are not yet fully developed.[14-17]

In this study, in order to clarify the fracture behavior of silicon nitride and silicon carbide under biaxial stresses, two types of tests with different stress modes were performed. Moreover, we investigated the fracture criterion applicable to the prediction methodology for the fracture strength of ceramic components under biaxial stresses.

EXPERIMENTAL PROCEDURES

Uniaxial strength tests and biaxial strength tests were performed. Size effect was assessed in uniaxial strength tests, furthermore stress state effect was assessed in biaxial strength tests. All specimens were tested at room temperature and in the air.

Test Materials

The materials investigated in this study are two kinds of sintered silicon nitride (SSN-A and SSN-B), and one kind of HIPed silicon carbide (SiC). Mechanical properties of these materials are shown in Table 1. The sintering additives were: SrO, MgO, CeO_2 and ZrO_2 for SSN-A, and Y_2O_3, MgO and ZrO_2 for SSN-B, respectively. In the case of SSN-A, two different kinds of lot (SSN-A1 and SSN-A2) were used. The sintering additives for SiC were B and C.

Table 1. Mechanical properties of SSN-A1, SSN-A2, SSN-B and SiC

Material	Density (g/cm^3)	Fracture toughness ($MN/mm^{3/2}$)	4-point bending strength (MPa)	Young's modulus (GPa)	Poisson's ratio	Hardness (GPa)
SSN-A1	3.27	6.2	740	280	0.27	15
SSN-A2	3.27	6.2	1010	280	0.27	15
SSN-B	3.25	6.1	970	300	0.27	17
SiC	3.17	2.6	610	420	0.15	24

Fracture toughness : Chevron notch method
Young's modulus : JIS R 1602
Hardness : Knoop, 300g

Uniaxial Strength Tests

Tensile strength tests and 4-point and/or 3-point bending tests were performed.

The machine drawing of the specimen for tensile test is shown in Figure 1-(a). The size at the gauge position of the specimen is 6mm in diameter and 24mm in length. For SSN-A1, tensile strength tests were performed using another three types of specimen. The sizes at the gauge position of specimen are 10mm, 14mm and 20mm in diameter and 40mm, 56mm and 80mm in length, respectively. The specimen gauge position was finished by a #600 diamond grinding wheel.

A 25 ton tension test apparatus was used, which was provided with a collet chuck, supported by a thrust bearing over self-aligning spherical bearings, to accurately align the specimen to the testing apparatus.[18] Three strain gauges were attached to each specimen at the gauge length center to measure strains at the outer peripheral section, to help confirming accurate alignment. The bending stress components of all the specimens tested were 5% or less. The test was carried out at a cross-head speed of 0.5mm/min.

The specimens for bending tests were cut into the shape specified by JIS R-1601 from the specimens which were prepared for tensile and torsional tests. The surface was finished by #800 diamond grinding wheel and edges were chamfered. 4-point and/or 3-point bending tests were conducted according to JIS R-1601.

Biaxial Strength Tests

Two types of tests with different stress modes were performed. Ball-on-ring and ring-on-ring tests were performed for the stress region for tension/tension biaxial stresses tests. Torsional, combined tensile/torsional and combined compressive/torsional tests were conducted for the stress region for tension/compression biaxial stresses tests.

(1)Tension/Compression Biaxial Stresses Tests[1,2]

The machine drawing of the specimen for torsional, combined tensile/torsional and combined compressive/torsional test is shown in Figure 1-(b). The pecimen for torsional, combined tensile/torsional and combined compressive/torsional tests were provided with parallel planes at both ends, so that tensile and compressive stresses could be applied simultaneously. The size at the gauge position of the specimen is 6mm in diameter and 24mm in length. The specimen gauge position was finished by a #600 diamond grinding wheel.

A tensile load was applied to the collet chuck, supported by a thrust bearing over self-aligning spherical bearings, and at the same time a torsional load was applied to the parallel planes at the both ends. The tensile/torsional test jig was developed, which enabled two different tests to be carried out simultaneously. The jig was fitted to the tensile/torsional test apparatus (model 1321, by Instron Corp.), which is capable of controlling the maximum tensile or compressive load of 5ton and the maximum torque of 100kg.m. Two 3-axial strain guages were attached to the specimen gauge position, to measure strains in the direction 45° to the specimen axis. Torsional, combined

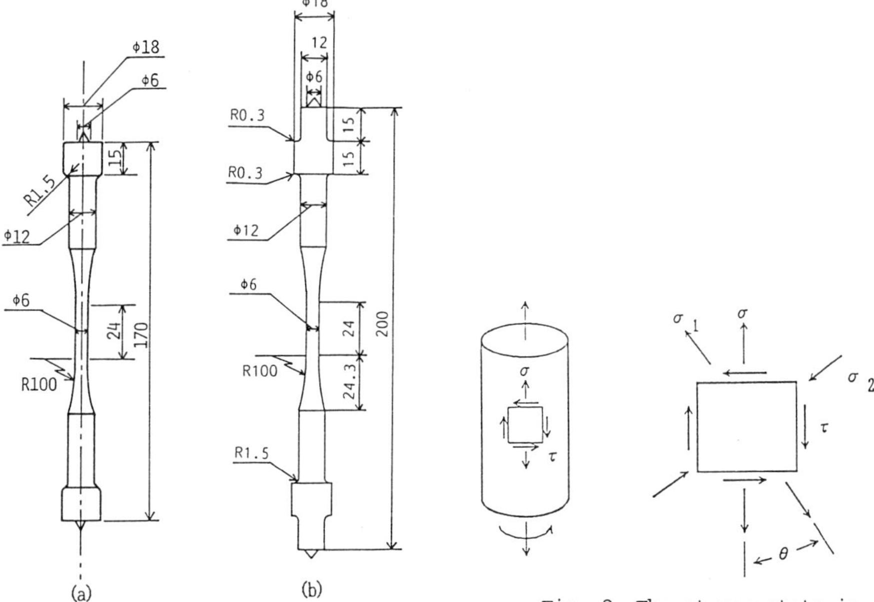

Fig. 1. Geometries and dimensions of specimens. (a) for tensile test, (b) for torsional, combined tensile/torsional or combined compressive/torsional test.

Fig. 2. The stress state in gauge position of caused by combined tensile/torsional loading.

tensile/torsinal and combined compressive/torsional tests were carried out at controlled stress speeds in constant τ/σ ratio : 0.5, 1, 2, 0, -2 and -1 for SSN-A1, 0.5, 1, 0, -2 and -1 for SSN-B, 1, 0,-2 and -1 for SiC, where τ is shear stress caused by a torsional load and σ is tensile or compressive stress.

The stress state in the gauge position of specimen caused by combined tensile/torsional loading is shown in Figure 2. Tensile stress σ is caused by tensile loading, and shear stress τ is caused by torsional loading. Maximum principal stress σ_1 and minimun principal stress σ_2 are expressed in equation (1) and (2). The fracture strength was expressed by σ_1 and σ_2.

$$\sigma_1 = \frac{\sigma}{2} + \sqrt{\frac{\sigma^2}{4} + \tau^2} \tag{1}$$

$$\sigma_2 = \frac{\sigma}{2} - \sqrt{\frac{\sigma^2}{4} + \tau^2} \tag{2}$$

(2)Tension/Tension Biaxial Stresses Tests[13]

Disk specimens were used for ball-on-ring and ring-on-ring tests. The dimensions of specimen were 40mm, 44mm and 70mm in diameter and 3mm, 3mm and 5mm in thickness for SSN-A2 and SSN-B, 44mm in diameter and 3mm in thickness for SiC. The tension face of disk specimen was finished by a #800 diamond grinding wheel.

Schematic views of ball-on-ring and ring-on-ring tests are shown in Figure 3. Ball-on-ring and ring-on-ring tests were carried out at a cross-head speed of 0.5mm/min. For SSN-A2, tests with six different volumes were made. The ring diameters of ball-on-ring tests were 30, 40, and 60mm. The inner and outer ring diameters of ring-on-ring tests were 10 and 30, 20 and 40, 30 and 60mm. For SSN-B, tests with five different volumes were made. The ring diameters of ball-on-ring tests were 30 and 40mm. The inner and outer ring diameters of ring-on-ring tests were10 and 30, 20 and 40, 30 and 60mm. For SiC, tests with two different volumes were conducted: The ball-on-ring test where the ring diameter is 40mm, and the ring-on-ring test where the inner and outer ring diameters are 20 and 40mm.

The stress distributions in disk specimens for SSN-A2 at ball-on-ring and ring-on-ring tests were calculated by finite element method (FEM). Calculated stress distributions for the ball-on-ring test where the ring diameter is 60mm

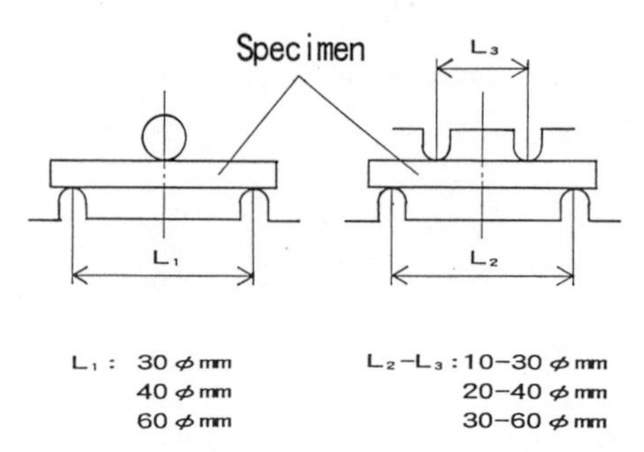

(a) Ball-on-ring test (b) Ring-on-ring test

Fig. 3. Schematic view of ball-on-ring and ring-on-ring tests.

and the ring-on-ring test where the inner and outer ring diameters are 30 and 60 mm are shown in Figure 4. The applied loads for the FEM analysis are 1200kg for ball-on-ring test and 3000 kg for ring-on-ring test. The fracture strength of ball-on-ring and ring-on-ring tests was calculated using the results of FEM analysis.

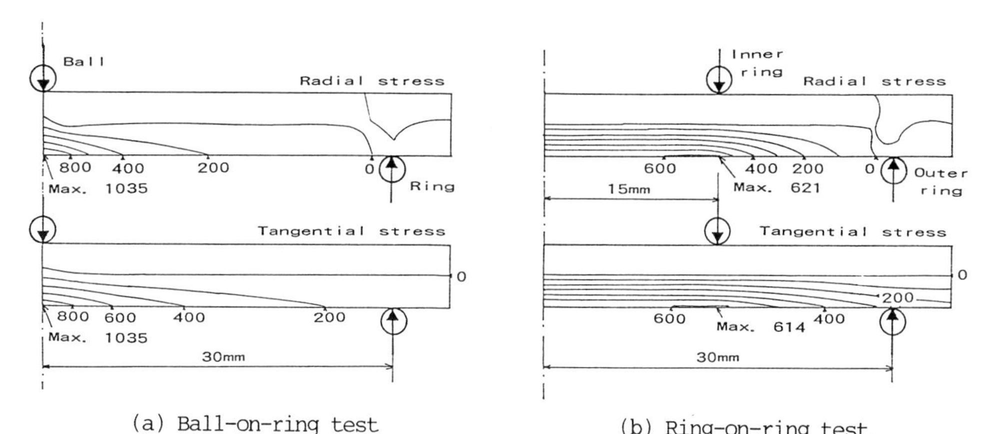

(a) Ball-on-ring test (b) Ring-on-ring test

Fig. 4. The stress distributions in the disk under the ball-on-ring and ring-on-ring tests analized by FEM. (a)ball-on-ring test of the ring diameter 60mm; (b)ring-on-ring test of the inner and outer ring diameters 30 and 60mm.

EXPERIMENTAL RESULTS AND DISCUSSIONS

In an attempt to develop strength distribution functions useful for reliability evaluation for ceramic components under multiaxial stresses, the predicted strength based on the fracture criteria already proposed was compared with experimental results.

Strength Evaluation by Uniaxial Distribution Function

The uniaxial Weibull distribution function, considering only the maximum principal stress in a ceramic component, has been used to explain the effects of component sizes and stress distributions on strength and fracture strength distribution when analyzing test results where specimens are subjected to a uniaxial stress.[19,20] In this study, at first the uniaxial distribution function was applied to the biaxial test results. The uniaxial Weibull distribution functionis represented by the following equation, using effective volume V_{ep}:

$$P_f=1-\exp\left\{-\left(\frac{\sigma_{1\max}}{\sigma_0}\right)^m V_{ep}\right\} \tag{3}$$

$$V_{ep}=\int_V\left(\frac{\sigma_1}{\sigma_{1\max}}\right)^m dV \tag{4}$$

where, P_f is the probability of failure, σ the scale factor, m the Weibull modulus, dV the differential check volume, σ_i the maximum principal stress at each position of the component, and σ_{1max} the highest value of the maximum prncipal stress σ_1 in the component, which is referred to as the strengh of the component. The expected value of σ_{1max} is given by:

$$E[\sigma_{1max}] = \sigma_0 V_{ep}^{-1/m} \Gamma(1+1/m) \quad (5)$$

where Γ is the gamma function. Relationship between effective volume and mean strength is as follows:

$$\left[\frac{V_{ea}}{V_{eb}}\right]^{1/m} = \frac{\sigma_b}{\sigma_a} \quad (6)$$

where V_{ea} and V_{eb} are the effective volume, σ_a and σ_b are the mean strength for components of a and b, respectively.

The ralationships between mean strength and effective volume for SSN-A1, SSN-A2, SSN-B and SiC under uniaxial stress are shown in Figure 5. Each effective volume was calculated using the uniaxial Weibull distribution function and taking into account only the maximum principal stress. Weibull moduli were 15 for SSN-A1, SSN-A2 and SSN-B, and 12 for SiC, which were obtained by 4-point bending tests. Linear relationships between the log of the mean strength and log of the effective volume for SSN-A1, SSN-A2, SSN-B and SiC were observed. Strength under uniaxial stress can be explained by size effect. The Weibull moduli predicted by the least square method using the relationships shown in Figure 5 were 15 for SSN-A1, SSN-A2 and SSN-B, and 11 for SiC. These values were used as the Weibull moduli of SSN-A1, SSN-A2, SSN-B and SiC during this study.

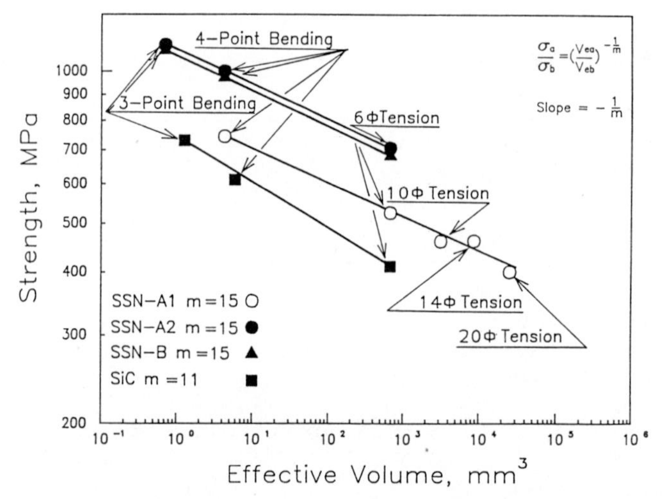

Fig. 5. The ralationships between mean strength and effective volume for SSN-A1, SSN-A2, SSN-B and SiC under uniaxial stress. ○ represents SSN-A1, ● represents SSN-A2, ▲ represents SSN-B and ■ represents SiC experimental result.

Strength Evaluation by Multiaxial Distribution Function

Matsuo has proposed a multiaxial distribution function to be applied to arbitrarily selected fracture criteria, where random crack directions are taken into consideration.[21] A multiaxial distribution function for fracture caused by internal defects may be represented by the following equation:

$$P_f = 1 - \exp\left\{ -\frac{2}{\pi\Omega} \int_V \int_0^{\pi/2} \int_0^{\pi/2} \left(\frac{Z}{\sigma_0}\right)^m \cdot \sin\phi \, d\phi \, d\theta \, dV \right\} \quad (7)$$

where, ϕ and θ are the angles between the line normal to crack surface and the principal axes of stress (see Figure 6). In the equation (7), Ω is introduced as a normalization constant to reconcile failure probability with that calculated by the uniaxial Weibull distribution function for a component subjected to a uniaxial, uniform tensile stress. Z is the equivalent stress, determined by unstable brittleness propagation conditions around a crack under a combined stress mode. It varies depending on the fracture criteria. An effective volume under multiaxial stresses is represented by equation (8). The multiaxial distribution function can be treated as the uniaxial Weibull distribution function by using effective volume represented by equation (8).

$$V_{em} = \frac{2}{\pi\Omega} \int_V \int_0^{\pi/2} \int_0^{\pi/2} \left(\frac{Z}{\sigma_{1max}}\right)^m \sin\phi \, d\phi \, d\theta \, dV \quad (8)$$

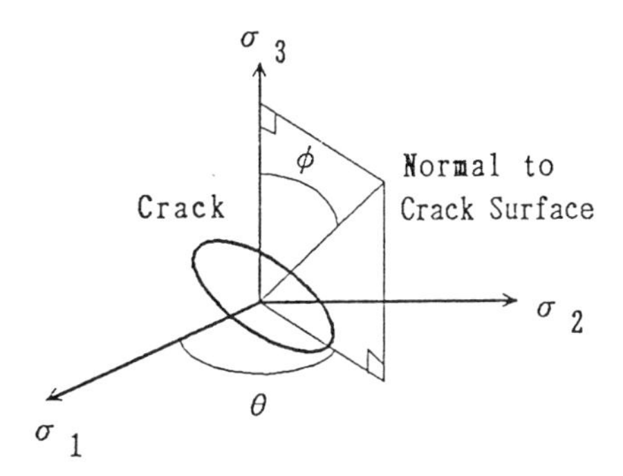

Fig. 6. Angles of crack surface.

Fracture Criteria

In this study, Weibull's multiaxial theory, G-Criterion for coplanar crack extension and non-coplanar crack extension, σ_θ-Criterion and modified G-Criterion were investigated.

(1)Weibull's multiaxial theory

Weibull attempted to predict the risk of rupture under multiaxial stresses, taking into consideration the vertical stresses in all directions.[19] Equivalent stress Z derived from this theory is the stress σ_n which is vertical to the crack surface, and is represented by the following equation:

$$Z = \sigma_n \quad (9)$$

No shear stress around a defect is considered in this equation.

(2)G-Criterion for coplanar crack extension

Batdorf considers that fracture occurs when the strain energy release rate G is beyond a critical level G_c in a model where fracture occurs unstably in a straight line over an extension of the crack, and adopts this theory as the fracture criterion.[22] Strain energy release rate G is given by the following equation:

$$G=\frac{\kappa+1}{8\mu}(K_{\mathrm{I}}^2+K_{\mathrm{II}}^2)+\frac{1}{2\mu}K_{\mathrm{III}}^2 \tag{10}$$

where, K_{I}, K_{II}, K_{III} are the stress intensity factors under stress modes I, II, III, $\mu=E/2(1+\nu)$, $\kappa=3-4\nu$ (plane strain condition) or $(3-\nu)/(1+\nu)$ (plane stress condition), E the Young's modulus, ν the Poisson's ratio. Equivalent stress Z under this condition is given by the following equation, assuming a penny-shaped crack:

$$Z=\sqrt{\sigma_n^2+\frac{\tau_n^2}{(1-0.5\nu)^2}} \tag{11}$$

where, σ_n is the normal stress vertical to the crack surface, and τ_n is the shear stress parallel to the crack surface.

(3)G-Criterion for non-coplanar crack extension

Lamon and Evans consider that fracture occurs when strain energy release rate G is beyond a critical level G_c in a model where fracture occurs unstably of a two-dimensional crack runs at an angle to the crack surface under a combined stress modes I and II, and adopts this theory as the fracture criterion.[16] Strain energy release rate G is given by the following equation, using a J-integral:[23]

$$G=\frac{\kappa+1}{8\mu}(K_{\mathrm{I}}^4+6K_{\mathrm{I}}^2K_{\mathrm{II}}^2+K_{\mathrm{II}}^4)^{1/2} \tag{12}$$

Equivalent stress Z is given by the following equation, using the stress intensity factors with respect to a two-dimensional crack:

$$Z=(\sigma_n^4+6\sigma_n^2\tau_n^2+\tau_n^4)^{1/4} \tag{13}$$

(4) σ_θ-Criterion

Erdogan and Sih consider that a crack propagates in direction in which a tensile stress σ_θ working on the crack tip is the largest value,[1] where θ is an angle formed by the stress direction with the crack direction. The angle θ_m in which the maximum σ_θ value is directed is given by the following equation:

$$\theta_m=\cos^{-1}\left[\frac{3\alpha^2+\sqrt{8\alpha^2+1}}{9\alpha^2+1}\right] \tag{14}$$

where, α is the ratio of the stress intensity factors with respect to a two-dimensional crack, $\alpha=K_{\mathrm{II}}/K_{\mathrm{I}}$, σ_θ under mode I is given by $K_{\mathrm{I}}/2\pi r$, and σ_θ under combined modes I and II is given by the following equation, if it attains a maximum:

$$\sigma_\theta = \frac{K_I}{\sqrt{2\pi r}} \times \cos\frac{\theta_m}{2}\left[\cos^2\left(\frac{\theta_m}{2}\right) + \frac{3}{2}\alpha\sin\theta_m\right] \tag{15}$$

Assuming that fracture results when the σ_θ value reaches a critical level, equivalent stress Z is given by the following equation:

$$Z = \sigma_n \times \cos\frac{\theta_m}{2}\left[\cos^2\left(\frac{\theta_m}{2}\right) + \frac{3}{2}\alpha\sin\theta_m\right] \tag{16}$$

(5)Modified G-Criterion

The authors proposed modified G-Criterion as fracture criterion for the strength prediction under multiaxial stresses.[12] Equivalent stress Z is given by the following equation:

$$Z = \sqrt{\sigma_n^2 + (\beta\tau_n)^2} \tag{17}$$

The characteristic of modified G-Criteion is that the influence of shear stress on a fracture can be varied by changing the β value. In this study, β was determined so that the predicted strength agreed with experimental results under tension/compression biaxial stresses. Then the strength under tension/tension biaxial stresses was predicted using a determined β.

The authors calculated an effective volume for each test using each fracture criterion, and predicted strength under biaxal stresses.

Comparison between Predictions and Experimental Results

Under Tension/Compression Biaxial Stresses

The experimental results and theoretical predictions based on various fracture criteria for SSN-A1 are shown in Figure 7. The experimental results and the predictions are normalized so that tensile strength is unity. The predictions based on maximum principal stress theory and Weibull's multiaxial theory are good agreement with experimental results.

The experimental results and theoretical predictions based on various fracture criteria for SiC are shown in Figure 8. The prediction based on G-Criterion is relatively good agreement with experimental results. The fracture criterion which leads the prediction to be is good agreement with experimental results is different, depending on each ceramic material.

β values for SSN-A1, SSN-B and SiC were determined so that the predictions based on modified G-Criterion were best agreement with experimental results. The experimental results and theoretical predictions based on modified G-Criterion for SSN-A1, SSN-B and SiC are shown in Figure 9. It can be understood that there are differences in the fracture behavior among SSN-A1, SSN-B and SiC. When β equals 0.65 for SSN-A1, 0.8 for SSN-B, 1.1 for SiC, the predictions are good agreement with expermental results. Therefore, there are differences in the contribution of shear stress to fractureamong various non-oxide ceramics, such as SSN-A1, SSN-B and SiC.

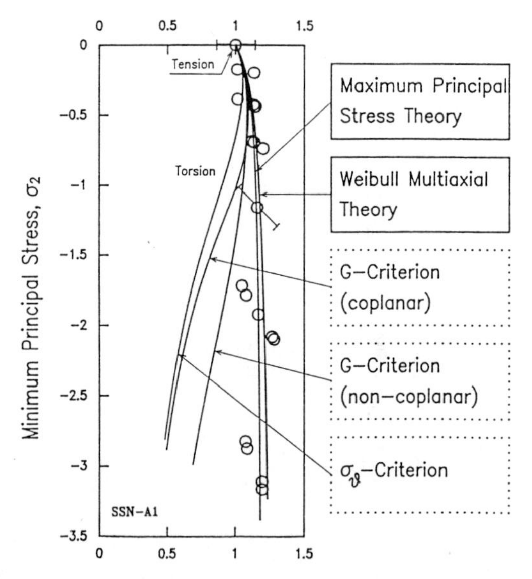

Fig. 7. Comparison of the experimental results and predictions based on various fracture criteria for SSN-A1 under tension/compresion biaxial stresses. Plots represent experimental results, lines represent predictions.

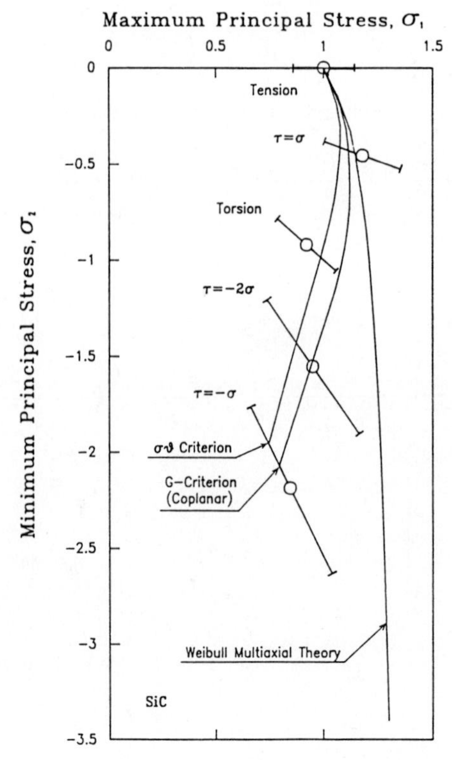

Fig. 8. Comparison of the experimental results and predictions based on various fracture criteria for SiC under tension/compresion biaxial stresses. Plots represent experimental results, lines represent predictions.

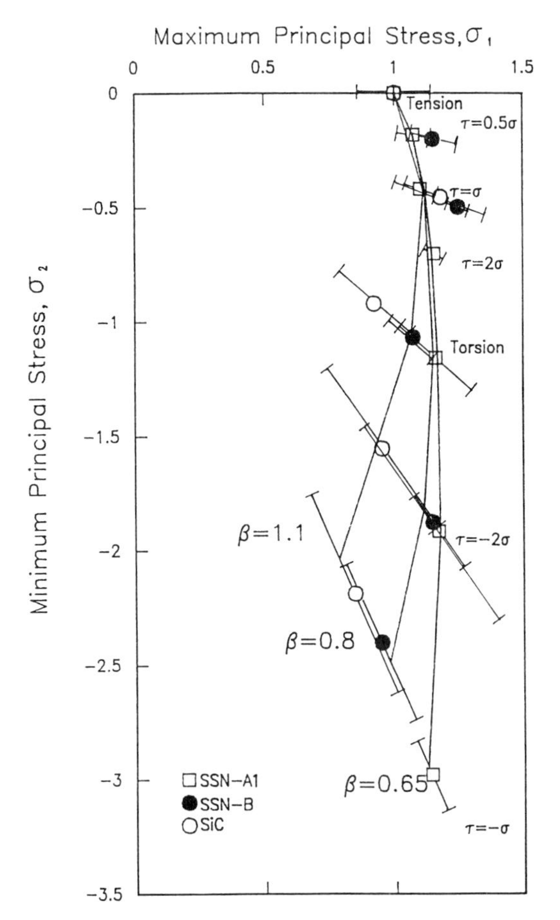

Fig. 9. Comparison of the experimental results and predictions based on modified G-Criterion for SSN-A1, SSN-B and SiC under tension/compression biaxial stresses. □ represents SSN-A1, ● represents SSN-B and ○ represents SiC experimental results, lines represent predictions.

The transmission electron micrographs of SSN-A, SSN-B and SiC are shown in Figure 10. SSN-A has a microstructure wherein pillar-shaped crystals are complexly interlocked with one another, such as composite materials; and the grain boundaries among crystals are in glassy phases. SSN-B has the same microstructure as SSN-A; and the grain boundaries among crystals are in crystalline phases. SiC has a microstructure where plate-shaped crystals are combined monolithically; and the grain boundaries among crystals hardly exist. It should be considered that the possible reason for the contribution of shear stress to fracture being different among SSN-A, SSN-B and SiC is probably due to differences in microstructure. Therefore, modified G-Criterion should be regarded as one of the best fracture criteria for various non-oxide ceramics under tension/compression biaxial stresses.

Under Tension/Tension Biaxial Stresses

Ball-on-ring and ring-on-ring tests were carried out for the tests under tension/tension biaxial stresses. Because the tests were conducted in different volume, the comparisons between predictions and experimental results were performed according to the following method.

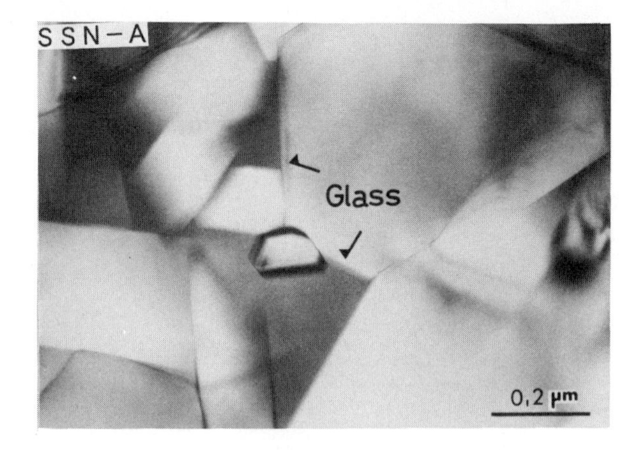

(a)SSN-A

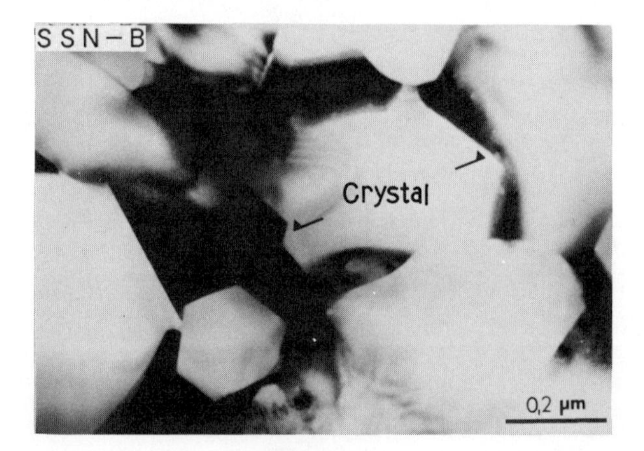

(b)SSN-B

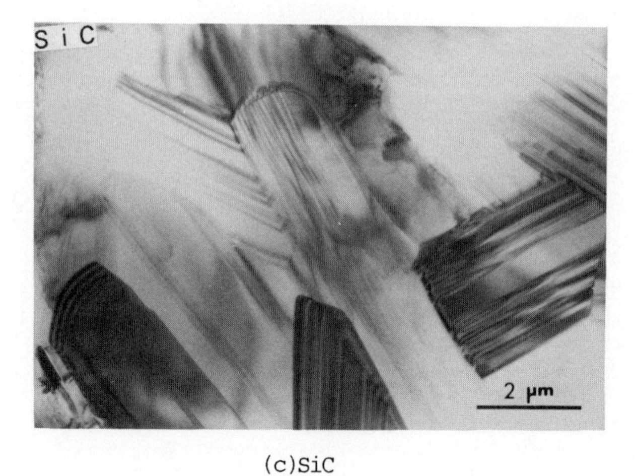

(c)SiC

Fig.10. Transmission electron micrographs of SSN-A, SSN-B and SiC.

(1)Prediction Method

Effective volume V_{e1} of an elemental body loaded with tension/tension biaxial stresses was calculated in accordance with each fracture criterion. Effective volume under tension/tension biaxial stresses is calculated continuously, by changing the ratio of loaded biaxial stresses. The strength σ_{T1} of an elemental body under uniaxial stresses was calculated using tensile strength. Then the prediction strength σ_{PR} of an elemental body was calculated using σ_{T1} and V_{e1} within equation (6).

(2)Calculation Method of Experimental Results

Effective volume V_{eD} of each test was calculated based on each fracture criterion. The strength σ_{D1} of an elemental body was calculated from each experimental results, V_{eD} and V_{e1} within equation (6).

Finally, the applicable fracture criterion for the strength prediction under tension/tension biaxial stresses was investigated by comparing σ_{PR} and σ_{D1}.

The experimental results and theoretical predictions based on various fracture criteria for SSN-A2 are shown in Figure 11. The experimental results and the predictions are normalized so that tensile strength is unity. The horizontal axis and the vertical axis represent principal stresses, σ_1 and σ_2. The prediction based on σ_θ-Criterion is good agreement with experimental results. The other predictions based on maltiaxial fracture criteria, such as Weibull's multiaxial theory, G-Criterion, modified G-Criterion using β of 0.65 are smaller than experimental results. Therefore, the predicted strength based on multiaxial fracture criteria could be considered to be good or low-risk for design of ceramic components. The prediction based on maximum principal stress theory, that is the uniaxial distribution function, is larger than experimental results. Therefore, the predicted strength could be considered to be high-risk.

The experimental results and theoretical predictions based on various fracture criteria for SSN-B are shown in Figure 12. The predictions based on multiaxial fracture criteria are smaller than experimental results. Therefore, the predicted strength could be considered to be low-risk for design of ceramic components. The prediction based on maximum principal stress theory is good agreement with experimental results.

The experimental results and theoretical predictions based on various fracture criteria for SiC are shown in Figure 13. The predictions based on σ_θ-Criterion is almost good agreement with experimental results, in spite of the fact that the prediction is a little smaller than experimental results. The other predictions based on multiaxial fracture criteria are smaller than experimental results. Therefore, the predicted strength could be considered to be low-risk for design of ceramic components. The prediction based on maximum principal stress theory is larger than experimental results. Therefore, the predicted strength could be considered to be high-risk for design of ceramic components.

Modified G-Criterion is expected to be applicable to low-risk prediction of fracture strength for various non-oxide ceramics with differences in microstructure under tension/tension biaxial stresses.

In both tension/compression and tension/tension biaxial stresses, modified G-Criterion is expected to be one of the fracture criteria applicable to various non-oxide ceramics.

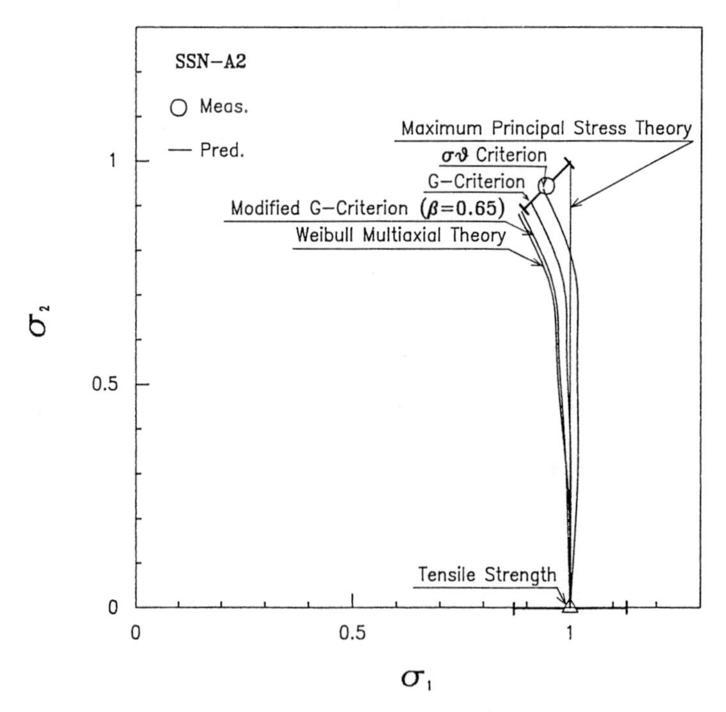

Fig.11. Comparison of the experimental results and predictions based on various fracture criteria for SSN-A2 under tension/tension biaxial stresses. Plots represent experimental results, lines represent predictions.

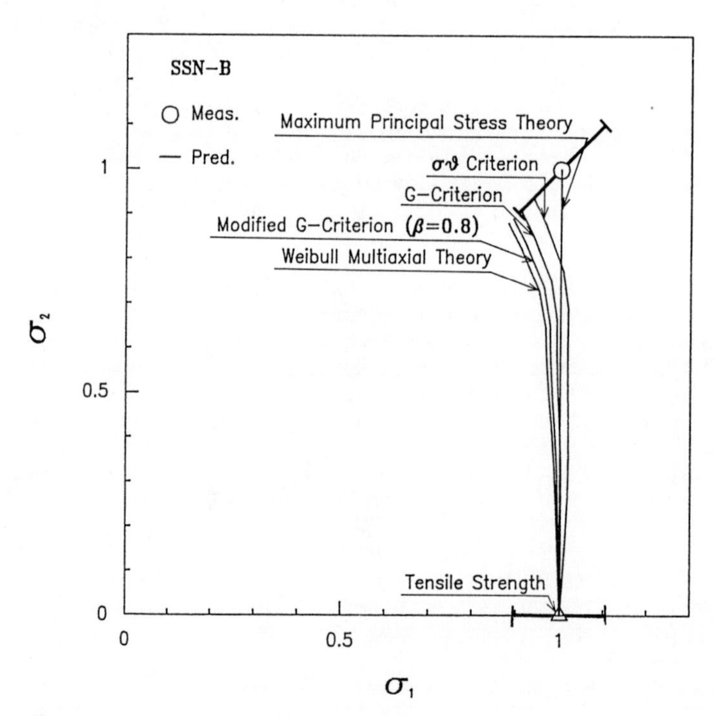

Fig.12. Comparison of the experimental results and predictions based on various fracture criteria for SSN-B under tension/tension biaxial stresses. Plots represent experimental results, lines represent predictions.

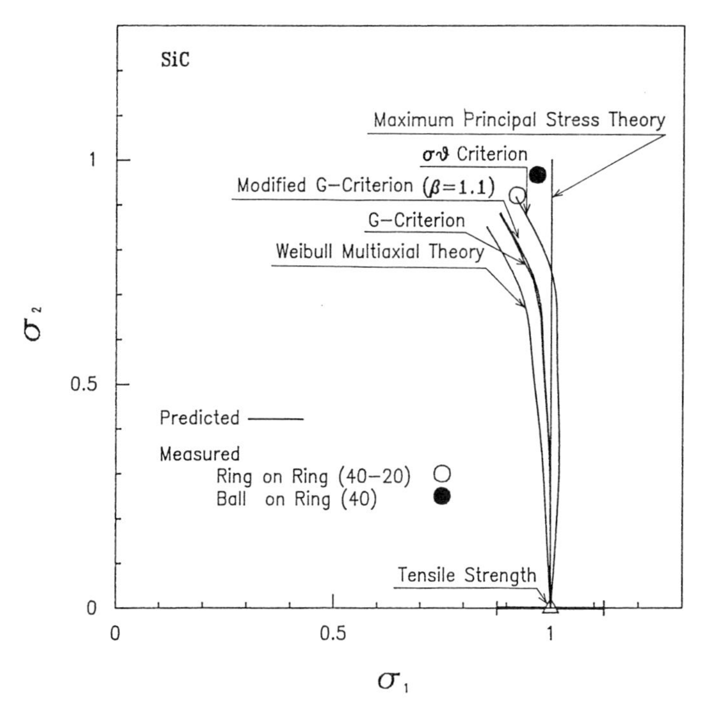

Fig.13. Comparison of the experimental results and predictions based on various fracture criteria for SiC under tension/tension biaxial stresses. Plots represent experimental results, lines represent predictions.

CONCLUSIONS

(1)Torsional, combined tensile/torsional, and combined compressive/torsional tests were conducted to clarify the fracture behavior of non-oxide ceramics under tension/compression biaxial stresses. There are differences in the fracture behavior among various ceramics, probably depending on microstructures.

(2)Ball-on-ring and ring-on-ring tests were performed to clarify the fracture behavior of non-oxide ceramics under tension/tension biaxial stresses. Most predicted strength according to the multiaxial fracture criteria is lower than experimental data, and most predicted strength according to maximum principal stress theory, that is the uniaxial distribution function, is larger than experimental data.

(3)The authors propose modified G-Criterion, $Z=(\sigma^2 + (\beta\tau)^2)^{1/2}$, as fracture criterion. β determined by tension/compression experimental data are 0.65 for SSN-A1, 0.8 for SSN-B, and 1.1 for SiC. Modified G-Criterion is expected to be one of the fracture criteria applicable to various non-oxide ceramic under biaxial stresses.

ACKNOWLEDGEMENT

This work was performed under the management of the Engineering Research Association for High Performance Ceramics as a part of the R&D Project of Basic Technology for Future Industries supported by NEDO (New Energy and Industrial Technology Development Organization).

The authors thank Dr.Junichi Hamanaka, Dr.Akihiko Suzuki and Mr.Seijiro Hayashi of Research Institute of Ishikawajima-Harima Heavy Industries Co. Ltd. for their valuable advice.

REFERENCES

1. F.Erdogan, G.C.Sih, J.Basic Eng., 85, 519-27(1963).
2. M.A.Hussain, S.L.Pu and J.Underwood, ASTM STP 560, 2-28(1974).
3. R.J.Nuismer, Int.J.Fract., 11, 245-50(1975).
4. G.C.Sih, Int.J.Fract., 10, 305-21(1974).
5. M.Ichikawa, and S.Tanaka, Int.J.Fract., 18, 19-28(1982).
6. K.Watanabe and H.Shiomi, Transactions of The Japanese Society of Mechanical Engineers-A, 49, 19-28(1983).
7. J.J.Petrovic, J.Am.Ceram.Soc., 68, 348-55(1985).
8. H.Awaji and S.Sato, Tras.ASME, J.Eng.Master. & Technol., 100, 175-82(1978).
9. G.Tappin, R.W.Davidge and J.R.Mclaren, "Fracture Mechanics of Ceramics,Vol.3", Plenum Press, New York, 435-49(1978).
10. M.G.Stout and J.J.Petrovic, J.Am.Ceram.Soc., 67, 14-23(1984).
11. D.K.Shetty, Trans.ASME, J.Eng.Gas Turbines & Powder, Vol.109, 282-89(1985).
12. I.Oda, M.Matsui, T.Soma, M.Masuda and N.Yamada, Nippon Seramikkusu Kyokai Gakujutsu Ronbunshi, 96 [5] 539-45(1988).
13. H.Tsuruta, N.Yamada, Y.Nakasuji, M.Masuda and M.Matsui, Nippon Seramikkusu Kyokai Gakujutsu Ronbunshi, 99 [7] 574-81(1991).
14. D.L.Hargosoch, Trans.ASME, 85-GT-100, 1-12(1985).
15. J.P.Gyekenyesi, Trans.ASME, J.Eng.Gas Turbines & Powder, 86-GT-34, 1-7(1985).
16. J.Lamon and A.G.Evans, J.Am.Ceram.Soc., 66, 177-82(1983).
17. J.Hamanaka, Y.Hashimoto, M.Itoh and N.Watanabe, Transactions of The Japanese Society of Mechanical Engineers-A, 52, 2187-93(1986).
18. T.Soma, M.Matsui and I.Oda, Proceedings of the International Conference of Non-Oxide Technical and Engineering Ceramics, Ireland, 361-74(1985).
19. W.Weibull, Ingeniors vetenskaps akademien Handlingar, Nr.151(1939).
20. D.G.S.Davis, Proc.Brit.Ceram.Soc.,22,429-52(1973).
21. Y.Matsuo, Transactions of The Japanese Society of Mechanical Engineers-A, 46, 605-12(1980).
22. S.B.Batdorf and H.L.Heinisch,Jr., J.Am.Ceram.Soc., 61, 355-58(1978).
23. T.K.Hellen and W.S.Blackburn, Int.J.Fract., 11, 605-17(1975).

BIAXIAL FLEXURE TESTING: ANALYSIS AND EXPERIMENTAL RESULTS

William F. Adler and Dennis J. Mihora

General Research Corporation
5383 Hollister Avenue
Santa Barbara, CA 93111

INTRODUCTION

Characterization of the damage in ceramics after exposure to particulate environments involves measurements of the loss of mechanical strength when impacted by waterdrops, hail, and sand.[1] The most desirable measure of strength degradation for particulate impacted specimens is a biaxial flexure test. The advantage of this test is that the central portion of the specimen is subjected to a uniform biaxial state of stress. This loading system removes subjectivity concerning the location and orientation of the most severe flaw generated by the particulate collisions.

The hydraulic pressure burst test and the ring-on-ring biaxial flexure test offer the most desirable configurations for determining the residual strength of particulate impacted specimens from among the biaxial flexure tests that have been described in the literature. Since elevated temperatures and high-strength ceramics are additional considerations, the ring-on-ring test became the preferred choice for some of the erosion studies at General Research Corporation (GRC), however both test configurations are being used.

The experimental erosion investigations at GRC place constraints on the dimensions of the specimens which can be used. The imposition of these restraints raised questions concerning the validity of the analytic expression typically used to evaluate the strength of specimens subjected to biaxial flexure in the ring-on-ring test. A general finite element analysis was formulated to remove the restrictions inherent in this analysis.

Significant differences in the stress distributions in the specimens for different ring materials and geometries used for the concentric ring biaxial flexure test are shown to exist. The results from these finite element analyses of the test configuration were used to design a concentric ring test setup which provides a uniform state of biaxial stress within the area of the specimen bounded by the load ring for all specimen dimensions and minimizes the stress amplification effects inherent in the typical concentric ring biaxial flexure test.

Fracture Mechanics of Ceramics, Vol. 10
Edited by R.C. Bradt *et al.*, Plenum Press, New York, 1992

The results presented are still incomplete but the computational capability is available to evaluate arbitrary specimen and fixture geometries in order to broaden the range of application of the concentric ring biaxial flexure test. The detailed finite element computations do not support the findings of Ritter and coworkers[2] concerning the amplification of the circumferential stress component at the load ring. In the present analysis both the radial and circumferential stress components are amplified to nearly the same extent in the vicinity of the load ring. However the radial stress component is usually greater on both the tension and compression sides of the specimen.

BACKGROUND

The biaxial flexure strength of ceramic materials using the ring-on-ring test configuration is typically determined from the theory of bending of thin elastic plates. The governing equations for the loading condition illustrated in Fig. 1 can be found in Roark and Young[3] or Timoshenko and Woinowsky-Krieger[4] when the circular plate is simply supported along its boundary. Vitman and Pukh[5] introduced a term which accounted for the overhang of the plate (the difference R-a in Fig. 1) that tends to stiffen the central region of the plate. Vitman and Pukh indicate that when (R-a) ≥ 6h the strength measurements no longer depend on the extent of the overhang for glass plates. The influence of the overhang will be included in the evaluation of the stress components for the ring-on-ring configuration.

The radial and circumferential stress components are equal in the region bounded by the loading ring (diameter 2b in Fig. 1) and achieve their maximum value throughout this region.

$$\sigma_{max} = \sigma_{rr} = \sigma_{\theta\theta} = \frac{3P}{4\pi h^2}\left[2\,(1+\nu)\,\ln\left(\frac{a}{b}\right) + \frac{(1-\nu)\,(a^2-b^2)}{R^2}\right] \qquad 0 \leq r \leq b \tag{1}$$

where P is the axial load applied by the load ring
h is the thickness of the test specimen (usually a disk)
ν is Poisson's ratio
b is the radius of the load ring
a is the radius of the support ring
R is the radius of the disk

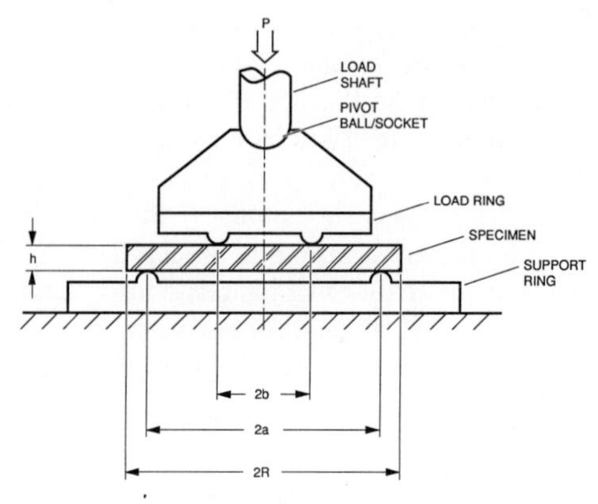

Fig. 1. Schematic of typical ring-on-ring biaxial flexure test configuration.

The approximate theory of bending upon which Eq. (1) is based is valid as long as the deflections of the specimen are small in comparison with its thickness. The assumptions for the approximate theory are

- that there is no deformation in the middle plane of the plate
- that points initially on a normal to the middle plane of the plate remain on the normal to the middle plane of the plate
- that normal stresses transverse to the middle plane of the plate can be ignored

The first assumption eliminates membrane stresses from the analysis. The second assumption eliminates transverse shear stresses from the analysis. The third assumption eliminates highly concentrated loads from being applied to the plate.

These assumptions are not always satisfied in the selection of specimen geometries for the concentric ring biaxial flexure test, so the use of Eq. (1) for evaluating the specimen's strength may not be correct. The current finite element capabilities can provide detailed analyses of considerably more representative models of the ring-on-ring test, however it does not appear that this capability has been applied to this test configuration. Preliminary computations for this purpose are presented in this paper. Prior work which compared the predictions from Eq. (1) with more general analyses will be outlined.

Kao, Perrone, and Capps[6] considered the case of a thin plate undergoing large deflections. They were able to demonstrate their large deflection calculations compared favorably with experimental measurements for a chemically strengthened glass. Kao and coworkers only provided calculations for the radial stress component, σ_{rr}, which showed the development of a nonuniform stress distribution within the area bounded by the load ring as the deflections/load increased. There was significant stress amplification at the load ring. The nonuniformity in stress is due to the membrane stresses which become more significant as the deflections increase. As indicated previously mid-plane stretching of the specimen is not included in the approximate theory upon which Eq. (1) is based.

Kao, Perrone, and Capps[6] used an 11 inch (27.94 cm) diameter glass plate which was 0.106 inch (0.269 cm) thick in their experiments. Metal rings with toroidal surfaces were used. The support ring was 10 inches (25.4 cm) in diameter and 5 inch (12.7 cm) and 2 inch (5.08 cm) diameter load rings were used.

Both Ritter, et al.[2], and Shetty, et al.[7], pointed out there is a stress amplification effect in the vicinity of the load ring radius on the tension side of the specimen. Ritter and co-workers carried out a two-dimensional finite element analysis for thin elastic plates for the ring-on-ring and piston-on-ring biaxial flexure tests. They found a 15 percent increase in the circumferential stress component, $\sigma_{\theta\theta}$, on the tension face of the specimen at the radius of the load ring for the concentric ring test. The radial stress component did not display this effect. Shetty and co-workers performed a two-dimensional, axisymmetric finite element analysis for the ball-on-ring test based on linear elasticity assigning the Hertzian loading function to the contact zone between the ball and the plate. Again they found the values of the stress components deviated from the analytical expressions for this case outside the contact zone of the ball.

Shetty and co-workers[7] also note that a stress magnification effect was found in the annular region directly below the load ring in the ring-on-ring configuration being used for lens fracture tests (Duckworth, et al.[8]). In order to evaluate this effect they used a strain-gaged 51 mm diameter by 2.54 mm thick 4340 steel disk for the test specimen. A ball bearing race 38 mm in diameter was used for the support ring and a ball bearing race 18 mm in diameter was used for the load ring. They found the circumferential stress component was 21 percent greater than the prediction from Eq. (1) at the load ring. The radial stress component did not exhibit this effect. Shetty and co-workers comment that their experimental result is consistent with the finite element predictions of Ritter and his co-workers[2]. Ritter's calculations were for a soda lime glass plate 38.10 mm in diameter and 3.18 mm thick. The support ring was 28.58 mm in diameter and the load ring was 14.30 mm in diameter.

It is evident from these investigations that the extent of the region in which the uniform biaxial stress state prevails may not be as large as determined from Eq. (1). It is advantageous to use a wide range of specimen dimensions including those which would be considered to be thick plates in the particulate impact investigations at General Research Corporation. In order to provide this flexibility in the ring-on-ring test a finite element analysis of this biaxial flexure test was undertaken. The results from this analysis will be described.

BIAXIAL FLEXURE TESTING ANALYSIS

The NIKE3D finite element program developed at Lawrence Livermore National Laboratory was used to calculate the stresses in the specimen as it is loaded in the ring-on-ring configuration. NIKE is an implicit, large deflection, time dependent, finite element (FE) code. The static stress results are late-time solutions of a slow ramp-up in force to a constant value. The mass and stiffness matrices of the structure are saved during the computations.

The limitations that exist with the closed form analysis in Eq. (1) are not present in the NIKE finite element analysis. There is no restriction placed on the diameter to thickness ratio of the disk. Large shear strains are included in the modeling which is especially significant for the thicker ceramic disks. Membrane stresses (large deflections) are combined with the bending stresses. The contact mechanics between the load ring and the disk allow localized deformations and sliding between the two members. The limitations of the Hertzian half-space model that assumes very thick plates is not an issue since stress through finite thickness plate is automatically calculated. The analyses also accounts for plastic flow and arbitrary distortions at the contact zone between the specimen and the load ring.

Figure 2 shows the FE layout of the two body contact problem that was used in the parametric calculations. The hash marks on the top of the load ring correspond to the applied pressure. The two circles on the lower side of the disk are the locations of lower surface displacement constraints. The highest density elements exist at the contact region and at the tension and compression faces of the disk. The merging of rectangular and pie-shaped elements to form the disk in Fig. 2 eliminates the usual singularities in hexahedron elements. Reduced accuracy occurs when 8-node hexahedrons degenerate to 6-node solids. The distribution of 8-node elements in the disk is shown in this three-dimensional perspective. As many as 30 azimuthal and 20 thickness elements have been used in various mesh concentrations in order to increase the accuracy of the computations.

To enhance the viewing resolution and accuracy of the calculations the finite elements are reduced in size in the contact zone. The number of elements were optimized to achieve maximum resolution within the usable 50 Mbyte memory size available on the Stellar 2000

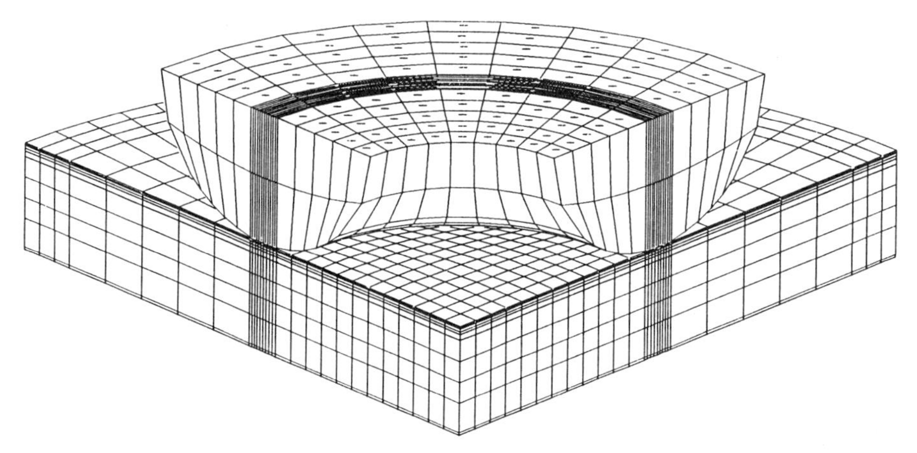

Fig. 2. Finite element mesh for a quarter model of a curved load ring and specimen.

computer used for these computations. Although the UNIX operating system operates in a virtual memory mode, the NIKE code requires that the mass and stiffness matrices fit into active memory. Approximately 16,000 equations (nodal degrees of freedom) can be solved on the Stellar.

The use of two slide-surfaces between the load ring and the specimen disk provides the mechanism for normal force (pressure) transfer between the contacting bodies. The master slide surface is coincident with the face of the load ring. The slave slide surface is coincident with the face of the disk. The slide surface algorithm properly transmits the normal force between the different nodal densities of the load ring and disk. Both slide surfaces are friction free.

The stresses through disks with different thicknesses were calculated using high-fidelity finite element modeling and the results were compared with the classic closed-form solution in Eq. (1). Thin disks ($2R/h > 20$) experience combined membrane and bending stresses. Very thick disks ($2R/h < 10$) introduce bending and shear stresses. The range of validity of the analytic formulation can therefore be determined.

The FE modeling yielded very favorable results. The analytic models are usable over a broad diameter to thickness range. The analytic results depart from the FE results for thin plates with substantial centerline deflection. A centerline deflection of ten percent of plate thickness introduces substantial membrane stress which increases the tension face stress. The analytic models applied to very thick plate overpredict the tensile and compressive stresses at the disk centerline.

Very large compressive contact stresses may be present directly under the load ring. For thick plates a load ring may act as an indenter introducing both compressive failure and nearby tensile cracking. The compressive contact stresses act on the compression side of the specimen similar to an infinite half-space. For very thin plates the load contact stresses are mirrored so tensile perturbations also exist on the tension side of the specimen. The tensile stress at the load ring radius is nearly as large as the compressive contact stress for thin plates.

An example of the effect of the contact stresses is provided for the rounded steel load ring shown in Fig. 2. The specimen is a 2.75 inch (7 cm) diameter by 0.108 in (0.27 cm) thick zinc sulfide disk. In the calculations which follow the value of Young's modulus for the specimen is 11.3 x 10^6 psi (77.9 GPa) and Poisson's ratio is 0.29. This is the specimen used for the strain measurements reported in a subsequent section. The curved steel load ring has a 0.125 inch (3.2 mm) radius.

The full thickness of the disk is shown in Fig. 3. The stress contour and stress magnitudes seem to differ for the radial and circumferential stress components in the vicinity of the contact zone. The maximum radial stress component under the load ring is 3.71 times the centerline stress of 16,600 psi (114.5 MPa). Directly under the load ring the ratio of the radial stress component and the circumferential stress component is 1.275. On the opposite face of the specimen, the tension side, this ratio is reduced to 1.182. This ratio is dependent on the plate thickness. For thicker plates the radial to circumferential stress ratio approaches unity, but for thin plates the localized compressive stresses on the compression face of the specimen influence the stresses on the tension side of the specimen.

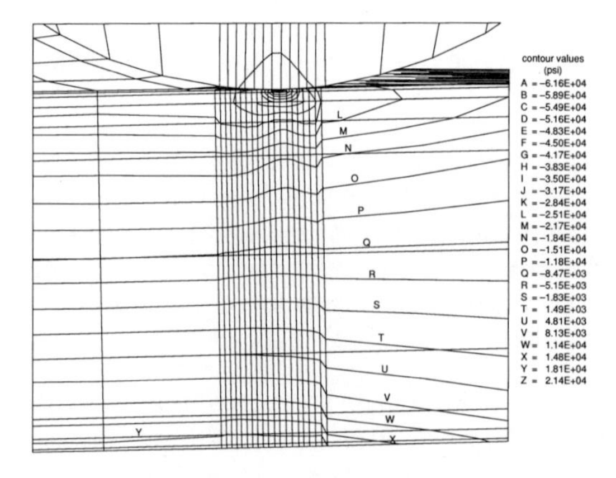

Radial stress component, σ_{rr}

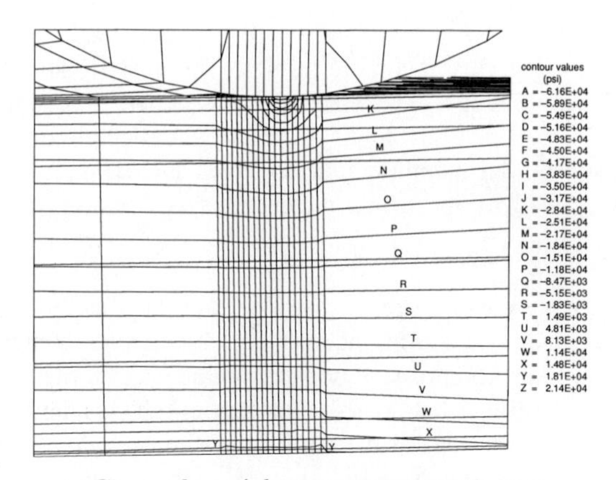

Circumferential stress component, $\sigma_{\theta\theta}$

Fig. 3. Radial and circumferential stress contours under a 0.125 in (3.2 mm) radius steel load ring for a 2.75 in diameter by 0.108 in (7 cm by 0.27 cm) zinc sulfide disk.

A method for eliminating large contact stresses is to substantially increase the contact area of the load ring and secondarily to reduce the modulus of elasticity of the load ring. With proper load ring design compressive stresses in the radial and circumferential directions will be virtually constant from the load ring to the disk centerline. Tensile stresses in the vicinity of the load ring radius on the tension side of the specimen can also be mitigated with a large contact area.

Since the finite element analyses are not restricted with respect to the geometry of the test configuration, a polymeric load ring (low modulus of elasticity) with a rectangular cross section was considered. The finite element layout for this case is shown in Fig. 4. The obvious advantage of the layout is the substantial increase in the contact area of this load ring over a curved load ring. The height to width ratio of the load ring cross section is 2:1. The outermost radius of the ring is constrained by a steel collar which is not shown. There is no constraint on the inner radius of the ring.

Finite element calculations were carried out for this and other load ring geometries. The rectangular load ring is made of Delrin. Delrin is the trade name for an acetal polymer manufactured by the E.I. DuPont de Nemours Company.

The specimen used throughout these computations has a Young's modulus of 14.5 x 10^6 psi (100 GPa) and a Poisson's ratio of 0.27: representative values of these parameters for germanium. The Delrin has an elastic modulus of 0.5 x 10^6 psi (3.5 GPa) and a Poisson's ratio of 0.30. The specimen for the initial computations is 0.788 inch (2 cm) in diameter and 0.078 inch (0.2 cm) thick.

The main concern with a rectangular cross section is the possibility of localized stress concentrations at the corners of the square load ring. Figure 5 displays the σ_{rr} and σ_{zz} stresses directly under the rectangular load ring. Only about half of the rectangular cross section is shown. The maximum σ_{rr} stress under the load ring is 10,800 psi (74.5 MPa) compression compared to the 11,200 psi (77.2 MPa) compressive stress at the specimen's

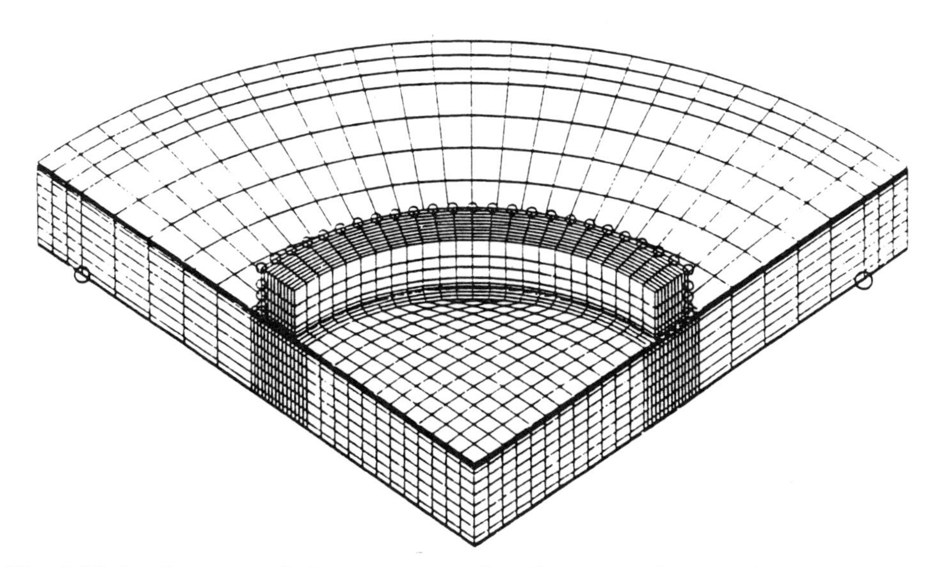

Fig. 4. Finite element mesh for a quarter model of a rectangular load ring and specimen.

centerline. This difference is negligible so the magnitude of the radial stress is not perturbed by the soft rectangular load ring. It will be shown that the steel load rings do produce a greatly elevated stress in this region. The 10,800 (74.5 MPa) compressive radial stress extends over 80 percent of the Delrin contact area. The σ_{zz} maximum stress is approximately 3,680 psi (25.4 MPa) compression located near the outboard radius of the rectangular indenter. There is a smooth stress gradient in the disk at the base of the square Delrin load ring. The edges of the indenter do not have stress peaks despite the deflection and angle change of the ceramic disk with the application of load to the load ring.

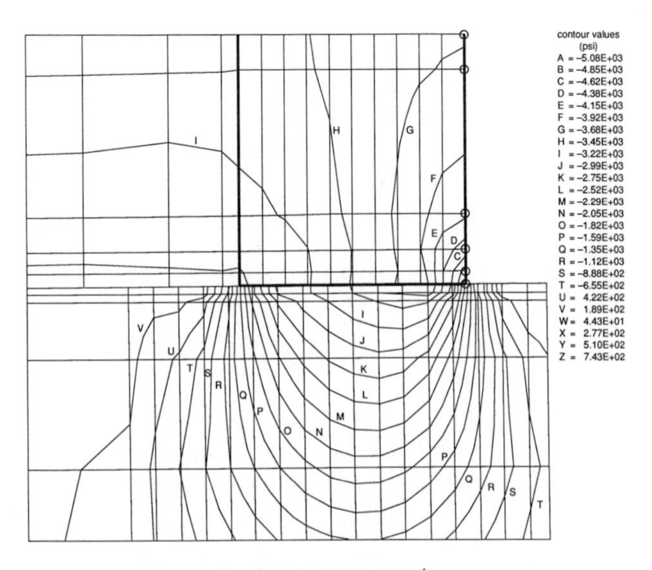

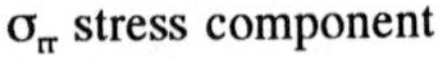

σ_{rr} stress component

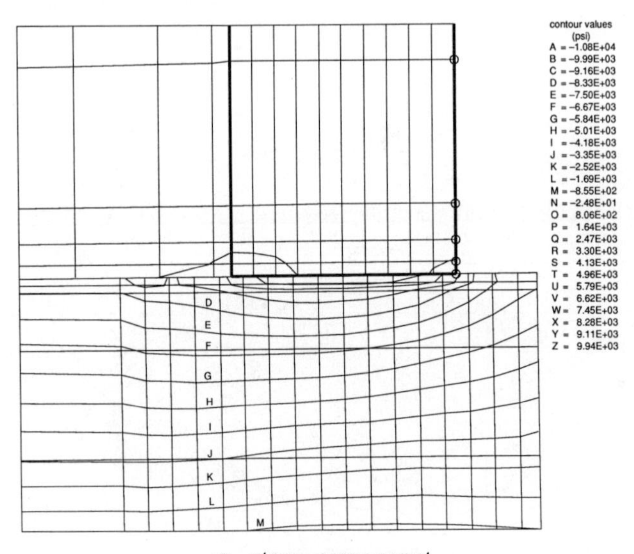

σ_{zz} stress component

Fig. 5. Stress distributions under the rectangular Delrin load ring evaluated when the centerline stress is 11.2 ksi (77 MPa) in compression.

The σ_{zz} stress in the disk directly under the load ring shown in Fig. 5 is relatively small. The σ_{zz} stress under the square Delrin load ring is a factor of three smaller than the σ_{rr} stress. The non-symmetry in the stress under the plate is a consequence of the deflection constraint and stabilization on the outer radius of the Delrin load ring by the steel collar.

Table 1 summarizes the FE results for five different load rings for comparison of the types of load rings that could be used in the biaxial flexure test. The results compare peak compressive stress in the specimen disk directly under the load ring to the centerline bending stress denoted by $\sigma(0)$. An axial load is applied to each ring to produce a nominal 10,000 psi (69 MPa) centerline bending stress in the disk. In these examples the same 0.788 inch (2 cm) in diameter by 0.079 inch (0.2 cm) thick specimen is used with the properties given previously for germanium. The three steel load rings have different radii of curvature. The local radius of curvature of the load ring, c, at the contact point is proportioned to disk thickness, h, the steel load rings have a Young's modulus of 30 x 10^6 psi (207 GPa) and a Poisson's ratio of 0.30.

The load ring geometries have a major influence on the maximum compressive and tensile stresses in the test specimen. The compressive contact stresses for thick disks ($2R/h < 10$) and the tensile stresses in the thin disks at the point opposite the contact face are major concerns in biaxial flexure testing. The localized, spatially amplified stresses can produce premature failures at the load ring. Failure here is not representative of the disk strength near the centerline where the biaxial stresses are constant. It is also very difficult to ascertain the stress that caused failure in a region near a steel load ring where values are rapidly changing. It is equally difficult to adjust the statistical failure data base when samples fail both in the center region and in a stress amplified region.

Figure 6 compares both the maximum tension/compression contact stress and bending stress in the radial direction from the finite element calculations for 3.0 inch (7.62 cm) diameter germanium disks of varying thickness. The single variable in these results is the disk thickness. The geometry of the load ring and the loading is held constant. The load ring has a radius of curvature of 0.125 inch (3.2 mm) at the contact point; a 147 lb (654 N) axial force is applied to the disk. The steel load ring is made of hardened steel with a modulus of elasticity of 30 x 10^6 psi (207 GPa) and a Poisson's ratio $\nu = 0.30$. The disk has the properties of germanium with an elastic modulus of 14.5 x 10^6 psi (100 GPa) and a Poisson's ratio $\nu = 0.27$.

Table 1. The Peak Compressive Stresses Under Various Load Ring Designs.

	$\sigma_{rr}/\sigma(0)$	$\sigma_{zz}/\sigma(0)$
Steel Curved Load Ring		
$c/h = 1.0$	3.92	5.12
$c/h = 1.6$	2.5	2.66
$c/h = 4.0$	1.6	1.6
Delrin Curved Load Ring		
$c/h = 1.0$	1.87	1.69
Rectangular Delrin Load Ring		
$c/h = \infty$	0.96	0.33

The thin plate bending theory of Vitman and Pukh[4] is shown by the dashed line in Fig. 6. The NIKE results are denoted with the triangle, square and circle symbols. The triangles with the "T-C" identifier are for the maximum tensile and compressive radial stresses at the disk centerline. The contact stress under the load ring asymptotes to the classic Hertzian infinite half-space theory for thick plates.

Membrane stresses in the disk are important for disks below 0.15 inch (3.8 mm) thick as they account for the difference in the centerline tension-compression stresses. For thin disks the radial tensile stress under the load ring on the tension face of the specimen can be almost as large as the compression contact stress. This tensile stress can be substantially larger than the maximum centerline tensile stress. However, the tensile and compressive bending stresses are less than for the linear theory. Membrane terms in the governing equations alter the linear bending stresses. For thicker plates, h > 0.2 inch (5 mm) in this case, the Vitman and Pukh model matches the NIKE3D centerline radial stress.

Two specimen dimensions were used in the biaxial flexure tests for germanium to be described: the small diameter disk used in the previous calculations and a 3.0 inch (7.62 cm) diameter by 0.244 inch (0.62 cm) thick disk. The stresses at the centerline and the

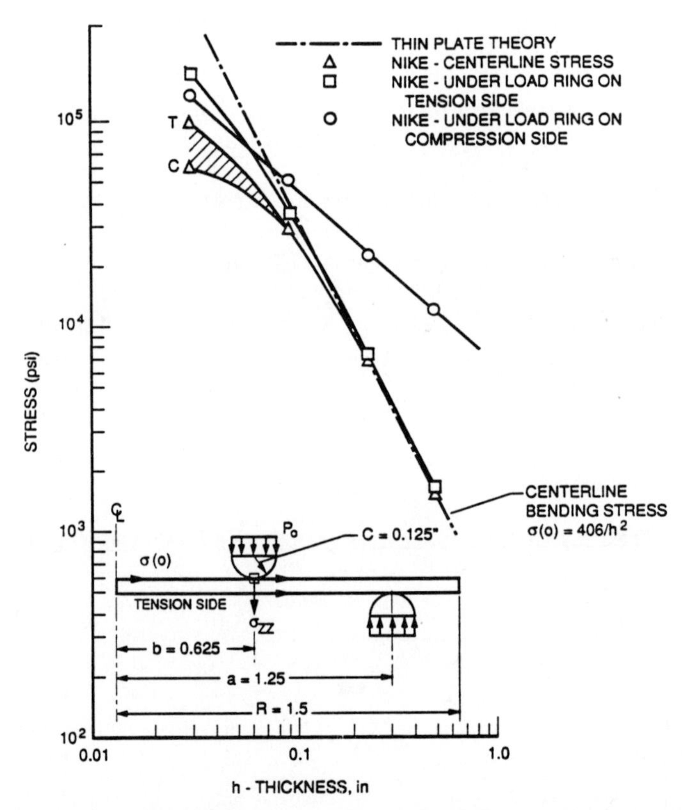

Fig. 6. Comparison of the radial stress in a 3 inch (7.62 cm) diameter germainum specimen determined from the analytic and finite element calculations using steel load rings.

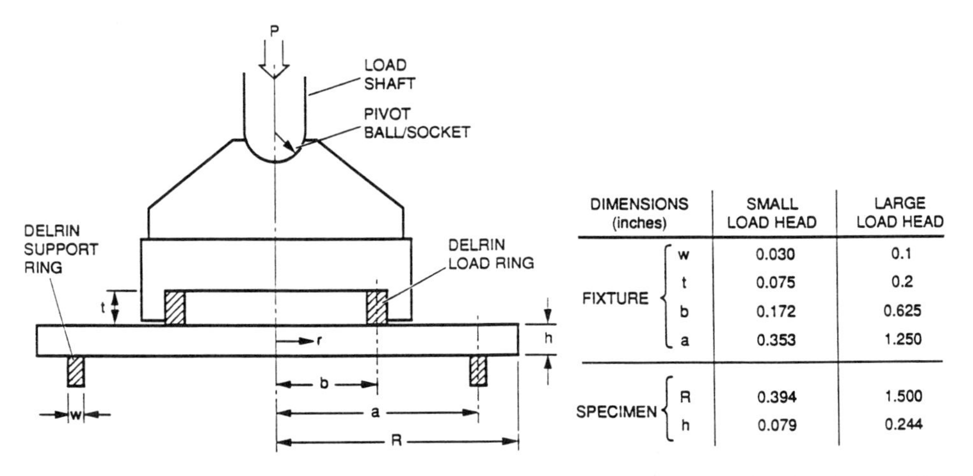

DIMENSIONS (inches)		SMALL LOAD HEAD	LARGE LOAD HEAD
FIXTURE	w	0.030	0.1
	t	0.075	0.2
	b	0.172	0.625
	a	0.353	1.250
SPECIMEN	R	0.394	1.500
	h	0.079	0.244

Fig. 7. The GRC biaxial flexure test fixture used to evaluate the strength of polycrystalline germanium disks.

load ring are compared for these two specimen dimensions for the rectangular Delrin load ring. The dimensions for the biaxial flexure test fixture and these specimen dimensions are provided in Fig. 7.

A load P was applied to the biaxial flexure test fixture to produce about 20,000 psi (138 MPa) stress levels in each specimen. The finite element computations were carried out as summarized in Table 2 along with the analytic results from Eq. (1).

The magnitude of the centerline tensile stresses in Table 2 are larger than the magnitude of the centerline compressive stresses. This difference in stress at the two surfaces of the disk are caused by membrane stresses in the disks which are not included in the analytic solution. In the FE approach inclusion of the membrane and bending stresses is intrinsic in the modeling. The membrane and bending contributions become more significant for configurations with larger deflections and for thinner disks.

Table 2. Stress Relations for Germanium Disks Using Rectangular Delrin Load Rings.

Specimen	Applied Load (lb)	σ_{max} from Eq. (1) (ksi)	Finite Element Results (ksi)			
			$\sigma(0)$ compression side	$\sigma(0)$ tension side	$\sigma(b)$ tension side	Centerline Deflection (in)
Small Disk	274.5	23.9	23.9	24.4	25.1	0.00648
Large Disk	2493.6	21.4	21.8	22.6	24.7	0.0049

Although the differences represented in Table 1 are not very great for the specimen geometries used here, the centerline tensile stresses evaluated from the FE predictions should be the reference stress defining the failure stresses for these disks. In cases where the disk fails away from the centerline, the applied tensile stresses are slightly larger. Since it is often difficult to accurately determine the initial location for fracture, the centerline stress is probably the best reference value which is a conservative prediction of the failure stress.

Since the FE calculations carried out so far were for specific geometries, a complete parametric characterization of the range of application of the linear bending theory has not been delineated. Due to the large number of test configurations that can be used for the biaxial flexure test it is necessary to consider the applicability of the linear bending theory on a case by case basis.

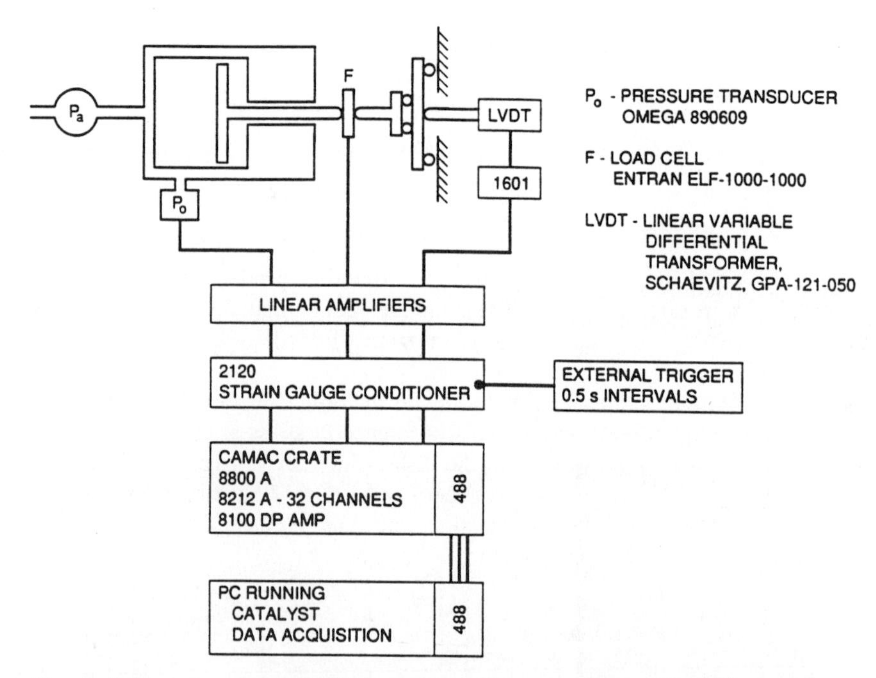

Fig. 8. Schematic of the GRC biaxial test fixture and data acquisition system.

BIAXIAL FLEXURE TEST FIXTURE

A schematic of the GRC ring-on-ring biaxial flexure test fixture and data acquisition hardware is shown in Fig. 8. A pneumatic piston was selected for load application which provides smooth administration of the load. The fixturing can accommodate different size disks by using various size load rings.

A pressure gage, pressure transducer, and load cell are used to record the time-dependent load on the specimen. A linear voltage differential transducer (LVDT) is used to measure the displacement at the center of the disk. Three channels of digitized data and one channel of analog data are acquired during the fracture test. The digital data acquisition equipment has 12 bit resolution. The LVDT has a repeatability of 0.000025 inch (0.000635 mm). The pressure transducer has a resolution of 0.1 psi and the load cell has a resolution of 0.24 lb. The data is acquired at time increments controlled by an external trigger which is typically every 0.5 sec. A CAMAC crate houses the ADC boards which transfer data to a PC via an IEEE 488 board. The stored data is available for later recall and plotting.

The finite element computations are used to review the loading conditions the specimens are experiencing, to determine the area over which the uniform stress state is distributed, and to evaluate the magnitude of the failure stress.

A thick germanium specimen disk is installed in the fixture in Fig. 9. An annular ring which is integral within the cylindrical head supplies a force to the disk. The axial force present in the cylindrical head originates at the pneumatic cylinder at the left end of Fig. 9. Contained within the cylindrical head is a spherical socket into which the axial shaft is positioned. The ball-in-socket design of the cylindrical load head yields a self-centering loading fixture which compensates for misalignment and variation in disk thickness. Both the pressurized cylinder and ring-on-ring fixturing have incorporated materials which minimize friction. The piston in the pneumatic cylinder rides on teflon sliders. Vacuum grease is applied to the face of the load rings to minimize friction.

An alignment fixture is used to center the disk on the support ring. The alignment fixture is removed after the components are properly positioned. The test specimen is then unconstrained along the perimeter during fracture testing.

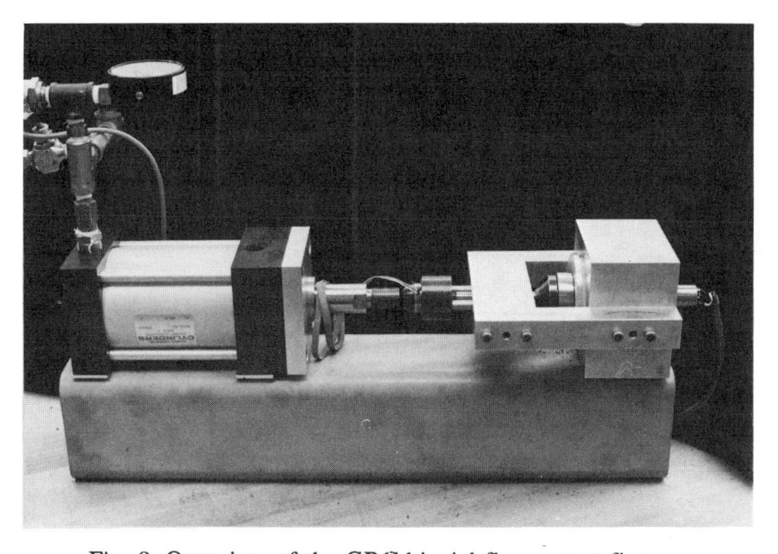

Fig. 9. Overview of the GRC biaxial flexure test fixture.

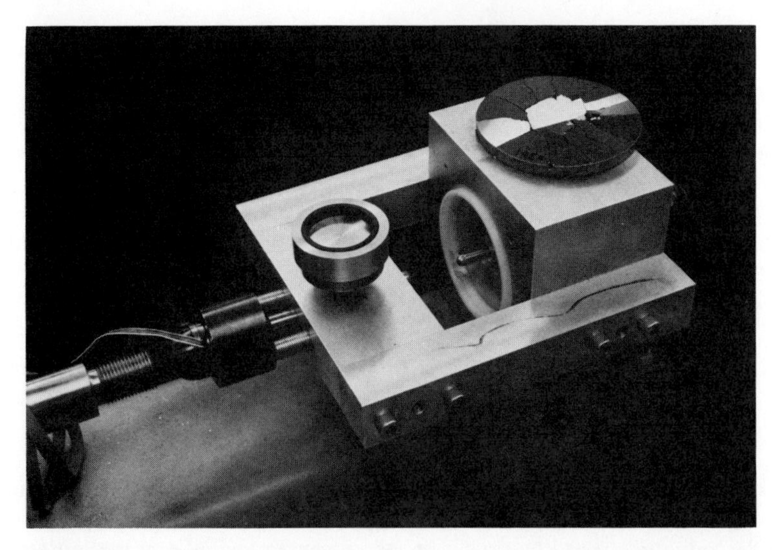

Fig. 10. Fractured germanium disk and load ring removed from the GRC biaxial test fixture showing the load and support rings and LVDT which contacts the tension face of the specimen.

Controlling the applied force to the specimen is substantially easier than controlling specimen deflection. The conventional testing machine does not have the resolution or accuracy of a small fraction of 0.001 inch (0.025 mm) which is the full scale failure condition for certain ceramic disks. A load of 10 to 3000 lb generated by the pneumatic cylinder is applied independent of disk deflection. The pressure in the pneumatic cylinder is ramped-up over a 2-to-4-minute period until the specimen fails. High accuracy is achieved in load application to the specimen by vernier control of the pressure source. The pressure, load, and deflection are continuously monitored through specimen failure.

Figure 10 shows the load and support rings when the fractured disk is removed from the biaxial flexure fixture. The outer support ring is white Delrin and it is press-fitted into the base. The inner load ring is black Delrin and it is press-fitted into the cylindrical load head. Two screws are used to limit the motion of the concentric rings after disk failure in order to avoid unnecessary specimen fracturing. The extended deflection probe of the LVDT is seen at the center of the support ring. This spring loaded probe measures centerline disk deflection.

BIAXIAL FLEXURE TEST RESULTS

The biaxial flexure test fixture with the flat Delrin load ring was used to evaluate the biaxial flexure strength of 30 0.788 inch diameter germanium disks and 30 3 inch diameter germanium disks. The germanium was polycrystalline and was polished to an 80/40 scratch to dig ratio. This is only a moderate level of surface quality. The finite element analysis provides an accurate determination of the failure stresses in each of these specimens.

The biaxial flexure test results are presented in terms of Weibull statistics in Fig. 11. All of the specimens failed within the load ring. The large diameter specimens failed at a moderately lower average stress, $\bar{\sigma}$, (27 percent less) than the small diameter specimens. From classical flaw statistics it is expected that the large specimens with a greater number of surface defects would statistically fail at a lower stress. The standard deviations, σ_n, are also listed in Fig. 11 for these tests.

For the present analysis the Weibull cumulative probability for failure is reduced to

$$P_f = 1\text{-}\exp(-\sigma/\sigma_o)^m \qquad (2)$$

The Weibull parameters σ_o and m are determined from the cumulative probability of failure as a function of the failure stress curves in Fig. 11. The Weibull fit to the test results is

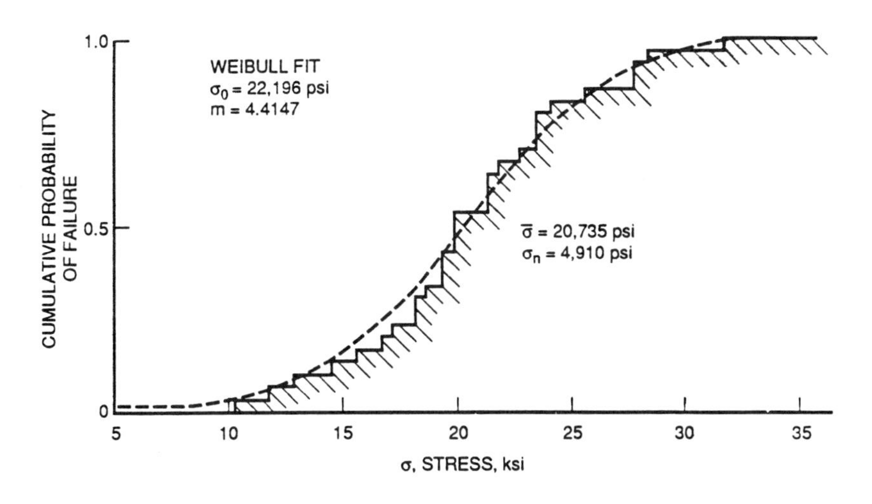

0.788 inch diameter disks

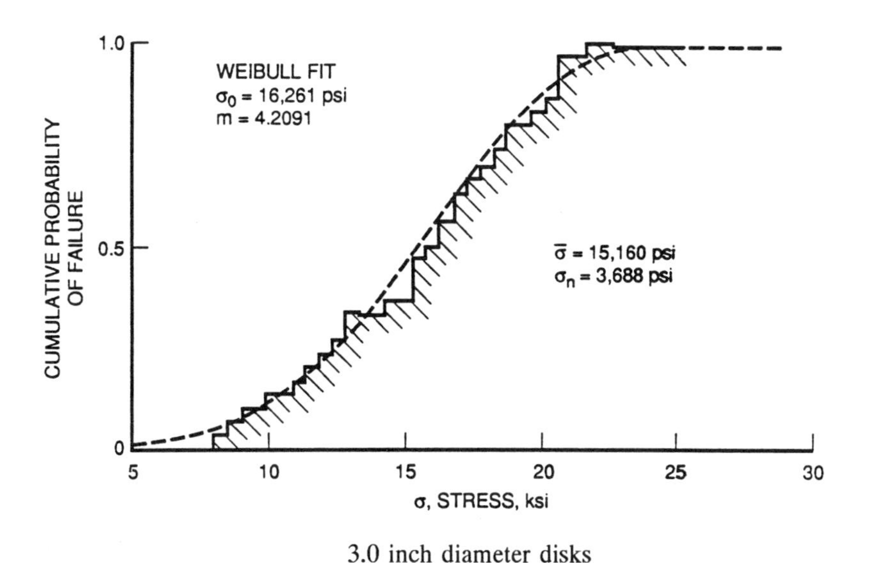

3.0 inch diameter disks

Fig. 11. Probability of failure curves for polycrystalline germanium specimens.

indicated by a dotted line. The values of σ_o for the large and small disks in Fig. 11 are in the same proportion as the average stresses: the value of σ_o for the large disks is 27 percent lower than the value for the small disks.

Statistical failures can be related to the dimensions of the test specimen. When the failure statistics are not dependent on the size of the specimen, the critical flaws present are independent of the sample size. The following relationships will demonstrate that the failure probabilities for the two sizes of germanium disks are almost independent of the specimen dimensions.

The failure stress is related to the size (length, area, volume) of the specimen by the relation (Batdorf and Heinisch[9]),

$$\sigma_1/\sigma_2 = (\text{dimension 2/dimension 1})^{1/m} \tag{3}$$

In the calculations which follow σ_1 is the failure stress for the small disk, σ_2 is the failure stress for the large disk, and m is the average of the values of m listed in Fig. 11. The ratio of the experimental failure stresses for the 0.788 inch and 3 inch diameter germanium disks is 22196/16261 = 1.366.

Specimen Dimensions	σ_1/σ_2
	1.366 (experimental)
thickness 2/thickness 1 = 3.09	1.299
diameter 2/diameter 1 = 3.54	1.341
area 2/area 1 = 12.54	1.798
volume 2/volume 1 = 38.75	2.335

This comparison indicates the flaw distribution is only slightly specimen size dependent. It is clearly seen that the area and volume are not the size factor which control the population of critical defects. A linear dimension appears to define the correlation with the statistically evaluated stresses: the most likely scaling parameters is the diameter of these disks. However it is somewhat difficult to provide a physical reason for this to be the case.

BIAXIAL FLEXURE TESTING - STRAIN MEASUREMENTS

Measurements of the effects of different load rings on ceramic disks were obtained by direct measurements of the strain at selected locations on a thin disk. These tests used a 2.75 by 0.108 inch thick zinc sulfide disk which was adequate in size to acquire good spatial resolution of the strains. This disk provides adequate room for bonding the strain gauges on the tension face of the specimen as seen in Fig. 12. One gauge is mounted at the disk's centerline. Gauges 2, 3, and 4 are centered directly under the load ring in the radial and circumferential directions. The active area of the selected Micro Measurements strain gauge, model GEA-06-125UN-120, is 0.030 by 0.030 inch (0.76 mm square). The Measurement Group Type 2120 instrumentation and signal conditioner were used with each strain gauge. The analog signals were routed to the CAMAC hardware for digital conversion and storage in a PC.

A statistical data base was acquired on disk strain for 35 test cases in which the angular orientation of the disk or the load ring with respect the test fixture were changed. Several disks were used in these tests. The large Delrin rings shown in Fig. 10 were used to load the zinc sulfide disks. The centerline stress varied from 5.0 to 6.0 ksi over the 35 cases. The circumferential position and orientation of the disks were slightly different for each case.

The statistics for these measurements indicated the radial stress ratio $\sigma_{rr}/\sigma(0) = 1.028 \pm 0.067$ and the circumferential stress ratio $\sigma_{\theta\theta}/\sigma(0) = 0.978 \pm 0.086$. The differences seen in these results are attributed to the test fixturing. There are variations in the flatness of the support and load rings. There is a small asymmetry of the loading around the load ring which is introduced at the ball/socket interface of the load shaft coming from the pneumatic cylinder. Anomalies of 0.1 mil in these components are significant enough to introduce the measured spatial errors.

Experiments were also performed to assess the spatial stress gradients under the load ring for the steel and Delrin rings. Instead of remounting numerous strain gauges, the entire disk with attached strain gauges was moved laterally relative to the fixed load and support rings. This lateral displacement allows a measurement of strains in the disk in close proximity to the load ring. This approach does introduce a non-symmetric configuration but it has only minor influence on the peak stress near the load ring. For small lateral displacements the data is representative of the case where many strain gauges are applied in a very small region.

Figure 13 shows the normalized radial tensile stress as a function of distance away from the contact point for both the Delrin and the steel load rings. The inserted sketch shows the dimensions of the test configuration and the direction associated with the lateral displacements. The nominal centerline stress for these cases was about 4.7 ksi (32.4 MPa). The steel ring had a local curvature of 0.125 inches and the rectangular Delrin ring had a width of 0.1 inch or 7.3 percent of the disk radius. The average ratios of σ_{rr}/σ_o and $\sigma_{\theta\theta}/\sigma_o$ obtained from 35 strain measurements as mentioned previously are shown in Fig. 13 for comparison with the tests summarized in Fig. 13.

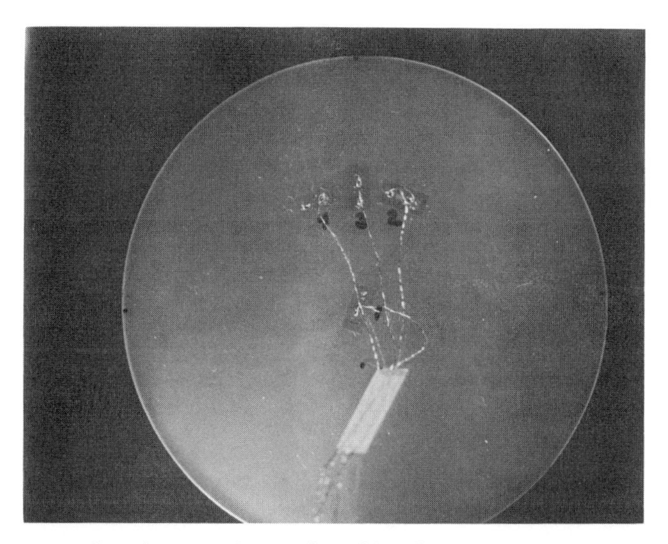

Fig. 12. Strain gauge locations on the tension side of the zinc sulfide disk. Gauge 1 is at the center and gauge 3 is at the load ring to measure radial strains. Gauges 2 and 4 are at the load ring and measure the circumferential strains.

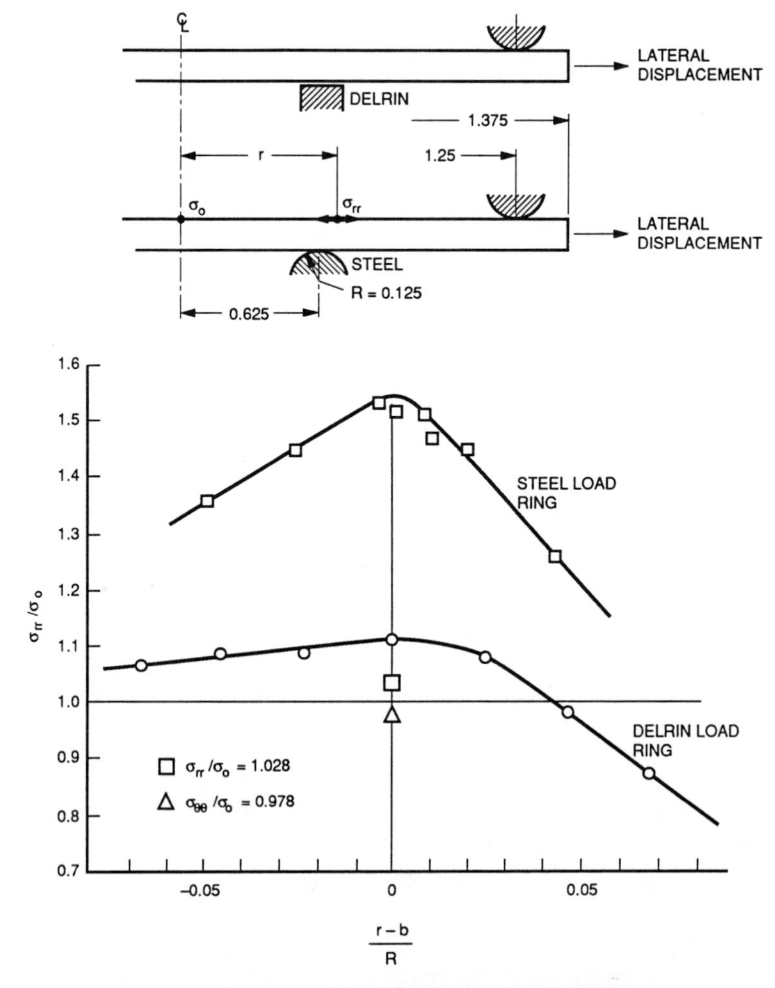

Fig. 13. Tension surface stresses for a 2.75 by 0.108 inch (7.0 by 0.27 cm) zinc sulfide disk loaded in biaxial flexure using curved steel and rectangular Delrin load rings.

The disk experiences a large amplitude tensile stress when using the steel load ring. The maximum failure stress can be 55 percent larger that the expected value when using the centerline stress as the reference state. Nearly 10 percent of the disk radius experiences a stress at least 30 percent larger than the centerline value. The Delrin ring imparts a much broader and uniform loading. The circumferential strains are not shown in Fig. 13 but they are always less than the radial strains over this lateral displacement range.

The primary purpose of the strain measurements was to compare the effectiveness of the Delrin over the steel load rings. A significant enhancement in performance is measured using the Delrin load rings which minimized the localized, large amplitude tension-side contact stresses. The secondary purpose was to quantitatively assess the relative magnitude of radial and circumferential stress variations from the centerline to the load ring. It was determined that the radial stress component is larger than the circumferential stress component at the position directly under the load ring where the maximum stresses occur.

CONCLUSIONS

The biaxial flexure test has been used by several investigators for ceramic materials, however there has been relatively little work evaluating the stress distributions in the test specimen which are responsible for its failure. A detailed finite element analysis was formulated for evaluating the stresses in a selection of specimen dimensions for the ring-on-ring biaxial flexure test. Significant radial stress amplification due to the contact stresses generated by conventional load rings can occur on the tension side of the specimen at the radius of the load ring. This result is exactly opposite to the available calculations of this effect carried out by Ritter, et al.[2], where the magnitude of the circumferential stress component at the load ring is amplified and there is no change in the radial stress component compared to the radial stress component at the center of the disk. The new finite element results were confirmed experimentally.

REFERENCES

1. J.E. Field, D.A. Gorham and D.G. Rickerby "High-Speed Liquid Jet and Drop Impact on Brittle Targets" *Erosion: Prevention and Useful Applications*, ASTM STP 664, W.F. Adler, ed, American Society for Testing and Materials, (1979) pp. 198-319.

2. R.J. Roark and W.C. Young, *Formulas for Stress and Strain*, 5th ed., McGraw-Hill New York (1975).

3. S. Timoshenko and S. Woinowsky-Krieger, *Theory of Plates and Shells*, 2nd ed., McGraw-Hill, New York (1959) p. 51.

4. F.F. Vitman and V.P. Pukh, "A Method for Determining the Strength of Sheet Glass", *Zavod Lab.*, 29:863 (1963).

5. R. Kao, N. Perrone, and W. Capps, "Large-Deflection Solution of the Coaxial-Ring-Circular-Glass-Plate Flexure Problem", *J. Am Ceramic Soc,* 54:566 (1971).

6. J.E. Ritter, K. Jaleus, A. Batakis and N. Bandyopadhyay, "Appraisal of Biaxial Strength Testing", *Int. Cong. on Glass, XII*, R.H. Doremus, W.C. Lacourse, J.D. Mackenzie, J.R. Varner and W.W. Wolf, ed., North Holland Publishing Co., New York (1980) pp. 419-24.

7. D.K. Shetty, A.R. Rosenfeld, P. McGuire, G.K. Bansal and W.H. Duckworth, "Biaxial Flexure Tests for Ceramics", *Ceramic Bulletin,* 59:1193 (1980).

8. W.H. Duckworth, A.R. Rosenfeld, S.H. Gulati, R.A. Rieger and K.E. Hoekstra, "Basic Principles of Lens Fracture Testing", *Am. J. Optometry and Physiol. Opt.,* 55:751 (1978).

9. S.B. Batdorf and H.L. Heinisch, Jr., Fracture Statistics of Brittle Materials with Surface Cracks, *University of California Report, UCLA-ENG-7703*, January 1977.

BENDING FRACTURE STRENGTH OF SINTERED SILICON NITRIDE DISKS WITH SHOULDER FILLET AT ROOM TEMPERATURE

Seijiro HAYASHI and Akihiko SUZUKI

Research Institute
Ishikawajima-Harima Heavy Industries Co., Ltd.
1-15 3-chome Toyosu, Koto-ku
Tokyo 135-91, Japan

ABSTRACT

Bending tests were performed on sintered silicon nitride disks with shoulder fillet at room temperature. Experimental fracture probability were compared with analytical predictions obtained by a structural reliability evaluation program, CCPRO. Results showed that; (1)the effective volume concept was applicable to biaxially loaded components, and (2)fracture criteria compared in this analysis gave little effect to fracture predictions.

INTRODUCTION

Ceramic materials exhibit a substantial dispersion in the fracture strength. This is attributed to microscopic flaws which inherently exist in the material. Microscopic flaws are random in size and orientation. As a consequence, the strength distribution of ceramic components in multiaxial stress states is a function of entire field solution of the stresses and cannot be determined by a maximum stress only.

Current design methodology is based on the Weibull theory and linear elastic fracture mechanics. In other words, it is required for assessing component reliability to use fracture strength distribution of the material and fracture criteria under multiaxial stress state.

Although many attempts were made toward precise reliability analysis, these were only verified with rather simple specimens, such as uniform pressure loading of disk(1), ball-on-ring or ring-on-ring test of disk(2), and smooth specimen under torsion and compressive load(2). Test results obtained from these specimens yielded some implications concerning the applicability of design methodologies and fracture criteria.

However, these implications lacks discussions on the influences of stress concentration to the reliability analysis.

Fracture Mechanics of Ceramics, Vol. 10
Edited by R.C. Bradt *et al.*, Plenum Press, New York, 1992

In this paper, bending fracture strength of pressureless sintered silicon nitride disks with shoulder fillet at room temperature is discussed. The experimental fracture probability is compared with predictions obtained by a structural reliability evaluation computer code.

Two objectives of this study are to assess the varidity of validity of reliability analysis method to multiaxially stressed specimens with stress concentration and to examine the influences of fracture criteria to the reliability analysis.

TEST MATERIAL AND SPECIMEN

Material

Material used in this study is pressureless sintered silicon nitride. Sintering aids are Y_2O_3, and ZrO_2. Table 1 shows the mechanical properties of the material provided by the manufacturer.

Table 1 Mechanical properties of the material.

Flexural Strength (4-point)	970 MPa
Young's Modulus	300 GPa
Fracture Toughness	6.4 MPa$\sqrt{m}$
Poisson's Ratio	0.27
Density	3.23 g/cc

Specimen

Specimen geometries are shown in Fig.1. Two series of disks, the first series with fillet radius of R=1mm and second with R=3mm, were employed for the experiment. Eight disks were tested at room temperature for each series.

All disks were machined in circumferential direction. Surface roughness of Rmax was 1.6S. After machining, all disks were tempered at 1100°C for 1 hour in the air. To ensure removal of the machining damage, toe of the fillet was polished circumferentially by SiC powder (average grain size was 0.001mm).

The four-point bend specimens, which geometry is based on the JIS (Japanese Industrial Standard) R1601, were machined from the same billet of the disks.

EXPERIMENTAL PROCEDURE

The fixture are shown in Fig.2. Out-of-plane bending load was applied via ball-on-ring jig which developed a biaxial stress state on the surface of the disk. Loading condition is shown schematically in Fig.1. Crosshead speed was 0.5mm/min. for all disks.

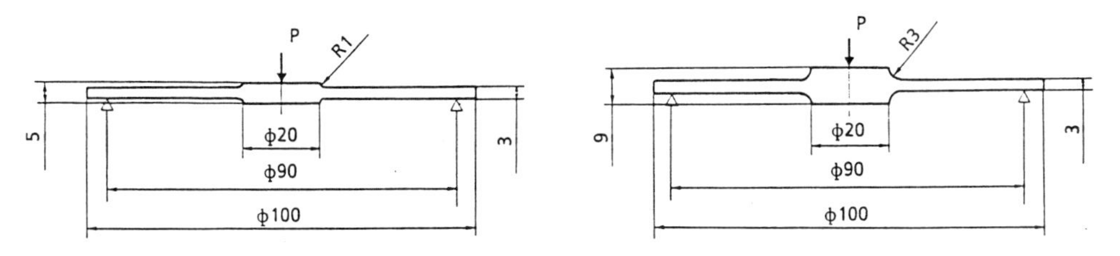

Fig. 1 Specimen geometries and loading condition.

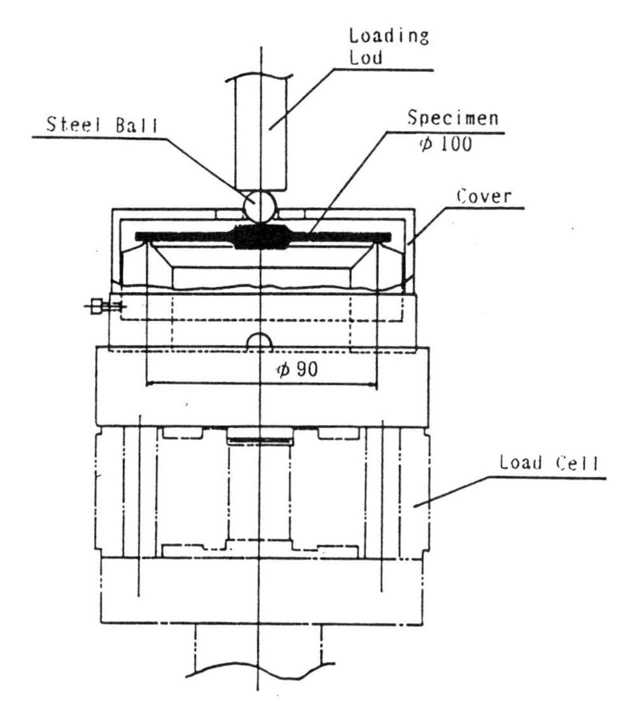

Fig. 2 Testing-fixture used in out-of-plane bending.

Since contact condition between the disk and the supporting ring affect test results, a pressure measuring film was used to confirm uniform loading between the disk and the supporting ring. After applying load of 49 N, a pressure measuring film revealed that no abnormally contact line was detected and the diameter of the supporting ring was 90mm.

Strain distribution around a toe of the fillet was measured by strain gages statically for each series of the disks to evaluate the boundary condition. The results were compared with the Finite Element (FE) analysis.

The inner and outer span of four-point bending test was 10mm and 30mm respectively (JIS R1601). A total of fourty specimens were fractured under four-point bending load. Twenty of those were tested with loading rate of 0.5mm/min. and the rest with 1.0mm/min.

RESULTS

Bending Strength Distribution of the Material

Figure 3 shows the strength distribution of four-point bending strength of the material. Average strengths and the Weibull moduli are summarized in Table 2. No effect of loading rate on the fracture strength and the distribution was obtained within the range tested in this study.

Fracture surfaces of all specimens were examined by Scanning Electron Microscope (SEM). Nineteen specimens fractured at the surface and 21 specimens from the inner pore. No preferential edge failure were observed.

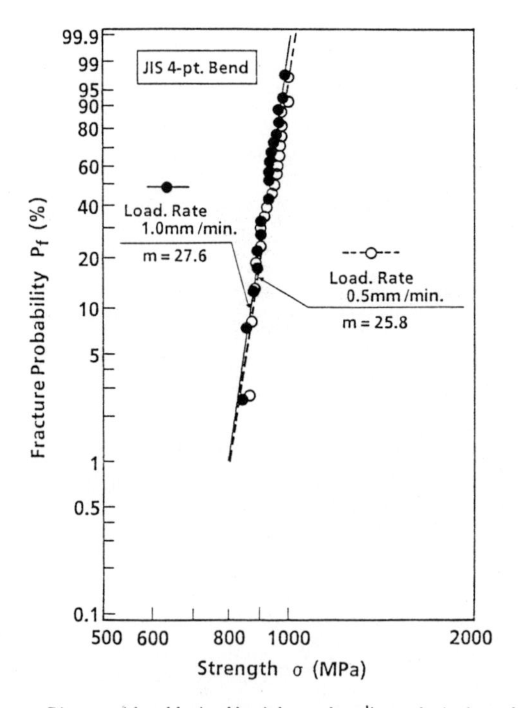

Fig. 3 Strength distribution in 4-point bending test.

Fracture Appearance of the Disks

Typical fracture patterns of the disks are presented in Fig.4. Fracture patterns were the same irrespective of the fillet radius.

All fracture initiated at the toe of the fillet. Crack propagated from the origin in both direction of left and right along the toe of the fillet. Multiple crack branching toward radial direction occured during the crack propagation along the toe. That resulted in fracture pattern shown in Fig.4.

Examination of fracture origin for all disks were conducted by SEM. The results are presented in Table 3.

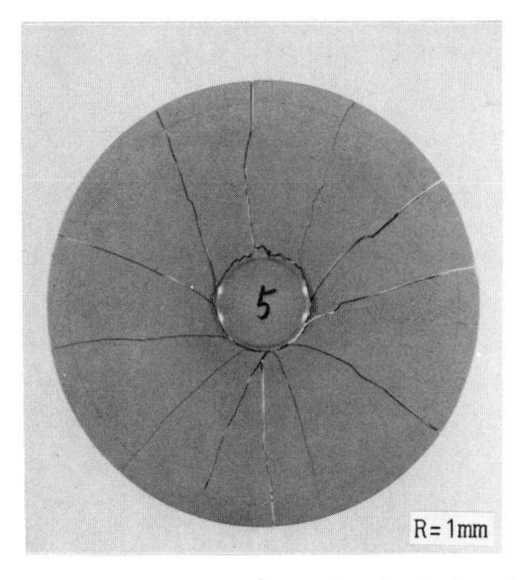

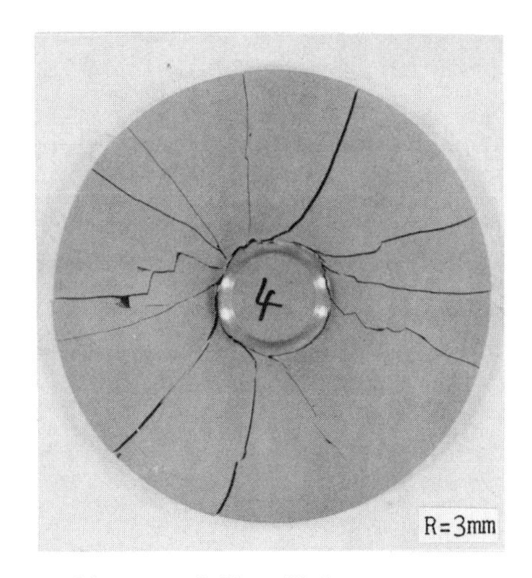

Fig. 4 Typical fracture patterns of the disks.

Table 2 Average strength and the Weibull moduli measured in different loading rate.

Load. Rate (mm/min.)	Average Strength (MPa)	Weibull Modulus	Sample Size
0.5	938	25.8	20
1.0	925	27.6	20

Table 3 Fracture origin.

<table>
<tr><th rowspan="2">Test Temperature</th><th rowspan="2">Fillet Radius</th><th colspan="2">Fracture Origin</th></tr>
<tr><th>Surface</th><th>Inner Pore</th></tr>
<tr><td rowspan="2">Room Temp.</td><td>1mm</td><td>3</td><td>5</td></tr>
<tr><td>3mm</td><td>4</td><td>4</td></tr>
</table>

Stress Analysis

Stress distribution of the disks were analyzed by the Finite Element Method. Stress distribution obtained by FEM gave good agreement with the experimental data as shown in Fig.5.

Figure 5 indicates that the boundary condition of the disk was the same as that of FEM model. Hence the tests were conducted correctly.

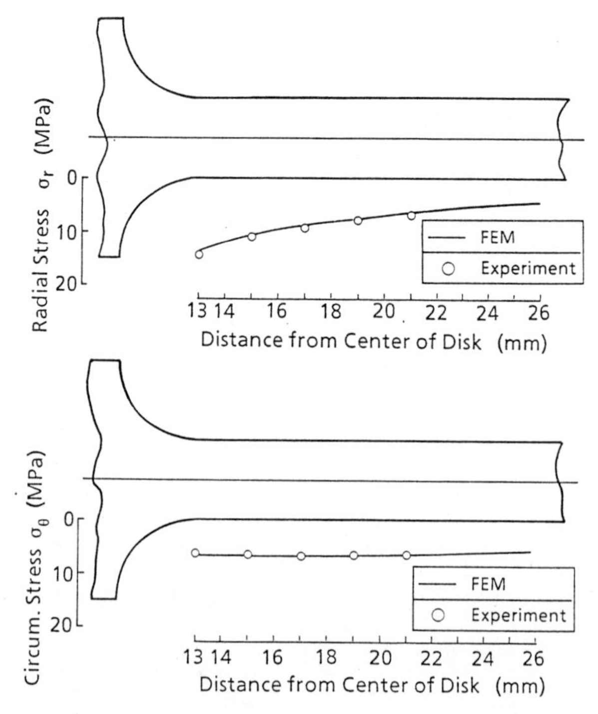

Fig. 5 Stress distribution measured around shoulder fillet compared to elastic prediction.

Fracture Probability Analysis

Program used in this study. Fracture probability analysis of the disks was predicted by using a computer program CCPRO (Ceramic Creep PROgram). CCPRO is a postprocessor of MSC/NASTRAN which can evaluate the reliability of ceramic components under various loading using stress analysis results from MSC/NASTRAN.

CCPRO is also capable to assess the effect of stress redistribution caused by creep on the reliability of ceramic components.

In the present study, only the elastic fast fracture module of the program was used.

CCPRO is based on the following assumptions.

1) The number of flaws present in the material is large.

2) The strength distribution which is considered to be the reflection of the flaw size distribution can be approximated by two-parameter Weibull distribution.
3) Penny-shaped cracks are assumed to be existing in the material.
4) There is no flaw interaction.
5) The flaws are uniformly distributed in both position and orientation.
6) A crack will propagate when a specific stress exceeds a critical value which is determined by fracture criteria.

Based on the assumptions described above, fracture probability under multiaxial stress state was obtained by

$$P_f(\sigma_N)=1-exp\left\{-\frac{1}{\Omega}\left(\frac{1}{m}!\right)^m\left(\frac{\sigma_N}{\overline{\sigma}_{ref}}\right)^m\left(\frac{1}{V_{ref}}\right)\right.$$

$$\left.\frac{\pi}{2}\int_V\int_0^{\pi/2}\int_0^{\pi/2}\left(\frac{Z}{\sigma_N}\right)^m y(\sigma_R\cdot 0)sin\phi d\phi d\theta dvol\right\} \tag{1}$$

where Ω : A factor adjust uniaxial fracture probability evaluated by any fracture criteria to tensile fracture probability of the material

m : Weibull modulus

σ_N : Characteristic strength

σ_{ref} : Average tensile strength of the material

V_{ref} : Volume of tensile specimen

Z : Equivalent normal stress which depends on fracture criteria (see Eq.2~Eq.6)

$y(\sigma_R\cdot 0)$: Heavyside step function

Fracture criteria compared in this program and corresponding equivalent normal stresses are described below.

a) Strain energy release rate criterion (G criterion)[3]

$$Z=\sqrt{\sigma_R^2+\left(\frac{\tau}{1-0.5\nu}\right)^2} \tag{2}$$

b) Modified G criterion[4]

$$Z=\sqrt{\sigma_R^2+\left[\beta\frac{\tau}{1-0.5\nu}\right]^2} \tag{3}$$

β: Shear sensitivity parameter

c) Maximum stress criterion (σ_θ criterion)[5]

$$Z=\sigma_R cos\frac{\theta_m}{2}\left[cos^2\frac{\theta_m}{2}-\frac{3}{2}\left(\frac{2}{2-\nu}-\alpha\right)sin\theta_m\right] \tag{4}$$

$$\theta_m = 2\arctan\left[\frac{1}{\alpha}\left\{\frac{1-\sqrt{1+8\alpha^2}}{4}\right\}\right]$$

$$\alpha = \frac{\tau}{\sigma_R}$$

d) Non-coplanar G criterion[6]

$$Z = \left[\sigma_R^4 + \frac{24}{(2-v)^2}\sigma_R^2\tau^2 + \frac{16}{(2-v)^4}\tau^4\right]^{\frac{1}{4}} \qquad (5)$$

e) Shetty's mixed mode equation[7]

$$Z = \frac{1}{2}\left[\sigma_R + \sqrt{\sigma_R^2 + \left\{\frac{4\tau}{C(2-v)}\right\}^2}\right] \qquad (6)$$

C: Shear sensitivity parameter

Here σ_R and τ are given by the following equations. The definition of crack orientation is shown in Fig.6.

$$\sigma_R = \sigma_1 \ell_1^2 + \sigma_2 \ell_2^2 + \sigma_3 \ell_3^2 \qquad (7)$$

$$\tau^2 = \sigma_1^2 \ell_1^2 + \sigma_2^2 \ell_2^2 + \sigma_3^2 \ell_3^2 - (\sigma_1 \ell_1^2 + \sigma_2 \ell_2^2 + \sigma_3 \ell_3^2)^2 \qquad (8)$$

$$\ell_1 = \sin\phi \cos\theta$$

$$\ell_2 = \sin\phi \sin\theta$$

$$\ell_3 = \cos\theta$$

Weibull modulus of the material. Fracture probability analysis requires the Weibull modulus, the strength distribution of the material. The Weibull modulus has to be obtained not only from fracture data of

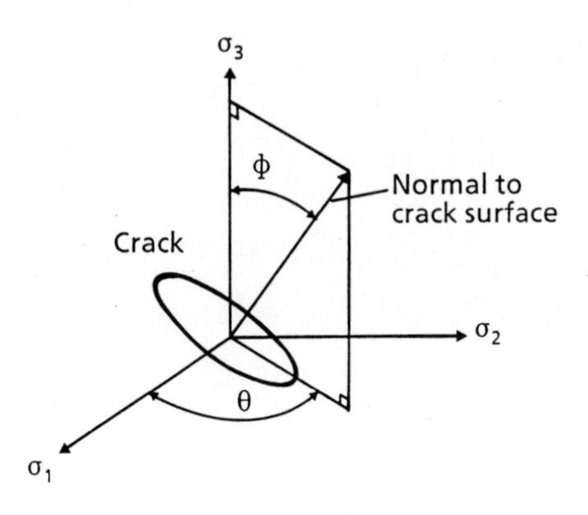

Fig. 6 The definition of crack orientation.

specimens of fixed geometry but also from data of specimens of various geometry under various loading including multiaxial stress state fracture data.

Matsui[8] obtained the relationship between average strength and effective volume of the material as shown in Fig.7.

They measured 4-point bending strength, tensile strength, and torsional strength and found that the results gave good agreement with the Weibull theory, Eq.(9).

$$\frac{\sigma_a}{\sigma_b}=\left(\frac{V_{e,a}}{V_{e,b}}\right)^{-\frac{1}{m}} \tag{9}$$

From the slope of the line in Fig.7, it was found that the Weibull modulus(m) of the material is 15. Since m=15 was obtained from many fracture data, it is more reliable than a value from 4-point bending fracture data.

All failure probability of the disks was done by using m=15 in this study.

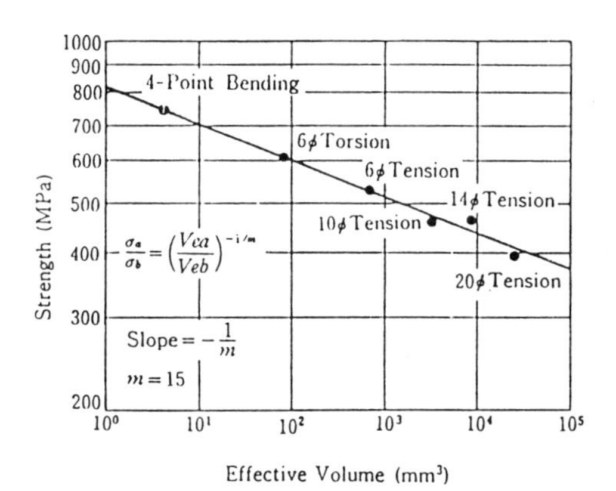

Fig. 7 Relationship between the average strength and the effective volume of the material.

Fracture probability of disks. Fracture probability for each disk in a test series was defined by

$$P_f=\frac{i-0.5}{N} \tag{10}$$

where i is the rank of a specimen in increasing order of fracture strength, and N is the sample size.

The strength distribution was expressed in terms of two-parameter Weibull distribution function, Eq.(11).

$$P_f=1-exp\left\{-\left(\frac{\sigma}{\sigma_0}\right)^m\right\} \tag{11}$$

The Weibull modulus m and the characteristic strength σ_0 were determined by the least square method[9]. The following regression curves were obtained for each series.

$$P_f = 1 - exp\left\{-\left(\frac{\sigma}{98\,0}\right)^{15.8}\right\} \qquad ;R=1mm \tag{12}$$

$$P_f = 1 - exp\left\{-\left(\frac{\sigma}{89\,0}\right)^{21.1}\right\} \qquad ;R=3mm \tag{13}$$

Figure 8 shows comparison of experimental fracture probabilities with predicted fracture probabilities for disks with shoulder fillet radius of 1mm. All fracture predictions by various fracture criteria yield good agreement with experimental data. Experimental results for disks with shoulder fillet radius of 3mm are plotted in Fig.9 along with various predictions. Although failure predictions are slightly conservative, the agreement between failure predictions and experiments is good.

It is interesting to note that the difference of fracture criteria used in this study give little effect to the predictions for both cases. This is because of the factor Ω in Eq.(1). The factor Ω is determined so as to fit uniaxial fracture probability evaluated by any fracture criteria with the tensile fracture probability of the material. Consequently, as far as the present model is applied to the reliability analysis, calculated fracture reliability is not affected by the fracture criteria largely in the case where stress state in the component is not so different from uniaxial stress state in the stress space.

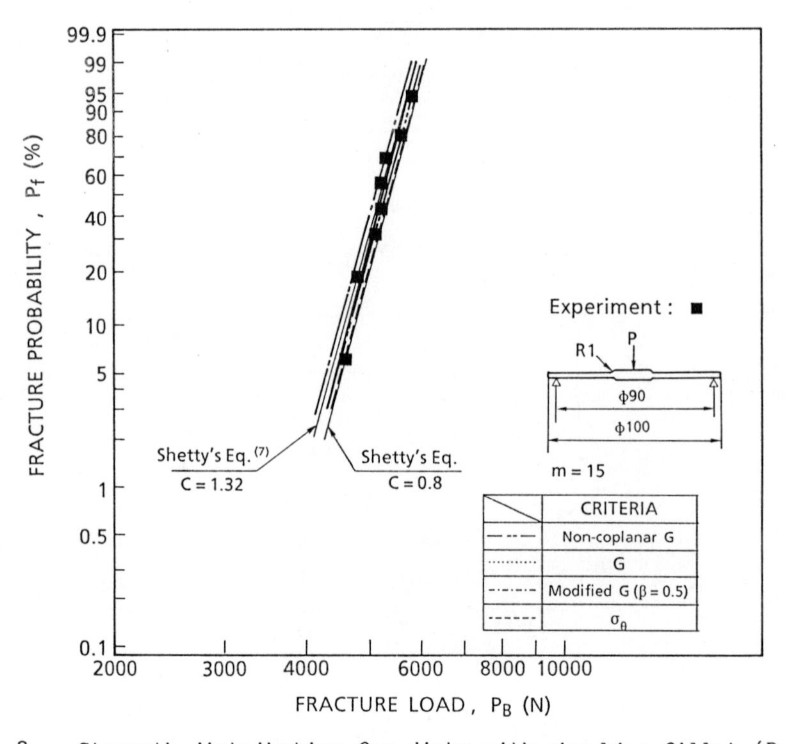

Fig. 8 Strength distribution for disks with shoulder fillet (R=1mm)

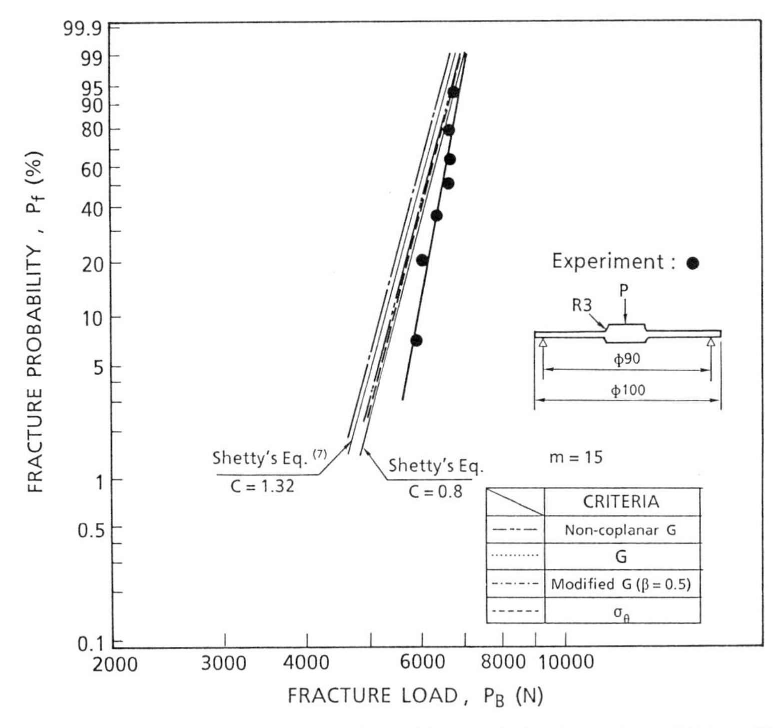

Fig. 9 Strength distribution for disks with shoulder fillet (R=3mm)

Relation between average strength and effective volume. The effective volume of a component under multiaxial stress state is given by the following equation.

$$V_e = \frac{2}{\pi}\frac{1}{\Omega}\int_V \int_0^{\pi/2}\int_0^{\pi/2}\left(\frac{Z}{\sigma_N}\right)^m y(\sigma_R \cdot 0) sin\phi d\phi d\theta dvol \tag{14}$$

Figure 10 is shown the relationship between the average strength and the effective volume of the disks with shoulder fillet. The vertical axis is the average strength. The average strength expressed in radial stress at a toe of the fillet is plotted for each series.

The equivalent stress, Z, depends on fracture criteria and crack shape. Hence, it is clear from Eq.(14) that the effective volume is a function of fracture criteria. Since little effect of fracture criteria to reliability analysis was observed, the effective volume, the horizontal axis of Fig.10, was calculated by the maximum stress criterion.

The dotted line in Fig.10 represents the prediction of Eq.(9). For the calculation of Eq.(9), the Weibull modulus m=15 and the average strength of 4-point bending specimens σ=940 MPa were used. It is clear in Fig.9 that the measured average strength are in close agreement with the prediction of Eq.(9). Fig.10 confirms the validity of the Weibull theory for treating the fracture reliability of components with stress concentration under biaxial stress state.

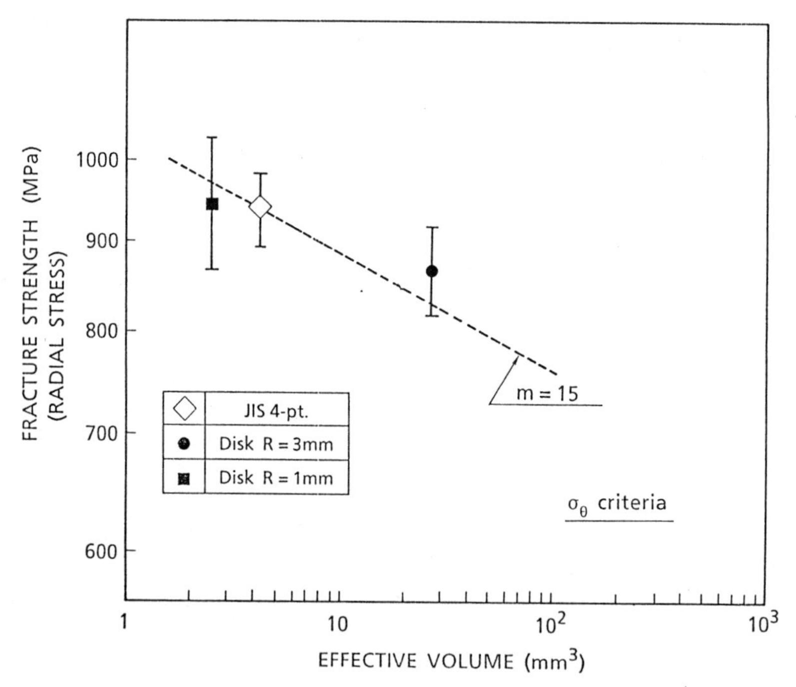

Fig. 10 Relation between fracture strength vs effective volume.

CONCLUSIONS

(1) Fracture probability of pressureless sintered silicon nitride disks with shoulder fillet subjected to biaxial flexure loading is accurately predicted by fracture mechanics based statistical theory.

(2) Fracture criteria used in this study give little effect to fracture reliability predictions.

ACKNOWLEDGEMENT

This work was performed under the management of Engineering Research Association for High Performance Ceramics as a part of The Fine Ceramics R & D Project for Future Industries supported by New Energy and Industrial Technology Development Organization.

REFERENCES

(1) D. K. Shetty, et al., "A Biaxial-Flexure Test for Evaluating Ceramic Strength", J. Am. Ceram. Soc., 66[1] 36-42(1983).

(2) Y. Nakasuji, et al., "Fracture Behavior of Non-Oxide Ceramics under Biaxial Stresses", in this volume.

(3) G. R. Irwin, "Fracture", Encycropedia of Physics, Chapter VI, Springer Verlag, Berlin (1958).

(4) Ref. 2.

(5) F. Erdogan and G. C. Sih, "On the Crack Extension in Plates Under Plane Loading and Transverse Shear", J. Basic Eng., Trans. ASME, 519-527(1963).

(6) T. K. Hellen and W. S. Blackburn, "The Calculation of Stress Intensity Factors for Combined Tensile and Shear Loading", Int. J. Fract., 11[4] 605-617(1975).

(7) D. Singh and D. K. Shetty, "Fracture toughness of polycrystalline Ceramics in Combined Mode I and Mode II Loading", J. Am. Ceram. Soc., 72[1] 78-84(1989).
(8) M. Matsui et al., "Tensile Strength of Sintered Silicon Nitride", Proc. Int. Conf. Non-Oxide Technical and Eng. Ceramics, p.361-374(1985).
(9) K. Trustrum and A. De S. Jayatilaka, "On Estimating the Weibull Modulus for a Brittle Material", J. Mater. Sci., 14[5] 1080-1084(1979).

EVIDENCE OF PLASTIC DEFORMATION ON SLOW CRACK GROWTH OF BORATE GLASS

Jun Matsuoka*, Kazuyuki Hirao and Naohiro Soga

Department of Industrial Chemistry
Faculty of Engineering, Kyoto University
Sakyo-ku, Kyoto 606, Japan

* Present Address : Department of Chemistry for Materials, Faculty of Engineering, Mie University
Tsu, Mie 514, Japan

INTRODUCTION

Slow crack growth of oxide glasses has widely been studied in various liquid environments, and this environmental effect on the K_1-v relations was discussed in terms of the liquid properties such as hydroxyl ion concentration,[1] cation species in aqueous solutions,[2] molecular structure,[3] molecular size,[4] water activity, viscosity, dielectric constant,[5] and testing temperature.[6,7] But most of these studies were limited to silica and silicate glasses, besides silica free oxide glasses.[8] Moreover, the silica containing glasses have been treated as brittle materials in which linear fracture mechanics is applicable. Although this concept that the glasses are brittle material is widely believed nowadays, the certification has been done only for soda-lime-silicate glass.[9]

Recently, we found that the large inelastic deformations were present on the mode-I fracture of sodium borate glasses at room temperature,[10] although the atmosphere was inert and fractured surfaces were optically flat. In this study, we show the fact that this large inelasticity is not due to the viscous flow but caused by the plastic deformation, that is, the magnitude of inelastic deformation energy is only a function of applied force field, but not a function of loading time.

CRITERION FOR NONLINEAR FRACTURE ANALYSIS

Energy release rate has commonly been used in linear fracture mechanics. In order to analyze the quasi-elastic nonlinear fracture phenomenon, we modified this concept as follows. Consider an inelastic specimen with a sharp crack, whose load-displacement curve is schematically illustrated in Fig.1. Here, P is load, U_t is total displacement, U_p is residual inelastic displacement, and $U_t - U_p$ is considered as elastic displacement U_e.

Energy release rate J_1 is used to estimate the linear part of fracture energy. J_1 is defined as the release of the elastically stored energy per crack extension of unit area,

$$J_1 = \frac{P^2}{2}\frac{d}{dA}\left(\frac{U_e}{P}\right)$$

$$= \frac{P^2}{2}\frac{d}{dA}\left(\frac{U_t - U_p}{P}\right) \qquad (1)$$

where A is the crack area.

The inelastic energy dissipation rate I_1 is defined as the work which is used to increase the inelastic displacement per unit area extension of the crack. Then, I_1 is expressed as

$$I_1 = \frac{P^2}{2}\frac{d}{dA}\left(\frac{U_p}{P}\right) \qquad (2)$$

J_1 and I_1 defined above are equal to the limit values of G_c and ϕ_p defined by Sakai et al[11] when the increase of crack area, ΔA, in their paper is extrapolated to be infinitely small. Total work used at the crack propagation of unit area, which is called crack growth resistance, R, is expressed by Eq.(1) and Eq.(2) as

$$R = J_1 + I_1 \qquad (3)$$

Stress field around the crack tip is proportional to the apparent stress intensity factor K'_1 defined as

$$K'_1 = (R \cdot E')^{1/2} \qquad (4)$$

where $E' = E/(1-\mu^2)$, and E and μ are the Young's modulus and the Poisson's ratio, respectively. Here, we assume that the magnitude of inelastic deformation is a function of K'_1. Although stress field exactly near the crack tip is a function of $(J_1 \cdot E')^{1/2}$ and the value is lower than that of K'_1 because of the stress relaxation around the crack tip, K'_1 was used for the analysis of the relationship of stress field versus crack velocity to make the analysis more convenient in comparing with other glasses.

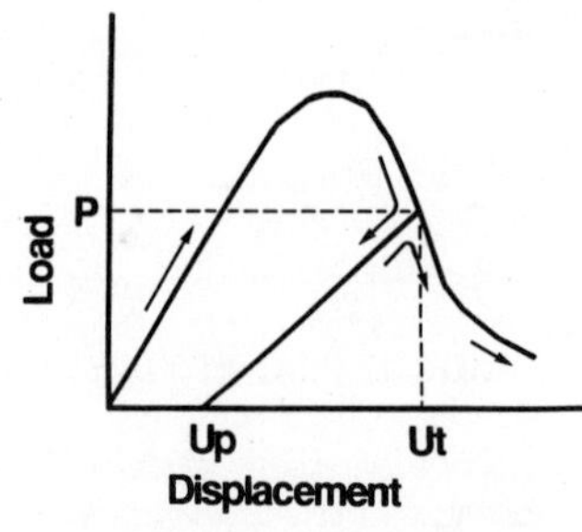

Fig.1 Load vs displacement on the fracture of inelastic specimen.

EXPERIMENTAL PROCEDURE

Glass sample of $20Na_2O{\cdot}80B_2O_3$ was prepared from reagent grade B_2O_3 and Na_2CO_3. About 100 g batches were melted at 1100 °C in a platinum crucible, poured into a steel mold of 3.5 x 4.5 x 1.5 cm, annealed immediately at 500 °C for 30 min, and then cooled slowly to room temperature in furnace. These glass blocks were machined with a 0.6 mm width diamond wheel saw to make the chevron notched short bar specimens shown in Fig.2(A), where $B=9.8\pm0.3$ mm, $W=27.0\pm0.2$ mm, $2H=40.00\pm0.03$ mm, $Sw=3.0\pm0.2$ mm, $Sh=10.05\pm0.02$ mm, and $\theta=60\pm1°$.

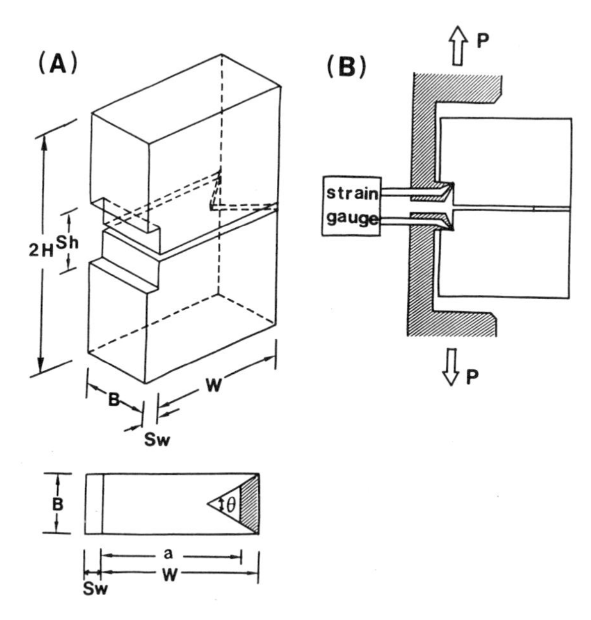

Fig.2 Short bar specimen used in this study (A), and the view of the sample mounted to testing machine jigs (B).

The specimen was set into a Instron testing machine by holding the slot of the specimen on loading knife edges as shown in Fig.2(B), and then loading-unloading procedures were performed. The crosshead speed of the machine was set to 0.001 mm/min for usual loading, and changed itself to 0.1 mm/min for unloading to zero load and for reloading to the former displacement.

Experiment was performed at room temperature (18 - 20 °C) with immersing the specimen in three different liquids such as kerosene, tetrahydrofuran (THF) and ethanol for the purpose of changing the K_1-v relations. Reagent grade kerosene dehydrated by sodium metal was employed as inert liquid against glass specimens. Reagent grade THF and 99.5 % ethanol containing water as impurity were used for corrosive liquid. A commercial soda-lime-silicate glass with the same size was also tested in kerosene to compare with the borate glasses.

Load point displacements were measured by a displacement gauge which was connected with the specimen directly by pressing with small springs. Inelastic displacements were measured more than ten times for one experiment by unloading to zero load, and the relation obtained between the total displacement and the inelastic displacement was interpolated to a smooth curve for analysis. The difference between the load point displacement and the corresponding inelastic displacement was considered as the elastic displacement.

The crack length was estimated by compliance method.[12] The crack length was determined by comparing the observed specimen compliance (elastic displacement per applied load) to the calibration curve which relates to the crack length and the dimensionless compliance. The calibration curve was obtained by measuring compliances of soda-lime-silicate glass specimens with several lengths of simulated cracks introduced by a 0.25 mm width diamond wheel saw. The calibration curve used in this study was

$$C' = 28.0 \ (1 - a^{*})^{-1.88} \ \{ 1 + 100 \ (1 - a^{*})^{3.5} \} \qquad (5)$$

$$C' = C \cdot B \cdot E \ / \ (1 - \mu^{2}) \qquad (6)$$

where C is the compliance, C' the dimensionless compliance, a^{*} the dimensionless crack length of a/W, B the specimen broadness. E and μ are Young's modulus and Poisson's ratio, respectively. The maximum error of this expression is evaluated to be less than 5 % in the case of 9.5 mm $< B <$ 10.1 mm and $a^{*} < 0.96$.

RESULTS

Fig.3 (a) and (b) show the relationship between P and U_t for borate glass of $20Na_2O \cdot 80B_2O_3$ and commercial soda-lime-silicate glass fractured in kerosene, respectively. In Fig.3 (a) and (b), the relationship between U_t and U_p is also shown. The larger inelastic displacement is shown for borate glass. On the other hand, the inelastic displacement of soda-lime-silicate glass was negligibly small. The inelastic displacement U_p of borate glass was nearly zero at first, but it increased rapidly when crack propagation occurred, and reached more than 25% of the total displacement.

Fig.4 shows the relation between the apparent stress intensity factor, K'_1, and the crack propagation velocity, v, obtained in this work. K'_1-v relations of borate glass fractured in both ethanol and in THF were shifted to the lower K'_1 side from that in kerosene. This shift is probably due to the stress corrosion caused by a small amount of water in both ethanol and THF. Crack growth resistance of both borate and soda-lime-silicate glasses in kerosene were nearly the same of about 7 J/m^2.

Fig.5 shows the magnitude of inelastic energy dissipation as a function of K'_1. It is clear that almost thirty percent of fracture energy was used to produce the inelastic deformation in borate glasses, which is much larger than that in soda-lime-silicate glass. The ratio of inelastic energy to total fracture energy also increased with an increase of stress intensity factor.

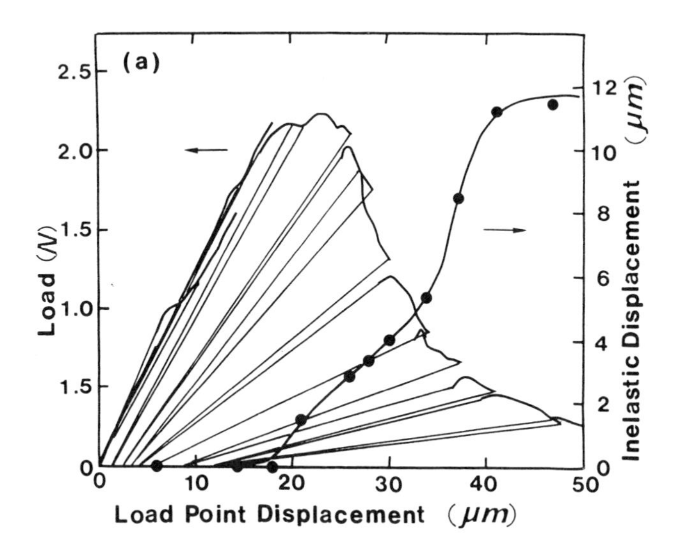

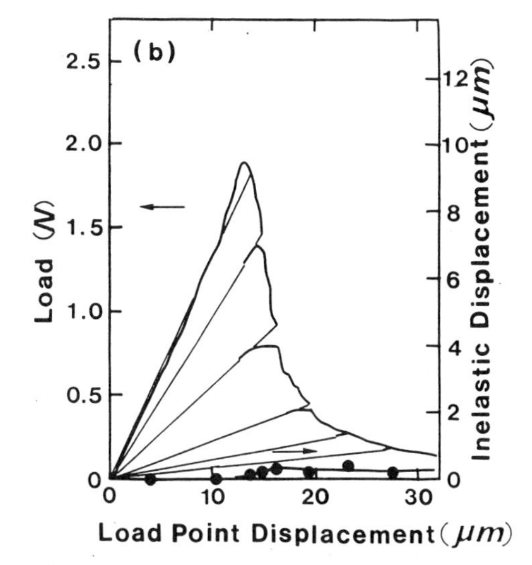

Fig.3 The relation between load and load point displacement, and the relation between inelastic displacement and load point displacement of $20Na_2O{\cdot}80B_2O_3$ glass (a), and commercial soda lime silicate glass (b) fractured in kerosene.

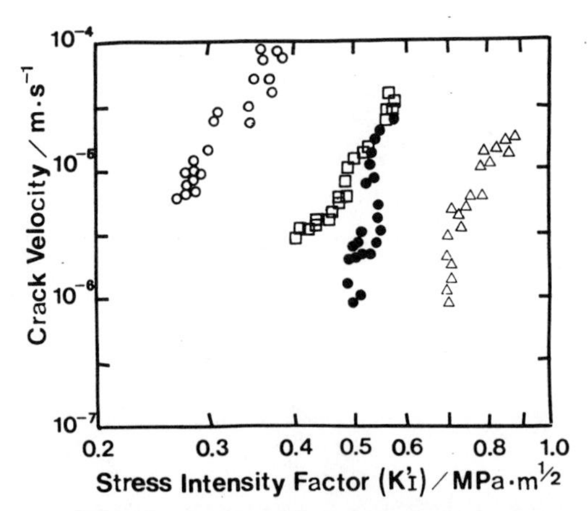

Fig.4 Relation between stress intensity factor and crack propagation velocity of $20Na_2O{\cdot}80B_2O_3$ glass fractured in kerosene (●), tetrahydrofurane (□) and ethanol (○), and that of commercial soda-lime-silicate glass fractured in kerosene (△).

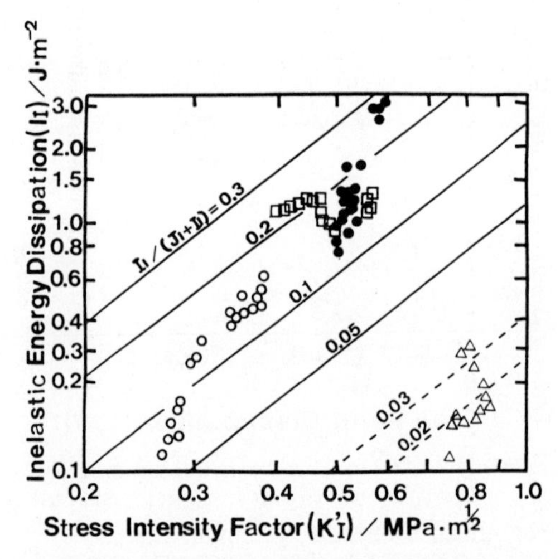

Fig.5 Relation between stress intensity factor and inelastic energy dissipation at slow crack growth. Symbols are equal to those in Fig.4.

DISCUSSION

Inelastic deformation at slow crack growth occurred only in borate glass, but not in soda-lime-silicate glass. The origin of this inelastic behavior is considered to be different from that of visco-elastic fracture occurred above or near the glass transition temperature, because the test was done at room temperature, which is much lower than the transition temperatures for the borate glass (450 °C) and soda-lime-silicate glass (540 °C). Moreover, borate and silicate glasses examined in this study have so different magnitude of inelasticity though the difference of the glass transition temperature of two glasses is not so large.

This inelastic deformation occurred not only in corrosive or water containing atmospheres, but also in inert atmosphere. This means that the inelasticity is not responsible for the softening of glass caused by the diffusion of water molecules into the glass structure.

In order to estimate whether the nature of this inelasticity is viscous or plastic, in another words the nature is time dependent or time independent, the loading time dependence of inelastic energy dissipation is investigated next. Fig.6 shows the dependence of inelastic energy dissipation on the stress intensity factor. Although crack growth velocity changes more than ten times in different atmospheres, the relation of K'_1 versus I_1 lies on one curve. Therefore, the inelastic energy dissipation is independent upon the crack growth velocity, which indicates that the magnitude of inelastic energy dissipation is independent upon the loading time at the crack tip. It is, therefore, concluded that the inelastic behavior of borate glass at slow crack growth is due to the plastic deformation, but not the viscous flow.

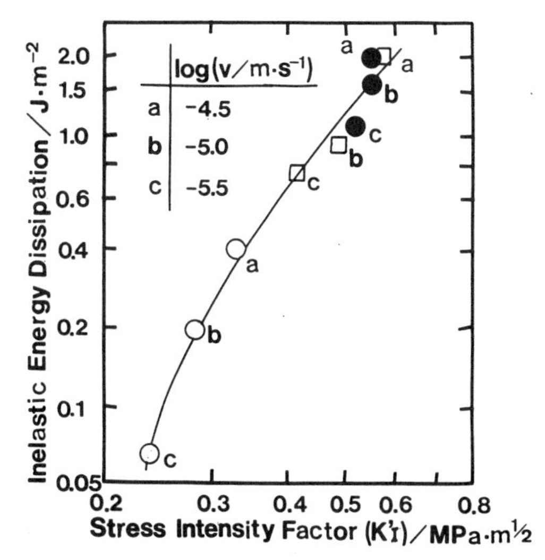

Fig.6 Dependence of inelastic energy dissipation on stress intensity factor of $20Na_2O{\cdot}80B_2O_3$ glass with some crack propagation velocities. Symbols are equal to those in Fig.4. The point at K'_1 = 0.23 MPa $m^{1/2}$ is extrapolated from the data.

By fitting the power law to the relationship of I_1 versus K'_1, or $I_1 = A\ {K'_1}^n$, the value of $n = 3$ was obtained.

According to this study and our previous work in Ref. 10, the plastic behavior occurred uniquely in borate glasses. Since there exists no inhomogeneity such as phase separation in this borate glass, it is probable that the nature of inelasticity is related to the atomic scale structure of borate glass. According to the structure for borate glass,[13] both three-fold coordinated and four-fold coordinated borons exist in glass network structure. This peculiar structure may cause the plastic deformation with a mechanism described below, which is an analogy of plastic deformation of crystals caused by the lattice defect migration under stress. Namely, the plastic deformation in borate glass is likely caused by the exchange of four-fold and three-fold coordinated boron sites under shear stress, which is schematically illustrated in Fig.7. One boron atom changes its coordination number from four to three by shear stress, and cooperatively, another boron atom changes its coordination number from three to four. In this mechanism, the plastic deformation can occur even without changing the potential energy from the initial state to the final state, because both of states have one four-fold boron and two three-fold borons in a deformation unit and no boron is connected with nonbridging oxygens. The transition state of this deformation mechanism is constructed by one three-fold boron connected with one nonbridging oxygen and two three-fold borons with no nonbridging oxygens. It is known that three-fold boron with one nonbridging oxygen exist stable as a structure unit of high alkali containing borate glasses, then the energy to create this structure unit might not be so large. This is why the activation energy of this deformation mechanism in the borate glass is small enough to cause the larger plastic deformation.

In silicate glasses, since only four-fold coordinated silicons are exist, the plastic deformation mechanism with low activation energy discussed above is not possible. Plastic deformation may only be possible by exchanging nonbridging sites to bridging sites in the highly modifier ion containing silicate glasses. Even in this case, since the transition state must contain a three-fold coordinated silicon or a five-fold coordinated silicon in a deformation unit, then the activation energy of this deformation mechanism must be very high. Therefore, the plastic deformation through this mechanism will be negligibly small in silicate glasses.

The plastic deformation occurred only in borate glass but not in silicate glass in this study. From the viewpoint of coordination exchange mechanism of plastic deformation proposed above, it is suggested that the glasses where the network-forming cations take two kinds of coordination states, such as borate, germanate and fluorozirconate glasses, have the potential to show the large plastic deformation at slow crack growth.

CONCLUSIONS

Large inelastic deformation was observed at the slow crack growth of sodium borate glass. The magnitude of inelastic energy dissipation increased rapidly with an increase of stress intensity factor, but it does not depend on crack growth velocity. This plastic deformation may be caused by the exchange of coordination state of three-fold and four-fold coordinated borons under shear stress.

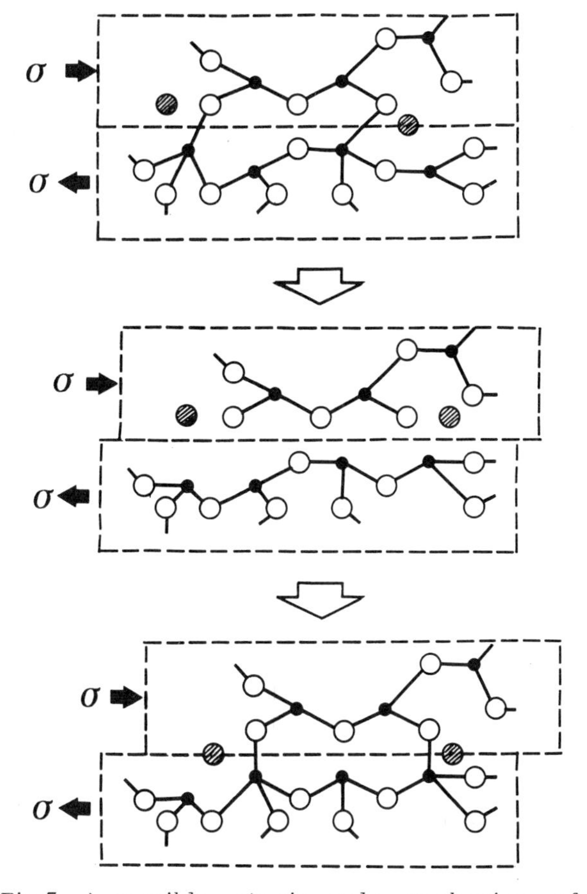

Fig.7 A possible atomic scale mechanism of plastic deformation of borate glass under shear stress around crack tip at slow crack growth. Open circles are oxygen anions, closed circles are boron cations and shadowed circles are sodium cations.

REFERENCES

1. S. M. Wiederhorn and H. Johnson, Effect of Electrolyte pH on Crack Propagation in Glass, J. Am. Ceram. Soc.,56:192 (1973)
2. G. S. White, S. W. Freiman, S. M. Wiederhorn, and T. D. Coyle, Effect of Counterions on Crack Growth of Vitreous Silica, J. Am. Ceram. Soc., 70:891 (1987)
3. T. A. Michalske and S. W. Freiman, A Molecular Mechanism for Stress Corrosion in Vitreous Silica, J. Am. Ceram. Soc., 66:284 (1983)
4. T. A. Michalske and B. C. Bunker, Steric Effect in Stress Corrosion Fracture of Glass, J. Am. Ceram. Soc., 70:780 (1987)
5. S. M. Wiederhorn, S. W. Freiman, E. R. Fuller, Jr., and C. J. Simmons, Effect of Water and Other Dielectrics on Crack Growth, J. Mat. Sci., 17:3460 (1982)

6 S. M. Wiederhorn, H. Johnson, A. M. Diness, and A. H. Heuer, Fracture of Glass in Vacuum, J. Am. Ceram. Soc., 57:336 (1974)

7 Y. Hibino, S. Sakaguchi, and Y. Tajima, Crack Growth in Silica Glass under Dynamic Loading, J. Am. Ceram. Soc., 67:64 (1984)

8 E. Gehrke, C. Uliner, M. Hänert, and G. Berger, Vergleichende Experimentelle Untersuchungen zum Risswachstum und zur Dynamischen Ermüdung einiger Phosphat- und Boratgläser, Silikattechnik, 33:238 (1982)

9 M. Inagaki, K. Urashima, S. Toyomasu, Y. Goto, and M. Sakai, Work of Fracture and Crack Healing in Glass, J. Am. Ceram. Soc., 68:704 (1985)

10 K. Hirao, J. Matsuoka, and N. Soga, Inelastic Deformation and Structure of Borate Glasses, J. Non-Cryst. Solids, 112:336 (1989)

11 M. Sakai, K. Urashima, and M. Inagaki, Energy Principle of Elastic-Plastic Fracture and Its Application to the Fracture Mechanics of a Polycrystalline Graphite, J. Am. Ceram. Soc., 66:868 (1983)

12 T. Nishida, Y. Takei, and T. Nishikawa, Slow Crack Growth of Reaction Bonded Si_3N_4 at High Temperature, *Yogyo-Kyokai-Shi* (J. Ceram. Soc. Japan), 90:254 (1982)

13 See, for example, articles in: "Borate Glasses," L. D. Pye, V. D. Fréchette and N. J. Kreidl, ed's, Plenum, New York, (1977)

PRECISION CRACK-OFF OF CERAMICS

Y.M. Chen, T.N. Farris*, and S. Chandrasekar

School of Industrial Engineering and
School of Aeronautics and Astronautics*
Purdue University
West Lafayette, IN 47907

INTRODUCTION

The "pinch-off" or crack-off of brittle materials was discovered in the early 1900's by P. W. Bridgman (Bridgman, 1912; 1949) while examining the rupture of solids at high pressure. In Bridgman's experiments, a glass rod was exposed to lateral fluid pressure while the two ends were kept free from pressure by using packing seals as in figure 1. At a certain pressure the rod broke and the divided parts were pushed out of the pressure chamber violently. The fracture surface formed was smooth and perpendicular to the axis of the rod. For brittle materials such as glass, the rupture was found to occur on a "perfectly" clean plane perpendicular to the axis. On the other hand, for ductile materials such as soft steel, copper, or aluminum, there was a significant contraction at the plane of breakage similar to the "necking" instability observed in tensile test specimens. The pressure at which this instability occurred under lateral pressure was found to be exactly the same as the stress at which necking occurred in a uniaxial tension test. The general features of the rupture phenomenon are well described by Bridgman who describes the failure during "pinching-off" as arising due to the axial tensile strain created in the solid by the lateral pressure on the curved surface. However, this does not address the question of dependence of the pressure at fracture on the surface state of a brittle solid, a fact pointed out by Bridgman himself. After Bridgman's discovery, the crack-off effect was little investigated until the 60's when Jaeger and Cook (1963) investigated this phenomenon in the context of the drilling of rocks. In the mid-70's and 80's there were several studies of this effect mainly directed at developing it as a process for slicing brittle materials, see for example Sato and Naoyuki (1975), Chandrasekar and Shaw (1988) and Santhanam (1989). Some Russian studies into this effect are briefly summarized in Cherepanov (1979).

In this paper we systematically investigate the fracture of borosilicate (BS) glass (Pyrex) and fine-grained polycrystalline aluminum oxide when subjected to lateral

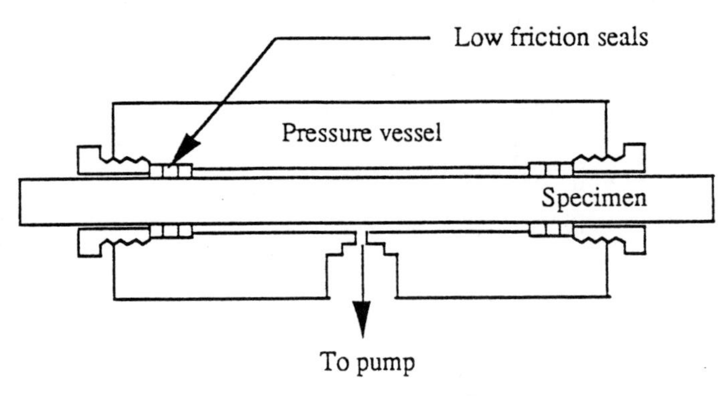

Figure 1. Bridgman's pinch-off apparatus.

pressure and characterize the cleaved surfaces thereby generated in terms of their surface finish and residual stresses. In all the fracture experiments conducted, the point of initiation of fracture is controlled by Vickers or Knoop indentation and its associated microcrack pattern. The pressure at which crack-off occurs is sensitively dependent on the indentation crack size and this inter-relationship is examined both experimentally and analytically. A fracture mechanics analysis, which incorporates the experimentally observed indentation cracking, is developed to calculate the pressure at fracture in the crack-off experiments in terms of the fracture toughness, indentation crack size, and indenter geometry. The predictions are in excellent agreement with the experimental results validating the crack-off model that has been developed. The effect of the pressure at fracture on the surface finish of the cleaved fracture surface of the brittle solid is investigated experimentally and explained qualitatively.

Our results indicate that the precision crack-off process is a potentially promising technique for cutting ceramics and glass. A discussion is undertaken of the critical issues involved in the development of this technique as a manufacturing process for slicing glass and ceramics. It is potentially an exciting and economical alternative to conventional diamond slicing processes which are costly, time consuming, and cause material loss.

INDENTATION FRACTURE

Crack-off of brittle solids has been typically observed to occur from a flaw on the surface of the specimen (Sato and Naoyuki 1975, Chandrasekar and Shaw 1988). Therefore, a pre-crack can often be used to control the origin of fracture in the crack-off process. In the present series of crack-off experiments on borosilicate glass and aluminum oxide, a pre-crack was introduced by indenting with a Vickers or a Knoop indenter. Typically, the indentations were made at normal loads greater than or equal to 5 kgf. The broad features of indentation cracking generated by pointed indenters, such as a Vickers pyramid, in this load range have been well studied; for recent reviews see Chaudhri and Phillips (1990) and Cook and Pharr (1990). We shall briefly discuss the nature of indentation cracking at loads $\geq 5kgf$.

When soda-lime glass is indented with a Vickers indenter at loads $\geq 5kgf$, full-penny shaped planar median cracks form under the indenter during loading; the cracks

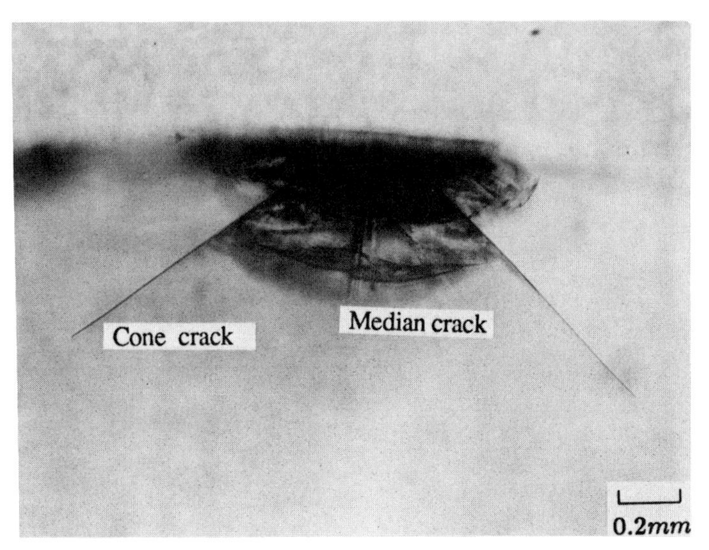

Figure 2. Side view of Vickers indentation in borosilicate glass showing median crack confined by the surfaces of a cone crack. The indentation was made close to the specimen edge and photographed in transmitted light, load = 490 N. Also the cone crack remains open after unloading.

are in a plane containing the loading axis and the indenter diagonal(s) (Lawn and Wilshaw 1975). The full-penny shaped median crack develops into a half-penny by breaking open to the surface either during the loading or unloading process (Chaudri and Phillips, 1990). Furthermore, during unloading lateral cracks with surfaces somewhat parallel to the indented surface grow in a shallow region (at less than or equal to indentation depth) near the surface (Lawn and Swain 1975). The indentation cracking produced by a Knoop indenter in soda-lime glass is similar to what is observed underneath a Vickers indenter except that a half-penny median crack forms instead of a full-penny shaped median crack during loading. In borosilicate glass, besides a median crack, a cone crack forms beyond a critical load when indented with Knoop or Vickers indenters. The median crack does not develop into a full-penny or even a half-penny shape, but is instead confined by the sides of the cone crack as in figure 2. The crack pattern in figure 2 was made by Vickers indentation close to the edge of borosilicate and photographed in transmitted light before sectioning with the crack-off process. The indentation load was 490 N. The fracture surface produced by the crack-off process is a cross-section through the indentation median and cone crack. This type of sectioning of the indentation using the crack-off process has been used in the study to make observations of the median crack and thereby make crack length measurements.

CRACK-OFF EXPERIMENTS ON BOROSILICATE GLASS AND ALUMINUM OXIDE

This section is devoted to a description of the experiments on the fracture of borosilicate glass and aluminum oxide rods when subjected to lateral fluid pressure, viz crack-off. Two principal questions are addressed : what is the effect of the pre-cursor flaw size on the fracture process in crack-off and how is the surface finish of the fracture surface influenced by the pressure at fracture?

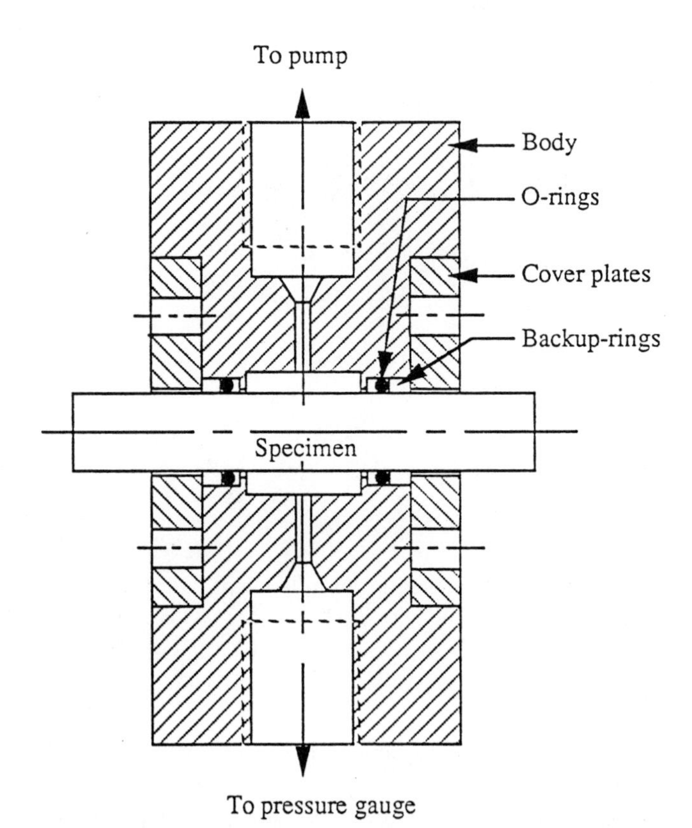

Figure 3. Pressure vessel used in the crack-off experiments.

Crack-off Apparatus

The pinch-off apparatus consists of three major components : the pressure pump system, the pressure vessel, and a digitized pressure indicating system. Figure 3 shows a schematic diagram of the pressure vessel used in the pinch-off experiments. This design is a variant of an earlier design by Chandrasekar and Shaw (1988); it is capable of sustaining fluid pressure up to 350 MPa. The sealing of the fluid pressure is provided by a combination of O-rings and Teflon back-up rings which are squeezed by the cover plates. The length of the specimen which is subjected to fluid pressure on the lateral faces is equal to the distance between the O-rings, which is about 25 mm. The ends of the specimen are open to the atmosphere as shown in figure 3. While cylindrical rods were the specimens used in the experiments, the design of the pressure vessel can be modified to accommodate non-axisymmetric specimens by simply changing the geometry of the O-rings and back-up rings. By heat-treating the steel pressure vessel, the same design may be used for pressures up to 500 MPa with minor changes in the sealing procedure. The central chamber of the pressure vessel is connected to the pressure pump and the pressure gauge as in figure 3.

A high pressure pump system (PS-150 Pumping System from High Pressure Equipment Co.) was used to generate the fluid pressure in the pinch-off pressure vessel. This pump is capable of generating pressures up to 150 ksi (1050 MPa) by means

Table 1. The physical properties of borosilicate glass and aluminum oxide.

Property	Borosilicate glass	Aluminum oxide
Young's modulus E (GPa)	63.0	352.0
Shear modulus G (GPa)	26.2	144.3
Poisson's ratio (ν)	0.20	0.22
Fracture toughness K_{IC} ($MPa\sqrt{m}$)	0.76	3.0 $\sim$ 4.0
Hardness (Knoop, kg/mm^2)	418	1523
Average grain size (microns)	-	10 -12

of an air operated hydraulic pump which is coupled to a high pressure intensifier. The pump supplies two stages of pressure levels — a low level ranging from 0 to 15 ksi (105 MPa) and a high level ranging from 0 to 150 ksi (1050 MPa). The pressurizing fluid used in the experiments was a low viscosity fluid consisting of one part of kerosene and two parts of a synthetic fluid, Diala - AX, manufactured by Shell Oil Co. A digitized pressure indicating system manufactured by Precise Sensors Inc. was used to measure the pressure in the pinch-off vessel. This pressure gauge can read pressures up to 100 ksi (700 MPa) with a resolution of ± 10 psi which is sufficiently accurate for our studies. Furthermore, it provides a recall of the peak pressure reached in a pressurizing cycle.

Experimental Procedures

The pinch-off experiments on aluminum oxide and borosilicate glass rods having nominal diameter of 12.7mm were carried out in the pressure vessel described in figure 3. Table 1 gives the relevant mechanical properties of these two materials. The specimens were indented under controlled loading conditions in a Vickers hardness tester to introduce a pre-crack – the median crack which served as the origin of fracture in pinch-off. Most of the specimens were indented with a Vickers diamond indenter whose diagonal was oriented perpendicular to the rod axis to within 5°. Some specimens were indented with a Knoop indenter whose long diagonal was perpendicular to the specimen axis to within 5°. In all the experiments reported, at least 5 specimens were indented at any given loading condition.

Several precautions were observed in the pinch-off tests. The location of the indentation was kept approximately mid-way between the sealing rings to ensure that the stress field in the vicinity of the indentation was not perturbed by the seals. It was usually observed that if the specimen fractured near the seals, the fracture surface was irregular. The pinch-off tests were always carried out within two days of making the indentation to minimize strength degradation effects arising out of post-indentation crack growth in borosilicate glass and aluminum oxide. Such crack growth has been observed by Gupta and Jubb (1981). In pinch-off experiments conducted on specimens which had been indented and preserved for 40 days, the fracture pressures were

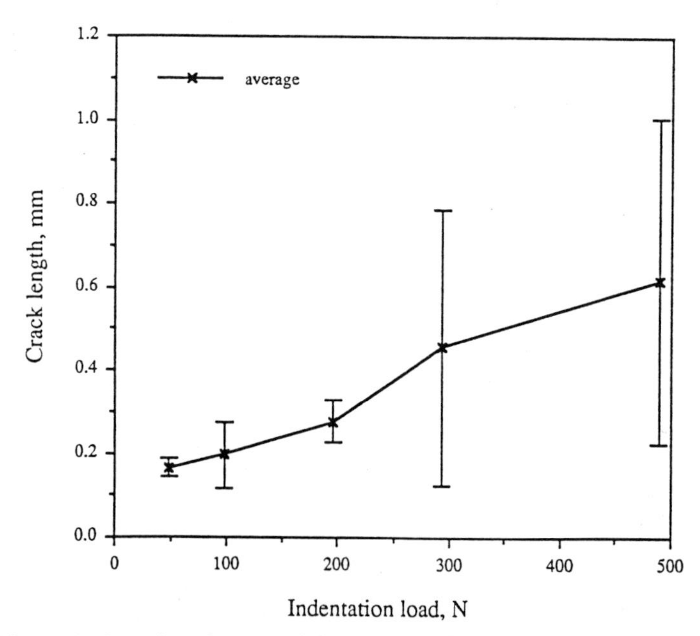

Figure 4. The variation of median crack length with Vickers indentation load in borosilicate glass measured by optical microscopy.

approximately 25 % lower than on specimens fractured within two days of indenting. Post-indentation crack growth is the obvious cause for this reduction in fracture pressure.

An imprint of the median crack generated by a Vickers or Knoop indenter in borosilicate glass is left behind on the fracture surface after crack-off as shown in figure 2. Using such a natural sectioning of the indentation, the length of the median cracks in borosilicate glass was measured as a function of the Vickers indentation load using optical microscopy. Figure 4 shows the experimentally measured variation of the median crack length with indentation load for the Vickers indenter. Crack lengths were measured in at least 5 different specimens for each indentation load.

In aluminum oxide, optical microscopy of the specimens which were indented and subsequently fractured in pinch-off was alone insufficient to identify the median crack boundary. This was due to a lack of contrast between the median crack zone and the remainder of the specimen surface. Petrovic (1983) found a similar problem with regard to the location of the median crack boundary in fracture surfaces of a glass ceramic which had been indented before fracture. However, if the fracture surface was coated with a thin layer of gold, the median crack boundary was clearly demarcated in the glass ceramic (Petrovic 1983). Following Petrovic (1983), the indented aluminum oxide specimens were coated with gold after being fractured in crack-off; this enabled a clear resolution of the median crack boundary by optical microscopy in specimens which had been indented at loads $\geq 30kgf$. The median crack lengths could then be measured. No cone cracks were observed in the alumina specimens indented with Vickers or Knoop indenters. In alumina specimens indented at loads $\geq 30kgf$ and fractured in crack-off, scanning electron microscopy (SEM) of the fracture surface was carried out to obtain the length of the median crack. Table 2 gives the average measured values of the median crack lengths in alumina using the above two techniques. The crack lengths are the average of measurements made over 5 specimens.

Fracture Pressure

After indentation at different loads a number of alumina and borosilicate glass rods were fractured in the crack-off apparatus of figure 3. Figure 5 shows the variation of the fracture pressure with indentation load and median crack size, obtained from these experiments. Every data point in figures 5(a) and 5(b) is the average of fracture pressures of at least 5 specimens. The fracture pressure decreases with increasing indentation load and median crack size as would be expected. The aluminum oxide rods fracture at higher pressures compared to borosilicate glass rods indented under similar conditions. This is due to the higher fracture toughness of aluminum oxide in comparison to glass, see table 1. The quantitative variations in the fracture pressure under different indentation conditions will be analyzed using a linear elastic fracture mechanics model of the crack-off process in the next section.

Table 2. The measured average length of median cracks in aluminum oxide.

Indenter (load)	Crack length (mm)
Vickers (30 kg)	0.400
Vickers (50 kg)	0.569
Knoop (30 kg)	0.480
Knoop (50 kg)	0.600

Characteristics of the Fracture Surface

The fracture surface of borosilicate glass and aluminum oxide produced by crack-off were analyzed using a surface profilometer (Talysurf 10) and the SEM. Residual stresses on the cut surface of aluminum oxide were measured using X-ray diffraction.

Surface Roughness The Talysurf measures the profile of the surface by traversing a diamond stylus of tip diameter $\sim$ 1 micron along a straight line on the surface. Its maximum vertical magnification is 50,000 while the maximum horizontal magnification is 100. When operating at its maximum vertical magnification, the profilometer can resolve a 0.05 micron change in the surface profile. Three parameters were used to characterize the surface roughness on the pinched off surfaces – the peak-to-valley height (R_q), the arithmetic average or root mean square value (R_a), and the flatness.

Figures 6(a) and 6(b) show typical Talysurf traces of the fracture surface of glass and aluminum oxide rods broken in pinch-off. The glass surface is very smooth with a peak-to-valley roughness R_q of less than 0.1 micron as in figure 6(a). However, there is some waviness. The peak-to-valley roughness R_q is usually defined as the distance between the highest peak and the lowest valley of the surface profile over a fixed traversing distance of the stylus. The aluminum oxide surface shows considerable roughness in the trace of figure 6(b) compared to the borosilicate glass. Based on surface roughness profiles such as those shown in figures 6(a) and 6(b), the peak-to-valley height R_q was obtained for fracture surfaces generated under different pinch-off conditions. The value of R_q was obtained in our experiments were obtained by averaging the 5 highest peak-to-valley numbers over a traversing distance of 10 mm.

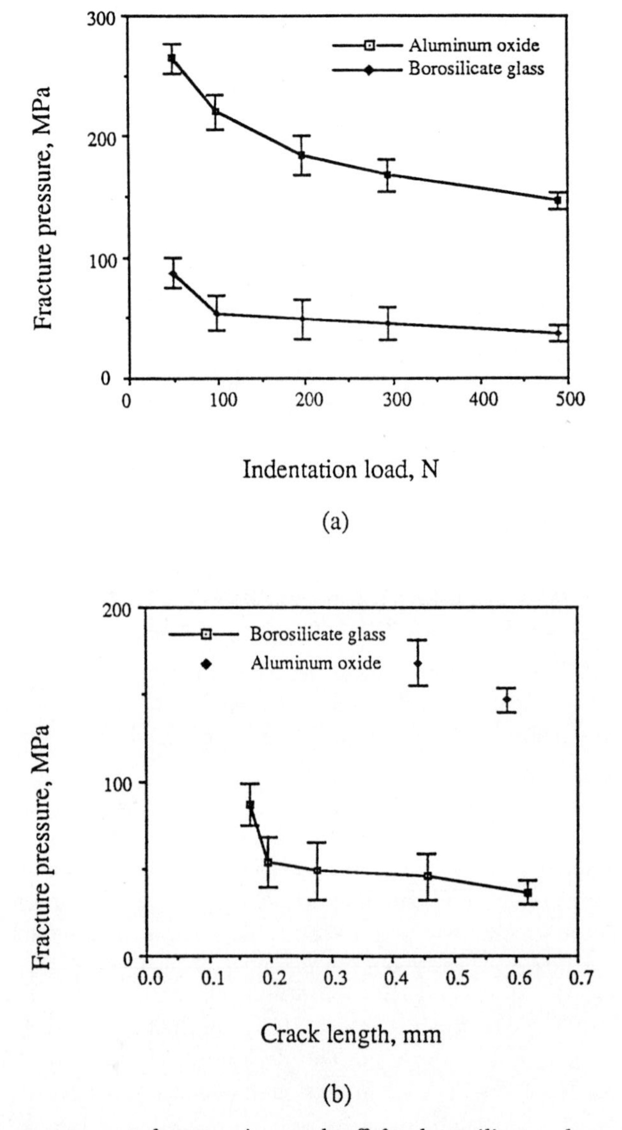

Figure 5. The pressure at fracture in crack-off for borosilicate glass and aluminum oxide rods of 12.7mm diameter (a) Variation of the pressure with Vickers indentation load (b) Variation of the pressure with median crack length.

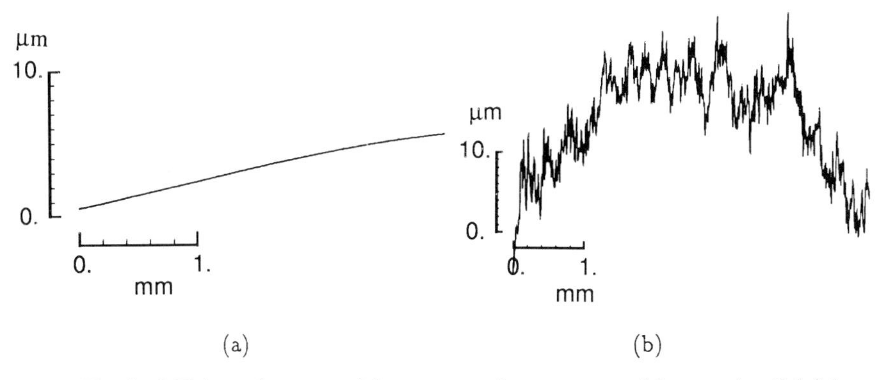

Figure 6. Typical Talysurf traces of fracture surface generated by crack-off (a) borosilicate glass (b) aluminum oxide.

Figure 7 shows R_q in the direction of crack propagation and perpendicular to the direction of crack propagation for cracked-off aluminum oxide surfaces. The peak-to-valley roughness values are similar in both of these directions, with $R_q \sim 15\ \mu m$. This roughness is similar to what is observed on a machined or coarse ground ceramic surface. R_q shows a decreasing trend with increasing fracture pressure. This means that fracture surfaces of aluminum oxide which broke at higher pressures were smoother than those which fractured at a lower pressure. A similar trend is observed in the arithmetic average (R_a) values of the surface roughness of aluminum oxide shown in figure 8(a). R_a decreases from $\sim$ 2.2 μm at fracture pressures of $\sim$ 160 MPa to $\sim$ 1.7 μm at pressures of $\sim$ 260 MPa. However in the case of R_a, its value in the direction perpendicular to the direction of crack propagation is greater than the value in the direction of crack propagation. The R_a values of fractured glass surfaces in figure 8(b) also show a general decrease with increasing fracture pressure, except for surfaces generated at the highest pressure. The peak-to-valley roughness, R_q, of glass is however little influenced by the fracture pressure and remains fairly constant at about 0.05 – 0.1 micron over the range of fracture pressure studied.

Figure 9 shows the variation in the flatness of cracked-off surfaces of borosilicate glass with fracture pressure. The flatness values given in the figure are obtained as the maximum deviation of the profile from the chordal length over a stylus traverse distance of 5 mm in the central region of the sample. From figure 9, it is clear that the variation in the flatness with fracture pressure for borosilicate glass is similar to the variation of R_a with fracture pressure. An interesting observation concerning the profile of the fracture surface of borosilicate glass and aluminum oxide is that the two mating fracture surfaces generated by pinch-off form a concave-convex combination. In all the pinch-off surfaces of borosilicate glass and aluminum oxide, the shorter in length of the two fracture pieces generated has a convex surface while the longer of the two pieces has a concave fracture surface.

Residual Stresses Residual stresses were measured on cracked-off surfaces of aluminum oxide using X-ray diffraction. Cr-Kα radiation with a spot of diameter 2.5 mm was used. The measured surface stress in aluminum oxide was compressive and about 8.3 MPa. Since this stress value is of the same order as the error in the measurement technique, it is concluded that the cracked-off surfaces produced by pinch-off are stress free. A similar observation was made in cracked-off Ni-Zn ferrite. That cracked-

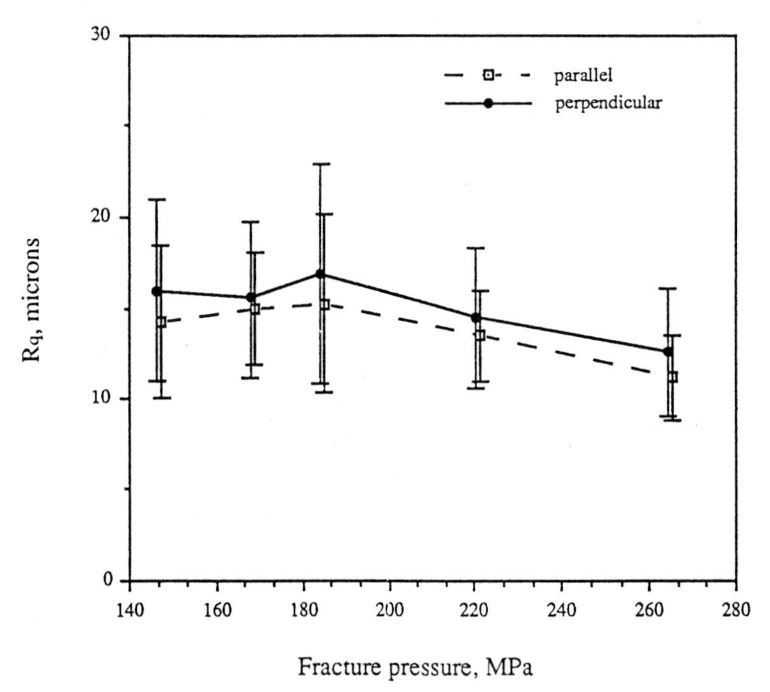

Figure 7. The peak-to-valley roughness (R_q) on the fracture surface of aluminum oxide measured parallel and perpendicular to the direction of crack propagation.

off surfaces are stress free is not surprising, since the cleavage fracture of brittle solids should involve very limited plastic deformation and negligible thermal effects over near-surface regions. In contrast residual stresses measured by X-ray diffraction on ceramic surfaces sliced using a diamond wheel are about 40 MPa compression in alumina and about 30 MPa compression in Ni-Zn ferrite. Both these values are significantly higher than on cracked-off surfaces because plastic deformation and temperature influences are significant during diamond slicing.

CRACK-OFF ANALYSIS

We now analyze the initiation of the fracture that leads to surface generation by the crack-off process. Our analysis is deterministic using linear elastic fracture mechanics; this is because pre-cracks due to microindentation exist in the specimen before it is subjected to lateral fluid pressure in the crack-off pressure vessel. This analysis is different from Matsuo's probabilistic analysis (Matsuo 1979) of the pinch-off process, wherein some arbitrary distribution of flaws is assumed to exist in the specimen. This analysis requires the use of two floating parameters to describe the flaw size distribution and is anyway not applicable to the present crack-off configuration wherein a well defined pre-crack is introduced by indenting. A two-dimensional finite element analysis has been conducted. It was found that the effect of the indentation crater or notch on the crack-off pressure is negligible for notch-to-crack length ratios less than 0.1 (Chen, 1990). A simplified three-dimensional model that approximates the median crack as a surface breaking half-penny crack is therefore discussed. Also, a three-dimensional boundary integral analysis is given to estimate the effect of the indentation crack shape on the pinch-off pressure (Murakami, 1985).

In both the two and three-dimensional analyses it is assumed that the fluid pen-

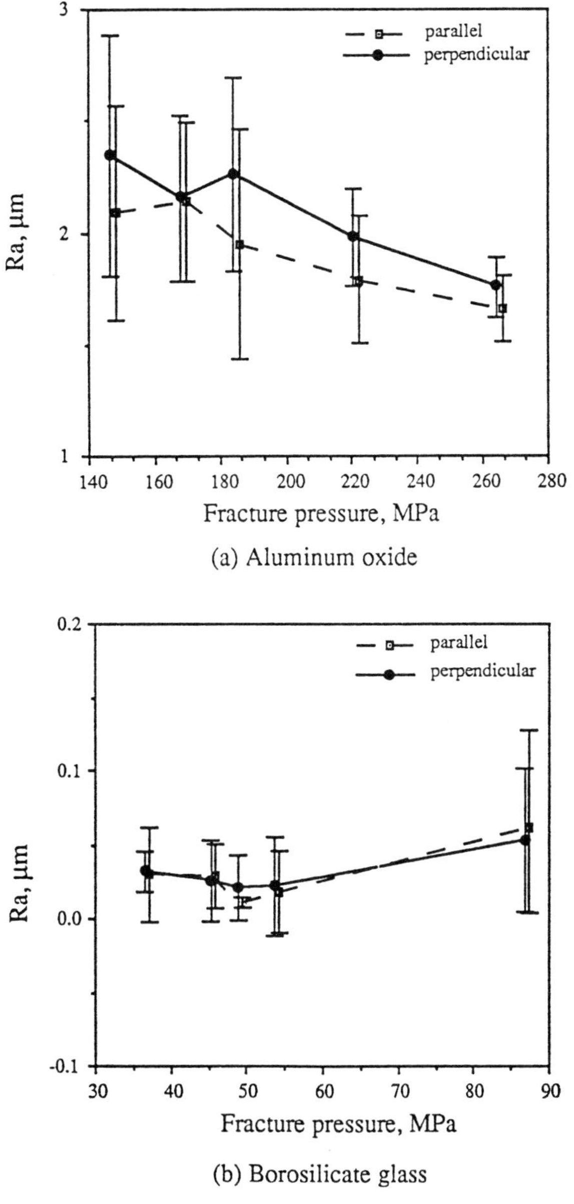

(a) Aluminum oxide

(b) Borosilicate glass

Figure 8. The arithmetic average, R_a, roughness value of the fracture surface measured parallel and perpendicular to the direction of crack propagation (a) aluminum oxide (b) borosilicate glass.

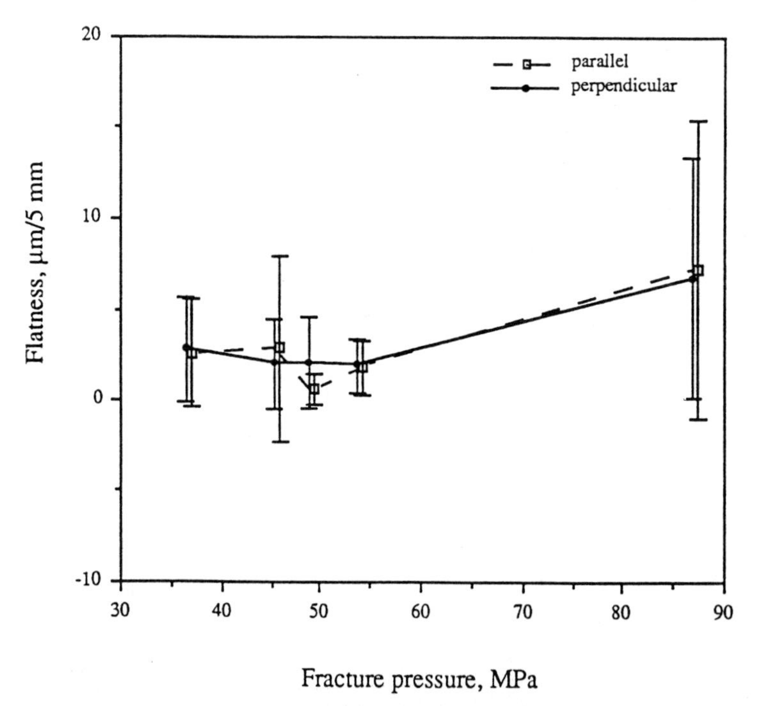

Figure 9. The flatness of borosilicate glass cleaved at different fracture pressures measured parallel and perpendicular to the direction of crack propagation.

etrates through the indentation crater and loads the faces of the median crack. This results in pure mode I loading for the 90° median crack oriented in the plane perpendicular to the rod axis. For mode I loading, the effect of the O-rings on the fracture pressure is analyzed in terms of two bounding cases namely perfect axial constraint and zero axial constraint loading of the rod by the O-rings.

Simplified Three-Dimensional Model

In general, the cracks produced by Vickers or Knoop indentation are much smaller than our specimen diameter of 12.7 mm. The crack size was always less than 5 % of the dimension of the cross-section of the specimen. Therefore, as a first approximation, the cracks are modeled as a single half-penny crack when the Knoop indenter is used (figure 10), and two orthogonal half-penny cracks when the Vickers indenter is used. These surface breaking cracks are in a semi-infinite media and are subjected to a uniform pressure acting on the crack surfaces. The stress intensity factors for the pressure acting on the crack surfaces are the same as that for remote uniform tension in a semi-infinite media given by Lee et al (1987) as

$$K_I^b \cong 0.666p\sqrt{\pi c}. \tag{1}$$

at the bottom of the crack and

$$K_I^t \cong 0.744p\sqrt{\pi c}. \tag{2}$$

at the top of the crack.

Table 3 shows the average pinch-off pressure, median crack depth, and the calculated stress intensity factor corresponding to the fracture pressure (equations (1)

Table 3. Fracture toughness of aluminum oxide estimated using precision crack-off.

Aluminum oxide					
Pinch-off	Crack	Free		Constrained	
pressure (MPa)	length (mm)	K_I^b (Eq.1) $(MPa\sqrt{m})$	K_I^t (Eq.2) $(MPa\sqrt{m})$	K_I^b (Eq.1) $(MPa\sqrt{m})$	K_I^t (Eq.2) $(MPa\sqrt{m})$
167.95	0.400	3.965	4.430	2.220	2.481
146.95	0.569	4.138	4.622	2.317	2.589
154.66	0.480	4.000	4.468	2.240	2.503
194.26	0.600	5.617	6.275	3.146	3.514

and (2)) for aluminum oxide. These K_I's are labeled as plane stress or "free" in table 3. The stress intensity factor at fracture of borosilicate glass ranges from 0.947 to 1.131 $MPa\sqrt{m}$ in our experiments and is larger than the published values of fracture toughness for this material, $K_{IC} = 0.76 MPa\sqrt{m}$ (Arora et al, 1979; Wiederhorn, 1969). The calculated stress intensity factor for aluminum oxide in table 3 is generally in good agreement with the reported value of $K_{IC} = 3.9 MPa\sqrt{m}$ for AD999 aluminum oxide (Anstis et al, 1981), except at the highest pinch-off pressure.

One explanation for the stress intensity factors being higher than the published values of K_{IC} is due to the effect of the O-rings being neglected in our calculation of K_I (plane strain). If we assume that the O-rings completely prohibit axial strain in the rod, which is called plane strain, the effective crack driving pressure becomes

$$p^* = (1 - 2\nu)p \tag{3}$$

The K_I at fracture calculated using p^* is also shown in table 3 in the columns labeled constrained.

In practice the loading effect due to the O-rings will be intermediate between plane stress and plane strain. Using published values of fracture toughness, $K_{IC} = 0.76$ $MPa\sqrt{m}$ for borosilicate glass and $K_{IC} = 3.50$ $MPa\sqrt{m}$ for aluminum oxide (manufacture's value for AD998), and the measured indentation crack size, the fracture pressure in pinch-off is calculated using equations 1 and 2. The calculations are carried out for both plane stress (no axial constraint) and plane strain loading (axial constraint by O-ring) of the rod. The analytically calculated fracture pressures and the corresponding experimentally observed values are given in figure 11 (a) and 11 (b).

DISCUSSION AND CONCLUSION

Fracture Pressure

In the previous section, a fracture mechanics analysis of the crack-off configuration was carried out. This analysis assumed, as its basis, that fluid penetrates into the median crack generated beneath an indentation and exerts a pressure normal to the crack faces as in figure 10. At a critical value of the pressure, when the stress intensity factor at the median crack tip reached a value equal to K_{IC}, fracture was predicted to occur in pinch-off. Comparing the experimentally observed fracture pressure versus median crack length variation with the analytically predicted fracture pressure in figures 11 (a) and 11 (b), it is clear that there is reasonably good agreement. The measured fracture pressures for the Vickers indentation generally fall within the two bounds predicted

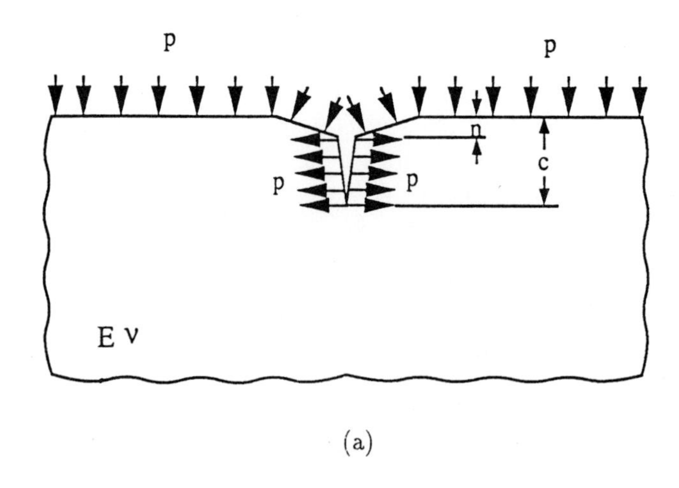

(a)

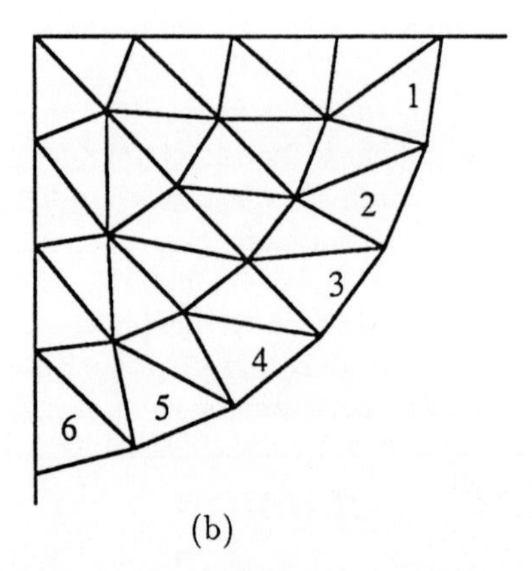

(b)

Figure 10. Schematic of the model used in the crack-off fracture mechanics analysis showing the discretization used in the body force method.

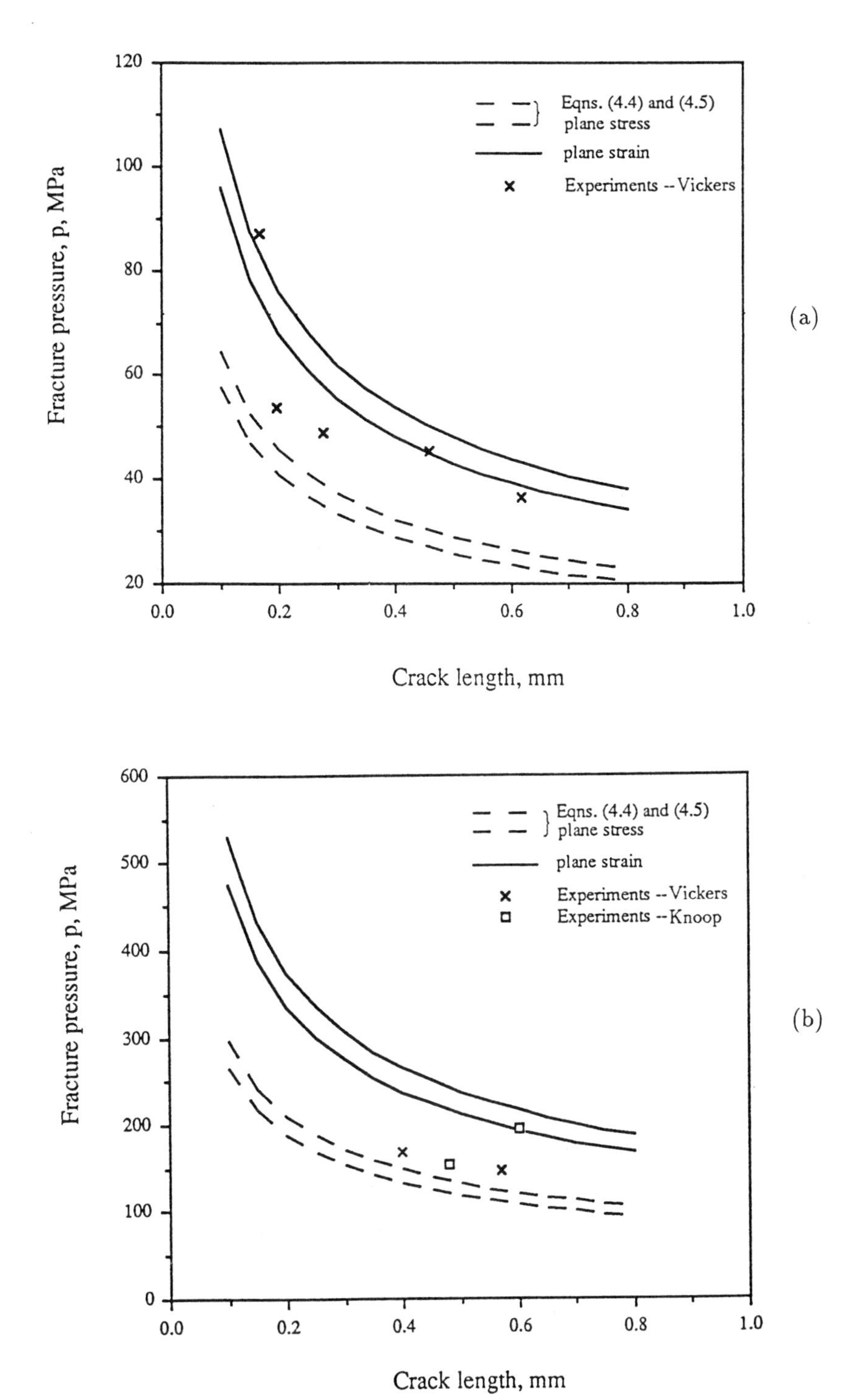

Figure 11. Comparison of the experimental crack-off fracture pressures for (a) borosilicate glass and (b) aluminum oxide with the analytically calculated pressures (full and zero axial constraint). (The solid curves in the figure are obtained by dividing the pressures corresponding to the dotted curved by ($1 - 2\nu$))

analytically using plane stress and plane strain loading of the rod. The latter type of loading occurs if the O-rings exert a compressive stress on the rod preventing any specimen elongation in the axial direction as a consequence of squeezing. In practice, the compressive stress exerted by the O-rings on the rod is not expected be large enough to cause a state of complete plane strain; the actual stress state will in all probability be somewhere between plane stress and plane strain. It must be pointed out that it is assumed that the median crack is the crack which eventually propagates and that the cone crack which is present, for example in borosilicate glass, does not affect significantly the stress intensity factors at the median crack tip. This assumption would be all right for the fracture of aluminum oxide which does not contain a cone crack but is probably not completely valid for the fracture of borosilicate glass. A fracture mechanics analysis of the interaction between the cone crack and the median crack is currently being attempted.

Surface Roughness

Figures 7 to 9 show that the surface roughness of borosilicate glass and aluminum oxide surfaces produced by crack-off generally improved with increasing crack-off pressure. Moreover, cracked off surfaces have a roughness comparable at least to coarse ground surfaces and are certainly orders of magnitude better in finish than the fracture surface produced by scribing and subsequent bending. In the case of cracked off borosilicate glass, the surface is mirror-like with a finish close to that obtained by polishing. In order to understand these characteristics of cracked-off surfaces let us qualitatively follow the crack propagation process.

SEM micrographs of crack-off surfaces of ceramics such as these shown in figure 12 (a) and 12 (b) indicate that the crack propagation is, to a large extent, through the grains (transgranular). This is seen in the numerous flat faces of the grains in figure 12. Furthermore by comparing the fracture surface of ferrite shown in figures 12 (a) and 12 (b) it is also clear that the crack propagating at a higher pressure shows more transgranularity, thereby leading to a better finish. On the other hand, intergranular crack growth, usually observed extensively on bending fracture surface causes the crack to weave in and out of plane along the grain boundaries, leading to a higher surface roughness. Another factor which contributes to a good finish in pinch-off is the pressure on the lateral curved faces of the rod. This lateral pressure produces a significant compression on planes other than the one normal to the rod axis thereby causing the crack to be confined to this plane. It is possible, however, that when the fracture pressure exceeds a certain critical value crack branching could occur (Lawn and Wilshaw, 1975b) leading to a roughening of the surface at high fracture pressure. This could well be the reason for the slight deterioration in the surface roughness of cracked-off glass observed at the highest fracture pressure for glass reported in the present experiments (figure 8 (b)). "Beach marks" observed on fractured specimens of borosilicate glass are thought to arise out of stress wave reflections from the end faces of the rod interacting with the crack propagation process. These "beach marks" were seen in almost all of the crack-off specimens of borosilicate glass near the exit end of the fracture.

Additional Comments, Observations, and Future Work

The foregoing tests have demonstrated that the crack-off process is a potentially promising manufacturing process for the slicing of brittle solids. The process provides a fast slicing method and is nowhere near as expensive as conventional slicing processes using

Figure 12. SEM micrograph of the fracture surface of cracked-off Ni-Zn ferrite generated at (a) fracture pressure of 48.3 MPa and (b) fracture pressure of 117.25 MPa [5].

diamond wheels which are currently being used for the slicing of ceramics. Under the conditions investigated in this study, the capability of the process to produce surfaces of aluminum oxide and glass having a good surface finish and devoid of residual stress has been demonstrated. The latter aspect makes it particularly attractive for cutting electronic ceramics.

The use of crack-off as a slicing process for ceramics would be significantly enhanced if advances are made in three areas :

(1) Improvement in surface finish of fractured surfaces to approach finishes comparable to what exists on fine ground surfaces.

(2) Ability to generate wafers as thin as 0.5 mm rapidly. Applications of this are in the slicing of silicon wafers.

(3) Slicing of rods of small diameter, say 1-2 mm or less so as to be applicable for optical fiber cutting.

With regard to the first mentioned area, we plan to investigate the process in higher pressure regimes, typically around 500 to 700 MPa. We have recently demonstrated, albeit in an academic fashion, the capability to make a wafer of glass as thin as 0.9 mm. This was done by gluing two rods of glass together end to end, making an indentation within a millimeter of the glued zone and propagating an indentation crack in crack-off. Subsequent ungluing of the rods resulted in a wafer about 0.9 mm thick.

It also appears that chipping and local surface roughening near the indented region on the cut surface could be minimized if one makes an "ideal" pre-crack. The "ideal" pre-crack would be a median crack which is exactly aligned perpendicular to the specimen axis and confined to this perpendicular plane; it will have no lateral or cone cracking associated with it. Ion-etching and wedge indentation show promise in this regard.

There are several fundamental aspects of the fracture process in crack-off which need to be studied further. A larger class of materials need to be investigated experimentally and analytically with the indentation pre-cracking being quantified more precisely in terms of its geometry. Here the analysis must include the interaction between the two median cracks in Vickers indentation and the cone and median cracks in borosilicate glass. Only then can the accuracy of the present model be put to its fullest test and completely verified. The effect of pressure on the surface finish needs further probing and understanding. Lastly, studies with chemically active fluids, i.e. fluids which interact chemically with the crack tip material, are sure to provide new insights into the process. They also offer hope for developing other interesting variations of this process.

Acknowledgment This work was partially supported by the National Science Foundation (NSF) through grants MSS 9057082 (Surface Engineering and Tribology Program, Dr. J. Larsen-Basse, Director). and DDM 9057916 (Manufacturing Processes Program, Dr. B. Kramer, Director)

References

[1] Anstis, G. R., Chantikul, P., Lawn, B. R. and Marshall, D. B., 1981, A critical evaluation of indentation techniques for measuring fracture toughness: I, direct crack measurements, *Journal of the American Ceramic Society*, **64**, 533-538.

[2] Arora, A., Marshall, D. B. , Lawn, B. R. and Swain, M. V., 1979, Indentation deformation/fracture of normal and anomalous glasses, *Journal of Non-Crystalline Solids*, **31**, 415-428.

[3] Bridgman P. W., 1912, Breaking tests under hydrostatic static pressure and conditions of rupture, *Philosophical Magazine*, **24**, 63-68.

[4] Bridgman P. W., 1931, *The Physics of High Pressure*, Macmillan, New York, also G. Bell, London, 78-98.

[5] Chandrasekar S. and Shaw, M. C., 1988, Precision crack-off of brittle materials, *ASME Journal of Enginering for Industry*, **110**, 187-191.

[6] Chaudhri, M. M., and Phillips, M. A., 1990, Quasi-static indentation cracking of thermally tempered soda-lime glass with spherical and Vickers indenters, *Philosophical Magazine, A*, **62**, 1-27.

[7] Chen, Y. M., 1990, M. S. Thesis, School of Aeronautics and Astronautics, Purdue University, W. Lafayette, Indiana.

[8] Cherepanov G. P., 1979, *Mechanics of Brittle Fracture*, McGraw-Hill Book Co., 878-880.

[9] Cook, R. F. and Pharr, G. M., 1990, Direct observation and analysis of indentation cracking in glasses and ceramics, *Journal of the American Ceramic Society*, **73**, 787-817.

[10] Gupta, P. K., and Jubb, N. J., 1981, Post-indentation slow growth of radial cracks in glasses, *Journal of the American Ceramic Society*, **64**, 112-114.

[11] Jaeger J. C. and Cook, N. G. W., 1963, Pinching-off and disking of rocks, *Journal of Geophysical Research*, **68**, 1759-1765.

[12] Lankford, J. 1981, Threshold microfracture during elastic-plastic indentation of ceramics, *Journal of Materials Science*, **16**, 1177-1182.

[13] Lawn, B. R. and Swain, M. V., 1975, Microfracture beneath point indentations in brittle solids, *Journal of Materials Science*, **10**, 113-122.

[14] Lawn B. R. and Wilshaw, T. R., 1975a, Indentation fracture: principles and applications, *Journal of Materials Science*, **10**, 1049-1081.

[15] Lawn, B. R. and Wilshaw, T. R., 1975b, *Fracture of Brittle Solids*, Cambridge University Press.

[16] Lee, J. C., Farris, T. N. and Keer, L. M., 1987, Stress intensity factors for cracks of arbitrary shape near an interfacial boundary, *Engineering Fracture Mechanics*, **27**, 27-41.

[17] Matsuo, Y., 1979, A probabilistic treatise on a lateral fluid pressure test of brittle materials, *Bulletin of Japan Society of Mechanical Engineering*, **22**, 1053-1058.

[18] Murakami, Y., 1985, Analysis of stress intensity factors of modes I, II and III for inclined surface cracks of arbitrary shape, *Engineering Fracture Mechanics*, **22**, 101-114.

[19] Petrovic, J.J., 1983, Effect of indenter geometry on controlled-surface-flaw fracture roughness," *Journal of the American Ceramic Society*, **66**, 277-283.

[20] Santhanam, S., 1989, Ph.D. Thesis, Department of Mechanical Engineering, Arizona State University.

[21] Sato Y. and Naoyuki, S., 1975, The cutting of brittle materials with lateral fluid pressure," *International Journal of Mechanical Science*, **17**, 705-710.

[22] Wiederhorn, S. M., 1969, Fracture surface energy of glass, *Journal of American Ceramic Society*, **52**, 99-105.

EFFECT OF DEFECTS AND GRAIN SIZE ON STRENGTH OF MULLITE CERAMICS

Yoshiaki Yamade, Yoshiaki Kawaguchi, Nobou Takeda*,
and Teruo Kishi*

Sumitomo Metal Industries, Research and Development Division
Fuso, Amagasaki, Hyogo, 660, Japan
Research Center for Advanced Science and Technology
The University of Tokyo
4-6-1, Komaba, Meguro-ku, Tokyo, 153, Japan

INTRODUCTION

Many studies have been conducted on the strength and fracture toughness K_{IC} of ceramics. However, more extensive studies are still necessary to explain the relationship between defects and grain size[1], and its effects on the strength and fracture toughness. Recent experimental observations have revealed the following interesting phenomenon on the defect/grain size effect[2]. If the defect size was smaller than 30 times of the grain size, the fracture toughness value [K_{IC}(defect)] calculated from the inherent defect size became smaller than that [K_{IC}(SEPB)] measured using a fracture toughness specimen with an artificially-introduced initial crack. These observations were reported on several engineering ceramics including silicon nitride, silicon carbide and alumina. Based on these observations, alumina with fine grain was developed to improve the strength. In the study of alumina[3], the fracture toughness increased with increasing grain size(d), and showed the maximum value when d=100μm. It seems to be impossible to increase both strength and fracture toughness values simultaneously by controlling the grain size.

The micro-fracture process initiating from the inherent defects, and the interaction of these defects and grain sizes should be clarified to understand the difference in fracture toughness values of K_{IC}(defect) and K_{IC} (SEPB). The AE technique is one of the most effective method to detect the microcracks in the non-destructive evaluation[4].

In this study, the objective is to investigate the effect of the grain size on the initial fracture process by using the AE technique. The detailed AE analysis has revealed the defect growth to explain the difference between K_{IC} (defect) and K_{IC} (SEPB) when the defect/grain size ratio is below 30.

EXPERIMENTAL PROCEDURE

Specimen

Commercial high-purity mullite powder (Al_2O_3=71.80%, SiO_2=28.05%)

was sintered using the fabrication process as shown in Fig. 1. The grain size was controlled by changing the sintering condition (Table 1). Three types of specimens (Fig. 2) were machined for four-point bend, single-edge precracked beam bending (SEPB) and double cantilever beam (DCB) tests. The surface of the specimens were polished using diamond paste up to the surface roughness of 0.8μm.

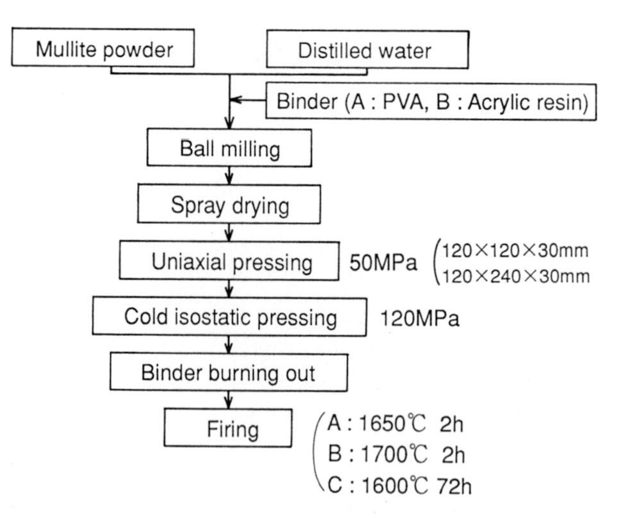

Fig. 1 Fabrication process of mullite.

Table 1 Sintering conditions and material properties.

Material	Sintering temp. (°C)	Sintering time (h)	Porosity (%)	Grain diameter (μm)	Binder
A	1650	2	2.9	0.95	PVA
B	1700	2	3.2	1.36	acrylic resin
C	1600	72	1.3	3.82	PVA

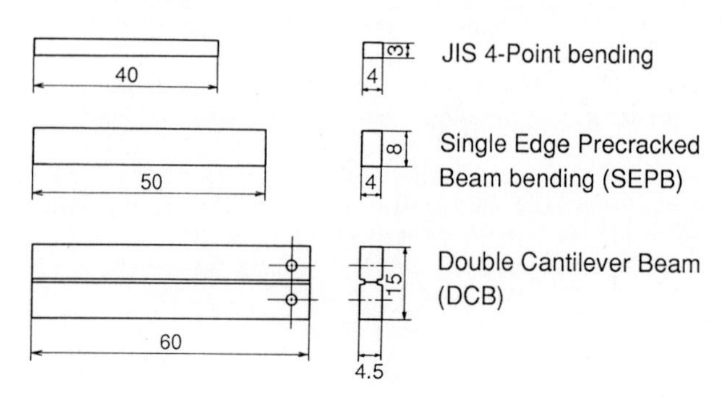

Fig. 2 Specimen configurations.

Microstructure

The material microstructure was observed using a scanning electron microscope (SEM) after polishing and thermal etching. The etching temperature was 50°C lower than the sintering temperature. The microstructure consisted of columnar grains. The grain size was calculated as an average of the long-diameter and short-diameter of each grain. The average grain size was obtained by measuring 200 grains for each type of mullite. Three-dimensional grain size is approximately 1.5 times as large as two-dimensional one in a case of cubic-shaped grain. In this study, the grain was columnar and the fracture surface mainly showed the intergranular fracture. Therefore, two-dimensional grain size was used for the following analysis.

Four-Point Bending Test

The 4-point bending test specified by JIS R1601-1981 was carried out. The specimen size was 3 x 4 x 40mm with the inner-span of 10mm and the outer-span of 30mm. The cross-head speed was chosen at 0.05, 0.5, 5.0, 50.0 and 500.0mm/min to study the strain rate dependence of strength. Twenty specimens were tested for each type of mullite at each cross-head speed.

Fracture Toughness

Fracture toughness was measured by SEPB(Single Edge Precracked Beam) method. In this method, a pre-crack was introduced by bridge-indentation in the specimens with the size of 4 x 8 x 50mm, and the 3-point bending was carried out with the span of 32mm and the cross-head speed of 1mm/min.

Evaluation of Slow Crack Growth (SCG) Behavior

The crack extension behavior was studied by using the double cantilever beam (DCB) specimens[7]. Mullite did not show any hysterisis loop during loading and unloading, and, therefore, the K_I-V (crack growth rate) curve was calculated from the load-COD curve by using a simple beam equation.

AE (Acoustic Emission) Measurement

AE signals were measured with the system shown in Fig. 3 during the 4-point bending test with the strain rate of 0.05mm/min[5]. In this measurement, the newly developed sensors were used for highly-sensitive AE measurement[6]. The highly sensitive piezoelectric device was directly connected to a pre-amplifier and contained in one sealing case to reduce the noise. This sensor structure improved the signal to noise ratio by approximately 13dB compared with conventional AE sensors.

These high sensitive sensors were attached to both longitudinal ends of the specimen. The AE processor (AE9600, NF Circuit Block) was used for triggering. The detected waveforms were stored in a digital waveform memory (AE9620), transferred to the computer (HP310, Hewlett-Packard) and recorded into a floppy disk. This waveform memory has sampling time of 50 ns, voltage resolution of 10 bits, recording length for one event of 2000 words and dead time (data processing time in the memory) of 10 ms. The total gain of amplifier is 80 dB. The trigger level is 5.6×10^{-6} V at the input level of the built-in amplifier.

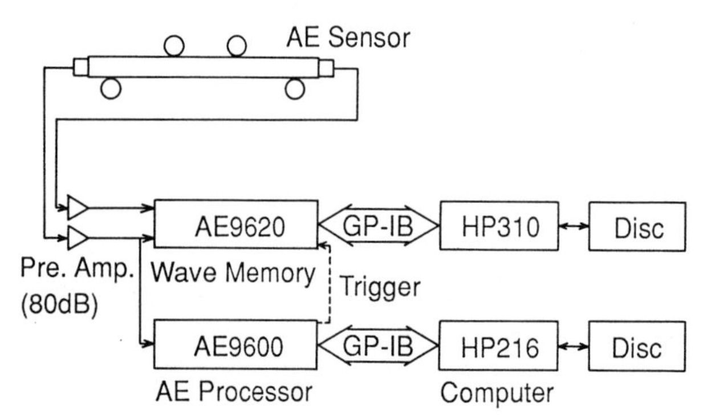

Fig. 3 Block diagram of 2-channel acoustic emission detection and recording system.

AE Waveform Analysis

Noises and AE signals were discriminated by directly observing the detected waveforms. Figure 4 shows the typical AE signal and noise waveforms. Pseudo-noises were recorded by striking the surface of a test piece set on the bending equipment, or by rubbing against the jig.

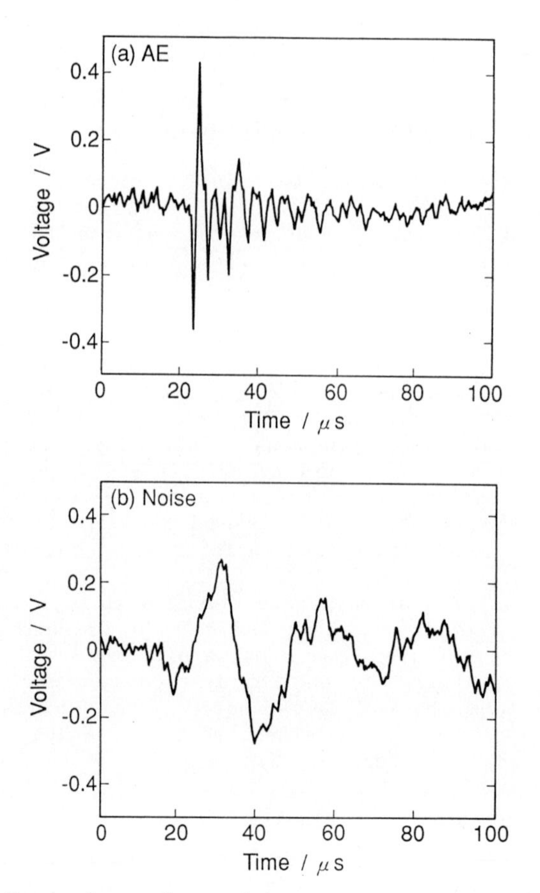

Fig. 4 Typical waveforms;(a) acoustic emission (b)noise.

The initial arrival of AE signals could not be distinguished easily because of the noise. The first peak should appear with a constant delay from the initial arrival resulting from the resonance of the sensors. The noise effect is the smallest because of the largest amplitude at the peak point. In this experiment, AE signal location was determined by the arrival-time difference of these peaks as shown in Fig. 5 and the P-wave velocity in each material. This method is superior to the conventional one, which depends on the threshold. The location error was estimated +0.8mm to -0.8mm for the specimen of 40mm in length.

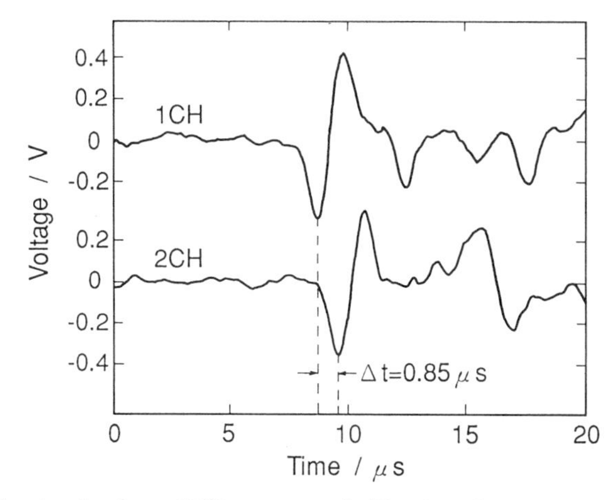

Fig. 5 Arrival-time difference of AE signals. Time difference obtained by using the first peaks.

RESULT

Microstructure of Sintered Mullite

Sintering temperature and sintering time have an appreciable effect on the porosity and the grain size as shown in Table 1. The aspect ratio of grains is approximately 1.6 for all types of mullite. The pores are found mainly at the triple point of granules (Fig. 6). The diameter of the pore is approximately 50μm.

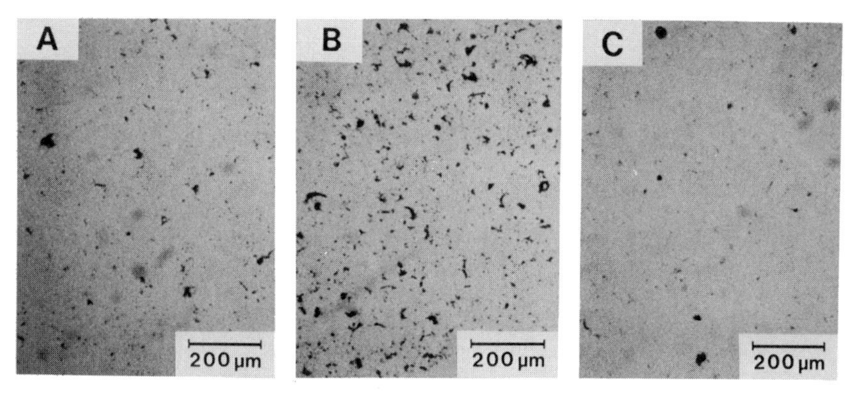

Fig. 6 Optical micrographs of polished surface of three types of mullite.

The glass phases are observed at regions near the triple point of grains by using a transmission electron microscopy (TEM) (Fig. 7). The glass phase contains more Si than the stoichiometric mullite (inside of grains), and the glass seemed to be SiO_2-rich glass.

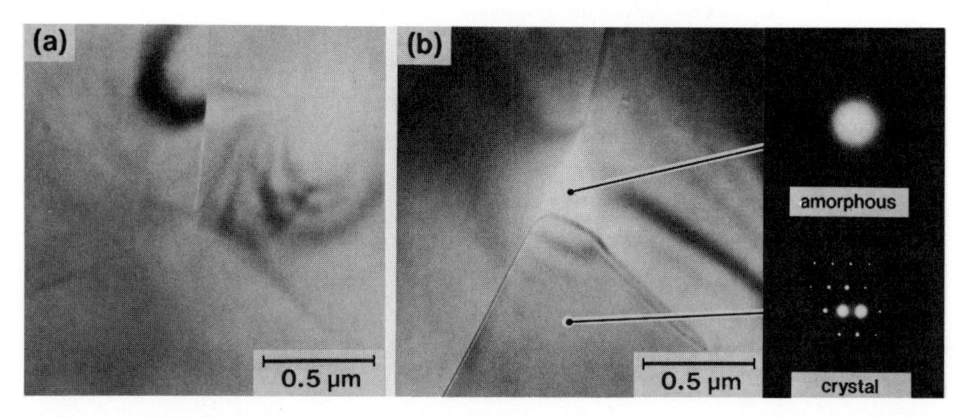

Fig. 7 TEM photographs near triple point in sample C; (a) non-glass (b)SiO_2-rich glass phase. The electron-diffraction patterns of amorphous (glass) and mullite crystal are also shown.

Effect of Grain Size on Fracture Toughness

The increase in the grain size increases the fracture toughness K_{IC} as shown in Fig. 8. This is an opposite tendency to the effect on the strength, as will be shown later.

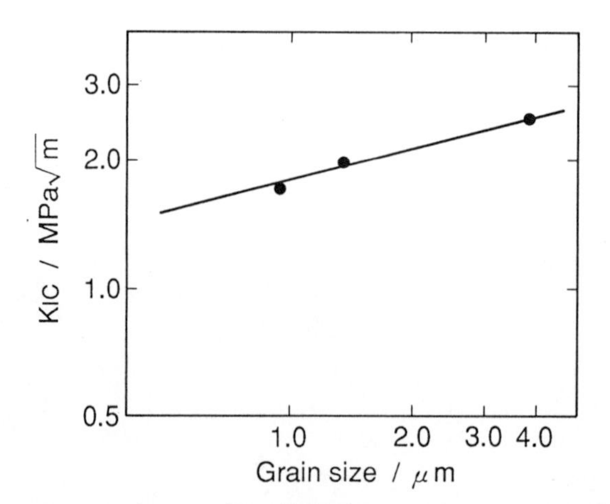

Fig. 8 Fracture toughness (K_{IC}(SEPB)) as a function of grain size.

K_I-V Diagram of Mullite

K_I-V diagram in N_2 gas, in the air and in the water are shown in Fig. 9. The plateau region of the K_I-V curve in the air (relative humidity= 65%) indicates that mullite shows stress corrosion cracking caused by water vapor.

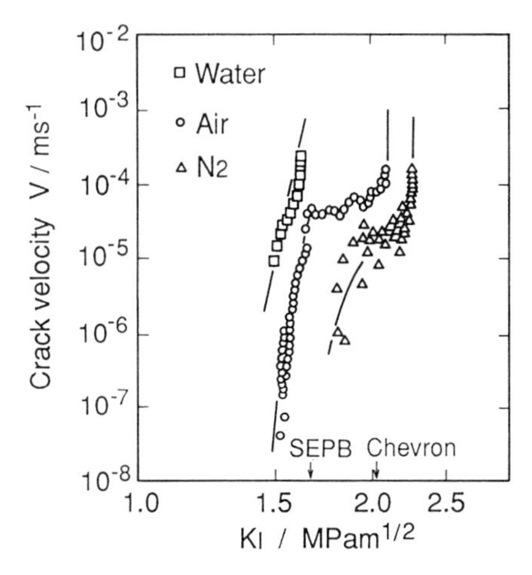

Fig. 9 K_I-V diagrams of sample C in different environments.

Effect of Loading Rate on Bending Strength

The effect of loading rate on the 4-point bending strength is shown in Fig. 10. The loading rate dependence of the bending strength seemed to be due to the stress corrosion cracking which was shown in the K_I-V diagram. The strength value changes linearly, which possibly means the same mechanism prevails in all cases of fracture. Therefore, it can be concluded that all tests carried out with other cross-head speeds can be treated as the same fracture phenomenon.

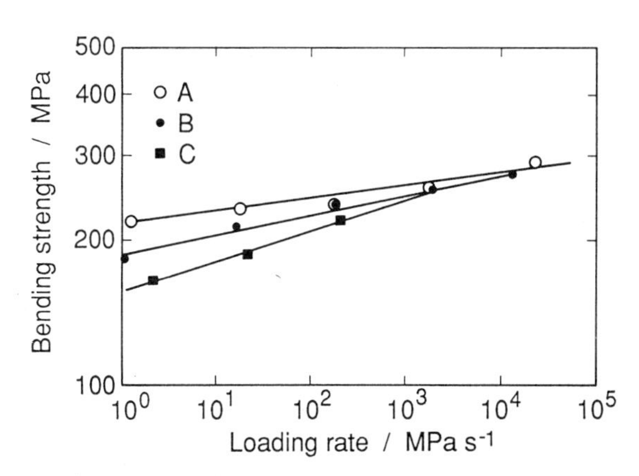

Fig. 10 Loading-rate dependence of 4-point bending strength.

Effect of Grain Size on Bending Strength

The decrease in the grain size increases the bending strength (Fig. 11). Through the SEM observations, the radius of the fracture origin is estimated as 20μm to 30μm when the defect is approximated as an ellipse or a semi-ellipse (Fig. 12). The increase in the grain size decreases the K_I value [denoted by K_{IC}(defect)] which is calculated from the size of fracture origin and the bending strength (Table 2). The effect of defect/grain size ratio on the K_{IC}(defect)/K_{IC}(SEPB) ratio is shown in Fig. 13. In the defect/grain size range smaller than 30, the K_{IC}(defect) value is found smaller than the K_{IC}(SEPB) value.

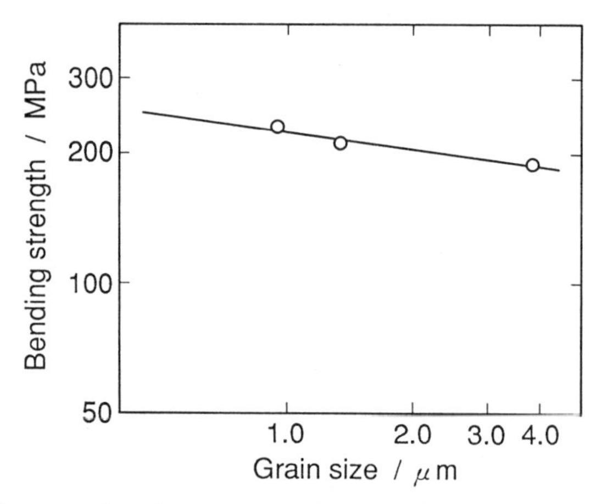

Fig. 11 4-point bending strength as a function of grain size. Cross-head speed was 0.5mm/min (JIS).

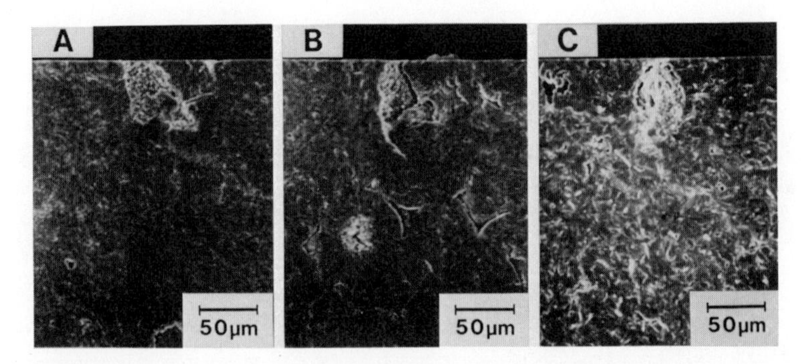

Fig. 12 SEM photos of fracture origins observed on fracture surfaces.

Table 2 Calculation of K_{IC}(defect).

Material	Bending strength (MPa)	Defect length (μm)	K_{IC}(defect) (MPa√m)	K_{IC}(defect) (Average) (MPa√m)	K_{IC}(SEPB) (MPa√m)	Grain diameter (μm)
A	250	25	1.41	1.46	1.73	0.95
	253	27	1.48			
	251	25	1.41			
	230	35	1.53			
B	231	25	1.31	1.19	1.98	1.36
	232	25	1.31			
	208	23	1.13			
	219	18	1.04			
	230	20	1.17			
C	194	20	0.97	1.10	2.50	3.82
	199	30	1.23			
	183	20	0.92			
	198	18	0.95			
	181	50	1.43			

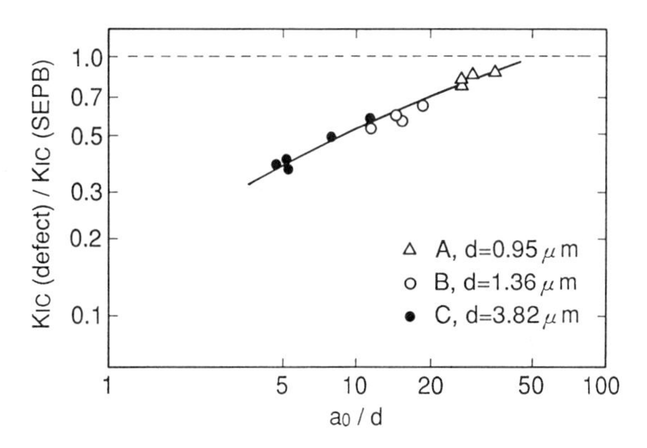

Fig. 13 K_{IC}(defect)/K_{IC}(SEPB) ratio as the function of a_0/d. Cross-head speed was 0.5mm/min.

Evaluation of Defect Growth by AE Technique

Figure 14 shows the AE location and cumulative AE event counts for three types of mullite with different grain sizes. AE signals begin to

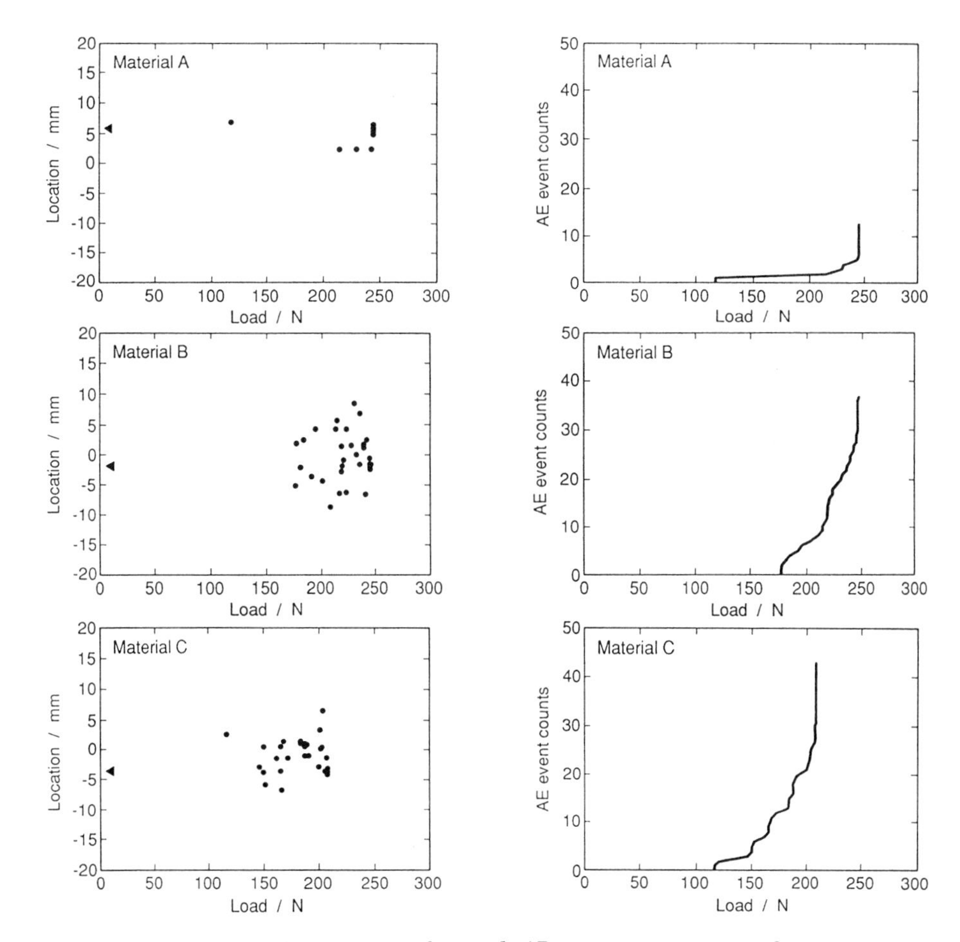

Fig. 14 AE source location and total AE event counts as function of load for three types of mullite. The 4-point bending test was carried out with 0.05mm/min of cross-head speed.

occur at approximately a half of the fracture strength. The increase in grain size increases the number of AE event counts: The average AE event counts were 8.5 for material A, 21.6 for material B and 28.3 for material C. The AE events generated from the fracture origins can be selected using AE location data.

It seems that the difference in the distribution of the AE location for each type of mullite results from the difference in the pore distributions (Fig. 6). It should be noted that the increase in grain sizes results in the increase in the amplitude of detected AE signals (Fig. 15). The sensitivity of adhered sensors was calibrated by using a breaking pencil lead[8]. According to the result, these sensors showed the almost similar response in calibration tests. This suggests that the amplitude of detected signals during bending tests is closely related to the size of the microcracks.

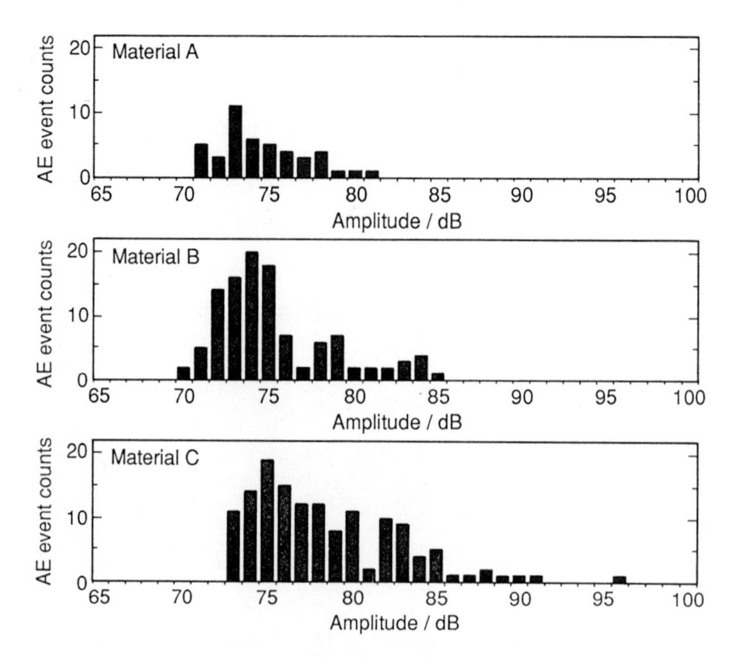

Fig. 15 AE signal amplitude distributions for three types of mullite. The total of 5 specimens is shown for each type.

Evaluation of Crack Growth from Artificial Defects

The AE measurement results for specimens with artificial defect are shown in Fig. 16. Initial artificial defect sizes on the fracture surface are shown in Table 3. The artificial defects introduced the concentration in AE generation from artificial defects. The stress level for the first AE detection (σ_{AE}) and the final failure stress are lower than those for specimens with natural defects. The stress concentration at the artificial defects are larger than that at natural defects. Then the failure from artificial defects occurred before the failure from natural defects. The increase in the indentation load increases the size of initial median-cracks, decreases the σ_{AE} value and increases the total AE event counts. The total AE event number is considered to be proportional to the crack extension area. There is no one-to-one correspondence between the growth ratio (final defect size-initial defect size) and the growth area (Fig. 17). The residual stress beneath of the indentation mark was not considered in the present analysis.

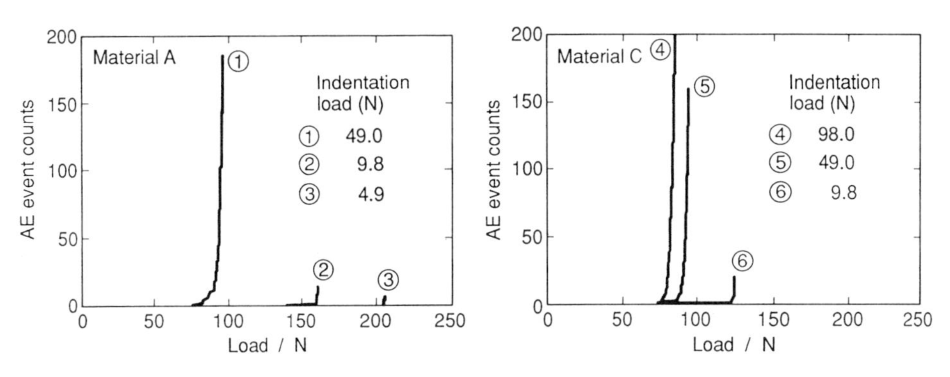

Fig. 16 AE event counts on the bending test with artificial defects. The cross head speed was 0.05mm/min.

Table 3 Summery of results on bending test with artificial defects.

Material	Vickers indentation load(N)	Initial crack length(μm)	K_{IC} (SEPB) (MPa$\sqrt{m}$)	Bending strength (MPa)	Final crack length(μm)	AE event counts	Crack area/ AE event ($\times 10^{-10}m^2$)
	4.9	25	1.73	215	51	7	4.4
A	9.8	40	1.73	159	93	14	7.9
	49.0	135	1.73	96	255	180	4.1
	9.8	54	2.50	123	327	18	91
C	49.0	154	2.50	94	556	154	29
	98.0	323	2.50	85	675	200	28

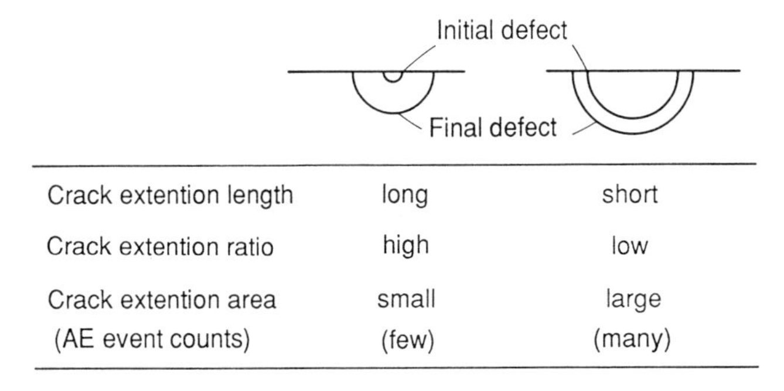

Crack extention length	long	short
Crack extention ratio	high	low
Crack extention area (AE event counts)	small (few)	large (many)

Fig. 17 Crack growth ratio and growth area.

Crack Growth Behavior During Cyclic Loading Test

Material C was used in this study. When several AE signals were detected during bending, the load was released immediately and the surface crack was observed using some penetrant. The AE results of two samples were shown in Fig. 18 and the photographs of specimen surfaces are shown in Fig. 19.

Two specimens show quite different types of AE behavior. In the case of C-1, AE signals begin to occur before the load reaches the previous maximum load. On the contrary, in the case of C-2, AE signals occurs just after the load is larger than the previous maximum load. Surface observations suggest that the AE sources in the C-1 specimen are surface cracks and those in C-2 specimen are inner cracks.

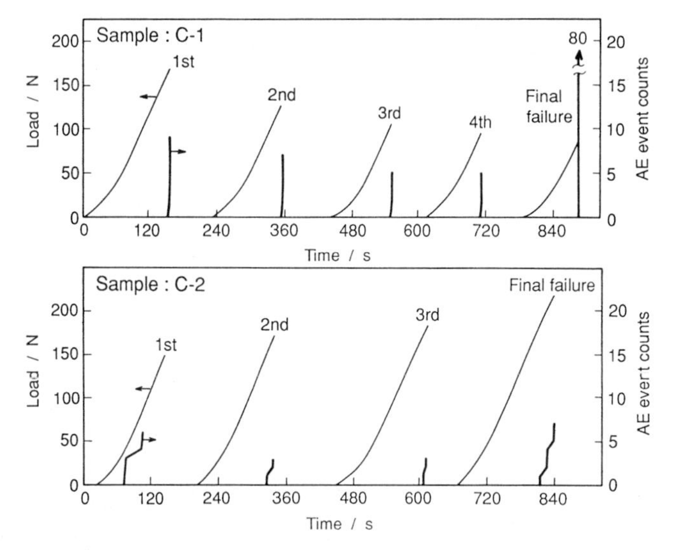

Fig. 18 AE event counts in cyclic loading test. The cross head speed was 0.05mm/min.

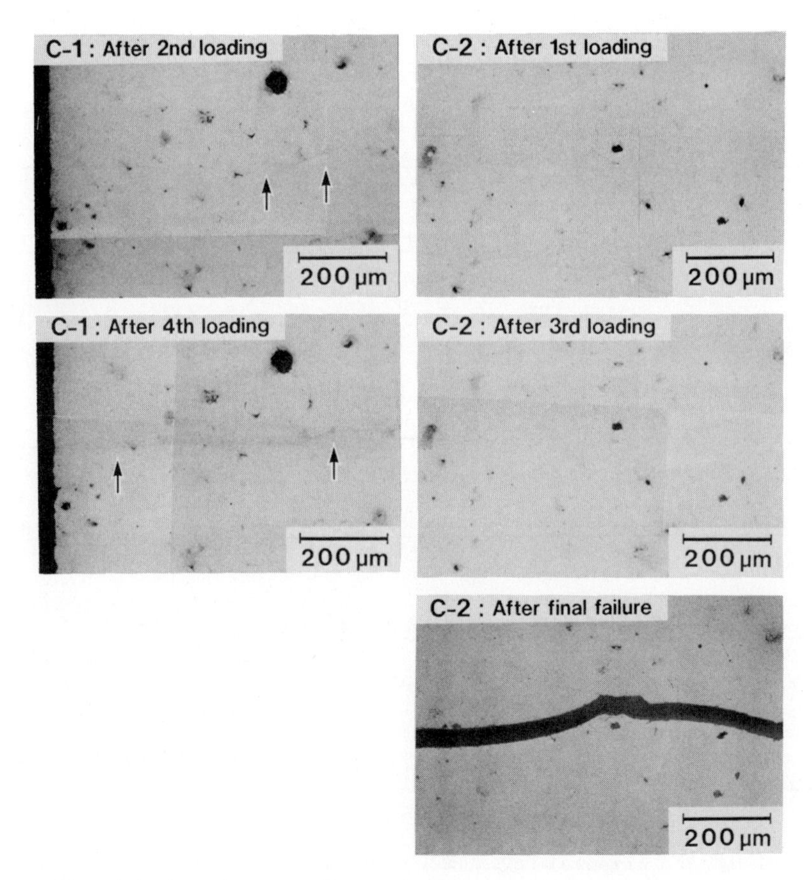

Fig. 19 Specimen surfaces during cyclic bending tests. (C-1)Surface crack (C-2)Inner crack.

DISCUSSION

Grain Size Effect on Fracture Toughness

The fracture toughness of mullite showed clear dependence on the grain size. The increase in the grain size increased the fracture toughness value as shown in Fig. 8. The mechanism of the fracture toughness increase could not be fully explained only with the present experiments, but some qualitative explanation is possible as follows: There are many studies on the grain-size dependence of fracture toughness for several ceramics. The studies on alumina should be the reference data to examine the behavior of mullite.

Rice et al.[3] showed that the increase in grain size increased fracture energy of alumina. The fracture energy showed the maximum value at 100μm in grain diameter. They suggested that fracture energy increment resulted from the microcracking mechanism. On the other hand, Nose and Kubo[9] proposed the other mechanism. The energy release rate was explained by the following equation. $\Delta G_{IC}=c\sigma_c{}^2 d/E$, $K_{IC}=K_0+c'\sigma_c d^{1/2}$ where c and c' are constants. This equation holds true under the assumption that the grain size does not change the fracture mode and the stored energy is consumed only to create change new surfaces.

The results of this experiment are qualitatively explained by the above two theories.

Decrease in Fracture Toughness for Specimens with Small Defects.

In the experiments conducted by other researchers, the grain size of ceramics were usually fixed and artificial defects of different size were introduced in the specimens. In this experiment, on the other hand, the defect sizes were approximately fixed and the grain size were changed by controlling the sintering condition. In both cases, the defect/grain size ratio were changed.

The decrease in fracture toughness for specimens with small defects has been reported by Evans and Langdon.[10], Rice et al.[13], and Miyata et al.[11] for other ceramics. The present study seems to detect the same phenomenon. The decrease in K_{IC} (defect) is observed if the defect size is smaller than 30 times of the grain size. Some researchers explained the phenomenon by the limit of the fracture mechanics equation applying for the small crack. If the crack size is close to the grain size, the ceramic material does not act as a homogeneous elastic medium, and the simple fracture mechanics equation cannot be applied directly because the final failure is supposed to occur from initial defects. On the other hand, the present AE measurement suggests the generation of microcracks in the ceramics before final failure. Their experiments were carried out in the air and the testing speed were 0.1mm/min or 0.05mm/ min, and the SCG behavior is expected in many ceramics[13] in such test conditions.

On the other hand, Evans and Langdon[10] suggested that the subcritical crack extension should be relatively larger (compared to the initial flaw size) for small flaws than for large ones, when slow crack growth occurs. However, he did not mention on the contradiction of the linear fracture mechanics equation. And the extension area were not estimated.

In the present experiment, crack growth areas are estimated qualitatively using AE measurement data, and the above phenomena were reasonably explained.

Estimation of Crack Growth by Using AE Results

Crack growth areas are estimated by using AE results with an assumption that microcracks grow continuously from defects.

The median-cracks extending from the Vickers indentation mark are semicircular as shown in Fig. 20(a). The two cracks are approximately of the same size and an arrow in Fig. 20(b) indicates an initial crack tip. Crack extension areas were estimated with assumption that the specimen fractured at the K_{IC} value measured in fracture toughness test and summarized in Table 3. These growth areas are divided by the total AE event count number to estimate the area for one AE event. The real area for one AE event should be smaller than this value, because there are some AE signals not detected during loading, but this crack growth area per one AE event is found close to the result from the surface observation of cyclic loading test in the previous section.

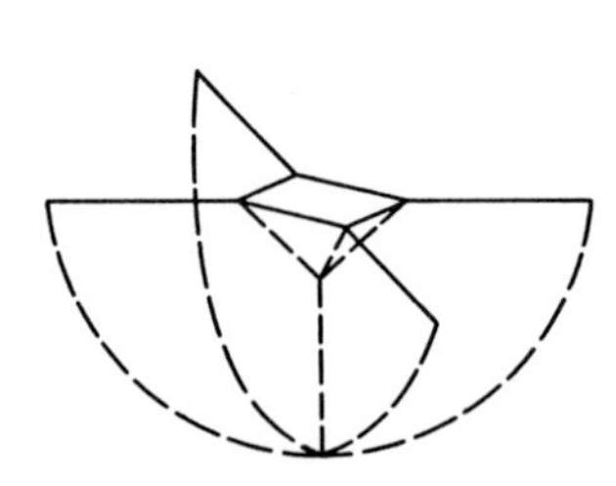

(a) Shape of median-crack

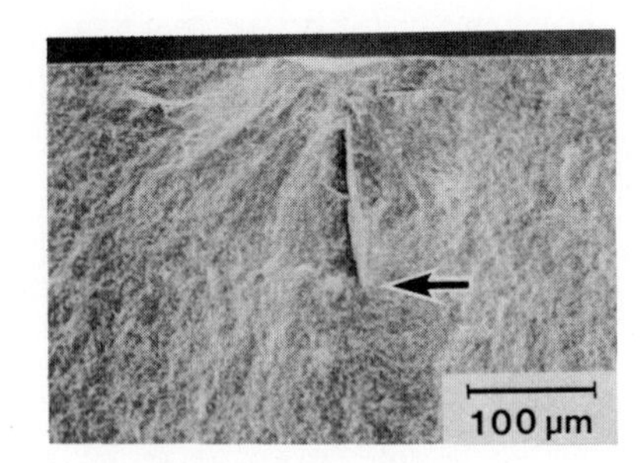

(b) Initial crack tip

Fig. 20 Median-crack introduced with a Vickers indentation.

The crack growth areas initiating from the natural defects are, then, estimated by multiplying the present AE event count number (generated from failure site ±0.8mm) by the crack growth area per one AE event obtained above. The results are shown in Table 4. The K_{IC}(final) value [K_{IC} (defect) value after crack growth correction] becomes closer to K_{IC}(SEPB) than that without considering crack growth. The increase in the defect/grain size ratio increases the change in the K_{IC} (defect) value. The increase in grain size increases the microcrack size and crack growth area.

Table 4 Estimation of defect growth and correction of K_{IC}(defect) by using AE measurement results.

Material	Initial defect length (a_0) (μm)	Bending strength (MPa)	K_{IC}(defect) / K_{IC}(SEPB)	AE event counts	Estimated Final defect length(a_f) (μm)	a_f / a_0	K_{IC}(final) / K_{IC}(SEPB)
A	50	185	0.85	4	62	1.25	0.95
	50	204	0.94	7	70	1.41	1.12
	60	219	1.10	1	63	1.05	1.13
	35	221	0.84	7	61	1.73	1.12
C	65	165	0.60	45	381	5.9	1.45
	42	173	0.51	16	228	5.4	1.18
	15	160	0.28	13	202	13.5	1.03
	22	192	0.41	9	169	7.7	1.13
	15	188	0.33	10	178	11.9	1.13

The Fracture Model Based on the AE Measurement

The results obtained in the present study are summarized as follows:

(1) Mullite generates AE signals approximately at a half of the maximum load during bending test. AE sources seem to be microcracks generated around the pore-defects.
(2) Continuous generation of AE signals is measured near the failure site just before the final failure. From these results, it is suggested that defects grow continuously before failure.
(3) There is little clear difference in crack growth and AE behavior between materials A and C with a Vickers mark in bending test. In the case of the fracture from natural defects, on the other hand, material C with a large grain size generates more AE events.

From the above results, it is concluded that the fracture of mullite ceramics from natural defects grows before final failure and changes from pore-shape to crack-shape. The ratio of initial defect/grain size ratio much affects on the growth rate, and the decrease in the ratio increases the crack growth area. Therefor, the difference between K_{IC}(defect) and K_{IC} (SEPB) is explained by considering the crack growth before final failure.

It is necessary to consider the difference in fracture toughness of single crystals and polycrystals[10,13]. The defects should grow up to the size which can be treated as a large crack when the final failure occurs. So the fracture toughness seems to become close to polycrystalline value.

The model of defect growth is proposed schematically in Fig. 21. In the range of small defect/grain size ratio, the crack growth area is large and the final failure occurs at a smaller stress. On the other hand, in the range of large a_0/d, crack growth is small and final failure almost occurs from the initial defect size.

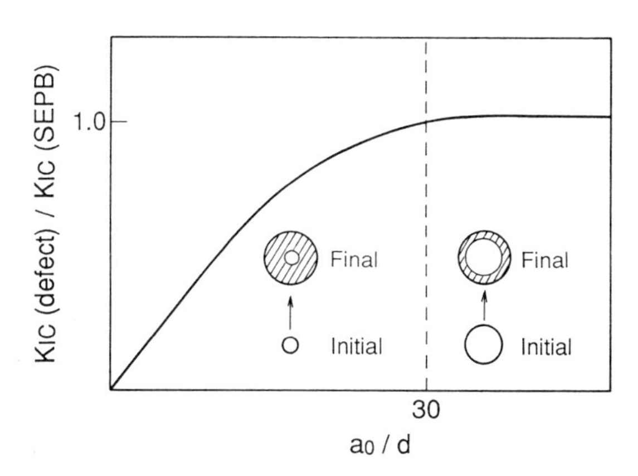

Fig. 21 Difference in crack growth behavior in different ranges of a_0/d. This explains the difference between K_{IC}(defect) and K_{IC}(SEPB).

In this experiment, mullite which shows SCG was used for the experiment. The source of microcracking is not necessarily due to the SCG mechanism. AE generation from the inner crack in cyclic loading tests is one of such evidences. The quantitative evaluation of microcracks is necessary for solving these problems. AE source analysis should be a most effective in the non-destructive evaluation.

CONCLUSION

In 4-point bending tests of mullite, the AE signals were detected before the final failure. The initial defect were extended before the final failure occurred. The number of AE signals increased with increasing grain size. In the case of large grains, defect growth seemed to be active.

The crack growth area in the bending test was estimated by the AE measurement, and the final failure seemed to occur after the K_{IC}(defect) value became close to K_{IC} (SEPB) value. If the defect growth is well considered, the final failure in the case of small size defect should be described by the LFM equation.

REFERENCES

1 H. P. Kirchner, R. M. Gruver and W. A. Scotter, Mater. Sci. and Enging., 22, 147-56 (1976)

2 I. Takahasi, S. Usami, K. Nakakado, H. Miyata and S. Shida, Yogyo-Kyoukai-Shi, 93 (4) 186-94 (1985) (in Japanese)

3 R. W. Rice, S. W. Freiman and P. F. Becher, J. Am. Ceram. Soc., 64 (6) 350-54 (1981)

4 M. Enoki and T. Kishi, "Acoustic Emission:Current Practice and Future Direction", ASTM STP 1077, ASTM, Philadelphia (1991) 47-66

5 Y. Yamade and T. Kishi, J. Ceram. Soc. Jpn inter. Ed., 98, 284-91 (1990)

6 M. Shiwa, H. Inaba and T. Kishi, Hihakai kensa, 39 (5) 374-82 (1990) (in Japanese)

7 Yoshiaki Yamade, Yoshiaki Kawaguchi, Nobuo Takeda and Teruo Kishi, Nippon Seramikkusu Kyoukai Gakujyutsu Ronbunshi, 99 (1148) 467-71 (1991)

8 N. N. Hsu, J. Simmons and H. C. Hardy, Materials Evaluation, 35, 100-06 (1977)

9 T. Nose and H. Kubo, Proc. Ann. Meeting Cer. Soc. Japan (1991) 455 (in Japanese)

10 A. G. Evans and T. G. Langdon, Prog. Mater. Sci., 21, 171-441 (1976)

11 H. Miyata, K. Nakakado and H. Kimoto, Nihon Zairyokyodo Gakkaishi, 21 (2) 45-59 (1986) (in Japanese)

12 S. M. Wiederhorn, "Fracture Mechanics of Ceramics, Vol.2", Plenum Press, New York (1973) 613-46

13 R. W. Rice, S. W. Freiman and J. J. Mecholsky, JR., J. Am. Ceram. Soc., 63 (3-4) 129-36 (1980)

EVOLUTION OF ELASTIC AND MECHANICAL PROPERTIES OF SILICA AEROGELS DURING GEL-GLASS TRANSFORMATION

F. Pernot, T. Woignier, and J. Phalippou

Laboratoire de Science des Matériaux Vitreux
C.N.R.S., U.R.A. 1119, U.S.T.L. Place E. Bataillon
34095 Montpellier Cédex 05, France

ABSTRACT

Pure silica aerogels are obtained from alcogels by hypercritical evacuation of the solvent, then partially densified by appropriate thermal treatments. Materials with relative densities ranging between 0.18 and 1.00 are thus prepared.

Their elastic and mechanical properties are then measured. It is shown that Young's modulus, strength and fracture toughness increase with increasing relative density; These evolutions are compared with various theoretical models.

Fracture energy does not show very significant variations as a function of densification.

INTRODUCTION

Silica alcogels obtained from alkoxide compounds are diphasic materials constituted by a solid and a neighbouring liquid. The solid part represents between 0.5 to 30% of the whole volume. It is extremely tenuous and brittle. Mechanical properties of silica alcogels were previously reported[1-4]

Alcogels dehydrated under usual conditions shrink and transform into xerogels, the mechanical properties of which were studied[2]. The measurements showed that these properties increased up to those of silica glass as a function of the sintering temperature.

Alcogels can also be dried under hypercritical solvent evacuation; the resulting materials, called aerogels, can be obtained free of cracks

Fracture Mechanics of Ceramics, Vol. 10
Edited by R.C. Bradt *et al.*, Plenum Press, New York, 1992

and are thus choice precursors for synthesis of very high purity silica glass[5].

The aerogel to glass transformation proceeds by a set of heat-treatments in the temperature range 500 to 1200°C. However, when suitable heat-treatments are used, partially densified aerogels (P.D.A.) can be obtained over a wide range of porosity (70 to 0%)[6,7].

In the literature most works dealt with the measurements of elastic and mechanical properties of silica aerogels[8-10]. A more complete study was therefore necessary on the elastic and mechanical behaviour of these materials during the aerogel-glass transformation. The experimental results are related to the textural properties of P.D.A..

MATERIALS AND EXPERIMENTAL DETAILS

Samples preparation

The gels are formed under neutral conditions by hydrolysis and polycondensation reactions of tetramethoxysilane [T.M.O.S. - $Si(OCH_3)_4$] in presence of methanol. The volume percentage of T.M.O.S. is equal to 46% and the H_2O/T.M.O.S. molar ratio is equal to 4. Alcogels are transformed into aerogels under hypercritical conditions for the solvent evacuation[5]. After hypercritical drying, the residual organic compounds are removed by an oxydation heat-treatment at 500°C. The aerogels are then sintered at 1100°C for different durations, thereby adjusting the final bulk density within the range 0.4 to 2.2 g/cm^3.

Methods of measurements

The bulk density, ρ_b, is measured from the weight of samples the dimensions of which are well known. The relative density, ρ, is obtained by the ratio of the bulk, ρ_b, to skeletal, ρ_s, density. The porosity, P, i.e. the fraction of pore space in the material is simply $(1 - \rho)$.

Flexural strength, σ_R, and Young's modulus, E, are measured on 1.5 x 3 x 18 mm bars. They are tested in three point bending with a 12 mm span in an "INSTRON" machine operating at a loading rate of 0.5mm/mn. All measurements are made in air at room temperature. Fracture bending stress and Young's modulus are calculated to be :

$$\sigma_R = \frac{3\ F\ L}{2\ b\ w^2} \qquad (1)$$

$$E = \frac{F L^3}{4 b w^3 \delta} \qquad (2)$$

where w is the sample thickness, b the sample width, L the span and δ the displacement for the applied load F.

With 3 x 6 x 38 mm bar specimens, fracture toughness K_{IC} is determined using the three point bend test of a single edge notched beam specimen (S.E.N.B.) over a 24 mm span. A notch as thin as possible of about 1.5 mm depth is machined at the midpoint of one 38 mm edge of each specimen. Samples are tested at a cross-head speed of 0.5 mm/mn. All measurements are carried out in air at room temperature. Fracture toughness is calculated from specimen dimensions, notch depth, a, and fracture load F [2,16] using the equations :

$$K_{IC} = \frac{3 F L}{2 b w^2} Y a^{1/2} \qquad (3)$$

w = sample height

$$Y = 1.93 - 3.07 \frac{a}{w} + 14.53 \left[\frac{a}{w}\right]^2 - 25.11 \left[\frac{a}{w}\right]^3 + 25.80 \left[\frac{a}{w}\right]^4 \qquad (4)$$

As shown previously[11,12], such conditions allow us to reach "true" initiating K_{IC} factor[11,13,14].

The knowledge of K_{IC} and σ_R allows calculation of other mechanical characteristics : critical flaw size, a_c, and fracture energy, Γ.

Critical flaw size (the size of the defect from which the fracture starts) may be obtained from[15] :

$$a_c = \left[\frac{Z}{Y} \frac{K_{IC}}{\sigma_R}\right]^2 \qquad (5)$$

Y : dimensionless term that depends on the crack depth and the test geometry[11],

Z : another dimensionless quantity that depends on the configuration of the crack[15].

Γ is given by[16] :

$$\Gamma = \frac{K_{IC}^2}{2 E} \qquad (6)$$

E : Young's modulus.

Since Γ is directly calculated from K_{IC} determined by S.E.N.B. method, it is also an initiation energy. In the calculation, Poisson's ratio is not taken into account since it has been pointed out that the stress state at the crack tip in porous materials is closer to plane stress than plain strain[17].

RESULTS AND DISCUSSION

Results of measurements are given in Table 1 : elastic and mechanical properties increase when increasing relative densities.

Table 1. Properties of partially densified aerogels

Bulk density (g/cm^3)	Relative density	Porosity (vol%)	Fracture stress (MPa)	Young's modulus (GPa)	Fracture toughness ($MPa\ m^{1/2}$)	Fracture energy ($J\ m^{-2}$)	Critical flaw size[a] (μm)
0.39	0.18	82	2.4	0.43	0.03	1.05	41.1
0.61	0.28	72	5.8	1.3	0.08	2.46	50.0
0.75	0.34	66	7.8	1.75	-	-	-
0.88	0.40	60	10.7	3.6	0.175	4.25	70.4
1.06	0.48	52	21.8	8.3	0.20	2.40	22.1
1.33	0.60	40	38.2	13.9	0.35	4.40	22.1
1.66	0.75	25	66.0	33.0	0.47	3.35	13.3
1.88	0.85	15	80.0	36.3	0.47	3.04	9.1
2.07	0.94	6	94.3	47.9	0.59	3.63	10.3
2.20	1.00	0	95.0	70.5	0.80	4.54	18.7

[a]Assuming a Griffith flaw : $(Z/Y)^2 = 1/1.21\,\pi$ in equation 5[18,19].

A number of equations have been proposed to relate elastic and mechanical properties to porosity of polycrystalline brittle solids[20]. Of these, a relationship has been proposed recently[21] :

$$Q = Q_s\,(1 - P)^n \quad \text{or} \quad Q = Q_s\,(\rho_b/\rho_s)^n \qquad (7)$$

where Q is the property studied (σ_R, E, K_{IC}, Γ), Q_s the same property for the solid. In contrast with classical semi-empirical relationships (linear or exponential approximation[20,21]), it obeys the boundary condition : Q = 0 for P = 1. This equation has been successfully applied to σ_R and E. The parameter "n" is defined as an empirical exponent in the literature previously mentionned[21]. However, it depends on the microstructure and has been calculated for cellular solids. The theoretical models proposed for such materials are then discussed and compared to experimental results.

Gibson and Ashby[22] define a cellular solid as a material made up of an interconnected network of solid struts or plates which form the edges and faces of cells. The simplest structure is a two-dimensional array of polygons, which pack to fill a plane area (*honeycombs*). More commonly, the cells are polyhedra, which pack in three dimensions to fill space (*foams*). If the solid of which the foam is made is contained in the cell edges only, the foam is said to be open-celled. If the faces are solid too it is said to be closed-celled.

The single most important structural characteristic of a cellular solid is precisely its relative density (ρ_b/ρ_s); it has been calculated as a function of the cell edge-length, ℓ, and the cell wall thickness, t; thus :

$$\frac{\rho_b}{\rho_s} \propto \left[\frac{t}{\ell}\right]^m \qquad (8)$$

where the exponent "m" depends both on the dimentionality of the structure (honeycomb versus foam) and on whether (if a foam) it has open or closed cells[22].

From these equations, expressions of elastic and mechanical properties of cellular solids have been proposed; for a regular hexagonal honeycomb[22,23] (Fig. 1) :

$$\frac{E}{E_s} \propto \left[\frac{\rho_b}{\rho_s}\right]^3 \qquad (9\ a)$$

and :

$$\frac{K_{IC}}{K_{ICs}} \propto \left[\frac{\rho_b}{\rho_s}\right]^2 \qquad (9\ b)$$

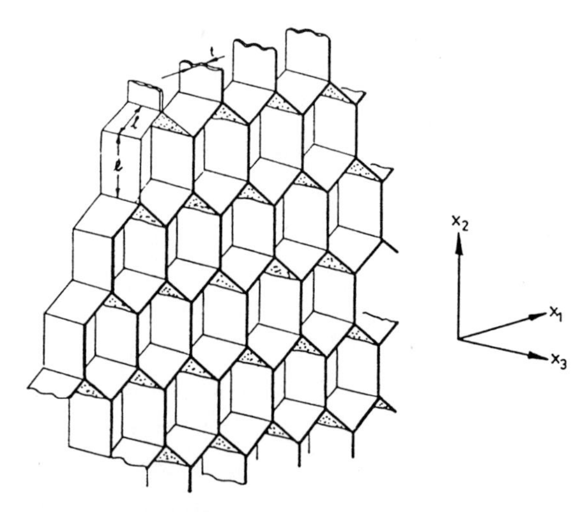

Fig. 1 A honeycomb with regular hexagonal cells from Gibson and Ashby[22]. Properties in the direction X_3 are not the same than in the plane X_1-X_2*.

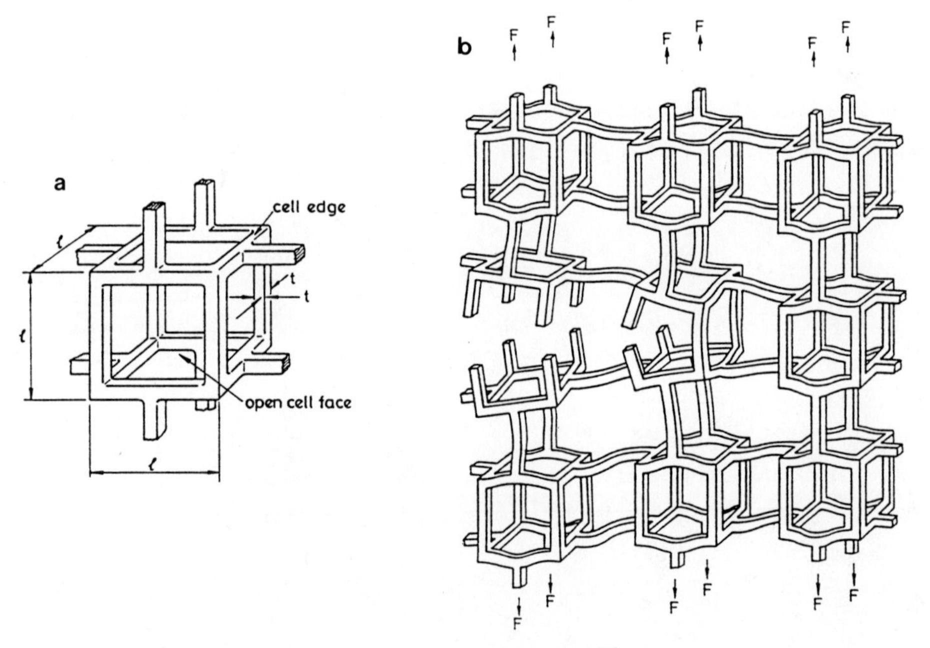

Fig.2 A cubic model for an open-cell foam[22] : a) a cubic cell showing the edge length, ℓ, and the edge thickness, t; b) propagation of a crack through a brittle cubic open cell foam.*

* Reprinted with permission of (L.J. Gibson, M.F. Ashby, "Cellular solids"), Copyright (1988), Pergamon Press PLC.

For an open-cell foam made of cubic polyhedra (Fig. 2 a) :

$$\frac{E}{E_s} \propto \left[\frac{\rho_b}{\rho_s} \right]^2 \qquad (10\ a)$$

and :

$$\frac{K_{IC}}{K_{ICs}} \propto \left[\frac{\rho_b}{\rho_s} \right]^{3/2} \qquad (10\ b)$$

This last relationship, due to Maïti[23] assumes that a crack in a foam extends in a discrete way, by successive failure of cell-edges, every step corresponding to a cell length (Fig. 2 b).

The analysis of the results obtained with partially densified aerogels on a LogLog scale shows that E varies as $(\rho_b/\rho_s)^{3.01 \pm 0.11}$ (Fig. 3 a) and K_{IC} as $(\rho_b/\rho_s)^{1.77 \pm 0.08}$ (Fig. 3 b).

At first sight, one might think that aerogels acts as regular honeycombs (see eq. 11 a & b); however, no structural studies of these materials sustain that they show an anisotropy (Fig. 1). This hypothesis must therefore be left definitely and a comparison with a three dimensional system is more suitable.

The analogy between partially densified aerogels and the cubic polyhedra open-cell foam is rather poor : only the exponent of K_{IC} (1.77 ± 0.08) show a fair agreement with that calculated by Maïti[23] (1.5); no agreement is noticed for the exponent of E (≈ 3.0) instead of 2 in the Gibson and Ashby model[22]. These discrepancies could be due to three reasons at least :

(i) All these models are deduced from the relationships between relative density and the ratio of the cell-edge length, ℓ, to the cell-edge thickness, t, which assume that t is much smaller than ℓ; in other words, the relative density should not be higher than 0.3. This is obviously not the case for most materials studied, the relative density of which reaches up to 1 (that of totally dense silica glass). During densification aerogels are no more foams in the sense of Gibson and Ashby[22].

(ii) When the density increases, porosity becomes partially closed; mixed models between open-cell and closed-cell foam both for E and K_{IC} would be perhaps more accurate. However such models, which would introduce more complex functions of (ρ_b/ρ_s), are not traduced by the experimental results, since each LogLog plot gives a single very good fit over the whole range of porosity tested (Fig. 3).

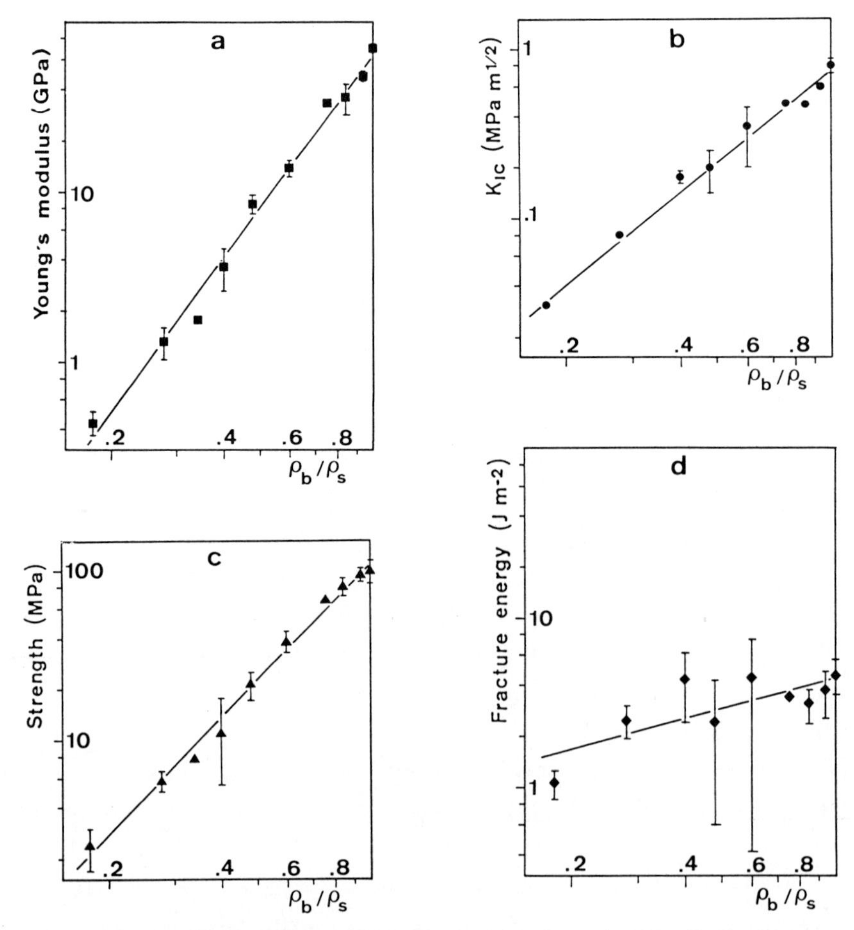

Fig. 3 LogLog plots of : a) E, b) K_{IC}, c) σ_R, d) Γ, as a function of (ρ_b/ρ_s)

(iii) More generally, the aerogel structure is very complex : SANS[24] and Raman[25] studies showed that the network could be described as a random of clusters, whose constitutive units have a size close to 1 nm. Such a structure could introduce unknown parameters, which in turn have not yet been introduced in theoretical models.

The evolution of flexure strength will be only briefly discussed; there, the LogLog plot (Fig. 3 c) also gives a good correlation; however, the exponent (2.28 ± 0.08) cannot be compared directly with the models, which consider the problem of brittle fracture in tension in term of K_{IC} only. Models discuss on crushing stress[22] [$(\sigma_{cr} \propto (\rho_b/\rho_s)^{3/2}$], which is equivalent to flexure strength only when critical flaw size is equal to a cell-length, ℓ. This is usually not true : critical flaw is more likely produced by fracture of several cells (see Fig. 2 b); anyway, the values of a_c (Table 1), which are not lower than 9 μm, obviously correspond to more than one cell-length.

For fracture energy, the LogLog plot (Fig 3 d) shows a slight increase with relative density; however, in view of measurement uncertainties, it is concluded that this increase is not very significant.

CONCLUSION

The evolution of elastic and mechanical properties of silica aerogels as a function of their porosity are satisfactorily traduced by power laws : σ_R varies as $(\rho_b/\rho_s)^{2.28 \pm 0.08}$, E as $(\rho_b/\rho_s)^{3.01 \pm 0.11}$, K_{IC} as $(\rho_b/\rho_s)^{1.77 \pm 0.08}$.

For K_{IC} and E the results show high discrepancies with previously proposed theoretical models. Some reasons have been suggested to explain these discrepancies; The complexity of aerogel structure could introduce unknown parameters so that new models should be proposed.

Simple models normally assume that all the solid material is connected and therefore contributes to the load, which is the case of Gibson and Maïti cubic polyhedra. In practice, the fraction of matrix solid that is connected is more likely not constant valued but increases as increasing density. It has been shown[24,25] that the structure evolves during the sintering. The size of the units increases and their spacial arrangement in the clusters is more dense. As a result, the stiffness of these materials does not only depend on the relative density but also varies as a function of the degree of interconnectivity of the solid network. Introducing this parameter in the equations would likely lead to better agreements between theory and experience.

REFERENCES

1. J.F. Quinson, J. Dumas, J. Serughetti, J. Phys. 46 C8 : 473 (1985).

2. M.J. Murtagh, E.K. Graham, C.G. Pantano, J. Amer. Ceram. Soc., 69 : 775 (1986).

3. G.W. Scherrer, S.A. Pardenek, R.M. Swiatek, J. Non-Cryst. Solids, 107 : 14 (1988).

4. T. Woignier, J. Phalippou, H. Hdach, G.W. Scherrer in : "Better Ceramics through Chemistry IV", "Mat. Res. Soc. Symp. Proceedings", vol. 180, B.J.J. Zelinski, C.J. Brinker, D.E. Clark, D.R. Ulrich, ed., Materials Research Society (1990) p. 1087.

5. J. Phalippou, T. Woignier, M. Prassas, J. Mater. Sci., 25 : 3111 & 3118 (1990).

6. J. Bouaziz, T. Woignier, D. Bourret, R. Sempere, J. Non Cryst. Solids, 82 : 225 (1986).

7. D. Bourret, R. Sempere, A. Sivade, J. Bouaziz, Rev. Phys. Appl., C 4 : 71 (1989).

8. M. Gronauer, A. Kadur, J. Fricke, "Aerogels" in : "Proceeding in Physics", J. Fricke, ed., Springer Verlag, Berlin, (1986) p. 6.

9. R. Calemczuk, A.M. De Goer, B. Salce, R. Maynard, A. Zarembovitch, Europhysic Letters, 3 : 1205 (1987).

10. T. Woignier, J. Phalippou, R. Vacher, J. Mater. Res. 4 : 688 (1989).

11. W.F. Brown, J.E. Srawley, in : "A.S.T.M. Special Technical Publication" n° 410, Baltimore, (1967), p. 1.

12. J.L. Chermant, F. Osterstock, G. Vadam, Verres Réfract. 33 : 843 (1979).

13. R.W. Davidge, G. Tappin, J. Mater. Sci. 3 : 165 (1968).

14. A.G. Evans, in : "Fracture Mechanics of Ceramics", vol. 1, R.C. Bradt, D.P.H. Hasselman, F.F. Lange, ed., Plenum Press, New York, (1974) p. 17.

15. A.G. Evans, G. Tappin, Proc. Br. Ceram. Soc. 23 : 275 (1972).

16. N. Miyata, H. Jinno, J. Mater. Sci. 16 : 2205 (1981).

17. N.A. Fleck, R.A. Smith, Powder Metall. 24 : 126 (1981).

18. J.L. Chermant, F. Osterstock, J. Mater. Sci. 11 : 1939 (1976).

19. R. Pabst, Z. Werkstofftech. 6 : 17 (1975).

20. R.W. Rice, in "Treatise on Materials Science and Technology", vol.11, R.K. MacCrone, ed., Academic Press, New-York, (1977) p. 199.

21. K.K. Phani, S.K. Niyogi, A.K. Maitra, M. Roychaudhury, J. Mater. Sci. 21 : 4335 (1986).

22. L.J. Gibson, M.F. Ashby, "Cellular Solids - Structure & Properties", Academic Press, New-York, (1988).

23. S.K. Maïti, M.F. Ashby, L.J. Gibson, Scripta Metal., 18 : 213 (1984).

24. R. Vacher, T. Woignier, J. Phalippou, J. Pelous, E. Courtens, J. Non Cryst. Solids, 106 : 161 (1988).

25. T. Woignier, J.L. Sauvajol, J. Pelous, R. Vacher, J. Non Cryst. Solids, 121 : 206 (1990).

A NEW THEORY OF NON-DESTRUCTIVE INSPECTION BASED ON FRACTURE MECHANICS AND FRACTURE STATISTICS

Y.Matsuo*, K.Kitakami** and S.Kimura*

*Department of Inorganic Materials, Faculty of Engineering
Tokyo Institute of Technology, Ookayama,Meguro-ku,Tokyo
**Canon Research Center, Atsugi,Kanagawa,Japan

ABSTRACT

Since ceramics is a typical brittle material, it is very important how to guarantee its strength reliability. In these years, Weibull's statistics, the proof test and the non-destructive inspection have been used for the above purpose. In this work, the statistical theory of non-destructive inspection (NDI) is presented considering the distribution functions of the fracture-location and the flaw-size at fracture origin, which are derived from the statistical theory of fracture-location (Oh-Finnie's theory) combined with competing risk theory and fracture mechanics. It is pointed out that one can select a screening-size only if the assumed stress field in N.D.I. is the same as the one in real loading. Furthermore, N.D.I. should be carried out within a definite region, called as a screening region, bordered by "N.D.I.-line" calculated from the joint cumulative distribution function.

1.INTRODUCTION

In order to assure the strength-reliability of ceramics, we must adopt a screening method, e.g., a proof test or a non-destructive inspection, for eliminating those components which involve large defects inside themselves or on their surfaces. The theories of proof test have been developed by many authors[1-3] and the applications to the practical uses are being performed [3]. However, it is necessary to make great efforts to perform the proof test, and there still remaines a doubt whether the components passed through the proof test are weakened or not.

On the other hand, since N.D.I. gives no damage on the components, it has been successfully adopted for metalic materials from early stage. Concerning ceramics, the technical development of N.D.I. has been achieved recently; for instance, the defects of 30-50μm in size can be detected by microfocus X-ray or ultrasonic microscope[5]. However, no valid theory has been presented for N.D.I. based on the distribution of the location and the size of the defects. Also there is no theory of N.D.I. in which the effects of stress states and the component configulation are taken into consideration.

The authors conclude that in order to establish the theory of non-destructive-inspection for ceramic materials it is essential to know the distributions of the location, the size and the orientation of the fracture origins.

In the following, first, we derive a general theory of brittle fracture by combining the statistical theory of fracture location[4] based on the competing risk theory[6] with the multiaxial Weibull distribution function[8]. On the basis of these analysis we suggest a new theory of non-destructive inspection for ceramics and show how to determine the screening flaw size and the screening region.

2. THE THEORY OF FRACTURE LOCATION COMBINED WITH COMPETING RISK THEORY AND THE FRACTURE MECHANICS

Suppose that a body invoving n kinds of fracture causes is subjected to a standard stress in a range (σm, $\sigma m+d\sigma m$) and fails by a cause Ci(the weakest flaw, called as a fracture origin) located at a small region (ξ, $+d\xi$) within the existing region Ai (i=1~n). Also suppose that the orientation of the flaw at the fracture origin lies in a range ($\alpha,\alpha+d\alpha$). Then the simultaneous distribution function for fracture is given by [9]

$$h_A(\sigma_m,\xi,\alpha)=\prod_{i=1}^{n} R_i\cdot\sum_{j=1}^{n}\lambda_j \tag{1}$$

where, Ri and λj are the reliability function and the failure rate with respect to fracture cause Ci, respectively, which are given as follows;

$$R_i=1-\int_0^{\sigma_m}\int_{\xi_t}\int_{\alpha_t} h_{A_i}(\sigma_m,\xi,\alpha)d\alpha d\xi d\sigma_m,\qquad \lambda_j=h_{A_j}(\sigma_m,\xi,\alpha)/R_j \tag{2}$$

The subscript A means the total existing region of individual fracture causes given by the direct sum of Ai as

$$A=A_1\oplus A_2\oplus\cdots\oplus A_n \tag{3}$$

The joint probabirity density function hAi($\sigma m,\xi,\alpha$) for fracture cause Ci, which is sharp enough that it is regarded as a crack, is given by

$$h_{A_i}(\sigma_m,\ \xi,\ \alpha)d\alpha d\xi d\sigma_m=\exp(-B_i)\frac{\partial}{\partial\sigma_m}(G_i)d\alpha d\xi d\sigma_m,\quad B_i=\int_{\xi_t}\int_{\alpha_t}G_i d\alpha d\xi,$$
$$G_i=\frac{2}{\pi}\left(\frac{z_i-\sigma_{u_i}}{\sigma_{oi}}\right)^{m_i}\cdot Y(z_i,\ \sigma_{ui})\left(\frac{d\tilde{A}}{d\alpha}\right)\left(\frac{dA_i}{d\xi}\right). \tag{4}$$

where the subscript t means the total region of each variable; Bi is a risk of rupture; mi,σoi and σui are Weibull's parameters; $d\tilde{A}$ is a surface element of a unit sphere; Y(,) is Heaviside's step function; Zi is an equivalent normal stress[10] defined by

Zi=K1c/Yi$\sqrt{\pi c}$, K1c: fracture toughness, Yi: geometric factor,
c: a representative dimension of a crack

In general, α is a 2-dimensional vector which has two independent components, θ and Φ, related to the space angles (see Fig.1). Equation (1) is valid for an arbitrary stress state and arbitrary types of fracture origins.

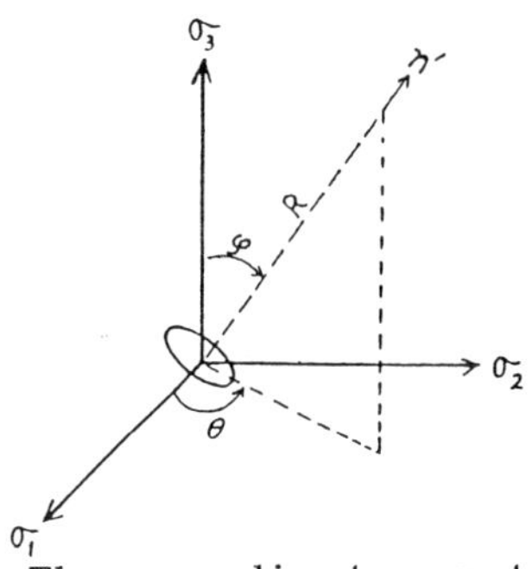

Fig.1 The coordinate systems aroud a crack

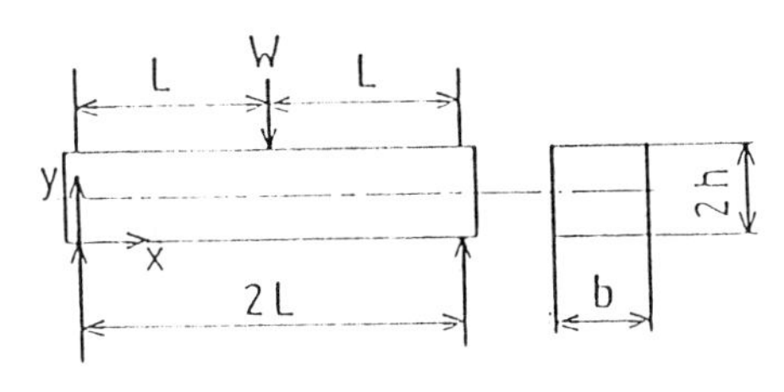

Fig.2 The coordinate systems in the 3-point bending test

3. ANALYSIS OF 3-POINT BENDING TEST

Using the general equation derived in the above, we analyse the 3-point bending test of a rectangular cross-sectioned ceramic specimen as a typical example (see Fig.2). For simplicity, we assume that there is only one type of fracture cause, namely, penny shaped crack (i=1). Also we assume that the location parameter $\sigma u1$ is equal to zero. As the first step, we employ the shear insensitive criteria for mixed mode fracture of a crack.

The equivarent normal stress Z1 is given by

$$Z1=\sigma\cos^2\Phi,\ \sigma=\sigma m\cdot x(h-y)/Lh,\ \sigma m=\sigma max=3WL/4bh^2 \qquad (5)$$

where σmax is the maximum principal stress acting in a specimen. Then, we obtain the joint probability density function for 3-point bending test as

$$h_{A_1}(\sigma_m,\xi,\alpha)d\alpha d\xi d\sigma_m = 2bm_1\left(\frac{2}{\pi}\right)\sigma_m^{m_1-1}\left\{\frac{x(h-y)}{\sigma_{01}Lh}\cos^2\phi\right\}^{m_1} \times exp\left\{-\frac{V_{e0}}{(2m_1+1)}\left(\frac{\sigma_m}{\sigma_{01}}\right)^{m_1}\right\}\sin\phi d\phi d\theta dx dy d\sigma_m \qquad (6)$$

where Veo is a non-dimensional effective volume. In Eq.(6) it is assumed that the crack plane is parallel to the thickness direction of a specimen. Therefore, $d\alpha=d\Phi$.

3.1 THE DISTRIBUTION FUNCTION OF FRACTURE LOCATION

The distribution function of fracture location along the specimen axis x is obtained as a marginal with respect to x from Eq.(6) as

$$H_{A_1}(x)=\left(\frac{x}{L}\right)^{m_1+1} \qquad (7)$$

The distribution function of the fracture location along y axis is derived below in a similar fashion as a marginal with respect to y,

$$H_{A_1}(y)=1-\left(\frac{h-y}{h}\right)^{m_1+1} \qquad (8)$$

Equations (7) and (8) exactly coincide with the ones proposed in the previous work[6,7] in which the basic equation was derived from ordinary Weibull function (uniaxial function). The varidity of these two equations were already proved in the previous work[6,7] using the experimental results carried out by Ito et al.[11].

3.2 THE DISTRIBUTION FUNCTION OF FLAW-ORIENTATION

The probability density function of flaw-orientation is obtained as a marginal with respect to flaw-orientation angle Φ as

$$h_{A_I}(\Phi)=(2m1+1)\cos^{2m1}\Phi \tag{9}$$

Figure 3(a) shows the calculated flaw-orientation probability density function. The density function shifts toward low angle and becomes sharp as m1 value increases. This means that the smaller the scatterness of fracture strength becomes, the smaller the flaw-orientation angle at fracture origin does. The calculated results mentioned above were derived from shear insensitive criteria. It goes without saying that we can adopt any other criteria for mixed mode fracture, for example, co-planar extension condition (Gc criteria) or maximum hoop-stress criteria. Figures 3(b) and (c) show the effect of the fracture criteria on the probability density function of the flaw-orientation. The density function is strongly

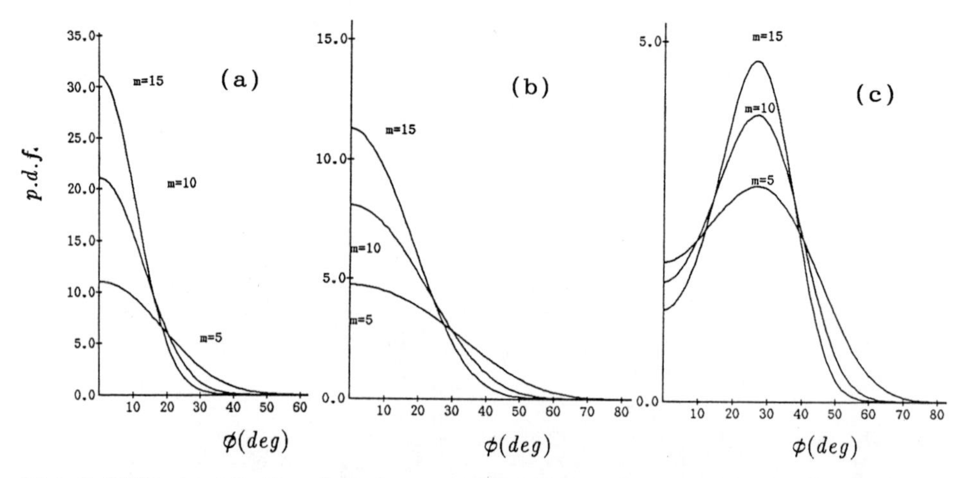

Fig.3 Effects of the fracture criteria on the probability density function of flaw-orientation. (a) shear insensitive (b) co-planar extension condition (c) maximum hoop-stress

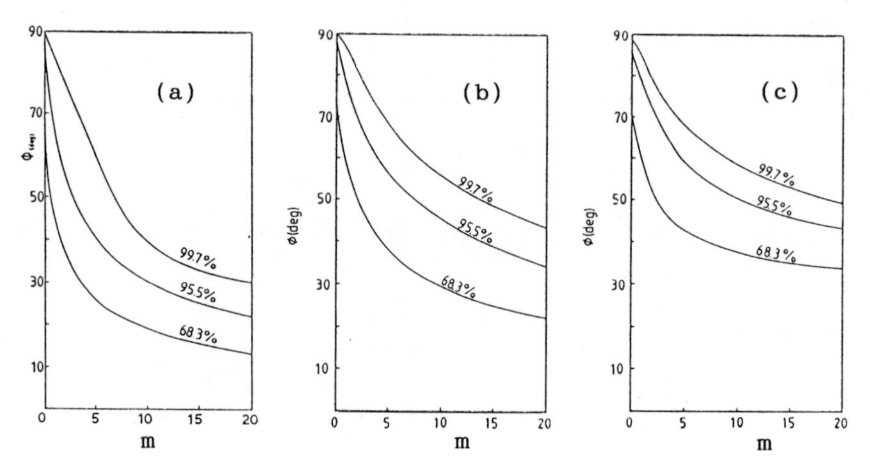

Fig.4 The angle which gives constant fracture probability (constant probability curve) (a) shear insensitive (b) co-planar extension condition (c) maximum hoop-stress

influenced by the fracture criteria. As long as the results shown in Fig.3, the maximum hoop-stress criteria shows the most wide distribution, namely, the most conservative estimate.

We calculated the angle which gives the constant fracture probability as a function of $m1$ by integrating the probability density function. We call this as a constant probability curve. Figure 4 shows the constant probability curves for three different types of fracture criteria. The curve is also strongly influenced by the fracture criteria. In Fig.4(a) (shear insensitive criteria), when $m1=10$, 99.7% of the specimens will fail within the angle less than 40 degree; on the contrary, 56 degree for co-planar extension condition and 58 degree for maximum stress criteria, respectively. These results may play an impotant role on non-destructive inspection. In the next paper, we will deal with this problem in detail.

3.3 THE DISTRIBUTION FUNCTION OF THE FLAW-SIZE AT FRACTURE ORIGIN

By changing the variable σm into the critical flaw-size dc through following two relations,

$$\sigma m=\sigma c\cdot Lh/x(h-y),\quad K1c=2/\pi\cdot\sigma c\sqrt{\pi dc/2}\quad(\Phi=0) \tag{10}$$

the joint distribution function with respect to the flaw size dc, x and y are given by

$$h_{A_I}(d_c,x,y) = bm_1\left(\sqrt{\frac{\pi}{2}}\frac{K_{IC}}{\sigma_{01}}\right)^{m_1}\cdot d_c^{-(m_1+2)/2} \times exp\left[-V_{e0}\left\{\sqrt{\frac{\pi}{2}}\frac{K_{IC}}{\sigma_{01}}\frac{Lh}{x(h-y)\sqrt{d_c}}\right\}^{m_1}\right] \tag{11}$$

The distribution function of the flaw size is calculated from this equation as a marginal of dc. Using Eq.(11), we analyse the three-point bending test data for HP-Si3N4 performed by Ito et al.[11]. The sample size of them is 415; 326 of them were failed by inner flaws, 77 were failed by surface flaws and 12 were unknown. The Weibull parameters used in the following calculations were estimated by multi-maximum likelihood method together with improved E-M algorithms[12,13] (see Table 1). Figure 5 shows the histogram of inner flaw-size at fracture origin observed by Ito et al. The theoretical values (open circles connected by solid lines) are almost coincide with the experimesntal data but slightly different from the ones in the region of small flaw-size. However, when we adopt the dependency rule of K1c value on the ratio between the flaw size and the grain size of the ceramics[14], the estimated values (closed circles connected by dotted lines) are improved in sufficient coincidence with the experimental results as seen in Fig.5.

Table 1 Estimated Weibull Parameters, K_{IC} and the dimensions of the HP-Si_3N_4 3-point-bend-specimen

Estimated Weibull Parameters		K_{IC} and Dimensions	
m_1	15.79	K_{IC}	4.06(MPa$\sqrt{m}$)
$\sigma_{\theta 1}$	959.9(MPa)	h	1.5(mm)
m_2	12.73	b	3.0(mm)
$\sigma_{\theta 2}$	1295(MPa)	L	10.0(mm)

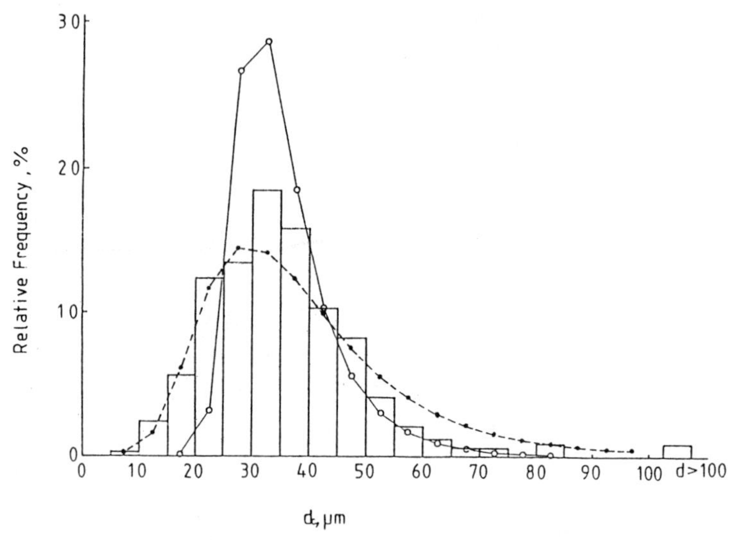

Fig.5 Histogram of flaw-size in 3-point bending of HP-Si3N4. (o signs: K1c=4.06 MPa$\sqrt{m}$, o signs: K1c is a function of the ratio between the flaw-size and the grain size)

4. THEORY OF NON-DESTRUCTIVE INSPECTION AND DISCUSSIONS

In the following sections a new theory of non-destructive inspection for brittle materials like ceramics is suggested on the basis of the analytical results obtained in the above sections. In this theory it is essential how to determine the screening the screening flaw-size (the threshold value) as well as the screening region.

4.1 DETERMINATION OF SCREENING FLAW-SIZE[15]

The first step to perform non-destructive inspection on ceramic materials is to determine the screening flaw-size (the threshold value). According to our work[14], the distribution function of the flaw-size at the fracture origin is not unique; it varies as the stress state and/or the dimensions of the specimen changes as we can easily see from Eq.(11). Figure 6 shows the probability density functions of the flaw-size at the fracture origin in case of 3-point bending and uniaxial tension. We see from this figure that the mode value in uniaxial tension is about 20μm larger than that of 3-point bending. Also we see that the volume of the specimen affects the probability density function strongly. Therefore, in determining the screening flaw-size, we should take the above results into consideration.

Now we suggest a method for determining the screening flaw-size as follows: The screening flaw-size should be selected from the flaw-size density function, for instance, dpu in tension as shown in Fig.6. The screening region is taken in the whole volume of the specimen. If the specimens which passed through the non-destructive inspection were subjected to 3-point bending, almost all the specimens would fail from the flaws less than dpu, as is easily seen from Fig.6. Therefore, in this case, the screening is invalid. On the contrary, if we determine the screening flaw-size dpl from the flaw-size density function in 3-point bending, and make screening in the whole body, almost all the specimens may be rejected, because almost all the specimens include the flaws larger than dpl.

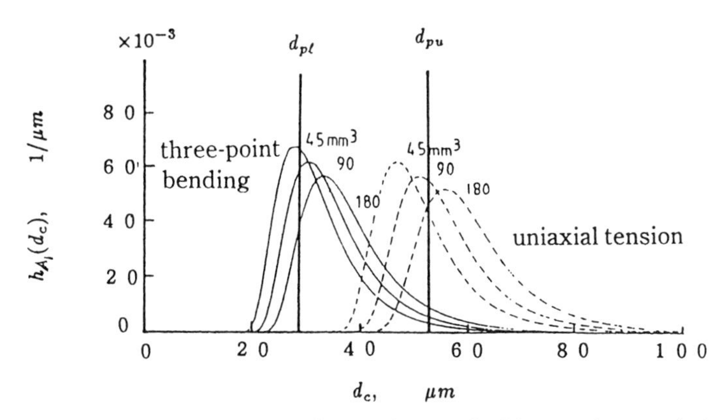

Fig.6 Probability density function of flaw-size. dpl and dpu are screening sizes from 3-point bending and from uniaxial tension, respectively.(m1=15.79, σo1=959.9)

There is one case which is valid. First let's select a screening flaw-size, dpl, from the flaw-size density function for 3-point bending, and make screening over the proper "screening region", which is defined in the following section; after that impose 3-point bending load to the survivals. Then non-destructive inspection is effective.

In the following section we suggest that the non-destructive inspection is valid only if the screening region is specified in a specific region determined by the joint probability density function of fracture location which depends on the stress state and the dimensions of the specimen.

4.2 DETERMINATION OF SCREENING REGION[16]

We also analyse 3-point bending problem as an example. The same treatment stated below is avalable for arbitrary state of stress and arbitrary configulation of a component. The joint probability density function of the fracture location (x,y) for inner flaw is derived from Eq.(6) as

$$h_{n1}(x,y) = \frac{(m_1+1)^2}{Lh}\left(\frac{L-x}{L}\right)^{m_1}\left(\frac{h-y}{h}\right)^{m_1} \tag{12}$$

here, for convenience of analysis, the origin of the coordinate axis is taken at the point A in Fig.2 where maximum stress arises. Figure 7 shows the calculated joint probability density function of Eq.(12) using Weibull parameters and the dimensions of a specimen shown in Table 1. It is seen that the neighbourhood of the maximum stressed point A takes high value of probability density. Figure 8 shows the joint equal-probability density curves (p.d.f.) drawn on x-y plane. In the neighbourhood of the point A, p.d.f. takes high value, and it becomes higher as m1 value increases.

The joint cumulative distribution function is obtained by integrating the density function (Eq.(12)) along the equi-probability density line as

$$H_{n1}(x,y) = \left\{1-\left(\frac{L-x}{L}\right)^{m_1+1}\right\}\left\{1-\left(\frac{h-y}{h}\right)^{m_1+1}\right\} \tag{13}$$

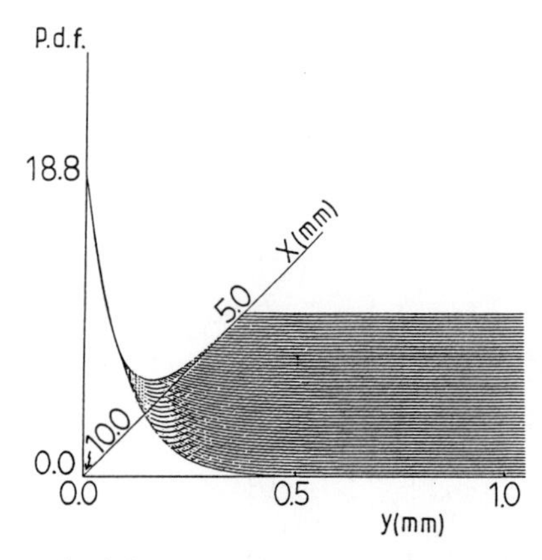

Fig.7 Joint probability density function of x and y

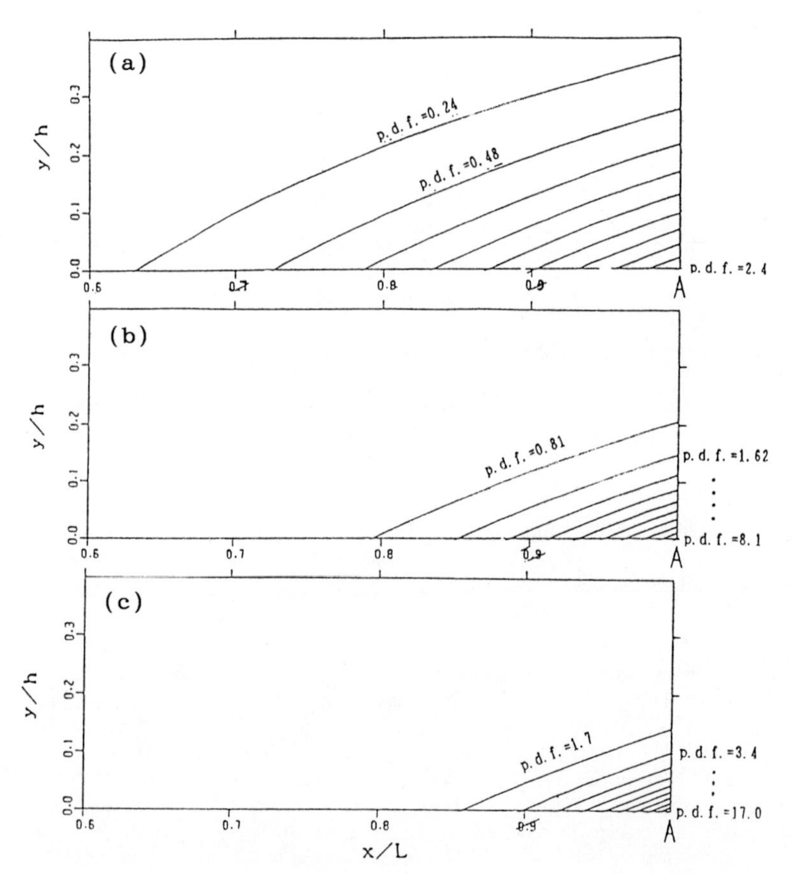

Fig.8 Joint equal-probability density curves drawn on xy plane (a) m1=5, (b) m1=10, (c) m1=15.

The solid lines in Fig.9 show the profiles of the particular regions which give constant fracture probabilities, say, 68.3%, 95.5% and 99.7%. The calculated lines tend to expand as the shape parameter m1 decreases. The screening region should be determined from these figures. If 0.3% fracture probability is permitted, we can determine the screening region within the solid line which gives 99.7% fracture probability. Therefore, we call these lines as NDI-lines.

It is usefull if NDI lines are drawn in normarized coordinates for given fracture probability and shape parameter. Figure 10 shows the linearized NDI-lines, which we call from now on as "screening diagram". From these figures we can find the screening region neccessary in non-destructive inspection.

The new method for non-destructive inspection suggested in the above is a "sufficient condition". In other words, it gives safety screening standard. If the screening is performed in neccessary and sufficient conditions, the non-destructive inspection should exactly coincide with so-called proof test[17].

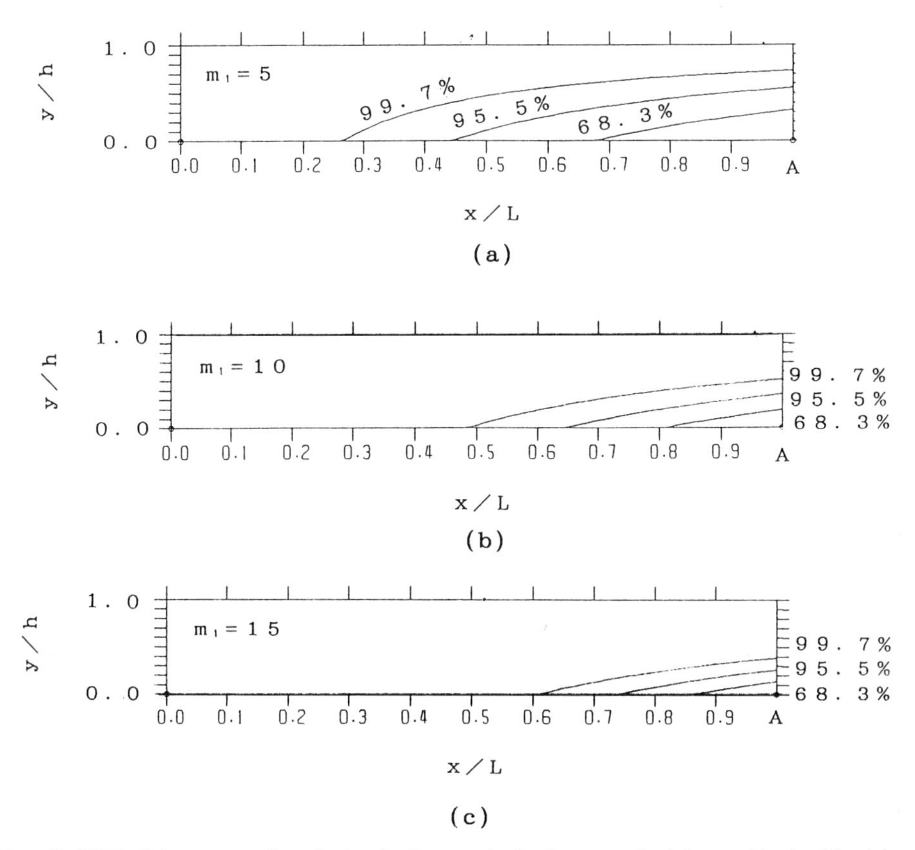

Fig.9 NDI-lines calculated from joint cumulative distribution function, Eq.(13). (a) m1=5, (b) m1=10, (c) m1=15.

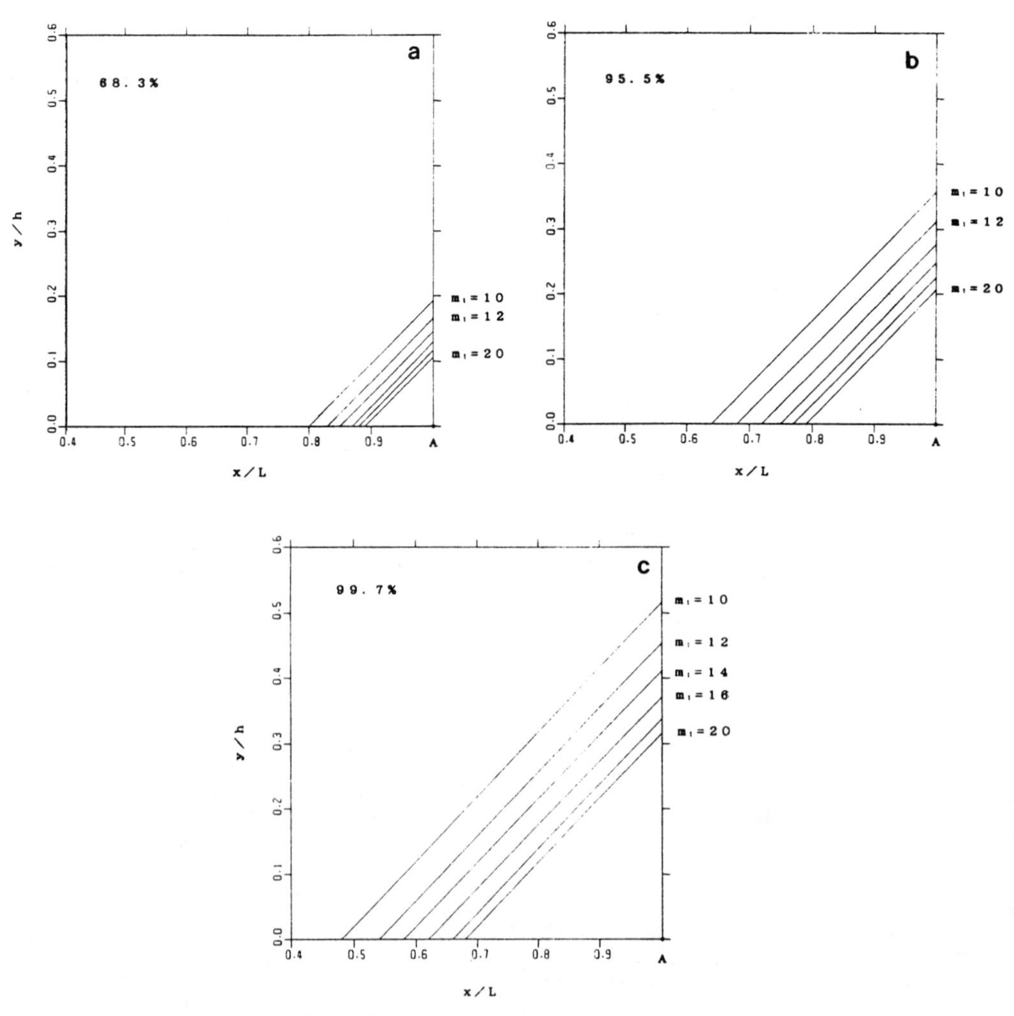

Fig.10 Screening diagrams.
(a) 68.3% ($\pm 1\sigma$)
(b) 95.5% ($\pm 2\sigma$)
(c) 99.7% ($\pm 3\sigma$)

5.CONCLUSIONS

A new theory which involves three independent random variables (fracture stress σm, fracture location !, and fracture orientation α) is formulated by combining the previous work with the multiaxial distribution function. From this theory, we derived the distribution function of the fracture location, flaw-orientation and flaw-size. The angles which give a constant fracture probability are calculated.

Based on these results, we suggested a new theory of non-destructive inspection, namely;

(a) The screening flaw-size should be determined from the probability density function and/or the cumulative distribution one for the "actual loading condition", since the flaw-size distribution is not unique but depends on the stress state as well as the specimen configurations.

(b) The screening region should be calculated from the cumulative distribution function for fracture location (x,y). For this purpose, we suggested NDI-lines and screening diagrams.

References

[1] Wiederhorn,S.M., in Fracture Mechanics of Ceramics (Eds. Bradt, R.C.,et al.),2(1974),613, Plenum Press.
[2] Matsuo,Y., J.Soc.Mat.Sci.Japan,33-370(1984),857.
[3] Hamanaka,J.,et al.,Trans.JSME.,52-481,A(1986),2187.
[4] Oh,H.L. and Finnie,I., Int.J.Fracture,6(1970),287.
[5] Munz,D. and Rosenfelder,O.,in Fracture Mechanics of Ceramics,Vol.7 (Eds.Bradt,R.C. et al.),(1986),265, Plenum Press.
[6] Matsuo,Y. and Kitakami,K., Trans.JSME.,51-471,A(1985),2575.
[7] Matsuo,Y. and Kitakami,K., in Fracture Mechanics of Ceramics,Vol.7 (Eds.Bradt,R.C.,et al.),(1986),223,Plenum Press.
[8] Matsuo,Y., Engineering Fracture Mechanics,14(1981),527.
[9] Matsuo,Y.,Kitakami,K. and Kimura,S., in Sintering '87, Vol.2, Proc. Int.Institute Sci. Sintering Symp.(Eds. Somiya,S. et al.),(1988), 802,Elsevier Appl.Sci.(The probability density function of flaw -orientation appeared in the reference is incorrect. It should be devided by the term "sin Φ" like Eq.9 in this paper.)
[10] Matsuo,Y.,Trans.JSME,46-406,A(1980),605.
[11] Ito,S.,et al., J.Soc.Mat.Sci.,Japan,30-337(1981),1019.
[12] Murata,H.,et al.,Trans.JSME,52-473,A(1986),27.
[13] Matsuo,Y.,in Ceramic Databook'86,(1987),134, Gordon and Breach Science Publishers.
[14] Matsuo,Y.,Kitakami,K. and Kimura,S.,J.Cer.Soc.Japan,94 (1986),711.
[15] Kitakami,K.,Matsuo,Y. and Kimura,S.,ibid.,99-5(1991),361.
[16] Kitakami,K.,Matsuo,Y. and Kimura,S.,ibid.,99-6(1991),457.
[17] Matsuo,Y.,Kitakami,K. and Kimura,S.,J.Mat.Sci.,22(1987),2253.

A STATISTICAL ANALYSIS OF LIFETIME PREDICTIONS FOR CERAMICS EXHIBITING A TRANSITION IN FATIGUE CRACK GROWTH BEHAVIOR

Toshio Ogasawara, Yoshio Akimune and Tohru Akiba

Central Engineering Laboratories, Nissan Motor Co. Ltd.
1, Natsushima-cho Yokosuka, 237, Japan

ABSTRACT

Reliability analyses for structural ceramic components are indispensable for determining the distribution and time-dependency of strength. Many kinds of ceramics exhibit a transition in fatigue crack growth behavior in a $K_I - V$ diagram resulting from crack growth acceleration due to cyclic loading and the reaching of the fatigue limit. Based on the Weibull distribution function, a statistical analysis was made of lifetime predictions for ceramics exhibiting a transition in fatigue crack growth behavior. The lifetime distribution was defined by a composite Weibull distribution function. Therefore, a change in the slope of the Weibull plot of the fatigue lifetime was seen for ceramic materials exhibiting a transition in fatigue crack growth behavior. The apparent Weibull shape parameters of fatigue lifetime distributions changed as a function of the maximum applied stress in cyclic loading.

INTRODUCTION

Reliability analysis methods based on fracture mechanics and Weibull statistics have been studied by various researchers for application to structural ceramic components because of their wide variability in strengths. For fast fracture, statistical analyses based on fracture mechanics and Weibull statistics have been performed under multi-axial stress states [1]-[8]. These theories have often been utilized in the design of ceramic components as a post processor program of a commercial finite element method [9].

In the most cases, scatter in the strength of ceramics is a consequence of the scatter in the size and distribution of the most critical cracks in components. This suggests that the fatigue lifetime of ceramics will invariably show considerable distribution. The most commonly used theory for lifetime prediction of ceramics is based on a power function of the applied stress intensity factor for subcritical crack growth caused by environment assisted stress corrosion cracking [10]-[15]. This theory agrees with the experimental results reported for glass, porcelain and some kinds of ceramics under

Fracture Mechanics of Ceramics, Vol. 10
Edited by R.C. Bradt *et al.*, Plenum Press, New York, 1992

constant and cyclic loading [11]. It has also been applied to estimate the strength and the lifetime of glass and ceramics after a proof test [10] [12]-[17].

Experimental results have been obtained recently for the lifetime of ceramics in cyclic fatigue tests. The acceleration of fatigue crack velocity of Al_2O_3, ZrO_2 and Si_3N_4 ceramics in cyclic fatigue crack propagation tests using Compact Tension (CT) specimens was observed by Reece *et al.* [29],Dauskardt *et al.* [27] [28] and Kishimoto *et al.* [30] [31] [32]. Kawakubo *et al.* [26] and Niwa *et al.* [35] investigated the fatigue behavior of Si_3N_4 ceramics in which a pre-crack was generated with a Vickers indenter. They confirmed accelerated degradation of fatigue lifetime by cyclic loading.

Fatigue failure from internal defects [34][37] and fatigue crack growth behavior under compressive cyclic loading [24] [25] have also been observed. These experimental results suggest the existence of another fatigue crack growth mechanism other than time dependent subcritical crack growth.

The evaluation of the fatigue behavior of ceramics from applied stress vs. lifetime diagrams (SN or ST diagrams) is also important in predicting lifetime of components. However, it is impossible to understand the fatigue behavior by simple extrapolation from SN or ST diagrams because of scatter in the lifetime data of ceramics. This makes it necessary to conduct a statistical analysis of fatigue lifetime data in the same manner as fast fracture strength analyses.

Calculations have been made of the fatigue lifetime distribution function based on a power function of fatigue crack growth and Weibull statistics [18]-[23]. The calculations were based on a simple power function in a $K_I - V$ diagram and a transition in fatigue crack growth behavior was not taken into consideration. However, fatigue crack velocity is not always represented by such a simple power function in a $K_I - V$ diagram. It is necessary to conduct a statistical analysis for ceramics exhibiting a transition in fatigue crack growth behavior in order to predict the lifetime of ceramic components and to estimate the fatigue behavior of the ceramic material from pre-existing cracks.

In this work, a new fatigue lifetime distribution function based on a power function of fatigue crack growth and Weibull statistics was calculated for ceramics exhibiting a transition of fatigue crack growth behavior. The applicability of the theory was verified by making comparisons with data in the literature.

STATISTICAL ANALYSIS FOR LIFETIME DISTRIBUTION OF CERAMICS

For ceramics without a transition in $K_I - V$ diagram

Lifetime distribution functions for ceramics in which crack velocity was represented by a simple power function were calculated by Olshansky *et al.* [19], Helfinstine [20], Hu *et al.* [22] and Matsuo *et al.*[21]. The most widely accepted statistical theory for analyzing the fast fracture strength of ceramics is the Weibull distribution function [1]. A two parameter form under nonuniform stress states is given for the surface cracks controlling inert strength σ_i :

$$F(\sigma_i) = 1 - \exp\{-(\sigma_i/\sigma_0)^m A_{eff}\} \tag{1}$$

where $F(\sigma_i)$ is the failure probability, m is the Weibull shape parameter and σ_0 is the Weibull scale parameter. A_{eff} is the effective surface area defined by

$$A_{eff1} = \int_A g(\mathbf{r})^m dA$$

where $\mathbf{r}$ is a position vector in the component, $g(\mathbf{r})$ is a non-dimensional scalar function representing the stress at the position vector $\mathbf{r}$, and $g(\mathbf{r})$ is normalized by the maximum stress in the component and the range of its value is $0 < g(\mathbf{r}) < 1$.

Critical crack size a_i of the fracture origin and fast fracture strength σ_i are related on the basis of fracture mechanics.

$$K_{Ic} = Y\sigma_i\sqrt{\pi a_i} \tag{2}$$

where K_{Ic} is the fracture toughness and Y is a constant related to the loading and crack geometry. Since R-curve behavior is not taken into consideration, K_{Ic} is independent of crack size.

It is assumed that cyclic fatigue failure of ceramics is caused by fatigue crack growth of pre-existing cracks. Then, the cyclic fatigue crack velocity da/dN is generally expressed as a power function of the applied maximum stress intensity factor K_{Imax}, so that

$$V = da/dN = AK_{Imax}^n \tag{3}$$

$$K_{max} = Y\sigma_{max}\sqrt{\pi a_i}$$

where A and n are fatigue crack growth parameters which depend on the material, environment and stress states.

The maximum stress in cyclic loading, σ_{max}, is used in Eq.(3), because the average crack velocity in one cycle under a cyclic stress condition is taken into consideration in this equation.

On the other hand, crack velocity caused by subcritical crack growth is given by

$$V = da/dt = AK_I^n \tag{4}$$

Although Eq.(3) is similar to Eq.(4), an exact definition of Eq.(3) is different from that of Eq.(4). This is because the fatigue parameters n, A in Eq.(3) include the effects of cyclic loading conditions such as the stress rate, frequency and stress wave form. When σ_{max} is uniform in a cycle, it is static fatigue. It seems reasonable to assume Eq.(3) for the fatigue crack growth velocity under cyclic loading.

The fatigue lifetime distribution function is calculated by applying the preceding assumptions [19] [20] [22].

$$F_f(N_f) = 1 - \exp\{-\int_A \left(\frac{N_f}{N_0}\right)^M dA\} \tag{5}$$

$$= 1 - \exp\{-\left(\frac{N_f}{N_0}\right)^M A_{eff0}\}$$

where A_{eff0} is the surface area in uniform stress condition, and M and N_0 are the shape and the scale parameters of the Weibull distribution function respectively:

$$M = m/(n-2)$$

$$N_0 = \frac{B}{\sigma_{max}^n} \sigma_0^{n-2}$$

where B is a fatigue crack growth parameter defined by

$$B = \frac{2}{(n-2)AY^2K_{Ic}^{n-2}\pi}$$

Matsuo *et al.* calculated a general fatigue lifetime distribution function $F_f(N_f)$ under nonuniform stress states [21].

$$F_f(N_f) = 1 - \exp\{-\int_A \left(\frac{N_f}{N_0}\right)^M dA\} \tag{6}$$

$$= 1 - \exp\{-\left(\frac{N_f}{N_0}\right)^M A_{eff1}\}$$

$$A_{eff1} = \int_A g(\mathbf{r})^{nM} dA$$

Eq.(6) coincides with Eq.(5) for uniform stress states except for the effective surface area.

The calculation of volume controlled fatigue failure is similar to the preceding calculation. In this case, the effective surface area A_{eff1} is changed an effective volume V_{eff1}.

For ceramics with a transition in $K_I - V$ diagram

Fig.1(a) shows a typical $K_I - V$ diagram of stress corrosion cracking in glass in which a subcritical crack growth behavior was observed [14]. In this case, trimodal $K_I - V$ curves were obtained and there were two transition points in fatigue crack growth behavior.

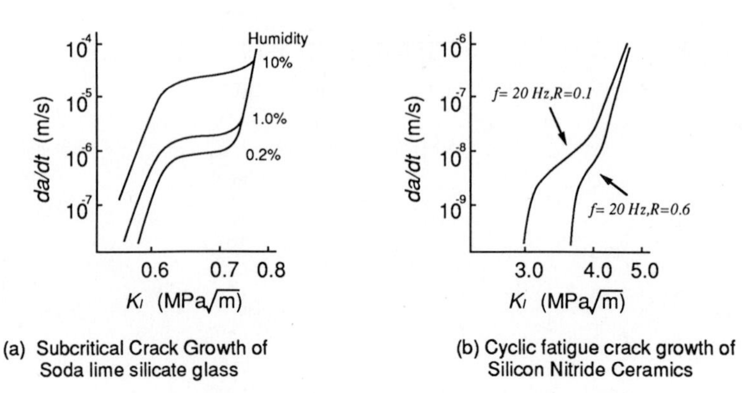

(a) Subcritical Crack Growth of Soda lime silicate glass

(b) Cyclic fatigue crack growth of Silicon Nitride Ceramics

Fig.1 Typical K_I-V diagram of glass and Ceramics

In a cyclic loading condition, Kawakubo *et al.* estimated a similar $K_I - V$ curve from fatigue lifetime data of pre-cracked Si_3N_4 ceramics [26]. Kishimoto *et al.* and Tanaka *et al.* observed accelerated crack velocity in the region of a low applied stress intensity factor in cyclic fatigue crack growth tests [30]-[32] [33]. Fig.1(b) shows a schematic $K_I - V$ diagram of fatigue crack growth in Si_3N_4 ceramics under cyclic loading by Tanaka *et al.*. Trimodal $K_I - V$ curves were also observed.

It is necessary to calculate a fatigue lifetime distribution function for ceramics exhibiting a transition in fatigue crack growth behavior. This theory is applied to make an estimation of the fatigue behavior of ceramic materials from pre-existing cracks and to predict the lifetime of ceramic components.

We assume a simplified $K_I - V$ diagram for ceramics with a transition in fatigue behavior, which has two gradients (see Fig.2). In this model, fatigue crack velocity is given by

$$V = da/dN = A' K_{Imax}^{n'} H(K_{Itrs1}, K_{Imax}) + A K_I^n H(K_{Imax}, K_{Itrs1}) \tag{7}$$

where K_{Itrs1} is a stress intensity factor at the transition point where fatigue crack growth parameters A', n' change to A, n.

$H(\alpha, \beta)$ is a Heaviside step function defined by

$$H(\alpha, \beta) = \begin{cases} 1 & \alpha > \beta \\ 1/2 & \alpha = \beta \\ 0 & \alpha < \beta \end{cases} \tag{8}$$

Fatigue parameters n, n', A and A' are related through the continuous condition at $K_{Imax} = K_{Itrs1}$.

$$A' = A K_{Itrs1}^{n-n'} \tag{9}$$

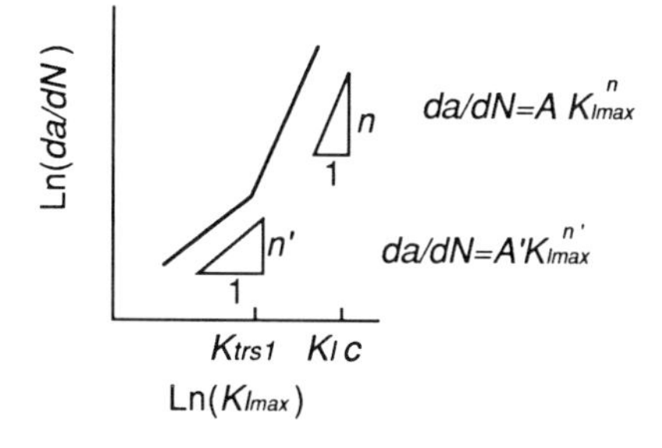

Fig.2 Simplifed K1-V Diagram with a transition

When the maximum stress σ_{max} is applied to the component, crack size a_i, which becomes a fracture origin, is related to the maximum stress intensity factor of the crack.

$$K_{Imax} = Y\sigma_{max}\sqrt{\pi a_i} \tag{10}$$

The following expressions are obtained from Eqs.(2) and (10) :

$$K_{Imax} = (\sigma_{max}/\sigma_i)K_{Ic}$$

$$\sigma_i = (K_{Ic}/K_{Imax})\sigma_{max} \tag{11}$$

Substituting Eq.(11) into Eq.(1) gives the distribution function $F_K(K_{Imax})$ for the stress intensity factor K_{Imax} of the fracture origin under maximum applied stress σ_{max}.

$$F_K(K_{Imax}) = 1 - \exp\{-\left(\frac{K_{Ic}\sigma_{max}}{K_{Imax}}\right)^m A_{eff}\} \tag{12}$$

The probability at $K_{Imax} = K_{Itrs1}$ is calculated by Eq.(12) and is equal to the fatigue failure probability at $N_f = N_{trs1}$.

$$F_K(K_{Itrs1}) = F_f(N_{trs1}) \tag{13}$$

where F_f is the fatigue lifetime distribution function. When the fatigue crack growth behavior changes at $N_f = N_{trs1}$, the fatigue lifetime distribution function is represented by a composite Weibull distribution function, because the probability function for $N_f > N_{trs1}$ is different from that for $N_f < N_{trs1}$.

$$F_f(N_f) = \left\{ \begin{array}{ll} 1 - \exp\{-G_1\} & N_f < N_{trs1} \\ 1 - \exp\{-G_2\} & N_f > N_{trs1} \end{array} \right. \tag{14}$$

where G_i is the risk of rupture in a Weibull distribution.

In the case of $N_f < N_{trs1}$ ($K_{Imax} < K_{Itrs1}$), fatigue crack velocity is given by

$$da/dN = A(Y\sigma_{max}\sqrt{\pi})^n \times$$

$$\left\{\left(\frac{K_{Itrs1}}{Y\sigma_{max}\sqrt{\pi}}\right)^{n-n'} \sqrt{a}^{n'} H(K_{Itrs1}, K_{Itrs1}) + \sqrt{a}^{n} H(K_{Itrs1}, K_{Itrs1})\right\} \tag{15}$$

$$A(Y\sigma_{max}\sqrt{\pi})^2 \int_0^{N_f} dN$$

$$= \int_{a_i}^{a_{trs1}} \left(\frac{Y\sigma_{max}\sqrt{\pi}}{K_{Itrs1}}\right)^{n-n'} \sqrt{a}^{n'} da + \int_{a_{trs1}}^{a_f} \sqrt{a}^{n} da \tag{16}$$

where a_i and a_f are the initial and critical(final) crack size respectively, a_{trs1} is the crack size at $K_I = K_{Itrs1}$ and N_f is the component lifetime. Then

$$K_{Itrs1} = Y\sigma_{max}\sqrt{\pi a_{trs1}} \tag{17}$$

Generally $\sqrt{a_f}^{2-n} << \sqrt{a_{trs1}}^{2-n}$ since for ceramics $n > 10$.

$$\sqrt{a_i}^{2-n'} = \sqrt{a_{trs1}}^{2-n'}\frac{n'-n}{2-n} + \frac{n'-2}{2}A'\sigma_{max}^{n'}(Y\sqrt{\pi})^{n'}N_f \tag{18}$$

Substituting Eqs.(2) and (17)into (18) yields the following equation:

$$\sigma_i^{n'-2} = \frac{1}{B'}\sigma_{max}^{n'}(N_f - N_u') \tag{19}$$

where

$$N_u' = \left(\frac{n'-n}{n-2}\right)\left(\frac{K_{Ic}}{K_{Itrs1}}\right)^{n'-2}\left(\frac{B'}{\sigma_{max}^2}\right) \tag{20}$$

$$B' = \frac{2}{(n'-2)A'Y^2K_{Ic}^{n'-2}\pi} \tag{21}$$

The surface of the component is divided into k infinitesimal surface elements ΔA_j, and a uniform stress state is assumed. Then, stress of a surface element ΔA_j is given by

$$\sigma_{max}(\mathbf{r}) = \sigma_{max}g(\mathbf{r})$$

where $\mathbf{r}$ is a position vector of ΔA_j, and $g(\mathbf{r})$ is a non-dimensional scalar function at $\mathbf{r}$, previously described. The distribution function of fast fracture strength σ_i at ΔA_j is expressed as

$$\langle F(\sigma_i)\rangle_j = 1 - \exp\{-\Delta A_j(\langle\sigma_i\rangle_j/\sigma_0)^m\} \tag{22}$$

The probability density function is given by partial differentiation of Eq.(22), such that

$$\langle f(\sigma_i)\rangle_j = \partial\langle F(\sigma_i)\rangle_j/\partial\sigma_i \tag{23}$$

$$= m\Delta A_j\langle\sigma_i\rangle_j^{m-1}/\sigma_0^m \exp\{-\Delta A_j(\langle\sigma_i\rangle_j/\sigma_0)^m\}$$

A coefficient J for a change of variable from $\langle\sigma_i\rangle_j$ to $\langle N_f\rangle_j$ is given by

$$\langle f(\sigma_i)\rangle_j = \langle f_f(N_f)\rangle_j \mid J \mid \tag{24}$$

$$\mid J \mid = \partial\langle N_f\rangle_j/\partial\langle\sigma_i\rangle_j = (n'-2)B'\sigma_i^{n'-3}/\{g(\mathbf{r})\sigma_{max}\}^{n'}$$

Probability function $F_f(N_f)$ is given by

$$\langle F_f(N_f)\rangle_j = 1 - \exp\{-\langle G_2\rangle_j\} \tag{25}$$

where

$$\langle G_2\rangle_j = \Delta A_j g(\mathbf{r})^{M'}\left(\frac{\langle N_f\rangle_j - N_u'/g(\mathbf{r})^2}{N_0'}\right)^{M'} \tag{26}$$

$$M' = m/(n'-2)$$

$$N_0' = \frac{B'}{\sigma_{max}^{n'}}\sigma_0^{n'-2}$$

Thus, the risk of rupture of the whole surface can be obtained by summing up all $\langle G_2\rangle_j (j = 1, 2, ..., k)$ and setting the limit as $\Delta A_j \rightarrow 0$:

$$G_2 = \int_A g(\mathbf{r})^{n'M'}\left(\frac{N_f - N_u'/g(\mathbf{r})^2}{N_0'}\right)^{M'} dA \tag{27}$$

The fatigue lifetime distribution function for $N_f < N_{trs1}$ can be expressed by an approximate three-parameter Weibull distribution function using location parameter N'_u. In the case of $n' < n$, Eq.(27) is not an exact three-parameter Weibull distribution function because $N'_u < 0$. It is difficult to integrate Eq.(27) analytically, because the location parameter is a function of $g(\mathbf{r})$. Therefore, a numerical integration is effective to use the equation in many cases.

In the case of uniform stress states, Eq.(27) is written in a very simple form because $g(\mathbf{r}) = 1$:

$$F_f(N_f) = 1 - \exp\{-\int_A \left(\frac{N_f - N'_u}{N'_0}\right)^{M'} dA\} \tag{28}$$

$$= 1 - \exp\{-\left(\frac{N_f - N'_u}{N'_0}\right)^{M'} A_{eff0}\}$$

where M' and N'_0 are the Weibull shape and scale parameters:

$$M' = m/(n' - 2)$$

$$N'_0 = \frac{B'}{\sigma_{max}^{n'}} \sigma_0^{n'-2}$$

According to Eq.(13), we obtain following equations:

$$(N_{trs1}/N_0)^M = \left(\frac{K_{Ic}\sigma_{max}}{\sigma_0 K_{Itrs1}}\right)^m$$

and

$$N_{trs1} = \left(\frac{K_{Ic}\sigma_{max}}{K_{Itrs1}\sigma_0}\right)^{n-2} N_0 \tag{29}$$

$$N'_u = \left(\frac{n' - n}{n' - 2}\right) N_{trs1} \tag{30}$$

$$N'_0 = N_0 \frac{n-2}{n'-2} \left(\frac{K_{Ic}\sigma_{max}}{K_{Itrs1}\sigma_0}\right)^{n-n'}$$

The failure probability function for $N_f > N_{trs1}$ is given by Eq.(6), so the fatigue lifetime distribution function for ceramics exhibiting a transition in fatigue crack growth behavior can be expressed by the following composite Weibull distribution function:

$$F_f(N_f) = \begin{cases} 1 - \exp\{-G_1\} & F_f(N_f) < F_f(N_{trs1}) \\ 1 - \exp\{-G_2\} & F_f(N_f) > F_f(N_{trs1}) \end{cases} \tag{31}$$

$$G_1 = \left(\frac{N_f}{N_0}\right)^M A_{eff1} \tag{32}$$

$$G_2 = \int_A g(\mathbf{r})^{n'M'} \left(\frac{N_f - N'_u/g(\mathbf{r})^2}{N'_0} \right)^{M'} dA \tag{33}$$

For ceramics with multiple transitions in $K_I - V$ diagram

A fatigue lifetime distribution function for ceramics that exhibits multiple transitions in fatigue crack growth behavior can be calculated by a similar method. For example, in the case of ceramics with two transition points as shown in Fig.3, the distribution function is given by following equation:

$$F_f(N_f) = \begin{cases} 1 - \exp\{-G_1\} & F_f(N_f) < F_f(N_{trs1}) \\ 1 - \exp\{-G_2\} & F_f(N_{trs1}) < F_f(N_f) < F_f(N_{trs2}) \\ 1 - \exp\{-G_3\} & F_f(N_{trs2}) < F_f(N_f) \end{cases} \tag{34}$$

where

$$G_3 = \int_A g(\mathbf{r})^{n''M''} \left(\frac{N_f - N''_u/g(\mathbf{r})^2}{N''_0} \right)^{M''} dA \tag{35}$$

$$M'' = m/(n'' - 2)$$

$$N''_0 = \frac{B''}{\sigma_{max}^{n''}} \sigma_0^{n''-2}$$

$$N''_u = \left(\frac{n'' - n'}{n' - 2} \right) \left(\frac{K_{Ic}}{K_{Itrs2}} \right)^{n''-2} \left(\frac{B''}{\sigma_{max}^2} \right) + N'_u$$

$$B'' = \frac{2}{(n'' - 2) A'' Y^2 K_{Ic}^{n''-2} \pi}$$

Under uniform stress states, we obtain:

$$((N_{trs2} - N'_u)/N'_0)^{M'} = \left(\frac{K_{Ic}\sigma_{max}}{\sigma_0 K_{Itrs2}} \right)^m$$

$$N_{trs2} = \left(\frac{K_{Ic}\sigma_{max}}{K_{Itrs2}\sigma_0} \right)^{n'-2} N'_0 + N'_u$$

$$N''_u = \left(\frac{n'' - n'}{n'' - 2} \right) N_{trs2} + \left(\frac{n' - n}{n'' - 2} \right) N_{trs1}$$

$$N''_0 = N'_0 \frac{n' - 2}{n'' - 2} \left(\frac{K_{Ic}\sigma_{max}}{K_{Itrs2}\sigma_0} \right)^{n'-n''}$$

In this manner, the fatigue lifetime distribution function of ceramics exhibiting more than three transition points can be calculated using a composite Weibull distribution function.

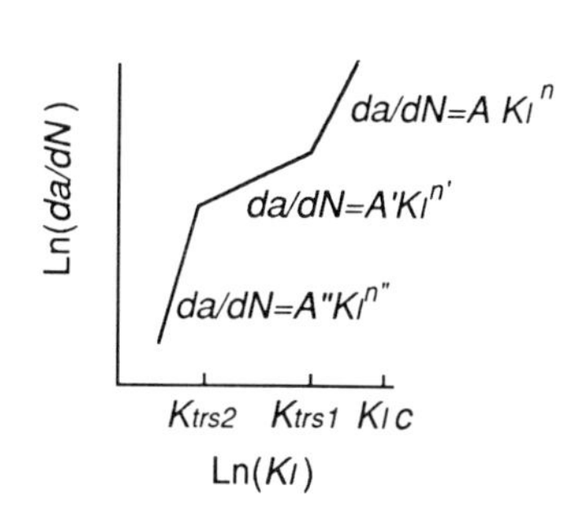

Fig.3 Simplifed K1-V Diagram with two transitions

MAPPING OF $K_I - V$ DIAGRAM ONTO WEIBULL PLOT OF FATIGUE LIFETIME DISTRIBUTION

This section discusses the mapping of a $K_I - V$ diagram onto a Weibull plot of the fatigue lifetime distribution, which is an important operation for application of this theory, since the $K_I - V$ diagram exhibits an one-to-one correspondence with the Weibull plot.

For example, Kawakubo *et al.* estimated a $K_I - V$ diagram as shown in Fig.4 from fatigue lifetime data of pre-cracked Si_3N_4 ceramics [26], where a region of $K_{Imax} = 5.2 \sim 6.0$ indicates an acceleration of crack velocity by cyclic loading, and a region of $K_{Imax} < 5.2$ corresponds to the fatigue limit behavior. A fatigue lifetime distribution was calculated using the parameters listed in Table 1, which were based on their experimental results. As the Weibull parameters were unknown, $m = 15$ and $\sigma_{max} = 650$ were used, and the effective surface area or volume was ignored, and an uniform stress state was assumed for the calculation. In their paper, crack velocity was represented as da/dt instead of as da/dN. When da/dt is converted into da/dN, it is necessary to integrate the cyclic loading time [11]. In this calculation, a square stress wave form as shown in Fig.5 was assumed, because the relation between da/dt and da/dN can be written in the following simple form:

$$\frac{da}{dt} = 2f\frac{da}{dN}$$

where f is frequency. In this square stress wave, stress ratio R $(= \sigma_{min}/\sigma_{max})$ is $R = 0$.

Fig.6 shows the $K_I - V$ diagram that was obtained when frequency was $f = 10Hz$. Weibull plots of the lifetime distribution calculated with Eq.(31) are shown in Fig.7. Changes in the slope of the Weibull plots appear at $N_f \approx 10^3$ and 10^4. The plots are upward curve in the short lifetime region $(N_f < 5 \times 10^3)$ and the large slope (large M) region in the Weibull plots corresponds to a small n region in the $K_I - V$ diagram. As lifetime increases, a gradual change in slope is seen observed, which is generally related to the fatigue limit behavior.

Fig.8 shows an SN diagram calculated with this model. The three curves indicate an equivalent failure probability of 10, 50 and 90% respectively. The pattern in Fig.8 agrees with the experimental results of Kawakubo *et al.* [26].

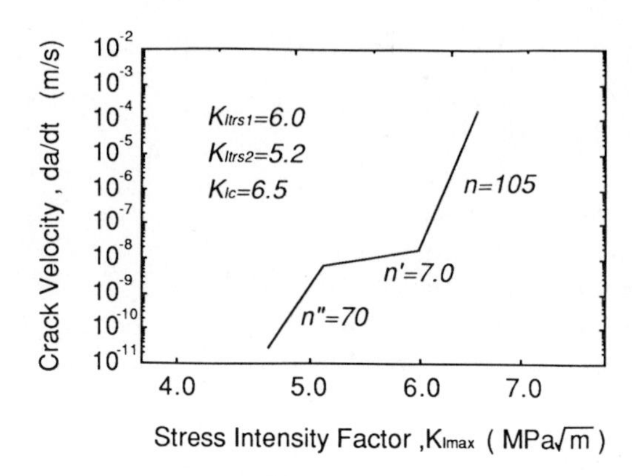

Fig.4 KI-V(da/dt) diagram of Si_3N_4 ceramics

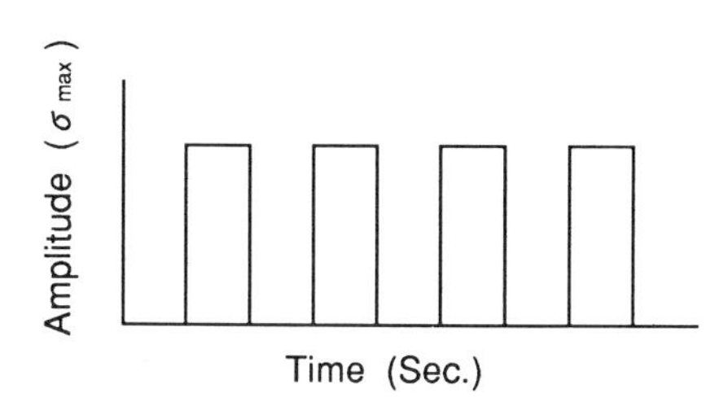

Fig.5 Square Wave Form
(f=10Hz,R=0, Duty ratio=1:1)

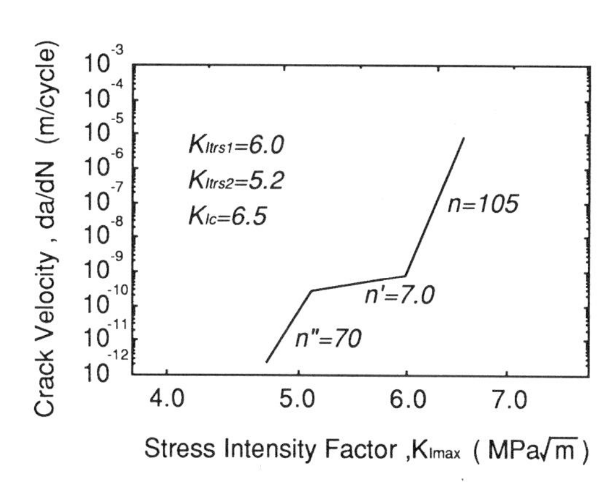

Fig.6 K_I-V(da/dN) diagram of Si_3N_4 ceramics under cyclic loading

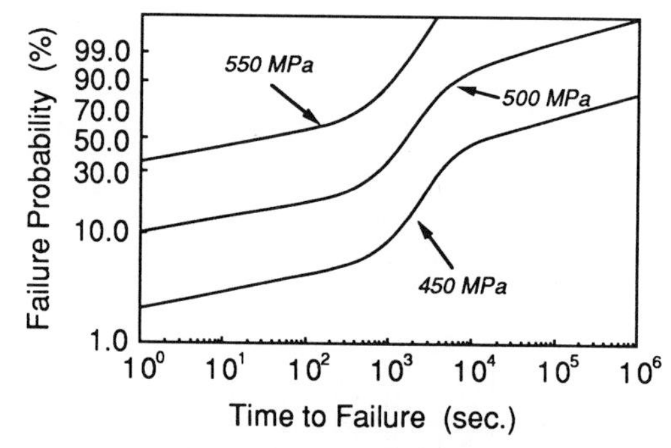

Fig.7 Weibull plot of lifetime distribution calculated by this theory

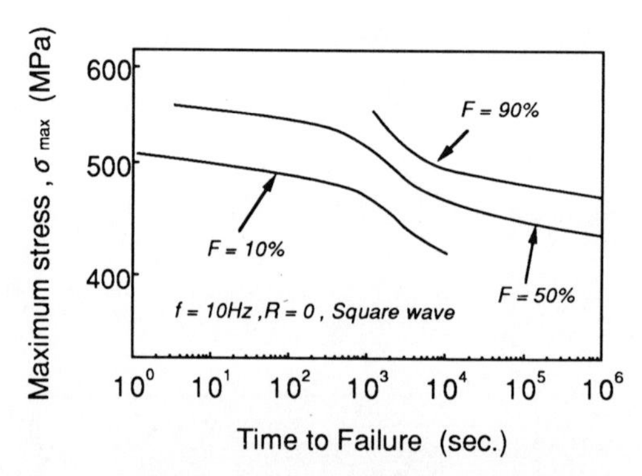

Fig.8 Estimated Stress - Time (ST) diagram from KI-V diagram shown in Fig.6

Table 1. Weibull and fatigue parameters

Weibull parameters	m		15.0
	σ_o MPa		650
Fracture toughness , K_{Ic} MPa$\sqrt{m}$			6.5
Transition points	K_{Itrs1} MPa$\sqrt{m}$		6.0
	K_{Itrs2} MPa$\sqrt{m}$		5.2
Fatigue parameters [m,Mpa,sec]	n	n	105
		n'	7.0
		n''	70
	A	*Static*	3.93×10^{-90}
		Cyclic *	2.00×10^{-91}

* *f = 10 Hz , R= 0 , Squar Wave*

Since a $K_I - V$ diagram is mapped onto a Weibull plot directly, it is possible to estimate a $K_I - V$ diagram from the schematic shape of a Weibull plot. If the experimental data are treated as a simple Weibull distribution function, the apparent Weibull shape parameter M of the fatigue lifetime distributions will change as a function of the maximum applied stress during cyclic loading. Such experimental results were reported by Matsuo *et al.* [21] and Sakai *et al.* [36].

CONCLUSIONS

A statistical analysis of the lifetime prediction for ceramics exhibiting a transition in fatigue crack growth behavior was conducted based on the Weibull distribution function. As a result, a lifetime distribution function was defined by a composite Weibull distribution function. A distribution function was also calculated in a similar manner for ceramics exhibiting multiple transitions.

The mapping of a $K_I - V$ diagram to a Weibull plot of a fatigue lifetime distribution was also discussed. Changes in the slope of a Weibull plot of fatigue lifetime data were seen for the ceramics exhibiting transitions in fatigue crack growth behavior. When the experimental data were treated as a simple Weibull distribution, apparent Weibull shape parameters of the fatigue lifetime distributions changed as a function of the maximum applied stress during cyclic loading. The theory presented here was shown to be useful for estimating the fatigue behavior of ceramics based on pre-existing cracks and for predicting the lifetime of ceramic components.

References

[1] W.A.Weibull, *J.Appl.Mech.*, **18**,293-297(1951)

[2] S.B.Batdorf and J.G.Crose, *J.Appl.Mech.*, **41**,459-464(1974)

[3] S.B.Batdorf, *Fracture Mechanics of Ceramics , Vol.3, Ed by R.C.Bradt et al.*, Plenum,New York(1974),pp1-30

[4] A.G.Evans, *Fracture Mechanics of Ceramics , Vol.3, Ed by R.C.Bradt et al.*, Plenum,New York,31-49(1987)

[5] Y.Matuso, *Engineering Fracture Mechanics*, **14**,527-538,(1981)

[6] Y.Matsuo, *Bulletin of JSME (in Japanese)*, **A46**,1203-1212,(1980)

[7] J.Lamon and A.G.Evans, *J.Amer.Ceram.Soc.*, **66**,177-182(1983)

[8] A.C.Rufin,D.R.Samos and R.J.H.Bollard, *AIAA Journal*, **22**,135-140(1984)

[9] J.P.Gyekenyesi, *Journal of Engineering for Gas Turbines and Power*, **108**,540-546(1986)

[10] A.G.Evans and S.M.Wiederhorn, *Int.J.Fract.*, **10**,379-92(1974)

[11] A.G.Evans, *Int.J.Fract.*, **16**,485-498(1980)

[12] A.G.Evans and E.R.Fuller, *Metall Trans.*, **5**,27-33(1974)

[13] A.G.Evans and E.R.Fuller, *Mater.Sci.Eng.*, **19**,69-77(1975)

[14] S.M.Wiederhorn, *Fracture Mechanics of Ceramics, Vol.2, Edited by R.C.Bradt et al.*, Plenum ,New York,(1974),pp613-646

[15] J.E.Ritter,Jr. and J.A.Meisel, *J.Amer.Ceram.Soc.*, **59**,478-481(1976)

[16] J.E.Ritter,Jr., *Fracture Mechanics of Ceramics , Vol.4, Edited by R.C.Bradt et al.*, Plenum,New York,(1978),pp667-686

[17] D.F.Jacobs and J.E.Ritter,Jr., *J.Amer.Ceram.Soc.*, **59**,481-487(1976)

[18] K.Jakus, D.C.Coyne and J.E.Ritter,Jr., *J.Mater.Sci.*, **13**,2071-80(1978)

[19] R.Olshansky and R.D.Maurer, *J.Appl.Phys.*, **47**,4497-4499(1976)

[20] J.D.Helfinstine, *J.Amer.Ceram.Soc.*, **63**,C113(1980)

[21] M.Matsuo, T.Oida, K.Jinbo, K.Yasuda and S.Kimura, *J.Ceram.Soc.Japan (Nippon Seramikkusu Kyokai Gakujutsu Ronbunshi) (in Japanese)*, **97**,136-142(1989)

[22] X.Hu, Y.Mai and B.Cotterell, *J.Mater.Sci.*, **24**,3118-3122(1989)

[23] T.Fett,G.Martin,D.Munz and G.Thun, *J.Mater.Sci.*, **26**,3320-3328(1991)

[24] S.Suresh and J.R.Brockenbrough, *Acta metall.*, **36**,1455-1470(1988)

[25] S.Suresh, L.X.Han and J.J.Petrovic, *J.Amer.Ceram.Soc.*, **71**,C158-161(1988)

[26] T.Kawakubo and K.Komeya, *J.Amer.Ceram.Soc.*, **70**,400-405(1987)

[27] R.H.Dauskardt ,D.B.Marshall and R.O.Ritchie, *J.Amer.Ceram.Soc.*, **73**,893-903(1990)

[28] R.H.Dauskardt ,W.C.Carter ,D.K.Veirs and R.O.Ritchie, *Acta metall.*, **38**,2327-2336(1990)

[29] M.J.Reece,F.Guiu and M.F.R Sammur, *J.Amer.Ceram.Soc.*, **72**,348-352(1989)

[30] H.Kishimoto, A.Ueno and H.Kawamoto, *J.Soc.Mater.Sci.Japan (in Japanese)*, **36**,1122-1127(1987)

[31] H.Kishimoto,A.Ueno,H.Kawamoto and Y.Fujii, *J.Soc.Mater.Sci.Japan (in Japanese)*, **38**,1212-1217(1989)

[32] H.Kishimoto,A.Ueno and H.Kawamoto, *Bulletin of JSME (in Japanese)*, **A56**,50-56,(1990)

[33] T.Tanaka, N.Okabe and Y.Ishimaru, *J.Soc.Mater.Sci.Japan (in Japanese)*, **38**,137-143(1989)

[34] M.Masuda,T.Soma and M.Matsui, *J.European.Ceram.Soc.*, **6**,253-258(1990)

[35] T.Niwa,K.Urashima,Y.Tajima and M.Watanabe, *J.Ceram.Soc.Japan (Nippon Seramikkusu Kyokai Gakujutsu Ronbunshi) (in Japanese)*, **99**,296-299(1990)

[36] T.Sakai and K.Fujitani, *J.Soc.Mater.Sci.Japan*, **35**,1371-1377(1986)

[37] T.Ogasawara, Y.Akimune and K.Yoneda, *J.Ceram.Soc.Japan (Nippon Seramikkusu Kyokai Gakujutsu Ronbunshi) (in Japanese)*, **99**,489-494(1990)

McCLINTOCK'S STATISTICS AND STRENGTH

SAFETY FACTOR FOR CERAMICS

Sergej M. Barinov and Vladimir Ya. Shevchenko

High Tech Ceramics Scientific Research Centre
Academy of Sciences of the USSR
Ozernaya 48, Moscow, 119361, USSR

ABSTRACT

An approach is proposed to substantiate a strength safety factors for ceramic materials accounting the strength statistics. Using a statistical distribution of the strength based on a model of occasional cracked grain boundaries, the correction coefficients for safety factors usually specified in practical engineering are evaluated. The dependence of correction coefficients on the admissible level of failure probability, component size / grain size ratio and the probability of existence of single grain-boundary crack is evaluated.

INTRODUCTION

Large-scale application of ceramics as a structural material is restricted by its brittleness and, as a result, its low reliability. In order to provide service reliability ceramic components are used in practice with high safety factors considerable superior to those of metallic structural materials.[1,2] For example, the recommended value of safety factor for glass which is an extremely brittle material is 8 - 10.[1] The necessity to specify margins seems to be caused by three factors: (i) a considerable "inert strength" scattering, (ii) a tendency of ceramics to delayed fracture in active media when the applied stress level is lower than the "inert strength" and (iii) a low energy consumption during crack propagation process resulting in catastrophic failure of ceramic components at overloads. A simple basis for providing a required reliability of ceramic components is to introduce the correction coefficients taking into account the necessity of increase the safety factors as result of above effects. Criteria of choosing operational stress taking into account the kinetics of slow crack growth are considered in[3,4], besides, Ref.[4,5] also allow the effects produced in ceramics defects statistics and loading conditions (static, dynamic, fatigue, creep). An attempt to take into consideration the effect of R-curve behaviour on reliability of ceramics is undertaken in[6]. The present paper makes an attempt to substantiate the ceramics safety factors specification on the basis of combination of an approach generally approved by engineering practice and strength statistics.

McCLINTOCK'S STATISTICS

The McClintock's theory is based on a statistical model of occasional cracked grain boundaries in a brittle polycrystal.[7] It is assumed that there are a number of independent grain boundary segments in structure, and that for each there is a finite probability of being cracked. Using the probability for an element of structure to consist of a crack that is longer than predetermined value, the exponential form of cumulative distribution function is derived. Taking into account the relation between stress intensity factor and crack length, the cumulative distribution function of strength per unit area, and the frequency distribution per unit area are obtained. According to above model, the cumulative extreme value distribution, for strength below s, is of the form

$$P(s) = 1 - \exp(-\frac{Aa}{S} \exp(-b/(s/s_o)^2)) \qquad (1)$$

where P(s) is the failure probability; A is the area of component; S is the area of an element of material's structure; a and b are the distribution parameters connected with the probability w the existence of crack with the dimension l; $s_o = K_{Ic}(2l)^{-1/2}$ is the strength of material containing a single crack with the length l. The expression for parameters a and b will be as follows:

$$a = w^{1/2}(0.5 - 1/\ln w)^{-1/2}$$

$$b = -\ln w$$

As it follows from (1)

$$s(P) = s_o(b/\ln((Aa/S)/(-\ln(1-P))))^{1/2} \qquad (2)$$

The main advantage of McClintock's approach as compared to the Weibull's theory is in the clear physical background for "weak link" nature, and in the possibility to predict the effects of component size / grain size ratio. Nevertheless, the correlation between estimates based on both approaches should be expected.

SAFETY FACTOR CORRECTIONS

The safety factor n for metallic structural materials is usually specified in practical engineering as a strength expectation $\bar{s}$ to operational stress s_a ratio. As to brittle ceramic materials an admissible level of failure probability P should be taken into account. Therefore, n is specified by s(P) to s_a ratio, or

$$n = s(P)/s_a = k(\bar{s}/s_a) \qquad (3)$$

where

$$k = s(P)/\bar{s} \qquad (4)$$

If $\bar{s}$ in (4) is equal to the distribution median s(0.5) then

$$k = (\ln((Aa/S)/(-\ln 0.5))/\ln((Aa/S)/(-\ln(1-P))))^{1/2} \qquad (5)$$

Thus, k depends on the Aa/S ratio, probability w of a single crack existence and failure probability P.

To calculate k for component with the surface area A being exposed to uniformly distributed stress s it is necessary to know the parameters a and, also, S which depend on w. The value S depends on the material grain size. In a first approximation we can consider that $S = l^2(1/w)^{1/2}$, as it is assumed in [7]. If $l \approx d$ where d is the grain size then

$$S \approx d^2(1/w)^{1/2} \qquad (6)$$

Thus, only w is the non-determined parameter of calculations. It can be determined by experiment, for example, on the basis of structural investigations of flaw detection as well as mechanical testing. In fact the w value for engineering ceramics may vary from 0.01 to 0.0001. If grain size is 300 μm, then, using (6), estimations give that above w values correspond to detection of a single crack at the area from 1 to 10 mm^2.

On the basis of the above representations we can calculate the value of k for different values of P, w and A/S using formula (5) and data given in Ref.[7] The obtained results are given in Table 1.

In order to determine experimentally the correction coefficient k there should be tested a series of samples with the surface area A made of a material with average grain dimensions d. The obtained results show the samples failure probability related to stress and it possible to determine strength representative values s(0.25), s(0.5) and s(0.75). The ratio r = s(0.75)/s(0.25) can be expressed using (2) in the following way:

$$r = \left(\frac{\ln((Aa/S)/(-\ln 0.75))}{\ln((Aa/S)/(-\ln 0.25))} \right)^{1/2} \qquad (7)$$

As it is follows from the above expression

$$Aa/S = F(r) \qquad (8)$$

where $F(r) = \ln(\ln 4/\ln 4/3))^{1/(r^2-1)}$.
By means of formulae (6) and (8) the value of w can be found on the basis of the known values A/d^2 and r:

$$f(w) = F(r)/(A/d^2) \qquad (9)$$

where $f(w) = w/(0.5 - 1/\ln w)$. The approximate estimation $f(w) = 2w$ can be assumed when the value of w is small. The k coefficient corresponding to the predetermined failure probability P is calculated using the found value w and the A/d^2 ratio on the basis of data given in the Table 1.

Ceramic materials are characterized by values r varying from 1.08 to 1.32. Respective values F(r) necessary to calculate w by formula (9) are given in the Table 2.

The Weibull's function is generally accepted to describe the statistic strength distribution. A considerable database of such distribution parameters has been accumulated. Therefore, it would be interesting to determine the aforesaid distribution parameters and McClintock's distribution parameters relationship. Structural ceramics are characterized by the value of Weibull's distribution function modulus m varying within the interval of 8 - 20. As it follows from Weibull's distribution function:

$$s(0.75)/s(0.25) = (\ln 0.25/\ln 0.75)^{1/m} \qquad (10)$$

Taking the logarithm of (10) where s(0.75)/s(0.25) = r we receive:

$$m = 0.6825/\lg r.$$

The m/r relationship is illustrated by the following data (Table 3).

Table 1. Dependence of k on P, w and A/S

P	w	A/S	k	P	w	A/S	k
10^{-2}	0.001	10^8	1.13	10^{-6}	0.001	10^8	1.37
		10^{10}	1.10			10^{10}	1.29
	0.00001	10^8	1.14		0.00001	10^8	1.41
		10^{10}	1.11			10^{10}	1.33
10^{-4}	0.001	10^8	1.25	10^{-8}	0.001	10^8	1.47
		10^{10}	1.20			10^{10}	1.37
	0.00001	10^8	1.28		0.00001	10^8	1.52
		10^{10}	1.23			10^{10}	1.43

Table 2. F(r) and r values

r	1.08	1.12	1.16	1.20	1.32
F(r)	$1.76.10^4$	$6.8.10^2$	$1.3.10^2$	34.7	11.5

Table 3. The m/r relationship

r	1.08	1.10	1.12	1.14	1.16	1.24	1.32
m	20.44	16.49	13.87	12.00	10.58	7.31	5.66

These data help to calculate r using Weibull's distribution function modulus to determine the coefficient k.

Thus, values of the coefficient k vary in the interval of 1.1 - 2 and increase with the increasing of the required material service reliability, grain dimension in relation to the component dimensions and defectness of material.

CONCLUSION

Estimates are given for ceramics strength safety factors based on the McClintock's inert strength statistics. It is shown that the strength safety factor for ceramics may considerable exceed that of structural metals. The correction coefficient k vary in the interval of 1.1 - 2 and increases with the increasing of the required material service reliability, grain dimension in relation to the component dimensions, and defectness of material. The advantage of the McClintock's statistical approach is in possibilities to predict the component/grain size effects on the reliability.

REFERENCES

1. R. Morrell, Design Criteria for Ceramics, in: Proceedings of Materials Selection and Design, Online Publ., Pinner, London (1985).
2. A. R. Hyde, Multidirectional Fibre Reinforced Ceramic Matrix Composites, Ceram. Ind. International 98:32 (1989).
3. S. M. Barinov, V. Ya. Shevchenko, Testing Methods for Lifetime Predictions for Ceramic Materials, Industrial Laboratory (USSR) 10:44 (1990).
4. Xiao-Zhi Hu, Yiu-Wing Mai, and B. Cotterell, A Statistical Theory of Time-Dependent Fracture for Brittle Materials, Philos. Mag. A, 58:299 (1988).
5. Xiao-Zhi Hu, Yiu-Wing Mai, and B. Cotterell, Lifetime Prediction of Ceramic Materials Subjected to Static Loads, J. Mater. Sci. Lett. 6:462 (1987).
6. Yu. L. Krassulin, S. M. Barinov, On the Reliability and Safety Factors for Ceramics, USSR Acad. of Sci. Reports (USSR) 245:883 (1985).
7. F. A. McClintock, Statistics of Brittle Fracture, in: "Fracture Mechanics of Ceramics, Vol. 1", R. C. Bradt, D. P. H. Hasselman, and F. F. Lange, ed., Plenum, New York - London (1974).

CREEP DAMAGE MECHANISMS IN STRUCTURAL CERAMICS

D. S. Wilkinson
Department of Materials Science and Engineering
McMaster University
Hamilton, Ontario L8S 4L7, CANADA

ABSTRACT

The mechanisms responsible for fracture in structural ceramics at elevated temperatures are reviewed. It is seen that some features of the process are common to a wide range of materials, with quite different microstructures. These include a sensitivity to flaws or microstructural inhomogeneities which nucleate microcracks. However, it is often observed that the flaws which control failure under creep conditions are different from those responsible for fast fracture at ambient temperatures. Another common feature is the development of cracks through a process of gradual damage accumulation. There are however, differences which depend on the nature of microstructure. In pure materials and some multi-phase alloys with high volume fractions of the binder phase, the heterogeneities responsible for fracture initiation are far apart, and extensive crack growth occurs prior to failure. Such materials are crack growth dependent. In multi-phase alloys in which the binder is a minor phase, the high degree of constraint produces many sites for damage to nucleate. However, the low ductility and damage tolerance of such materials leads to failure after only a modest amount of crack propagation. In such materials, modified Monkman-Grant behaviour is observed.

INTRODUCTION

The utilization of structural ceramics at elevated temperatures is ultimately limited by their creep resistance and damage tolerance. Typical operating lifetimes for high temperature components (e.g. in heat engine applications) are in the range of several thousand hours. Thus, reliable data on the structural integrity of materials over long periods of time is required if ceramics are to become widely used at elevated temperatures. However, creep testing is tedious and expensive. Moreover, materials

Fracture Mechanics of Ceramics, Vol. 10
Edited by R.C. Bradt *et al.*, Plenum Press, New York, 1992

with a strong potential for application at high temperatures are still being developed. It is premature therefore, to engage in extensive materials testing programs at this time, although a number of prototypical programs have been carried out (Govila, 1980; Quinn and Quinn, 1983; Tanaka et al, 1992). Nonetheless, materials developers need guidance as to the microstructural features which promote or degrade creep fracture lifetime and creep deformation resistance. It is therefore essential that a deeper understanding be developed related to the mechanisms responsible for creep and creep fracture in structural ceramics.

In this paper, our current understanding of high temperature fracture in a range of structural ceramics will be reviewed.

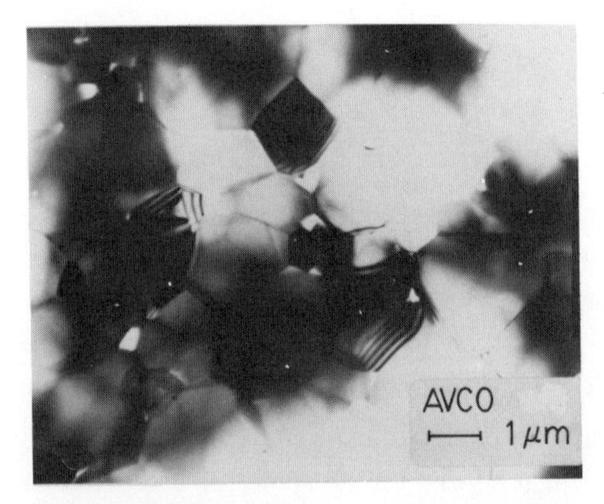

Figure 1. A typical microstructure of a pure ceramic material, in this case hot-pressed alumina, showing a uniform, high density microstructure. Some grain boundary porosity is evident.

MICROSTRUCTURES FOR HIGH TEMPERATURE MATERIALS

Ceramic materials of interest at high temperatures generally exhibit microstructures that fall into one of three main categories. The first of these are the single-phase materials, of which pure alumina has been most extensively studied (Fig. 1). These materials are processed so as to obtain clean, "dry" grain boundaries with essentially no glass. Many of these materials (pure alumina, for example) are not serious contenders as high temperature structural materials. Because of the simplicity of their microstructures however, they are ideal for model studies of creep fracture (Porter et al, 1981; Hsueh et al, 1981; Dalgleish et al, 1985; Robertson et al 1991: Wilkinson et al, 1991). Moreover, these materials are often potential matrix materials for ceramic matrix composites. Thus their study provides valuable baseline data for understanding more complex systems.

The second class of microstructures of interest involves the use of vitreous bonding (Jakus et al, 1986; Wiederhorn et al, 1987; Page et al, 1987). These materials are sintered using additives that enhance densification through the formation of a liquid phase, which, upon cooling, forms a glass (generally about 10% by volume). Thus the microstructure consists of crystalline grains embedded in a continuous glass matrix. The glass typically exists both as thin (nanometre-scale) films along grain boundaries, and as larger (micron-scale) pockets distributed amongst the grains. The most studied example of this class of material is sintered silicon nitride (Fig. 2).

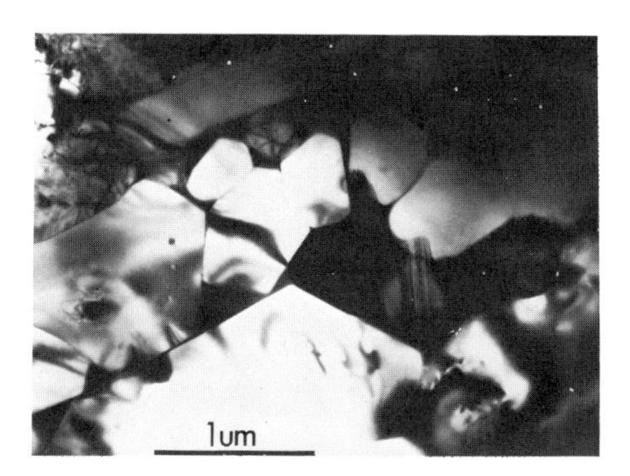

Figure 2. The microstructure of sintered silicon nitride consists of elongated grains of β-Si_3N_4 grains bound together by a glassy phase, typically about 10% by volume.

The third class of microstructure consists of multi-phase crystalline ceramics, typically made by adding discontinuous reinforcement (particles, whiskers or platelets) to a matrix phase. A wide range of such materials has been studied with the aim of improving their fracture toughness at ambient temperatures. Only a limited amount of high temperature data is currently available (Choksi and Porter, 1985; Han and Suresh, 1989). A typical example of such a material is SiC whisker reinforced alumina.

An additional material, which has been extensively studied at elevated temperatures but which does not fit neatly within this classification scheme is siliconized SiC (Carroll and Tressler, 1985; Carroll et al, 1986; Wiederhorn et al, 1986, 1988; Quinn, 1987; Carroll and Tressler, 1989). This is a reaction bonded material whose microstructure consists of SiC grains embedded, not in glass, but in a silicon matrix (see Fig. 3). The volume fraction of the silicon is usually quite large (in the range 20 to 40%), much greater than that of the glass in vitreous bonded materials.

Not surprisingly, the creep damage mechanisms leading to failure at elevated temperatures are somewhat different in these different classes of materials. However, they also share some features in common. In particular, all materials exhibit the growth of cavities under appropriate conditions. Moreover, all materials exhibit some sensitivity to flaws as initiation sites for the nucleation and development of microcracks.

CAVITY DEVELOPMENT

Creep fracture in ceramic materials occurs exclusively along grain boundaries. This is in contrast with metals and their alloys in which transgranular failure can be induced if a sufficiently high stress is used (Ashby et al, 1979; Gandhi and Ashby, 1979).

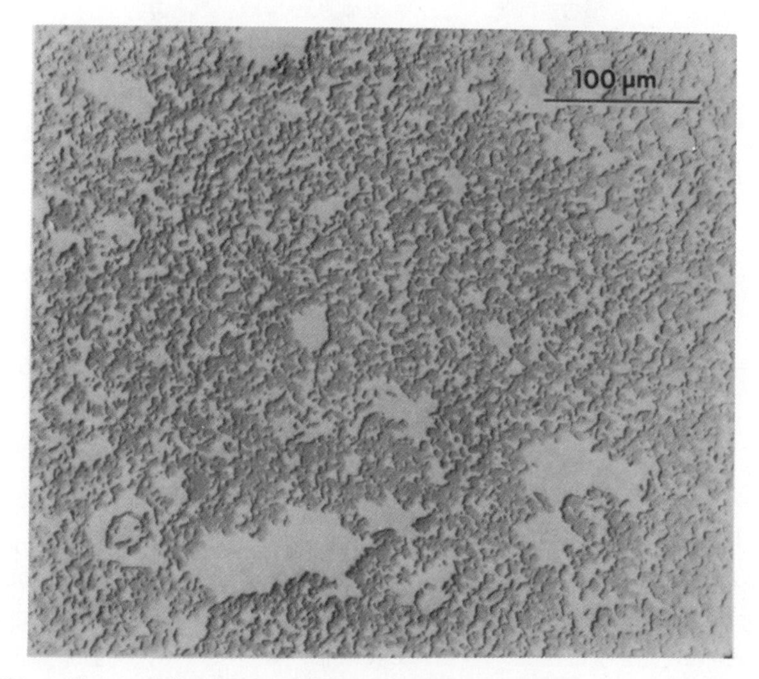

Figure 3. Siliconized silicon carbide exhibits a two-phase microstructure with an abnormally high volume fraction of the binder phase (Wiederhorn et al, 1986).

Ceramic materials differ in their microstructure and behaviour from metals in two important aspects. Firstly, the grain size is much smaller (by up to two orders of magnitude), and secondly, dislocation plasticity is of much less importance. Both of these factors promote the grain boundary failure process. Because of the fine grain size of most ceramic materials, the dominant site for intergranular cavity nucleation is at grain boundary triple junctions, although cavity nucleation on grain boundary facets has been noted, both in single phase materials (Choksi and Porter, 1987) and in vitreous bonded alumina (Wiederhorn et al, 1987). Indeed, work by Choksi and Porter (1987) have shown that this particularly in an alumina with a highly bimodal grain size distribution, in which case mid-facet is found in leger grained regions. Models for cavity nucleation (Argon et al, 1980; Wang et al, 1985) suggest that substantial stresses are required, much greater than those typically applied to ceramic (or metallic) materials in service. This has puzzled researchers for many years. One suggestion has been that transient stresses are generated at grain boundaries due to the stochastic nature of grain boundary sliding (Argon et al, 1980). These rapid bursts of sliding can produce high stresses at inclusions lying along the grain boundaries and at triple junctions. This concept is attractive for creep resistant metallic alloys where large grains are involved, and grain boundaries decorated with many fine particles are common. This is not the case in ceramics. It appears more likely however that, in single phase materials, incipient porosity provides the initiation sites for most cavities. For example, in a hot-pressed alumina, with a density of over 99.9%, transmission electron microscopy has shown that roughly one out of every 4 triple junctions contains a small cavity (Robertson, 1989; Robertson et al, 1991). In glass containing ceramics, the most probable mechanism of cavity nucleation involves the generation of a negative pressure in the larger glass pockets (Thouless and Evans, 1984), as modelled by Tsai and Raj (1982) and Marion et al (1983). Evidence of such behaviour has been found by Clarke (1985) in a vitreous bonded alumina. Similar behaviour is also seen in siliconized SiC, in which cavities form preferentially in silicon at the Si/SiC interface (Wiederhorn at al, 1988).

The role of microstructural inhomogeneity on cavity nucleation has been extensively studied in hot-pressed alumina. Such a material can be expected to have a relatively uniform microstructure. However, inhomogeneities do exist. Their effect on fracture is found to depend on the temperature and stress. At high stresses, failure occurs by slow crack growth from a single dominant crack. These cracks tend to nucleate at one of several characteristic defect types. These include: processing flaws in the shape of penny shaped cracks normal to the hot pressing direction, regions of much larger than normal grain size resulting from an inhomogeneous distribution of MgO, and glassy deposits on the surface of the bend bar (Fig. 4 a and b). Cracks grow from these regions until K_{Ic} is reached. Calculations of K_{Ic} based on measurements of the crack dimensions at the point of failure and a knowledge of the applied stress yield values similar to those obtained by direct measurement (Wilkinson et al, 1991).

Figure 4. Typical processing flaws which lead to fast failure by crack growth include a) large grained regions, and b) flat penny shaped cracks.

a

b

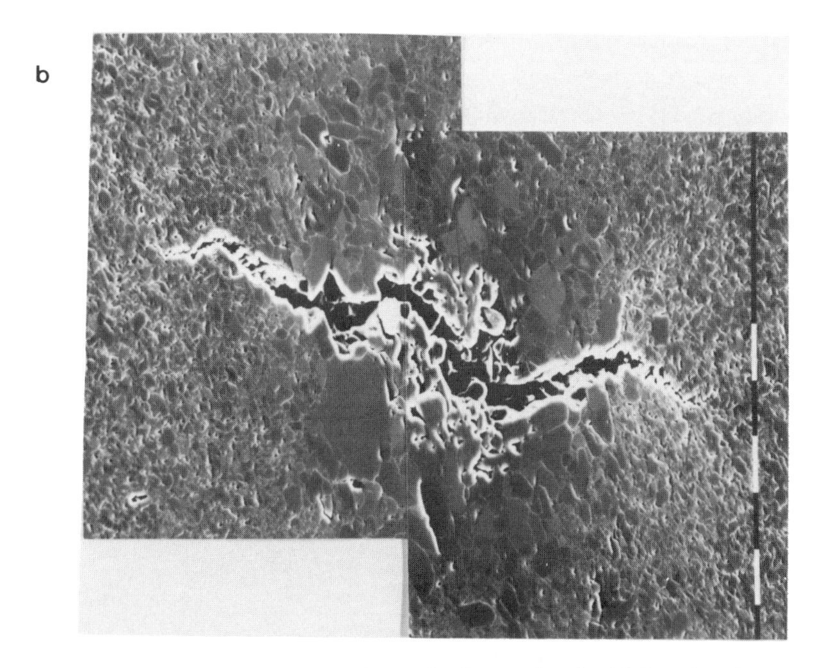

Figure 5. a) The tensile surface of a flexure bar in Al_2O_3, showing an extensive network of large microcracks, and b) the nucleation of microcracks at a large grained region.

As the stress is lowered and the temperature increased, a more generally distributed damage pattern is observed, involving both large microcracks and grain-sized cavities (Hsueh and Evans, 1981, Robertson et al, 1991). The microcracks are the result of cavity linkage. They tend to nucleate at large grained regions or on single large grains (Fig. 5 a and b). As the stress is lowered even further however, the density of microcracks starts to decrease again. This is a manifestation of the role of strain rate in building stress concentrations required for the nucleation of damage. As the stress is lowered individual large grains no longer act as nucleation sites. Only the large grain clusters are able to concentrate sufficient stress. Thus very few microcracks are seen and the material begins to exhibit behaviour characteristic of superplasticity.

The picture that emerges from this study is that even under near-ideal conditions, failure initiation in ceramics is related to processing flaws. At elevated temperatures, inhomogeneities which concentrate stress are most damaging. In pure alumina, large grained regions are of primary concern. In siliconized silicon carbide, regions in which SiC particles are clustered together also help to concentrate stress. Thus cavities tend to form on Si/SiC interfaces, but only in these regions (Wiederhorn et al, 1986). This is clearly seen in Coors SCRB 210 (Fig. 6) which has a bimodal distribution of large (50 micron) and small (3 micron) SiC particles (Wiederhorn et al, 1988). The cracks which develop in these materials follow the large grains, and damage nucleates most readily wherever two large grains are in near contact. In glass-containing ceramics, a detailed examination of the distribution of creep damage is more difficult. However, it is clear that cavities tend to form along in the glass pockets at multiple grain junctions, and propagate along the thin amorphous films which lie between the grains(Wiederhorn et al,

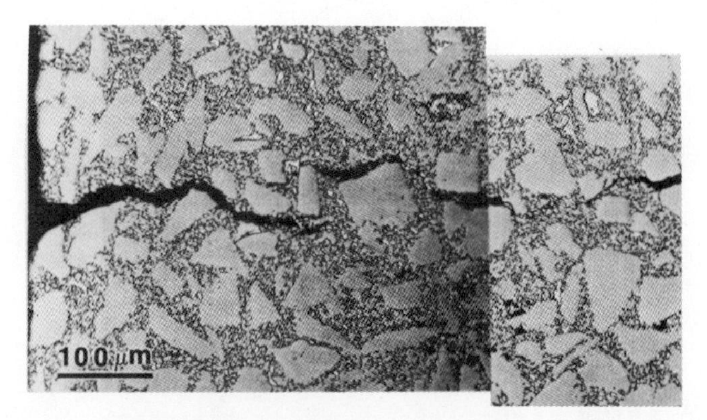

Figure 6. A microcrack growing through a siliconized SiC composite containing both large and small SiC particles (Wiederhorn et al, 1986)

1991; Clarke, 1985). It appears likely that damage will tend to occur preferentially in regions of abnormally low glass content. As in the Si-SiC case, the grain boundary phase acts to shed load, although by a different mechanism, and regions with little glass are less effective at doing that. Indirect evidence for such behaviour can be seen in sintered silicon nitride containing about 10 vol.% of an Al_2O_3-SiO_2-Y_2O_3 glass (Chadwick and Wilkinson, 1990, 1992). If such a material is annealed so as to crystallize about half of the glass, the rate of deformation at a given stress is largely unaltered; however, the time to failure and the ductility are dramatically reduced (Fig. 7), indicating that the glass plays an important role in developing a damage-tolerant microstructure.

A further complicating factor is related to the stability of these grain boundary glasses at elevated temperatures. If devitrification occurs simultaneously with the creep process, then the interpretation of creep data can be ambiguous. In particular, since both creep and devitrification may involve diffusion on a scale equivalent to the grain size, the processes can be expected to exhibit similar kinetics. Thus, the microstructure will evolve over the same time scale as creep testing. A simple model calculation (Wilkinson, 1988) suggests that this leads to abnormally high estimates of the stress exponent for creep, and will also make interpretation of time to failure data difficult.

Several important conclusions arise from an assessment of the role of inhomogeneity on high temperature failure on ceramics. The first of these is that microstructure always plays an important role. We also note however, that the flaws

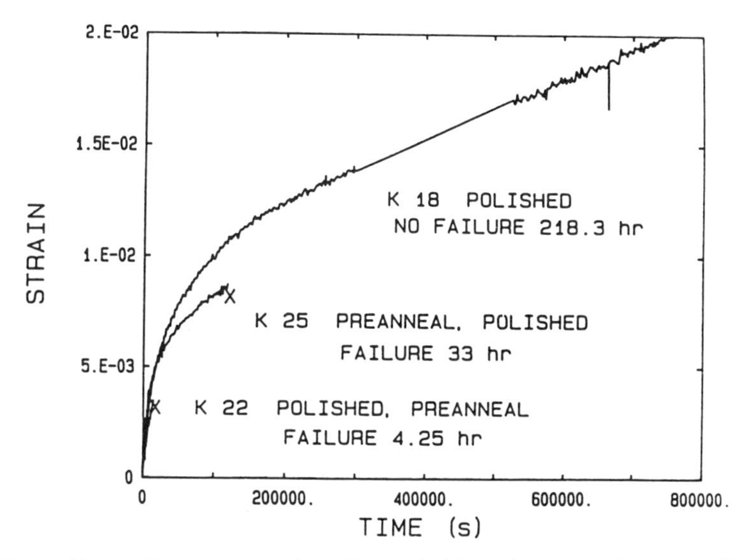

Figure 7. Creep curves for silicon nitride before and after annealing so as to crystallize about half of the intergranular glass.

which are important after long exposures at high temperatures are often very different from those which control the strength of the same materials at ambient temperatures, or in short term testing at elevated temperatures. Indeed, in some instances the flaws which are responsible for high temperature do not exist in the as-fired body, but develop during service. One example of these are the oxide pits which can develop on the surface of nitride ceramics (Fig. 8). Thus ambient temperature testing or rapid high temperature testing cannot be relied upon to help optimize materials for extended service under creep conditions.

DAMAGE ACCUMULATION AND LIFETIME

Except for the high stress regime, in which failure occurs by the growth of a single crack, high temperature failure occurs by a process of gradual damage accumulation. From the preceding discussion, it is clear that such damage is nucleated preferentially at specific sites within a microstructure. The extent of general damage accumulation will therefore depend on the distribution of the strength-controlling inhomogeneities, and on the general damage tolerance of the material. In many multi-phase ceramic materials, that damage tolerance is relatively low (Fig. 6). This is because damage tends to nucleate at the interface between the refractory phase and the softer binder phase. It then grows through the binder phase. As discussed above, the

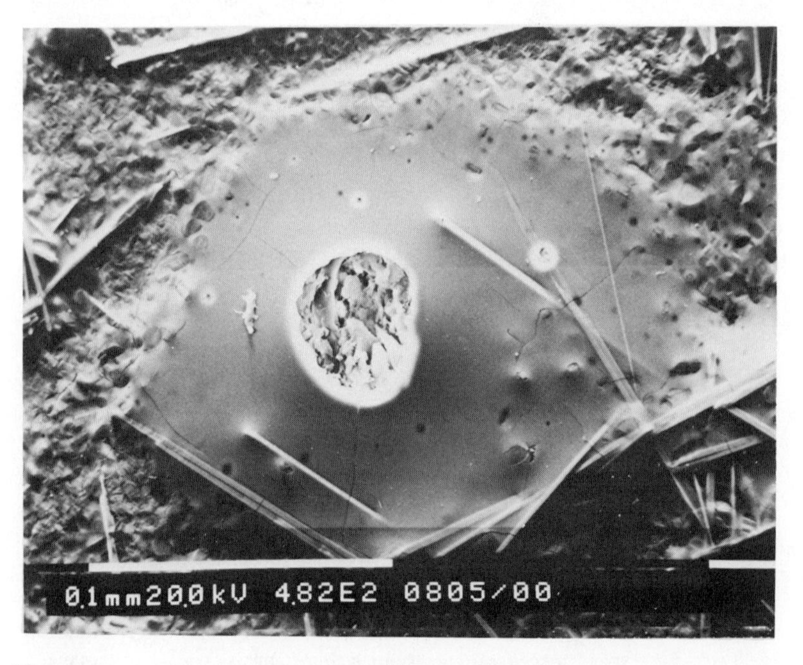

Figure 8. An example of an oxide pit which develops on the surface os sintered silicon nitride during high temperature exposure in air.

binder phase is often found in thin layers between refractory grains (e.g. vitreous bonded ceramics). Cracks grow readily along such interfaces because of the strong degree of constraint which is present there. Larger pockets of the binder phase can be effective however, in blunting such cracks, increasing the damage tolerance of the structure. In particular, materials such as some of the Si-SiC materials which contain large volume fractions of the binder phase (in excess of 30 vol.%) exhibit considerably greater damage tolerance and creep ductility.

The question arises as to what controls the actual lifetime of such materials - cavity nucleation, cavity growth, or crack growth and the final damage coalescence process. No definitive answer can be given to this question. However, there is considerable evidence that the creep lifetime is strain controlled. This is indicated by the existence of a modified Monkman-Grant relationship for such materials, in which the minimum creep rate $\dot{\epsilon}$ is related to the time to failure t_f by a relationship of the form

$$t_f \, \dot{\epsilon}^m = C \tag{1}$$

where C is a constant independent of applied stress and temperature. Wiederhorn and co-workers have shown that this relationship is valid for two grades of siliconized silicon carbide with m values of 1.45 and 2.39 (Wiederhorn et al, 1991), and for SiC whisker reinforced silicon nitride with an m of 1.69 (Hockey et al, 1991). Similar data is available for silicon nitride and vitreous bonded alumina (Ferber et al, 1990). A value of m in excess of one indicates a decreasing failure strain as the stress increases. The source of this behaviour is not clear. However, it suggests that regions under stress concentrations are less able to redistribute load as the far-field strain rate increases. In addition, Wiederhorn et al (1991) have suggested a model based on microcrack growth and coalescence. The model is essentially statistical, and assumes that cracks grow by the random attachment of cavities. It therefore neglects the role of a stress concentration at the head of a crack to enhance the driving force for crack propagation. Such a simplification is likely to be useful only if the cracks themselves are not much larger than the cavities that attach to them. The cavities are approximately one grain diameter in size. First order calculations (Wilkinson, 1983) indicate that until a crack attains a length of several times the grain diameter, the crack tip stress field does not extend beyond the next grain, and the far-field stress is the dominant parameter. Wiederhorn et al applied their model to a Si-SiC for which the critical crack length at failure was 22 μm, which is within the limits just described.

In single phase materials, the situation is quite different. The large grained regions (Fig. 5b) which provide the most important sites for microcrack nucleation are isolated. Thus cracks must nucleate and grow through large regions of homogenous

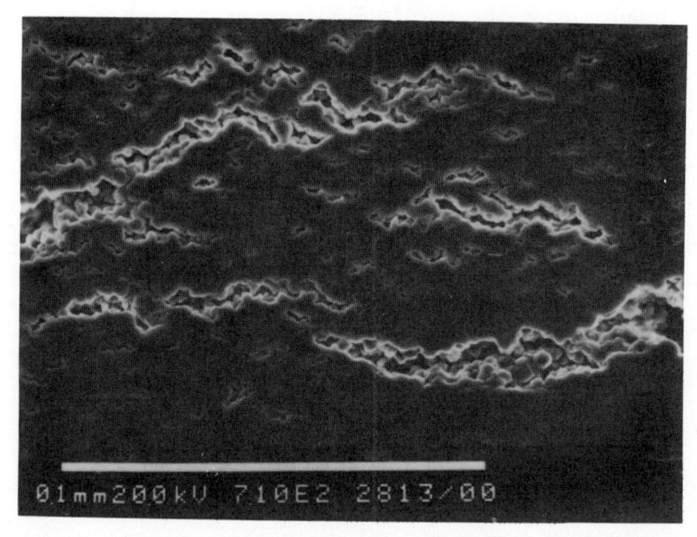

Figure 9. The growth of microcracks occurs by a damage process involving grain boundary cavitation. Under these conditions (100 MPa at 1150°C in alumina), cracks are quite sharp.

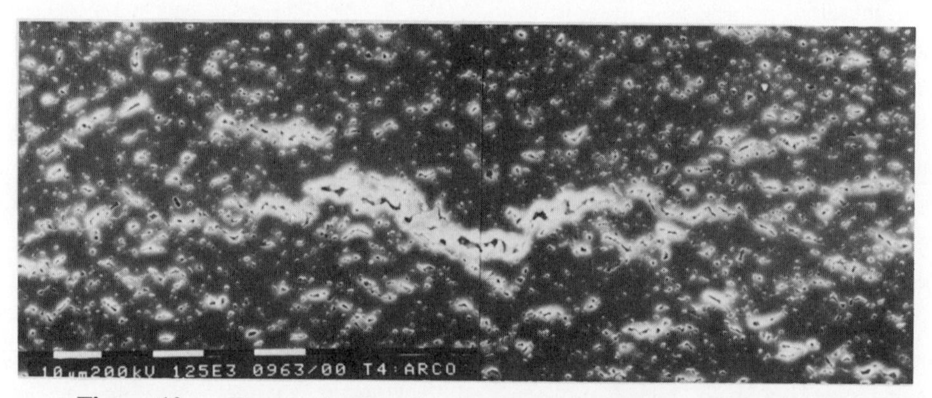

Figure 10. Creep damage in hot-pressed alumina showing the evolution of microcracks from individual facet cavities (bar = 10 microns).

material before failure. A detailed study of alumina (Dalgleish et al, 1985; Robertson et al, 1991, Wilkinson et al, 1991) has shown that the damage accumulation process leads to the development of much larger cracks prior to failure than is common in multi-phase materials. This shown in Fig. 5a for an alumina specimen tested in flexure. Many cracks have developed on the tensile surface of the bar with lengths ranging up to about 0.5 μm. A detailed examination of such cracks (Fig. 9 and 10) indicates that these cracks grow by a process of cavitation; i.e. cavities nucleate ahead of a crack, grow to facet size, and eventually link up with the main crack. The pattern of cavitation, with a high density of cavities ahead of each microcrack indicates a strong influence of the concentrated stress field of the crack on cavity nucleation and growth (Fig. 11). In the stress regime corresponding to the most dense microcracking, cavities are observed to develop as sharp facet cracks. This suggests that their growth is limited by the rate of surface diffusion. However, using standard models of creep lifetime due to surface diffusion controlled cavitation, and all reasonable values of the surface diffusivity, the lifetime is always underestimated compared to that measured experimentally (Wilkinson et al, 1991). Thus, it appears that creep life is controlled by the rate of crack propagation across the specimen. Because of the extensive creep prevalent in these materials a crack tip parameter which accounts for the stress relaxation produced by creep is required (Kromp and Pabst, 1980; Wilkinson et al, 1991). Thus a model has been developed for crack growth controlled by a C^* stress field (Wilkinson et al, 1991). The time to failure is determined by integrating the rate of crack propagation over time, until the crack size exceeds that for catastrophic crack propagation (i.e the critical stress intensity factor is reached). Models predict that the crack growth rate scales as $(C^*)^{n/(n+1)}$ where n, the stress exponent for creep, is equal to 2 for alumina. Moreover, C^* scales as the stress times the strain rate, i.e. as σ^{n+1}. Therefore one might expect a simple σ^{-2}

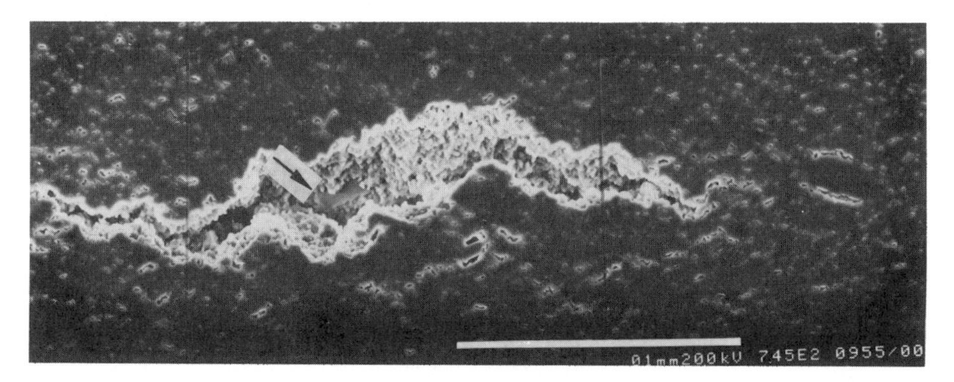

Figure 11. The growth of a microcrack in alumina, showing a large grain (arrowed) at the centre of the crack, and another ahead of the crack, decorated with cavities.

dependence of the time to failure on the stress. However, the net section stress changes as the crack grows, thus altering the value of C^* applied to the crack. Moreover, the critical condition for final failure is also stress dependent. Thus a more detailed analysis is required (Wilkinson et al, 1991) which leads to the prediction of a $\sigma^{-2.4}$ dependence of the time to failure for this case which is excellent agreement with the experimentally measured values.

One of the more interesting results of this investigation is the large difference in lifetime for tensile and flexure specimens tested at the same nominal stress and temperature. Flexure tests last longer by more than a factor of 5. This is a commonly observed phenomenon in ceramics containing a viscous phase and in siliconized silicon nitride. There it can be explained on the basis of the difference in creep response of the material in tension and compression. It has been suggested that either creep cavitation occurs more readily in tension, thus contributing to the creep behaviour (Wiederhorn et al, 1986), or that the glass contributes to creep by viscous flow, which has been shown to lead to a difference in the creep response in tension and compression (Dryden et al, 1989; Chadwick et al, 1992). However, these phenomena do not apply to single phase microstructures. Moreover, there is no measurable difference in the creep resistance of alumina in tension compression or flexure. Only the lifetime is affected by the stress state. It therefore appears that differences in the damage morphology are responsible. For example, crack linkage is known to involve the development of bands of cracking on planes inclined to the stress axis during flexural testing (Dalgleish et al, 1985, Wilkinson et al, 1991). This phenomenon, which has been called shear band formation by Dalgleish et al (1985), does not occur in tension (Wilkinson et al, 1991). The development of shear bands is apparently related to the development of a rotational hinge

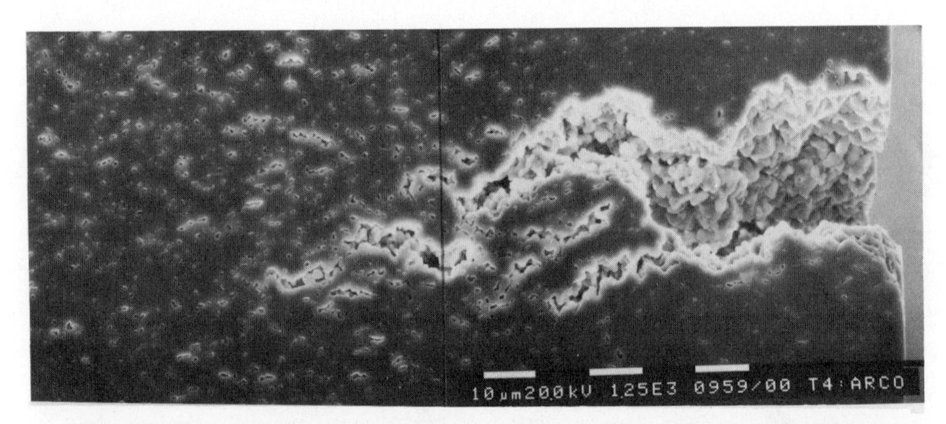

Figure 12. At low stresses, the cracks in alumina start to blunt and bifurcate. This suggests a dramatic increase in damage tolerance and is accompanied by a large ductility increase.

in flexure specimens which constrains the final failure process. Further evidence for this is offered by calculations of the fracture toughness made by measuring the final crack lengths at failure. The values so measured from tensile specimens are quite reasonable. However, those estimated from flexure specimens are much higher (by about a factor of 3). Thus cracks have not propagated catastrophically, even when K_{Ic} is exceeded because of the constraint effects. Flexure testing is therefore unsuitable for developing accurate data on lifetime under creep conditions. Indeed, it overestimates the life and thus leads to non-conservative predictions.

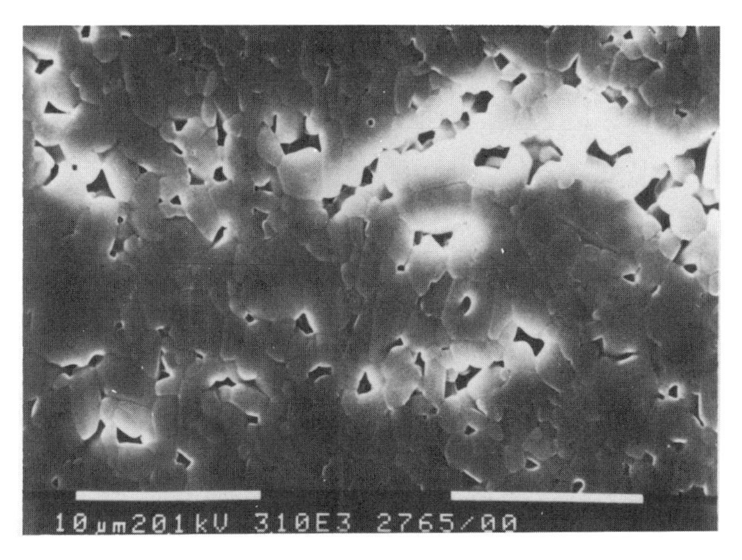

Figure 13. At low stresses in alumina facet sized cavities become angular in shape, the result of growth by grain boundary sliding.

Another interesting phenomenon is observed in alumina as the stress is lowered. The creep ductility increases rapidly and the material exhibits characteristics typical of superplastic flow. Large crack opening displacements (about 30 times the grain size) are observed. The crack tips, while still discernable as such, become bifurcated and disconnected (Fig. 12). Moreover, the morphology of individual cavities also changes (Fig. 13). They are no longer crack-like, but become quite angular. These are formed by cavities growing across individual facets and then blunting by grain boundary sliding. Such cavities are very resistant to coalescence with each other. Moreover, the local strains produced by this process are considerable, and are responsible for the large crack opening displacements observed.

SUMMARY

A brief review of the mechanisms responsible for high temperature failure in a range of structural ceramics has been presented. It is clear that there are some similarities between the processes involved for a wide range of microstructures. However, there are a number of features which are microstructurally dependent. In almost all cases, the failure origin can be traced to a microstructural heterogeneity. These may be the result of processing flaws. They may however, be generated in service (e.g. through an oxidation process). Those microstructural flaws which are most damaging are those which help to concentrate stress, such as large grained regions or the development of surface pits. As the stresses are lowered, the materials is more able to relieve the stresses around these inhomogeneities, and superplasticity will eventually become evident.

REFERENCES

Argon, A. S., Chen, I.-W. and Lau, C. W., 1980, Intergranular cavitation in creep: theory and experiments, in *Creep-Fatigue-Interactions*, eds. R. M. Pelloux and N. S. Stoloff, The Metallurgiacal Society of AIME, pp. 46-85.

Ashby, M. F., Gandhi, C., and Taplin, D. M. R., 1979, Overview no. 3, Fracture mechanism maps and their construction for f.c.c. metals and alloys, Acta Metall., **27**, 699-729.

Carroll, D. F. and Tressler, R. E., 1985, Time-dependent strength of siliconized silicon carbide under stress at 1000°C and 1100°C, J. Amer. Ceram. Soc. **68**, 143-46.

Carroll, D. F. and Tressler, R. E., 1987, Time dependent mechanical behaviour of silicon carbide ceramics at elevated temperatures, in *High Tech Ceramics*, ed P. Vincenzini, Elsevier, Amsterdam, 1335-44.

Carroll, D. F., R. E. Tressler, R. E., Tsai, Y. and Near, C., 1986, High temperature mechanical properties of siliconized silicon carbide composites, in *Tailoring Multiphase and Composite Ceramics*, eds. R. E. Tressler, G. L. Messing, C. G. Pantano and R. E. Newnham, Plenum, 775-88.

Carroll, D. F. and Tressler, R. E., 1989, The effect of creep damage on the tensile creep behaviour of a siliconized silicon carbide, J. Amer. Ceram. Soc., **72**, 49-53.

Chadwick and Wilkinson, 1990, Effect of annealing on creep and creep fracture in Si_3N_4, Proc. on 4th Intl. Conf. on *Creep and Fracture of Engineering Materials and Structures*, eds. B. Wilshire and R. W. Evans, Institute of Metals, pp.377-87.

Chadwick, M. M. and Wilkinson, D. S., 1992, The creep behaviour of sintered silicon nitride, to be published.

Chadwick, M. M., Wilkinson, D. S. and Dryden, J. R., 1992, Creep due to a non-Newtonian viscous phase, to be published.

Choksi, A. H. and Porter, J. R., 1985, Creep deformation of an alumina matrix composite reinforced with silicon carbide whiskers, J. Amer. Ceram. Soc., **68**, C144-45.

Choksi, A. H. and Porter, J. R., 1987, Cavity development during creep deformation in alumina with a bimodal grain size distribution, J. Amer. Ceram. Soc., **70**, 197-203.

Clarke, D. R., 1985, The high temperature deformation of a polycrystalline alumina containing an intergranular glassy phase, J. Mater. Soc., **20**, 1321-32.

Dalgleish, B. J., Slamovich, E. B. and Evans, A. G., 1985, The duality in the creep rupture of a polycrystalline alumina, J. Amer. Ceram. Soc., **68**, 575-81.

Dryden, J. R. Kucerovsky, D. Wilkinson, D. S. and Watt, D. F., 1989, Creep deformation due to a viscous grain boundary phase, Acta Metall., **37**, 2007-15.

Ferber, M. K., Jenkins, M. G. and Tennery, V. J., 1990, Comparison of tension, compression and flexure creep for alumina and silicon nitride ceramics, Ceram. Eng. Sci. Proc., **11**, 1028-45.

Gandhi, C. and Ashby, M. F., 1979, Overview no. 5, Fracture mechanism maps for materials which cleave: f.c.c., b.c.c. and h.c.p. metals and ceramics, Acta Metall., **27**, 1565-1602.

Govila, R. K., 1980, Ceramic life prediction methods, AMMRC report no. TR80-18, Army Materials and Mechanics Research Center, Watertown, Mass.

Han, L. X. ans Suresh, S., 1989, High temperature failure of an alumina-silicon carbide composite under cyclic loads: meachanisms of fatigue crack tip damage, J. Amer. Ceram. Soc., **72**, 1233-37.

Hockey, B. J., Wiederhorn, S. M., Liu, W., Baldoni, J. G. and Buljan, S.-T., 1991, Tensile creep of whisker-reinforced silicon nitride, J. Mater. Sci., **26**, in press.

Hsueh, C. H. and Evans, A. G., 1981, Overview 14, Creep fracture in ceramic polycrystals - II. Effects of inhomogeneity on creep rupture, Acta Metall., **29**, 1907-17.

Jakus, K., Wiederhorn, S. M. and Hockey, B. J., 1986, Nucleation and growth of cracks in vitreous-bonded aluminum oxide at elevated temperatures, J. Amer. Cer. Soc., **69**, 725-31.

Kromp, K. and Pabst, R. F., 1980, Uber die ermittlung con J-integralwerten bei keramischen werkstoffen im hochtemperaturbereich, Materialpruf., **22**, 241-45.

Marion, J. E., Evans, A. G., Drory, M. D. and Clarke, D. R., 1983, Overview no. 28, High temperature failure initiation in liquid phase sintered materials, Acta Metall, **31**, 1445-57.

Page, R. A., Lankford, J., Chan, K. S., Hardman-Rhyne, K. and Spooner, S., 1987, Creep cavitation in liquid-phase-sintered alumina, J. Amer. Ceram. Soc., 137-45.

Porter, J. R., Blumenthal, W. and Evans, A. G., 1981, Overview 14, Creep fracture in polycrystals - I. Creep cavitation effects in polycrystalline alumina, Acta Metall. **29**, 1899-1906.

Quinn, G. D. and Quinn, J. B., 1983, Slow crack growth in hot-pressed silicon nitride, Proc. Intl. Conf on *Fracture Mechanics of Ceramics, Vol. 6*, eds. R. C. Bradt, A. G. Evans, D. P. H. Hasselman and F. F. Lange, pp. 603-36.

Quinn, G. D., 1987, Static fatigue of a siliconized silicon carbide, AMMRC report no. MTL TR87-20, Army Materials and Mechanics Research Center, Watertown, Mass.

Robertson, A. G., 1989, Swelling and creep damage accumulation in hot-pressed alumina, PhD. thesis, McMaster University, Hamilton, Ontario, Canada.

Robertson, A. G., Wilkinson, D. S. and Caceres, C. H., 1991, Creep and creep fracture in hot-pressed alumina, J. Amer. Ceram. Soc., **74**, 915-21.

Tanaka, T. Nakayama, H., Okabe, N., Yamamoto, S. and Fukui, S., 1992, Creep rupture map of engineering fine ceramics, in this volume.

Thouless, M. D. and Evans, A. G., 1984, Nucleation of cavities during creep of liquid-phase-sintered materials, J. Amer. Ceram. Soc., **67**, 721-727.

Tsai, R. L. and Raj, R., 1982, Creep fracture in ceramics containing small amounts of a liquid phase, Acta Metall, **30**, 1043-58.

Wang, J. S., Stephens, J. J. and Nix, W. D., 1985, A statistical analysis of cavity nucleation at particles in grain boundaries, Acta Metall., **33**, 1009-21.

Wiederhorn, S. M., Hockey, B. J. and Krause, R. F. Jr., 1987, Influence of microstructure on creep rupture, in *Ceramic Microstructures '86: Role of Interfaces*, J. A. Pask and A. G. Evans, eds. Plenum Press, New York, pp. 795-806.

Wiederhorn, S. M., Chuck, L. Fuller, E. R. Jr. and Tighe, N. J., 1986, Creep rupture in siliconozed SiC, in *Tailoring Multiphase and Composite Ceramics*, eds. R. E. Tressler, G. L. Messing, C. G. Pantano and R. E. Newnham, Plenum, pp. 755-73.

Wiederhorn, S. M., Roberts, D. E., Chuang, T.-J. and Chuck, L., 1988, Damage enhanced creep in siliconized silicon carbide, J. Amer. Ceram. Soc., **71**, 602-8.

Wiederhorn, S. M., Hockey, B. J. and Chuang, T.-J., 1991, Creep and creep rupture of structural ceramics, Proc. NATO Advanced Study Workshop on *Toughening Mechanisms in Quasi-Brittle Materials*, S. P. Shah, ed., Kluwer Academic Publishers, Dordrecht, pp. 555-76.

Wilkinson, D. S., 1983, Steady state crack growth in creeping solids, Proc. Intl. Conf. on *Advances in Life Prediction Methods*, ASME, pp. 259-67.

Wilkinson, D. S., 1988, Effect of devitrification on creep deformation in glass-containing ceramics, J. Amer. Ceram., Soc., **71**, 562-65.

Wilkinson, D. S., Caceres, C. H. and Robertson, A. G., 1991, Damage and fracture mechanisms during high temperature creep in hot-pressed alumina, J. Amer. Ceram. Soc., **74**, 922-33.

HIGH TEMPERATURE FRACTURE MECHANISM OF GAS-PRESSURE SINTERED SILICON NITRIDE

N.Kohler[1], Y.Ikuhara[2], H.Awaji[3] and K.Funatani[4]

[1] Fachhochschule Rhineland-Pfalz Koblenz
[2-4] Japan Fine Ceramics Center
Mutsuno 2-4-1, Atsuta-ku, Nagoya 456, JAPAN

INTRODUCTION

Silicon nitride has been adopted as automotive engine components in mass-production scale. It is a very hopefull candidate material for future Ceramics Gas Turbine (CGT) components. However, the CGT is to be operated at about 1,400℃, requiring more intensive researches and developments to make materials durable even at the temperature. For that purpose it is substancial to investigate the mechanism of high temperature fracture behavior of these materials.

The high temperature mechanical strength of a GPSSN (Gas Pressure Sintered Silicon Nitride), fracture behavior and the microstructure are studied in this research. The mechanical properties of the GPSSN measured are the flexural strength and the fracture toughness. The high temperature fracture behavior is correlated to the results of microstructure observed by SEM (Scanning Electron Microscope) and TEM(Transmission Electron Microscope). The discussion is mainly focussed on the deformation and the fracture behavior at high temperatures which is accompanied by a slow crack growth phenomenon. The investigation of microstructure at different temperatures gives further information to considering the high temperature mechanical behavior of GPSSN.

MATERIALS

Test specimen were machined from two step gas pressure sintered block (37 x 41 x 15 mm) of GPSSN(NTK). The density measured in water is 3.266 g/cm^3. The relative theoretical density is 99.9 %. The chemical composition(wt %) includes:

Al_2O_3=4.34, Y_2O_3=4.50, CaO = 0.001,
MgO = 0.0003, Fe_2O_3 = 0.0069

The X-ray analysis (Phillips PW-1700) showed that the diffraction pattern is in a good agreement with the ASTM-index. The microstructure is observed by Secondary electron image of plasma-etched surface (Fig.1) and Reflective electron image (Fig.2). The photograph of Figure 2 clearly recognize that the present material is composed of two phases. The secondary phase forms a grain boundary network in which elongated silicon nitride grains are inbedded. A quantitative inspection of the microstructure of a plasma-etched surface yielded the following characteristic averaged grain dimensions;

Grain length = 1,57 μm, Grain diameter = 0.27 μm
Aspect ratio ≒ 5.81.

Fig.1 SEM-micro-graph of a plasma-etched surface.

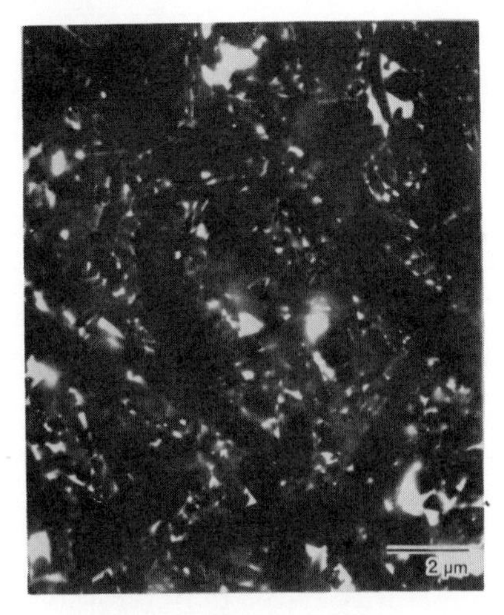

Fig.2 Reflecting electron image of a mirror surface.

The Vickers hardness (10 kgf) of the material has the average value of $Hv_{(10)}$ = 17 GPa.

MECHANICAL TESTS

Flexural Strength

The flexural strength was measured from room temperature to 1,500 ℃ with a particular emphasis on the high temperature strength.

Thus the test was conducted from 1,000 to 1,500℃ in each 50 ℃, using JIS R (1987) and the cross head speed (CHS) of 0.5 mm/min. The flexural test was carried out on an Instron-1185 Machine in a nitrogen atmosphere to prevent oxidation and the decomposition of the silicon nitride.

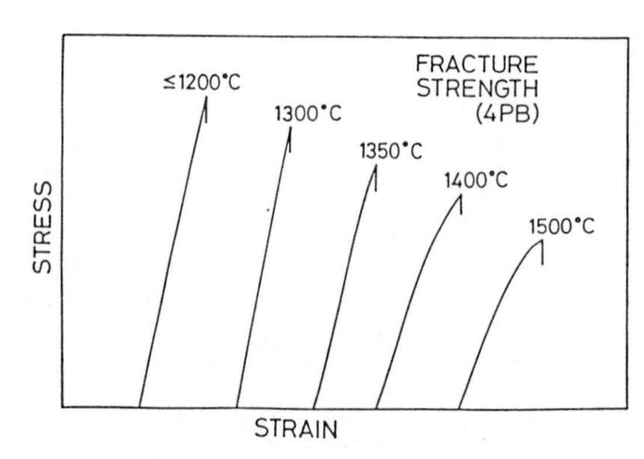

Fig. 3 Stress-Strain diagram of flexural tests, (0.5mm/min.)

Figure 3 shows a schematic difference of the stress-strain curve for different temperatures. There are no significant changes in the stress-strain curves below 1,200 ℃. However, at temperature higher than 1,300 ℃ the curves indicates plastic deformation that seems to be related to short range dislocations and slips which increase as the temperature becomes higher. At 1,500℃ much larger strain is recorded which can be related to the grain boundary cavity formation as confirmed by SEM and TEM observations.

Figure 4 shows the typical change of strength with temperature. The curve has an S-shape in the temperature range from 1,100 to 1,300 ℃, which have been related to a large variation in the inherent defect distribution of the materials[1]. The high temperature fracture strength is understood to be largely affected by the Slow Crack Growth (SCG) behavior. The S-curve behavior could be resulted from the SCG at lower temperature and from the defect distribution[2,3]. However, the detailed discussion should be given with the results of microstructural observations.

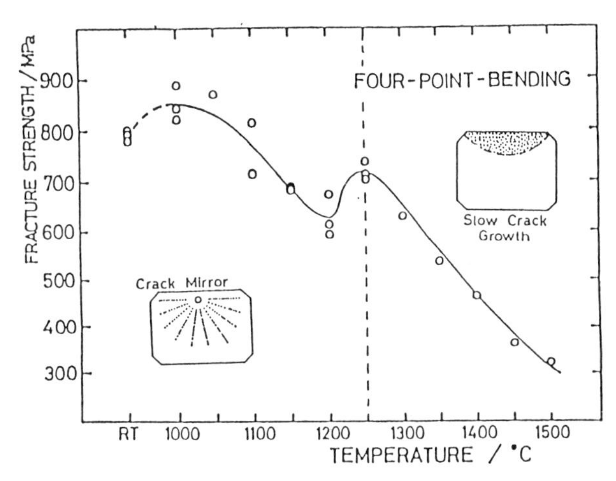

Fig.4 Relation between temperature and fracture strength (four point bending)

Fractographs

Fractographic observations by SEM are conducted after flexural strength tests, which demonstrated two different characteristic fracture surfaces. From room temperature to 1,200 ℃, the fracture surface is rather rough, while at higher temperatures fracture surface at the tension side becomes more flat. Both of the fractographs can be divided into two regions; the crack starting region and the unstable catastrophic fracture zone. The fractographs show a so-called fracture mirror in the crack starting region of the specimens teted below 1,200 ℃ as in Figure 5. This can not be observed in the at higher temperature fractographs. The fractographs of 1,250 to 1,500 ℃-specimens have a characteristic crescent shaped region due to the SCG which starts from the tension side surface and propagates into as shown in the upper side of Figure 6. The lower zone in Figure 6 shows a rather flat surface attributed to the catastrophic fracture. At the center of the fracture mirror of the speciment tested at room temperature and 1,200 ℃, a semi-circular-shaped defect was observed as the fracture origin. There are different types and sizes of the defects in the near surface tension side, which are ranging from about 40 to 150 μm. They are inclusions and pores with large grown crystals as shown in Figure 7, and Y-rich spot with fine needle or whisker like crystals grown within the cavity as shown in Figure 8.

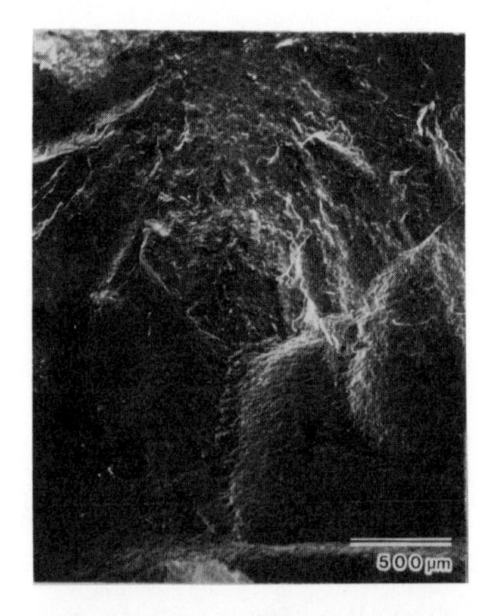

Fig.5 Fractograph of R.T. testing.

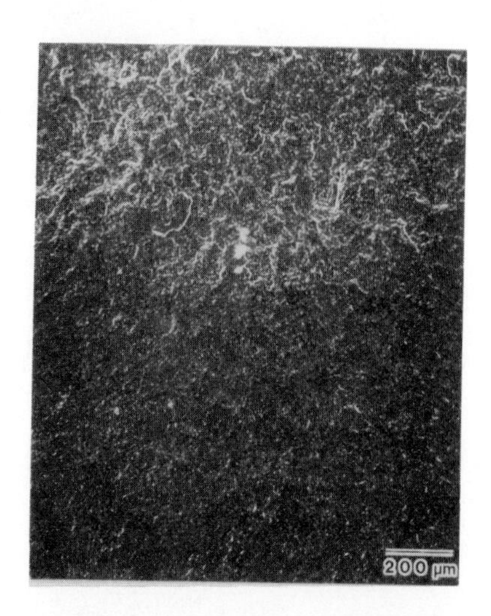

Fig.6 Fractograph of 1,300℃ testing.

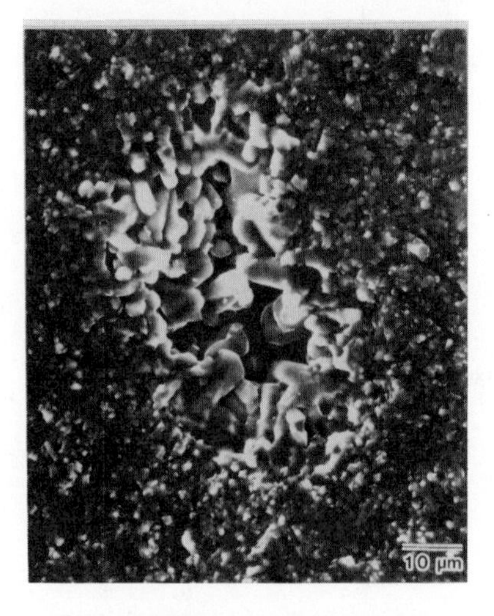

Fig.7 SEM micrograph of a pore defect.

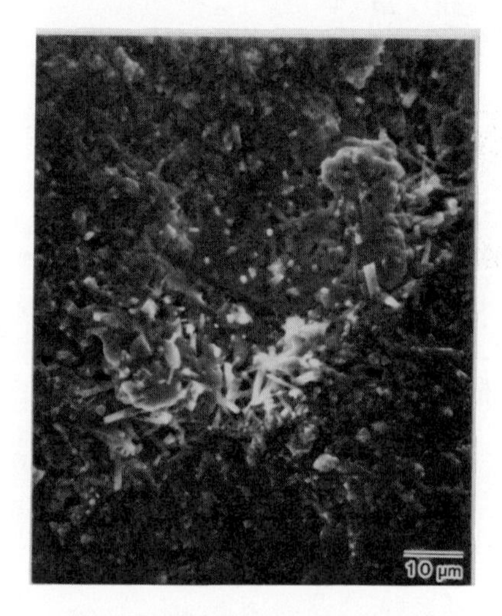

Fig.8 SEM micrograph of a Y-rich defect.

Fracture Toughness

The fracture toughness was measured by the SEPB- and the SEVNB-configurations.

a) SEPB method

The SEPB fracture toughness test method is composed of Vickers indentation, pre-cracking and three point bending test. The pre-crack length was measured by optical microscope. The cross head speed (CHS) of the test was 0.5 mm/min at room temperature and 1,000 ℃ and 5.0 mm/min., at 1,300 and 1,400 ℃.

The critical intensity factor K_{1C} is calculated by following equation accounting the dimension of specimen size and crack depth.

$$K_{1C} = \sigma_f \cdot a^{1/2} \cdot Y$$

Table 1 Fracture toughness (SEPB) of GPSSN

Part 1 : Cross Head Speed; v = 0.5 mm/min.

Tempeature T (℃)	Crack Length a (mm)	Average Fracture Toughness K_{1C} (MPa·m$^{1/2}$)
R.T.	2.08	6.19 ± 0.12
1,000	2.01	5.42 ± 0.08

Part 2 : Cross Head Speed; v = 5.0 mm/min.

Temperature T (℃)	Crack Length a(mm)	Average Fracture Toughness K1C (MPa·m1/2)
1,300	1.77	4.86 ± 0.88
1,400	1.96	6.15 ± 0.65
(SCG)	(2.03)	(6.47 ± 0.77)

(The values in the brackets include the slow crack growth region)

The test results are listed in Table 1. Even at CHS of 5.0 mm/min., fracture toughness values and fractographs reveal some effects of SCG at 1,400 ℃.

b) SEVNB method

The SEVNB (Single Edge V-Notched Beam) fracture toughness test specimens are mechined by V-shaped diamond wheel to form a single V notch. The SEVNB test is conducted by three point bending jig.

By the effect of V-notch, calculation of the K_{1C} value should take two factors into account and modification of the equation. They are the notch radius "r" and the average grain size "d". Those two parameters affect the stress distribution at the "process zone" and unstable crack propagation behavior[4,5].

The SEVNB tests are conducted from room temperature to 1,500 ℃ by CHS of 0.5 mm./min. The modified K_{1C} values are calculated by the factor "1.19" using following values[6] and listed in Table 2.

V-notch radius: r = 38 μm
Average grain size: d = 0.92 μm

Table 2 SEVNB test results from R.T. to 1,500 ℃ and modified value

Temperature T (℃)	Average Crack Length a = mm	Average Fracture Toughness K_{1C} (MPa·m$^{1/2}$)	Modified Fracture Toughness K_{1C} (MPa·m$^{1/2}$)
R.T.	1.45	6.82 ± 0.25	5.73 ± 0.21
1,000	1.44	6.12 ± 0.14	5.15 ± 0.12
1,100	1.48	6.10 ± 0.10	5.12 ± 0.08
1,200	1.45	5.85 ± 0.09	4.75 ± 0.08
1,250	1.49	5.55 ± 0.13	4.66 ± 0.11
1,300	1.49	5.55 ± 0.12	4.67 ± 0.10
1,400	1.47	7.69 ± 0.13	6.46 ± 0.11
" (SCG)	(1.78)	(9.54 ± 0.12)	(8.02 ± 0.10)
1,500	1.48	6.05 ± 0.17	5.09 ± 0.14
" (SCG)	(2.41)	(13.40 ± 0.81)	(11.26 ± 0.68)

(The value in brackets include the slow crack graowth region.)

The temperature dependent S-curve behavior of the fracture strength is not observed in the K_{1C} test results. The fracture toughness shows a steady decrease with temperature up to 1,300 ℃ as shown in Figure 9. The S-curve behavior, therefore, seems to be dictated by the stress gradient in flexural deformation made as well as the defects within the sintered

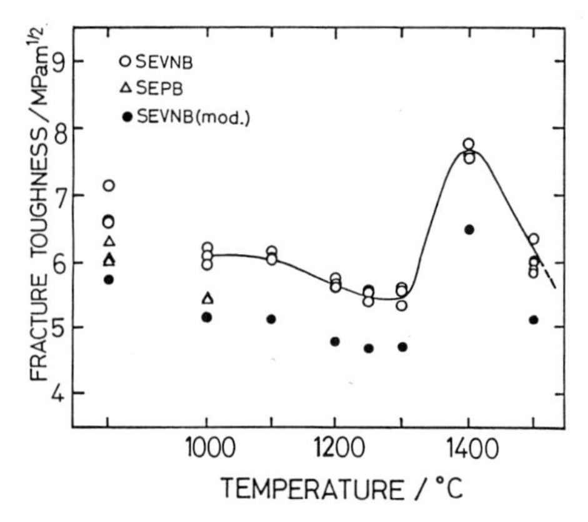

Fig.9 Relation between fracture toughness and temperature.

body. On the other hand, fractographs of specimens tested above 1,300℃ showed influence of SCG, and the modified value which calculated from the area of final steady fracture show quite sharp increase of toughness as shown in Figure 10.

Effect of CHS on SCG Behavior

The effect of SCG behavior on the fracture process and mechanisms of the silicon nitride was investigated by changing the cross head speed of, SEVNB-specimens at temperatures of 1,300, 1,400 and 1,500℃. The CHS was changed in the range of 0.5, 1.0, 5.0, and 50 mm/min. The test result at 1,300 ℃ is not affected by CHS, but the K_{Ic} value decreases steadily as the CHS increases at 1,400 ℃. On the other hand, the K_{Ic} value at 1,500℃ becomes the highest at 1.0 mm/min., and then decreases steadily at higher CHS.

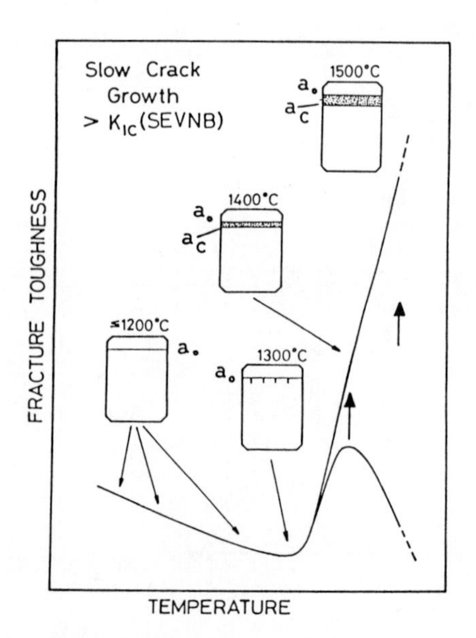

Fig.10 Relation between temperature and fracture toughness influenced by SCG behavior.
(a_o is the original V-notch depth, a_c is the critical crack length after SCG.)

The test results show that the effect of SCG decreases and then disappears with increasing CHS as shown in Figure 11. The fracture toughness value at higher temperature is largely affected by SCG behavior. The critical crack length after SCG decreases by elevating temperature, and apparent fracture toughness value calculated from origainal V-notch depth become far smaller than the corrected value calculated from actual critical crack length.

There are four different parameters which significantly influence on the K_{1C}-determination:

- the maximum load F_{max}
- the load at fracture initiation F_c
- the original crack length a_0
- the slow crack extension by crack growth a_c

The combination of those parameters yields are:

- K_{1Cmax} obtained by F_{max} and $a_{max} = a_0 + a_C$
- K_{1C} with F_{max} and a_0
- K_{1C}^0 calculated from F_C and a_{max}

Table 3 Fracture toughness (SEVNB) at different CHS-values

Temperature T (℃)	Cross Head Speed v (mm/min.)	Crack Length a (mm)	Average Fracture Toughness K_{1C} (MPa·m$^{1/2}$)
1,300	0.5	1.49	5.55 ± 0.12
	1.0	1.49	5.59 ± 0.05
	5.0	1.49	5.27 ± 0.09
	50.0	1.49	3.02 ± 0.11
1,400	0.5	1.47	7.69 ± 0.13
	(SCG)	(1.78)	(9.54 ± 0.12)
	1.0	1.49	7.23 ± 0.18
	(SCG)	(1.88)	(8.13 ± 0.36)
	5.0	1.49	5.45 ± 0.07
	50.0	1.50	2.99 ± 0.10
1,500	0.5	1.48	6.05 ± 0.17
	(SCG)	(2.48)	(13.40 ± 0.81)
	1.0	1.20	7.88 ± 0.27
	(SCG)	(1.89)	(10.88 ± 1.20)
	5.0	1.22	8.81 ± 0.29
	(SCG)	(1.32)	(8.99 ± 0.28)
	10.0	1.21	5.61 ± 0.00

(The value in bracket include the slow crack growth)

The value of K_{1C} and K_{1Cmax} are obviously important in explaining SCG behavior. The value of K_{1Cmax} becomes larger at higher temperatures. This suggests the limitation of the conventional K_{1C} because of the prominant stress relaxation and plastic deformation. The effect of SCG is also apparent in fractographic observation as shown in Figure 12. The effect of CHS is lager at low speed but it can be eliminated by increasing CHS.

MICROSTRUCTURE AND FRACTOGRAPHY STUDIES BY TEM

Microstructure

The characterization of the microstructure of specimens tested at room temperature, 1,200, 1,350, and 1,500 ℃ is conducted through the microstructural observation by TEM. The microstructure of 1,200℃-sample shows

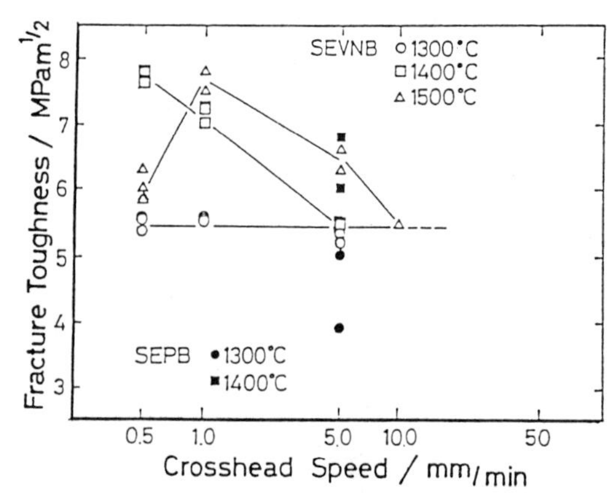

Fig.11 The effect of CHS on the high temperature fracture toughness.

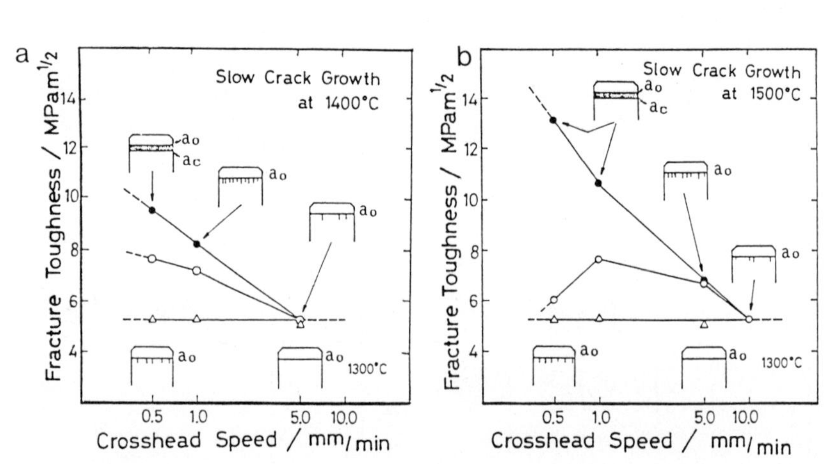

Fig.12 The effect of SCG on the fracture toughness at a)1,400 and b)1,500 ℃.

no fundamental changes compared with that of R.T.. In contrast, the microstructure of 1,350 and 1,500 ℃-specimens is strongly infuenced by the deformation and important difference is recognized. The stress/strain-diagram of the samples tested at 1,350 ℃ is the first diagram which shows deformation, and is characterized by a lot of "lattice dislocations" and some of small cavities in microstructure as shown in Figure 13. The amount of deformation and cavities increases as testing temperature become higher.

Deformation and Cavity Formation

The specimen tested at 1,500 ℃ which exhibits an intensive plastic deformation in the stress/strain-diagram shows a high amount of cavity formation and only a few lattice dislocations. At this temperature, the grain boundary slip governs the formation of numerous cavities as observed in Figure 14-a). In specimens tested at 1,350 and 1,500 ℃, some oriented grains or slip like patterns lying completely in plain are observed in Figure 14-b).

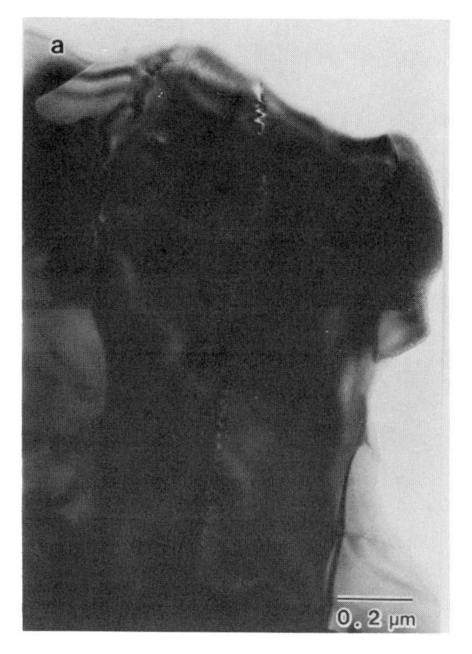

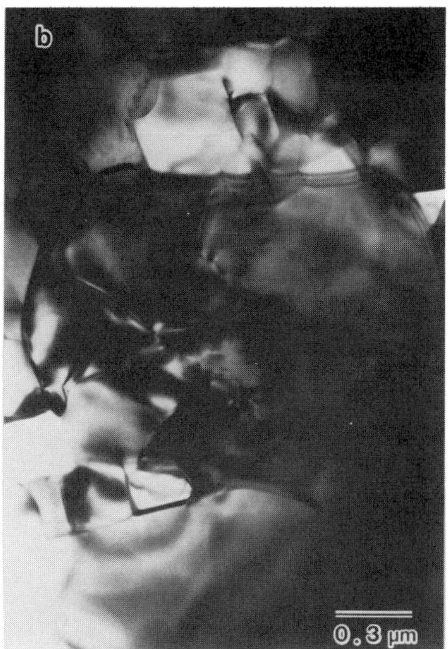

a) Dislocation wake   b) Dislocation network

Fig.13 Dislocations observed in a specimen tested at 1,350 ℃.

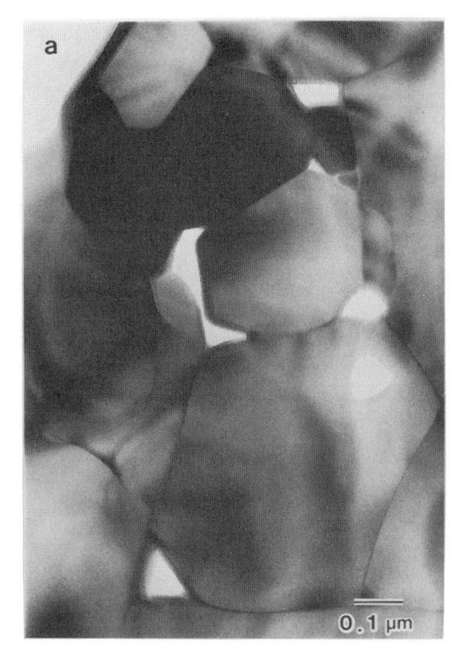

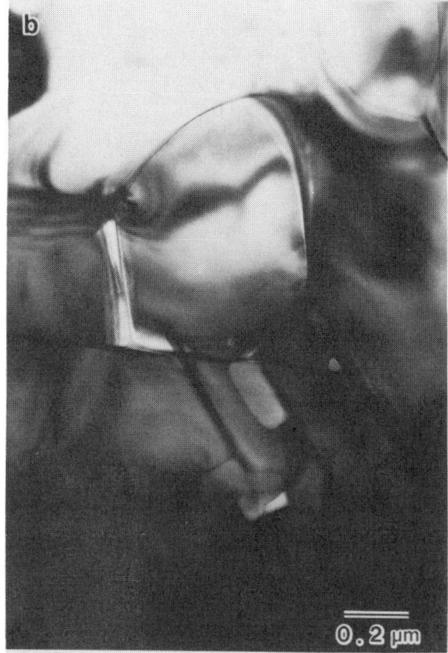

a) Grain boundary cavities   b) Slip like plain

Fig.14 High temperature fracture behavior in samples tested at 1,500 ℃.

CONCLUSION

The high temperature strength and the fracture behavior of a GPSSN are investigated through flexural testing, fracture toughness testing, microstructure and fractographic observations by SEM and TEM. Following important fracture mechanisms are made clear. They seem to give important suggestions for future material development.

1) Fracture originates from an internal defect of the specimen at temperature below 1,250 ℃, but slow crack growth behavior becomes important at higher temperature, and the crack starts from the tension surface of the flexural test specimen.

2) Flexural strength is strongly affected by the temperature of measurement. It shows an S-curve behavior with a peak at about 1,250 ℃, which may be caused by the stress relaxation around defects such as Y-rich spots and pores.

3) Fracture toughness values at higher temperature by the SEPB and the SEVNB methods have a finite difference because of the slow crack growth behavior. Higher cross head speed should be applied to measure exact fracture strength at temperature higher than 1,250 ℃.

4) Stress/Strain diagrams at temperature higher than 1,350 ℃ exhibit a yielding behavior, which may be related to the short range dislocations and slips. At 1,500 ℃, however, many grain boundary cavities are observed with a few re-oriented crystals, being caused by grain boundary deformations and transgranular slips.

5) Stress relaxation behavior at higher temperatures before forming grain boundary cavities seems to have important effects on the improvement of the high temperature strength of ceramic materials through designing and controlling the phase and microstructure.

REFERENCES

1. R. F. Pabst, Bedeutung und Besonderheiten des Zusammenhang von Herstellung Festigkeit und Konstruktion bei keramischen Werkstoffen, Fortschrittsberichte der DKG, Band 1 (1985), p. 1-8
2. G. D. Quinn, Fracture Mechanism Maps for Hot-Pressed Silicon Nitride, Ceramic Engineering and Science, Proceedings, V.5, No.7, p. 596-602
3. G. D. Quinn, J. B. Quinn, Slow Crack Growth in Hot Pressed Silicon Nitride, Fracture Mechanism of Ceramics, V.6, (1981), p. 603-636
4. H. Awaji, T. Watanabe, T. Yamada, Evaluation of Fracture Toughness by a Single-Edge V-Notched Beam Method, Transaction of Japanese Society of Mechanical Engineering, V. 56, No.525, (1990-5), p. 1148-1153
5. A. G. Evans, K. T. Faber, Crack-Growth Resistance of Microcracking Brittle Materials, Journal of American Ceramic Society, V. 67, No.4, (1984), p. 225-269
6. D. Munz, T. Fett, Mechanische Verhalten keramischer Werkstoffe, Versagensverlauf, Werkstoffauswahl, Dimensionierung, Springer Verlag, (1989)

APPENDIX

High temperature fracture mechanisms are affected by dislocations and slips before large deformation become observable. The behavior of processing zone and deformation caused by dislocation and slips seems to play a very important role even in ceramics materials. It is very important to understand fracture mechanisms through direct observation of the behavior in ceramic materials.

In-situ observation of crack propagation in Alumina at room temperature and 800 ℃ are performed by Dr. Y. Ikuhara and Mr. T. Suzuki (JFCC).

The results obtained from the observations show the state of the size and stress pattern of processing zone and dislocation wake as shown in followings photographs copied from the Video tape record.

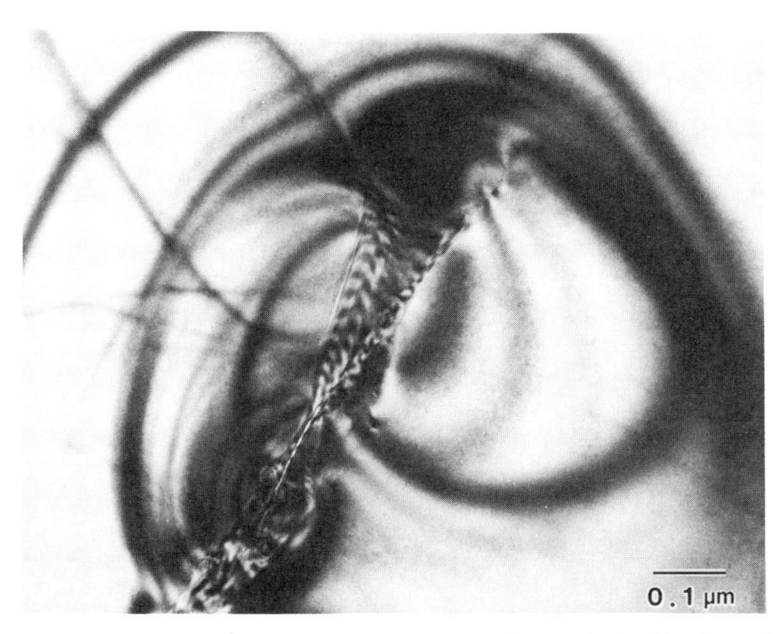

Fig.1 Processing zone and dislocation wake observed in Alumina at room temperature.

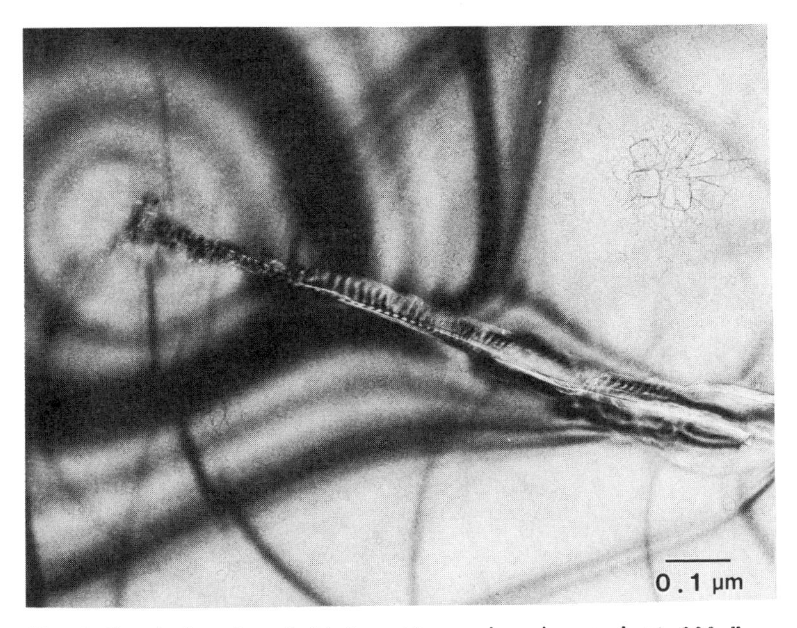

Fig.2 Crack front and dislocation wake observed at 800 K.

CHARACTERIZATION OF THE FLEXURAL STRENGTH DEGRADATION OF A COMMERCIAL HOT-PRESSED SILICON NITRIDE IN A HIGH-TEMPERATURE SULPHIDIZING ENVIRONMENT

C. Saraiva Martins, M. Steen, J. Bressers, and L. Guerra Rosa*

Institute for Advanced Materials, Joint Research Centre
Petten Establishment, C.E.C., the Netherlands
and *Instituto Superior Técnico, Lisbon, Portugal

ABSTRACT

In this paper the results of residual strength tests on a hot-pressed silicon nitride after corrosive exposure in a simulated coal gasification environment are presented. It is shown that the weight gain of the specimens stabilizes after 200 hours exposure at 1300 °C, and that it does not depend on the surface to volume ratio of the samples. For room temperature testing, an air environment can be used for reference purposes. At higher temperatures, however, the oxidation occurring in air may obscure the true effect of the corrosion exposure, and tests at high temperatures are therefore performed in nitrogen. For the as-received material, the Weibull modulus remains constant from room temperature up to 1300 °C, and bulk flaws cause failure at all temperatures. In the residual strength tests at room temperature, the same value of the modulus is obtained and failure is again caused by flaws in the bulk of the material. This indicates that the exposure does not affect the intrinsic failure behaviour of the material and that the corrosion induced small strength degradation can be taken into account through a reduction in the load bearing section caused by the presence of a corroded surface layer.

Fracture Mechanics of Ceramics, Vol. 10
Edited by R.C. Bradt *et al.*, Plenum Press, New York, 1992

INTRODUCTION

Research to bring hot-pressed silicon nitride ceramics to an exploitable status is currently receiving much attention. For high temperature load-bearing applications the major problems to be solved are reliability and corrosion resistance in reducing gaseous environments containing sulphur, such as in the chemical and petrochemical process industries and in coal gasification technology. Environmental and economical factors justify a closer look at the coal gasification process which, together with combined cycle power generation, enables low NO_x and SO_x emissions to be achieved. In order to have an economically viable process, however, gas turbine inlet temperatures above 1250 °C are needed[1]. Using today's gas turbine technology, it is impossible to reach these temperatures, and the potential economical benefit of coal gasification is offset by the costs associated with the necessary cooling of the turbine blades and cleaning of the gas.

Previous studies[2] in gas mixtures similar to those which are likely to be encountered in coal gasifiers have reported that severe corrosive attack of hot-pressed silicon nitride can be prevented by either reducing the temperature to below 1200 °C, or by slightly humidifying the gas. Under these conditions, the extent of degradation has been found to depend on the partial oxygen and sulphur pressures of the environment. The corrosive exposure conditions of this investigation (i.e. gas composition, absolute pressure and temperature) have been chosen to simulate real coal gasification environments[3] as closely as possible.

In previous studies, mainly the aspects of weight gain and microstructural analysis have been considered. In the current investigation, the effect of corrosion exposure is additionally assessed by residual mechanical strength measurements.

MATERIAL CHARACTERIZATION

Commercially available hot-pressed silicon nitride from CERADYNE Inc. (USA), is used in this study. Detailed information on the chemical composition is lacking, but Y_2O_3 and Al_2O_3 are claimed to be used as sintering aids during processing. Determination of the density by immersion in water and in mercury yields an average value of 3.26

Mg/m^3, which proves to be uniform between different plates and different locations within the same plate. The total porosity level is less than 0.1 percent. X-ray diffraction analysis on as-received material reveals the presence of β-Si_3N_4 and of a minor amount of $Y_{10}N_2(SiO_4)_6$. The typical microstructure of a polished surface observed in SEM consists of grey and white phases. The former is identified by EDS as containing Si (β-Si_3N_4), while the latter is an intergranular yttrium containing phase. The grain size ranges from 1 to 10 μm, with an aspect ratio between 1 and 10.

SPECIMEN PREPARATION

Bend specimens with dimensions (45±0.5)x(4±0.05)x(3±0.05) mm^3 are cut perpendicular to the hot-pressing direction from a 200x200x10 mm^3 plate. All long faces are ground parallel to the longitudinal axis using a coarse and fine grinding sequence. Final finishing is performed with a D46 diamond grinding wheel at a stock removal rate not exceeding 0.025 mm per pass for the last 0.1 mm. Edge chamfering is not applied. A batch of 20 specimens has been measured with a Perth-O-Meter to evaluate the surface condition, yielding a roughness parameter R_a = 0.43±0.03 μm. Surface preparation of the other specimens used in this study (cubes 6x6x6 mm^3 and rectangular samples representing the gauge section of tensile specimens 30x8x6 mm^3) is performed in a similar way to the bend test specimens.

EXPERIMENTAL PROCEDURE

The composition of the corrosive gas for this investigation is chosen to simulate real coal gasification environments as closely as possible, while still maintaining experimental simplicity. The decisive parameters for the simulation gas are the partial pressures of sulphur and oxygen, which should be the same as under service conditions. The equilibrium partial pressures of these species in the gas for the experiments, as well as those in the real environment are calculated using an in-house computer programme CORGA, based on the minimization of the free energy of the gas system[4]. Compared to actual coal gasifier conditions, these calculations yield a higher temperature and a lower total pressure for the simulated environment, whose composition is set as 0.4% H_2S + 0.75% H_2O + H_2 bal.

Prior to all exposures the samples are cleaned ultrasonically in acetone and ethanol for one hour and dried overnight in an oven at 80 °C. They are subsequently weighed using a balance with a 0.02 mg reproducibility. Exposure takes place in an autoclave at 1200 and 1300 °C under an absolute pressure of 1.5 bar. The specimens are supported by a high purity alumina disk with drilled holes to minimize the contact surface between the specimens and the support disk. After sealing, the autoclave is filled with flowing active gas and heated at a rate of 10 °C/min. After reaching the exposure temperature, the specimens are held there for a specified amount of time under a gas flow rate of 17.5 l/h. The water vapour content of the gas is measured at the outlet of the autoclave with a dewpoint-meter. To prevent leakage of the active gas to the laboratory environment, the active part of the autoclave is surrounded by a protective argon blanket.

A critical issue in corrosion testing is the surface to volume ratio of the specimen, giving rise to a size effect in the dependence of weight gain on exposure time. In order to investigate the influence on the corrosion kinetics, samples with different area to volume ratios, A/V, have been exposed. The values chosen for this ratio correspond to those used for conventional corrosion testing (6x6x6 mm^3 cubes, A/V=1.0), for the four-point flexure specimens (45x4x3 mm^3, A/V=1.2), and for the gauge length of uniaxial tensile specimens (30x8x6 mm^3, A/V=0.65). At least two specimens of each of these types are exposed to the active gas mixture and held at the exposure temperature for specified periods of time. After each exposure step, the specimens are cooled down to room temperature under gas flow and removed from the rig for weight change determination.

The flexural tests of this study are performed in the four-point bending mode with outer and inner spans of 40 and 20 mm, respectively. For room temperature tests a stainless steel jig is used, while tests at 1200 and 1300 °C are performed on a silicon carbide jig. In both jigs, the rollers can rotate freely in their supports. In order not to mask the influence of the corrosive exposure, the high temperature tests are carried out in an inert nitrogen atmosphere. A constant displacement rate of 0.5 mm/min is used for testing. The load is measured by a 2 kN load cell with an accuracy of 0.5 %, and recorded versus cross head displacement and time.

RESULTS AND DISCUSSION

Comparative testing in air and in nitrogen at 1300 °C for 2 hours has revealed a completely different surface morphology. SEM investigations show that air testing results in the formation of surface pores and cracks. Also yttrium rich crystals are found at the surface. The XRD spectrum shows a surface composition of 68% Si_3N_4, 16% $Y_2Si_2O_7$ and 16% SiO_2. The weight gain in air is 0.066 mg/cm^2. After nitrogen testing a much smoother surface is obtained, silica does not seem to be formed, and the weight gain is only 0.009 mg/cm^2. However, qualitative elemental analysis of the surface, using EDS with different beam intensities, reveals a high percentage of yttrium and aluminium, suggesting that the intergranular phase tends to emerge at the outer surface. On the basis of these results, the strength determinations at temperatures of 1200 and 1300 °C are performed in a nitrogen atmosphere.

The results of the flexure tests on the unexposed as-received material are shown in the Weibull plot of figure 1. Twenty specimens have been tested at room temperature and at 1200 and 1300 °C. The parameters of the Weibull distribution (i.e. characteristic strength σ_0 and modulus m) have been determined by conventional linear regression analysis using $p_i=(i-0.5)/n$ as an estimator, where i and n represent the rank and total number respectively. At room temperature a characteristic strength of 917 MPa and a modulus of 15.8 are obtained. These values agree well with those given in the sales information of the manufacturer, viz. $\sigma_0=925$ MPa and m=15 to 25. At 1200 °C in nitrogen σ_0 decreases to 778 MPa, while the value m=17.1 falls within the limits of the scatterband on the room temperature value. At 1300 °C in nitrogen, the characteristic strength drops to 440 MPa which is less than half the value obtained at room temperature. The value of the Weibull modulus, however, remains constant at 15.3. The high temperature strength values observed in this investigation are smaller than those mentioned in the manufacturer's data sheet, where characteristic strength values in air of 895 MPa at 1200 °C and 500 MPa at 1400 °C are quoted. Oxide strengthening, which is not present in the current experiments may explain this discrepancy.

The similarity in the values of the Weibull modulus at the three temperatures suggests that at each temperature the same flaw population is governing failure. This is corroborated by fractographic analysis which reveals that all flexure specimens fail from bulk flaws. A representative example of the fracture surfaces of a specimen failed at room temperature is shown in figure 2a at low magnification (x32). A dark black spot can easily be seen in the region of failure initiation.

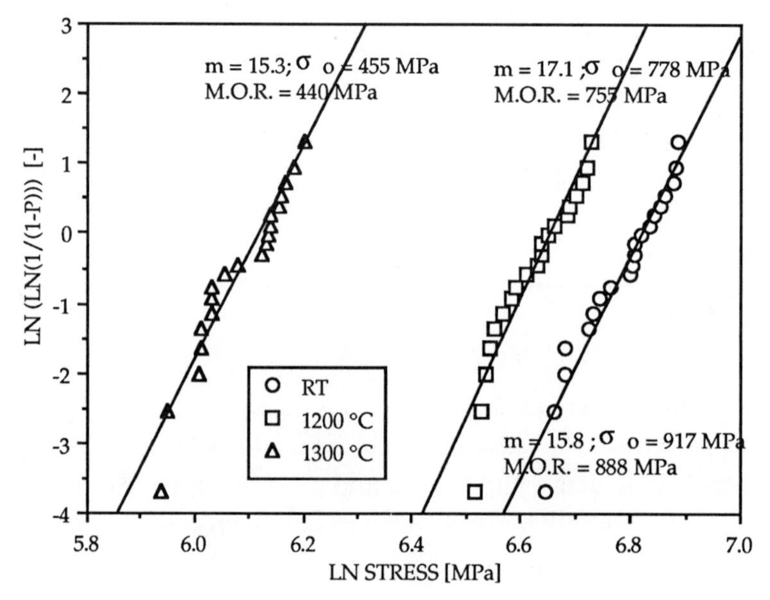

Fig. 1. Weibull plot of bend tests on as-received material

SEM investigations at higher magnification reveal that in all cases the failure causing defect is situated in the immediate vicinity of this dark spot, but does not necessarily coincide with it. At higher temperatures, dark spots do not occur so prominently but failure is still due to bulk flaws. At all temperatures, Fe and Mn inclusions as well as large grains (fig. 2b) act as failure initiation sites.

The weight change per unit surface area as a function of exposure time at 1300 °C is shown in figure 3 for different surface to

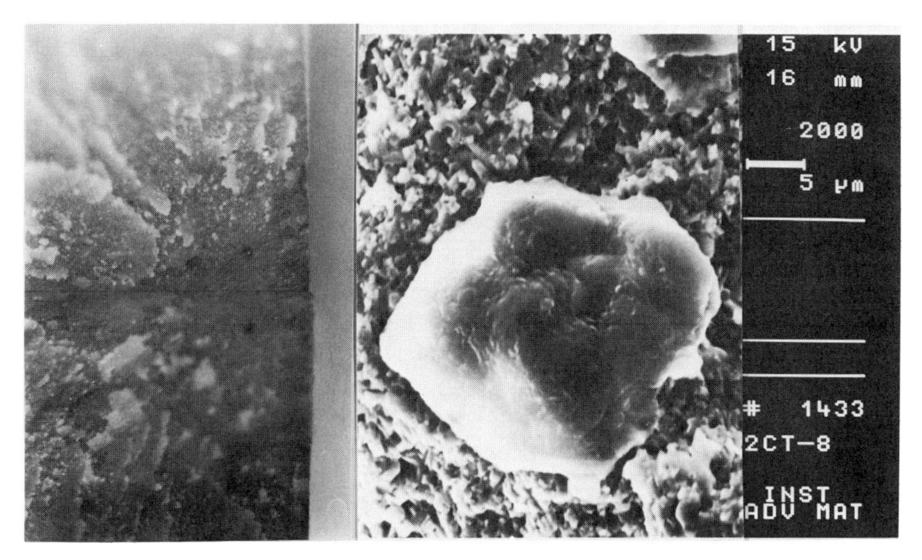

Fig. 2. Macro- and micrograph of fracture surface of specimen tested at room temperature

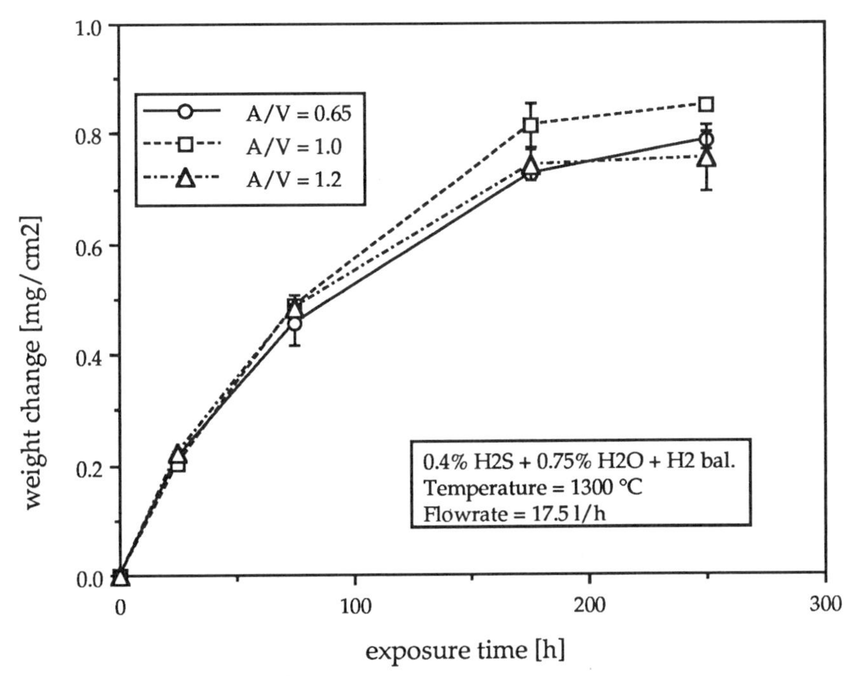

Fig. 3. Weight gain during exposure at 1300 °C

volume ratios. A large effect of the A/V ratio is not apparent. When the weight change is expressed per unit volume, however, a definite dependence on the A/V ratio is found, indicating that the former representation is to be preferred and that within the A/V ratios covered in this investigation, the change in weight during corrosive exposure is a surface controlled phenomenon. Figure 3 also shows that the weight gain seems to level off after exposure times of about 200 hours. On this basis, the exposure of the bend test specimens is set at 1300 °C for 200 hours.

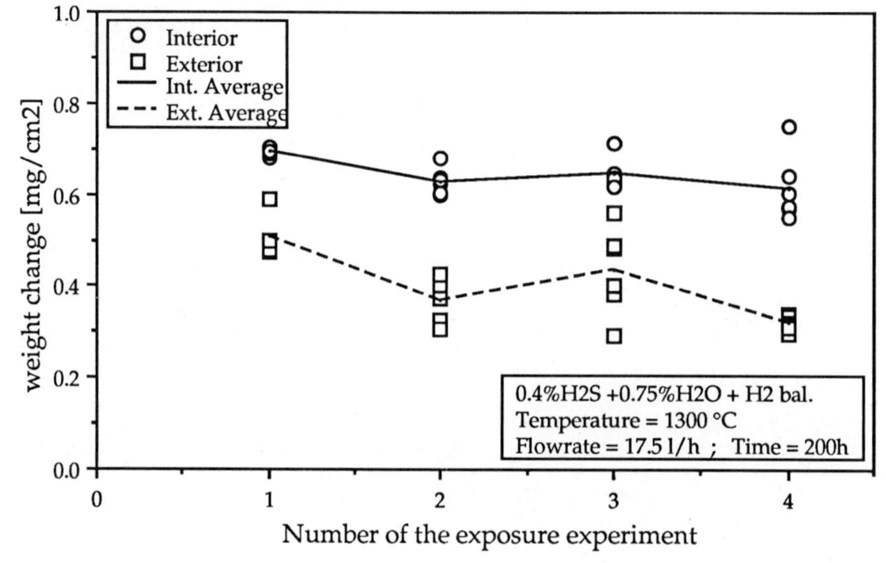

Fig. 4. Effect of specimen position in the jig on weight gain

A peculiar feature noticed during the course of the exposure experiments is that the weight gain of the specimens, and thus the corrosion rate, depends on the position in the supporting jig. As indicated in figure 4, specimens located near the centre of the support jig consistently display a larger increase in weight than those situated nearer to the outer circumference. This difference has not been investigated to a larger extent, but is believed to be caused by a difference in the local gas flow rate around the specimens.

Table 1. Surface roughness parameters of as-received and exposed samples

		R_a [μm]	R_{3z} [μm]	R_t [μm]
transverse	as-received	0.43	2.20	3.60
	exposed	0.87	4.10	7.40
longitudinal	as-received	0.30	1.48	2.14
	exposed	0.90	4.60	7.60

The effect of corrosive exposure at 1300 °C on the surface roughness is summarised in Table 1. As expected, a notable increase is observed and the original difference between the longitudinal and transverse directions has disappeared.

For the time being, residual strength measurements have only been performed at room temperature. The results are shown in the Weibull plot of figure 5, where the data on as-received material are repeated for comparison purposes. The modulus remains approximately at the value for material in the as-received condition, suggesting that the same flaw population is causing failure in both the unexposed and exposed conditions. This is confirmed by fractographic analysis which shows that failure also initiates from bulk flaws in the exposed material. The presence of the corroded outer surface layer therefore does not seem to affect the intrinsic failure behaviour of the material at room temperature. Pre-corrosion and subsequent mechanical loading can thus be considered mutually independent, and the effect of corrosive exposure, which is exclusively surface-related (as also evidenced by the independence of the weight change per unit exposed surface area on the A/V ratio), can be fully decoupled from the volume-dominated strength properties. Assuming this is further substantiated by detailed microstructural investigations, the effect of pre-corrosive exposure on the residual mechanical strength can simply be taken into account geometrically by assuming effective values for the width and thickness of the bend specimens. The simplest approximation consists of neglecting the load bearing capacity of the corrosion layer. Using this approximation and the experimentally observed 16% decrease between the characteristic strength of the exposed and the as-received material (figure 5), the average thickness of the corrosion layer in the bend specimens is calculated as 90 μm, which agrees rather well with the mean thickness of the corrosion layer observed by light optical microscopy.

In the future, residual strength tests under flexure and in the uniaxial tensile mode will be carried out at high temperatures. These results will be complemented by microstructural investigations. Also, the effect of corrosive exposure on the time-dependent strength properties (creep, dynamic fatigue) and on fracture toughness will be investigated.

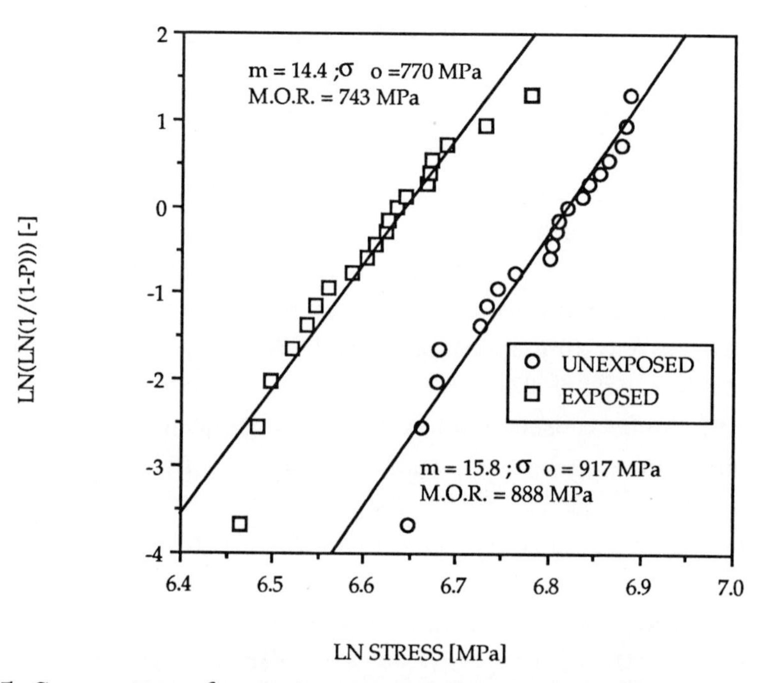

Fig. 5. Comparison of room temperature bend test results obtained on as-received and exposed specimens

CONCLUSIONS

This investigation has shown that the weight gain of samples of a hot-pressed silicon nitride exposed to a simulated coal gasification environment, in which the partial pressures of oxygen and sulphur are similar to those in reality, increases for exposure durations up to 200 hours at 1300 °C. When expressed per unit area of exposed surface, the weight increase does not seem to depend on the surface to volume ratio of the specimens, indicating that it is a mainly surface-related phenomenon. In the course of the investigation, it has been observed, however, that the increase in weight depends on the position of the samples in the specimen holder, suggesting that the local gas flow rate around the specimens is important. This aspect has not been investigated further.

In order to evaluate the corrosion induced mechanical strength degradation, flexure tests have been carried out on the material in the as-received and exposed condition at different temperatures. For room temperature testing, a laboratory air environment is adequate for reference purposes. At the higher test temperatures of 1200 and 1300 °C, however, testing in air results in an appreciable weight gain and a rough corroded surface. In order not to mask the real effect of corrosion exposure, the high temperature tests have been carried out in nitrogen.

Bend tests on the material in the as-received condition show that the Weibull modulus remains approximately constant from room temperature up to 1300 °C, indicating that the same flaw population is causing short-term failure at all temperatures. This is confirmed by fractographic analysis, which shows that bulk flaws act as failure initiation sites.

In the residual strength tests at room temperature after corrosive exposure, a slight reduction in characteristic strength is observed compared to the as-received material. The value of the Weibull modulus, however, is approximately the same, again suggesting that bulk flaws initiate failure. This is confirmed by fractographic analysis and indicates that the corrosion exposure does not affect the intrinsic failure behaviour of the material. The net effect of the exposure can thus be assimilated with a reduction in the load bearing section of the specimens, resulting in a slight decrease in the nominal characteristic strength value. Preliminary results show that this reduction simply corresponds to the decrease in specimen dimensions by an amount equal to the thickness of the corroded outer surface layer. Further microstructural investigations and complementary residual strength tests at high temperatures are currently underway.

ACKNOWLEDGEMENTS

The authors would like to thank their colleagues G. von Birgelen, V. Harrison and P. Young for their technical assistance. This work has been performed within the research and development programme of the Commission of the European Communities. One of the authors (CSM) wishes to thank the Commission for a grant.

REFERENCES

1. B. J. Davidson, D. B. Meadowcroft and J. Stringer, High temperature alloy requirements for coal fired combined cycles, *in*: "High temperature alloys for gas turbines and other applications", D. Reidel publishing company, Dordrecht (1986)
2. F. Costa Oliveira, Corrosion of silicon nitride in $H_2S/H_2O/H_2$ gas mixtures at 1200-1300 °C, *in* : "Proceedings of the first European ceramic society conference", Elsevier Applied Science, London and New York (1989)
3. D. R. Holmes and J. Stringer, Limitations of the use of alloys in coal gasifiers for power generation, *in* : "Proceedings of the international conference on materials to supply the energy demand", ASM publications (1980)
4. W. B. White, S. N. Johnson and G. B. Dantzig, *J. Chem. Phys.* 28:751 (1958)

HIGH TEMPERATURE FAILURE MECHANISMS OF SINTERED SILICON NITRIDE

A.K. Mukhopadhyay[1], D. Chakraborty[2], and S.K. Datta[2]

[1]Center for Advanced Materials Technology
Department of Mechanical Engineering
The University of Sydney
NSW 2006, Australia

[2]Central Glass and Ceramic Research Institute
Calcutta 700032, India

1. INTRODUCTION

A clear understanding of high temperature failure mechanisms of engineering ceramics is essential for their successful use as hot zone components in the turbine engine and for related structural applications. As an engineering ceramic the sintered silicon nitride (SSN) has shown tremendous potential for commercial exploitation in terms of thermomechanical properties, lower production cost, easy machinability, and fabricability to close tolerance of complex shapes. The various oxide additives used in sintering contribute to formation of a silicate rich liquid phase that aids densification, but remains on cooling from sintering temperature as an amorphous phase at the grain boundary of SSN[1-4]. Softening of the amorphous phase at high temperature causes strength degradation[1-5]. In contrast the use of a nitrogen rich liquid phase has recently led to superior thermomechanical behaviour of SSN at high temperatures[6-8]. In this paper we report the high temperature failure mechanisms of these SSN ceramics. In most of the previous works [1-5] emphasis have been focussed on the study of temperature dependence of flexural strength and the interdependence of flexural strength and fracture toughness have been only cursorily explored[7,8]. The major aim of present work has therefore been directed to evaluate the high temperature strength, fracture toughness and creep simultaneously, for SSN ceramics of given compositions with nitrogen rich grain boundary phase. For the purpose of comparison only similar tests were conducted also on SSN synthesised with a conventional silicate-

Fracture Mechanics of Ceramics, Vol. 10
Edited by R.C. Bradt *et al.*, Plenum Press, New York, 1992

rich liquid phase. Following these tests, extensive use of both scanning and transmission electron microscopy were made to elucidate the failure mechanims at ambient as well as at high temperatures of SSN.

2. EXPERIMENTAL WORK

The Si_3N_4 powder used for the synthesis of SSN was prepared in the Central Glass and Ceramic Research Institute laboratory. The samples A, B, B1, B2, B3 were synthesised from Y_2O_3-AlN-SiO_2 system. The samples C, C1, D, D1 were synthesised from MgO-AlN-SiO_2 system. The sample E synthesised from MgO-SiO_2 system was included in this study for the purpose of comparison only. Post sintering heat treatment led to the synthesis of samples B1, C1, D1 corresponding to the as sintered products B, C, D respectively. Data on composition and microstructural parameters are given in Tables 1 and 2.

Ground and polished specimens (45 mm by 4.5 mm by 3.5 mm) were used in the temperature range 30^0C to 1250^0C for the measurement of flexural strength and fracture toughness (K_{1c}) by four point loading with 10 mm moment arm and 1.25 $Nmin^{-1}$ loading rate in a commercial machine (422S, Netzsch, Germany). The single-edge-notched-bar technique (SENB) with notch tip radius of 120 um and normalized notch length of 0.3 was used for K_{1c} evaluation[9]. Typical K_{1c} value of SSN is about 5 $MPa.m^{1/2}$ and hardness is about 15 GPa[2,8]. Thus yield stress is about H/3 i.e 5 GPa. Therefore the plastic zone size is $(K_{1c}/\text{yield stress})^2$ i.e about 1 micron. Evidently the sample size strictly abides by the ASTM size restrictions for the plain strain crack toughness (K_{1c}) testing[7]. The recorded load deformation behaviour at ambient of both notched and unnotched bars exhibited linear elastic fracture characteristics except at high temperatures eg. at and above 1000^0C. Using similar unnotched bars in four point loading creep experiments in the stress range 75-350 MPa stress at 1100^0C - 1450^0C were conducted in a compressive creep apparatus (2390, Applied Test Systems, USA). The creep activation energy (Q) and creep stress exponent (n) were calculated from experimental data following the equations[10,11]:

$$Q = [\, R \ln (\dot{\epsilon}_2/\dot{\epsilon}_1) \,]/\{ (1/T_1)-(1/T_2) \} \qquad (1)$$

and

$$\dot{\epsilon} = A \, (\sigma)^n \exp (-Q/RT) \qquad (2)$$

where R is the universal gas constant, $\dot{\epsilon}_1$ and $\dot{\epsilon}_2$ refer respectively to steady state state creep strain rates at the creep test temperatures T_1 and T_2 measured in absolute scale, (σ) is the apllied stress and A is an empirical constant. Obviously, the equation (1) refers to constant stress-temperature change experiments for evaluation of Q and the equation (2) refers to isothermal stress-change experiments for calculation

Table 1. Composition of SSN.

Samples	Mol % Batch Composition				Sintering Liquid Composition	
	Y_2O_3	AlN	MgO	SiO_2	Wt. % (approx.)	N_2 Atom % (approx.)
A	10.0	45.0	-	45.0	11.9	18.6
B	7.7	43.6	-	48.6	12.3	16.1
B1	7.7	43.6	-	48.6	12.3	16.1
B2	7.7	43.6	-	48.6	12.3	16.1
B3	7.7	43.6	-	48.6	12.3	16.1
C	-	32.6	30.4	37.0	20.7	16.0
C1	-	32.6	30.4	37.0	20.7	16.0
D	-	40.0	10.0	50.0	23.1	13.8
D1	-	40.0	10.0	50.0	23.1	13.8
E	-	-	66.7	33.3	9.0	-

- not present in the powder batch as individual component.

Table 2. Microstructural parameters of SSN.

Samples	% Theoretical Density	Aspect Ratio
A	96.6	3.45
B	94.7	2.66
B1*	94.7	-
B2**	93.8	-
B3**	88.8	-
C	95.9	3.38
C1*	95.9	-
D	95.9	2.30
D1*	97.5	-
E	94.7	3.10

* subjected to post sintering heat treatment at 1200-1400°C in nitrogen atmosphere.
** starting composition same as that of B.

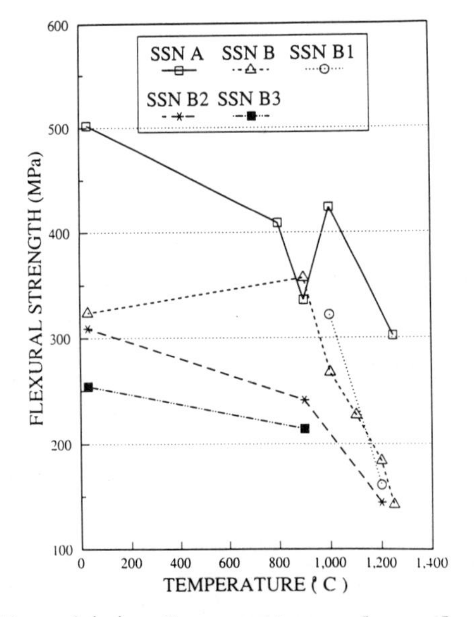

Fig. 1(a). Temperature dependence of Flexural Strength (SSN A, B, B1, B2, B3).

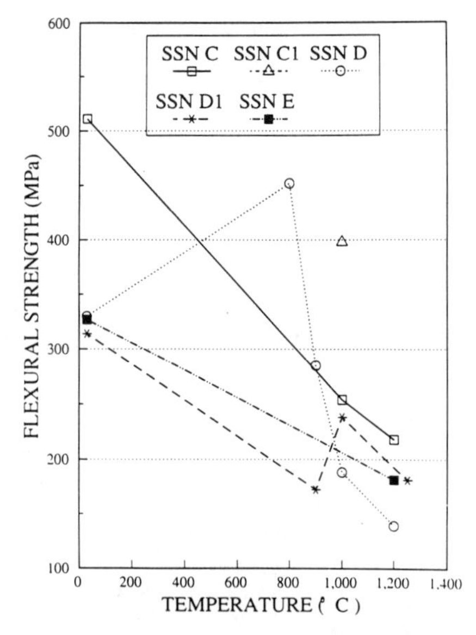

Fig. 1(b). Temperature dependence of Flexural Strength (SSN C, C1, D, D1, E).

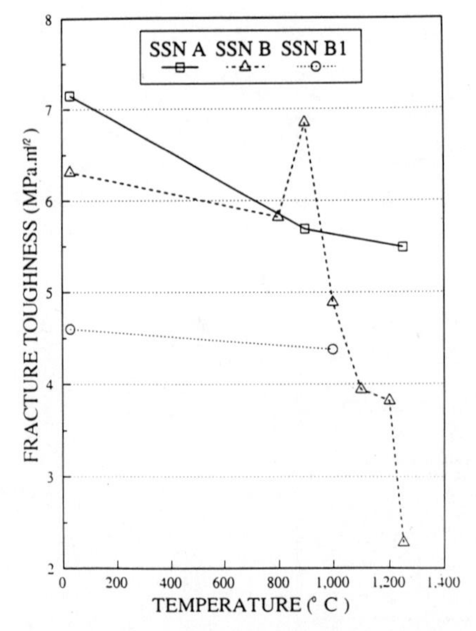

Fig. 2(a). Temperature dependence of Fracture Toughness (SSN A, B, B1).

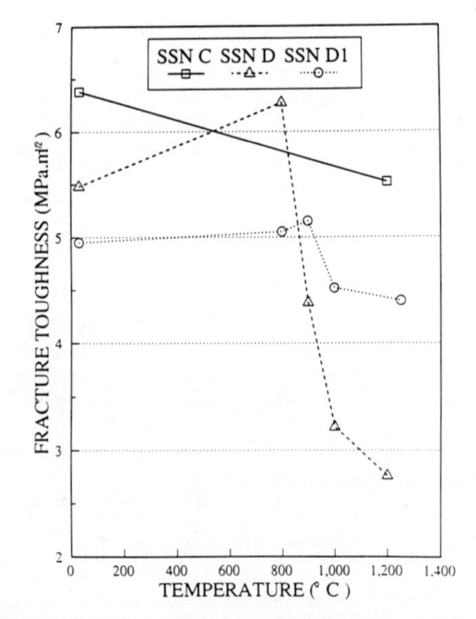

Fig. 2(b). Temperature dependence of Fracture Toughness (SSN C, D, D1).

of n.All fracture surfaces were examined by scanning and transmission electron microscopy (SEM and TEM) to identify failure processes[12].At least four to six specimens were used to obtain a representative value of each mechanical property. The scatter of data was less than 10% on average.

3.RESULTS and DISCUSSIONS

The results of flexural strength, fracture toughness, creep resistance and corroborative fractographic experiments are followed by comprehensive discussion to elucidate the failure mechanisms of SSN.

3.1 Flexural Strength

Data on temperature dependence of flexural strength are presented in Figures 1(a) and 1(b) for SSN synthesised from Y_2O_3-AlN-SiO_2 and MgO-AlN-SiO_2 systems respectively. In general there is a strength degradation above 900^0C. In particular the SSN D samples show a remarkable drop in the flexural strength in the temperature range $800-1200^0C$. The SSN samples B2,B3 show inferior strength to SSN B in spite of identical starting compositions.Post sintering heat treatment[7] improves the strength at 1000^0C of B1,C1,D1 by about 20%, 60%, 30%, respectively,and at 1250^0C of D1 by about 30% compared to those of parent SSN B,C,D at the same temperatures.The sample SSN E, synthesised with conventional silicate rich liquid from MgO-SiO_2 system, showed much inferior strength characteristics at high temperature to those of SSN A and SSN C [Figures 1(a) and 1(b)] . In the case of as sintered samples, the SSN A and C show the best strength characteristics with respect to others in their respective sintering liquid systems. All samples taken together, SSN A has the best strength behaviour for all temperatures.

3.2 Fracture Toughness (K_{1c})

Data on the temperature dependence of fracture toughness are presented in the Figures 2(a) and 2(b) for SSN synthesised from Y_2O_3-AlN-SiO_2 and MgO-AlN-SiO_2 systems, respectively. The trend of temperature dependence of K_{1c}, is similar to that of strength [Figures 1(a) and 1(b)] . Therefore, the K_{1c} of SSN decreases above 900^0C. At temperatures higher than 900^0C the load-deformation behaviour of all the present SSN materials started exhibiting non linearity. The degree of non-linearity was, however, found to be sensitive to composition of the starting powder batch.This aspect has been dealt with in further details later in this presentation. SSN samples A and C have the best K_{1c} characteristics in their respective sintering liquid systems. Amongst all the silicon nitrides,the SSN A has the best K_{1c} behaviour. Post sintering heat treatment led to considerable improvement in K_{1c} above 800^0C of SSN D1 relative to SSN D. However, SSN B1 was inferior to SSN B in fracture toughness.

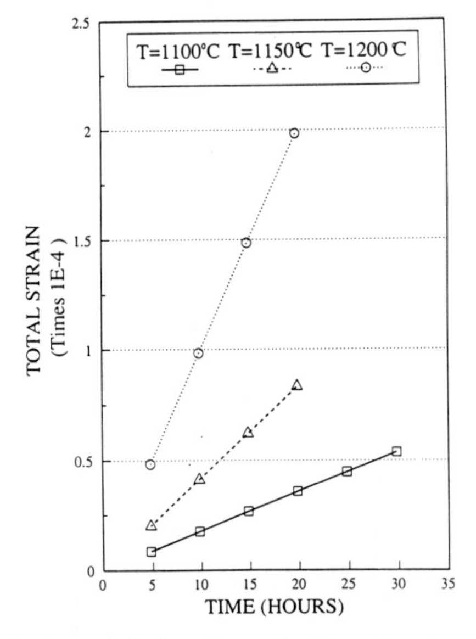

Fig. 3(a). Total Strain vs. Time of SSN D.

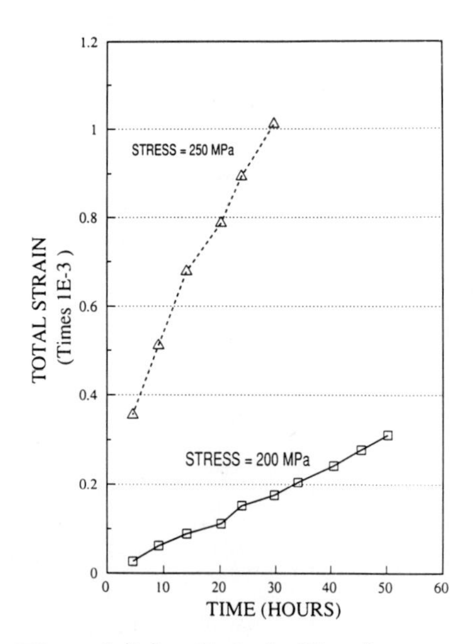

Fig. 3(b). Total Strain vs. Time of SSN A.

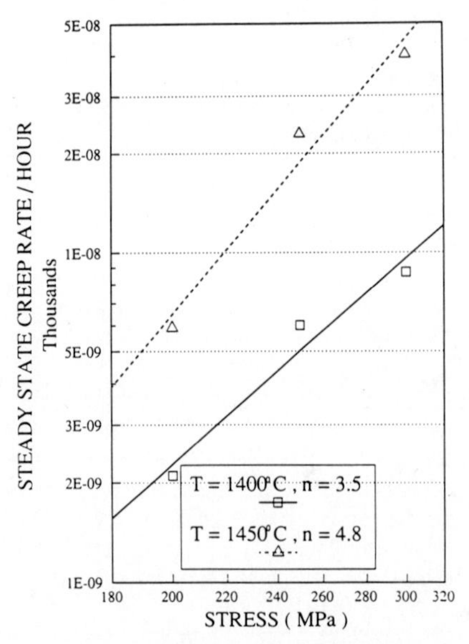

Fig. 3(c). Stress Dependence of Strain Rate (SSN A).

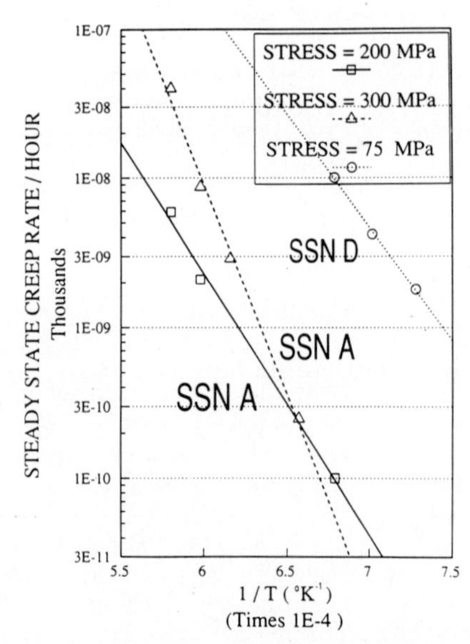

Fig. 3(d). Temperature Dependence of Creep Rate.

3.3 Creep Deformation

Data on typical variation of total strain with time are presented in Figures 3(a) and 3(b) for the SSN samples D and A respectively. For the creep tests at the 75 MPa stess and for the temperature range 1100-1200°C, the SSN D had a creep strain rate of $(1.8\text{-}10)\times10^{-6}$/hr. This type of material was the least creep resistant of the two types of SSN studied here. Attempts to carry out additional creep tests at higher temperatures and with higher stresses was rendered unsuccessful because of specimen rupture. As temperature of creep test was increased by 50°C from 1100°C, the total strrain increass 2.3 times on average of that at 1100°C. At 1200°C, however, the total strain increases by a factor of 5.6 in comparison to that at 1100°C. Although, at the given stress level the total strain of SSN D increases with both time and temperature, the influence of temperature on strain and in consequence, strain rate, appears to be of more predominant significance. In fact the strain rate increases by almost an order of magnitude as the temperature is increasd from 1100°C to 1200°C.

The stress-strain behaviour of most of the samples included a small elastic deformation upon initial loading [(not included in Figures 3(a) and 3(b)], a primary or transient creep period in which the strain rate decreased with time eg. about first five hours in the creep test at 250MPa stress and 1450°C of the sample SSN A, and a steady state creep period where the strain rate did not vary with time. The onset of tertiary creep was detected for the different SSN ceramics at different temperatures and stresses. In the case of SSN A the steady state creep strain rate was found to be[6] in the range of $(0.79\text{-}40)\times10^{-6}$/hr corresponding to the stess range 200-350 MPa at the test temperature range of 1200-1450°C. As shown in Figure 3(b) at the test temperature of 1450°C the total creep strain of SSN A suffers significant increase following the increment of stress from 200 MPa to 250 MPa. Consequently, the creep strain rate increases by a factor of about four. However, this stress and temperature levels were much higher than what could be used in the case of SSN D. In fact the SSN A ruptures after a period of 0.8 hr. in the creep tests conducted at 350 MPa stress at a temperature of 1450°C[6], indicating the more creep resistant nature of SSN A compared to SSN D. Availability of very limited amount of data in the case of SSN D precluded the experimental estimation of creep stress exponent.

Data on stress and temperature dependence of steady state creep strain rate are presented in the Figures 3(c) and 3(d), respectively, for the SSN A and D. The former is the most creep resistant and the later the least. The SSN A have stress exponent (n) of 3.5 in the creep tests at 1400°C under the stress range 200-300 MPa. In the same stress range the creep tests conducted at 1450°C of the same material n is 4.8 (Figure 3c). This result apparently indicates a temperature dependence of the strss exponent in the creep of SSN A.

Figure 3(d) shows that at a creep stress of 75 MPa at 1100-1200°C, the SSN D has an activation energy (Q) of about 74

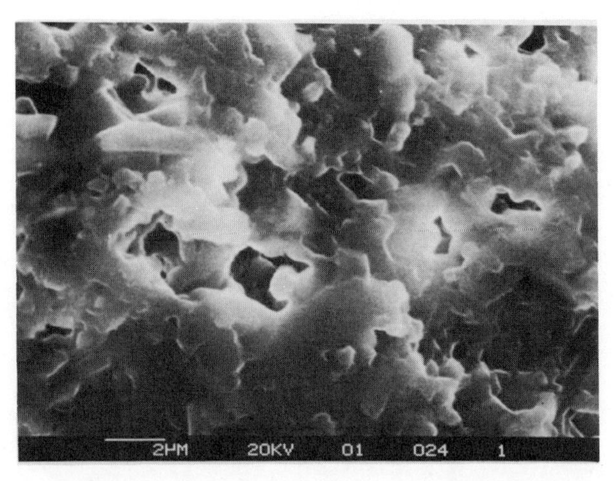

Fig. 4(a). Transgranular Fracture at Ambient of SSN A.

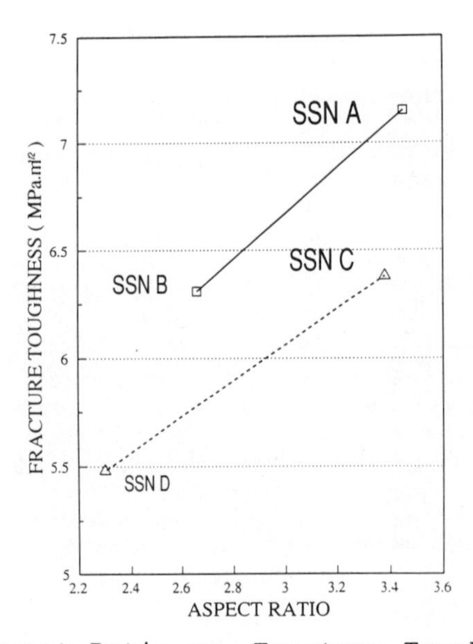

Fig. 4(b). Aspect Ratio vs. Fracture Toughness of SSN.

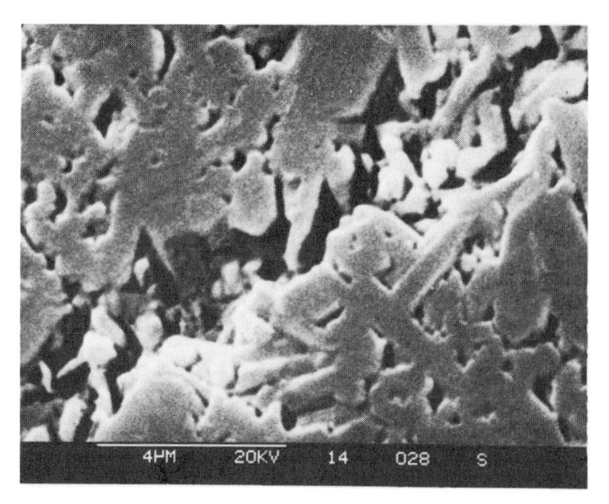

Fig. 4(c). Tortuous Crack Path of SSN C.

Fig. 4(d). Flatter Crack Path of SSN D.

KCal/mole. In the temperature range 1200-1450°C of SSN A under applied stress of 200 MPa, the creep activation energy is about 79 KCal/mole. However, at an applied stress of 300 MPa the Q value increases to about 132 KCal/mole corresponding to creep test temperature of 1250-1450°C for the SSN A. This result would apparently imply a stress dependence of activation energy.

3.4 Failure Mechanisms

The materials SSN A and C, containing the highest amount of nitrogen in the lowest amount of liquid phase in starting compositions of their respective sintering liquid systems [see Table 1], possess the highest values of flexural strength and fracture toughness at both ambient and high temperatures.The similar natures of temperature dependence of strength and fracture toughness possibly indicate that strength is most probably controlled by K_{1c}.For a given sintering liquid, toughness enhancement is related to two factors : (a) an increase in aspect ratio, (b) an increase in extrinsic grain boundary toughness. This is because, the SSN microstructure may be considered to be that of a composite material consisting of the beta-Si_3N_4 grains separated by the grain boundary oxynitride glass. From the compositional view point [see Table 1] the grain boundary glasses are of Y-Si-Al-O-N and Mg-Si-Al-O-N types for the Y_2O_3-AlN-SiO_2 and MgO-AlN-SiO_2 sintering liquids respectively[13]. From the works of Drew et al.[14] and independent work at the authors laboratory[15,16] it is known that K_{1c} of oxynitride glasses improve with nitrogen content. This would imply the presence of a relatively tougher grain boundary phase in SSN A than B or in SSN C than D.As shown in Figure 4(a), the fracture mode at ambient typically involves transgranular crack propagation as well as grain pull out. The higher aspect ratio of grains in SSN A than in B or in SSN C than D [see Table 2] would almost certainly involve more frictional energy dissipation in grain pull-out,thus contributing to toughening [Figure4(b)]. The crack follows a tortuous path in the tougher material [Figure 4(c)] but takes a relatively flatter course in the less tough material [Figure 4(d)] .

The samples SSN B2 and B3 with density lower than B have inferior strength than SSN B at elevated temperature although, they have identical starting compositions [see Tables 1 and 2]. The other important factor is that the material with tougher grain boundary phase possesses higher strength and toughness at higher temperatures. Further, as Y-Si-Al-O-N glass is more refractory than Mg-Si-Al-O-N glass, the SSN A shows the best strength and toughness behaviour at elevated temperature[14]. The increase in strength and toughness at high temperature of heat treated samples e.g. B1, C1, D1 may be linked to the increase in degree of crystallinity of the grain boundary phase as indicated by others[3,4]. However, SEM examination did not reveal any significant difference in fracture mode from as sintered ones which show predominantly intergranular fracture [Figure 5(a)]. The general trend of degradation in strength and toughness at and above 1000°C is definitely linked to some inelastic deformation process as indicated typically in the Figures 5(b)

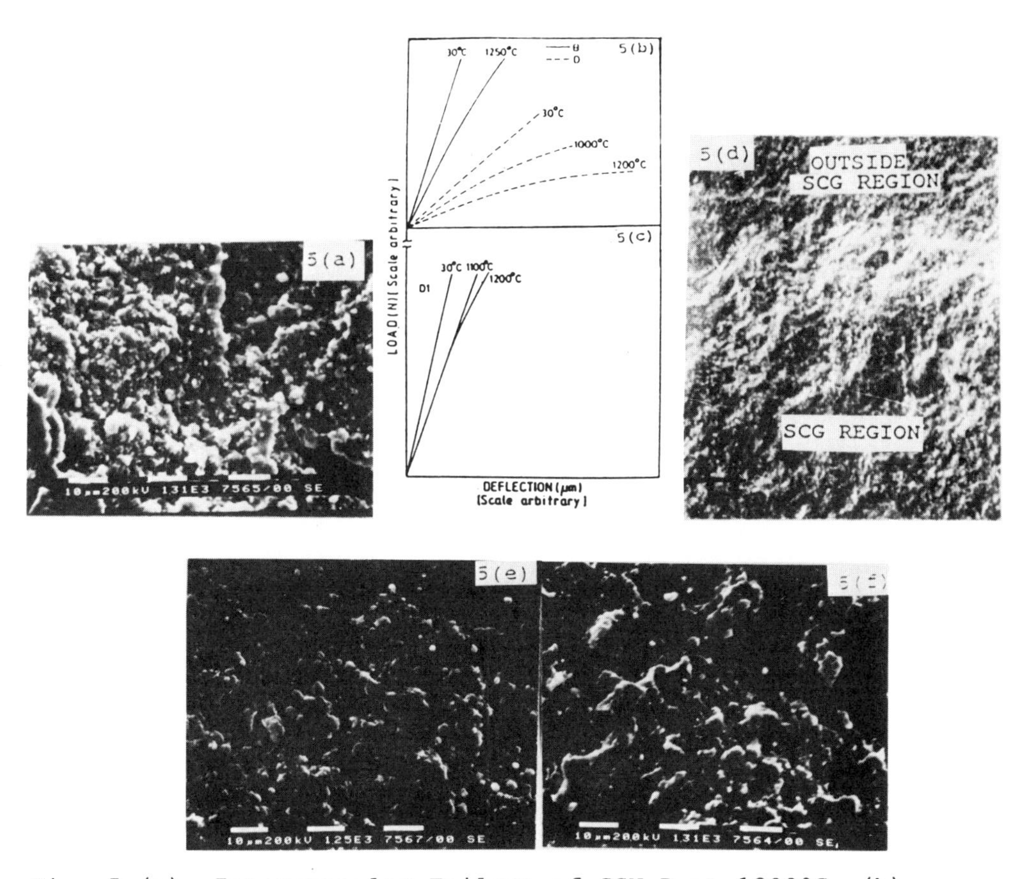

Fig. 5.(a). Intergranular Failure of SSN D at 1200°C; (b) Schematic Load vs. Deformation Behaviour of SSN; (c) Schematic Load vs. Deformation of SSN D1; (d) SCG at 1200°C of SSN A, Bar = 100μm; (e) Intergranular Failure Inside SCG Region; (f) Transgranular Fracture and Cavitation Outside SCG Region.

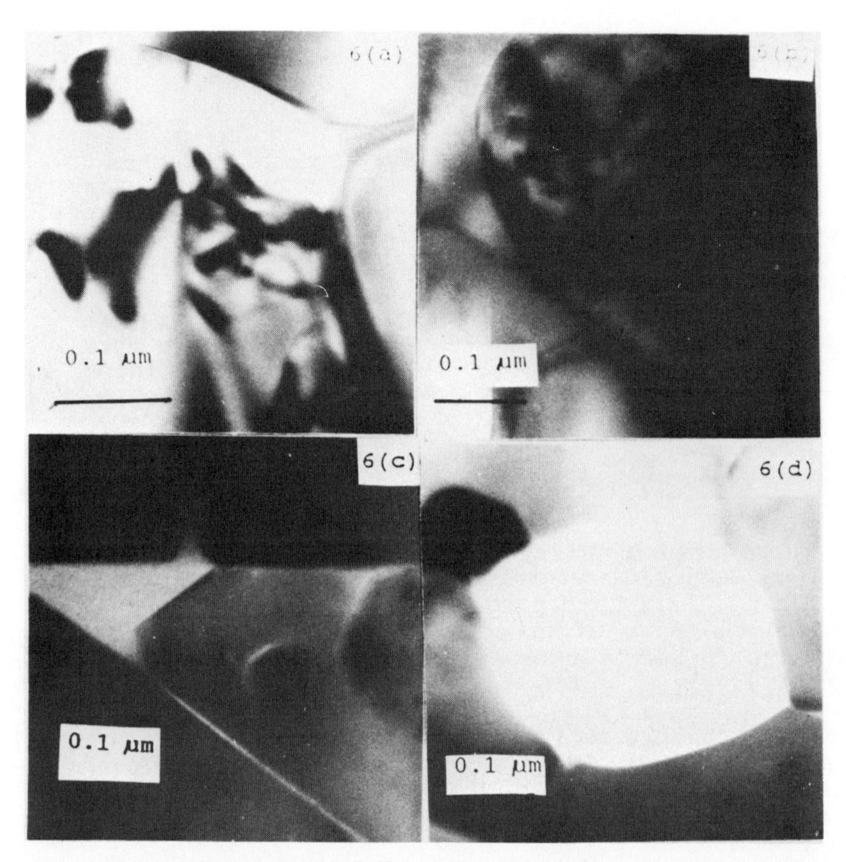

Fig. 6. (a). Dislocation Network in Grains of SSN D (75 MPa, 1200°C), Bar = 0.1 μm; (b) Dislocation Tangles in Grains of SSN D (75 MPa, 1200°C), Bar = 0.1 μm; (c) Grain Boundary Sliding in SSN D (75 MPa, 1200°C), Bar = 0.1 μm; (d) Creep Cavitation in SSN D (75 MPa, 1200°C), Bar = 0.1 μm.

and 5(c) for SSN B, D and D1. Similar phenomena have been reported for SSN[1] and sintered sialon[9] and have been attributed to the presence of slow crack growth (SCG). Fractographic evidence shows typical example of slow crack growth on the fracture surface of SENB specimen at 1200°C [Figure 5(d)]. Note the difference in fracture mode inside and outside the slow crack growth region [Figures 5(e) and 5(f)]. The occurrence of SCG in the material adjacent to the tip of the notch indicates the presence of a stress corrosion assisted process. In the fast fracture experiments at high temperature, the largest pre-existing crack may thus grow sub-critically until it has reached a critical size when fast fracture initiates through intergranular crack propagation.

As the SSN D sample was found to be the least creep resistant, transmission electron microscopic study has been done extensively on this material. Similar studies on the SSN A sample are yet to be finished, although the preliminary observations[17] indicate the presence of similar features. In the case of SSN D creep tested under 75 MPa stress at 1200°C the dominant features typically observable are : (1) dioslocation network and dislocation tangles in the grains [Figures 6(a) and 6(b)], (2) growth of wedge shaped voids in the two- and three- grain junctions [Figure 6(b)] , (3) grain boundary sliding leading to void growth [Figure 6(c)] and (4) extensive cavitation [Figure 6(d)] .

The deformation in dense silicon nitride undergoing slow fracture experiments should be explained in terms of (1) the flow characteristics of the of the amorphous grain boundary phase that bonds the silicon nitride grains and (2) the geometric constraints imposed by the system. The available model[18] for creep in a partially amorphous glass ceramic may in principle be applied to the case of sintered silicon nitride. According to this model, a given volume of the specimen will contain x_0 number of glassy areas of which $x(\sigma)$ contain voids under an applied stress(σ). If each void containing amorphous regions influence y nearest neighbours such that they have insignificant resistance to the deformation compared to that of the void-deficient regions, then, the material devoid of void must have to support a greater proprtion of the applied stress by a factor of (x_0) / $[\ x_0 - y * x(\sigma)\ ]$. If it is assumed that the grain boundary sliding and grain boundary fluid movement vary with stress, then, the creep rate of the composite material could be written as:

$$\dot{\epsilon} = \text{constant}*(\sigma)*x_0 \ / \ [\ (x_0) - y*x(\sigma)\ ] \qquad (3)$$

Unfortunately, however, the parameters x_0, y and $x(\sigma)$ are yet to be known for this material eg. SSN. Theoretical estimates of the stress exponent n range between 1 and 6 in the cases of glass ceramics and fire bricks treated using this model[18]. It is intersting to note that SSN A had n values of 3.5 and 4.8. The indication is that the void formation during creep might have modified the rate dependence of stress beyound that expected from a simple newtonian flow. Only where the viscous glass is free to flow and fill the voids as they are formed should we expect the pure newtonian creep to prevail. But that too would require the satisfactions of the conditions viz. (1) void formation must create a hydrostatic pressure gradient favouring viscous flow into the voids and (2) the system must have sufficient glass available to fill the voids from the grain boundaries. The observed n values, which are far from unity, thus imply the presence of a modified-newtonian deformation process in the creep of SSN. At time t=0, the entire stress is supported by viscous liquid and elastic grains. As time progress, the liquids flow from region of higher pressure to regions of lower pressure. As this flow occurs, we may anticipate,the grains must support a larger stress. Locally, the strss may be quite high enough to induce plastic deformation as indicated by the presence of dislocations [Figures 6(a) and 6(b)] .

A parallel possibility would be the openings of wedge cracks at the grain junctions to maintain the strain continuity during the non-accomodated grain boundary sliding[19,20]. Thus during the initial stages of creep thre would be successive nucleation of grain-boundary wadges spreading with time accross the specimen cross-section. During the steady state creep, these wadges may be assumed to grow at a rate proportional to the steady state creep strain rate. When strain reaches a limiting value, grain boundary separations might be expected to link to form unstable cracks which would continue to grow depending on stress and temperature of sample. This proposed rationale for the deformation process is in accordance with the transmission electron microscopic observations presented herein for SSN [Figure 6]. It is emphasised that local pastic deformation processes may be active simultaneously with the non-accomodated grain boundary sliding process in the deformation mechanism of SSN. The observed cavitation [Figure 6(d)] would help the pores at triple grain junctions to act as further potential source of failure.

4. CONCLUSIONS

(a) The SSN with highest density, aspect ratio,and nitrogen content in lowest amount of sintering liquid has highest values of strength and fracture toughness at both ambient and elevated temperatures.

(b) The toughening mechanism involves possibly frictional energy dissipation in pulling out of large aspect ratio grains, a tortuous crack path and a small contribution from grain boundary phase, which assumes more importance at temperatures above ambient.

(c) The failure mode changes from a predominantly transgranular one at ambient to a predominantly intergranular one at elevated temperature. At and above 1000^{0}C SCG dominates failure in fast fracture experiments. The slow fracture process involves on the other hand grain boundary sliding and/or cavitation.

5. ACKNOWLEDGEMENTS

The authors wish to thank Professor Y.W.Mai for many helpful discussions and suggestions in this work. A.K.Mukhopadhyay is supported by an Australian Research Council Postdoctoral Reasearch Fellowship tenable at the University of Sydney. The authors also greatfully acknowledge the experimental assistance of their colleagues at the University of Sydney and The Central Glass and Ceramic Research Institute, Calcuutta, India. The kind interest of Dr.B.K.Sirkar, Director, Central Glass and Ceramic Reasearch Institute, Calcutta, India, in this work is also greatfully acknowledged.

REFERENCES

1. R. K. Govila, " Strength characterisation of yttria-doped sintered silicon nitride " ; J. Mater. Sci. 20 : 4345 (1985).

2. A.Okada and N.Hirosaki, " Sub-critical crack growth in sintered silicon nitride exhibiting a rising R-Curve curve " J.Amer.Ceram.Soc. 73 : 2095 (1990).

3. L. A. Pierce, D. M. Mieskowski and W. A. Sanders, " Effect of grain boundary crystallisation on the high temperature strength of silicon nitride " ; J. Mater. Sci. 21: 1345 (1986).

4. T. Hayashi, H. Munakata, H. Suzuki and H. Saito, "Pressureless sintering of Si_3N_4 with Y_2O_3 and Al_2O_3 " J. Mater Sci. 21: 3501 (1986).

5. W. A. Sanders and D. M. Mieskowski, " Strength and microstructure of sintered Si_3N_4 with rare-earth-oxide additions " ; Am. Ceram. Soc. Bull. 64: 304 (1985).

6. D. Chakraborty and A. K. Mukhopadhyay, " Creep of sintered silicon nitride " ; Ceramics Int. 15: 237 (1989).

7. A. K. Mukhopadhyay, D. Chakraborty and J.Mukerji, in " High Technology Ceramics " ; P. Vincenzini. (Editor), Elsevier Science Publishers, Amsterdam, 1987.

8. A. K. Mukhopadhyay, D. Chakraborty and S. K. Datta, " On the Microhardness characteristics of Silicon Nitride and Sialon Ceramics " ; J. European Ceram. Soc. 6: 303 (1990).

9. A. K. Mukhopadhyay and D. Chakraborty, " High Temperature Fracture Toughness and Fractographic study of reaction sintered sialon and sintered sialon " ; Mater. Sci. Eng. A 104 :215 (1988).

10. R. M. Arrons and J. K. Tien, " Creep and strain recovery in hot-pressed silicon nitride " ; J. Mater. Sci. 15: 2046 (1980).

11. S. U. Din and P. S. Nicholson, " Creep deformation of reaction-sintered silicon nitride " ; J. Amer. Ceram. Soc. 58: 500 (1975).

12. D. Chakraborty and A. K. Mukhopadhyay, " Fractrography of silicon nitride and sialon ceramics " ; Central Glass and Ceramic Bull. 34: 59 (1987).

13. P. K. Das and J. Mukerji, " Dense Si_3N_4 by liquid phase sintering " ; Ind. J. Technol. 24: 209 (1986).

14. R. A. L.Drew, S. Hampshire and K. H. Jack, " Nitrogen Glasses " ; Proc. Brit. Ceram. Soc. 31: 119 (1981).

15. P. K. Das, J.Mukerji and D. Chakraborty, " Glasses and Melts in the system MgO-AlN-SiO_2 " ; Am. Ceram. Soc. Bull. 66: 1616 (1989).

16. D. Chakraborty and A. K. Mukhopadhyay, " Study on hardness and toughness of nitrogen glasses " ; Unpublished Work (1990).

17. A. K. Mukhopadhyay, " Fractographic observation and transmission electron microscopic study on creep of silicon nitride ceramics " ; Unpublished work (1991).

18. R. Morrel and K. H. G. Ashbee, " High temperature creep of lithium zinc silicate glass-ceramic " ; J. Matre. Sci. 8: 1253 (1973).

19. A. H. Cottrell, in " Structural Processes in Creep " ; Iron and Steel Institute, London, (1961).

20. C.Zener, in " Elasticity and Unelasticity " ; University of Chicago Press, (1948).

FRACTURE BEHAVIOR OF A SILICON NITRIDE CERAMIC CONTAINING A HIGH TEMPERATURE DUCTILE PHASE

Roger L.K. Matsumoto and Julie W. Pier

Hercules Incorporated
Hercules Research Center
c/o 1313 N. Market Street
Wilmington, DE 19894-0001

ABSTRACT

A silicon nitride ceramic composite containing dispersed cobalt silicide has been fabricated and evaluated. The cobalt silicide phase undergoes a brittle to ductile transition at approximately 900°C. Fractures below this temperature are predominantly brittle. The composite, which has a fracture toughness at room temperature of 10.5 MPa√m, is toughened by the crack front interacting with the dispersed particulates by several well known mechanisms. Since the silicide particulates become ductile around 900°C, and can undergo ligament stretching, an added high temperature toughening mechanism may be operative. Fractographic analyses have been performed on this material on surfaces fractured at several temperatures up to 1500°C. Surfaces investigated were from chevron notch beam samples, which contain areas of controlled fracture as well as areas of fast fracture. The results of the fractographic analyses are correlated with the fracture toughnesses obtained.

INTRODUCTION

The fracture of a composite ceramic is affected by the nature of the dispersed phase. The toughness of the composite can be enhanced over the base matrix. Toughening mechanisms will depend upon the second phase in the matrix.

A brittle second phase can yield increased toughness in the composite by several distinct mechanisms. Crack deflection and crack front bowing (crack pinning) [1,2] are examples. Interactions of the crack front with the stress field surrounding the particulates can also contribute to increased toughness [3]. The stress field is a function of the elastic moduli and thermal expansion coefficients of the matrix and dispersed phases. These interactions are enhanced when the dispersed phase has the lower modulus, the higher thermal expansion coefficient, or both. If the local stress is higher than the strength of either phase, then microcracking can occur. Microcracking can be an effective way to increase the toughness of composites [4].

Ductile inclusions offer crack bridging by ligament stretching as an added mechanism [5-10]. The contribution to toughening will depend upon the interfacial bond between the particulate and matrix. Unbonded particulates act as an independent phase and do not contribute to toughening. Well bonded particulates allow only limited deformation during fracture. Toughening will be greatest in systems where the interface is weak. In those cases, debonding, as well as ligament stretching, will contribute to the toughness.

It would be interesting to test these diverse mechanisms in a single ceramic system. The dispersed second phase would, however, have to be both brittle and ductile for this to be possible. It is possible in some systems for this requirement to be met. Specifically, several silicides undergo a brittle to ductile transition at an elevated temperature. Thus, they are brittle below that temperature and ductile above. Fracture behavior would thus depend upon the temperature at fracture, as different toughening mechanisms would be operative.

This paper will report on one such system where the dispersed phase is a mixture of cobalt silicides (Co_2Si and CoSi) and the matrix is silicon nitride. The surfaces of this material, fractured at temperatures from 25°C to 1500°C, have been examined and correlated with the fracture toughnesses obtained.

EXPERIMENTAL PROCEDURES

Details of the preparation of this material were given previously [11]. Some of the data obtained on the cobalt silicide phase are: thermal expansion coefficient (20° - 1200°C) of 17.8 X 10^{-6} /°C [11]; approximate brittle to ductile transition of 900°C [11]; melting points of silicides, Co_2Si = 1327°C, and CoSi = 1395°C [12].

Test samples were obtained by hot pressing a spray dried mixture of silicon nitride and cobalt metal powders. The cobalt metal reacts with the silicon nitride in-situ to form the two silicides. The material was densified at 1750°C at a pressure of 27.6 MPa for one hour. The pressed part had a bulk density of 3.5 g/cm^3 and a silicide volume of about 13%. These parts were cut into 3mm X 4mm X 50mm bars for four point bend strength testing and into 4mm X 5mm X 50mm bars for chevron notch beam samples for fracture toughness measurements.

As the brittle to ductile transition temperature for the silicide mixture was determined to be approximately 900°C, test temperatures selected were room temperature, 800°C, 1100°C, and 1500°C. Thus, temperatures below and above the brittle to ductile transition temperature were chosen and also a temperature above the melting point of the silicides. The crosshead speed of the testing machine was 0.2mm/min for all tests except for the strength test at 1500°C. Because the material exhibited superplastic behavior at this temperature, the speed was increased to 5mm/min in order to obtain fracture. Spans used were 20mm and 40mm for both chevron notch and strength testing.

Fracture toughness was determined using the method of Munz, et al. [13]. The dimensions of the notch are shown in Figure 1. The crosshead speed was 0.2 mm/min at all test temperatures.

Fracture surfaces were examined by scanning electron microscopy (SEM).

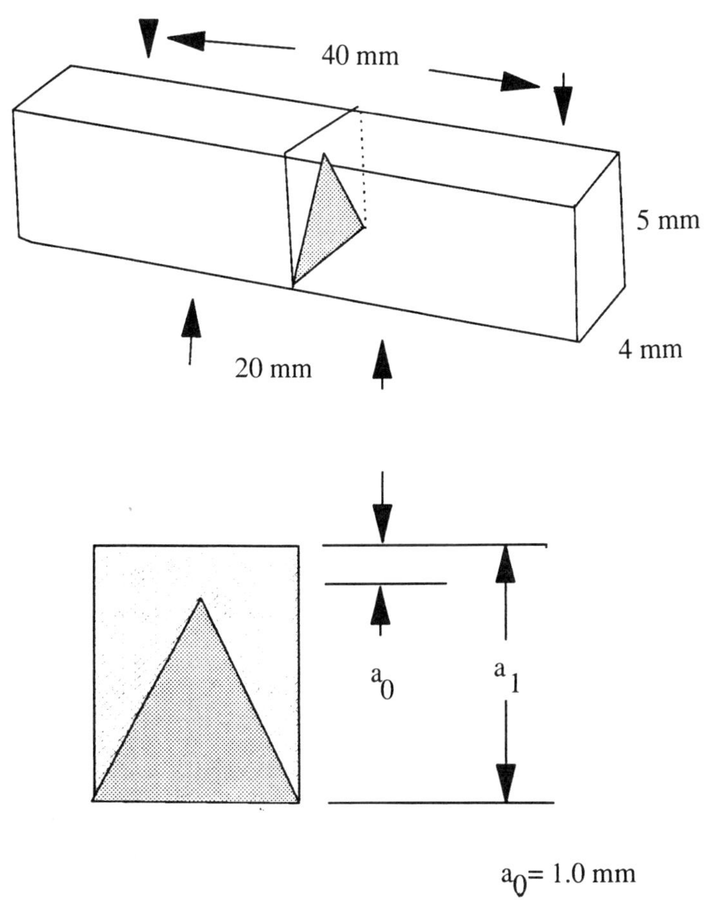

Figure 1 Dimensions of the chevron notch beam sample.

Surfaces were examined orthogonally to the surface as well as at a 45 degree angle. On each chevron notch fracture surface, the area which experienced controlled crack growth was compared with the area which experienced fast fracture.

Some representative materials were also investigated by transmission electron microscopy (TEM).

Crystalline phases were determined by X-ray diffraction (XRD).

RESULTS

Flexural strength, fracture toughness and elastic modulus values obtained at the four test temperatures are shown in Table I. Figures 2 and 3 are optical micrographs of fractured surfaces. SEM micrographs of fractured surfaces are shown in Figures 4 to 12. TEM micrographs are shown in Figures 13 and 14. XRD analysis showed that silicon nitride exists fully in the beta phase and that there is an approximately equal quantity of CoSi and Co_2Si.

Table I Mechanical property values

TEST TEMPERATURE (°C)	FLEXURAL STRENGTH (MPa)	ELASTIC MODULUS (GPa)	FRACTURE TOUGHNESS (MPa√m)
23	703 ± 83 *	248	10.5 ± 0.1 *
800	626 ± 44	213	12.2 ± 1.4
1100	484 ± 23	157	6.0 ± 0.2
1500	(140) **	(87) **	2.7 ± 0.5

Notes: * - one standard deviation
** - loading rate = 5.0 mm/ min., all other values obtained at 0.2 mm/min. Material behaves superplastically at 1500°C.

DISCUSSION

Flexural strength values obtained (703 MPa at room temperature, 626 MPa at 800°C, and 484 MPa at 1100°C) appear to be typical for silicon nitride ceramics. The decrease in strength at 1100°C is normally due to the softening of the intergranular phase. The apparent superplastic behavior at 1500°C may be due to the relatively fine grain structure of this material as seen in the SEM micrographs, such as Figure 4. Even at the much higher strain rate employed at 1500°C in order to fracture the test bar, considerable slow crack growth occurred prior to actual failure. This can be seen in Figure 2, where approximately half of the thickness of the bar had undergone slow crack growth prior to failure. Under these conditions, strength values reported have little real meaning.

The elastic modulus value at room temperature of 248 GPa is lower than the typical value of 310 MPa for silicon nitride. This is probably due to the presence of the cobalt silicide phase. The reduction in elastic modulus at the higher temperatures may be due to the increasing ductility of the silicides and the softening of the grain boundary phase. The value at 1500°C shown has little meaning, again because of the superplastic-like nature of the material at this temperature.

The fracture toughness value of 10.5 MPa√m obtained at room temperature is relatively high compared with a typical value for silicon nitride of 6 MPa√m. A toughness increase at 800°C is not usual for a ceramic material. The value of 6.0

Figure 2 Slow crack growth during fracture at 1500°C.

MPa$\sqrt{m}$ at 1100°C is typical for a normal silicon nitride at room temperature. The value reported at 1500°C is suspect because the elastic beam condition was not met by the superplastic-like behavior at this temperature. An explanation for the fracture toughness values obtained can be found by examining the SEM micrographs. In the chevron notch beam test geometry used, a stable crack front progresses from the tip of the chevron notch (which is located at a_o = 1.0mm or α = 0.2) to the point where α = 0.4, that is, 2.0 mm from the sample edge. Thus, there is a triangular section of up to 1.0 mm from the tip where the crack front velocity is related to the speed of the cross head of the test machine. Beyond that point, the crack front propagates unstably and rapidly to failure. Examination of the fractured surface in these two areas can reveal topographic features which may be a function of the crack velocity. Figure 3 is an optical micrograph of a chevron notch tested at 1500°C. The rough area is the region which experienced controlled crack growth, and is the region in which the toughness determination is made. The smooth area is where the crack front propagated unstably. Texture differences are less obvious on surfaces fractured at lower temperatures.

In Figures 4 to 12, parts A and C are SEM micrographs of the controlled fracture area, while parts B and D are from the uncontrolled fast fracture area of the chevron notch sample. Parts A and B are micrographs of the larger silicide particulates dispersed in the silicon nitride matrix, while parts C and D are micrographs of the matrix.

The surface created during fracture at room temperature is shown in Figure 4. An examination of the micrographs show that there is no difference between the controlled and uncontrolled fracture regimes with respect to surface morphology. The matrix shows a mixed fracture mode of intergranular and intragranular fracture. The elongated beta grains are clearly seen and are approximately 0.5 micrometers in diameter and from 1.0 to 3.0 micrometers long. The silicide phase exhibits brittle fracture. In the oblique angle view shown in Figure 5, the fracture plane through the silicide phase can be seen to be the same as the matrix.

The surface created during fracture at 800°C is shown in Figure 6. Again, the matrix shows no difference between the two fracture rate regimes and still has a mixed

Figure 3 Surface of chevron notch sample fractured at 1500°C.

mode brittle fracture. The silicide phase is now showing definite signs of ductility, even at this temperature. The micrograph of the controlled fracture (Figure 6-A) shows considerable ductility of the silicide, while Figure 6-B, that of the uncontrolled fracture region, shows a more brittle fracture of the silicide. Ductility is rate dependant and thus the controlled fracture rate is allowing the silicide to deform prior to fracture. An examination of the 45° oblique view, Figure 7, clearly shows the ductility of the silicide in part A. However, some ductility can also be seen in part B, in the rapid crack growth area. Large beta silicon nitride columns can be clearly seen in these silicide islands.

It thus appears that at 800°C, the fracture of this material has the added toughening mechanism of ductile ligament stretching. This mechanism may by limited at this temperature, since it appears that the silicide phase is well bonded to the matrix and little or no debonding has occurred. However, it does seem to explain the increase in the fracture toughness value over the value obtained at room temperature.

When fractured at 1100°C, the surface created in the matrix rapidly forms an oxidized glassy layer. This is shown in Figures 8 and 9. The silicides do not oxidize under these conditions, as seen in the same figures. The greater extent of ductility in the silicides is apparent in parts A and B of those figures. When the glass is removed, with an HF etch (Figures 10 and 11), the debonding between the silicide and matrix is revealed. The matrix appears to have fractured in a mixed mode. Thus, it appears that while contributions from other mechanisms have decreased due to the softening of the intergranular phase, the overall contribution to the fracture toughness by ductile ligament

Figure 4 Fracture surface created at room temperature.

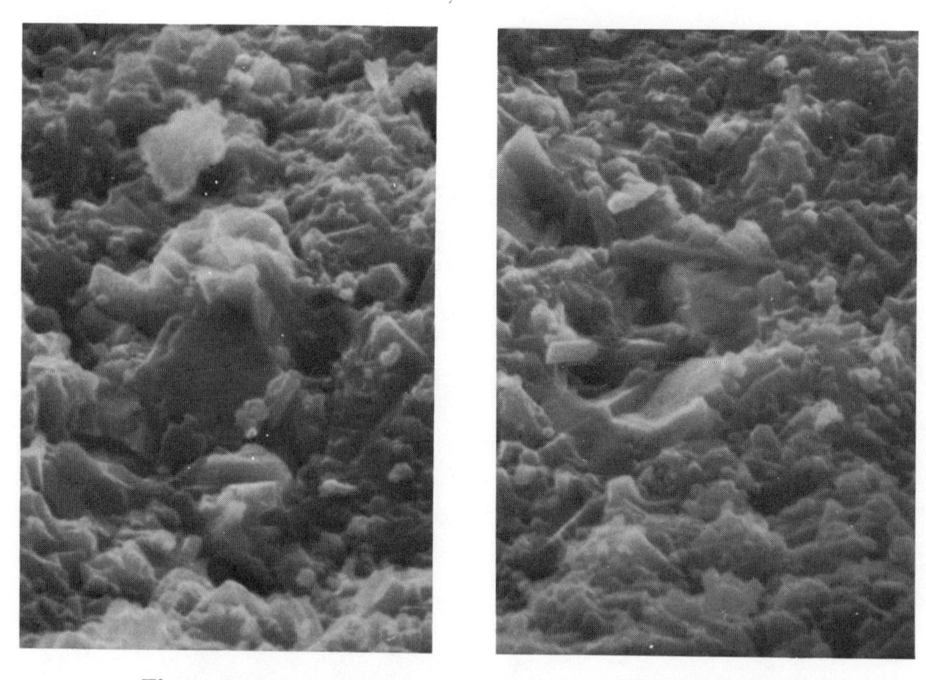

Figure 5 Oblique view of room temperature fractured surface.

debonding/stretching is increasing. Consequently, the fracture toughness at 1100°C is enhanced by the presence of the ductile silicide particulates, and thus is relatively high for a silicon nitride ceramic.

At 1500°C, the silicide phase is molten and the oblique view of the fracture surface shown in Figure 12 (after etching) reveals large silicide spheres. The matrix has fractured in an almost intergranular mode. This would be expected given that the intergranular phase would have been quite soft.

The micrographs of the fracture surfaces show that the ductility of the silicide phase at higher temperatures apparently contributes to the toughness of the material. The high toughness of this material at room temperature is not as well explained by the micrographs since the silicide phase is completely brittle. The large thermal expansion mismatch, 17.8×10^{-6} /°C for the silicides compared with about 3×10^{-6} /°C for silicon nitride, should have resulted in a large stress gradient around the silicide particles, especially since the bond to the matrix appears strong. This stress gradient should result in the crack front deviating away from the silicide particles, since the silicide has the higher thermal expansion coefficient and the lower elastic modulus. Instead, it appears that the crack front is not visibly affected by the presence of the silicides. A possible explanation may be found in TEM micrographs obtained on this material. Figure 13 shows the interface between silicide and matrix phases. Dislocation bands can be seen at the interface while the bulk of both phases appear to be unstressed. On a larger scale, as seen in Figure 14, cracks appear within the silicide phase. It thus seems likely that,

Figure 6 Surface fractured at 800°C.

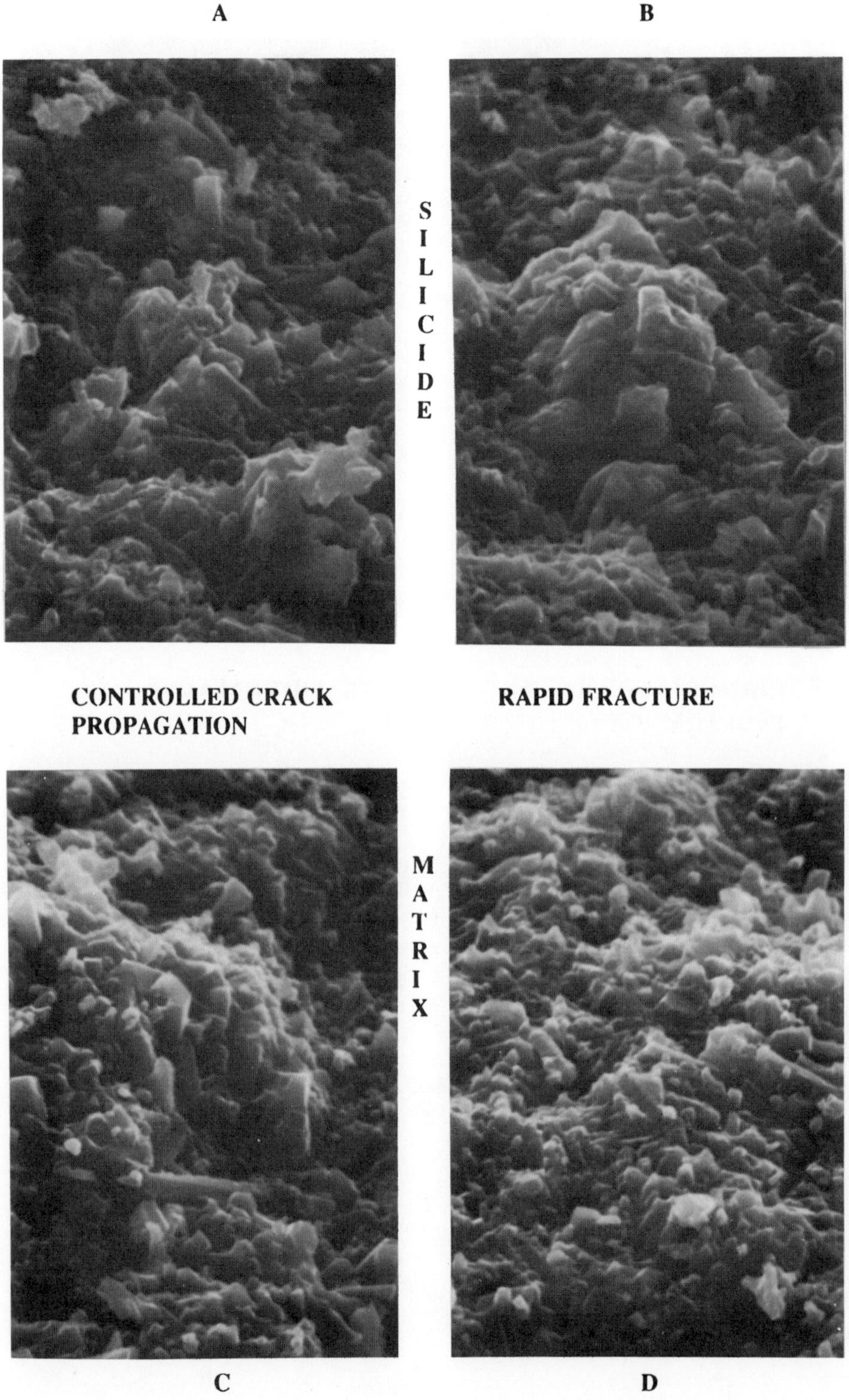

A B

S I L I C I D E

CONTROLLED CRACK PROPAGATION

RAPID FRACTURE

M A T R I X

C D

Figure 7 Oblique view of surface fractured at 800°C.

Figure 8 Surface fractured at 1100°C. Not etched.

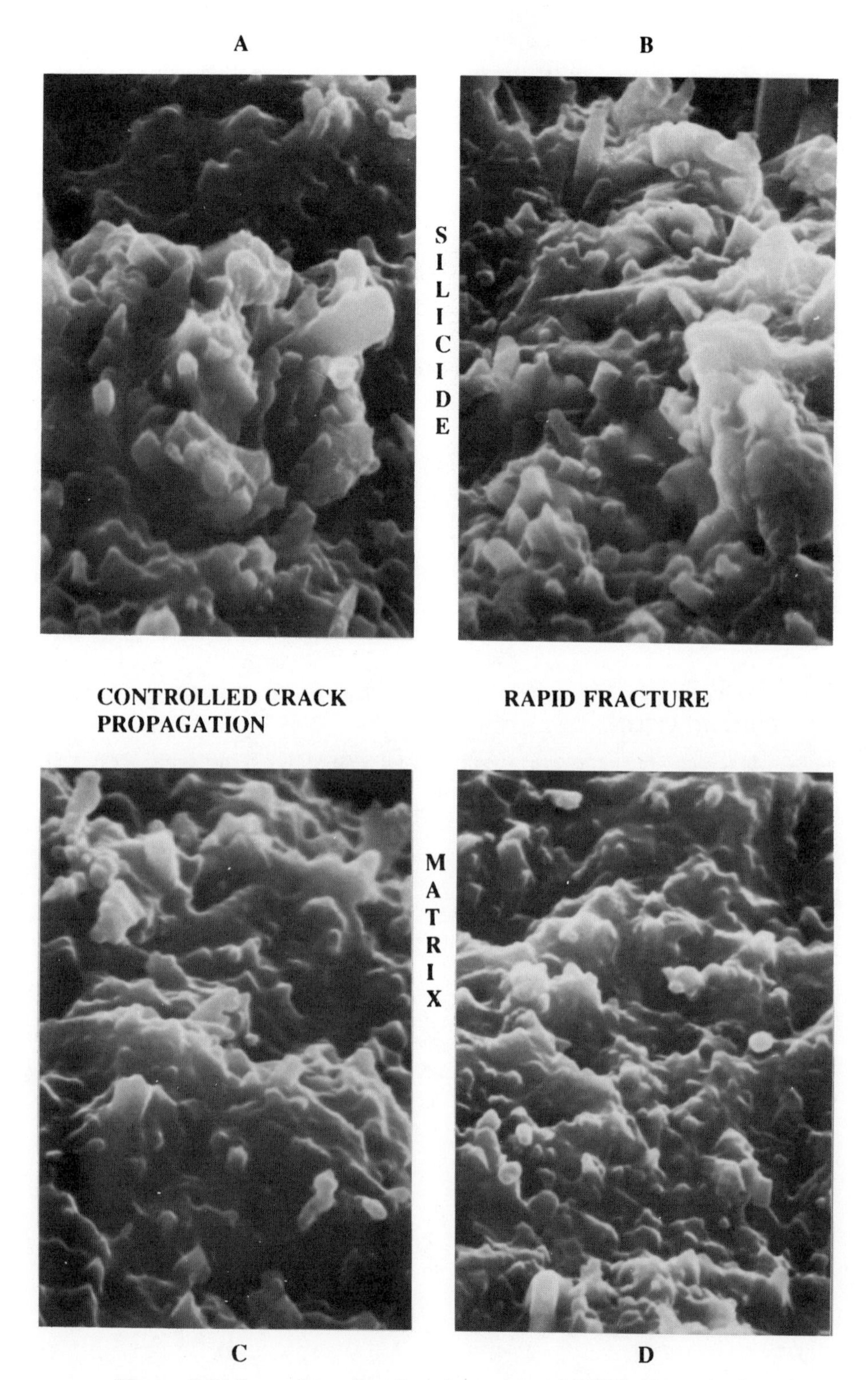

Figure 9 Oblique view of surface fractured at 1100°C. Not etched.

Figure 10 Surface fractured at 1100°C. Etched.

Figure 11 Oblique view of etched surface fractured at 1100°C.

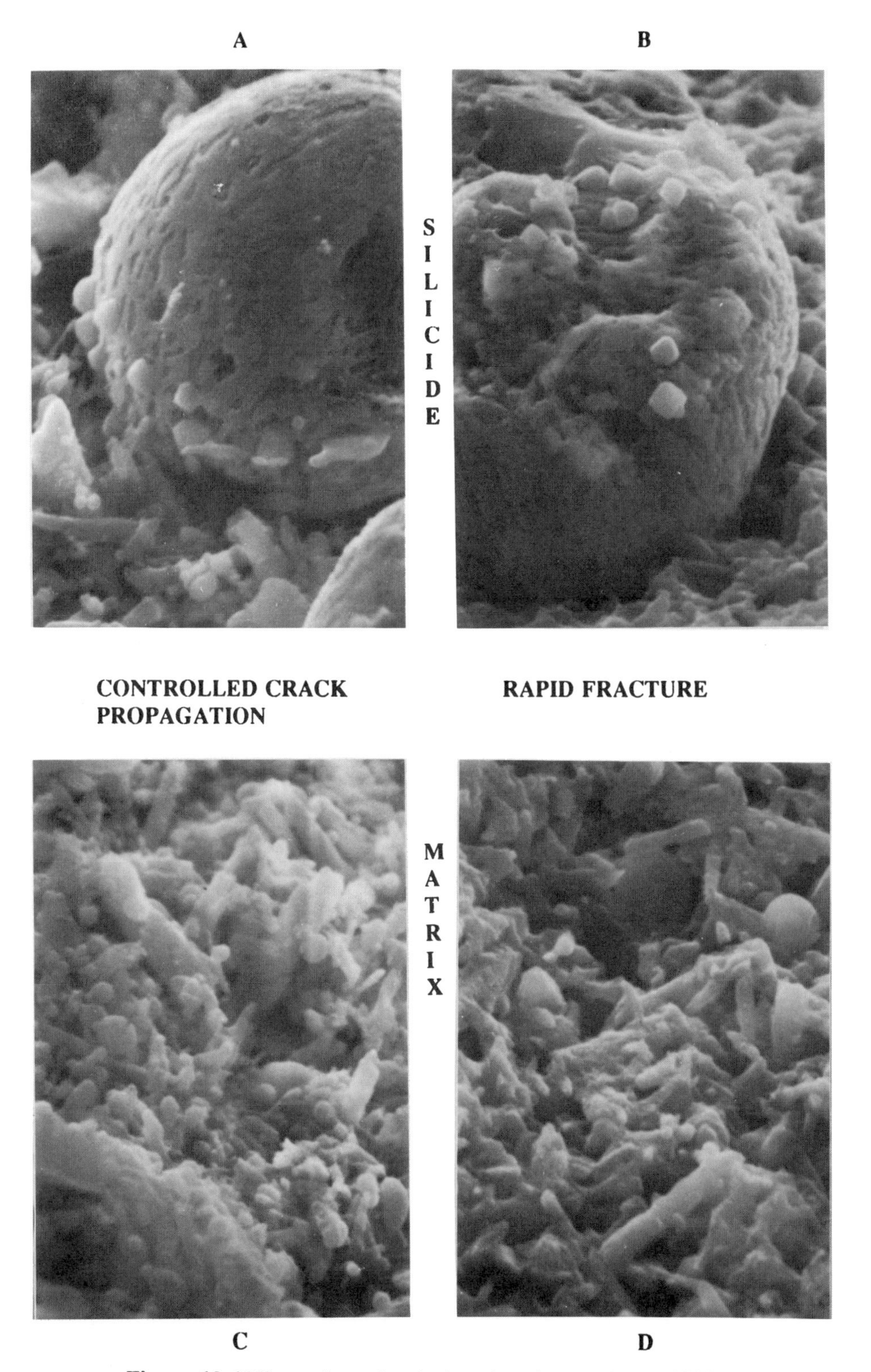

Figure 12 Oblique view of etched surface fractured at 1500°C.

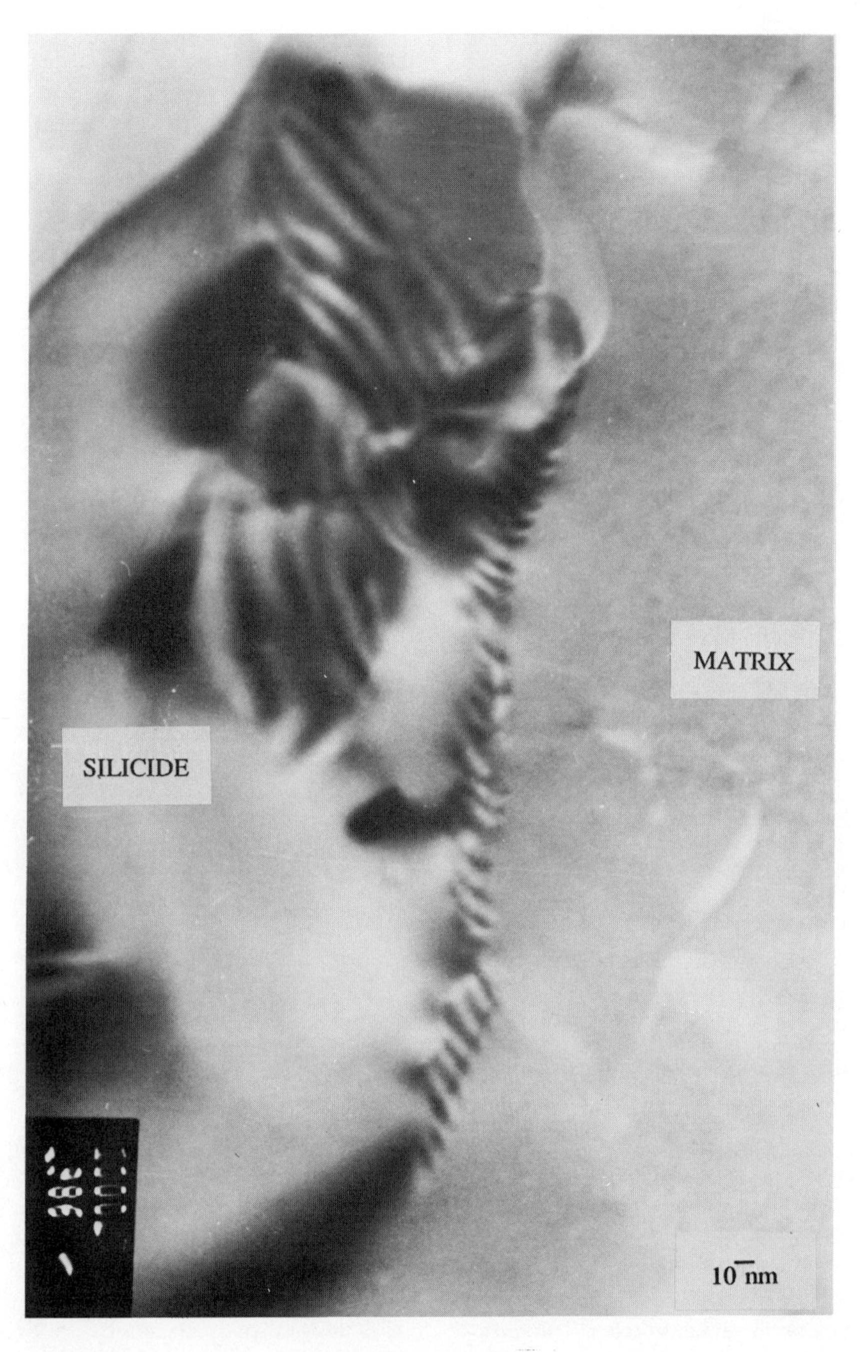

Figure 13 Interface between cobalt silicide and silicon nitride showing dislocation bands.

Figure 14 Cracks in the cobalt silicide phase within the silicon nitride matrix.

despite the large difference in thermal expansion coefficients, the two phases exist at room temperature in a largely unstressed state due to dislocations and microcracking in the silicide phase.

The relatively high fracture toughness of this material at room temperature may thus be due to several factors. The most obvious is the self reinforcement by the elongated beta silicon nitride grains. The size of the elongated grains is not as great as would be expected for a material of this toughness, however. The microcracks in the larger silicide phase particles may be another mechanism which increases the overall toughness in the composite. Other mechanisms such as crack pinning or bowing by the silicides are possible, but their effect would not be seen on a fracture surface.

SUMMARY

A silicon nitride ceramic matrix composite containing cobalt silicides, formed by reacting cobalt metal with silicon nitride, has been shown to be a material with high fracture toughness, 10.5 MPa√m at room temperature. This toughness has been shown to be due to the interlocking beta phase of the matrix and possibly to microcracking in the silicide phase. The elevated toughness at 800°C, 12.2 MPa√m, has been shown to be due to an additional toughening mechanism. The dispersed silicide phase becomes ductile at elevated temperatures due to the brittle to ductile transitions of the silicides. The ductile cobalt silicides appear to be influencing the toughness at 800°C. Higher levels of ductility of the silicides and their debonding from the matrix at 1100°C appear to be the primary reasons for the retention of a reasonable fracture toughness level, i.e., 6.0 MPa√m. Without these mechanisms, the toughness would probably be much lower as the intergranular glassy phase has softened considerably. The ductile nature of the silicides are clearly seen in the SEM micrographs for these temperatures. At 1500°C, the material behaves superplastically and is subject to slow crack growth, even at high strain rates.

ACKNOWLEDGEMENTS

Experimental assistance was provided by M.J. Centuolo, J.A. Mahoney, J.F. Brereton, T. Prindle, G.G. Weaver, N.P. Pyrros, and R. Kriedler.

TEM results were obtained by Professor D.A. Bonnell of the University of Pennsylvania in Philadelphia.

Valuable discussions with Professor A.V. Virkar of the University of Utah in Salt Lake City, were helpful with fracture toughness methodology.

Discussions with Professor A.G. Evans of the University of California at Santa Barbara were helpful with sorting out fracture mechanisms.

REFERENCES

1. R.W. Davidge and T.J. Green, "The Strength of Two-Phase Ceramic/Glass Materials, J.Mat.Sci., 3 629-634 (1968).
2. D.J. Green, "Fracture Toughness Prediction for Crack Bowing in Brittle Particulate Composites," J.Am.Ceram.Soc., 66 [1] C4-C5 (1983).
3. M. Taya, S. Hayashi, A.S. Kobayashi, and H.S. Yoon, "Toughening of a Particulate-Reinforced Ceramic-Matrix Composite by Thermal Residual Stress," J.Am.Ceram.Soc., 73 [5] 1382-1391 (1990).
4. N. Claussen, J. Steeb, and R.F. Pabst, "Effect of Induced Microcracking on the Fracture Toughness of Ceramics", Am.Ceram.Soc.Bull., 56 [6] 559-562 (1977).
5. D.J. Green, P.S. Nicholson, and J.D. Embury, "Fracture of a Brittle Particulate Composite; Part 2, Theoretical aspects," J.Mat.Sci., 14 1657- 1661 (1979).
6. A.K. Khaund and P.S. Nicholson, "Fracture of a Brittle Composite: Influence of Elastic Mismatch and Interface Bonding," J.Mat.Sci., 15 177-187 (1980).
7. V.V. Krstic, P.S. Nicholson, and R.G.Hoagland, "Toughening of Glasses by Metallic Particles," J.Am.Ceram.Soc., 64 [9] 499- 504 (1981).

8. F. Erdogan and P.F. Joseph, "Toughening of Ceramics through Crack Bridging by Ductile Particles," J.Am.Ceram.Soc., 72 [2] 262-270 (1989).
9. W.H. Tuan and R.J. Brook, "The Toughening of Alumina with Nickel Inclusions," J.Europ.Ceram.Soc., 6 31-37 (1990).
10. T.L. Jessen and D. Lewis III, "Fracture Toughness of Graded Metal-Particulate/Brittle-Matrix Composites," J.Am.Ceram.Soc., 73 [5] 1405-1408 (1990).
11. R.L.K. Matsumoto and G.G. Weaver, "High Temperature Properties of a Silicon Nitride - Intermetallic Silicide Composite", to be published in proceedings of the 15th Annual Conference on Composites and Advanced Ceramics.
12. Handbook of Chemistry and Physics, 64th edition, R.C. Weast, ed., CRC Press, Boca Raton, FL., p. B-88.
13. D. Munz, R.T. Bubsey, and J.L. Shannon, Jr., "Fracture Toughness Determination of Al_2O_3 Using Four-Point-Bend Specimens with Straight-Through and Chevron Notches," J.Am.Ceram.Soc., 63 [5-6] 300-305 (1980).

BRITTLE-TO-DUCTILE TRANSITION IN SILICON NITRIDE

Y. Mutoh, K. Yamaishi, N. Miyahara and T. Oikawa*

Department of Mechanical Engineering
Nagaoka University of Technology
Kamitomioka, Nagaoka-shi 940-21, Japan
*Advanced Technology Research Center, NKK Co.
Kawasaki-ku, Kawasaki-shi 210, Japan

INTRODUCTION

Since silicon nitride is hard to sinter, it is general to sinter with small amounts of sintering aids, such as Al_2O_3 and Y_2O_3. Therefore, very thin glass phase formed at the grain boundary has an influence on high temperature mechanical properties of sintered silicon nitride. Silicon nitride is one of major candidate materials for potential structural components in high-performance turbine application. In service, components in turbine application may be subjected to loading at elevated temperatures. It is, therefore, essential to gain detailed understanding of fracture behavior at elevated temperatures. From this viewpoint, the characteristics of high temperature strength[1-5] and creep strength[6-9] have been actively investigated lately for silicon nitride in comparison with other ceramic materials. However, the detailed fracture behavior of silicon nitride including the influence of intergranular glass phase at elevated temperatures has not always been obvious.

In this study, fracture toughness tests of HIP-sintered silicon nitride were carried out to investigate the temperature dependency of fracture toughness and the detailed fracture behavior at elevated temperatures. The transition of fracture mechanism and the role of glass phase in the transition behavior were also discussed.

EXPERIMENTAL PROCEDURES

Specimen

The material used is a HIP-sintered silicon nitride with additives of 2w%Y_2O_3 and 5w%Al_2O_3. Bending-type fracture toughness specimens, dimensions of which were 5× 10 × 55mm, were machined. A fatigue precrack was introduced into the specimen. The method for fatigue-precracking[10] was as follows. A through-thickness sharp crack was introduced by the Bridge Compression (BC) method[11], as shown in Fig.1, where several Vickers indentations were used as the crack initiator. The anvil with groove width of 8mm was

Fracture Mechanics of Ceramics, Vol. 10
Edited by R.C. Bradt *et al.*, Plenum Press, New York, 1992

used to obtain almost 3-4mm length of a through-thickness crack. Subsequently, a fatigue crack was extended from the BC precrack introduced. Fatigue precracking was conducted using a servohydraulic fatigue test machine under a three-point bend type of loading with stress ratio of 0.1 and frequency of 20Hz. The maximum applied load was controlled to make the maximum stress intensity factor K_{fmax} lower than 3.8 MPa$\sqrt{m}$. The resultant total length of precrack was in the range from 5 to 5.5mm. The reason for using not the BC-precracked specimen but the fatigue-precracked specimen in this study is as follows. The BC precrack is a sharp pop-in crack and is introduced under a high K-level equivalent to K_{Ic} or K_{Ia}. Therefore, the damage zone with microcracks, which sometimes induces the lower K_{Ic}-value, is often formed at the tip of the BC precrack[12]. On the other hand, the fatigue precrack can be introduced under a sufficiently low K-level. Therefore, the damage zone formed at the tip of the precrack is small enough to obtain the accurate K_{Ic}-value.

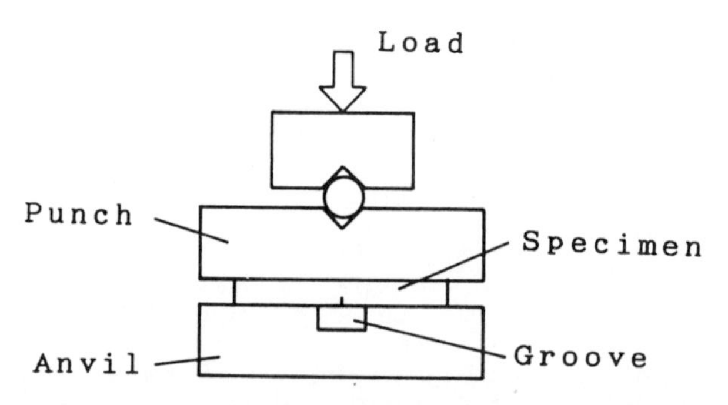

Fig.1 Schematic illustration of the BC method for precracking.

Fracture toughness test

High temperature fracture toughness tests in both air and vacuum were conducted using an Instron type universal test machine with electric furnaces. The pressure in a vacuum chamber was lower than 6mPa. Test temperatures were every 100℃ from 1000℃ to 1500℃. Fracture toughness tests, which were conducted in three point bending with span length of 40mm at the crosshead speed of 0.5mm/min, were started after keeping the attained test temperature for 10min. For comparison, a fracture toughness test at room temperature was also carried out in air.

Since a non-linear fracture behavior was observed above 1200℃, the elastic-plastic fracture toughness J_{Ic} tests based on the R-curve method were carried out at 1400℃ in both air and vacuum atmospheres. The accurate load-displacement curve is essential to obtain in the J_{Ic} test. In the present experiment, since fracture toughness tests were conducted in the electric furnace, it was difficult directly to detect the displacement of loading point of the specimen. The load-point displacement was obtained to subtract the displacement due to the loading system from the total displacement measured outside of the furnace. The complience of the loading system was obtained to measure the load-displacement curve for the case that the sialon block of 20× 30× 60mm in size, which was assumed to be a rigid body, was inserted instead of the fracture toughness specimen. For confirmation, the load-point displcement of the fracture toughness specimen without a precrack was obtained according to the above mentioned method. This obtained displacement was consisitent well with the calculated one based on the elastic beam theory. The specimens were loaded up to certain levels of displacement at 1400℃ and subsequently fractured at room temperature to measure the stable crack length according to the J_{Ic} test standard JSME S-001.

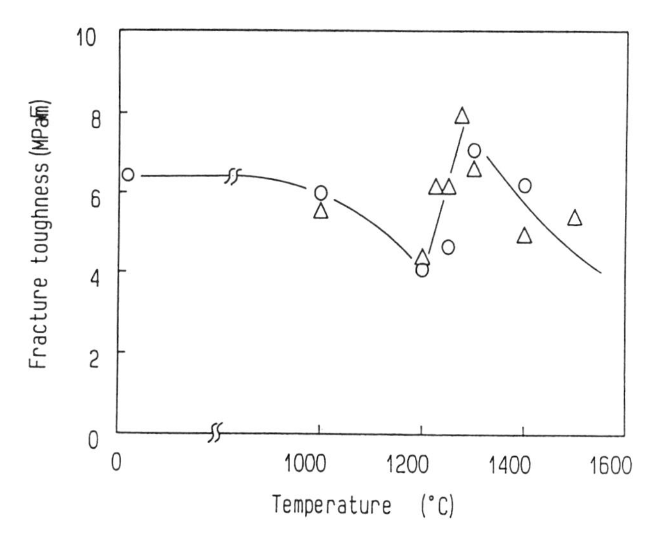

Fig.2 Relationship between K_{Ic} and temperature.

RESULTS AND DISCUSSION

Fracture toughness at elevated temperatures

The relationship between fracture toughness K_{Ic} and temperature is shown in Fig.2. Although the non-linear fracture behavior with stable crack growth in the temperature region higher than 1200℃ was found, the apparent K_{Ic}-values calculated using the maximum load were plotted in the figure. As can be seen from the figure, the fracture toughness K_{Ic} decreased with increasing temperature up to 1200℃ . However, the brittle-to-ductile transition was found in the temperature range from 1200℃ to 1275℃ : the K_{Ic}-value rapidly increased from 4MPa$\sqrt{m}$ to 8MPa$\sqrt{m}$. The fracture toughness again decreased with increasing temperature above the transition temperature. No significant difference in the temperature dependeny and transition behavior of fracture toughness was found between air and vacuum atmospheres.

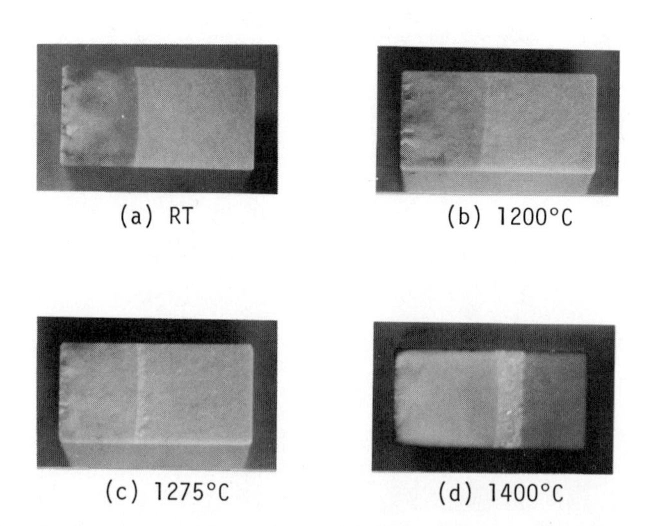

(a) RT (b) 1200°C

(c) 1275°C (d) 1400°C

Fig.3 Macroscopic view of fractured specimens.

<u>Fractographic observations</u>

On the basis of the macroscopic observations, unstable fracture directly occurred from the tip of fatigue precrack in the lower temperature region than 1200℃. On the other hand, the stable crack growth region was found between the fatigue precrack region and the unstable fracture region in the higher temperature region than 1275℃, as shown in Fig.3. The similar stable crack growth behavior of silicon nitride at elevated temperatures has been observed lately[2,9]. On the basis of microscopic observations, the intergranular fracture was dominant in the lower temeprature region than the transition temeprature. On the basis of the conventional observation, where the fracture surface was observed from the perpendicular direction to the surface, no significant difference in fracture surface morphology other than the adhesion of glassy phase on the stable crack growth region was found between the lower and higher temperature regions than the transition temperature. The fracture surface was observed from the near-parallel direction to the surface (10° ~ 15° upper direction from the

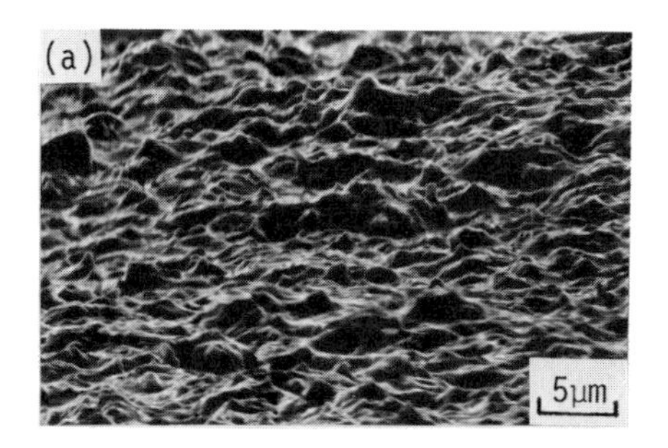

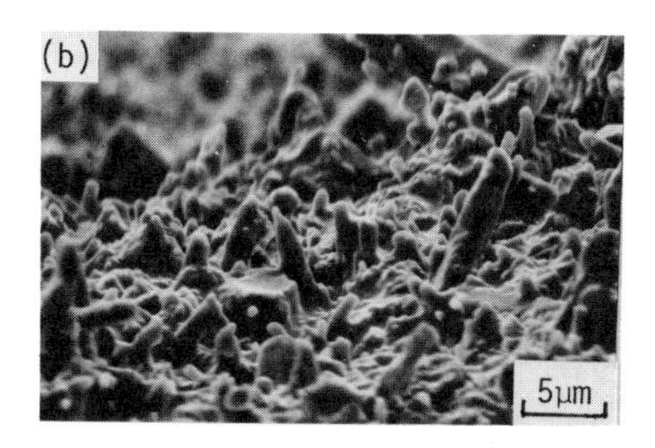

Fig.4 Fractographs of the specimen tested at 1400 ℃, where the fracture surface was observed from the near-parallel direction to the surface (10° ~ 15° upper direction from the parallel direction). (a)Fatigue precrack region, (b)Stable crack growth region, (c)Unstable fracture region

parallel direction) to find inherent morphology of stable crack growth region. From the results, many pulled-out β-Si_3N_4 particles, which could not been found in the fatigue precrack and unstable fracture regions, were observed in the stable crack growth region, as shown in Fig.4.

Observations of microstructures below the fracture surface

On the basis of the fractographic observations, the stable crack seemed to propagate with pulling-out of β-Si_3N_4 particles. To investigate the mechanism of stable crack growth further, the longitudinal cross-section of the fractured specimen, which was cut from the center of the specimen, polished and finally etched by using boiled phosphoric acid, was observed in detailed. The results are shown in Fig.5. As can be seen from the figure, there were many pulled-out β-Si_3N_4 particles in the stable crack growth region as observed on the fracture surface. Furthermore, the β-Si_3N_4 particles below the fracture surface seemed to be oriented in the tensile direction. Then, the inclined angles of β-Si_3N_4 particles from the fracture surface were measured in the region within 10μm depth from the fracture surface. The results are shown in Fig.6. It was found from the figure that the β-Si_3N_4

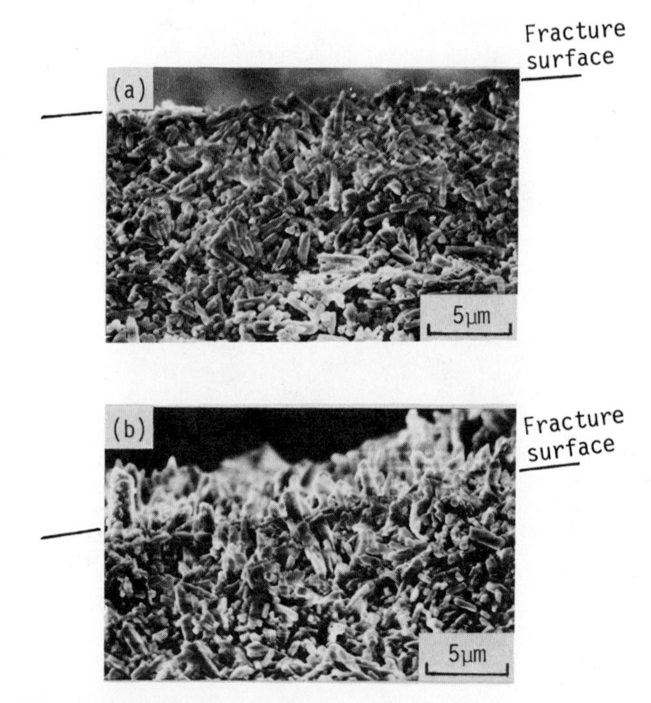

Fig.5 Microstructures below the fracture surface for the specimen tested at 1400 ℃. (a)Fatigue precrack region, (b)Stable crack growth region

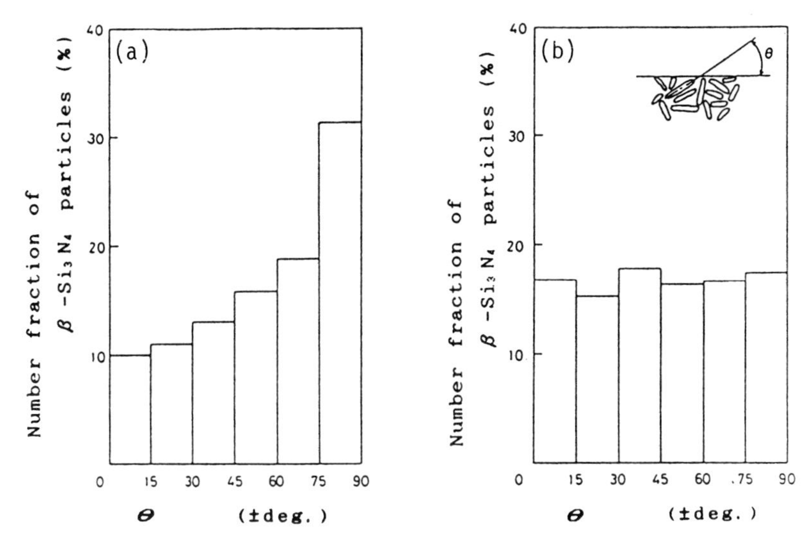

Fig.6 Histograms of the inclined angle θ of β -Si_3N_4 particles from the fracture surface. (a)Fatigue precrack region, (b)Stable crack growth region

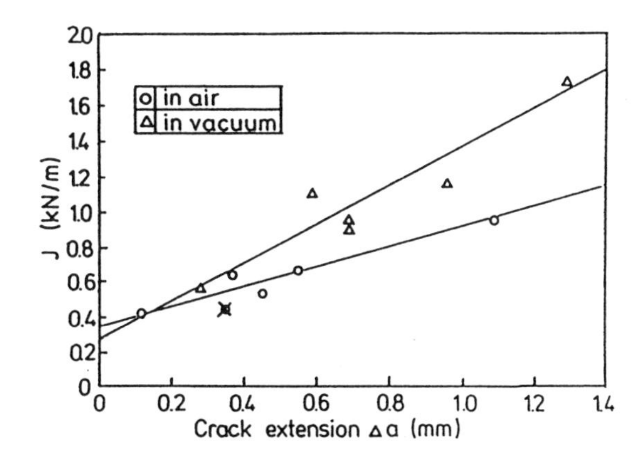

Fig.7 R-curves for the specimens tested at 1400 ℃ in air and vacuum.

particles were oriented in the perpendicular direction to the fracture surface in the stable crack growth region, while they were not oriented and randomly distributed in the fatigue precrack region. This result suggests that the stable crack propagates with large deformation including the orientation of β-Si_3N_4 at the crack tip.

R-curve for the stable crack growth at 1400℃

The relationship between J-value and stable crack length at 1400℃ is shown in Fig.7. As can be seen from the figure, the resistance of crack growth significantly increased with increasing the crack length. This increase in resistance of the stable crack growth seemed to induce the increase in fracture toughness in the brittle-to-ductile transition region as shown in Fig.2. The J_{IC}-values estimated based on the R-curve shown in Fig.7 were 0.37kN/m and 0.27kN/m for air and vacuum atmospheres, respectively. On the other hand, the slope of the resistance curve (the tearing modulus) in vacuum was steeper than that in air. However, further works are requested to discuss whether these differences in both atmospheres have significance.

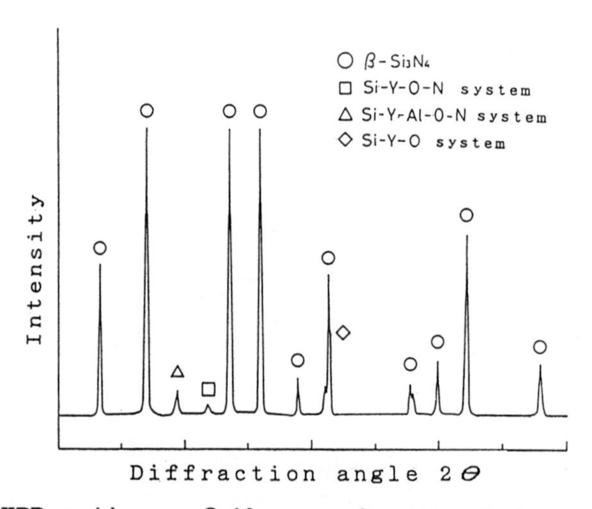

Fig.8 XRD pattern of the specimen tested at 1400℃.

Temperature dependency of viscosity of the intergranular glass phase

It is considered that fracture behavior of silicon nitride at elevated temperatures is related to characteristics of intergranular glass phase. The result of X-ray diffraction on the fracture surface of the specimen tested at 1400 ℃ in vacuum is shown in Fig.8. The crystallized glass phases of Si-Y-O-N, Si-Y-O and Si-Y-Al-O-N systems were found. These glass phases, which were initially amorphous, seemed to be crystallized during the heat cycle of the fracture toughness test. However, major part of glass phase was thought to remain amorphous and their composition could not be determined by the X-ray diffraction. Although the specification of exact composition of glass phase was difficult as mentioned above, considering together with the composition of source powders, it is expected to be the oxynitride glass of Si-Y-Al-O-N system. The temperature dependency of viscosity for oxynitride glass, which had not yet been clear, was estimated based on the relationship between viscosity and temperature for the similar system of glass which had been

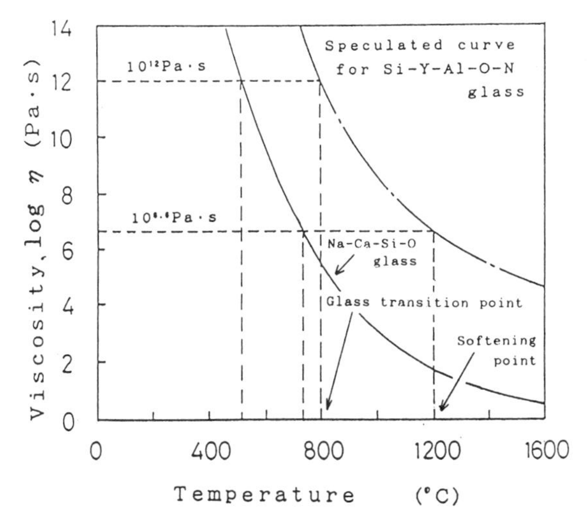

Fig.9 Relationship between viscosity and temperature for glasses.

reported. Figure 9 shows the relationship between viscosity and temperature for the Na-Ca-Si-O system of glass[13]. It is known that the similar temperature dependency of viscosity is observed for other glasses with different composition[14]. The viscosity is significantly increased with including nitrogen in the glass[15]. Since there are a lot of nitrogen around the intergranular glass phase in silicon nitride, large amounts of nitrogen must solute in the glass phase. Therefore, the viscosity-temperature curve for the intergranular glass phase is estimated to be located in the upper-right direction of that for the glass of Na-Ca-Si-O system shown in Fig.9. Refering the temperature dependency of fracture toughness shown in Fig.2 and that of bending

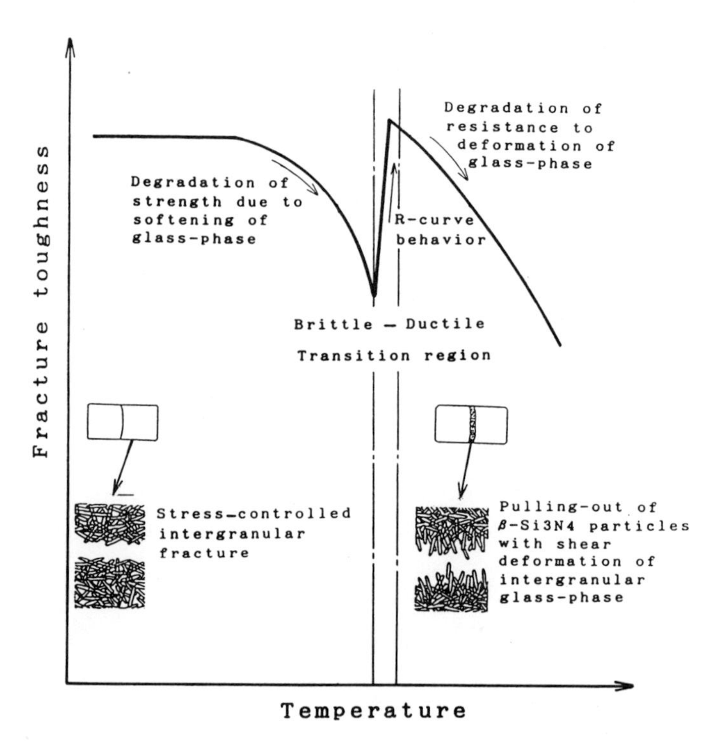

Fig.10 Schematic model for the temperature dependency of fracture toughness and the transition of fracture mechanism in silicon nitride.

strength which will be shown later, both K_{IC}-value and bending strength decreased above 800℃. Shifting the viscosity-temperature curve for the glass of Na-Ca-Si-O system in the upper-right direction to adjust the glass transition point T_g to 800℃, the one-dotted curve shown in Fig.9 was obtained. Based on the curve, the softening point T_s of the intergranular glass phase, at which the glass can not keep its shape without outer support, seemed to be around 1200℃. Since the softening point of the intergranular glass phase is consistent with the brittle-to-ductile transition temperature, it is thought that the orientation of β-Si_3N_4 particle is possible when the temperature attains the softening point of the glass phase. As the result, the stable crack seems to grow with bridging and pulling-out of the oriented β-Si_3N_4 particles.

Temperature dependency of fracture toughness and fracture mechanism

From the foregoing results and discussion, the characteristics of fracture toughness of silicon nitride at elevated temperature will be summarized as schematically shown in Fig.10. In the lower temperature region than the transition temperature, the intergranular fracture is dominant. Since the strength of the intergranular glass phase is reducing above 800℃ which corresponds to the glass transition point, the fracture toughness decreases with increasing temperature. When the temperature attains the softening point of the glass phase (about 1200℃) and consequently the intergranular glass phase becomes adequately soft, the oriented β -Si_3N_4 particlces are bridged and pulled out with shear deformation of the intergranular glass phase. This change of fracture mechanism results in the brittle-to-ductile transition of fracture. In the higher temperature region than the transition temperature, the decrease in viscosity is promoted and consequently the resistance of shear deformation of the intergranular glass phase decreases with increasing temperature. As the result, the fracture toughness reduces with increasing temperature.

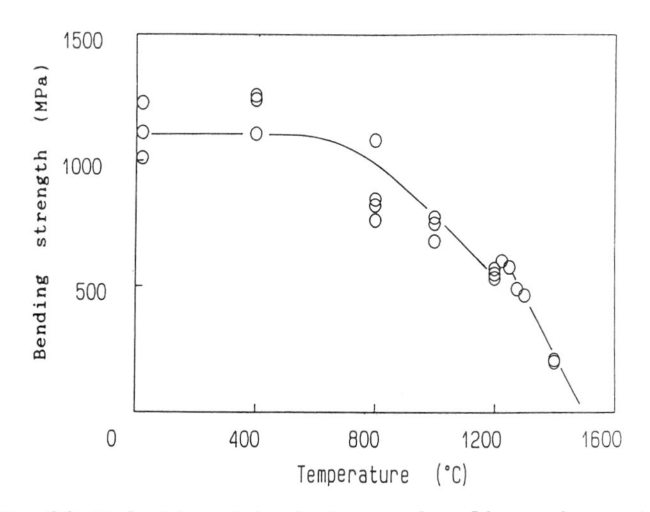

Fig.11 Relationship between bending strength and temperature for silicon nitride.

Transition behavior in bending strength test

It is speculated that the brittle-to-ductile transition behavior, which is observed in fracture toughness test and correlated with the temperature dependency of viscosity of the glass phase, is also found in bending strength test. The bending strength tests at elevated temperatures in vacuum were carried out using the specimens of 3× 4× 35mm in dimension, which were machined from the same material as used for the fracture toughness specimen. The bending test was conducted under three-point

Fig.12 Stable crack growth region on the fracture surface of the bending strength specimen tested at 1400 ℃ .

Fig.13 A fractograph of the stable crack growth region for the bend specimen tested at 1400 ℃ , where the fracture surface was observed from the near-parallel direction to the surface.

bending with span length of 30mm at the crosshead speed of 0.5mm/min. The relationship between bending strength and temperature is shown in Fig.11. Although the bending strength decreased with increasing temperature above 800℃ , it increased in the temperature range from 1200℃ to 1300 ℃ , where the increment of bending strength was not so remarkable compared with that of fracture toughness. From the macroscopic observations, the stable crack growth region was found for the specimens tested at higher temperatures than 1300℃ , as shown in Fig.12. Observing the fracture surface in the stable crack growth region from the near-parallel direction to the surface, the pulled-out β -Si_3N_4 particles were found, as shown in Fig.13. Therefore, although the transition of bending strength was not so remarkable compared with that of fracture toughness, the brittle-to-ductile transition of fracture mechanism was also found to occur in bending strength test.

CONCLUSION

Fracture toughness tests of HIP-sintered silicon nitride in air and vacuum were carried out to investigate the temperature dependency of fracture toughness and the fracture mechanism at elevated temeprature. The main results obtained are summarized as follows.

(1)Fracture toughness decreased with increasing temperature in the temperature range from 800℃ to 1200℃. The brittle-to-ductile transition was found in the temperature range from 1200℃ to 1275℃: the fracture toughness rapidly increased in the transition region. Above the transition temperature, the fracture toughness decreased with increasing temperature.

(2)The intergranular fracture was dominant in the lower temperature region than the transition temperature. It seemed that since the strength of intergranular glass phase decreased above the glass transition temperature (almost 800℃), the fracture toughness also decreased with increasing temperature above 800℃.

(3)Above the transition temperature, the stable crack growth with bridging and pulling-out of β-Si_3N_4 particles which were oriented in the tensile direction was observed. The brittle-to-ductile transition temperature seemed to be consistent with the softening temperature of the intergranular glass phase.

(4)The increase in fracture toughness in the transition region was induced by the increase in resistance of the stable crack growth.

(5)The similar brittle-to-ductile transition behavior was found in the bending strength test.

REFERENCES

1. F.F.Lange, High-Temperature Strength Behavior of Hot-Pressed Silicon Nitride: Evidence for Subcritical Crack Growth, J.Am.Ceram.Soc., 57:84(1974).
2. S.H.Knickerbocker, A.Zangvil, and S.D.Brown, High-Temperature Mechanical Properties and microstructures for Hot-Pressed Silicon Nitrides with amorphous and Crystalline Intergranular Phases, J.Am.Ceram.Soc., 68:C-99(1985).
3. R.K.Govila, Fracture Phenomenology of a Sintered Silicon Nitride Containing Oxide Additives, J.Mater.Sci., 23:1141(1988).
4. K.Tanaka, Y.Mutoh, and T.Niwa, Fracture Strength, Fracture Toughness and Their Scatter Characteristics at High Temperature in Silicon Nitrides Produced by Several Sintering Methods, Trans.JSME, Ser.A, 54:1086(1988).
5. T.Ohji, Y.Yamauchi, W.Kanematsu, and S.Ito, Tensile Strength and Fracture Defects Expanded by Subcritical Crack Growth of Silicon Nitride at High Temperatures, J.Mater. Sci.Letters, 9:1266(1990).
6. W.R.Cannon and T.G.Langdon, Creep of Ceramics, J. Mater. Sci., 23:1(1988).
7. T.Fett, K.Keller, and D.Munz, An Analysis of the Creep of Hot Presssed Slicon Nitride in Bending, J.Mater.Sci., 23:467(1988).
8. G.D.Quinn, Fracture Mechanism Maps for Advanced Structural Ceramics, J.Mater.Sci., 25:4361(1990).
9. J.Crampon and R.Duclos, Creep and Microstructure of Electrical Discharge Machinable Si_3N_4 composites, Acta Metall., 38:805(1990).
10. Y.Mutoh, K.Tanaka, and N.Miyahara, A Method for Evaluating the Fracture Toughness of Ceramics, Trans.JSME, Ser.A, 55:2144(1989).
11. T.Sadahiro and S.Takatsu, A New Precracking Method for Fracture Tougness Testing of Cemented Carbides, Modern Develop. Powder Metall., 12,13 and 14:561(1981).

12.T.Nishida and Y.Mutoh, Fracture Toughness Tests of Sintered and Melted High Speed Steels, in "Mechanical Behavior of Materials- VI ", M.Jono and T.Inoue ed., Pregamon Press, Vol.3:563(1991).
13.S.Sakuhana,"Oxynitride Glass", Uchidarokakuho, Tokyo(1989).
14.T.Ohta, F.Tsuchiya, K.Kawamura, S.Nakanishi, and J.Fukunaga, High and Medium Range Viscometers and Their Test with Some Alkali Silicate Glasses, Nihon-Seramikkusu-Kyokai-gakujutsuronbun-shi, 99:168(1991).
15.C.Schrimpf and G.H.Frischat, Some properties of Nitrogen-Containing Na_2O-CaO-SiO_2 Glasses, J.Non-crystal Solids, 52:479(1982).

A GENERIC MODEL FOR CREEP RUPTURE LIFETIME ESTIMATION ON FIBROUS CERAMIC COMPOSITES

Tze-jer Chuang

Ceramics Division
National Institute of Standards and Technology
Gaithersburg, Maryland 20899 U.S.A.

ABSTRACT

Because of their high strength and toughness at elevated temperatures, fiber reinforced ceramic composites (e.g. SiC_f/SiC, SiC_f/Si_3N_4) have become potential candidates for the next generation of turbine engine materials for aerospace applications. A generic model is proposed for assessing lifetime of this class of materials when subjected to long-term creep rupture conditions. This two-dimensional model consists of interfacial cracks growing between square grains and rectangular fibers in the direction normal to the principal tensile stress axis. Neglecting transient effects, the total lifetime is derived based on the criterion that rupture is due to coalescence of adjacent cracks. It is found that lifetime is inversely proportional to crack growth rate, volume fraction, and aspect ratio of the fibers; but extremely sensitive to the applied stress owing to the high power of the V-K_I law. This lifetime estimation seems to be in fair agreement with the creep rupture data of SiC_w/Si_3N_4 composite with 0 and 30 vol% reinforcement tested at 1250°C in air. Furthermore, TEM performed on the post crept specimens revealed that creep damage is predominantly in the form of microcracks at matrix/matrix as well as fiber/matrix interfaces, approximately in accord with the model simulation.

INTRODUCTION

In recent years, fiber reinforced ceramic matrix composites have attracted considerable interest because of their potential advantages over conventional materials in structural applications at elevated temperatures. Those advantages include increased strength, enhanced fracture toughness, and high creep and corrosion resistance in severe service environments. The design life of a typical structural member, regardless of its application, will always include an unsteady or cyclic thermal-mechanical loading phase as well as a sustained loading period. Therefore, if ceramic composites are to be used

Fracture Mechanics of Ceramics, Vol. 10
Edited by R.C. Bradt *et al.*, Plenum Press, New York, 1992

successfully and confidently in high temperature, load-bearing applications they must demonstrate adequate strength capability during the initial transient start-up, ***as well as*** maintain an acceptable level of reliability during the steady-state portion of the life cycle. As of now, the lack of design methodology for assessing service life is one of the leading hurdles discouraging widespread implementation of these emerging materials.

There are at least a few factors responsible for the current deficiency: (1) since this class of materials is still in the development stage, parameters describing physical and mechanical behaviors are ever changing; (2) collection of relevant data under sustained loading conditions for this new material is a formidable task as it is time consuming, and at the end the collected data can be obsolete and become irrelevant when the material is improved; and (3) absence of theories which allow extrapolation of short-term laboratory data to long-term service conditions. This last factor is especially critical because, for long term applications, theories will always be needed for any material to be adopted for load-bearing purposes.

The present paper addresses the last issue by presenting such a theory from which a rough estimation of the creep rupture lifetime can be made. Experimental observations on the resulting microstructure that has been exposed to high temperature, sustained loading conditions will be made first on a set of ceramic composites with different amounts of reinforcement. It will be shown that the formation and propagation of microcracks along interfaces between fiber/matrix and matrix/matrix constitute major damage, and final coalescence is responsible for creep rupture. Because those crack-like cavities were found predominantly at interfaces orientated in the direction normal to the principal tensile stress axis, the rôle that mass transport plays in void growth and void linkage must be an important aspect of the creep rupture process. Accordingly, mass diffusion-induced microcrack growth in both single phase (aimed at modeling matrix/matrix interfaces) and bimaterials systems (aimed at modeling fiber/matrix interfaces) will be considered next. Theoretical predictions of the crack growth velocity at steady state creep will be made for a given applied stress intensity and temperature. Finally, a generic creep rupture model invoking the crack growth rate will be presented from which a lifetime expression can be derived in terms of applied stress, temperature, fiber loading, aspect ratio and architecture, and other materials properties (e.g. diffusivities, surface free energies, elastic properties, etc.). Lifetime data are then used to verify the theoretical lifetime predictions. Implications are discussed in terms of the effects of applied stress, fiber concentration and other relevant parameters on the creep rupture lifetime. Good agreement of the stress dependence as well as volume fraction dependence suggest that this generic model is useful for rough estimates of lifetime by a design engineer as a means of assessing reliability of a structure or a structural component subjected to creep rupture conditions.

EXPERIMENTAL OBSERVATIONS

In order to elucidate the mode of rupture, microstructural studies using transmission electron microscopy (TEM) on the as-received and post-crept specimens were carried out to pinpoint the origin of failure [1]. The materials selected for current study are hot-pressed silicon nitride with 6.0 wt% yttria and 1.5 wt% alumina as sintering aids (commercially known as AY6[1]) and AY6 reinforced with 30 vol% SiC whiskers. Figure 1 is a TEM micrograph showing the microstructure typically observed in the as-received state. As can be seen, the material is virtually fully dense, containing a negligible amount of pores or cavities, although there is a sporadic distribution of glassy phase present at triple point junctions and interfaces. Tensile creep testing was performed on these two materials in the range of 50-100 MPa at 1250°C in air. Detailed test procedures are described elsewhere [2-3]. Data on both creep strain versus exposure time for a fixed stress and time-to-rupture versus applied stress were collected. Figure 2 presents a limited amount of data on the second set. These data were reported on a log-log scale. Thus, it is seen that creep rupture lifetime is extremely sensitive to the applied stress, regardless of whether or not the material is reinforced. Furthermore, addition of fiber or whisker reinforcement to the matrix ceramic drastically shortens the life of the material [1]. The implications of these two facts will be fully addressed in subsequent sections.

The microstructural changes induced by tensile creep were evaluated by TEM performed on the post-crept specimens. Sections were cut from the gauge section and examined under TEM. The results were then compared with sections taken from the unstressed part of the same specimen as well as from the as-received material (see Fig.1). In this way, the effects of tensile creep on the microstructure could be determined.

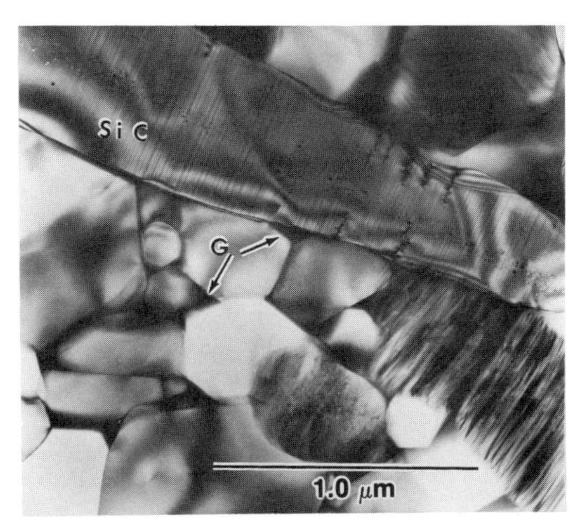

Figure 1. TEM micrograph showing the as-received state of the silicon nitride reinforced with 30 vol% SiC whiskers. (After Hockey, *et al.* [1])

[1]AY6 is a designation for a grade of Si_3N_4 made by GTE. Its use should not be construed as endorsement by the National Institute of Standards and Technology.

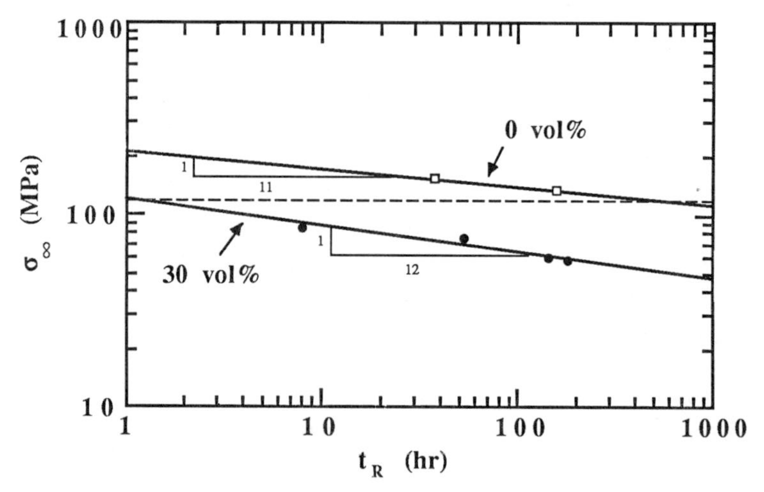

Figure 2. Time to rupture versus applied stress curves for 0 vol% and 30 vol% SiC reinforcement of silicon nitride at 1250°C.

A typical microstructure that was produced as a result of tensile creep rupture is shown in Figure 3 [1]. Figure 3a is a high magnification TEM micrograph showing the damage pattern on a local scale. Generally speaking, creep damage appears in a form of interfacial crack-like cavities. Both types of interface (i.e. matrix/matrix and fiber/matrix) are seen to be susceptible to creep cavitation. However, damage seems to be concentrated on those interfaces whose orientations are roughly normal to the principal tensile stress axis (Fig.3a). As a result, randomly distributed cracks parallel to

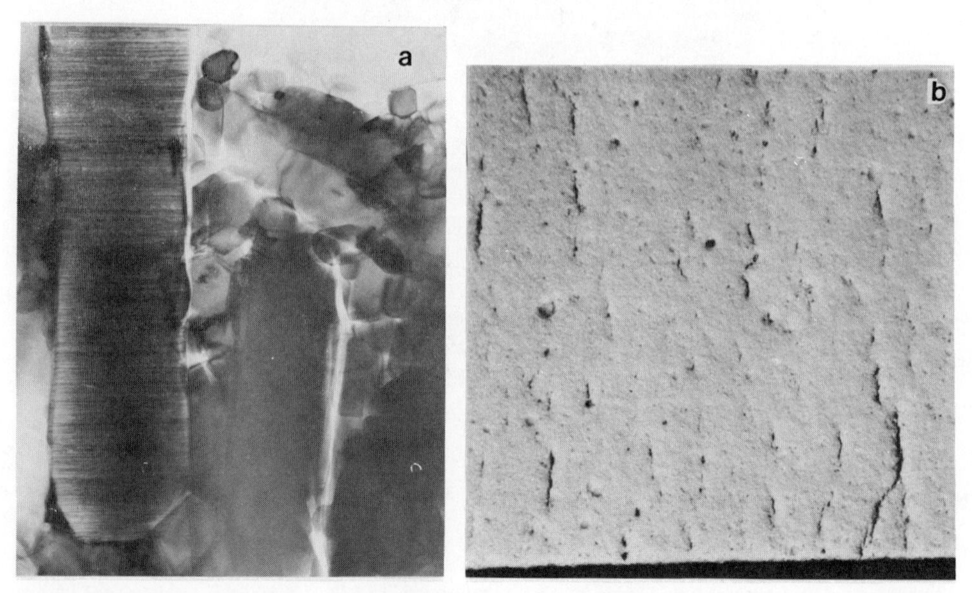

Figure 3. Micrographs showing microcracks at matrix/matrix and matrix/whisker interfaces. (a) high magnification taken in gauge section; (b) optical micrograph taken of surface.

each other on a global scaleare observed. Figure 3b is an optical micrograph obtained from the surface of a post-crept tensile specimen. From these pictures it is clear that interfacial crack growth is an important part of the creep rupture process and their linkage leads to final creep fracture. In the following sections, theoretical predictions of the crack growth rate in single phase and bimaterial systems will be formulated, followed by development of a generic crack growth model for lifetime estimation.

THEORY OF DIFFUSIONAL CRACK GROWTH

At elevated temperatures in excess of about one third of the homologous temperature, mass transport becomes activated along high diffusivity paths such as interfaces and internal free surfaces. Because it takes less energy to form a void at an interface than in the bulk, creep cavities are predominantly observed at interfaces, rather than inside grains. Moreover, of all boundaries, cavitation seems to favor those normal to the principal tensile stress axis. Those observations suggest that stress driven diffusion around the cavity periphery plays an important rôle in the cavity growth process. One possible and convincing mechanism leading to cavity growth involves a coupled process of transportation of species along cavity surfaces towards the tip via surface diffusion, and from there atoms are further driven away along the interface via grain-boundary diffusion. In this manner, the cavity tip is allowed to advance in a steady-state fashion. A direct proof showing this mechanism is indeed operative was provided by Varma and Dyson[4] for a nickel-base alloy. Based on the specific kinetics, many diffusional cavity growth models have been proposed [5-17] of which the diffusional crack growth model is particularly relevant to the present case. Two separate systems will be discussed, namely, crack growth in a single phase material, and in a bimaterial. The former aims at modeling matrix/matrix interfaces, whereas the latter aims at fiber/matrix interfaces, taking the dissimilarity into account.

Single Phase System

Let us consider a crack-like boundary cavity growing between two adjoining (homogeneous, isotropic and elastic) grains subjected to a remote applied tensile stress normal to the crack plane. Figure 4 is a TEM micrograph showing the morphology of such a crack observed in a polycrystalline alumina beam subjected to a flexure stress of 20 MPa at 1720°C [18]. The major goal is to solve the crack tip velocity as a function of applied stress and temperature. The problem can be divided into two parts. The first part involves a determination of the crack morphology developed by surface self-diffusion, which thereby provides the required boundary conditions at the moving crack tip in terms of the unknown crack velocity. The second part contains a stress analysis which relates the interfacial normal stresses at the far-field to that presented at the crack tip. The solution enables the unknown crack growth rate to be expressed in terms of the given applied stress or applied stress intensity factor which will be shown to be appropriate for the mechanics of diffusive crack growth.

For a single phase material, the crack shape is symmetric with respect to the tip, thus only the upper surface needs be considered. The near-tip shape on a nondimensional physical plane can be solved from a nonlinear second order ordinary differential equation derived from Fick's laws and steady-state conditions. It was shown

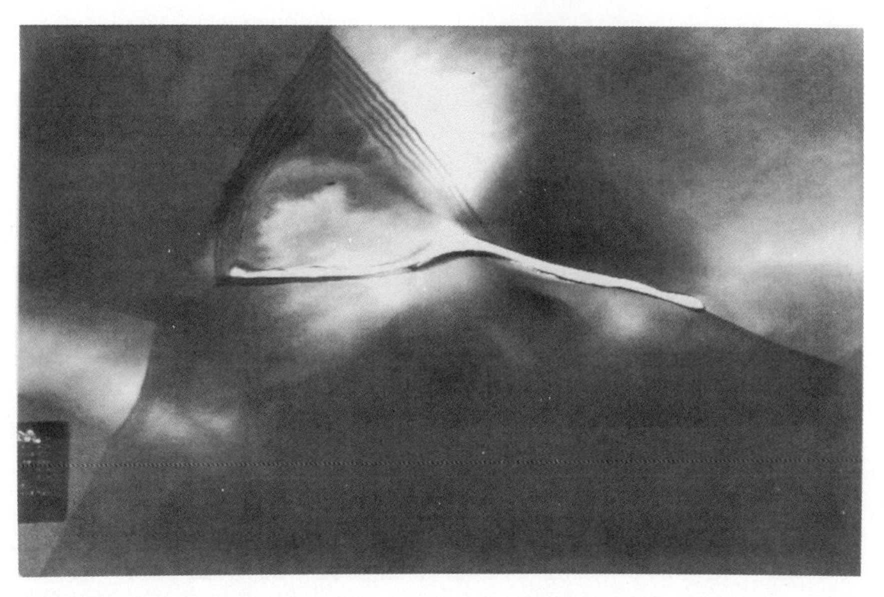

Figure 4. Diffusional crack growth in alumina [18]. The crack is ≈ 125 nm thick.

that a unique profile exists regardless of the applied stress if the lengths are scaled with the parameter ℓ [12]:

$$l = \left(\frac{D_s \gamma_s \Omega^{4/3}}{VkT}\right)^{\frac{1}{3}} \tag{1}$$

where D_s is the surface diffusivity, Ω the atomic volume, γ_s the surface free energy, V the crack-tip velocity, and kT has its usual meaning. Further, the near-tip profile can be described in closed form by a logarithmic function to within ±3%. Plotting of this solution indicates a constant crack thickness develops in agreement with what is observed in Fig.4. It is further predicted that a higher crack velocity (hence, a higher applied load) will lead to a thinner crack opening. The crack thickness is inversely proportional to $V^{1/3}$ according to Eqn.(1). With the near-tip shape solved, the surface curvature adjacent to the tip and the matter flux at the tip can be quantified using continuity in chemical potentials and conservation of mass, which, in turn, define the boundary conditions at the crack tip in stress in terms of the unknown V: $\sigma(x=0) \propto V^{1/3}$; $\sigma'(x=0) \propto V^{2/3}$ where x=0 refers to the crack tip and the prime designates the first derivative with respect to x.

In order to acquire a final solution relating crack velocity to applied stress intensity K, a precise analysis of the tensile stress distribution at the interface ahead of the moving crack tip must be performed. In the absence of diffusion, the applied K induces the well-known elastic stress field with a characteristic $x^{-1/2}$ type singularity. However, as matter diffuses from the crack surfaces and deposits along the interface, the so called "wedging" effect is produced which alleviates the stress concentration at the crack tip. Chuang [14] has formulated this problem using the concept of infinitesimal edge dislocations to evaluate the residual stresses induced by mass transport along the interface. The result is an integral equation for the unknown $\sigma(x)$, where x=X/L is a nondimensional boundary coordinate scaled by L :

$$L = \sqrt{\frac{\pi}{4} \cdot \frac{ED_b\delta_b\Omega}{(1-\nu^2)VkT}} \tag{2}$$

where E and ν are Young's modulus and Poisson's ratio respectively and $D_b\delta_b$ is the grain-boundary diffusivity. The magnitude of L is typically in the order of μm [18].

A complete solution for the integral equation was obtained numerically [14]. It was found that the stress distribution is dependent on a parameter α which, in turn, is a function of V. Typical stresses for α=0,1,5,10 and 20 respectively are plotted in Fig.5. For the sake of comparison, the elastic stresses for the case of α=20 are given by a dashed line. It is seen from this plot that the stress singularity at the crack tip has been eliminated by diffusion and the peak stresses now occur at around X $\approx$ 0.9 L instead. The sizes of the diffusion zone are estimated to be approximately 4L, beyond which the influence of diffusion becomes insignificant. The undisturbed far-field stresses are solely controlled by the applied K, which can then be related to $\sigma(0)$ and $\sigma'(0)$ at the crack tip by the numerical solution. This relationship can be cast in the following form:

$$K = 0.75\ \sigma(0)\ L^{1/2} + 0.60\ \sigma'(0)\ L^{3/2} \tag{3}$$

This is a significant result as this equation appears to have a widespread applicability, not only to the current model of diffusive crack growth in a single phase elastic system, but also to a future model of bimaterials system as will be discussed later. In addition, Cao *et al.* [19] applied this equation to a problem of high temperature stress corrosion cracking of ceramics in which corrosive liquid product filled up the crack-like cavity and is a major species in the mass transport process.

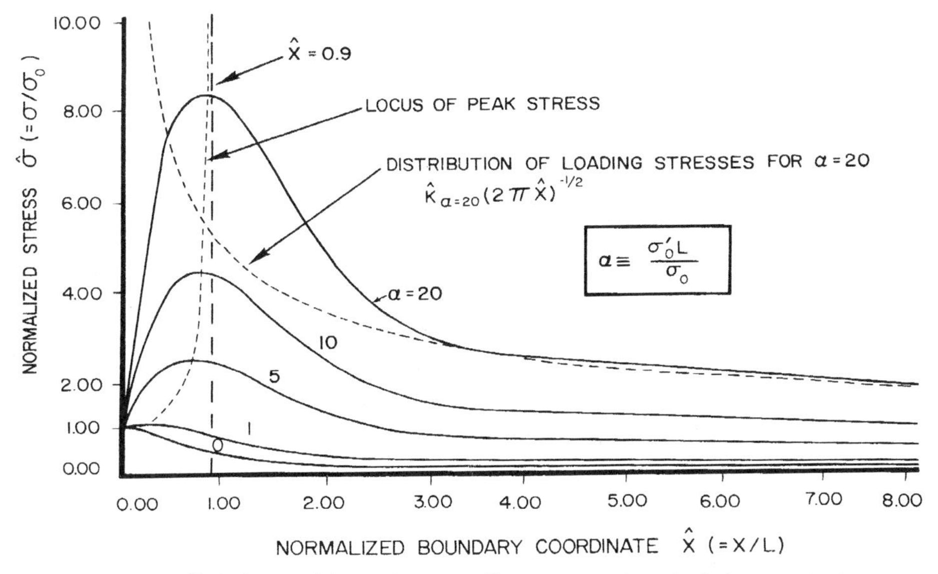

Figure 5. Solutions of boundary tensile stresses ahead of the crack tip.

The final goal of expressing K in terms of V or vice versa can be achieved by substituting the stress conditions at the crack tip (namely, $\sigma(0) \propto V^{1/3}$; $\sigma'(0) \propto V^{2/3}$) in Eqn.(3). It can be shown that if the physically inadmissible branch is discarded, a one-to-one relationship between the crack-tip velocity, V and the applied stress intensity, K_I in mode I can be established and it takes the form: $k=0.845\ (v^{-1/12} + v^{1/12})$ where k and v are nondimensional K_I and V normalizing by K_G and V_{min} where K_G[2] is the critical K-value based on the Griffith theory and V_{min} is a materials constant depending on temperature, diffusivity and elastic constants. Solving v in terms of k, one finally arrives at the following expression:

$$v = (0.59\,k + \sqrt{0.35k^2-1}\,)^{12} \tag{4}$$

It is seen that for a creep crack growing in a single phase ***elastic*** material by diffusion, a unique v-k relationship emerges, irrespective of materials species or temperature. However, as indicated in Eqn.(4), the v-k relationship, in general, can not be cast in a simple form of a power-law equation which is conventionally adopted as an empirical expression. Fig.6 plots the equation in a double log space, and a few remarks are in order. First, there exists a threshold stress intensity, $K_{th} = 1.69\ K_G$, below which the applied stress is not sufficient to drive the crack, and as a result, the crack ceases to propagate and sintering may actually occur. Secondly, at higher K values, say $K>4K_G$, a power-law equation, $V=(\text{constant})\cdot K^{12}$ becomes an asymptote and is a fair representation of Eqn.(4). Finally, for intermediate K-levels in the range $4K_G > K > 1.69K_G$, the theory predicts that the stress exponents vary from 12 to infinity.

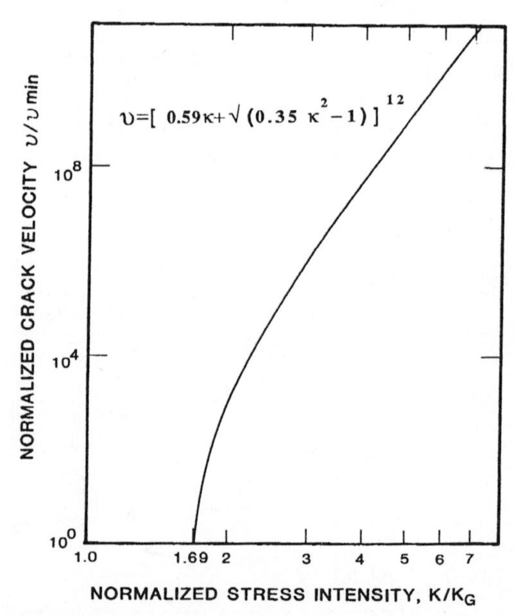

Figure 6. V-K_I relationship predicted by the diffusive crack-growth theory.

---

[2]K_G is related to true surface energy and should not be confused with K_{IC} which is, in general, about one order of magnitude higher.

Bimaterial System

In the case of a crack growing along a fiber/matrix interface under creep conditions, what the effects, if any, of dissimilarity between the fiber and the matrix in terms of differing physical and mechanical properties are on the crack growth behavior becomes an interesting subject and deserves a thorough investigation. As the geometry dictates (see Fig.3a), the crack appears to be much thinner than the grain size or fiber diameter. Plane-strain conditions should prevail and this case can be modeled as a two dimensional two-phase solid containing an interfacial crack, phase 1, representing the fiber and phase 2 the matrix. The problem now is to solve the crack growth rate for a given stress and temperature, and again this can be divided into two parts.

For the first part of the problem, Chuang *et al.* [20] have recently investigated the crack tip morphology that could be developed by surface diffusion controlled crack growth along a bimaterial interface. In contrast to the symmetric case developed in single phase systems, four asymmetric cases are possible for bimaterial systems depending on the degree of dissimilarity in surface free energy of the two adjoining phases. Excluding the physically inadmissible case which is not observable, Figure 7 presents schematic sketches for the three possible cases of crack tip morphology. As can be seen, the upper and lower cavity surfaces are no longer symmetrical with respect to the tip owing to distinctive properties of the dissimilar phases. A tip morphology map in the space of γ_1 versus γ_2 can be constructed to demonstrate prospective areas where each case applies. The near tip shape again can be uniquely described by a logarithmic function as in the case of single phase materials. However, the maximum half-thickness is now 2.0 ℓ instead of 1.41 ℓ. Those results yield a more complex expression for the matter flux and the root radii (or the surface curvatures) at the crack tip. These expressions involve physical properties of the two dissimilar phases. However, the velocity dependence remains the same, namely, $\kappa_{tip} \sim V^{1/3}$; $J_{tip} \sim V^{2/3}$. Accordingly, the boundary conditions at the moving crack tip have the following relationships: $\sigma(0) \sim V^{1/3}$ and $\sigma'(o) \sim V^{2/3}$. Of course, the proportionality constants will have lengthy expressions in terms of materials constants and temperature.

To formulate the desired K-V relationship, a stress analysis at the interface ahead of the moving crack tip must be carried out. The elastic stress field, in the

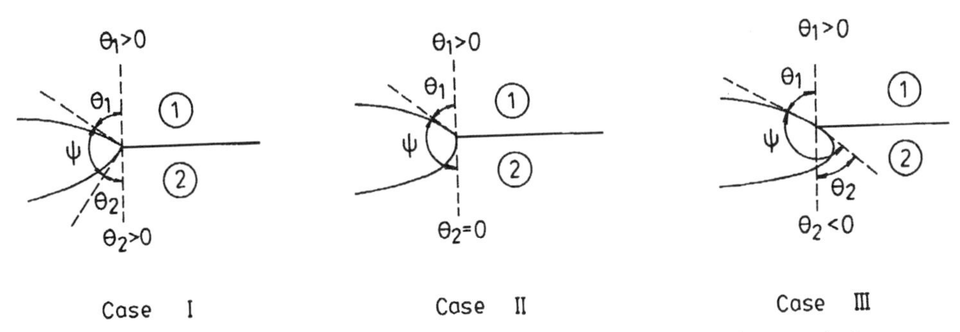

Figure 7. Schematic of the three asymmetric cases of crack tip morphology.

absence of creep, in a bimaterial containing an interfacial crack has been investigated by many authors [21-25]. The results show many characteristics distinct from the single phase case: (1) Uniaxial stressing results in a mix-mode response. Thus, a far-field applied tensile stress will produce shear stresses at the interface. (2) The normal stress distribution at the interface ahead of the crack tip induced by K_I has the following form: $\sigma(x)=K_I \cos(\varepsilon \cdot \ell n\ x)/\sqrt{2\pi x}$, where ε is a dissimilarity index defined by

$$\varepsilon = \frac{1}{2\pi} \ln\left(\frac{1-\beta}{1+\beta}\right) \tag{5}$$

Here β is a Dundur's parameter defined by [24]

$$\beta = \frac{\mu_1(\kappa_2-1)-\mu_2(\kappa_1-1)}{\mu_1(\kappa_2+1)+\mu_2(\kappa_1+1)} \tag{6}$$

where μ is shear modulus; $\kappa=3\text{-}4\nu$ for plane-strain conditions and the subscripts refer to phases. Plotting this stress distribution function shows the curve being oscillatory decaying with increasing distances from the crack tip. (3) The elastic crack opening displacements show an unrealistic interpenetration behind the crack tip, although the overlapping zone size is of atomic dimensions; and finally (4) the relationship between K_I and σ_∞ will involve ε in addition to the crack size a:

$$K_I = \sigma_\infty\sqrt{\pi a}\,[\cos(\varepsilon \ln 2a) + 2\varepsilon \sin(\varepsilon \ln 2a)] \tag{7}$$

The above elastic behaviors are indeed a fair representation of the short term materials responses. However, under long term creep conditions where mass transport processes are active, it can be argued from physical principles that surface diffusion will prevent interpenetration and grain-boundary diffusion will relax the shear stresses produced at the interface resulting in grain-boundary sliding. This latter phenomenon is widely observed during high temperature creep especially for ceramic systems where a liquid phase often exists between the grains. Based on the assumption that the interface ahead of the crack tip cannot resist shear, it was found after including the residual stresses induced by diffusion, that the tensile creep stresses no longer have an oscillatory character and have the same solutions as indicated in Fig.5, **if** a new length parameter is used as a scaling parameter [26]:

$$L = \sqrt{\frac{\pi^2\lambda(1-\beta^2)\langle D_b\Omega\rangle\delta_b}{VkT}} \tag{8}$$

where $\langle D_b\Omega\rangle = \frac{1}{2}(D_{b1}\Omega_1 + D_{b2}\Omega_2)$ is the average of the two-phase properties and λ is defined by

$$\lambda = \frac{\mu_1\mu_2[\mu_1(1+\kappa_2)+\mu_2(1+\kappa_1)]}{\pi(\mu_1+\mu_2\kappa_1)(\mu_2+\mu_1\kappa_2)} \tag{9}$$

As a result, the V-K_I law for bimaterial systems has the same form as the single phase systems depicted by Eqn.(4), except that the normalizing parameters V_{min} and K_G will have different expressions involving the materials constants associated with the two-phase properties [26].

It should be noted that when the properties of phase 1 are identical to that of phase 2, then $\beta=0$ according to Eqn.(6) and from Eqn.(9) $\lambda=2\mu/\pi(1+\kappa)=E/4\pi(1-\nu^2)$, thus L in Eqn.(8) reduces to L of the single phase case expressed in Eqn.(2).

LIFETIME ESTIMATION

In general, creep rupture lifetime can be expressed in the following functional form: $t_R = f(\Lambda, \ell, A_s, V_f, \upsilon_f, \upsilon_m)$ where t_R is time to rupture; Λ is a geometric factor depending on grain morphology and fiber architecture; ℓ is average fiber length; A_s is the aspect ratio of the fiber; V_f is the volume fraction of the fiber; υ_f is the crack velocity at the fiber/matrix interface and υ_m at the matrix/matrix interface. The first four parameters are geometry-dependent. So for a given composite configuration, these parameters are fixed. The stress and temperature dependences reside in the last two which are related to the thermal-mechanical forces driving the cracks to rupture. If an explicit function for t_R can be found, then lifetime can be estimated via the V-K_I relationship discussed in the previous section. Hence there is a need of constructing a creep rupture model for assessing lifetime.

The structure of fibrous composites can be classified into two classes: (1) continuous and (2) discontinuous or chopped fiber reinforcement. The continuous fiber reinforced composites are formed by 1D,2D or 3D laminates or hybrids depending on the degree of complexity in design. Chopped fibers, on the other hand, are usually randomly dispensed throughout the discontinuous fiber composites owing to the ease of processing. It is clear that a realistic model will demand a sophisticated statistical approach or a computer simulation method to handle the intrinsic complexity in microstructure.

For the sake of demonstration, we select the simplest case of a ceramic composite reinforced with unidirectional chopped fibers for modeling purposes. This generic model is sketched in Figure 8 in an attempt to simulate the real systems to the first order approximation. The global composite is envisioned to be composed of periodic cells. Each cell as sketched in Fig.8 contains rectangular grains and parallel fibers. The whole composite is subjected to a remote tensile stress σ_∞. Failure is supposed to occur when two neighboring microcracks grow to coalesce with one another. These model configurations and assumptions are in line with the experimental observations (see Figs.1 and 3). Neglecting the incubation time for crack formation and the transient creep, the time to rupture can be computed from fiber spacing divided by crack growth rate; the former quantity is derivable from the geometric factors such as

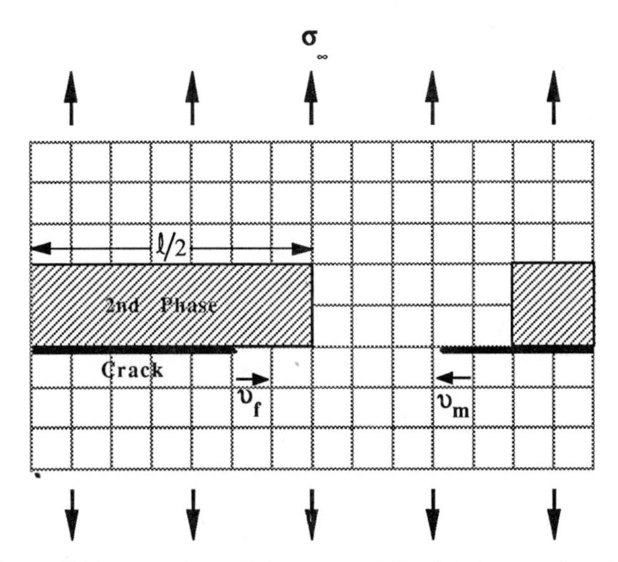

Figure 8. A generic model proposed for lifetime estimation.

volume fraction, aspect ratio, etc. Working out the details of the expression for the fiber spacing, it can be shown that t_R takes the following functional form:

$$t_R = (\upsilon_f^{-1} - \upsilon_m^{-1})\, l + \Lambda l A_s^{-1/3} V_f^{-1/3} \upsilon_m^{-1} \qquad \textbf{(10)}$$

Although this equation is only strictly valid for this simple model, it does contain the key parameters affecting the creep rupture life. It is seen, for example, that higher crack velocities lead to shorter lives as expected. Increasing the volume fraction of the reinforcement, on the other hand, will not prolong the life. The implication of this surprising finding will be discussed in terms of long-term versus short-term behaviors.

The stress dependence of the lifetime is implicit in Eqn.(10). In principle, the crack velocity obtained in the single phase theory can be applied to υ_m at the matrix/matrix interfaces and the predicted velocity in the bimaterial systems, Eqns.(7-9), can be adopted for υ_f at the fiber/matrix interface. Substituting these expressions in Eqn.(10), one could obtain a specific expression for the stress dependence on the service life. We will come back to address this point when we discuss the comparison between theory and experiment in the following section.

DISCUSSION

We have presented experimental observations of the microstructural development resulting from exposing the material to long-term creep rupture conditions. A theory of diffusional crack growth in both single phase and two-phase materials was presented, followed by a generic creep rupture model. This model, when combined with the crack growth theory, is capable of estimating the service life of any generic composite under given thermal-mechanical conditions if the physical and mechanical properties are well defined. There are two major issues which are interesting and they deserve further elaboration. These issues address how well do the theory and experiment correlate with one another as well as the contrast between long-term and short-term mechanical behaviors. They are discussed separately below.

Comparison between Theory and Experiment

The theoretical prediction of lifetime, Eqn.(10) involves a series of geometric factors, materials constants and crack growth rates. After substituting the crack velocities, a specific expression for lifetime in terms of applied K or applied stress and temperature can be obtained. Nevertheless, we do not write down the full expression for t_R here, because data on many materials constants are not available as of now. Thus, a direct comparison of the theory and the rupture data can not be made at this time.

However, it is still possible to make comparison of the stress dependence or the stress exponent, if a few reasonable assumptions are made. With the aim of comparing theoretical predictions with creep rupture data reported in Fig.2, we observe that those data probably lie in the higher stress range. Accordingly, it is assumed $K_I > 4K_G$ and Eqn.(4) then simplifies to $v = 7.53\ k^{12}$. Furthermore, Eqn.(7) relating K_I to the applied stress can be cast in the form: $K_I = Y(a)\ \sigma_\infty$. A combination of these two expressions results in $V = (\text{constant}) \cdot \sigma_\infty^{12}$, irrespective of single phase or bimaterial systems, although the pre-exponential constant is system-dependent. Because of a lack of fiber data at elevated temperatures, it is not possible to predict whether $\upsilon_m > \upsilon_f$ or vice versa. So to a first order of approximation, it is further assumed that the crack velocities are approximately equal. Then the lifetime in Eqn.(10) takes the following simple form:

$$t_R = \frac{\Lambda l}{A_s^{1/3} V_f^{1/3} \upsilon} = \frac{\Lambda' l}{A_s^{1/3} V_f^{1/3} \sigma_\infty^{12}} \tag{11}$$

It is predicted that rupture life can be expressed as a power law function of stress with a stress exponent of -12. Examination of the limited rupture data plotted on a log-log scale (see Fig.2) indeed shows a slope of -12 for the 30 vol% composite and -11 for the 0 vol% composite in close accord with the predicted value. Therefore, it can be said that a good agreement of the stress dependence on the lifetime between theory and experiment has been reached.

Equation (11) also predicts rupture lifetime to be proportional to the volume fraction to a power of -1/3. To make a rough comparison with rupture data reported in Fig.2, we take advantage of the following theoretical relationship

$$t_{R1} / t_{R2} = (V_{f2} / V_{f1})^{1/3} \tag{12}$$

Let t_{R1} represent rupture time for the 0 vol% curve and t_{R2} designate rupture time for the 30 vol% curve in Fig.2. At a stress of 120 MPa, data in Fig.2 show that $t_{R1} = 600$ hrs, whereas $t_{R2} = 1$ hr. Eqn.(12) predicts that the vol% of the curve 1 should be 1.4×10^{-7}% which is virtually 0 vol% in agreement with the volume fraction of the curve 1. Although this is only a rough estimation, it seems that the limited amount of data do support qualitatively the theoretical predictions of the volume fraction dependence.

Long-term Vs. Short-term Behaviors

The materials reponses in the long term often differ in character with their short-term counterparts. The contrasts are sometimes overlooked and this point should be borne in mind. Here we only emphasize a few physico-mechanical phenomena for the sake of discussion: (1) fracture mode; (2) effect of volume fraction; (3) strength versus rupture time; and finally (4) crack opening displacement.

When a brittle ceramic specimen is loaded, either quasi-statically or rapidly, during a short-term mechanical test for the purposes of obtaining data on strength or deformation characteristics, the specimen's response is largely elastic at room temperature. At elevated temperatures, time-independent plasticity might play some rôle. When the specimen fractures, it can be attributed to a critical crack size at a stress intensity $K_I = K_{IC}$. If the material is a fiber reinforced ceramic, the response remains elastic, except now the fracture mode is more complicated: the running crack, in order to advance, must either break the fibers (if they are relatively weak) or pass around them (if the fibers are stronger than the matrices as are the usual cases), thereby causing fiber pull-out as crack opening displacement increases. As a result, cracks are retarded and more energy is required to advance the crack, leading to an enhancement of strength and toughness. Based on this principle, the idea of adding the fibers as a way of reinforcing the matrix prevails in the ceramic community. In general, increasing fiber loading enhances the strength and toughness of the composite (see, for example, [27-29]). At high temperatures, fast fracture at higher loads is dominated by subcritical "creep" crack growth, but the mode of failure remains similar, i.e. via fiber pull-out as crack advances. Nair *et al.* recently modeled this problem by taking viscous sliding into account during fiber pull-out [30]. These events are pseudo-creep in appearance; but the whole process generally ends in a short period.

A completely different picture emerges when the composite is subjected to low, sustained stresses at elevated temperatures. It takes a long time to complete the testing cycle. During the long-term test, time-dependent processes, notably mass diffusion, are activated. The material's reponses are dominated by irreversible creep; elastic behavior, in general, becomes insignificant. The rupture mode is the linkage of interfacial microcracks as indicated in Fig.3. Rupture mechanisms such as fiber pull-out are seldom observed in the ***long-term*** creep rupture process.

The sharp contrast in rupture mode between long term and short term behavior has many implications. The first impact is the effect of fiber loading. It is now well established that increasing the volume fraction of fibers generally will enhance the strength and toughness of the composite. On the other hand, adding more fibers in the matrix may not necessarily improve the material's long-term performance. Creep rupture data in Fig. 2 demonstrate that, at 120 MPa, 1250°C, adding 30 vol% of reinforcement, while enhancing the strength, will shorten the lifetime from 600 hrs to one hour. It becomes a challenge for the process engineer to strike a balance between short-term gain and long-term loss.

Another noteworthy behavior is the crack opening displacement. In the short-term, the crack behaves elastically. From fracture mechanics principles, it is well-known that higher applied stresses lead to larger crack opening displacements, although the shape remains parabolic. On the other hand, in the long-term creep crack growth such as the case shown in Fig.4, the permanent crack opening remains constant far behind the crack tip (i.e. has a constant thickness instead of a parabolic shape). Furthermore, higher applied stresses lead to thinner cracks, as implied by Eqn.(1), in contrast to expectations from elastic crack behavior.

CONCLUSION

A generic model for estimating creep rupture lifetime has been proposed. The model entails three important ingredients: (1) the rupture mode entails coalescence of adjacent microcracks along interfaces consistent with microstructural observations; (2) for the purpose of assessing the stress and temperature dependences, a rigorous theory based on diffusive crack growth is presented in a single phase system as well as in a bimaterial system;and (3) simple grain and fiber morphologies are assumed for the investigation of the effect of fiber loading. The model predicts, in certain appropriate circumstances, rupture life to be inversely proportional to the applied stress to a power of 12; and also to be inversely proportional to the fiber volume fraction to a power of one third. Limited creep rupture data on SiC_w reinforced silicon nitride seem to support the theoretical predictions. Hopefully, as more and more high temperature creep rupture data on fibrous ceramic composites become available, this proposed model can be subjected to more stringent verifications to determine its validity and the scope of applicability.

ACKNOWLEDGEMENT

It is a pleasure to thank NASA Lewis Research Center, High Temperature Materials Program, Structural Integrity Branch, Dr. John P. Gyekenyesi, for support under Interagency Order No. C-82000-R. The author is grateful to B. J. Hockey for providing photomicrographs in Figs.1 and 3; and to G. Quinn and S. M. Wiederhorn for critically reviewing the manuscript.

REFERENCES

1. B. J. Hockey, S. M. Wiederhorn, W. Liu, J. G. Baldoni and S.-T. Buljan, "Tensile Creep of Whisker-Reinforced Silicon Nitride," **J. Mater. Sci.**, in press.
2. D. F. Carroll, S. M. Wiederhorn and D. E. Roberts, " Technique for Tensile Creep Testing of Ceramics," **J. Am. Ceram. Soc. 72** [9] 1610-14 (1989).
3. S. M. Wiederhorn, D. E. Roberts, T.-J. Chuang and L. Chuck," Damage-Enhanced Creep in a Siliconized Silicon Carbide: Phenomenology," **J. Am. Ceram. Soc. 71** [5] 602-608 (1988).
4. R.K. Varma and B.F. Dyson, " Metallographic Detection of Atom-plating due to Cavity Growth," **Scripta Metall. 16** [11] 1279-1284 (1982).

5. D. Hull and D.E. Rimmer,"The Growth of Grain-Boundary Voids under Stress," **Philos. Mag. 4** [42] 673-87 (1959).
6. M.V. Speight and J.E. Harris,"Kinetics of Stress-Induced Growth of Grain-Boundary Voids," **Metal Sci. J. 1** [1] 83-85 (1967).
7. R. Raj and M.F. Ashby, "Intergranular Fracture at Elevated Temperature," **Acta Metall. 23** [6] 653-666 (1975).
8. J.P. Hirth, "Nucleation of Void Sheets in Creep and Hydrogen Attack," **Res. Mechanica Letters 1** [1] 3-5 (1981).
9. R. Fuentes-Samaniego and W.D. Nix, "Steady-state diffusional growth of an axisymmetric cavity on a grain boundary," **Philos. Mag. 44** [3] 601-612 (1981).
10. W.D. Nix, "Introduction to the Viewpoint Set on Creep Cavitation," **Scripta Metall. 17** [1] 1-43 (1983).
11. R.J. Charles, "Diffusion-Controlled Stress Rupture of Polycrystalline Materials," **Metall. Trans. A 7** [8] 1081-89 (1976).
12. T.-J. Chuang and J.R. Rice, "The Shape of Intergranular Creep Cracks Growing by Surface Diffusion," **Acta Metall. 21** [12] 1625-28 (1973).
13. T.-J. Chuang, K.I. Kagawa, J.R. Rice and B. Sills, "Overview No.2: Non-equilibrium Models for Diffusive Cavitation of Grain Interfaces," *ibid.* **27** [2] 265-284 (1979).
14. T.-J. Chuang, "A Diffusive Crack Growth Model for Creep Fracture," **J. Am. Ceram. Soc. 65** [2] 93-103 (1982).
15. L.-E. Svensson and G.L. Dunlop, "Growth of Intergranular Creep Cavities," **Int. Metals Reviews** [2] 109-131 (1981).
16. V. Vitek, "A Theory of Diffusion Controlled Intergranular Crack Growth," **Acta Metall. 26** [9] 1345-56 (1978).
17. D. S. Wilkinson and V. Vitek, "The Propagation of Cracks by Cavitation: a General Theory," *ibid,* **30** [12] 1723 (1982).
18. T.-J. Chuang and N.J. Tighe, "Diffusional Crack Growth in Alumina," pp. 129-132, **Proceed. 3rd Int. Conf. Fundamental Fracture,** Irsee, Germany, June, 1989.
19. H.C. Cao, B.J. Dalgleish, C.-H. Hsueh and A.G. Evans, "High-Temperature Stress Corrosion Cracking in Ceramics," **J. Am. Ceram. Soc. 70** 257-264 (1987).
20. T.-J. Chuang, S.-T. Shiue and S. Lee, "Asymmetric Tip Morphology of Creep Microcracks Growing along Bimaterials Interfaces," **Acta Metall. Mater.** submitted.
21. M. L. Williams, "The Stresses Around a Fault or Crack in Dissimilar Media," **Bull. Seismol. Soc. America, 49**, 199-204 (1959).
22. A. H. England, "A Crack between Dissimilar Media," **J. Appl. Mech. 32** [6] 400-402 (1965).
23. J. R. Rice, "Elastic Fracture Mechanics Concepts for Interfacial Cracks," ***ibid,* 55** [3] 98-103 (1988).
24. J. Dundurs, "Elastic Interaction of Dislocations with Inhomogeneities," in *Mathematical Theory of Dislocations,* pp.70-115, ASME, New York (1968).
25. Z. Suo, "Singularities Interacting with Interfaces and Cracks," **Int. J. Solids Structures, 25,** [10] 1133-42 (1989).
26. T.-J. Chuang, J.-L. Chu and S. Lee, "Diffusive Crack Growth at a Bimaterials Interface," to be published.
27. B. Budiansky, J. W. Hutchinson and A. G. Evans, "Matrix Fracture in Fiber Reinforced Ceramics," **J. Mech. Phys. Solids, 34** [2] 167 (1986).

28. A. G. Evans and R. M. McMeeking, "On the Toughening of Ceramics by Strong Reinforcements," **Acta Metall. 34** [12] 2435 (1986).
29. D. B. Marshall and B. N. Cox, "Tensile Fracture of Brittle Matrix Composites: Influence of Fiber Strength," ***ibid 35*** [11] 2607 (1987).
30. S. V. Nair, K. Jakus and T. Lardner, "The Mechanics of Matrix Cracking in Fiber Reinforced Ceramic Composites Containing a Viscous Interface," **Mech. of Mater.**, in press.

SOME ASPECTS OF THE MORPHOLOGY AND CREEP BEHAVIOR OF A UNIDIRECTIONAL SiC_f-MLAS MATERIAL

Dominique Kervadec and **Jean-Louis Chermant**

Laboratoire d'Etudes et de Recherches sur les Matériaux, LERMAT, URA CNRS n°1317, ISMRa, 6 Bd du Maréchal Juin, 14050 CAEN Cedex, France

INTRODUCTION

Ceramic matrix composites (CMC) are today used in aeronautics and aerospace structures : for example, high thermal protections for Columbia or Hermès shuttles, parts for supersonic ramjets or combustion chambers, nozzles for liquid propulsion, turbojet parts, thrust chambers or apogee thrusters [1,2,3] . Information is now available on their mechanical behaviour and toughness (for example[4,5,6,7,8], the remaining problem being to know exactly what type of rupture parameters - G, J, K, R - the design department must utilize to calculate the structures in CMC! It requires also information on life time predictions in different environment. These works emphasize mainly on the role of the fiber/matrix interfaces, their change versus temperature and the poor chemical stability at high temperature of the Nicalon[R] SiC fibers as confirmed by Wiederhorn and Hockey[9] or Le Costumer et al.[10]. That is the reason why research is also going on new types of ceramic fibers, such as Fiberamic[R] SiCON (Rhône-Poulenc, Aubervilliers, France)[11].

To date only some investigations have been performed on the creep behaviour of ceramic matrix reinforced with long ceramic fibers : Brennan and Prewo[12], Abbé[13,14,15,16] and Holmes[17]. These references are concerning respectively with SiC_f-LAS, SiC_f-SiC and SiC_f-Si_3N_4 materials.

The present work concerns with first results on bending creep behaviour of a unidirectional SiC_f-MLAS composite under vacuum, in the temperature range 1073-1473K. These composites with a vitroceramic matrix are developed for uses at lower temperatures than composites with a SiC matrix[18].

MATERIALS AND TEST PROCEDURES

Ceramic material composites investigated in this work were unidirectional SiC_f-MLAS, developped by Aérospatiale Company (Saint-Médard en Jalles, France). Fibers used were

Nicalon NLM 202 (Nippon Carbon, Japan). The stackings of SiC fibers were infiltrated by a slurry, followed by an uniaxial pressing and a heating at 1273-1373K. The slurry was made of a mixture of aluminum, lithium, magnesium and silicium oxides : $MgO-Li_2O-Al_2O_3-SiO_2$[19]. The apparent density of such materials, with 34% of fibers (in volume), is 2.45, with 1% of open porosity. Figure 1 presents two optical micrographs of a SiC_f-MLAS. Compared to the chemical vapor infiltration process, used for example for C_f-SiC or SiC_f-SiC materials[20], slurry infiltration process leads to materials with a very small porosity. An other advantage of this last method is to produce CMC with a short process route (some hours).

Morphological and mechanical tests were performed on polished specimens, in using classical technics : grinding and diamond pastes (until 0.25 μm).

Morphological parameters were obtained using a Leitz Texture Analyzer System (TAS, Leitz, Wetzlar, Germany). Scanning electron micrographs were performed on a Jeol T 330 (Jeol, Tokyo, Japan), transmission electron micrographs on a Jeol 200CX (Jeol, Tokyo, Japan) on argon ion milled specimens (Brot, Paris, France) and microanalysis on a SEM Jeol GLS 840 with an EDS system from Tracor (Middeton, USA).

Creep experiments were conducted in three point bending under vacuum (~ 10^{-4} torr) in a Sesame opening furnace (VMDI, Paris, France) with an upper temperature limit of 2273K. The push rods are made of tungsten and are 20 mm in diameter. The knives are TiC cylinders 5 mm in diameter. The bending span is 14 mm. This furnace is located inside an Instron 1380 creep apparatus (Instron, Bucks, Great-Britain). The upper push rod is fixed to the upper crosshead of the Instron machine. The lower push rod is fixed to the moving lateral reaction plate (Fig. 2). All tests were performed at constant stress. Displacement was measured on the external part of the device using linear variable differential transducers (LVDT, Penny and Giles, Dorset, Great-Britain) and given by the difference between the values of 3 LVDT in equilateral position on the upper push rod and those on the lower push rod (the center of gravity corresponding exactly to the loading axis). With a such equipment, a displacement of less than one micron

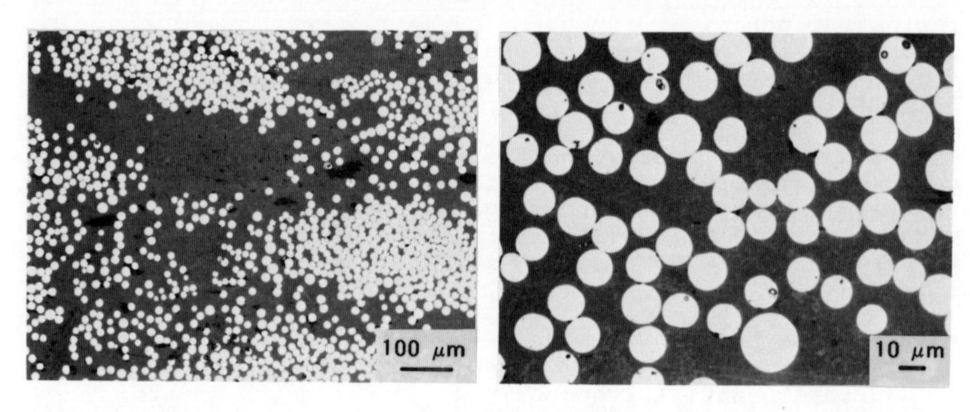

Fig. 1. Micrographs of an unidirectional SiC_f-MLAS composite material (Aérospatiale origin).

Fig. 2. Creep test equipment.

can be measured accurately. During all these experiments, particular attention was paid on the accuracy on the alignment and parallelism of the equipment.

The size of the polished specimens was : L = 18 mm, l = 4 mm and e = 2.5 mm. Creep tests were run at 1073, 1173, 1273, 1373 and 1473K, under a load range between 25 and 450 MPa, each load being maintained during a minimum of 50 hours. The experimental method used was the incremental stress at constant temperature, which is more easy to undertake than incremental temperature tests.

MORPHOLOGY AND MICROSTRUCTURE

At the macroscopic scale, such CMC materials are made of three apparent phases : fibers, matrix and pores. At the microscopic scale, there is to add the fiber/matrix interphase. The mechanical properties will depend mainly on these interphase.

The morphological parameters at the macroscopic scale (i.e. optical microscope) were measured using image analysis and mathematical morphology techniques[21,22,23]. Due to the initial fiber bundles, the distribution of the fibers is non homogeneous in the bulk material. The measured parameters were only :

- size distribution of the fibers : the mean diameter of the fibers is 13.5 μm (Fig. 3),
- surface area of the porosities on the surface of the specimens : $A_A(P) = 0.8\%$,
- mean contact number between fibers in a bundle : $n_c = 0.48$.

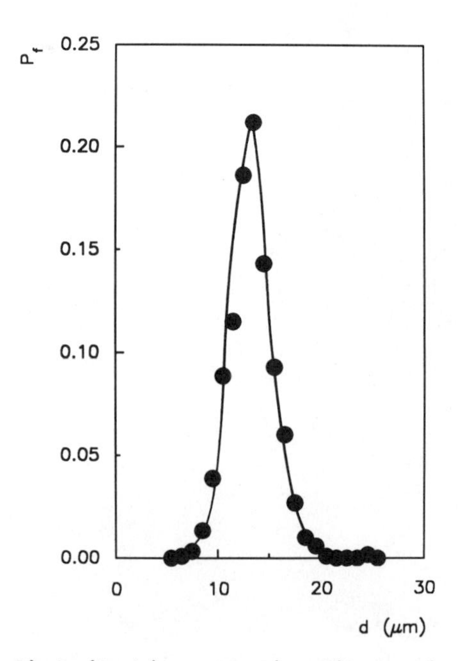

Fig. 3. Size distribution of SiC fibers in as-received SiC_f-MLAS material.

Figure 4 presents two transmission electron micrographs, for an as-received material showing circumferential cracks or decohesion appearing around the fibers and the interphase between the matrix and a SiC fiber. Such circumferential cracks have already been observed by Bischoff et al.[24] on similar as-processed state materials. Diffraction patterns confirm the cristallinity of the vitroceramic phase in MLAS matrix and that the SiC fibers are mainly amorphous with the presence of some SiC nanocrystals (probably β SiC[25]).

CREEP RESULTS

If we except the work of the team of Brennan and Prewo at UTRC, East Hartford (USA) (for example[12]), few mechanical strength results with temperature have been published on 1D or 2D SiC_f-LAS composites. For the MLAS matrix, only one paper were presented, by the team of Jamet at Aérospatiale, Saint Médard en Jalles (France)[2] : figure 5 reports the change in rupture stress in three point bending, σ_r, as a function of temperature, T, for SiC_f-MLAS materials. This result was the starting point of our creep investigation, in order to define the range of temperatures and stress levels for use in the creep experiments of this type of material.

On figure 6 is plotted the three point bending creep behaviour at 1173, 1373 and 1473K for a stress level of 200 MPa. Steady state creep is always observed and also the tertiary state for 1473K tests. For tests performed at 1073K no creep was observed, but only the failure of the specimens by a shear process at very low stress.

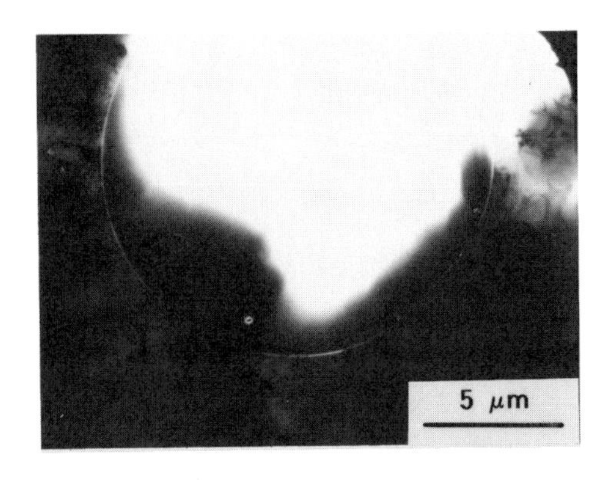

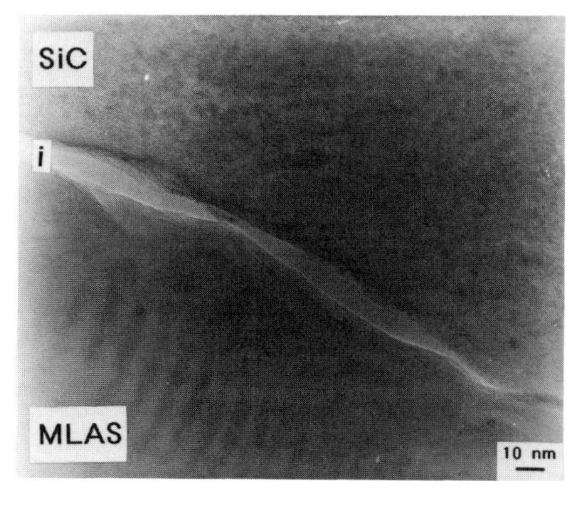

Fig. 4. Transmission electron micrographs of an as-received SiC_f-MLAS material.

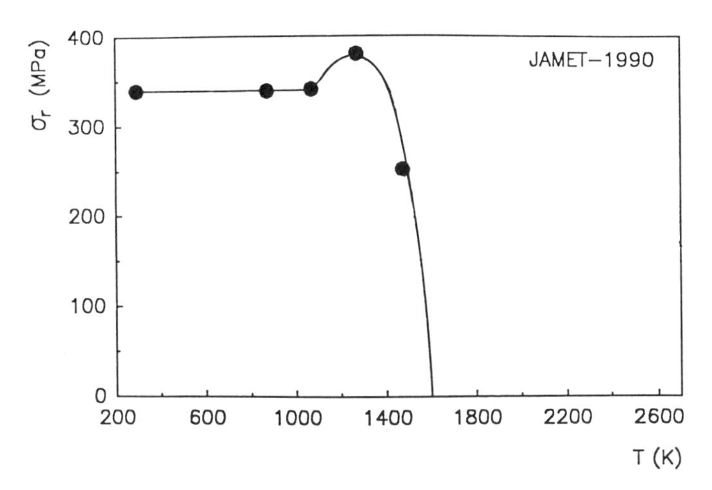

Fig. 5. Change in the rupture stress, σ_r, as a function of temperature, T, for SiC_f-MLAS.

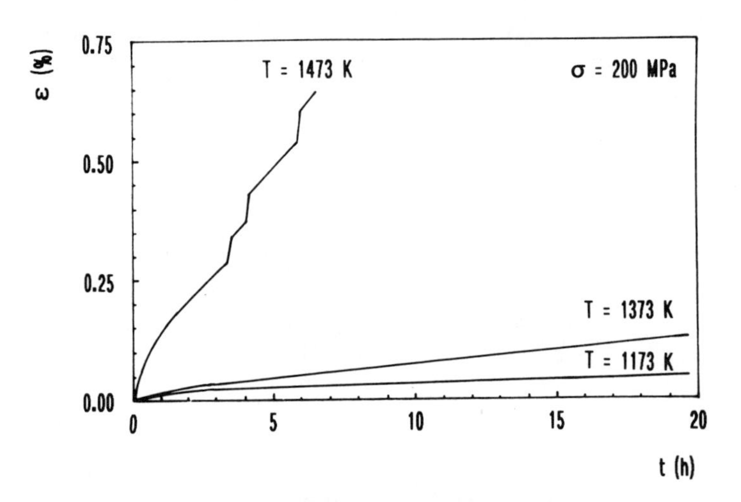

Fig. 6. Creep curves at three temperatures under a stress of 200 MPa.

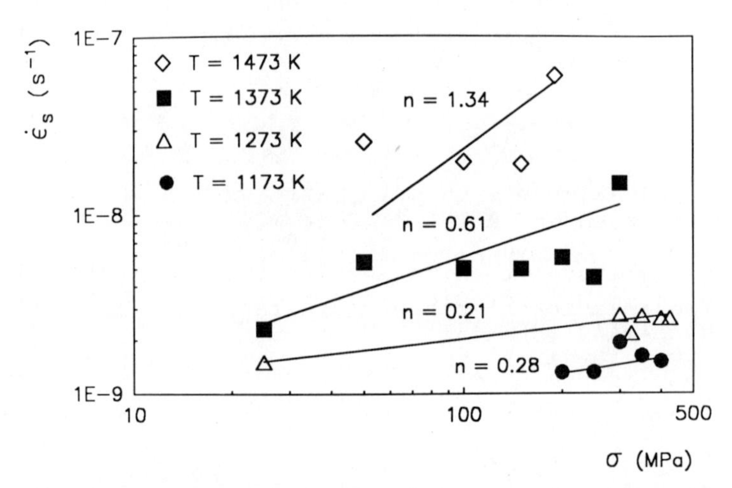

Fig. 7. Steady state creep rate, $\dot{\epsilon}_s$, as a function of stress, σ, for SiC_f-MLAS materials.

Steady state creep rate, $\dot{\epsilon}_s$, as a function of stress, σ, is shown on figure 7. From the classical creep equation (Norton or Dorn law) :

$$\dot{\epsilon}_s = A\,\sigma^n \exp - \frac{E}{kT}$$

with : $\dot{\epsilon}_s$ the creep rate in the steady state,
σ the applied stress,
n the stress exponent,
E the activation energy for creep,
k the Boltzmann's constant,
T the temperature,
A a constant.

the stress exponent can be calculated. The slope of the plots $\dot{\epsilon}_s$ as a function of σ gives access to n. From our results, the n values can be estimated at the different temperatures :

1173K : n = 0.3
1273K : n = 0.2
1373K : n = 0.6
1473K : n = 1.3

Figure 8 presents the corresponding Arrhenius plot.

DISCUSSION

These results concern the first ones regarding creep investigation of a new batch of CMC made with long silicon carbide fibers in a MLAS vitroceramic matrix.

The analysis of the Arrhenius plot indicates that in the temperature range 1073-1473K two mechanisms arise : at low temperature, 1173-1273K, a constant value of the apparent activation energy, E_a, of 80 $kJ.mol^{-1}$ is measured, whatever is the applied stress level ; for higher temperatures, 1373-1473K, there is a continuous increase of E_a with temperature, until 400 $kJ.mol^{-1}$: in that domain, E_a depends on both stress and temperature. Moreover, whatever is the test temperature, at low stress level (for example 25 MPa) only one mechanism is involved, which is the same that in the low temperature range.

No activation energy values are presently known for mechanisms related to MLAS vitroceramic matrix. We can only note on one hand, that the 80 $kJ.mol^{-1}$ value is of the same order of magnitude than the value found by Wiederhorn et al.[26], 69 $kJ.mol^{-1}$, for SiC microcracking. On the other hand, Bunsell et al.[27] found a value of 300 $kJ.mol^{-1}$ for the SiC fiber creep. Then, we can propose that for SiC_f-MLAS materials creep is governed by microcracking at low temperature ($\leq$ 1273K) and by creep of the SiC fibers at higher temperatures, at least until 1473K.

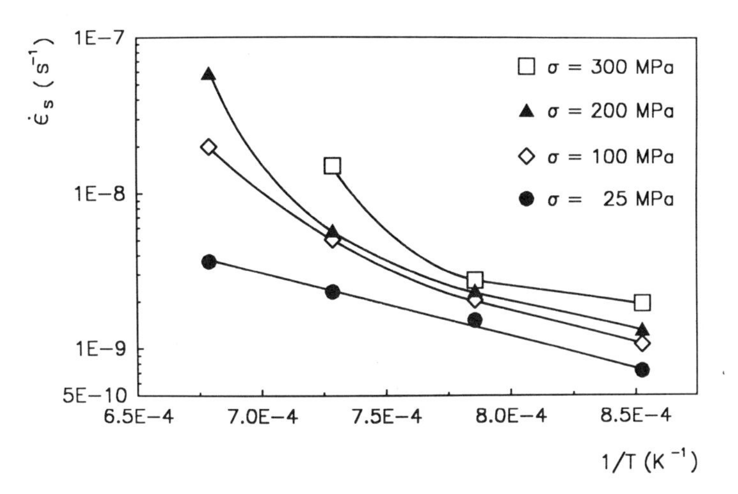

Fig. 8 . Arrhenius plot for SiC_f-MLAS materials.

But there is also to take into consideration the chemical change in the SiC fibers and fiber/matrix interfaces during such test exposures. This is due to the fact that the Nicalon[R] fibers NLM 202 contain also 12 O wt%, which interacts specially at the interfaces.

These results are confirmed also by the scanning electron microscope observations, paying attention to the change in the fiber aspects and fiber fracture features with temperature.

*** 1073-1173K**

At 1073-1173K, specimens fracture by shear stress (Fig. 9). This could be avoided with thinner specimen. The outer aspect of the fibers seems not to be modified after creep exposure.

*** 1273K**

From 1273K shear effect is accompanied by microcracking in the vitroceramic matrix and by the beginning of the fiber pull-out. No modification of the fibers and of their interfaces is observed (Fig. 10).

Fig. 9. Optical macrogaph of a specimen after a creep test at 1173K.

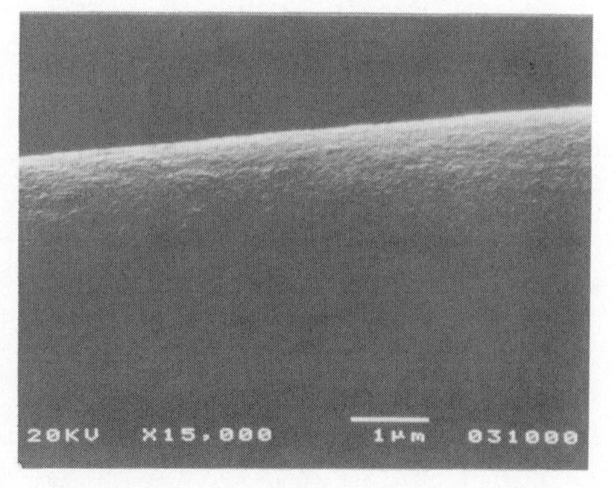

Fig. 10. Scanning electron micrographs of SiC_f-MLAS after a creep test at 1273 K : microcracks and surface of a fiber.

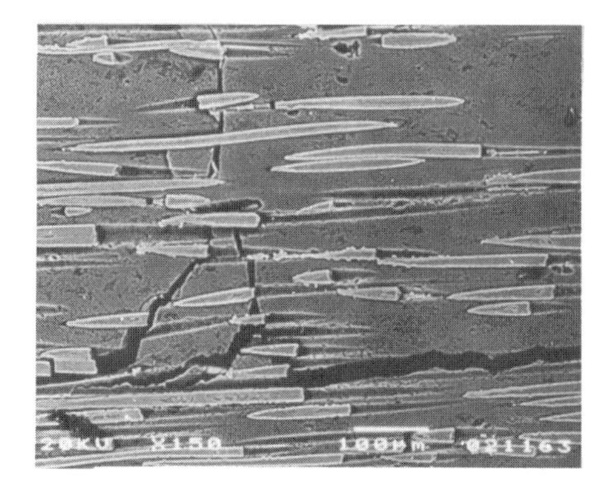

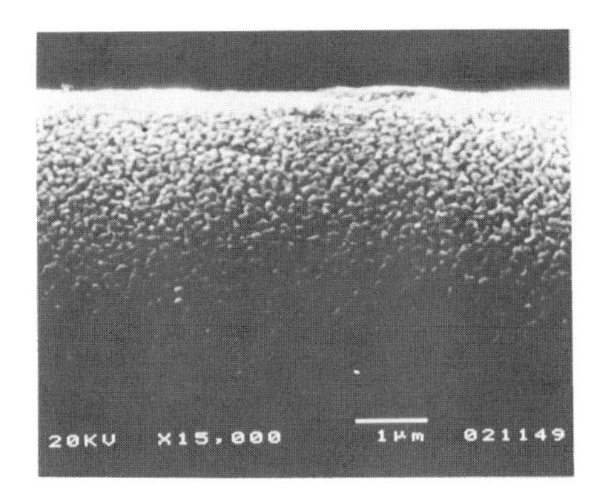

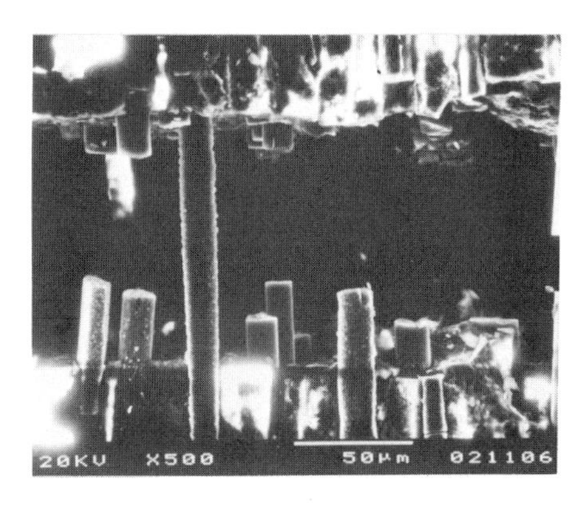

Fig. 11. Scanning electron micrographs of SiC_f-MLAS specimens after creep tests at 1373K : microcracks, surface of a fiber and fiber pull-out.

*** 1373K**

At 1373K, rupture arises on the surface in tension and with delamination perpendicular to the main fracture path. The beginning of the interface modifications due to an oxydation is observed for fibers closed to the surface of the specimens. At this temperature fiber pull-out is also involved. It can be estimated in the range 10 to 50 μm (Fig. 11).

*** 1473K**

At 1473K, same features are observed than at 1373K, but they are more pronounced : the modification of the fiber/interfaces is greater close to the surface and the fiber pull-out can reach some millimeters in length. Flat fractures of the fibers are also observed (Fig. 12). If a test is conducted until the tertiary stage, strain jumps are observed on the plot strain, ϵ, as a function of time, t, (Fig. 13). As the curves remain parallel, this is certainly due to the fracture of fiber bundles which were previously pulled-out (Fig. 14). Such observations were already made by Holmes[17].

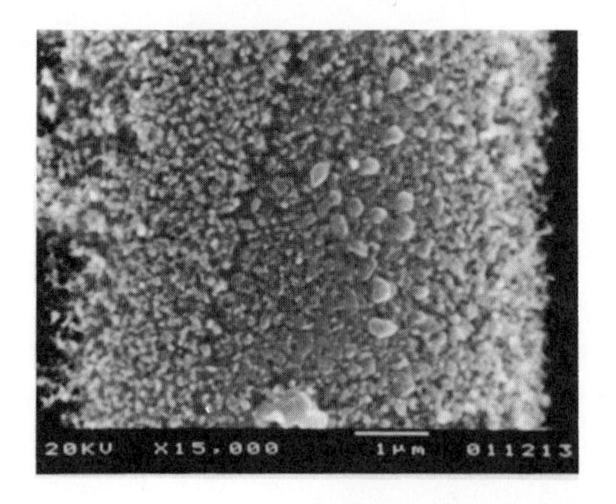

Fig. 12. Scanning electron micrographs of SiC_f-MLAS specimens after creep tests at 1473K : surface of a fiber, fiber pull-out and fiber rupture.

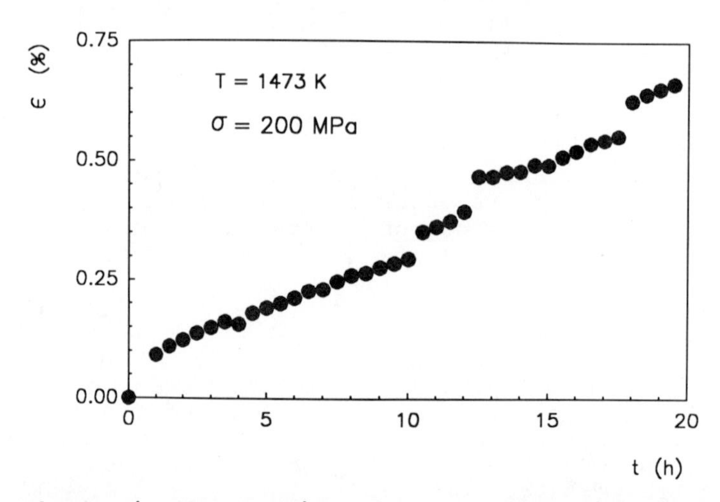

Fig. 13. Change in the strain, ϵ%, as a function of time, t.

These different observations raise the influence of the interfacial cohesion of the fibers with the MLAS matrix and the chemical change of the fibers after a long exposure at such temperatures under vacuum.

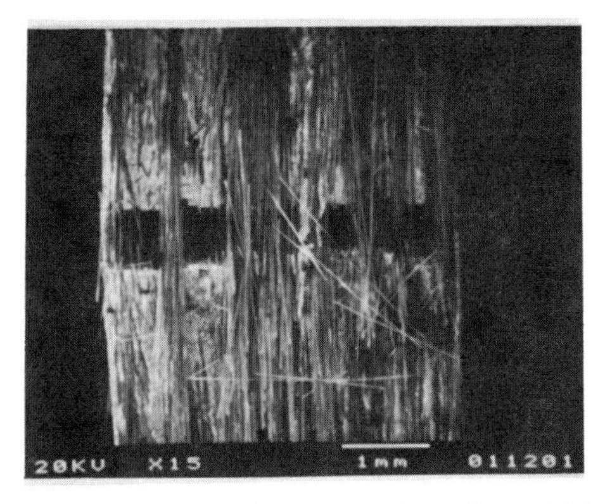

Fig. 14. Scanning electron micrograph of a SiC_f-MLAS specimen after creep test at 1473K. Tensile surface.

Creep behaviour of SiC_f-MLAS materials is different from SiC_f-SiC materials regarding the matrix behaviour and the fiber pull-out as a function of temperature which is opposite for these two types of CMC : pull-out increases with temperature for SiC_f-MLAS and decreases for SiC_f-SiC[15].

Microanalysis and TEM observations on crept specimens are the next step of this investigation.

CONCLUSION

In this work, first results on the creep behaviour of a unidirectional SiC_f-MLAS composite were presented in the temperature range 1073-1473K, tested in three-point bending under vacuum. It can be said that, on one hand, such materials are very sensitive to shear stress which interacts often at low temperature if specimens are too thick. On the other hand, based on analytical results and microstructural analysis of test specimens after creep deformation, in a first attempt it can be proposed that the creep behaviour of such materials is primarily due to MLAS matrix microcracking at temperatures lower than 1273K and to SiC fiber creep in the temperature range 1373-1473K.

The role of the fiber/matrix interfaces and the chemical change of the fibers during creep tests are the main parameters to control.

ACKNOWLEDGMENTS

This work was supported by Centre National de la Recherche Scientifique (CNRS), Centre National des Etudes Spatiales (CNES), Direction des Recherches, Etudes Techniques (DRET), and Aérospatiale Company (Saint-Médard en Jalles, France), Société Européenne de Propulsion (SEP, Saint-Médard en Jalles, France) and Société Nationale d'Etude et de Construction de Moteurs d'Aviation (SNECMA, Evry, France), under a

Groupement Scientifique (G.S.) on "Comportements thermomécaniques des composites céramique-céramique à fibres". The authors are grateful to Drs F. Albugues and J.F. Jamet, from Aérospatiale Compagny, for many valuable discussions, and to Dr P. Ruterana for observations in transmission electron microscopy.

REFERENCES

1. A. Melchior, M.F. Pouliquen and E. Soler - in "AIAA/SAE/ASME/ASEE", 23rd Joint Propulsion Conference, June 29 - July 2, 1987, San Diego, Cf, USA (1987).
2. J.F. Jamet - in "French Aerospace 90 Aeronautical and Space Conference", June 12-14, 1990, Washington, DC, USA (1990).
3. J.L. Chermant - Jap. J. Appl. Phys. Series 2, Lattice defects in ceramics, **S2** : 179 (1989).
4. M. Gomina, F. Abbé, P. Fourvel, M.H. Rouillon, J. Vicens, Chermant J.L. - Jap. J. Appl.Phys. Series 2, Lattice defects in ceramics, **S2** : 75 (1989).
5. M. Gomina, P. Fourvel and M.H. Rouillon - J. Mat. Sci., **26** : 1891 (1991).
6. H. Osmani, D. Rouby and G. Fantozzi - Comp. Sci. Tech., **37** : 191 (1990).
7. A.G. Evans and D.B. Marshall -"The mechanical behavior of ceramic matrix composites", presented at the 7th International Congress on Fracture, ICF7, March 20-24, 1989, Houston, Tx, USA.
8. L.P. Zawada, L.M. Butkus and G.A. Hartman - Ceram. Eng. Sci. Proc., **11** : 1592 (1990).
9. S.M. Wiederhorn and B.J. Hockey -"High temperature degradation of structural composites", presented at the 7th World Ceramics Congress, Montecatini Terme, Italy, June 24-30, 1990.
10. P. Le Costumer, M. Monthioux and A. Oberlin - "Matériaux composites pour applications à hautes températures", March 29-30, 1990, Bordeaux, France. Proceedings AMAC-CODEMAC, edited by R. Naslain, J. Lamalle and J.L. Zulian, p 43 (1990).
11. G. Perez and O. Caix - in "Developments in the science and technology of composite materials", ECCM4, Sept. 25-28, 1990, Stuttgart, Germany, p 573 (1990).
12. J.J. Brennan and K.M. Prewo - J. Mat. Sci., **17** : 2371 (1982).
13. F. Abbé, J. Vicens and J.L. Chermant- J. Mat. Sci. Let., **8** : 1026 (1989).
14. F. Abbé and J.L. Chermant - in "Creep and fracture of engineering materials and structures", edited by B. Wilshire and R.W. Evans, Proc. Fourth Int. Conferenrence, University College, Swansea, Great-Britain, April 1-6, 1990. The Institute of Metals, p 439 (1990).
15. F. Abbé - Thèse de Doctorat de l'Université de Caen, July 11, 1990.
16. F. Abbé and J.L. Chermant - in "7èmes Journées Nationales sur les Matériaux Composites", JNC7, Lyon, France, Nov. 6-8, 1990. Proceedings AMAC, edited by G. Fantozzi and P. Fleischmann, p 401 (1990).
17. J.W. Holmes - J. Mat. Sci., **26** : 1808 (1991).

18. D. Kervadec and J.L. Chermant - Rev. Comp. Nouv. Mat., **1**: 9 (1991).
19. J.F. Jamet - in "40th Congress of the International Astronautical Federation", Malaga, Spain, Oct. 8-13, 1989.
20. F. Christin, R. Naslain and C. Bernard - in Proc. "7th Int. Conference on CVD", edited by T.O. Sedwick and H. Lydtin, The Electrochemical Society, Princeton, p 499 (1979).
21. J. Serra - Mathematical morphology and image analysis. Academic Press (1982).
22. M. Coster and J.L. Chermant - Précis d'analyse d'images. Les Editions du CNRS (1985). Les Presses du CNRS (1989).
23. F. Abbé, L. Chermant, M. Coster, M. Gomina and J.L. Chermant J.L. - Comp. Sci. Tech., **37** : (1990).
24. R. Bischoff, M. Rühle, O. Sbaizero and A.G. Evans - J. Amer. Ceram. Soc. **72** : 741 (1989).
25. C. Ponthieu, M. Lancin and J. Thibault-Desseaux - Phil. Mag., **62A** : 605 (1990).
26. S.M. Wiederhorn, D. Ellis Roberts, T.J. Chuang and L. Chuck - J. Amer. Ceram. Soc., 71 : 602 (1988).
27. A. Bunsell, G. Simon, Y. Abe and M. Akiyama - in "Fibre reinforcements for composite materials", Composite materials series 2, Elsevier, p 427 (1988).

CREEP RUPTURE MAP OF ENGINEERING FINE CERAMICS

Tsuneshichi Tanaka[1], Hideaki Nakayama[2], Nagatoshi Okabe[3], Shigeru Yamamoto[1], and Satoshi Fukui[4]

[1]Faculty of Science and Engineering, Ritsumeikan University Kita-ku, Kyoto, 603, Japan
[2]Faculty of Junior College of Automobile Industry, Osaka Sangyo University, Daito-shi, Osaka, 574, Japan
[3]Heavy Apparatus Engng. Laboratory, Toshiba Corporation Tsurumi-ku, Kanagawa, 230, Japan
[4]Faculty of Mechanical Engineering, Technical College of Takamatsu Takamatsu-shi, Kagawa, 761, Japan

ABSTRACT

The main objective of this study is to clear the strength and strain behaviors of engineering fine ceramics in long term high temperature creep. For this purpose, a series of creep tests was carried out on a silicon nitride and a silicon carbide at high temperature levels up to 1723K and during long term up to 1000 hours by using a newly developed high temperature tensile creep testing machine. The results indicate that the relationship of applied stress vs. rupture time of both ceramics at each temperature level can be approximated by a straight line on full log paper with a little scatter. Then, discussion on the creep behaviors of the silicon nitride from the viewpoint of creep strain revealed that the minimum creep strain rate well governed the rupture time, and the creep rupture maps constructed from two different viewpoints indicated that the tertiary creep stage was a coalescence stage of multi-site microcracks.

INTRODUCTION

Engineering ceramics have excellent mechanical properties; high wear resistance, high oxidation resistance, high compressive strength [1] and high strength at elevated temperature [2], but oppositely these show extremely low fracture toughness which is a fatal defect as a structural element in the general concept of mechanical design [3]. In order to use these hard and brittle ceramics as engineering materials, a new design concept has to be constructed through a detail and precise evaluation of strength characteristics of these materials [4-8].

Though one of the major characteristics of fine ceramics is their superior high temperature strength, strength and strain behaviors in high temperature creep have not been systematized yet. Many experiments have been undertaken to systematize the creep behavior of engineering fine ceramics [9-17]. And, in most cases, the creep behaviors have been discussed on the basis of creep data obtained under bending and/or compressive load conditions as was indicated in the review by Cannon and Langdon [18]. For example, fracture mechanism maps covering wide ranges of stress, temperature and time to failure were constructed by Quinn [19], and Quinn and Braue [20] as results of their extensive experiments on a hot-pressed silicon nitride and a sintered silicon nitride under 4-point bending.

In order to evaluate the creep behavior of fine ceramics precisely from the fundamental viewpoint, however, it is preferable to discuss the creep behavior based on data obtained under uniaxial tensile load conditions as is suggested by Carroll and Tressler [13].

Fracture Mechanics of Ceramics, Vol. 10
Edited by R.C. Bradt *et al.*, Plenum Press, New York, 1992

For this purpose, a high temperature tensile creep testing machine for fine ceramics was newly developed together with a displacement-meter. To begin with, alignment of load transmission train of this machine was carefully checked and sufficient performance of this testing machine was confirmed. Then, a series of long term high temperature creep tests was carried out on two typical engineering fine ceramics; a sintered silicon nitride and a sintered silicon nitride.

Experimental results were discussed from several viewpoints as follows: (1) relationships of applied stress vs. rupture time for both ceramics; (2) modified life estimation method for the silicon nitride; (3) creep deformation curve and relationship between the minimum creep rate and the rupture time of the silicon nitride; (4) two types of creep rupture maps for the silicon nitride constructed from two different viewpoints, i.e., one is from rupture point on the creep deformation curve and another from rupture surface morphology, and furthermore the correspondence of the creep deformation curve and the rupture surface pattern.

SPECIMEN, TESTING MACHINE AND EXPERIMENTAL PROCEDURES

Experimental Materials

A sintered silicon nitride with additives of Al_2O_3 and Y_2O_3 and a sintered silicon carbide with additives of B and C were used as experimental materials. The shape and sizes of round bar type specimen are indicated in Fig.1. Sintered silicon nitride specimens were post heat-treated at the temperature of 1273 K during 40 hours after the grind-finishing to the shape indicated in Fig.1 and then supplied to the experiments. And the sintered silicon carbide specimens were used without any post heat-treatment.

High Temperature Tensile Creep Testing Machine

In order to carry out long term high temperature creep tests, at first, a high temperature creep testing machine for fine ceramics was newly developed, whose whole view is shown in Fig.2. As shown in this figure, this testing machine is a single lever type one with a lever ratio of 1:10, and has a loading capacity of about 30 kN. And, this testing machine is equipped with a high temperature electric furnace which has 4 heaters of Kanthal Super 33 connected in series. By using this machine, long term creep tests at temperature levels up to 1800 K can be carried out. As is shown, the whole structure is almost the same as creep tester for ordinary metallic materials. But two points are improved; one is a specimen chucking device to prevent the failure of brittle ceramics by the application of unexpected stress component such as bending stress, and another is a displacement-meter to pick up the elongation of specimen at high temperature.

Detailed structure of specimen chucking device is indicated in Fig. 3. As is shown in this figure, upper and lower gripping rods are connected to pull rods by knife edge pins. Such a structure of load transmission train eliminates the effect of bending load. Specimen is attached to gripping device by using a couple of semi-circular spacer rings. Spacer rings and specimen grips are made by a sintered silicon carbide and knife edge pins and pull rods are by stainless steel, JIS.SUS304.

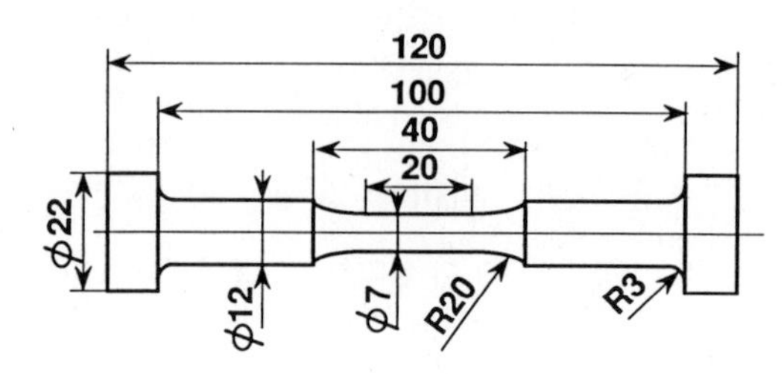

Fig.1. Shape and sizes of specimen.

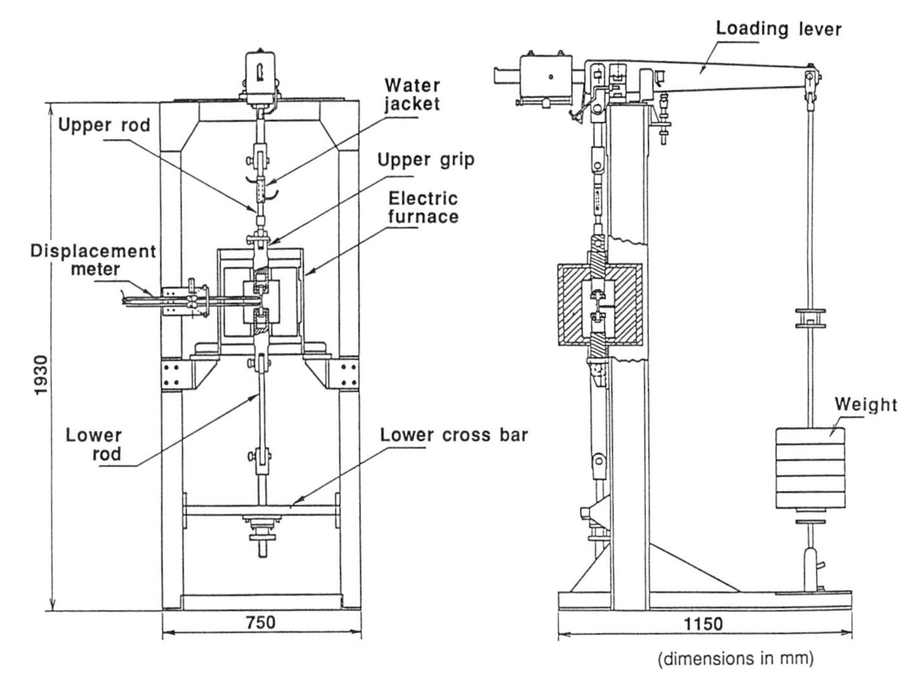

Fig.2. Whole view of newly developed tensile high temperature creep testing machine for fine ceramics.

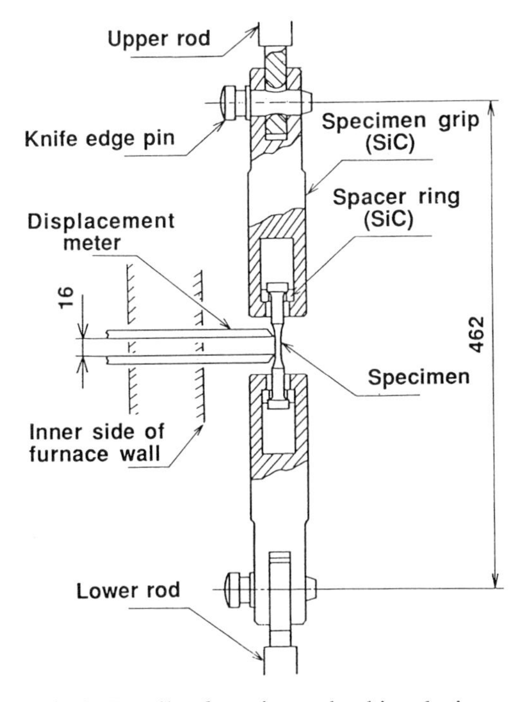

Fig.3. Details of specimen chucking device.

Figure 4 shows the displacement-meter developed in this study. A couple of rods to pick up the elongation of specimen directly is made by silicon carbide. The right side ends of these rods contact to specimen with an appropriate pressure, and an eddy current type displacement sensor is attached at the left side ends of these rods. The supporting point of these rods is made by INVAR to minimize the effect of ambient temperature fluctuation on the measurement accuracy. Creep strain was determined for the gauge length of 16mm in the specimen parallel part with the length of 20 mm. The resolution power of this device is about 1 μmeter.

This testing machine was set in a simple constant temperature room to reduce the measurement error of the displacement-meter. But the temperature in the room fluctuated in the range of ±2K, and the temperature change of 1K generates the analogue output of about 250 μstrain due to the thermal deformation of elements of the displacement-meter. Therefore, creep strain less than 1000 μstrain must be considered as experimental error of this device. Really, as shown in Fig.5 indicating a detailed variation of creep strain, creep strain fluctuates within a range of about 1000 μstrain with a period of a day. Other details of this testing machine are described in the previous report by the authors [17].

Experimental Procedures

Procedures to conduct a series of high temperature creep tests are as follows: At first, initial tensile stress of 13MPa was applied in room temperature, and at this stage, alignment of load transmission train was carefully checked. Then, the temperature was gradually raised

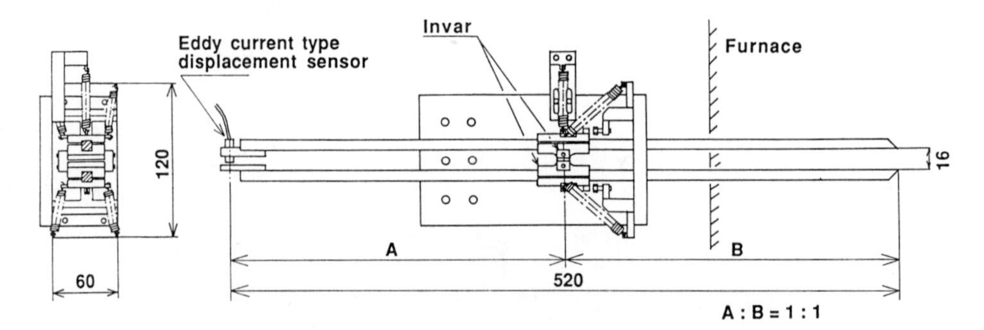

Fig.4. Structure of high temperature displacement-meter.

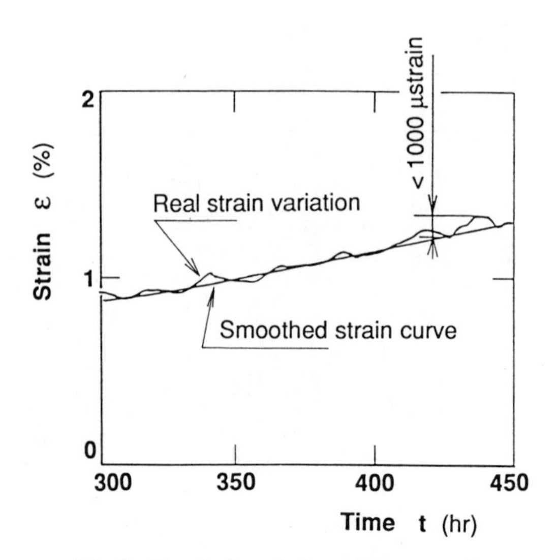

Fig.5. Detailed variation of creep strain.

at a rate of 10K/min to the desired temperature to prevent the fracture of silicon carbide elements in the furnace by thermal stress. And, the desired load was applied statically not to induce any transient influence after allowing the temperature to stabilize for 12 hours. It was confirmed that the temperature distribution on the specimen parallel part of 20mm long is within ±0.5% of the controlled value in the temperature range of this study.

An example of ruptured silicon nitride specimen is shown in Fig.6. A majority of the specimen ruptured inside the gauge length area, but a few specimens ruptured outside the area, whose results were excepted from the following discussion.

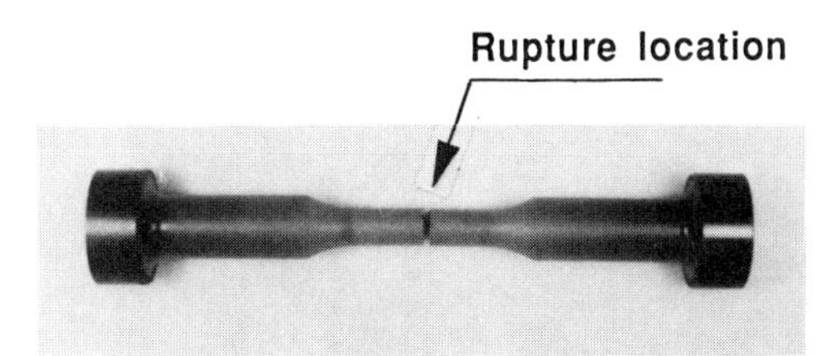

Fig.6. An example of ruptured silicon nitride specimen.

EXPERIMENTAL RESULTS AND DISCUSSION

Experimental results obtained in this study are summarized in Tables 1 and 2. As shown in Table 1, in the case of the silicon nitride, the specimen which did not rupture over the predetermined censoring time of 500 and/or 1000hours was supplied to creep test at the higher stress or the higher temperature. Data for asterisked specimens are the results by these used specimens. Then, only the relationships of the stress vs. the rupture time are listed in the case of the silicon carbide, for the creep strain of the silicon carbide was within the experimental error range of this displacement-meter (<1000 μstrain) even at the high temperature of 1723K. Therefore, discussion on the creep strain behaviors was made only for the case of the silicon nitride. In order to discuss the creep strain behaviors of silicon carbide, creep test data at elevated temperature levels over 1800K must be accumulated. The study on this point is now under investigation by the authors.

Creep Rupture Curve

Figures 7 and 8 show the creep rupture curves for both the silicon nitride and the silicon carbide. Relationships between the applied stress and the rupture time plotted on full log paper well fit straight lines with a little scatter in comparison with those of static and cyclic fatigue lives [21], and so, can be formulated by the following expression;

$$t_r \sigma^m = C \qquad (1)$$

where C and m are the constants. These two figures also indicate that the creep rupture strength of the silicon carbide is considerably higher than that of the silicon nitride.

Regression lines which were determined by the least square method in both figures give two constants, m and C. Properties of these two constants, m and C, were evaluated to establish a creep rupture life estimation method, especially from the viewpoint of their temperature dependence.

At first, dependence of the power constant m on the temperature of both ceramics is indicated in Fig.9. From this figure, it can be assumed that the power constant m is independent on the temperature. Therefore, the average values of m, $\overline{m}$, for both fine ceramics are indicated in the figure. The value of $\overline{m}$ for the silicon nitride is 6.99 and that for the silicon carbide is 21.79, respectively. Then, the temperature dependence of the constant C is indicated in Fig.10 for respective ceramics. In this figure, C' which is the logarithmic value of C is plotted by open circle and open square instead of C. And $\overline{C'}$ shown by solid circle and solid square is the average value of C' determined by using the average value of m.

Table 1. Results for the sintered silicon nitride.

Specimen number	Temperature T (K)	Stress σ (MPa)	Rupture time t_r (hr)	Rupture strain ε_r(%)	Minimum creep rate $\dot{\varepsilon}_{min}$ (%/hr)
2	1518	153	0.02		
3	1518	51	77.1		
4	1518	39	223.0		
5	1408	129	30.5		
6	1298	229	9.5		
7	1408	102	194.5		
8	1523	38	30.0	2.755	0.048
9	1473	50	152.0	1.989	0.013
10	1473	50	427.5	2.743	0.005
11	1523	30	129.0	4.285	0.014
12	1523	30	115.0	3.415	0.017
13	1523	25	733.0	10.014	0.008
14	1473	50	859.0	2.927	0.002
15	1523	25	305.5	4.455	0.008
16	1523	25	207.0	3.313	0.011
17	1473	70	52.0	1.562	0.015
18	1523	50	4.0	1.240	
19	1473	80	21.0	1.263	0.053
20	1473	90	7.0	0.914	
21	1423	100	92.0	1.119	0.007
22	1423	80	>500.0		
22*	1423	200	3.0	0.190	0.046
23	1423	130	45.5	1.501	0.018
24	1423	160	4.25	0.830	
25	1273	200	>1000.0		
25*	1298	200	>1000.0		
25**	1373	200	229.5	0.478	0.001
26	1373	200	186.0	1.083	0.004
27	1298	235	>1000.0		
27*	1348	235	301.5	1.136	0.001
28	1373	200	143.0	0.710	0.004
29	1373	200	227.0	0.872	0.004
30	1323	200	>500.0		
30*	1348	200	>1000.0		
30**	1348	300	0.033	0.097	
31	1373	100	>500.0		
31*	1373	160	>500.0		
31**	1473	40	>500.0		
31***	1523	15	>1000.0		

Table 2. Results for the sintered silicon carbide.

Specimen number	Temperature T (K)	Stress σ (MPa)	Rupture time t_r(hr)
1	1623	200	1.0
2	1623	180	12.0
3	1623	160	14.0
4	1623	160	39.5
5	1673	150	40.0
6	1673	130	>86.0
7	1673	200	0.061
8	1673	170	0.11
9	1723	150	0.62
10	1773	100	177.5
11	1773	130	19.5
12	1723	130	10.0
13	1773	130	0.67

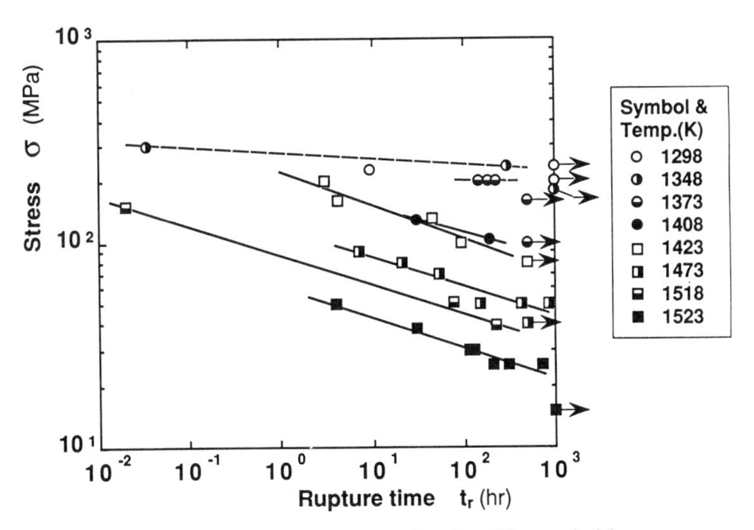

Fig.7. Creep rupture curves for the silicon nitride.

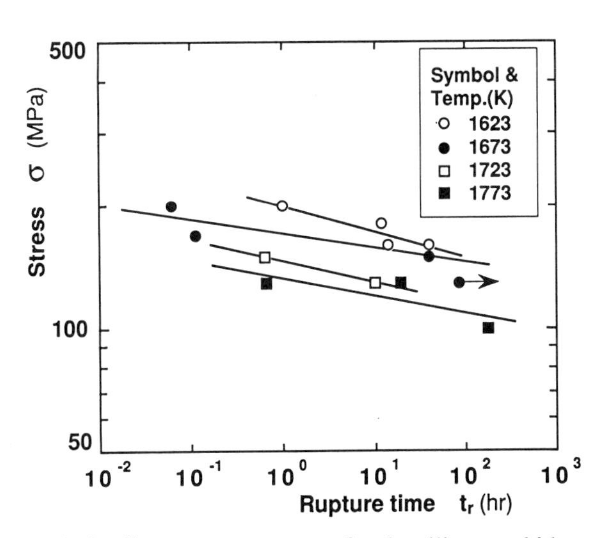

Fig.8. Creep rupture curves for the silicon carbide.

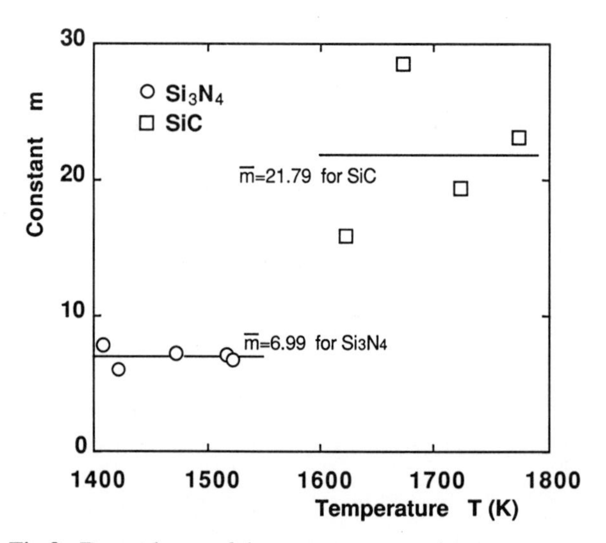

Fig.9. Dependence of the constant m on the temperature.

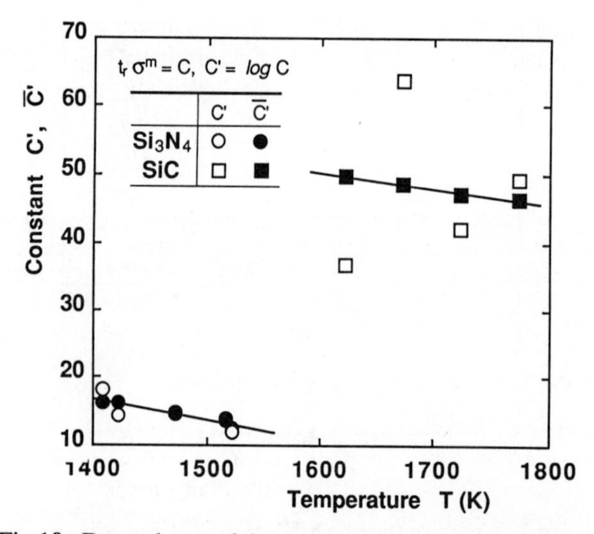

Fig.10. Dependence of the constant C' on the temperature.

Table 3. Values of constants m, C' and $\overline{C'}$.

Tempetature T (K)	m	C' (= *log* C)	$\overline{C'}$
Si_3N_4			
1408	4.89	18.14	16.28
1423	6.06	14.24	16.24
1473	7.21	14.92	14.52
1518	7.07	13.76	13.62
1523	6.72	12.00	12.40
	($\overline{m}$ = 6.99)		
SiC			
1623	15.95	36.63	49.79
1673	28.60	63.74	48.53
1723	19.43	42.08	47.14
1773	23.19	49.16	46.36
	($\overline{m}$ = 21.79)		

Values of the constants m, C' and $\overline{C'}$ at respective temperature levels are summarized in Table 3. Contrary to the case of the power constant m, it is reasonable to consider that the constant C, or $\overline{C'}$ is a function of the temperature, whose temperature dependence can be expressed as follows;

$$\overline{C'}(T) = 59.99 - 0.03T \quad \text{for silicon nitride}$$
$$\overline{C'}(T) = 87.62 - 0.02T \quad \text{for silicon carbide} \qquad (2)$$

At first, in Figs.11 and 12 are indicated the approximation lines for creep rupture life at respective temperature levels for both ceramics determined by using the values of the constants, $\overline{m}$ and $\overline{C'}$, in Table 3 . As shown in these figures, a group of the data points at each temperature level well fits the approximation line at corresponding temperature as a first approximation. And this gives validity for the above assumptions on the two constants; the power constant m is independent on the temperature and the constant C is a function of the temperature.

Then, by using the average value of m and the temperature dependence of the constant C given by Eq.(2), creep rupture life estimation equation can be established. But, as shown in Figs.9 and 10, the silicon carbide shows large scatter in the temperature dependence of two constants m and C'(=*log* C), presumably depending on the smallness of sample size. Therefore, the life estimation method is discussed only for the case of the silicon nitride. By substituting the value of $\overline{m}$ and Eq.(2) for the silicon nitride into Eq.(1), creep rupture life estimation equation for the silicon nitride is given as follows;

$$t_r = Exp\,[E\,(59.99 - 0.03T) - 6.69\,\ln\sigma]\,, \quad E = \ln 10 \qquad (3)$$

In Fig.13 are shown the modified life estimation curves for the silicon nitride determined by the above expression. In this figure, data points shown by open circle are rupture specimens, those by solid circle are run-out specimens and figures near open circles are rupture times in hour. And, squares are high temperature fast tensile fracture strength obtained by the silicon nitride specimen fabricated with the same processes as the specimen used in this study [2]. Equation (3) gives straight line relationship on the semi-log paper, so the upper parts of these estimation lines are curved not to exceed the fast tensile fracture strength shown by dashed line. As can be observed, data points well fit the modified life estimation curves drawn in the life range from 0.1 to 1000hours.

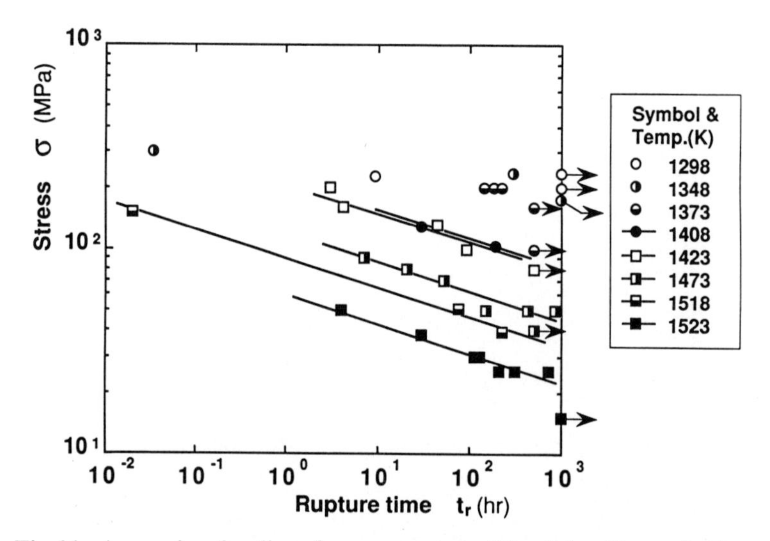

Fig.11. Approximation lines for creep rupture life of the silicon nitride.

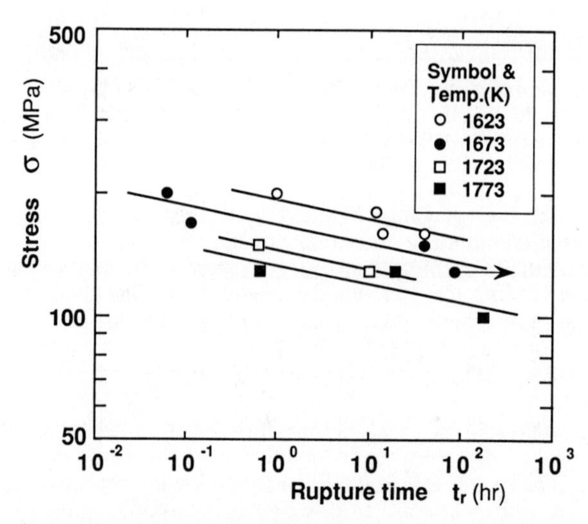

Fig.12. Approximation lines for creep rupture life of the silicon carbide.

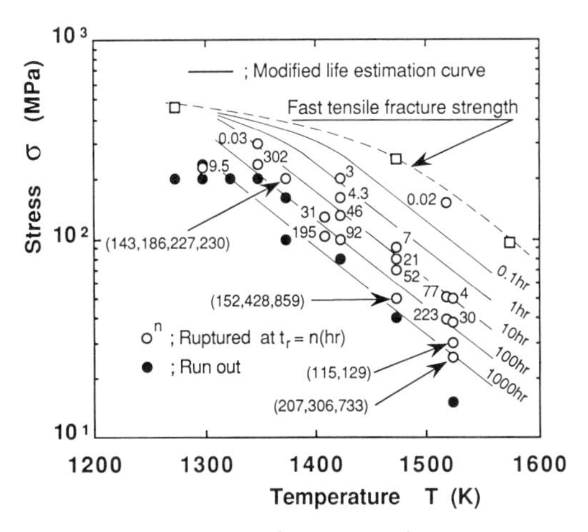

Fig.13. Modified life estimation curves for the silicon nitride.

Creep Deformation Behaviors

In Fig.14 are shown creep strain curves for the silicon nitride. In this figure, (a) shows the results in short term creep, (b) shows those in intermediate creep range and (c) is the results in long term creep. As is observed, several specimens show three-stage creep deformation processes analogous to the case of ordinary metallic materials [9]; transient creep stage, then steady creep stage, and last tertiary creep stage . And it is noticeable that the creep strain over 10% can be observed even in the case of brittle ceramics. The creep strain data obtained as analogue data from the eddy current type displacement sensor were converted to digital data and stored in a memory of a personal computer. Then, creep strain rate was determined by applying the 7-point polynomial method [22] to these digital data. Figure 15 shows thus obtained creep strain rate curves. This figure clearly shows the existence of steady creep stage, or the minimum creep rate. A majority of the specimen ruptured at the steady and/or the tertiary creep stage, and only three specimens ruptured at the transient creep stage.

Figure 16 shows the relationships between the rupture time and the minimum creep rate of the silicon nitride under several temperature levels. This figure first indicates that the minimum creep rate well governs the rupture time, and next the relationship of these two parameters can be well expressed by the following equation;

$$t_r \, \dot{\varepsilon}_{min}^{\;0.912} = 2.428\times10^{-2} \qquad (4)$$

As is well known, the minimum creep rate is as a function of applied stress and temperature, and expressed by the following Arrhenius type equation;

$$\dot{\varepsilon}_{min} = A\sigma^n \, exp\left(-\frac{Q}{RT}\right), \quad A;\ \text{constant} \qquad (5)$$

where Q is activation energy for creep deformation, R is gas constant with the value of 8.31J/mol•K, T is absolute temperature and n is the stress exponent. When the temperature T is constant, the above equation is reduced to the following form, which gives the stress exponent n.

$$\dot{\varepsilon}_{min} = B\sigma^n, \quad B;\ \text{constant} \qquad (6)$$

The relationships of the minimum creep rate and the stress at three temperature levels are shown in Fig.17, and slopes of these regression lines give the stress exponent n. For example, the result under the temperature of 1423K gives the stress exponent of 3.4, and the other two results give the values of 4.7 and 3.4, respectively.

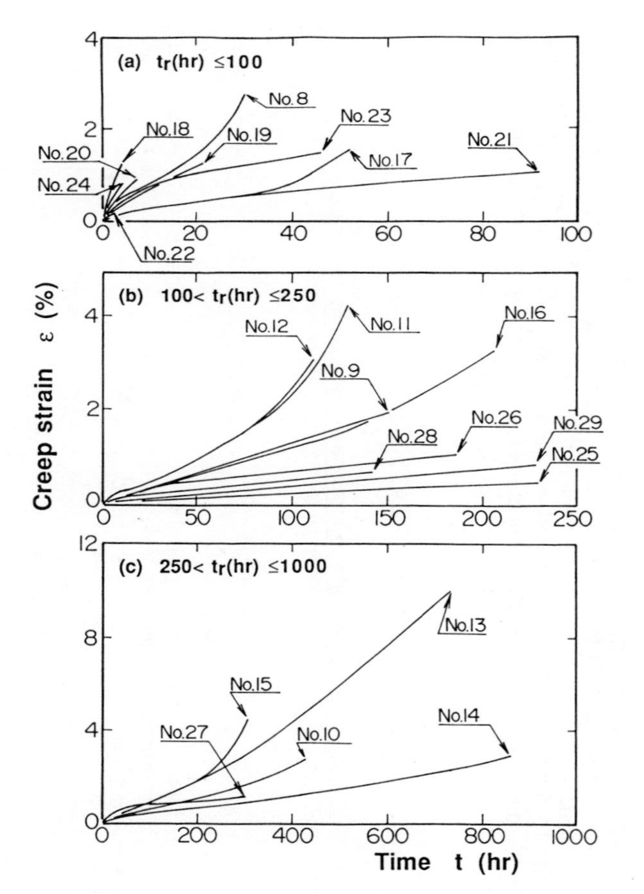

Fig.14. Creep strain curves of the silicon nitride.

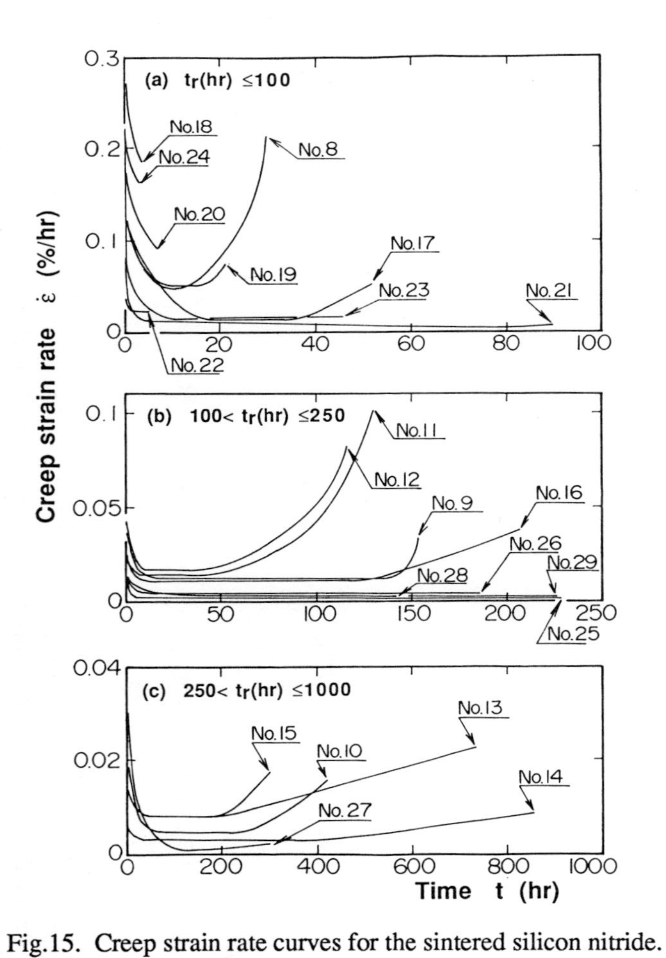

Fig.15. Creep strain rate curves for the sintered silicon nitride.

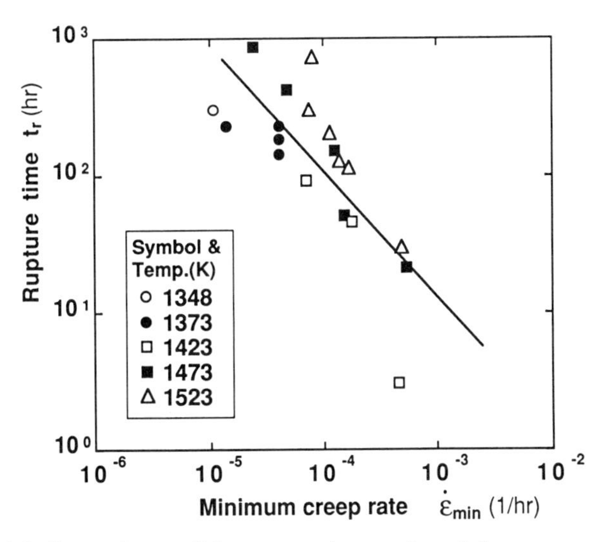

Fig.16. Dependence of the rupture time on the minimum creep rate of the silicon nitride.

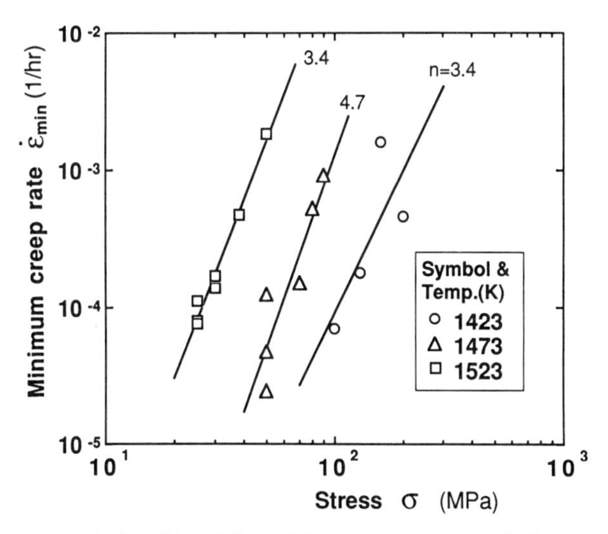

Fig.17. Relationship of the minimum creep rate and the stress, and the values of the stress exponent n (Si_3N_4).

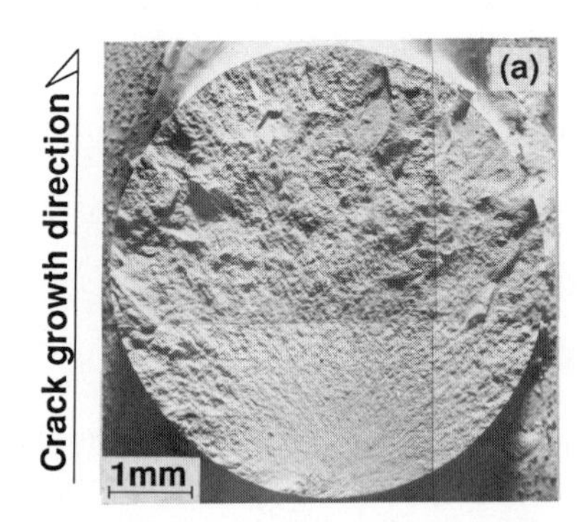

(a) T=1373K, σ=200MPa, t_r=143hours.

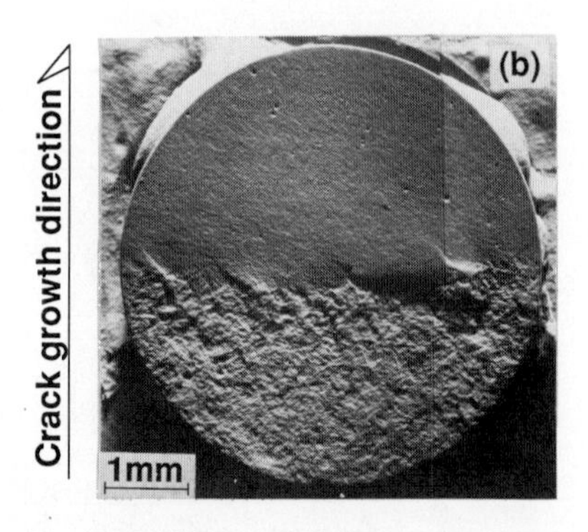

(b) T=1523K, σ=50MPa, t_r=4hours.

Fig.18 (a),(b). Creep rupture surfaces of the silicon nitride.

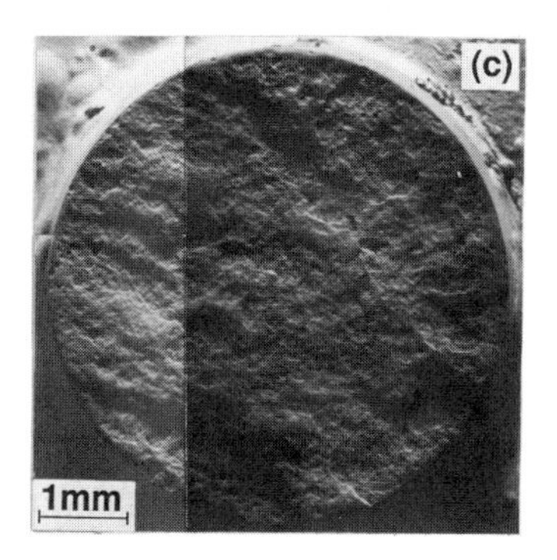

(c) T=1523K, σ=30MPa, t_r=129hours.

Fig.18 (c). Creep rupture surfaces of the silicon nitride.

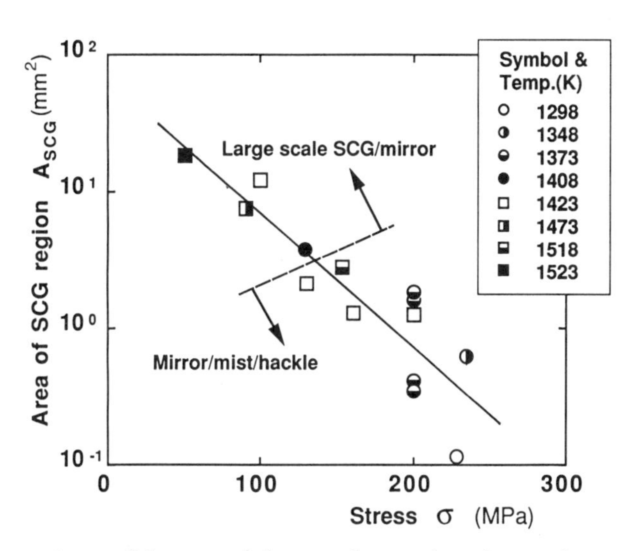

Fig.19. Dependence of the area of slow crack growth region on the stress (Si_3N_4).

Rupture Surface Morphology

Figure 18 shows the pictures of three different creep rupture surfaces of the silicon nitride. (a) shows the rupture pattern composed of mirror, mist and hackle, which well resembles that in fast tensile fracture. (b) shows the rupture surface composed of large scale slow crack growth region (the lower half region) and mirror (the upper half region), and the area of the large scale S.C.G. region increases with the decrease in the applied stress as shown in Fig.19. Figure 19 also shows the stress dependence of the area of S.C.G. region in rupture surfaces composed of mirror, mist and hackle. Figure 18 (c) shows a typical creep rupture surface, and all of the area is covered with glass layer. It is assumed that such a rupture surface pattern can be constructed by multi-site cracking and their developments.

An example of rupture surface of the silicon carbide is shown in Fig.20. In the case of the silicon carbide, all of the creep rupture surface showed the same pattern composed of mirror, mist and hackle as indicated in Fig.20.

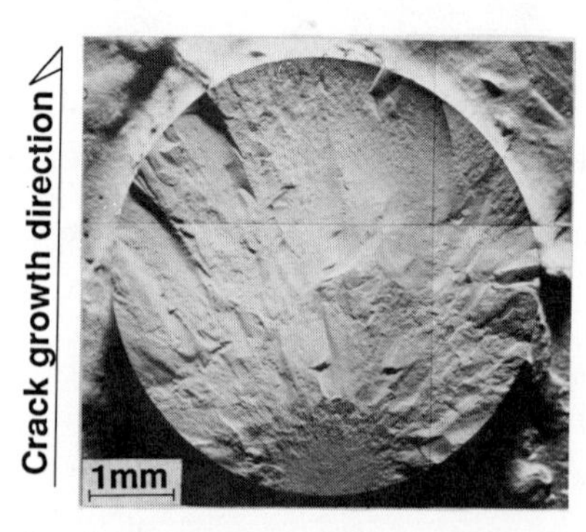

Fig.20. An example of creep rupture surface of the silicon carbide. (T=1623K, σ=160MPa, t_r=39.5hours)

Creep Rupture Maps

Figures 21 and 22 show creep rupture maps constructed from two different viewpoints together with the modified life estimation curves mentioned in previous section; one is constructed from the viewpoint of rupture point on creep deformation curve, and another is from the rupture surface morphology.

In Fig.21, symbols are classified into three groups depending on the rupture point on creep deformation curve, for example open circle means the specimen ruptured at transient creep stage. And figures near four data points indicate the number of data points superposed at the same location, and the specimens which are superposed on the same location ruptured at the same stage of creep deformation curve. Then, symbols in Fig.22 are also classified into three groups depending on the rupture surface morphology; data points shown by open circle mean that the rupture surfaces of these specimens consist of mirror, mist and hackle.

By comparing Figs.21 and 22, it is observed that locations of the solid circles in these figures completely coincide with each other excepting two points in Fig.22 whose creep deformation behaviors were not observed. This indicates that the typical creep rupture surface can be observed only for the specimen which ruptured at tertiary creep stage, in other words, tertiary creep stage is a coalescence stage of multi-site microcracks.

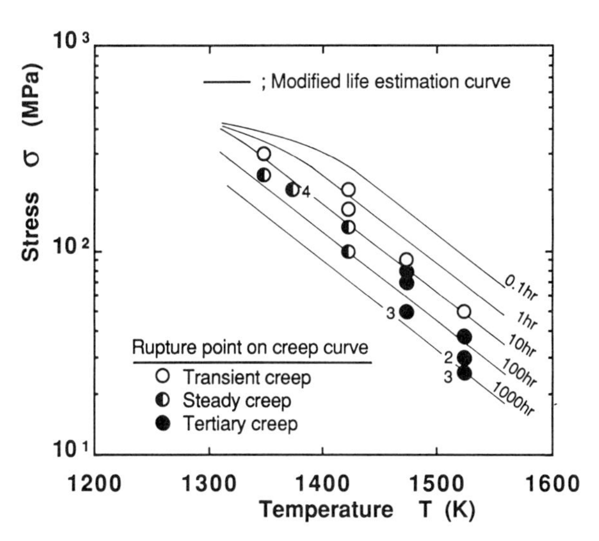

Fig.21. Creep rupture map constructed on the basis of rupture point on creep deformation curve.

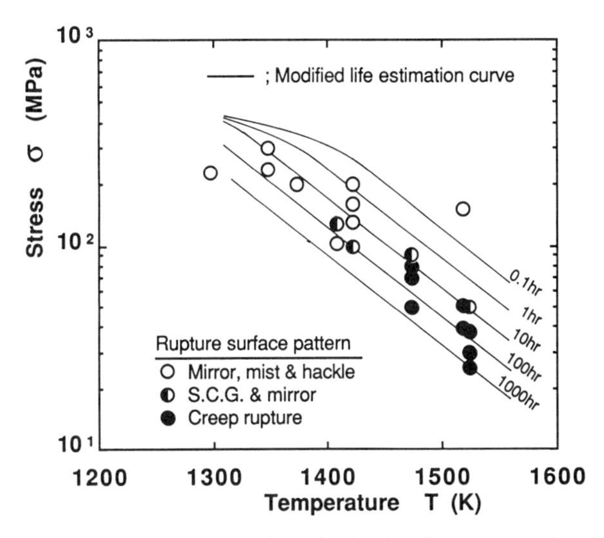

Fig.22. Creep rupture map constructed on the basis of rupture surface morphology.

CONCLUSIONS

Major conclusions obtained in this study are summarized as follows;

(1) Relationships between the applied stress and the creep rupture life of both the sintered silicon nitride and the sintered silicon carbide at several temperature levels well fitted straight lines on full log paper with a little scatter. And, life approximation lines for both ceramics could be determined on the assumptions that, in the expression of $t_r \sigma^m = C$, the power constant m was independent on the temperature and the constant C was a function of the temperature.

(2) Creep rupture lives of the sintered silicon nitride were well approximated by the modified life estimation curves determined by the following equation in the life range from 0.1 to 1000hours;

$$t_r = Exp\,[E\,(59.99 - 0.03T) - 6.69\,ln\,\sigma]\,,\ E = ln\,10.$$

(3) The sintered silicon nitride showed 3-stage creep deformation curve analogous to the case of metallic materials. And a majority of the specimen ruptured at steady creep stage and/or tertiary creep stage. On the other hand, the sintered silicon carbide did not show any meaningful creep deformation in the temperature range up to 1723K.

(4) The relationship of the minimum creep rate and the rupture time of the sintered silicon nitride could be approximated by a straight line on full log paper as follows;

$$t_r\,\dot{\varepsilon}_{min}^{\ 0.912} = 2.428\times10^{-2}$$

Therefore, the minimum creep rate well predicts the creep rupture time of this type of ceramics.

(5) In the case of the sintered silicon nitride, three types of creep rupture surface patterns were observed; (a) mirror, mist and hackle pattern, (b) large scale S.C.G. region and mirror pattern and (c) typical creep rupture surface which seemed to be constructed by multi-site cracking and their developments. On the other hand, all of the sintered silicon carbide specimens showed the same rupture surface composed of mirror, mist and hackle.

(6) Creep rupture maps for the sintered silicon nitride, which were constructed from two different viewpoints, indicated that the typical creep rupture surface could be observed only for the specimen ruptured at tertiary creep stage, in other words, tertiary creep stage was a coalescence stage of multi-site microcracks.

REFERENCES

1. T.Tanaka, S.Yamamoto, N.Okabe, H.Nakayama and Y.Fukuchi, On the Evaluation Method of Compressive Strength for Ceramics, *J. of the Society of Materials Science, Japan,* **38:** 437 (1989).
2. T.Tanaka, N.Okabe and A.Sakaida, Tensile Strength of Sintered Silicon Nitride at Room and High Temperatures, *J. of the Society of Materials Science, Japan,* **36:** 563 (1987).
3. T.Tanaka, N.Okabe, A.Sakaida and H.Nakayama, Fracture Toughness Tests of Sintered Silicon Nitride at Room and High Temperatures, *J. of the Society of Materials Science, Japan,* **38:** 261 (1989).
4. T.Tanaka, N.Okabe and Y.Ishimaru, Fatigue Crack Growth and Crack Closure of Silicon Nitride under Wedge Effect by Fine Fragments, *J. of the Society of Materials Science, Japan,* **38:** 137 (1989).
5. A.Ueno, H.Kishimoto, H.Kawamoto and M.Asakura, Crack Propagation Behaviors of Sintered Silicon Nitride under Cyclic Load of High Frequency and High Stress Ratio, *J. of the Society of Materials Science, Japan,* **39:** 1570 (1990).
6. H.N.Ko, Cyclic Fatigue Behavior of High-Strength Sintered Si_3N_4 under Rotating Bending, *J. of Materials Science Letters*, **10:** 545 (1991).
7. M.Okazaki, A.J.MacEvily and T.Tanaka, On the Mechanism of Fatigue Crack Growth in Silicon Nitride, *Metallurgical Trans., A,* **22A:** 1425(1991).
8. H.Nakayama, K.Ohira, N.Okabe and T.Tanaka, On the Fracture Strength of Fine Ceramics at High Temperatures under Impact 3-Point Bending Load, *J. of the Society of Materials Science, Japan,* **39:** 1630 (1990).
9. R.Kossowsky, D.G.Miller and E.S.Diaz, Tensile and Creep Strengths of Hot-Pressed Si_3N_4, *J. of Materials Science*, **10:** 983 (1975).

10. R.M.Arons and J.K.Tien, Creep and Strain Recovery in Hot-Pressed Silicon Nitride, *J. of Materials Science*, **15:** 2056 (1980).
11. R.K.Govila, High Temperature Uniaxial Tensile Stress Rupture Strength of Sintered Alpha SiC, *J. of Materials Science*, **18:** 1967 (1983).
12. A.Bauarroudj, P.Goursat and J.L.Besson, Oxidation Resistance and Creep Behavior of a Silicon Nitride Ceramics Densified with Y_2O_3, *J. of Materials Science*, **20:** 1150 (1985).
13. D.F.Carroll and R.E.Tressler, Accumulation of Creep Damage in a Siliconized Silicon Carbide, *J. of the American Ceramic Society*, **71:** 472 (1988).
14. R.K.Govila, Strength of Slip-Cast, Sintered Silicon Nitride, *J. of American Ceramics Society*, **73:** 1744(1990).
15. Z.C.Jou and A.V.Virkar, High-Temperature Creep and Cavitation of Polycrystalline Aluminum Nitride, *J. of American Ceramics Society*, **73:** 1928 (1990).
16. C.F.Chen and T.J.Chuang, Improved Analysis for Flexural Creep with Application to Sialon Ceramics, *J. of American Ceramics Society*, **73:** 2366 (1990).
17. T.Tanaka, N.Okabe, S.Yamamoto, H.Nakayama, A.Segawa and T.Fujii, Strength and Strain Behavior in High Temperature Creep of Fine Ceramics, *J. of the Society of Materials Science, Japan,* **39:** 1692 (1990).
18. W.R.Cannon and T.G.Langdon, Review: Creep of Ceramics (Part 1, Mechanical Characteristics), *J. of Materials Science*, **18:** 1(1983).
19. G.D.Quinn, Fracture Mechanism Maps for Advanced Structural Ceramics (Part 1, Methodology and Hot-Pressed Silicon Nitride Results), *J. of Materials Science*, **25:** 4361(1990).
20. G.D.Quinn and W.R.Braue, Fracture Mechanism Maps for Advanced Structural Ceramics (Part 2, sintered silicon nitride), *J. of Materials Science*, **25:** 4377 (1990).
21. T.Tanaka, N.Okabe, H.Nakayama, Y.Ishimaru and T.Imamichi, Effect of Stress Pattern on Fatigue Strength of Sintered Silicon Nitride, *J. of the Society of Materials Science, Japan,* **39:** 694 (1990).
22. Standard Test Method for Constant-Load-Amplitude Fatigue Crack Growth Rates above 10^{-8}m/cycle, ASTM Designation, E647-81: 765(1981).

OBSERVATIONS ON THE ROLE OF CRACKS IN THE NON-LINEAR DEFORMATION AND FRACTURE BEHAVIOR OF POLYCRYSTALLINE CERAMICS

D. P. H. Hasselman, K. Y. Donaldson and A. Venkateswaran

Department of Materials Engineering
Virginia Polytechnic Institute and State University
Blacksburg, VA 24061 USA

ABSTRACT

A review is presented of the authors' research on the role of cracks in the non-linear deformation and fracture behavior of polycrystalline structural ceramics. The presence of cracks can have two principal effects on the creep behavior, namely "crack-enhanced creep," which increases the rate of creep and "elastic creep by crack growth," due to the accompanying decrease in Young's modulus. These two mechanisms serve to explain observations on the anomalous creep behavior reported in the literature such as nonlinear creep at low stress levels, increasing creep rate with increasing grain size, anomalously high activation energies and diffusion coefficients and large differences in the creep rates in tension and compression.

Decreases in Young's modulus of a polycrystalline alumina subjected to displacement-controlled loading by as much as a factor of five confirms the significant role of elastic creep by crack-growth and crack-enhanced creep in the non-linear deformation of structural ceramics. Decreases in Young's modulus of another polycrystalline alumina by a factor of two at low displacement-controlled strain rates with no such decreases at high strain rates indicates the significant effect elastic creep can have on the loading rate-sensitivity of the strength of structural ceramics.

For an alumina with glassy grain boundary phase without and with artificial flaws very poor correlations were found for the stress dependence of the creep rate and the time-to-failure, but excellent correlations between creep rate and time-to-failure. These observations suggested the phenomenon of "crack-enhanced creep fracture."

INTRODUCTION

Many structural materials when subjected to mechanical loading can exhibit non-linear deformation (i.e., inelastic behavior). In ductile metals at ambient temperatures dislocation motion and multiplication represent the primary mechanisms for non-linear deformation [1-4]. For polycrystalline ceramics at elevated temperatures, non-linear

Fracture Mechanics of Ceramics, Vol. 10
Edited by R.C. Bradt *et al.*, Plenum Press, New York, 1992

deformation by diffusional creep is a generally well-accepted phenomenon [5-7]. Amorphous materials such as glasses and polymers deform non-linearly by viscous flow [8]. Such viscous flow can constitute the primary mechanism for non-linear deformation in polycrystalline ceramics with a glassy grain boundary phase [9-11]. In general, extensive experimental studies and analyses have shown that nonlinear deformation and its associated dominant mechanisms are highly dependent on stress (or stress rate) and temperature, as well as on microstructural variables such as grain size, porosity and crystallinity and/or the width or composition of the glassy grain boundary phase, if present.

The presence or formation of microcracks is increasingly recognized as a separate effect which can influence the deformation of brittle materials. In geological materials at ambient temperatures, the presence and/or growth of cracks can lead to nonlinear stress-strain behavior, creep, dilatancy, stress relaxation and stress-strain behavior which is a function of prior loading history [12-18]. Microcracking in geological materials also can lead to major reductions in elastic moduli [19-21]. During displacement controlled loading of materials such as rock and concrete, the formation of cracks and subsequent decrease in elastic moduli results in a reduction in stress, referred to as "strain softening" [22-26].

In brittle polycrystalline structural ceramics at ambient temperatures microcracks can form under the influence of internal stresses which arise from thermal anisotropy of the individual grains [27] or from thermal expansion mismatches in composites.

At elevated temperatures, microcrack formation in polycrystalline ceramics subjected to mechanical load can occur as the result of grain boundary sliding [28] or cavitation [29]. Viscous flow of a glassy grain boundary phase can lead to grain boundary separation. It is expected that if cracks can influence the mechaical behavior at ambient temperatures similar effects are expected to occur at elevated temperatures. Two such effects have been recognized.

The first, due to the transfer of stress from microcracked regions to regions without microcracks, results in an increase in the effective creep rate over the corresponding value found for the uncracked material. This mechanism, labelled "cracked-enhanced creep", was originally formulated by Weertman [30] for Griffith cracks and was extended to penny-shaped cracks by two of the authors of this paper [31].

For a plate with parallel non-interacting Griffith cracks of equal size, subjected to a tensile stress perpendicular to the plane of the cracks, the creep rate, $\dot{\epsilon}_c$, for linear creep is [30]:

$$\dot{\epsilon}_c = \dot{\epsilon}_o \left[1 + 2\pi N a^2\right] \quad (1)$$

where ϵ_0 is the creep rate in the absence of cracks, N is the number of cracks per unit area and a is the crack half-length. Of interest to note is that for any crack geometry and under conditions of linear creep, the relative increase in creep rate due to the presence of the cracks is equal to the relative decrease in Young's modulus. For a plate with parallel non-interacting Griffith cracks, the effective Young's modulus, E_{eff}, is given by [31,32]:

$$E_{eff} = E_0 \left[1 + 2\pi N a^2\right]^{-1} \quad (2)$$

where E_0 is Young's modulus of the crack-free plate. For this reason, in general, if a solution is available for the effect of the cracks on elastic behavior, the corresponding solution for the creep behavior is easily obtained. In order to maintain constancy of volume, this requires that the value for Poisson's ratio in the elastic solution is set equal to 0.5.

For non-linear creep, the rate of crack-enhanced creep is [30]:

$$\dot{\epsilon}_c = \dot{\epsilon}_0 \left[1 + 2\pi N a^2 n^{1/2}\right] \tag{3}$$

where n is the stress exponent for the underlying creep mechanism. Examination of eqs. (1), (2) and (3) show that the inverse of the relative effect of cracks on Young's modulus represents the lower bound on the relative effect of cracks on creep rate. For non-linear creep, the relative increase in the creep rate will always be higher than the corresponding relative change in specimen compliance.

From eq. (2), the elastic strain under conditions of an applied tensile stress, σ, is:

$$\epsilon_{el} = \sigma\left[1 + 2\pi N a^2\right] / E_0 \tag{4}$$

The crack density will be assumed to be constant, so that the elastic creep is the result of crack growth only. Differentiating with respect to time yields the rate of elastic creep by crack growth:

$$\dot{\epsilon}_{el} = 4\pi a \dot{a} \sigma N / E_0 \tag{5}$$

where $\dot{a}$ is the rate of crack growth.

The concepts of crack-enhanced and elastic creep were used effectively to explain several apparent anomalies in the creep and fracture behavior of structural ceramics [33], such as the transition from linear to power-law creep in uranium dioxide at levels of temperature and stress far below those predicted from theory [34], an anomalous grain size dependence of creep rate [35], discrepancies between diffusion coefficients inferred from creep data and those measured directly [36], an anomalously high activation energy for creep [37] and the relaxation of residual stresses in aluminum oxide at temperatures (850° C) far below those at which diffusional creep can make a significant contribution [38].

The purpose of this paper is to present a summary of the authors' observations (39-41) on the effect of elastic and crack-enhanced creep on the non-linear and fracture behavior of polycrystalline ceramics.

EXPERIMENTAL RESULTS, DISCUSSION AND CONCLUSIONS

A. Role of Cracks in the Stress-Strain Behavior of a Coarse-Grain Polycrystalline Aluminum Oxide [39]

Samples of a translucent polycrystalline aluminum oxide (VISTAL, Coors Ceramics Company) with a mean grain size of 38 μm consisted of circular rods about 50 mm long by about 6 mm in diameter. These samples were subjected to four-point bending in Argon at temperatures of 1650, 1700 and 1750°C at strain rates ranging from 0.002 to 0.217 min^{-1}. Figure 1 shows an SEM-micrograph of an as-fired surface. Figure 2 shows the observed stress-strain behavior for a number of samples deformed to fracture at 1700°C over a range of strain rates. Pronounced non-linear behavior, especially at the lower values of displacement rate, are evident. Note that at a displacement rate of 0.005 min^{-1} the stress-strain curve exhibits a negative slope. This effect is not expected from the usual mechanism for non-linear deformation, but is in accordance with the mechanism of strain-softening under displacement-controlled loading conditions, during which Young's modulus decreases faster than the corresponding increase in displacement.

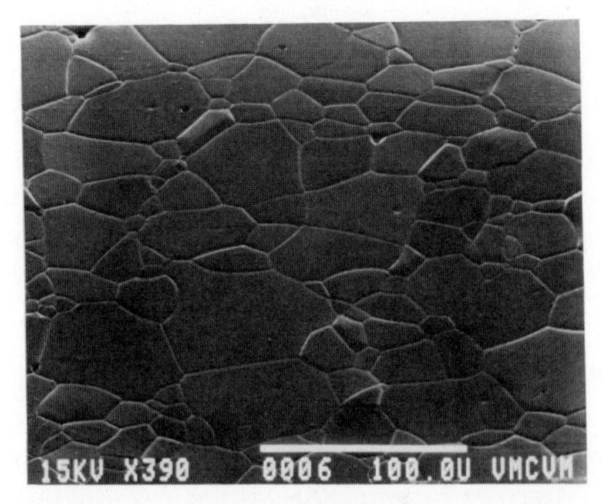

Figure 1. SEM-micrograph of as-fired surface of VISTAL alumina.

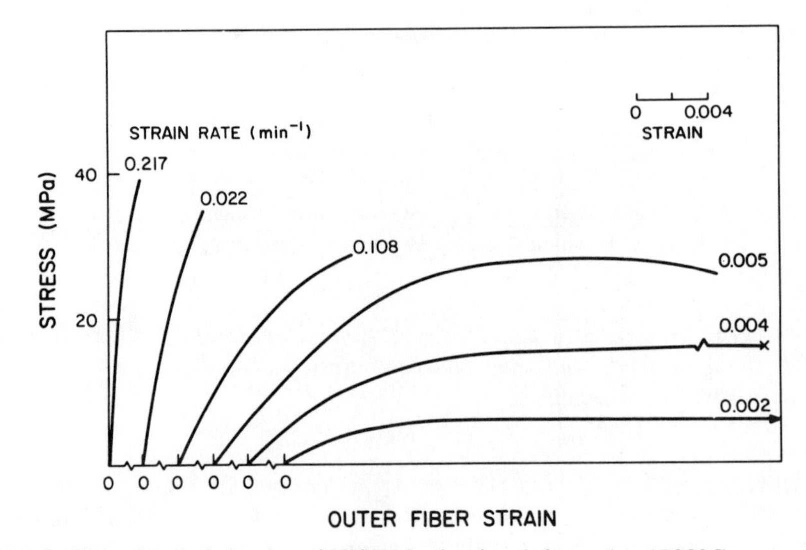

Figure 2. Stress-strain behavior of VISTAL alumina deformed at 1700°C over a range of strain rates.

A number of additional samples were loaded over a range of displacement rates to approximately 80 to 90% of the failure strain, after which Young's modulus was measured at room temperature. Figure 3 shows the results obtained for specimens deformed at 1700°C. Numerically almost identical results were obtained for samples deformed at 1650°C. Note that Young's modulus following deformation at a rate of 0.004 min^{-1} has a minimum value which is only about 20% of the value prior to deformation. At the lowest values of displacement rate the values of Young's modulus higher than this minimum result because the stress level does not reach a value sufficiently high for microcrack formation to occur. At the highest values of displacement rates, although the stresses are sufficiently high, the time period under load is too short for extensive microcrack formation to occur. Figure 4 shows a cross section near the position of maximum stress of a sample deformed at 0.0043 min^{-1} at 1650°C. Crack formation is clearly evident.

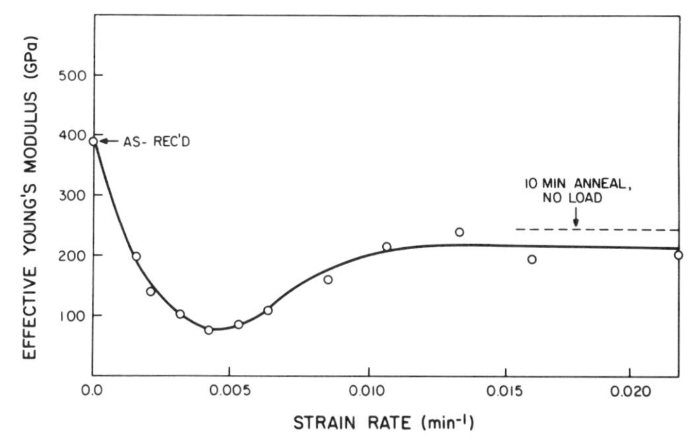

Figure 3. Young's modulus at room temperature of VISTAL alumina following deformation to approximately 80 to 90% of failure strain for a range of strain rates.

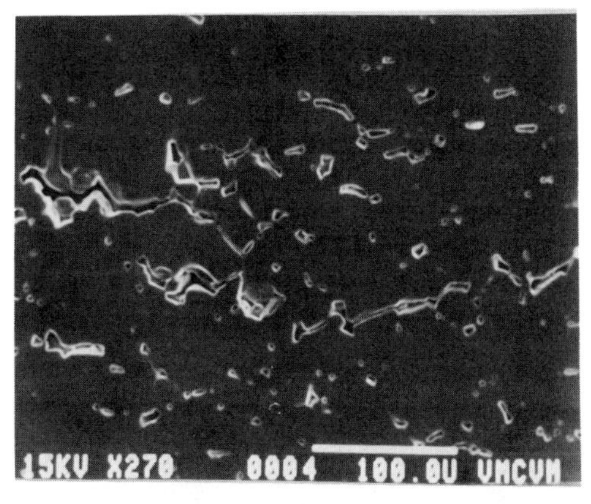

Figure 4. SEM-micrograph of polished section through VISTAL alumina deformed at 1650°C at a strain rate of 0.0043 min^{-1} near the position of maximum stress at the right-hand side of the micrograph.

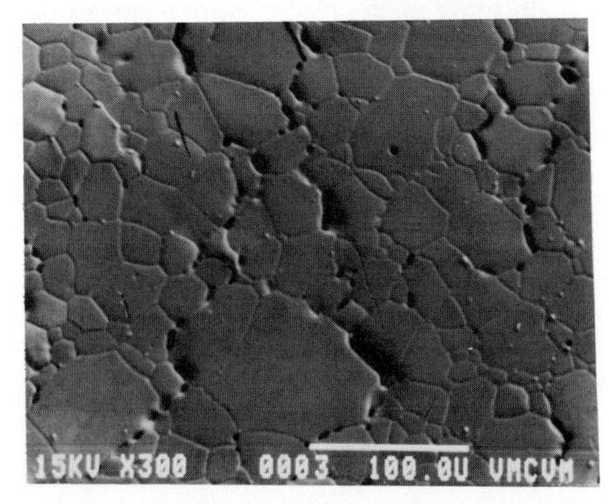

Figure 5. SEM-micrograph of surface of VISTAL alumina deformed at 1650°C at a strain rate of 0.0043 min^{-1} at position of maximum stress.

Figure 5 shows the corresponding crack formation at the surface. At the higher and lower displacement rates the density and dimensions of the microcracks were found to be lower than those shown in figs. 4 and 5. It should be noted that the observed crack density was not sufficient to explain the observed decrease in Young's modulus by a factor of five. It is speculated that the microstructure exhibits additional "damage", not observable by scanning-electron-microscopy. Theoretical modelling which included the effect of crack-enhanced and elastic creep by crack growth confirmed that the decrease in Young;'s modulus would be a maximum at intermediate displacement rates.

The above observations are pertinent to the interpretation of non-linear stress-strain curves. It should be recognized that the component of elastic strain due to the decrease in Young's modulus, assuming reversible crack closure, is recovered on unloading. Under load, however, the decrease in Young's modulus results in an elastic component of the strain, which is superposed on those strain components which arise from true non-linear mechanisms of deformation. This elastic component, under load, yields an "apparent" non-linear contribution to the overall stress-strain curve. It is the view of the present writers that many observations of the non-linear deformation of structural ceramics at elevated temperatures can be attributed, at least in part, to increases in elastic strain due to decreases in Young's modulus as the result of crack formation.

B. Role of Multiple Crack Formation and Associated Decrease in Young's Modulus in the High-Temperature Strain-Rate Sensitivity of the Failure Stress [40]

Samples for this study consisted of a relatively fine-grained polycrystalline aluminum oxide known as AlSiMag 838 made by the General Electric Company. The SEM-fractograph shown in fig. 6 indicates that this material is nearly fully dense with an average grain size of approximately 5 μm with an occasional grain as large as about 20 μm. Annealed specimens of this material in the form of circular rods with a length of about 50 mm and a diameter of approximately 4.8 mm were deformed to fracture in four-point bending over a range of strain rates at temperatures of 1350, 1400, and 1450°C under displacement-controlled loading conditions. Additional specimens were deformed to about 75 and 85% of the failure strain. Young's modulus was then measured at room temperature.

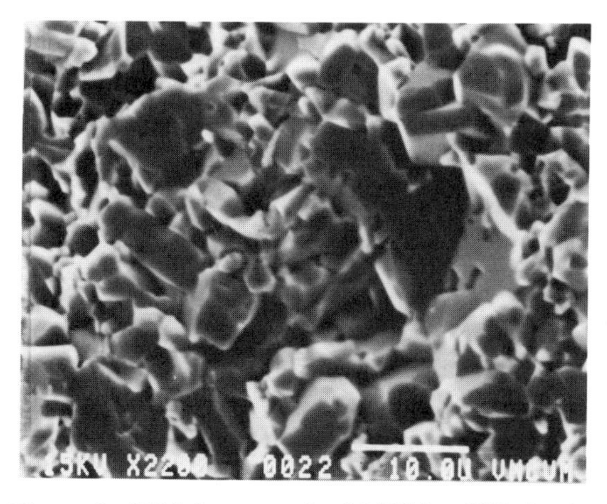

Figure 6. SEM-fractograph of AlSiMag 838 alumina

Figure 7 shows the stress-strain curves at 1400°C for a range of strain rates. Again, the deformation is highly non-linear. The decreasing stress with increasing strain due to the effect of "strain-softening" as the result of crack formation is clearly evident. Figure 8 is an SEM-micrograph of a polished cross section of a specimen deformed at 1400°C at a strain rate of 0.08 min^{-1} near the position of maximum tensile stress, which is at the right-hand edge of the photo-micrograph. Extensive crack formation is evident. The degree or extent of crack formation was found to decrease with increasing strain rate. This effect arises because at the lower strain rates, although the stresses are lower, more time is available for crack growth to occur.

Figure 9 shows the experimental data for the fracture stress as a function of strain rate at 1400 and 1450°C. A very strong strain-rate sensitivity is clearly evident.

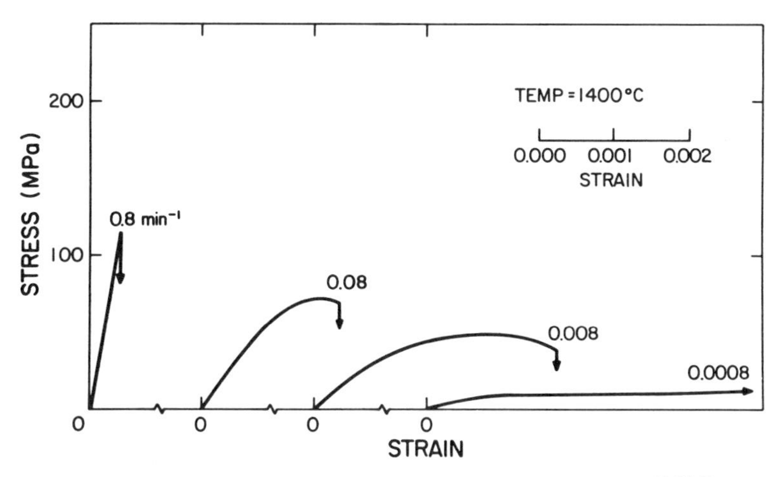

Figure 7. Stress-strain curves of AlSiMag 838 alumina deformed at 1400°C over range of strain rates.

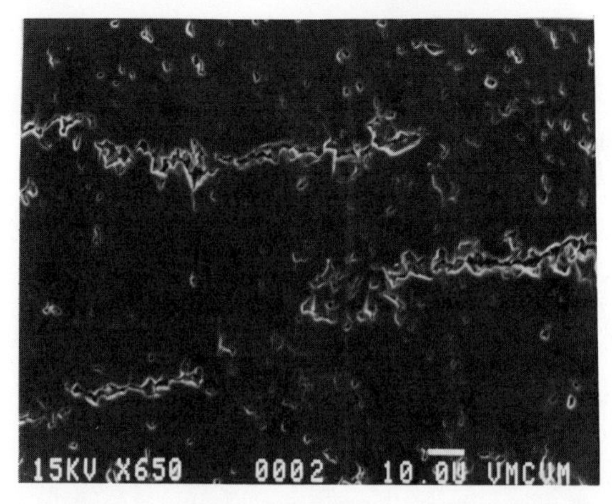

Figure 8. SEM-micrograph of polished section of AlSiMag 838 alumina deformed at 1400°C at a strain rate of 0.08 min^{-1} near the position of maximum tensile stress at the right-hand edge of micrograph.

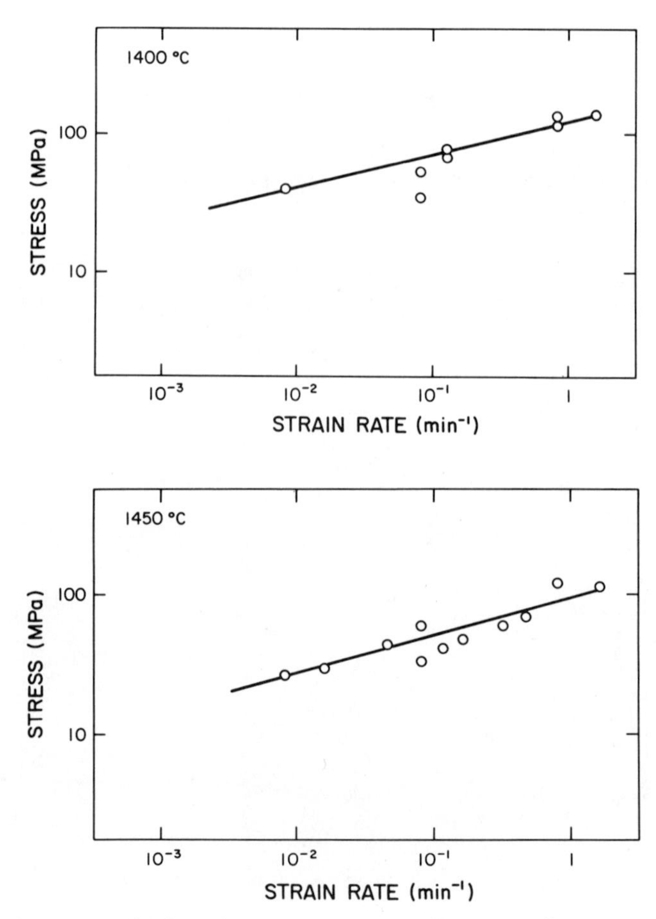

Figure 9. Strain-rate sensitivity of fracture stress of AlSiMag 838 alumina at 1400 and 1450°C.

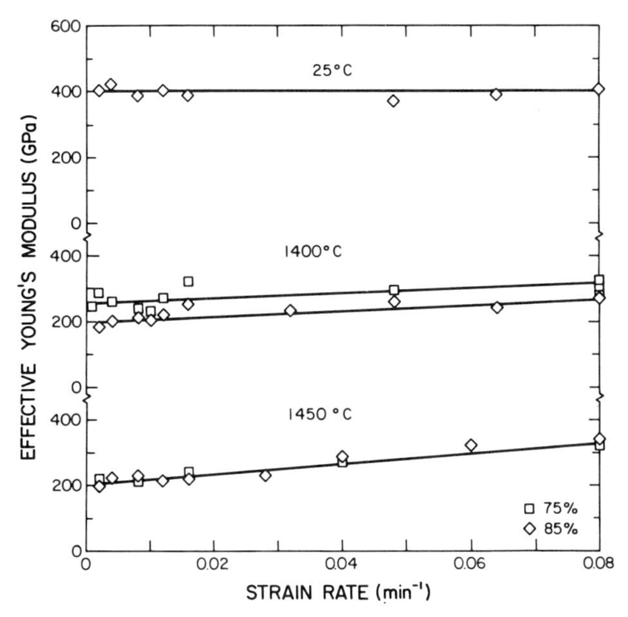

Figure 10. Young's modulus of 838 alumina at room temperature following deformation at 1400 and 1450°C to approximately 75% (□) and 85% (◇) as a function of strain rate.

Figure 10 shows the data for Young's modulus following deformation to approximately 75 or 85% of the failure strain as a function of strain rate at 1400 and 1450°C. Included in the figure are the corresponding data at 25°C, which shows no dependence on strain rate, as expected. For the samples subjected to deformation at 1400 and 1450°C, Young's modulus increases strongly with increasing strain rate or in other words, the decrease in Young's modulus was greater at the lower values of strain rate than at the higher values. This latter observation is in accordance with the observed decrease in crack density with increasing strain rate. Such a decrease in Young's modulus under conditions of displacement controlled loading has the effect of lowering the resulting stress at any given value of displacement from the value which would have resulted if Young's modulus had remained unaffected. Because this effect was observed to be a function of strain rate, it represents a variable which can affect the strain-rate sensitivity of the failure stress in addition to subcritical crack growth. Both effects need to be considered in assessing the strain-rate sensitivity of structural ceramics.

C. Crack-Enhanced Creep-Fracture of Structural Ceramics [41]

The ceramic chosen for this study consisted of a polycrystalline alumina with glassy grain boundary phase made by the Coors Ceramics Company, designated as AD-94. Figure 11 shows an SEM-micrograph, which indicates a mean grain size of approximately 5 to 6 μm. Figure 12 shows the distribution of Mg, which is the principal constituent of the glassy phase. Annealed specimens were subjected to a constant stress at 1225°C. Both the creep rate as well as the time-to-failure were noted. The creep rate to be reported corresponded to the minimum creep rate within the secondary creep regime. Figures 13a and b show the experimental data for the stress dependence of the creep rate and time-to-failure. Figure 13c shows the Monkman-Grant plot of the creep rate and time-to-failure.

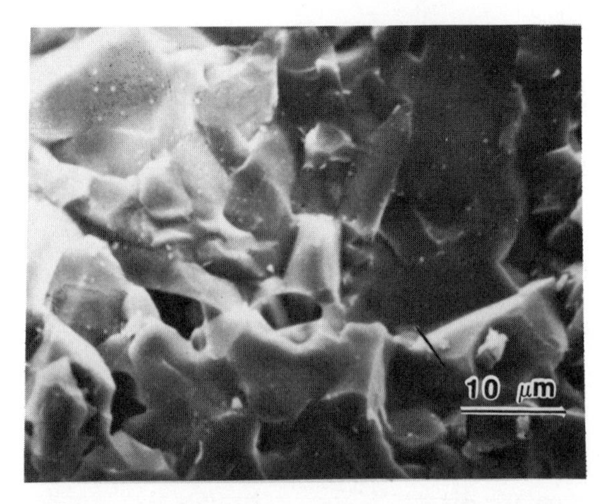

Figure 11. SEM-fractograph of AD-94 alumina.

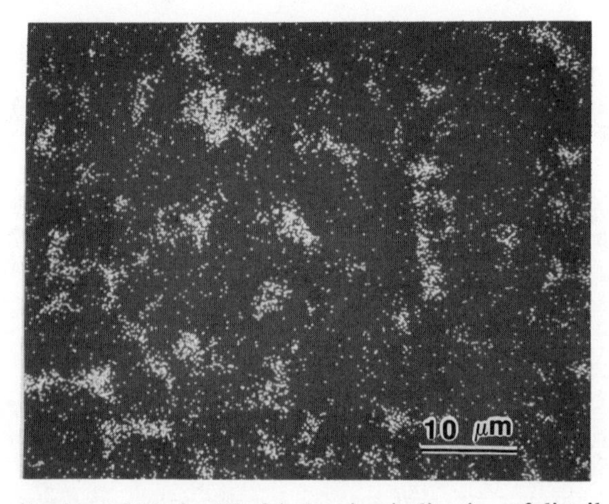

Figure 12. Distribution of Mg in AD-94 alumina indicative of distribution of glassy phase.

Noteworthy of the data shown in figs. 13a and b is that both sets of data for the creep rate and time-to-failure show very poor correlations with the magnitude of stress. However, much better direct correlation is obtained between creep-rate and time-to-failure. These differences in correlations have some interesting implications. The viscous deformation of the glassy grain boundary phase is expected to be the primary mechanism of creep. The time-to-failure is governed by the growth of cracks. At the temperature of these experiments, the pockets of glassy phase most likely represent the precursors of the cracks, which by the mechanism of cavitation grow along the glassy grain boundaries. At least in principle, the rate of creep deformation can be calculated from the viscosity of the glassy phase and microstructural information such as the width of the glassy phase along the grain boundary and other relevant details. Such a calculation of the creep rate, however, does not require information on the rate of crack growth. Conversely, the time-to-failure can be calculated on the basis of fracture-mechanical principles from information on the size of the crack precursor and data for the rate of crack growth as a

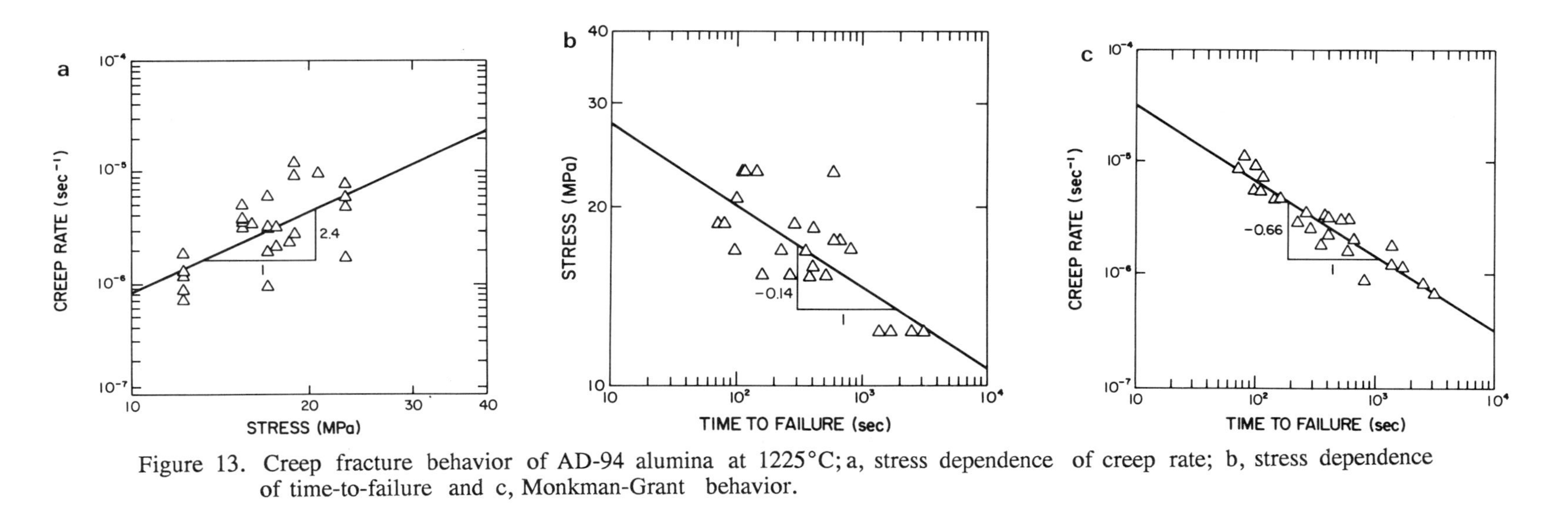

Figure 13. Creep fracture behavior of AD-94 alumina at 1225°C; a, stress dependence of creep rate; b, stress dependence of time-to-failure and c, Monkman-Grant behavior.

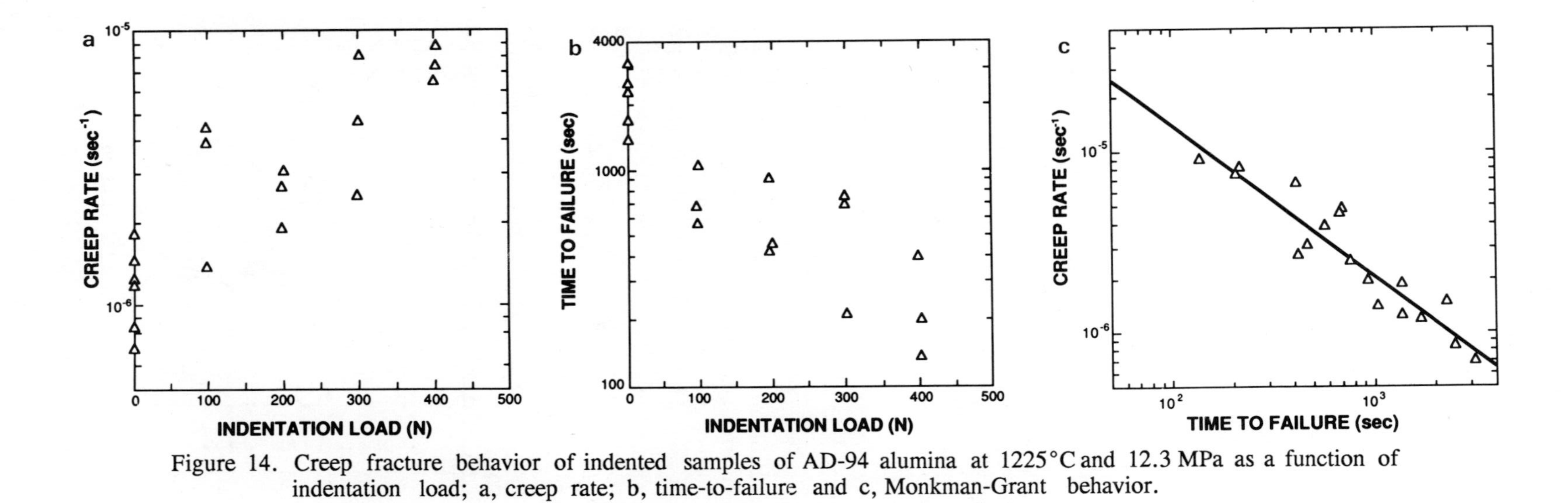

Figure 14. Creep fracture behavior of indented samples of AD-94 alumina at 1225°C and 12.3 MPa as a function of indentation load; a, creep rate; b, time-to-failure and c, Monkman-Grant behavior.

function of stress intensity factor. Again, such calculation does not require information on the kinetics of creep deformation. Clearly, both the rate of creep and the time-to-failure depend on the viscous flow of the glassy phase. For this reason, general correlations for data between the rate of creep and time-to-failure over a range of stress or temperature are to be expected. Nevertheless, different specimens may exhibit the same rate of creep, but widely differing times-to-failure and vice-versa. The data of fig. 13, however, suggest the existence of a direct specimen-to-specimen correlation. This, in turn, suggests that the presence of the failure-causing cracks plays a role in the rate of creep deformation. This hypothesis was confirmed by measuring the creep and creep fracture behavior of specimens which were pre-cracked by means of hardness indentations using a range of indentation loads. Independent measurements showed that indentation loads of 100, 200, 300 and 400 N resulted in indentation cracks with a radius of approximately 300, 550, 700 and 800 μm, respectively.

Figure 14 shows the dependence of the creep rate and time-to-failure at 1225°C and stress of 14.5 MPa as a function of indentation load together with the accompanying Monkman-Grant plot. The rate of creep increases rapidly with indentation load (i.e., crack size), accompanied by corresponding decreases in the time-to-failure. It is of interest to note that the creep rate and time-to-failure show similar poor correlations with the magnitude of stress as the corresponding data for the as-received (non-indented) specimens, reflective of the heterogeneous nature of the material of this study. The much better correlation for the Monkman-Grant behavior appears to be independent of the indentation load. The rapid increase in creep rate with indentation load (i.e., crack size) suggests that inherent cracks with dimensions only a small multiple of the grain size of the material of this study are sufficiently large to have a measurable effect on creep rate.

Figure 15 shows the experimental data for the dependence of the specimen compliance on indentation load. An increase of specimen compliance with indentation load is observed. This relative dependence, as discussed earlier, represents the lower bound on the corresponding relative dependence on creep rate. Comparison with the data shown in fig. 14 shows that the relative increase in creep rate with increasing indentation load is far greater than the corresponding increase in specimen compliance, as expected for non-linear creep. It may be added in passing that similar results for the creep-rupture behavior were obtained with samples with artificial cracks in the form of slots introduced by diamond-sawing [41].

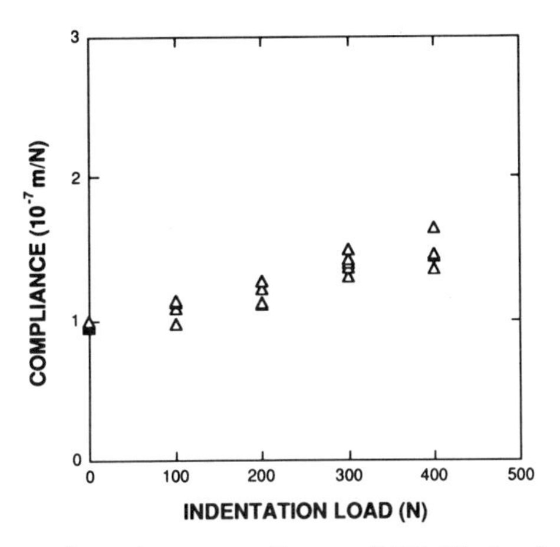

Figure 15. Dependence of specimen compliance of AD-94 alumina on identation load.

The data for the creep behavior shown in fig. 14 represents positive proof that the presence of a crack influences the kinetics of the creep process, i.e., supports the existence of crack-enhanced creep. This is in support of the interpretation given to the data presented in fig. 13. The associated decrease in the time-to-failure prompted the authors of this study to label the failure mechanism as "crack-enhanced creep fracture."

In general it is suggested that in the studies of the creep behavior of structural ceramics, measurement of the associated time-to-failure be included as well, as it can provide useful information for improved interpretation of the creep behavior.

GENERAL CONCLUSIONS

The overview of the experimental data presented in this paper offers strong evidence for the role of elastic creep by crack growth and crack-enhanced creep in the non-linear deformation, the strain-rate sensitivity and the creep fracture behavior of structural ceramics.

ACKNOWLEDGEMENT

The present paper was prepared with support provided by the Army Research Office under Contract No.: DAAL03-88-K-0073.

REFERENCES

1. A. H. Cottrell, "Dislocations and Plastic Flow in Crystals," Clarendon Press, Oxford (1956).
2. G. E. Dieter, "Mechanical Metallurgy," 2nd ed. McGraw-Hill, New York (1976).
3. W. D. Kingery, H. K. Bowen and D. R. Uhlmann, "Introduction to Ceramics," 2nd ed. Wiley, New York, NY (1976).
4. J. Weertman, Am. Soc. Met. Trans. Q., 61 (1968) 681.
5. F. R. N. Nabarro, p. 75 in "Report on a Conference on Strength of Solids," The Physical Society, London (1948).
6. C. Herring, J. Appl. Phys., 21 (1950) 437.
7. R. L. Coble, J. Appl. Phys., 34 (1964) 1679.
8. H. S. Y. Hsich, J. Mat. Sci., 15 (1980) 1194.
9. R. Morrell and K. H. G. Ashbee, J. Mat. Sci., 8 (1973) 1253.
10. R. Kossowsky, D. G. Miller and E. S. Diaz, J. Mat. Sci., 10 (1975) 1983.
11. F. F. Lange, pp. 361-81 in "Deformation of Ceramic Materials." Edited by R. C. Bradt and R. E. Tressler, Plenum Press, New York (1975).
12. W. F. Brace and E. G. Bombolakis, J. Geophys. Res., 68 (1963) 3709.
13. N. G. W. Cook and K. Hodgson, J. Geophys. Res., 70 (1965) 3709.
14. W. G. Brace, B. W. Paulding, Jr., and C. Scholz, J. Geophys. Res., 71 (1966) 3939.
15. C. H. Scholz, J. Geophys. Res., 73 (1968) 1947.
16. C. H. Scholz, J. Geophys. Res., 73 (1968) 3295.
17. D. J. Holcomb, J. Geophys. Res., 86 [B7] (1981) 6235.
18. P. L. Swanson, pp. 299-318 in "Fracture Mechanics of Ceramics," Vol. 8, edited by R. C. Bradt, A. G. Evans, D. P. H. Hasselman and F. F. Lange, Plenum Press, New York, NY (1968).
19. J. B. Walsh, J. Geophys. Res., 70 (1965) 381.
20. R. L. Salganik, Izv. Akad. Nauk SSR Mekh. Tverd. Tela, 8 (1973) 149.

21. B. Budiansky and R. J. O'Connell, Int. J. Solid Struct., 12 (1976) 81.
22. S. K. Ghosh and M. Z. Cohn, pp. 315-32 in "Inelasticity and Nonlinearity in Structural Concrete," ed. by M. Z. Cohn, University of Waterloo Press, Waterloo, Canada (1973).
23. G. Maier, A. Zanelani and J. Dotreppe, J. Eng. Mech. Div., Am. Soc. Civ. Eng., 102 [E2] (1984) 1015.
24. Z. P. Bazant, J. Eng. Mech. Div., Am. Soc. Civ. Eng., 102 [E2] (1976) 331.
25. Z. P. Bazant and B. H. Oh, J. Eng. Mech. Div., Am. Soc. Civ. Eng., 110 (1984) 1015.
26. E. P. Chen and L. M. Taylor, pp. 175-86 in "Fracture Mechanics of Ceramics," Vol. 7, ed. by R. C. Bradt, A. G. Evans, D. P. H. Hasselman, and F. F. Lange, Plenum Press, New York, NY (1986).
27. A. G. Evans, Acta. Metal., 26 (1978) 1845.
28. A. G. Evans, Acta Metal., 28 (1980) 1155.
29. T. J. Chuang, K. I. Kagawa, J. R. Rice and L. B. Sills, Acta Metal., 27 (1979) 265.
30. J. Weertman, Trans. Am. Soc. Met., 62 (1969) 502.
31. D. P. H. Hasselman and A. Venkateswaran, J. Mat. Sci., 18 (1983) 161.
32. A. Venkateswaran and D. P. H. Hasselman, J. Mat. Sci., 19 (1981) 1627.
33. D. P. H. Hasselman and A. Venkateswaran, pp. 525-45 in "Deformation of Ceramic Materials," ed. by R. E. Tressler and R. C. Bradt, Plenum Press, New York, NY (1984).
34. J. B. Ainscough, F. Rigby and S. A. Morrow, J. Am. Ceram. Soc., 64 (1981) 315.
35. B. Burton, G. L. Reynolds and J. B. Barnes, J. Mat. Sci., 8 (1973) 1690.
36. A. Venkateswaran and D. P. H. Hasselman, J. Am. Ceram. Soc., 67 (1984) C-144.
37. S. I. Warshaw and F. H. Norton, J. Am. Ceram. Soc., 45 (1962) 479.
38. Y. Tree, A. Venkateswaran and D. P. H. Hasselman, J. Mat. Sci., 18 (1983) 2135.
39. D. P. H. Hasselman, A. Venkateswaran and K. Y. Donaldson, J. Mat. Sci., 24 (1989) 671.
40. A. Venkateswaran, K. Y. Donaldson and D. P. H. Hasselman, J. Am. Ceram. Soc., 71 (1988) 565.
41. K. Y. Donaldson, A. Venkateswaran, D. P. H. Hasselman, J. Mat. Sci., (in review).

FRACTURE TOUGHNESS MEASUREMENT

BY INDENTATION FRACTURE METHOD AT ELEVATED TEMPERATURE

Shuji Sakaguchi, Norimitsu Murayama,
Yasuharu Kodama and Fumihiro Wakai

Government Industrial Research Institute, Nagoya
1-1, Hirate-cho, Kita-ku, Nagoya 462 JAPAN

ABSTRACT

Fracture toughness measured by indentation fracture method at elevated temperature up to 1200°C was investigated. Three Kinds of ceramic samples, such as silicon nitride, silicon carbide and alumina, were tested with using high temperature Vickers hardness tester. Vickers indent loads were 5kg and 10kg, and indent diagonal length and median crack length were measured. High temperature test was conducted in argon atmosphere. Chevron notched beam technique was used to obtain the values of fracture toughness for the reference. Young's modulus and internal friction were measured by the mechanical resonance.

Fracture toughness by indentation fracture method is usually obtained by the following equation, but the exponent, t, and the coefficient, A, were determined by the room temperature experiments.

$$K_{IC}=A\cdot(E/H)^{t}\cdot(P/c^{3/2})$$

Then, the exponent, t, for E/H is investigated for applying this equation to high temperature measurement.

The ratio of Young's modulus, E, and Vickers hardness, H, does not show large difference on each sample at room temperature, namely E/H is about 15 or 20. However, Vickers hardness decreases more rapidly than Young's modulus with increasing temperature, and E/H becomes about 40 on silicon carbide and alumina at 1200°C. Then, the exponent for E/H can be evaluated more precisely by the high temperature measurement than by the room temperature measurement. This exponent is not unique but decreases with increasing E/H value, and it becomes 0 when E/H is larger than 25, where it is 1.5 when E/H is smaller than 25.

On the other hand, internal friction was investigated for the determination of the limit temperature for applying this method. This method cannot be applied if the sample becomes plastic, because the residual stress by the Vickers indent should be different. Internal friction increases exponentially with increasing temperature over a certain temperature, and this temperature should correspond to the limit temperature for applying the indentation fracture method. On the present experiment, silicon nitride measured at 1200°C corresponds to this case, then the datum was neglected from the analysis.

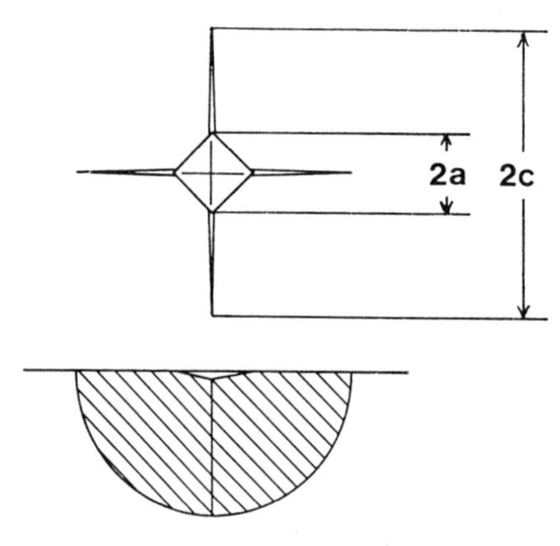

Fig. 1 Shapes of the Vickers indent and the median crack.

INTRODUCTION

Indentation fracture method is the most convenient technique to measure the fracture toughness of engineering ceramics.[1] A lot of studies were carried out to determine the empirical equation for this technique, and in Japanese Industrial Standard, JIS, it is standardized as JIS R1607.[2] However, these studies were carried out only at room temperature. There are very few researches that this technique is applied at elevated temperature,[3] or the empirical equation is discussed for the high temperature measurement.[4] In this study, the equation for the fracture toughness calculation by indentation fracture method is discussed with the results of the measurement at elevated temperature up to 1200°C.

Equation (1) is the general form of the equation for obtaining the fracture toughness, K_{IC}, by indentation fracture method.

$$K_{IC}=A\cdot(E/H)^{t}\cdot(P/c^{3/2}) \quad (1)$$

where A is a coefficient, E is Young's modulus, H is Vickers hardness, t is an exponent for the dimensionless value of (E/H), P is the load for Vickers indent and c is the crack length by the indent (Fig. 1). In the former researches, t is reported as 0,[5,6] 0.4[1,7] and 0.5.[2,8,9] However, these values are determined from the data by the room temperature tests, then the range of E/H is relatively small, namely 15 to 20 for engineering ceramics. WC/Co was used for widen the E/H range,[1] where E/H is about 44 for WC/Co, but it has two problems to analyze the results with engineering ceramics. First, as it has higher fracture toughness value compare to engineering ceramics, it is doubtful that the Vickers indent forms the appropriate median crack on this material. Second, if the indent is formed from the plastic deformation, the residual stress around the indent should be different from that on the ceramics, then the relation between the crack length and the fracture toughness should be different.

Three kinds of the experimental data are required to determine the coefficient, A, and the exponent, t, on the equation (1). (1) High temperature Vickers hardness test for the measurement of the indent diagonal, 2a, and the median crack length, 2c. (2) Fracture toughness measurement at elevated temperature for the reference. (3) Young's modulus measurement at elevated temperature. For fracture toughness measurement, chevron notched beam method was used. Young's modulus was measured by the mechanical resonance. Internal friction is obtained simultaneously with the Young's modulus measurement, and we take into account the result of the internal friction to the discussion.

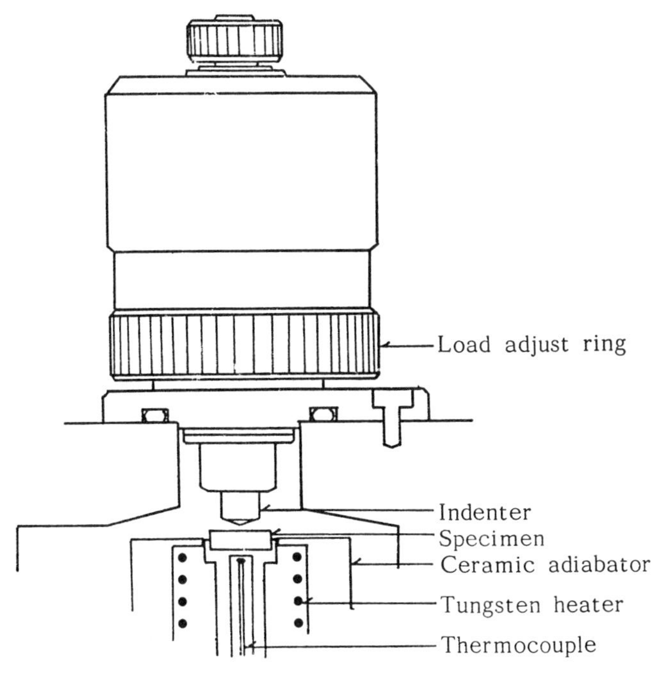

Fig. 2 Loading and indenting part of the high temperature Vickers hardness tester.

EXPERIMENTALS

The tested samples are hot-pressed silicon nitride with yttria and alumina as sintering additives, pressureless sintered silicon carbide with boron and carbon compound as additives and pressureless sintered 99.9% alumina. We also tested yttria stabilized tetragonal zirconia polycrystals (Y-TZP), but the data is not included for the analysis because the crack formation is different from other materials.

High Temperature Vickers Hardness Test

High temperature Vickers hardness test was carried out with using the equipment of Type AVK-HF hardness tester made by Akashi Co. The maximum temperature is 1200°C and the maximum load is 50kg. Fig. 2 shows the loading and indenting part of the equipment. Hardness test was carried out under argon atmosphere with using diamond Vickers indenter. Six indents were put continuously as quick as possible, and the first indent was neglected, because the indenter was not heated and the indent temperature is not stable at the first indent. The indent size 2a and the median crack length 2c are measured at the tested temperature. Applied loads are 5 and 10kg for silicon nitride, silicon carbide and alumina.

Chevron Notched Beam Test

Chevron notched beam method was used to obtain the reference values of fracture toughness at elevated temperature. Fig. 3 shows the plane section of the chevron notch. The calculation of the fracture toughness was carried out with using the equation (2) from the maximum load, P, during the flexural test.[10] This equation is obtained by the method of Munz,[11] which is based on the concept of slice model by Bluhm [12]

$$K_{IC}=(P/BW^{1/2})(2.91+4.97\alpha_0+11.64\alpha_0^2) \times[(S_1-S_2)/W](\alpha_1-\alpha_0)/(1-\alpha_0) \quad (2)$$

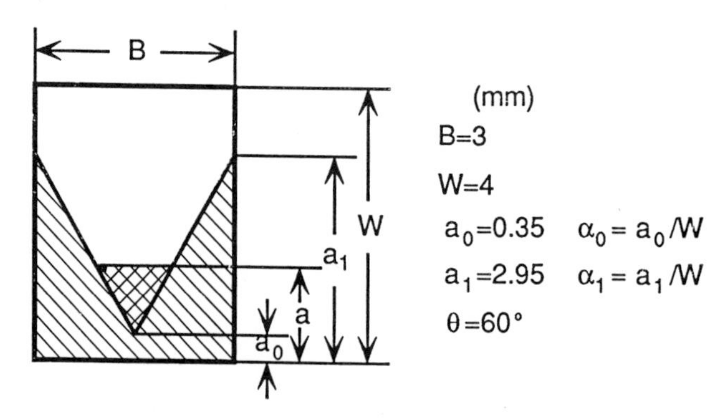

Fig. 3 Chevron notch plane section and the notations for the equation (2).

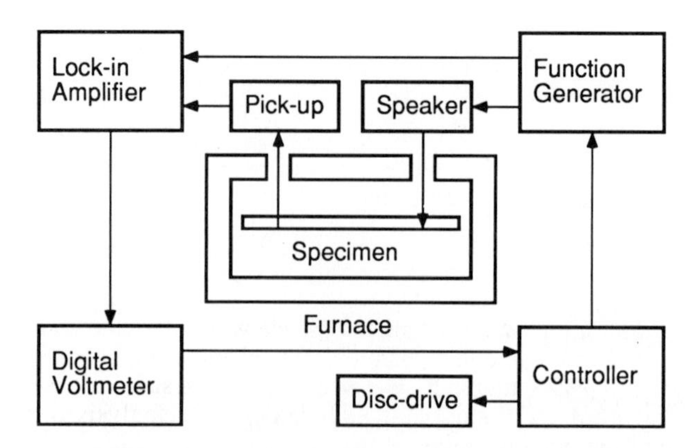

Fig. 4 Block diagram of the Young's modulus and internal friction measurement system.

where S_1 and S_2 are outer and inner spans, those are 30 and 10mm, respectively. Other notations and the size of the specimen are shown in Fig. 3. The thickness of the machined notch was 0.1mm. The cross-head speed of the flexural test was 0.05mm/min.

Young's Modulus and Internal Friction Test

Young's modulus and internal friction are measured by the mechanical resonance technique.[13-15] Fig. 4 shows the block diagram of the measurement system. Molybdenum silicide was used for heating element and the specimen is heated in air. Specimen was hanged into the furnace with mullite fiber. The specimen was a rectangular bar, and the size was 100x 20x 2mm.

Calculation of the Young's modulus and internal friction is as follows. First of all, the frequency of the vibration , f,is swept with using the function generator, and the signal ,S, of the vibration of the specimen is detected by the pick-up. The relation between f and S is approximated to the equation (3), which describes the forced vibration with damping, by the method of least squares. Then, the parameters f_0, k and I are determined.

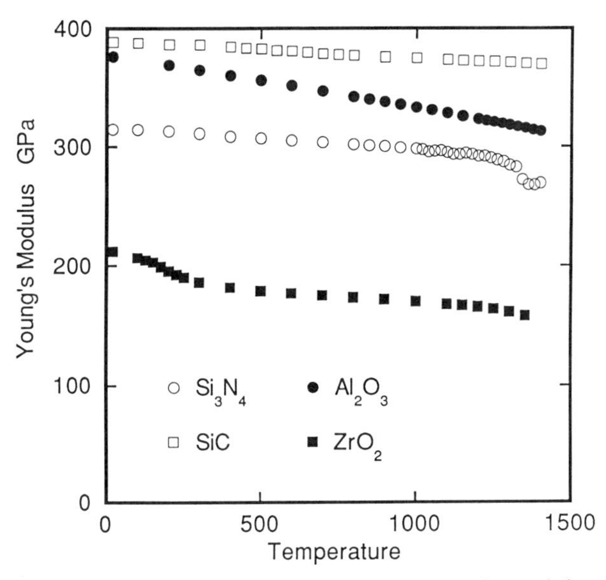

Fig. 5 Temperature dependence of the Young's modulus of each specimen.

$$S=\frac{I/4\pi^2 m}{\sqrt{(f^2-f_0^2)^2+k^2f^2/\pi^2}} \quad (3)$$

where m is mass of the specimen.

Young's modulus is obtained with equation (4), and internal friction, Q^{-1} is obtained with equation (5) with f_0 and k.

$$E=0.9465\frac{m\cdot f_0^2}{b}\left(\frac{l}{t}\right)^3\left[1+6.59\left(\frac{t}{l}\right)^2\right] \quad (4)$$

$$Q^{-1}=k/\pi f_0 \quad (5)$$

where l, b and t are the length, width and thickness of the specimen, respectively.

RESULTS

Fig.5 and 6 show the temperature dependence of the Young's modulus and Vickers hardness of each sample. The degradation of the Vickers hardness with elevating temperature is much larger than that of Young's modulus. Especially, on silicon carbide, the degradation of the Young's modulus is very small up to 1400°C, but the Vickers hardness at 1000°C is less than half of that at room temperature. Silicon carbide is well known by its high resistant property against high temperature on bending test. However, the Vickers hardness decreases rapidly. As, the hardness is related to the compressive strength,[16] the temperature dependence of the compressive strength of silicon carbide should be discussed.

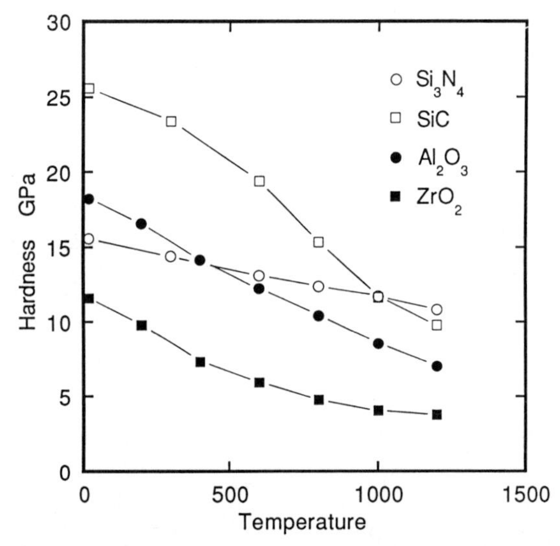

Fig. 6 Temperature dependence of the Vickers hardness of each specimen.

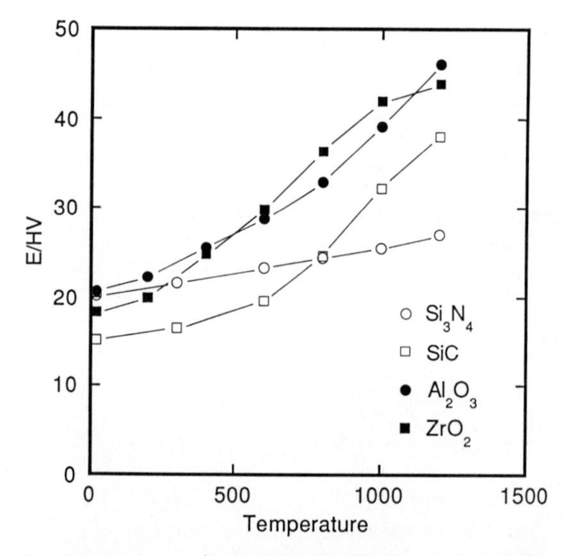

Fig. 7 Temperature Dependence of the ratio of Young's modulus and Vickers hardness. The E/H values are about 15 or 20 at room temperature, but the values increase with increasing temperature.

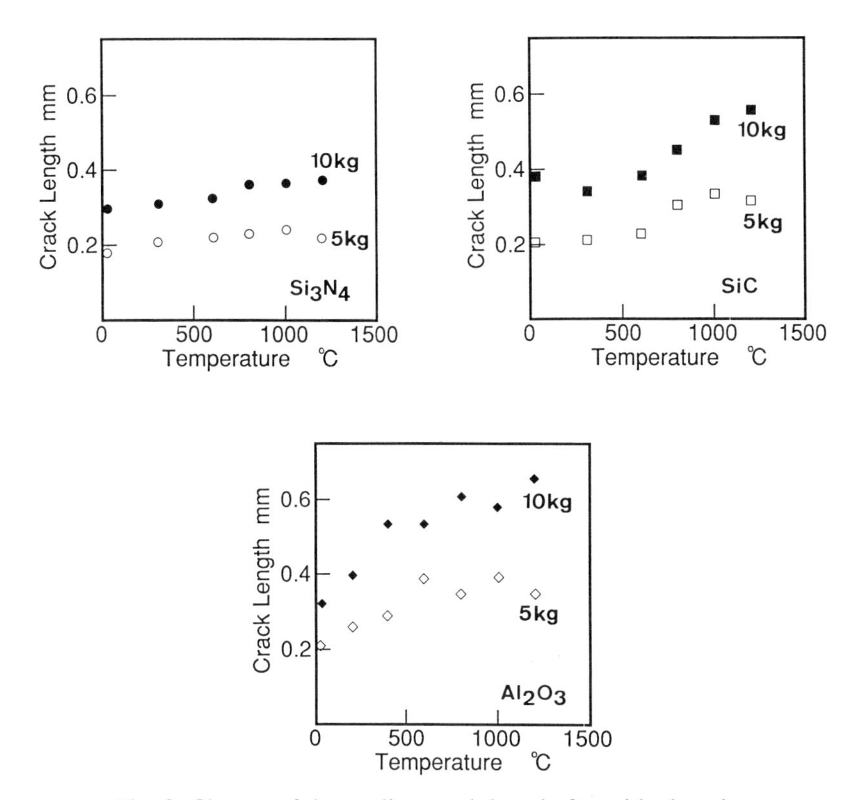

Fig. 8 Change of the median crack length, 2c, with changing temperature on silicon nitride, silicon carbide and alumina at two load levels.

Fig. 7 shows the temperature dependence of E/H value of each sample. At room temperature, all the samples show the E/H value about 15 or 20, but it increases with increasing temperature. It becomes more than 40 for alumina above 1000°C, then, these data are good for determining the exponent for E/H.

Fig 8 shows the change of the crack length of each material with increasing temperature at each load level. Silicon carbide shows little degradation of fracture toughness with increasing temperature, as it is shown afterwards, but the change of the crack length is observed because of the significant change of the Vickers hardness.

Fig. 9 shows the fracture toughness of each sample measured by chevron notched beam method. Each point shows the average of five samples. These curves are used as the reference of fracture toughness.

DISCUSSIONS

Internal Friction

Fig. 10 shows the temperature dependence of the internal friction of silicon nitride, silicon carbide and alumina. The increase of internal friction means the absorption of the

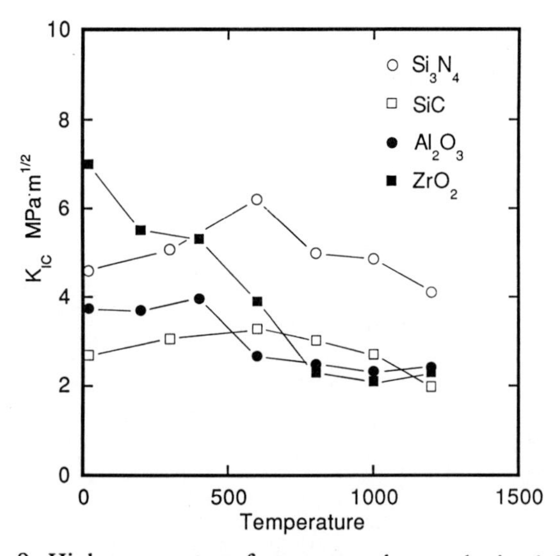

Fig. 9 High temperature fracture toughness obtained from chevron notched beam specimens.

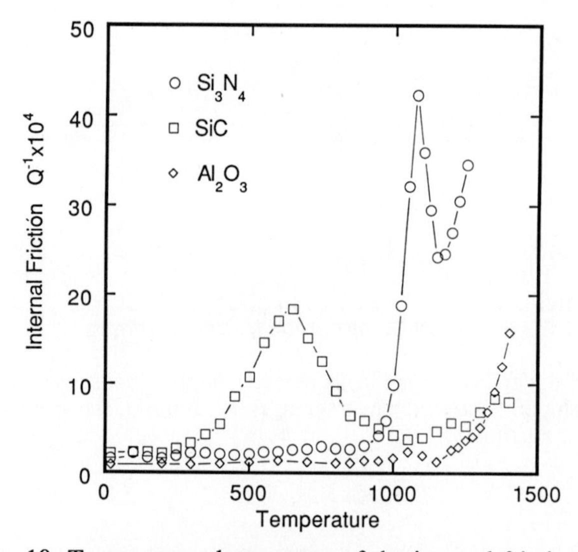

Fig. 10 Temperature dependence of the internal friction of silicon nitride, silicon carbide and alumina.

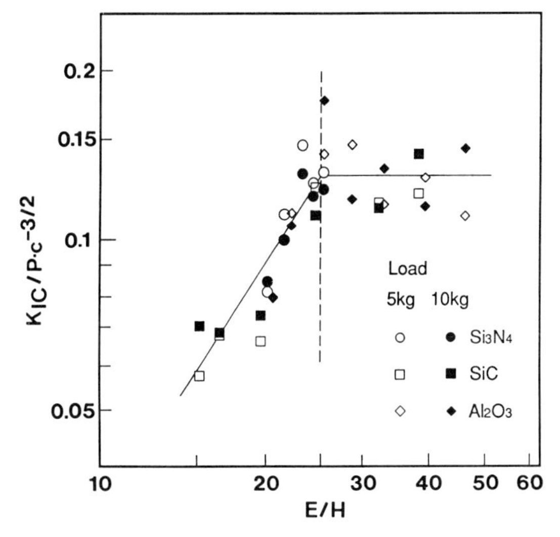

Fig. 11 The relation between E/H and $K_{IC}/(P/c^{3/2})$. The exponent, t, is 1.5 for E/H is smaller than 25, and 0 for E/H is larger than 25.

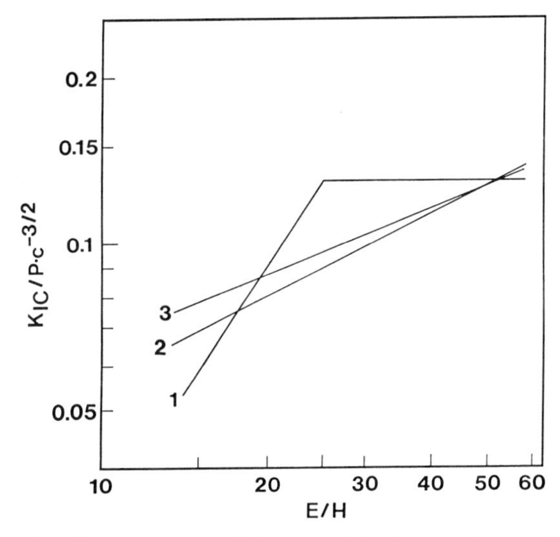

Fig. 12 Comparison of the empirical equations of the present study (line 1), Ref. 2 and 9 ($K_{IC}=0.018(E/H)^{1/2}(P/c^{3/2})$; line 2) and Ref. 7 ($K_{IC}=0.036E^{0.4}P^{0.6}a^{-0.7}(c/a)^{-1.5}$; line 3).

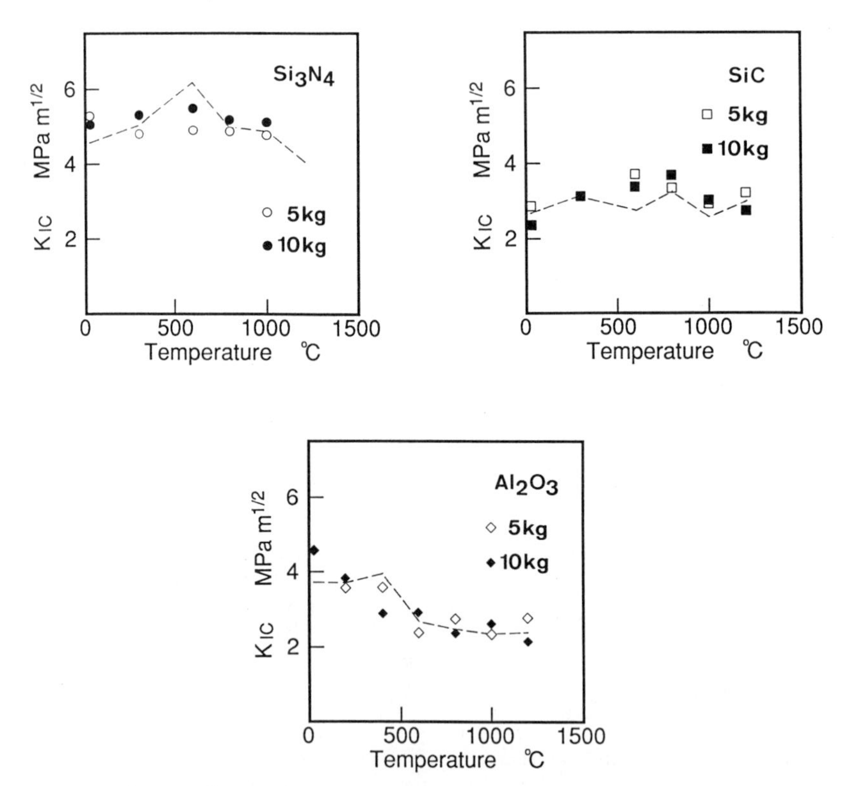

Fig. 13 Comparison of the fracture toughness obtained from chevron notched beam (dashed line), and from the present equations for indentation fracture method at two load levels on silicon nitride, silicon carbide and alumina.

energy of mechanical vibration. Especially, exponential increase of the internal friction with ascending temperature corresponds to the elastic-plastic transition of the material.

On the present specimens, silicon nitride shows the abrupt increase of the internal friction over 1050°C, then, the data of silicon nitride at 1200°C should be excluded from the analysis. On silicon carbide, internal friction shows a peak at around 600°C. However, this increase does not show exponential behavior, then it does not corresponds to the plastic deformation of the material. On alumina, it shows exponential increase over 1250°C. Therefore, on silicon carbide and alumina, the data up to 1200°C can be adopted to the analysis.

Empirical Equation

The relation between $K_{IC}/(P/c^{3/2})$ and E/H is investigated in order to determine the exponent, t, for E/H. Fig. 11 shows the log-log plot of these two values. As it is shown in this figure, the slope of the plot, which corresponds to the value of t, is not unique. The exponent t should be two values below or over 25 of E/H. In the region of E/H<25, t is obtained as 1.5, and for E/H>25 t is 0.

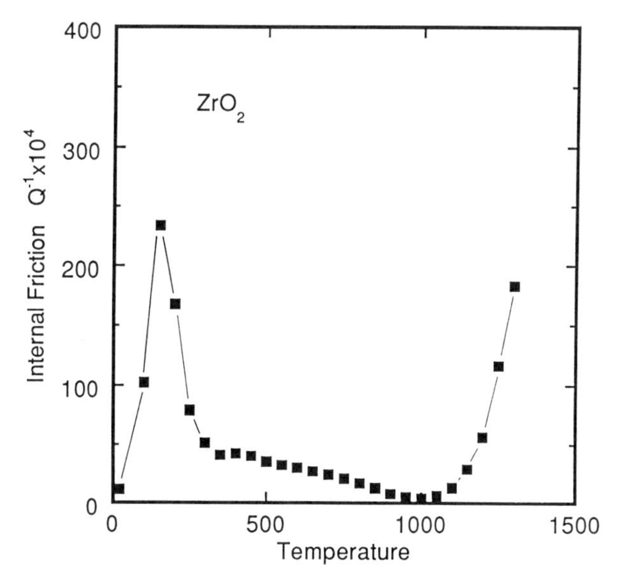

Fig. 14 Temperature dependence of the internal friction of tetragonal zirconia polycrystals.

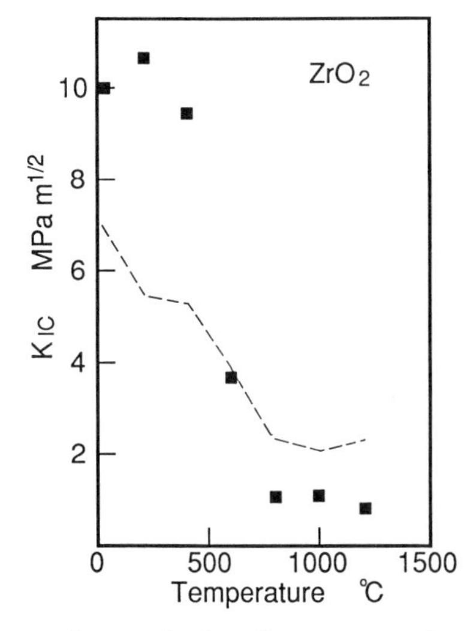

Fig. 15 Comparison of the fracture toughness obtained from chevron notched beam (dashed line), and from indentation fracture method on tetragonal zirconia polycrystals.

From these results, the coefficient A on the equation (1) is determined, and the empirical equation for each condition is obtained as follows.

$$K_{IC}=1.04\times10^{-3}\cdot(E/H)^{1.5}\cdot(P/c^{3/2}) \qquad (E/H<25) \qquad (6)$$

$$K_{IC}=0.130\cdot(P/c^{3/2}) \qquad (E/H>25) \qquad (7)$$

Fig. 12 shows the difference of present equations and former researches. It shows that the lines for present study and former studies are crossing on two points, that is at E/H is between 15 and 20, and about 50. It means that if we make an approximation with using the data from ceramic materials and WC/Co at room temperature, we also get the same line as the previous researches, because the typical E/H value of ceramics is 15 or 20, and the value of WC/Co is about 50. However, we show that it is not good estimation for intermediate E/H region from the data obtained at elevated temperature.

Fig. 13 shows the comparison between the calculated fracture toughness by equation (6) and (7), and the measured fracture toughness by chevron notched beam.

Measurement of Zirconia

TZP is also tested like other three materials. The Vickers load was changed from 50kg at room temperature to 1kg over 1000°C as the fracture toughness of TZP decreases very rapidly with increasing temperature. As shown in Fig. 5 to 7, E/H of TZP increases from 17.9 at room temperature to 43.4 at 1200°C. Fig. 14 shows the temperature dependence of the internal friction of TZP. It increases exponentially at around 1200°C

Fig. 15 shows the comparison between the fracture toughness by indentation fracture method, calculated with present equations, and the fracture toughness measured by chevron notched beam. At low temperature, fracture toughness measured by indentation fracture method shows higher values, because the crack formation is different, which is palmqvist type crack.[17] At 1200°C, these two values should be different because of the influence of the plasticity, that is detected by internal friction measurement. However, it shows different values at 800°C and 1000°C. TZP has its own toughening mechanism, so, it may have some other difficulties to measure the fracture toughness by indentation fracture method.

CONCLUSIONS

Fracture toughness measurement by indentation fracture method was carried out at elevated temperature up to 1200°C. By the high temperature measurement, the exponent, t, for E/H can be determined precisely as the range of the E/H becomes larger. The value of t is not unique but changing with the E/H, and t is obtained as 1.5 at E/H<25 and 0 at E/H>25. Form these results, Empirical equations for fracture toughness calculation is obtained as equation (6) and (7) for each condition of E/H.

The results of internal friction measurement was discussed. Abrupt increase of the internal friction with increasing temperature corresponds to the brittle ductile transition of the material. Then, it is closely related to the residual stress of the indentation. Therefore, it should be excluded from the analysis if the material shows the increase of internal friction. On this study, measurement of silicon nitride at 1200°C is for this case, then the datum is excluded from the analysis.

REFERENCES

1. A. G. Evans and E. A. Charles, Fracture toughness determination by indentation, J. Am. Ceram. Soc., 59:371 (1976).
2. JIS R1607, Testing methods for fracture toughness of high performance ceramics, Japanese Standards Association, Tokyo, (1990).

3. M. Shimada, M. Koizumi, A. Tanaka and T. Yamada, Temperature dependence of K_{IC} for high-pressure hot-pressed Si_3N_4 without additives, J. Am. Ceram. Soc., 65:C-48 (1982).
4. S. Sakaguchi, N. Murayama and F. Wakai, Fracture toughness of engineering ceramics measured by indentation fracture method at elevated temperatures, pp337, in, "MRS Int'l Mtg. on Adv. Mats. Vol. 5", M. Doyama, S. Somiya and R. P. H. Chang eds., Mat. Res. Soc., Pittsburgh, (1989).
5. K. Niihara, Micro-fracture of ceramics by indentation, Bull. Ceram. Soc. Jpn., 20:12 (1985).
6. K. Tanaka, Evaluation of elastic/plastic indentation stress to determine fracture toughness, J. Jpn. Inst. Metals., 48:1157 (1984).
7. D. B. Marshall and A. G. Evans, Reply to "comment on 'elastic /plastic indentation damage in ceramics: the median/radial crack system'", J. Am. Ceram. Soc., 64:C-182 (1981).
8. B. R. Lawn, A. G. Evans and D. B. Marshall, Elastic/plastic indentation damage in ceramics: The median/radial crack system, J. Am. Ceram. Soc.,63:574 (1980).
9. T. Miyoshi, N. Sagawa and T. Sassa, Research for the evaluation of fracture toughness of engineering ceramics, Trans. Jpn. Soc. Mech. Eng., A, 51:2489 (1985).
10. S. Sakaguchi, N. Murayama, Y. Kodama and F. Wakai, Equation for the evaluation of fracture toughness by chevron notched beam with using JIS type bending bar specimen, J. Ceram. Soc. Jpn., 99:47 (1991).
11. D. Munz, R. T. Bubsey and J. L. Shannon Jr., Fracture toughness determination of Al_2O_3 using four-point-bend specimens with straight-through and chevron notches, 63:300 (1980).
12. J. I. Bluhm, Slice synthesis of a three dimensional "work of fracture" specimen, Engng. Fract. Mech., 7:593 (1975).
13. S. Spinner and W. E. Tefft, A method for determining mechanical resonance frequencies and calculating elastic moduli from these frequencies, Proc. Am. Soc. Testing Mats., 61:1221 (1961).
14 S. Sakaguchi, F. Wakai and Y. Matsuno, Elastic properties of engineering ceramics at elevated temperature, J. Ceram. Soc. Jpn., 95:476 (1987).
15 S. Sakaguchi, N. Murayama and F. Wakai, Internal friction of Si3N4 at elevated temperatures, J. Ceram. Soc. Jpn., 95:1219 (1987).
16. R. W. Rice, The compressive strength of ceramics, pp195, in, "Materials Science Research Vol. 5: Ceramics in Severe Environments", W. W. Kriegel and H. Palmour III eds., Plenum, New York, (1971).
17. K. Niihara R. Morena and D. P. H. Hasselman, Evaluation of K_{IC} of brittle solids by the indentation method with low crack-to indent ratios, J. Mat. Sci. Lett., 1:13 (1982).

EFFECTS OF EXPOSURE TIME AND STRAIN RATE ON THE ELEVATED-TEMPERATURE TENSILE STRENGTH OF SiC MONOFILAMENTS

William S. Hong, Janet M. Sater, and Michael A. Rigdon

Institute for Defense Analyses
Alexandria, Virginia, 22311-1772 USA

Michael G. Jenkins and Mattison K. Ferber

Oak Ridge National Laboratory
Oak Ridge, Tennessee, 37831-6064 USA

ABSTRACT

Tensile tests of one production lot of SCS-6 were conducted at 20°C, 1000°C, 1200°C, and 1400°C. Strain rates of 0.2%, 0.6%, and 2.0%/min (155 mm gripped length) were employed. Fibres underwent isothermal soak times of 3 min and 30 min at elevated temperatures prior to testing. Results of the Yates algorithm analysis and analysis of variance of the empirical data did not show any coupled effects due to the test condition parameters but did indicate that the individual effects of parameters were significant. The empirical results revealed that higher test temperatures or longer soak times always decreased the observed tensile strengths, but that the strain rate effects tend to decrease with increasing temperatures.

INTRODUCTION

Design prediction of the mechanical behaviour of ceramic matrix composites (CMCs) requires accurate knowledge of the mechanical behaviour of the constituent materials, especially in the range of elevated temperatures at which the CMCs will be employed. Although ceramic fibres play an essential role in the performance of CMCs at elevated temperatures, recent assessments [1-2] of currently available ceramic fibres have concluded that these fibres lack either the required properties at elevated temperatures, or the ability to retain such properties for practical durations. Proper design utilization of the fibres has been further hampered by the lack of consistent test data

Fracture Mechanics of Ceramics, Vol. 10
Edited by R.C. Bradt *et al.*, Plenum Press, New York, 1992

related to their mechanical properties, especially at the anticipated use temperature.

Concurrent with these trends are greater interests in developing both a general elevated-temperature database for fibres and consistent testing methods that will enhance the utility of such a database. Although the mechanical properties of bare fibres at elevated temperatures are not necessarily a valid indication of how such fibres will perform in a composite matrix, the need to differentiate between competing fibres for design and comparative trade studies, as well as for generating quality assurance data, is real.

The general state of the elevated-temperature fibre properties database (consisting mostly of tensile strength with some modulus data) has been examined in detail [2]. Typically, the data show a lack of agreement in reported properties between different testing groups for the same type of fibre. A specific case can be seen in Fig. 1, which plots the tensile strength versus test temperature for Nicalon® Si-C-O ceramic filaments as reported by four test groups. Details of the test procedures used by the groups are summarized in Table 1. The use of different gage lengths (and how these are defined, i.e., gripped length or heated length) and strain rates may be expected to affect strengths. However, the use of varying thermal histories before testing may be the most important factor affecting the data as listed in Table 1. Additional factors such as pretest

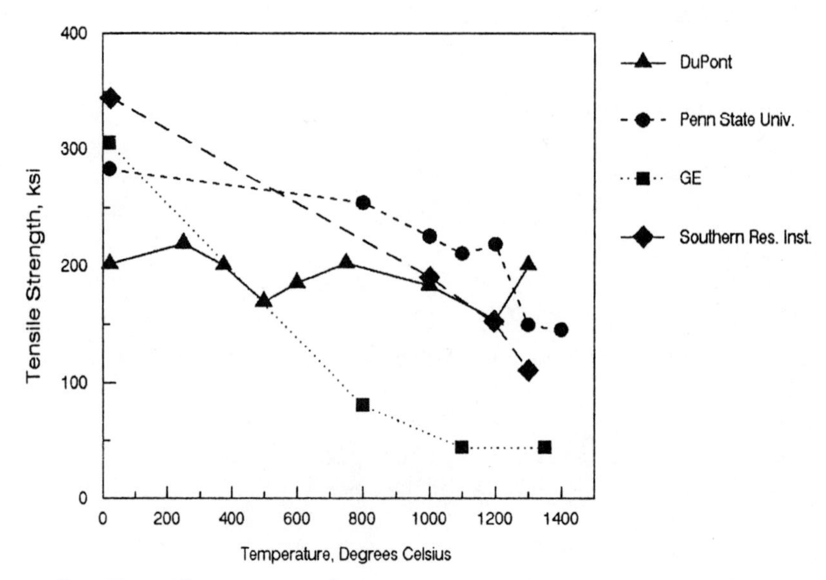

Figure 1. Tensile strength versus test temperature for Nicalon filaments. Data from various researchers as reported in Ref. 2.

Table 1. Summary of Elevated-Temperature Fibre Test Conditions and Procedures Used To Obtain Data in Fig. 1 as Given in Ref. 2

Organization	Atmosphere	Gage Length (cm)	Strain Rate (min^{-1})	Thermal History
Penn State	Ambient Air	7.5 (heated)	0.0066	Rapid heatup (1-2 min)
DuPont	Ambient Air	13.3 (gripped)	0.071	Rapid heatup (< 10 s)
Southern Research	He	5.8 (heated)	0.02	0.55°C /s; 5 min soak
GE	Ambient Air	20 (gripped) 2.5 (heated)	0.006	10°C /min; 10 min soak

specimen selection, sizing removal techniques, methods used to measure fibre diameters, etc., will also be expected to influence results, but their consideration is beyond the scope of the present study.

The exact reasons why test groups choose methods such as those shown in Table 1 are not known. Often, these procedures evolve in order to generate self-consistent test data for comparison and quality evaluation purposes within the testing organization. Thus, one method may not be more "correct" than any other. Nevertheless, difficulties will occur with any attempt to correlate or compare test data from one group to another.

While test procedures for fibres have direct bearings on the resulting data, few recent efforts are known to have been made that systematically varied certain test parameters, and then quantified their effects on elevated-temperature mechanical properties. The purpose of the present study was to make such an investigation, and demonstrate that such alterations in testing have measurable effects on the results. Because of this study's limited scope, the only test parameters that were systematically altered were test temperature, thermal history (soak time at temperature before testing), and initial strain rate. Eventually, the need for true comparability of ceramic fibre, elevated-temperature, mechanical property data must be addressed through more standardized testing and reporting procedures.

TEST MATERIAL

The commercially-obtained[a] SiC fibres were produced by chemical decomposition of a silane-hydrogen gas mixture on a resistively-heated, graphite-coated, carbon monofilament as it passed through a CVD reactor [3]. The result of the SiC deposition is a composite-like filament, roughly containing three regions as viewed in cross section [3]. The inner-most region is the graphite-coated carbon core (~37 μm diameter). The next region is a sheath (~40 μm I.D., ~80 μm O.D.) of carbon-rich, β-SiC with columnar grain sizes ~50 nm on average. The final or outer region is a sheath (~80 μm I.D., ~142 μm O.D.) of stoichiometric to silicon-rich β-SiC with columnar grains similar to those in the second region but with an average grain size of ~100 nm [3]. Finally a carbon-rich outer coating ~4 μm thick encircles the fibre for improved handling qualities and inteface control in any intended matrices.

The effect of this composite-like structure is that some fibre properties, such as fast-fracture tensile strength, may depend strongly on the carbon-rich coatings and the carbon core [3]. However, such properties as thermal expansion, elastic modulus, or thermally activated behaviour such as creep, or strength degradation due to exposure time, should be controlled by the SiC sheath. Previous studies [3-8] of this fibre under elevated-temperature conditions indicate that these observed mechanical behaviours of the SCS-6 are similar to those of standard SiC.

EXPERIMENTAL SETUP AND PROCEDURE

All testing was conducted on a screw-drive, electro-mechanical load frame[b] operable under load- or displacement-control. Tests were conducted in cross-head displacement (XHD) control which was related to quasi-strain control by assuming for long, gripped lengths, the measured XHD represented the fibre displacement. Thus, fibre strain, ε_f, was estimated as:

$$\varepsilon_f = XHD / L_o \quad (1)$$

where L_o was the total, original fibre length.

---

[a] SCS-6, Textron Specialty Materials, Lowell, Massachusetts.

[b] 1380 Low Cycle Creep Fatigue Machine, Instron Corporation, Canton, Massachusetts.

Fibres were gripped by water-cooled, pneumatic grips[c] employing flat-surfaced, titanium (Ti-6Al-4V) faces at the grip/fibre interfaces. The closure force at the grip faces, controlled by adjustable air pressure, was set so as to just prevent fibre slippage during room temperature tests. Total, free fibre length from grip to grip was 155 mm for all tests. The strain rates used were primarily 0.2 and 2.0% min^{-1}, with a limited series of tests also made at 0.6% min^{-1}. All crosshead speeds were based on the 155 mm gage length.

Elevated temperatures were provided by a two-zone, resistance-heated furnace[d]. The hot zone of the furnace was ~50 mm in length, the total height of the furnace refractory, 100 mm. Test temperatures were measured by Type S thermocouples at three spots in the hot zone of the furnace. The thermocouple readouts were subsequently used for closed-loop temperature control of the two sets of heating elements within the hot zone. All testing was conducted in ambient air (20°-25°C, 35%-55% RH). Furnace test temperatures were ambient, 1000°C, 1200°C, and 1400°C with elevated-temperature soak times of 3 or 30 minutes. A narrow vertical slot was cut in the front of the furnace insulation, allowing the roller-mounted furnace to be quickly pulled around gripped fibres, thus enabling fast and reproducible heatup of fibres prior to testing.

An AT-compatible personal computer with a analog/digital data acquisition system was connected to the test setup and used for both monitoring the test and for data acquisition. Load, displacement, furnace temperature, and time data were recorded in numeric (digital) format.

TEST RESULTS

The intent of these investigations was to show that particular test variables influence the fibre's elevated-temperature tensile strength. Originally only three variables were to be investigated at two levels in the form of a replicated 2^3 factorial experiment design. (Additional sets of test conditions were added later to fill in data points.). The three variables- temperature, time at temperature, and strain rate- were tested at high and low levels. Two sets of tests, in which each sample set contained 10 to 12 samples, were run for each of eight combinations of test conditions, thus providing replication of results. The measured load was then used to calculate tensile strength assuming a constant fibre diameter (~140 μm).

---

[c] Pneumatic Action Grips, Instron Corporation, Canton, Massachusetts.

[d] 1600°C Short Furnace, Instron Corporation, Canton, Massachusetts.

Measured strengths of the SCS-6 fibre are plotted as a function of test temperature in Fig. 2, with the data grouped by combinations of initial strain rate and thermal history. For clarity, only the averages of each test series are plotted, with most of the data points representing at least 10 individual tests; the only exception being the 1400°C tests for which only five tests were performed. The dependence of strength on test temperature is obvious. The room temperature average strength (not plotted) was 3679 MPa.

Other trends are discernible from Fig. 2 as well. At 1000°C, the strength is decreased by using either a lower strain rate or longer exposure to the test temperature. At 1200°C, the data points appear to be grouped by thermal history, with the two sets of tests exposed for 3 minutes lying above those heat treated for 30 minutes. Once again, the effect of slower strain rates (at a fixed temperature and exposure time) results in lower strengths, but the averages lie much closer to each other than at 1000°C. This would suggest that strain rate effects on strength at 1200°C are less clear, and with possibly little statistical significance.

Relative effects of strain rate can be seen more clearly in Fig. 3- the same average strengths are plotted versus nominal strain rates, grouped by thermal history (temperature and soak time). At 1000°C, the relative soak time and strain rate effects are evident. But at 1200°C, the minimal strength

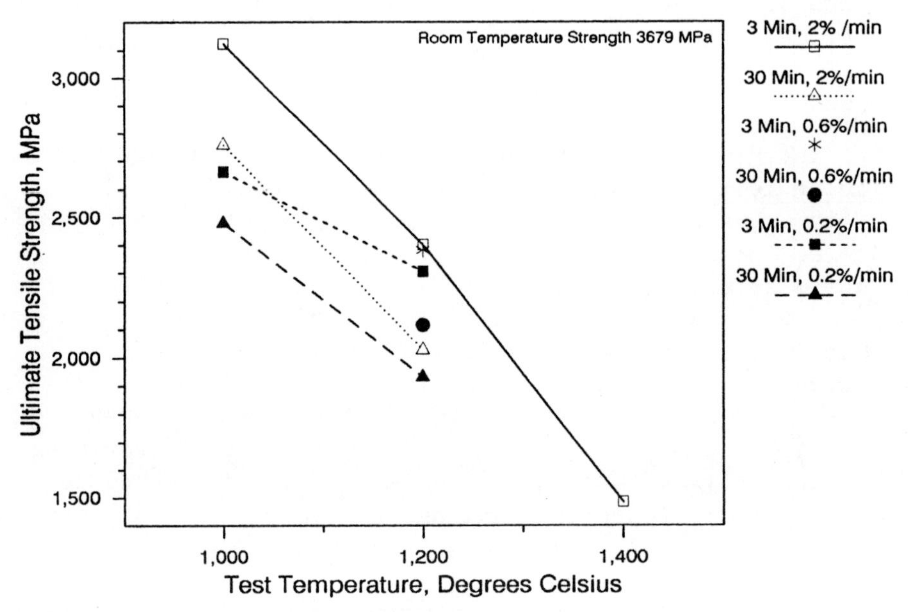

Figure 2. Tensile strength versus test temperature for SCS-6 monofilament fibres.

differences between 0.2% min^{-1} and 2.0% min^{-1} made it desirable to generate additional 1200°C tests using an intermediate strain rate of 0.6% min^{-1}. Considering expected statistical variability and the fact that these data were generated after the main series of tests, the intermediate strain rate test results do not contradict the apparent trend of reduced strain rate dependence for SCS-6 fibre at 1200°C.

In order to qualitatively assess changes to the SCS-6 fibres due to elevated-temperature exposures, fibre fragments were periodically recovered from the furnace and load stage for examination in the scanning electron microscope (SEM). Because the fibres shattered when they broke, it was not possible to recover the primary fracture surfaces responsible for failure. Fig. 4a is an SEM micrograph of an as-received (unheated) SCS-6 fibre fracture surface. Note the "flaking" of the carbon-rich outer layer, as well as the interface coating between the carbon core and the SiC sheath.

Figs. 4b and 4c are fracture micrographs of fibres exposed to 1000°C and 1200°C heat treatments (fibre sections that had been inside the furnace hot zone were distinguishable by a characteristic iridescence on their outer surfaces). The 1000°C specimen fracture surface was obtained by bending of the fibre, hence its "stepped" nature. Visible differences between the fibres in these micrographs are admittedly less than obvious. Note, however, that the fracture, or 'peeling' of the outermost

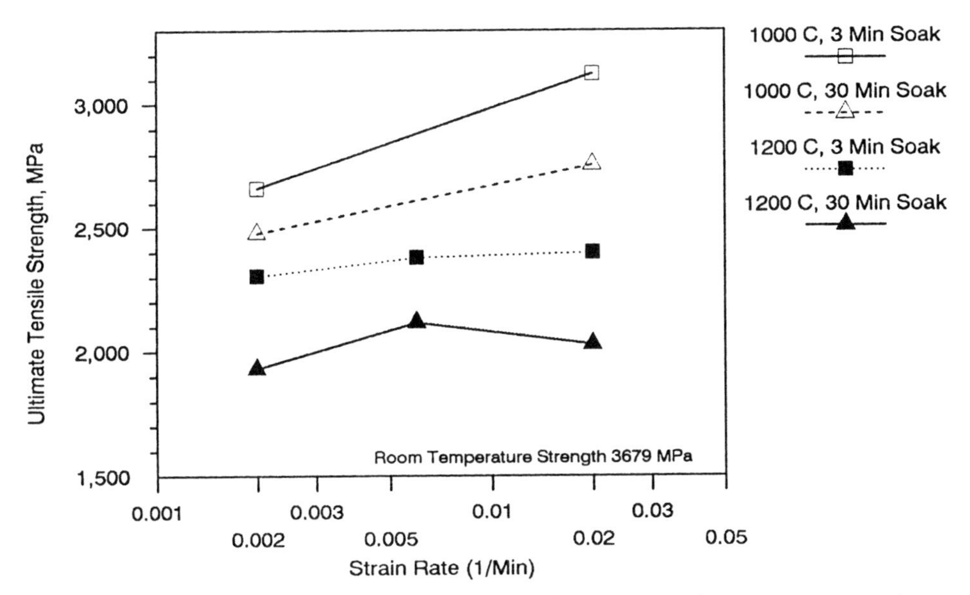

Figure 3. High temperature tensile strength versus strain rate for SCS-6 monofilament fibres.

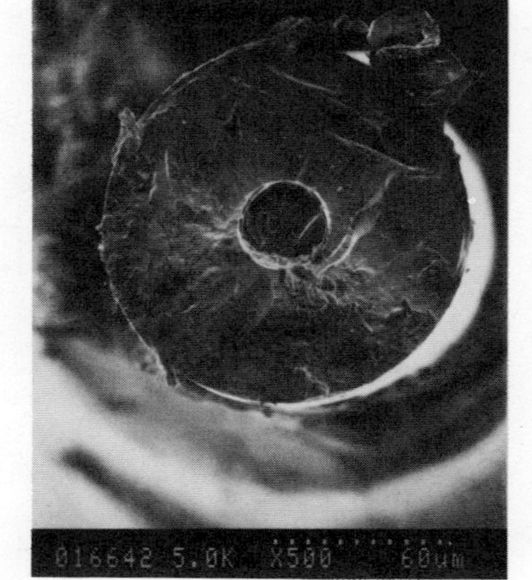

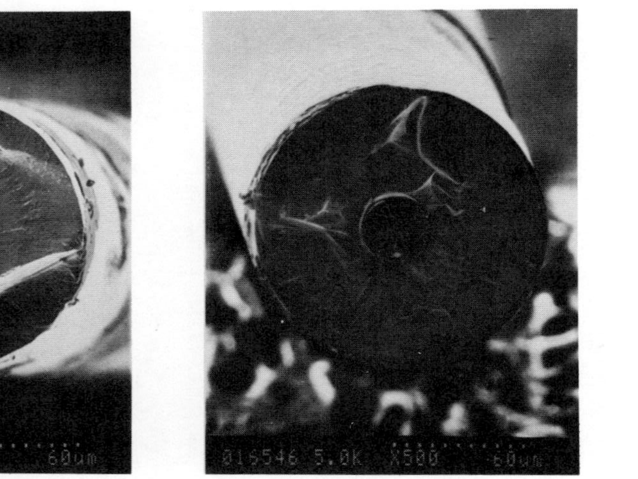

Figure 4. SEM micrographs of SCS-6 fibre fracture surfaces

a) SCS-6 fracture surface, tested at room temperature

b) SCS-6 fracture surface, tested at 1000°C

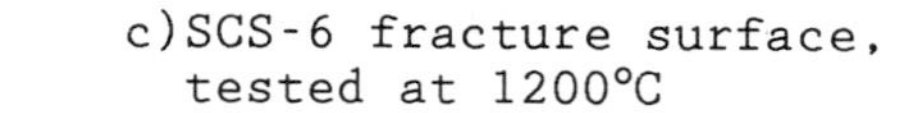

c) SCS-6 fracture surface, tested at 1200°C

layer in both Figs. 4b and 4c is of a more brittle nature, indicative of reactions that have taken place during elevated-temperature exposures. Higher magnification examination was also done on untested fibres that had been heat-treated in air for one hour at 1000°C and 1200°C, using a conventional box-type furnace. Some coarsening of the SiC grains due to the heat treatments could be discerned in these closer examinations.

STATISTICAL EVALUATION

An additional objective of these experiments was to estimate differences between the two experiment levels, if any, and to assess that estimate via some simple statistical techniques.

There are several statistical techniques that can be used to determine the relative effects of these different factors on the strength (y). These effects can be single such as temperature (X_T), time (X_t), or strain rate (X_e); coupled such as temperature and time (X_TX_t), temperature and strain rate (X_TX_e), or time and strain rate (X_tX_e); or higher order such as temperature, time, and strain rate ($X_TX_tX_e$). The Yates algorithm is one of the fastest, most simple methods for analyzing the results. Assumptions for using this approach include [9-11]:

1. A number of random samples of fibres have been selected from a population; experiments have been performed in a random order.

2. The samples are independent; each of the factors - temperature, soak time, and strain rate - are independent.

3. A normal distribution of results for each combination of test conditions is assumed.

The procedure is described elsewhere [9-11] and only the results are presented in Table 2.

Relatively speaking, the main effects appear to be more significant than the coupled effects and the higher order effect. The magnitudes of these effects are used in the next step of the analysis, which is to estimate the noise that is generated in the data by 'Mother Nature.' This can be done by analysis of variance, where variance is essentially the average of the squares of the distances from the mean. Assumptions for this calculation include those mentioned previously as well as the following [9, 10]:

1. Measurements/observations are random with respect to any conditions not systematically varied or removed from analysis by other methods.

Table 2. Results of Yates Algorithm from 2^3 Factorial Experiment Design for SCS-6 High Temperature Tensile Strength Data

Effect	Magnitude of Effect (MPa)
$\overline{y}$	2,416
X_T	587
X_t	324
X_e	232
$X_T X_t$	52
$X_T X_e$	140
$X_t X_e$	43
$X_T X_t X_e$	47

2. The sample average, standard deviation or standard error, and variance are best estimates of the population mean, standard deviation, and variance.

3. Means and variances are additive.

4. Experimental errors are independent and distributed normally about zero (the mean value).

5. Variances of the experimental errors for all combinations of test conditions are equal and have a common value about zero.

6. The previous assumption of a normal distribution allows use of Student's t statistics; a two-sided test is used to determine the confidence limits. Note that the t test is valid for any sample size. It provides a good approximation even though the population may be non-normal to a noticeable degree.

The principle is relatively simple and is shown by Eqn. (2) which basically states the following. The total variance of the fibre tensile strength data at elevated temperature consists of three parts: sums of the variances due to the sample average from test data; the variances due to the experimental error from the effects; and the variances due to noise generated by 'Mother Nature.'

$$\text{SSq(total)} = \text{SSq(average)} + \text{SSq(effects)} + \text{SSq(residual)}$$

$$\Sigma \overline{y}^2 = N\overline{y}^2 + \Sigma[N(\text{effect})^2/4] + \text{SSq(residual)} \qquad (2)$$

The SSq(residual) is the number that is not known but it can be determined. Once determined it can be used in other calculations to determine the confidence limits. These calculations are shown elsewhere [9-11] and relevant results are shown in Table 3.

The important result from this table is that from the estimated standard error of an effect the Student's two-sided t distribution for eight degrees of freedom can be used to determine the confidence limits which are ±169 MPa for 90% confidence limits and ±210 MPa for 95% confidence limits. Confidence limits can be plotted with the magnitudes of the effects (from Table 2) as illustrated in Fig. 5. It may be said with 95% confidence that the main effects of temperature, soak time, and strain rate have some influence on the fibre tensile strength since the magnitudes of the effects lie outside the noise level, i.e., any one of these factors will affect the tensile strength. There do not appear to be any coupled or higher order interactive effects based on the limited number of variables that were investigated in this study.

That this is a relatively simple set of experiments cannot be overstated. It is extremely important to note that other factors not considered here may also significantly affect the fibre tensile strength; that is to say that there may even be other coupled or higher order interactive effects. Examples of other factors include temperature gradients in the furnace, variations in fibre surface quality, variations in fibre diameter, and atmospheric conditions, among others.

DISCUSSION

Figs. 2 and 3 show that cumulative thermal history (test temperature and total exposure time at temperature prior to loading) affects the measured strength of SCS-6, as has been seen in other elevated-temperature studies of fibre properties [2]. The data also indicate, however, that as testing occurs above a certain temperature, strain rate effects are less noticeable. For SCS-6, faster strain rates result in higher measured strengths at 1000°C, but varied strain rates at 1200°C cause little change. These apparent findings were supported through the statistical evaluation made in the previous section, where strain rate effects were confirmed but found to be less important than temperature or soak time.

It is possible that the relative strain-rate insensitivity of SCS-6 at the higher temperatures is related to creep behavior, which has been shown to occur under similar thermal conditions in air [3]. The tensile failure test used in this study was carried out in a crosshead displacement controlled mode, and not designed to determine the onset of fibre creep.

Table 3. Results of Analysis of Variance Calculations from 2^3 Factorial Experiment Design for SCS-6 Elevated Temperature Tensile Strength Data

SSq(residual)	264,779, with 8 degrees of freedom
Estimated Variance about the Mean	33,097
Estimated Variance of an Effect	8,274
Estimated Standard Error of an Effect	91 MPa

What is clear from the present investigation is that the spread of measured fibre strengths at elevated temperature due to altered test procedures can be significant. This is evident even if average strengths are considered, as has been done here, without taking into further account statistical variability due to normal experimental factors. That the average strength of SCS-6 can be nearly the same at both 1000°C and 1200°C, by changing only some major test procedure variables, is a significant point when examining elevated-temperature mechanical property data of any fibre. An additional consideration is that these results were obtained with SCS-6, which has the reputation for being a robust fibre with good handling qualities. However, it also has an unusually large diameter compared to other ceramic fibres, many of which have diameters an order of magnitude smaller and are usually part of multi-filament tows.

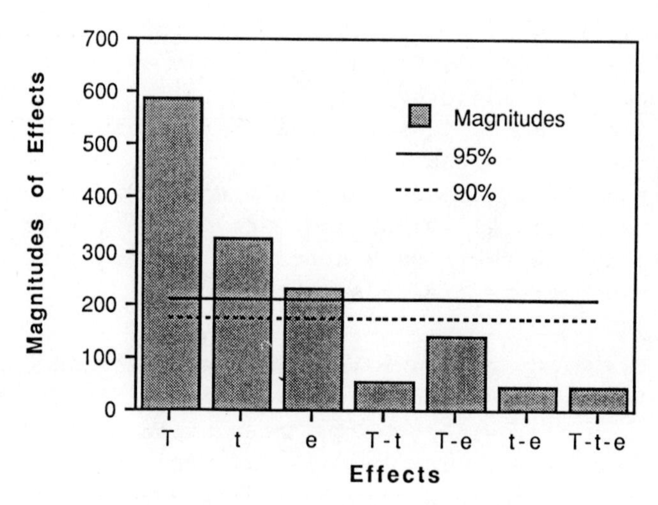

Figure 5. Bar graph with confidence limits illustrating effects of experimental variables.

Comparison [2] of the present study's strength results with those from other studies for the SCS-6, revealed similar trends of decreasing strength with increasing temperature as shown in Fig. 6. However, 'scatter' in the data at any one temperature approached 1.0 GPa which is about 1/3 of the average room temperature strength. Even when strength values were normalized with respect to the measured room temperature strength, making such direct comparisons is difficult. In many cases, reporting of the test results did not include such important test parameters as isothermal soak time, strain rate, gripped or gage length of the fibre, variation of fibre diameter over the length, and test environment. A need appears to exist for standards for both performing tests of ceramics fibres as well as reporting of the test data.

CONCLUSIONS

Salient conclusions are the following: 1) higher test temperatures or longer soak times always decreased the observed tensile strengths of SCS-6 fibres, but strain-rate effects tend to decrease with increasing temperatures; 2) the choice of testing procedures for ceramic fibres at elevated temperatures have potentially significant effects on the measured results; 3) procedures for testing and reporting results obtained from monofilament fibres need to be standardized if designs utilizing CMCs are to be exploited fully.

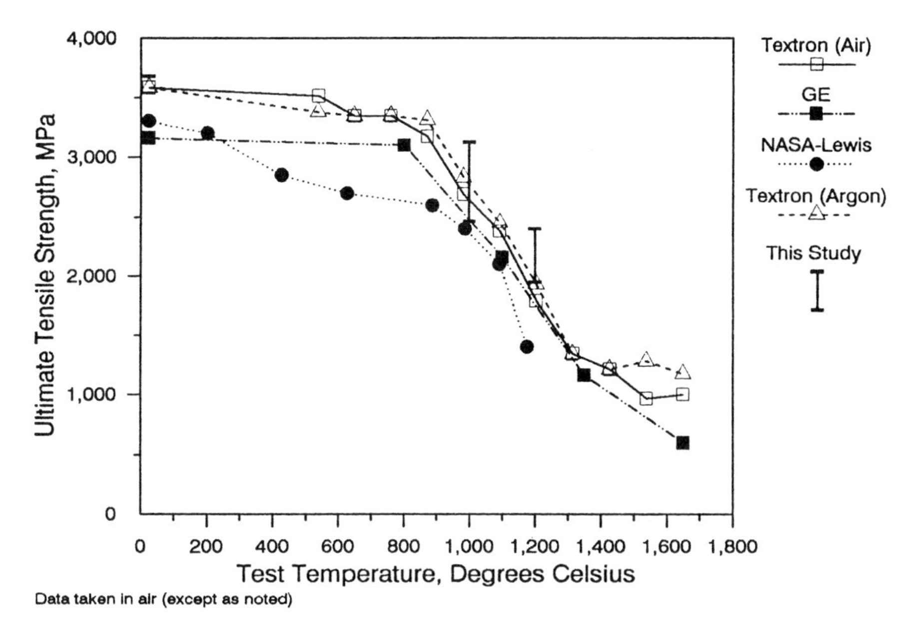

Figure 6. Tensile strength versus test temperature for SCS-6 monofilament fibres, comparing data from various researchers [2] with the range of averages obtained in this study.

ACKNOWLEDGMENTS

Research sponsored by the U.S. DOE, Assistant Secretary for Conservation and Renewable Energy, Office of Transportation Technologies, as part of the High Temperature Materials Laboratory Program, under contract DE-AC05-84OR21400 managed by Martin Marietta Energy Systems, Inc. Research also sponsored by the Institute for Defense Analyses as part of its Corporate Research Project program. Contribution of the SCS-6 fibre from Textron and assistance with scanning electron microscopy from Ms. D. Coffey of the HTML is gratefully acknowledged.

REFERENCES

1. T.I. Mah, M.G. Mendiratta, A.P. Katz and K.S. Mazdiyasni, Recent Developments in Fiber-Reinforced High Temperature Ceramic Composites, Amer. Ceram. Soc. Bull., 66[2]:304 (1987).
2. W.S. Hong, M.A. Rigdon, N.L. Fortenberry, "Reinforcement Options for High Temperature Composites and A Comparison of High Temperature Tensile Testing Results for Ceramic Fibers," IDA Paper P-2483, Institute for Defense Analyses, Alexandria, Virginia, (1990).
3. J.A. DiCarlo, Creep of chemically vapour deposited SiC fibers," J. Mater. Sci., 21:217 (1986).
4. J.A. DiCarlo, W. Williams, Dynamic Modulus and Damping of Boron, Silicon Carbide, and Alumina Fibers, Ceram. Eng. Sci. Proc., 1[7-8]:671 (1980).
5. R.T. Bhatt, M.D. Kraitchaman, "Environmental Effects on the Tensile Strength of Chemically Vapor Deposited Silicon Carbide Fibers," NASA Technical Memorandum 86981, NASA Lewis Research Center, Cleveland, Ohio (1985).
6. J.A. DiCarlo, High Temperature Properties of CVD Silicon Carbide Fibers, in "Whisker and Fiber-Toughened Ceramics, Proceedings of an International Conference," R.A. Bradley, D.E. Clark, D.C. Larsen, J.O. Stiegler, eds., ASM International, Materials Part, Ohio, pp 1-8 (1988).
7. J.A. DiCarlo, Thermal Stability of SCS-6 SiC Fibers, Paper #69 in "Advanced High Temperature Engine Materials Technology Program Review," NASA Conference Publication 10039,NASA Lewis Research Center, Cleveland, Ohio, (1989).
8. G.N. Morscher, J.A. DiCarlo, and J. Wagner, A Simple Test for Thermomechanical Evaluation of SiC Fibers, Paper #49 in "Advanced High Temperature Engine Materials Technology Program Review," NASA Conference Publication 10051,NASA Lewis Research Center, Cleveland, Ohio, (1990).
9. E.L. Crow, F.A. Davis, and M.W. Maxfield, "Statistics Manual," Dover Publications, Inc.,New York, New York (1960).
10. G.E.P. Box, W.G. Hunter, and J.S. Hunter, "Statistics for Experimenters An Introduction to Design, Data Analysis, and

Model Building," John Wiley & Sons, Inc.,New York, New York, (1978).
11. J.S. Hunter, "Statistical Design of Experiments," Society of Manufacturing Engineers, New York, New York, (1989).

STRESS-RELAXATION TESTS USED TO DETERMINE THE ELEVATED-TEMPERATURE CREEP PARAMETERS OF STRUCTURAL CERAMICS

Michael G. Jenkins and Mattison K. Ferber

Oak Ridge National Laboratory
Oak Ridge, Tennessee, 37831-6064 USA

ABSTRACT

Strain-controlled, stress-relaxation tests were conducted on aluminum oxide and silicon nitride tensile specimens for temperatures of 1000°C to 1260°C. In conjunction with a preliminary investigation of the use of these tests for the determination of elevated-temperature creep parameters, a phenomenological theory of plastic deformation was applied to the empirical results to characterize the deformation of the material in terms of a mechanical equation of state. Elevated-temperature creep parameters determined from single specimen tests and the mechanical equation of state approach were compared with the same creep parameters determined from multiple-specimen tests and conventional analysis approaches. The mechanical equation of state approach offers the potential of mechanistically characterizing elevated-temperature plastic behaviour of ceramics using fewer specimens over shorter test times.

INTRODUCTION

Creep and plastic deformation at elevated temperatures is one of many important considerations for the implementation of structural ceramics in advanced heat engine designs [1,2]. Some candidate ceramic materials, such as monolithic silicon nitride, which exhibit excellent room-temperature strengths and fracture resistances may show poor resistance to creep deformation at the intended design temperatures (1200°C to 1400°C) [3-5].

Traditionally [6] creep parameters have been determined from constant load (stress) tests in which total strain is recorded as a function of time, as shown in Fig. 1a. The generalized behaviour for steady-state creep can be described as a power-law dependence of the strain rate, $\dot{\varepsilon}$, on the applied stress, σ, as shown in Fig. 1b, such that [6]:

$$\dot{\varepsilon} = A\sigma^n \tag{1}$$

where n is the stress exponent and A is a constant incorporating such phenomenological terms as activation energy and diffusion parameters. The lengthy times involved in creep testing to final rupture require that many such tests be terminated when the investigator is satisfied that steady-state creep has been achieved. Accelerated methods for determining the mechanistic creep parameters would be useful, if information such as total strain to failure, total time to failure, or material degradation, are not desired.

Potentially, one such accelerated method involves load- (stress-) relaxation tests of various materials to obtain σ - $\dot{\varepsilon}$ relations in terms of a mechanical equation of state and the state of strain hardening in the material. Although previous tests [7-11] generally focused on metals at relatively low temperatures (20°C to 550°C), at least one study [10] involved an oxide ceramic (UO_2) at temperatures in the range of 1400°C to 1800°C.

This paper concerns the preliminary investigation of the application of such an approach to structural ceramics. Three ceramics were tested in single-specimen, constant-strain, stress-relaxation tests at temperatures from 1000°C to 1260°C. The analytical approach describing the mechanical equation of state and the state of strain hardening is discussed. The experimental procedure is detailed, followed by a listing of the materials evaluated. Finally, the application of the mechanical

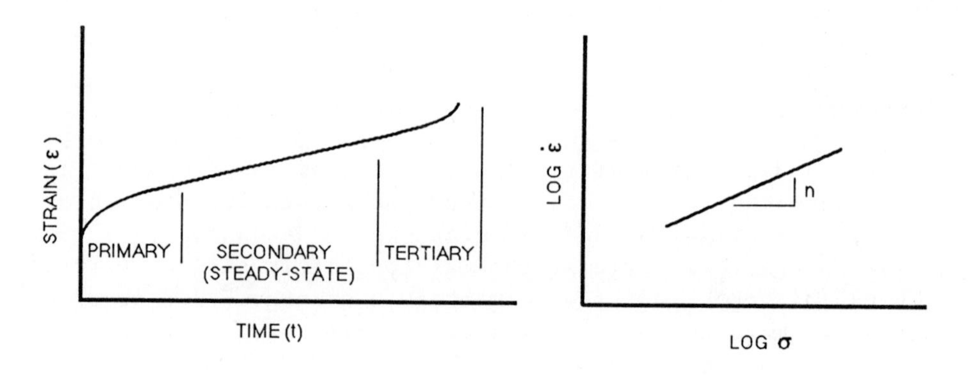

Figure 1. Characterization of elevated-temperature creep.

equation of state to the stress-relaxation test results is discussed and the creep parameters thus determined are compared to the same parameters determined from conventional creep tests of these same materials.

ANALYTICAL BASIS

A theory of the mechanical equation of state has been proposed [12] which hypothesizes that all deformation histories for a given strain-hardening process can be described incrementally by the differential form:

$$d \ln \sigma = \gamma \, d\varepsilon_p + \nu \, d \ln \dot{\varepsilon}_p \tag{2}$$

where σ is the tensile stress, ε_p is the plastic tensile strain, $\dot{\varepsilon}_p$ is the plastic tensile strain rate, γ and ν are coefficients related to strain-hardening and strain-rate sensitivity, respectively. The differential form of Eq. (2) links the smooth variation of any of the deformation variables in terms of the others. In general, the coefficients, γ and ν at any stage of the deformation process depend on the current values of at least two of the variables, σ, ε_p, and $\dot{\varepsilon}_p$, in addition to the previous deformation history. Given any deformation path, γ and ν can be identified as the expressions:

$$\gamma \equiv (\partial \sigma / \partial \varepsilon_p)_{\dot{\varepsilon}_p} \tag{3}$$

and

$$\nu \equiv (\partial \sigma / \partial \dot{\varepsilon}_p)_{\varepsilon_p} \tag{4}$$

Therefore, in the assumed mechanical equation of state, ν is the slope of curves of constant plastic strain, is not dependent on the deformation history, and is dependent only on the current values of σ and $\dot{\varepsilon}_p$.

The basis [12-14] for the theory is that the mechanical state of a material after some plastic deformation, depends on the deformation path rather than on the final strain state. Other cited attempts at descriptions of plastic properties [14-16] have included only ε_p as the prominent state variable thereby limiting the description of the strain hardening in a material to the amount of accumulated plastic strain. This is in contradiction to the common practice of assuming the attainment of a 'steady-state' regime which, by definition, means that no significant change will occur in the state of the material with further deformation [12-14]. The theory described here seeks to establish a description of the state of strain-hardening in general terms of a mechanical equation of state which is not solely dependent on a description of the accumulated strain and is also not dependent on whether or not a 'steady-state' deformation regime exists.

TEST MATERIALS

The mechanical equation of state is intended to be generally applicable to the isothermal deformation of 'well-behaved' polycrystalline materials (e.g. face-centered cubic metals) deforming by intragranular slip [15] although applications have been made to body-centered cubic metals [8] and some polycrystalline ceramics [10]. The intent of the present study was to apply the mechanical equation of state to structural ceramics which do not plastically deform in the usual sense of the term (i.e. slip system flow at the crystallographic level). Instead, the primary 'plastic' mechanisms in these ceramics appear to be due to intergranular rather than intragranular mechanisms.

Creep or plastic deformation behaviour of many structural ceramics is often directly related to intergranular diffusion of material or grain boundary sliding (GBS) [6] due to the presence of amorphous, grain boundary phases. The amounts of the amorphous materials, while usually small (~2%-8% by weight) are essential for the proper processing and densification of the ceramic. The glassy nature of these phases, limits the refractoriness of the materials at temperatures above ~800°C to 1000°C. Thus, the overall creep response of the ceramic at a given temperature may be controlled largely by the resistance of the intergranular phase to the sliding of the grains of the bulk ceramic or diffusion of material at the grain boundaries.

The materials tested in this study contained various types and amounts of intergranular phases as indicated in Table 1. The aluminum oxide (AD94) had been studied as a 'model' material in previous investigations [16]. The two silicon nitride materials represent candidate materials for advanced heat engine applications. The first silicon nitride (SN220M) is a 'low-temperature' silicon nitride developed for applications (e.g. cam followers) requiring high resistance to wear and good strength retention at moderate temperatures (1000°C to 1200°C). The second silicon nitride (PY6) is an advanced silicon nitride developed for applications (e.g. gas turbine components) requiring oxidation resistance and good strength retention at higher temperatures (1200°C to 1400°C).

TEST PROCEDURE

All testing was conducted in uniaxial tension in ambient air (room temperature = 20°C to 25°C at 15% to 35% RH) with elevated temperatures provided by a resistance-heated furnace with a 50 mm hot zone. Cylindrical, tensile test specimens (see Fig. 2) with total lengths of 160 mm, 16 mm diameter button-head

Table 1. Summary of test materials

Type of Material	Material Designator/ Supplier	Material Description
Aluminium Oxide (Al_2O_3)	AD94 / Coors Ceramic Co., Golden, Colorado	Slightly porous, 97% theoretical density, 94 wt% Al_2O_3, 5 wt% SiO_2, 12 μm average grain size
Silicon Nitride (Si_3N_4)	SN220M / Kyocera Corp., Kyoto, Japan	Dense; >99%, theoretical density, fine, equiaxed grains, hot pressed, 90.5 wt% β-Si_3N_4, 6 wt% Al_2O_3 , 3.5 wt % Y_2O_3
Silicon Nitride (Si_3N_4)	PY6 / GTE Laboratories Inc., Waltham, Massachusetts	Dense; >99% theoretical density, duplex microstructure, hot isostatically pressed, 94 wt% β-Si_3N_4, 6 wt% Y_2O_3

diameters, and gage sections of 6.35 mm diameter were fabricated from the as-received materials. Load was transferred into the button-head ends of the specimens through deformable collets which interfaced with water-cooled grip ends located outside the furnace. Self-contained, self-aligning, hydraulic couplers were incorporated into the load train at each grip end to minimize bending (< 3%) [16].

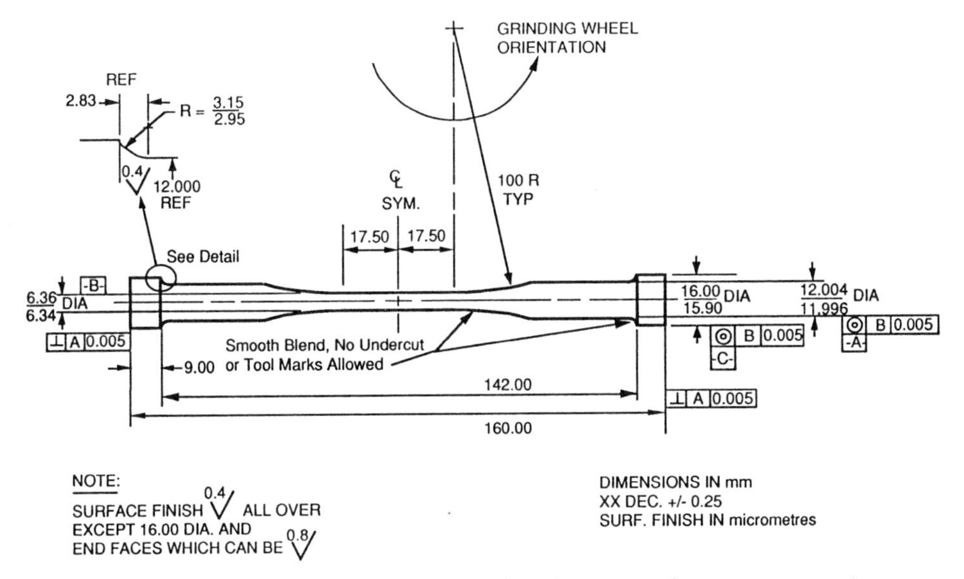

Figure 2. Cylindrical, button-head, tensile test specimen.

Testing was conducted on a commercial, electro-mechanical, single-screw load frame with analog control capabilities for both load or strain. A remote, electronic load cell incorporated into the load train was used to measure load. Strain was measured with a contacting extensometer which employed silicon carbide extensions with knife edge contacts at the specimen interface. Remote, air-cooled capacitive sensors detected changes in specimen strain transmitted through the ceramic extensions as shown in Fig. 3.

Stress-relaxation tests were conducted in strain-control using the contacting extensometer and a dedicated, analog, PID-type controller. Constant-stress, creep tests were conducted in load-control using the load cell and a similar dedicated, analog, PID-type controller. Specific test conditions for each material are shown in Table 2.

The stress in the gage section of the tensile specimen is defined as $\sigma = P/A_0$, where P is the applied (or measured) load and A_0 is the original cross-sectional area of the gage section. The total strain in the gage section is defined as $\varepsilon_t = \Delta l / l_0$, where Δl is the applied (or measured) displacement in the gage section over the original gage length, l_0, of the extensometer.

TEST RESULTS

Single specimens were used in the stress-relaxation tests in which successive stress-time (σ-t) curves were generated at

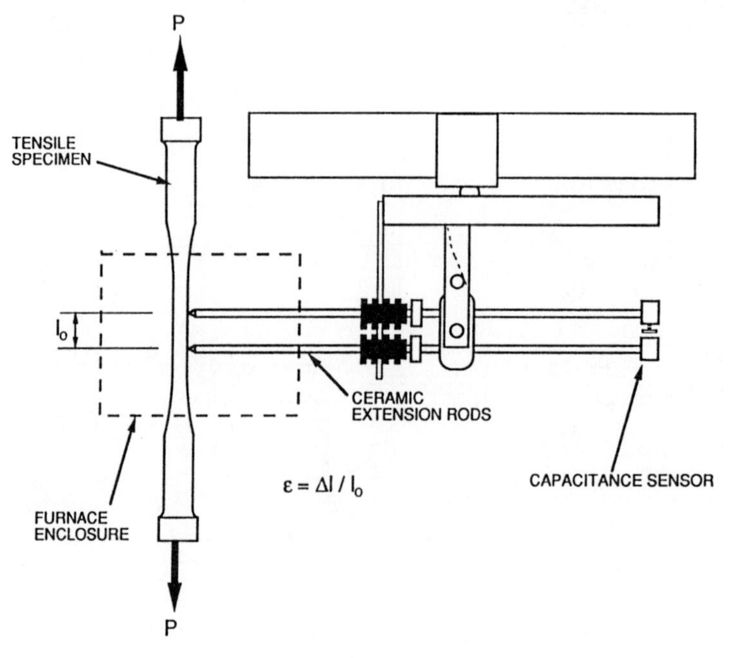

Figure 3. Schematic illustration of extensometer/test set-up.

Table 2. Summary of Test Conditions

Material	Temperature (°C)	Stress Relaxation[a] ε_t (µstrain)	Constant Stress[b] σ_o (MPa)
AD94	1000	360 320 240 160	21 [16] 28 [16] 33 [16] 39 [16]
SN220M	1000	580 Failed upon restraining	140 [16] 175 [16] 210 [16]
SN220M	1200	200 160 120 80	20 [16] 25 [16] 32 [16]
PY6	1260	640 560 480 400 320	100 [5] 113 [5] 125 [5] 138 [5] 150 [5] 175 [5]

[a] Single specimen; gage length = 25 mm

[b] Multiple specimens; cross sectional area of gage length = 31.7 mm^2.

decreasing, fixed total strain levels, ε_t, after establishing the initial level of plasticity. Figure 4 shows an example of a σ-t curve for the AD94 at 1000° C. Superposed on Fig. 4 is the constant, total strain maintained in the specimen during the entire test.

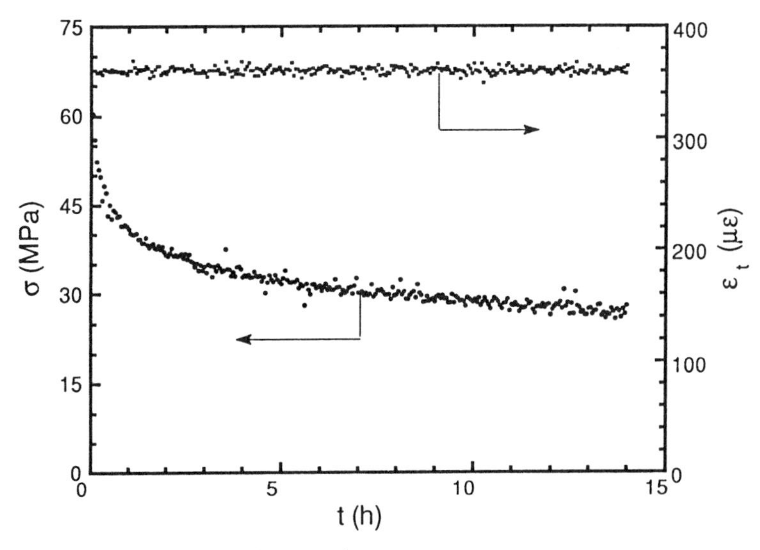

Figure 4. Stress and total strain versus time for AD94 at 1000°C for one initial strain.

The total strain rate, $\dot{\varepsilon}_t = d\varepsilon_t/dt$, is written as:

$$\dot{\varepsilon}_t = \dot{\varepsilon}_p + \dot{\varepsilon}_a \tag{5}$$

where $\dot{\varepsilon}_p$ is the unrecoverable plastic strain rate and $\dot{\varepsilon}_a$ is the anelastic strain rate which is the time rate of change of the recoverable strain, ε_a, such that at $t = 0$, $\dot{\varepsilon}_a = 0$. As an estimate, ε_a is approximated as σ_o/E, where E is the elastic modulus of the material at the test conditions. For constant-strain tests, where $\dot{\varepsilon}_t$ is zero, Eq. (5) can be arranged such that the unrecoverable plastic strain rate is determined as:

$$\dot{\varepsilon}_p = \dot{0} - \dot{\sigma}/E \tag{6}$$

Figure 5 shows multiple log-log plots of σ versus $\dot{\varepsilon}_p$ for the AD94 material at 1000°C in which each curve was generated from a single specimen but at a decreasing, constant values of ε_t. That is, after the completion of each stress-relaxation test, the remaining stress on the specimen was reduced to zero and a lower, constant value of ε_t was applied to the specimen so as to begin the next stress-relaxation test.

On each curve in Fig. 5 unique, σ - $\dot{\varepsilon}_p$ pairs exist at which the slopes ($d\sigma/d\dot{\varepsilon}_p$) of the tangents are all equal. Representative lines connecting such pairs on each curve for $d\sigma / d\dot{\varepsilon}_p = 0.18$ and 0.11 are shown in Fig. 5. Linear curve-fits of these points show parallel lines with slope, $\nu = 0.38$. The significance of ν is that it should approximate the inverse of the creep stress exponent, n, such that $\nu \approx 1/n$. Additionally,

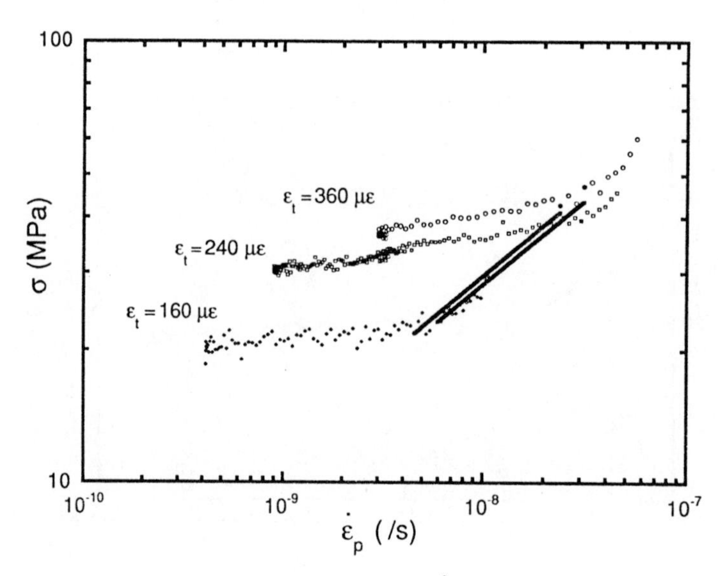

Figure 5. Stress - strain rate (σ - $\dot{\varepsilon}_p$) relations with fits of $d\sigma / d\dot{\varepsilon}_p$ data for AD94 at 1000°C.

ν can also be used as a scaling parameter to translate each separate $\sigma - \dot{\varepsilon}_p$ curve onto the initial $\sigma - \dot{\varepsilon}_p$ curve at which the level of plasticity was developed thus creating a master curve uniquely defining the $\sigma - \dot{\varepsilon}_p$ relation for each material at each test temperature.

For direct comparison with the customary form of creep test results, the inverses ($\dot{\varepsilon}_p - \sigma$) of either such master curves or initial curves are shown in Figs. 6-9 for the various materials and temperatures. Also shown in these figures are the results of constant-stress creep curves for these same materials. The results summary of these analyses are shown in Table 3.

As part of the present study, it is postulated that curve fits of individual, similar segments of the $\sigma - \dot{\varepsilon}_p$ curves for the constant-stress creep tests and the stress-relaxation tests should show similar power law relations between $\sigma - \dot{\varepsilon}_p$ with the exponent, ν^*, again related to the creep stress exponent as $\nu^* \approx 1/n$. These results are also shown in Table 3.

DISCUSSION

Apparent discrepancies are shown in Table 3 and Figs. 6-9 for the measured steady-state creep behaviour and the behaviour predicted by the mechanical equation of state as applied to the stress-relaxation tests. In many stress-relaxation tests of successive, decreasing values of ε_t, regardless of the material, the resulting $\sigma - \dot{\varepsilon}_p$ curves were not always easily recognized as being parallel to each other under the assumed constant levels of plasticity. In these cases it was not straightforward to

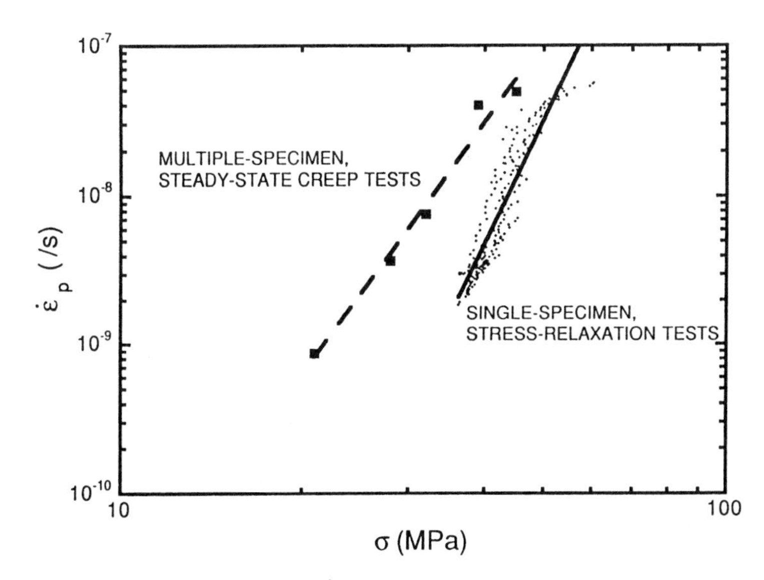

Figure 6. Comparisons of $\dot{\varepsilon}_p - \sigma$ relations for AD94 at 1000°C.

Table 3. Summary of test results

Material	Temperature (°C)	Stress Relaxation[a] ν	ν^*	Constant Stress 1/n
AD94	1000	0.38 ($n \approx 2.6$)	0.14 ($n \approx 8.0$) 0.08 ($n \approx 13$) 0.07 ($n \approx 14$) 0.07 ($n \approx 14$)	0.19 ($n = 5.4$) [16]
SN220M	1000	N/A	0.06 ($n \approx 16$)	0.19 ($n = 5.4$) [16]
SN220M	1200	0.14-0.66 ($n \approx 1.5$-7.1)	0.32 ($n \approx 3.1$) 0.15 ($n \approx 6.7$) 0.10 ($n \approx 9.9$) 0.15 ($n \approx 6.7$)	0.26 ($n = 3.8$) [16]
PY6	1260	0.40 ($n \approx 2.5$)	0.14 ($n \approx 7.2$) 0.15 ($n \approx 6.8$) 0.04 ($n \approx 28$) 0.03 ($n \approx 32$) 0.02 ($n \approx 42$)	0.18 ($n = 5.6$) [5]

[a]Listed in order of decreasing ε_t as shown in Table 2.
Note: Calculation of n as $n \approx 1/\nu$ and $n \approx 1/\nu^*$.

establish consistent $d\sigma/d\dot{\varepsilon}_p$ translation lines and hence the slope, ν. Consequently the agreement between the comparisons of the constant-stress creep behaviour and the stress-relaxation behaviour suffered. Whether this difficulty was due to statistical variations or problems with the test setup is difficult to say given the limited number of stress-relaxation tests conducted.

Note that the best agreement of test results occurred for values of ν^* determined from the σ - $\dot{\varepsilon}_p$ relations of the stress-relaxation tests conducted with the largest values of ε_t, i.e. tests which established the plasticity state. However in nearly all such cases the n value determined from $1/\nu^*$ was greater than the n value determined from the creep tests, which is

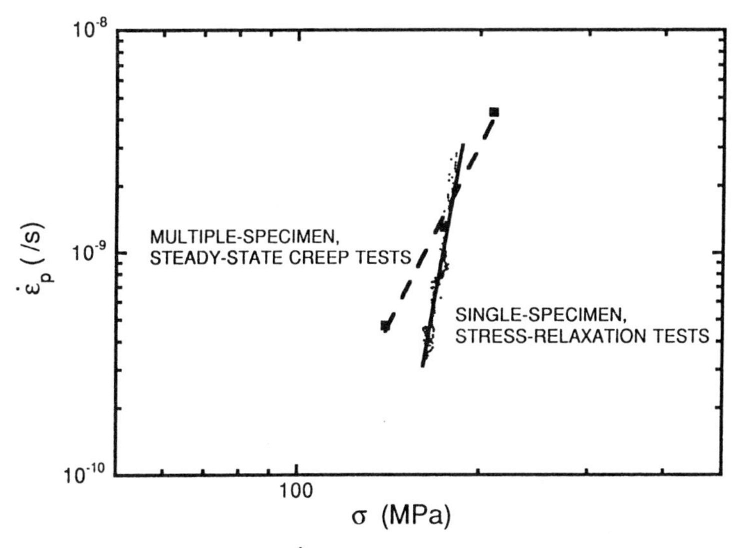

Figure 7. Comparisons of $\dot{\varepsilon}_p$ - σ relations for SN220M at 1200°C.

consistent with the observation that higher n values would be indicative of non-steady-state (transient) creep-type behaviour [15].

Figure 8 for the SN220M at 1200°C shows the best agreement between the results from the two test methods. The constant-stress, strain-time curves for this case [16] were sigmoidal in shape with very brief 'steady-state' regimes. Thus the good agreement of the $\dot{\varepsilon}_p$ - σ plots is reasonable since this brief type of behaviour would be similar to that exhibited by the transient-type of stress-relaxation tests.

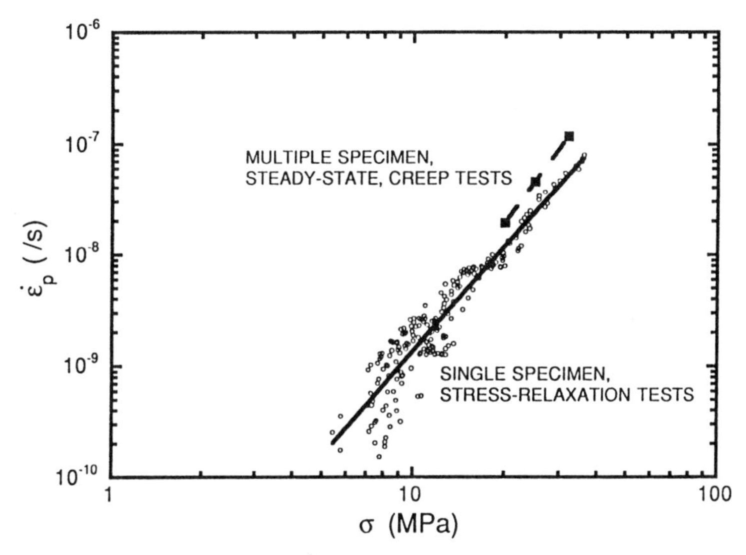

Figure 8. Comparisons of $\dot{\varepsilon}_p$ - σ relations for SN220M at 1000°C.

Figure 9 for the PY6 at 1260°C also shows good agreement between the results from the two test methods. The strain-time curves [5] displayed dominant 'steady-state' regimes, thus the agreement in the $\dot{\varepsilon}_p$ - σ plots is indicative of the attainment of a 'good' plastic state.

Other problems, such as lack of plasticity, susceptibility to slow crack growth, and residual porosity, which may have contributed to the lack of agreement between the two test methods are inherent in the materials themselves. Many structural ceramics even in more plastic states such as long-term creep deformation will seldom exhibit strain of more than 1.0 to 1.5% (10 000 to 15 000 μstrain). Thus, establishing a condition of plasticity under the relatively rapid straining of the stress-relaxation tests without abruptly failing the specimen may be impossible. Additionally, the repeated stressing and relaxing of the materials under the elevated-temperature environment may contribute to the initiation and propagation of small, environmentally-assisted cracks. Such slow crack growth behaviour may have contributed to the poor agreement in the comparison of the results in Fig. 7 as well as to the failure of the SN220M at 1000°C upon restraining the specimen after the initial stress-relaxation test.

Initially, the application of the mechanical equation of state [14] explicitly excluded the role of intergranular mechanisms such as grain boundary sliding (GBS) as opposed to intragranular slip mechanisms. Studies [8-10] which attempted to verify the original equation of state, showed that GBS could be accommodated provided that the region of interest did not

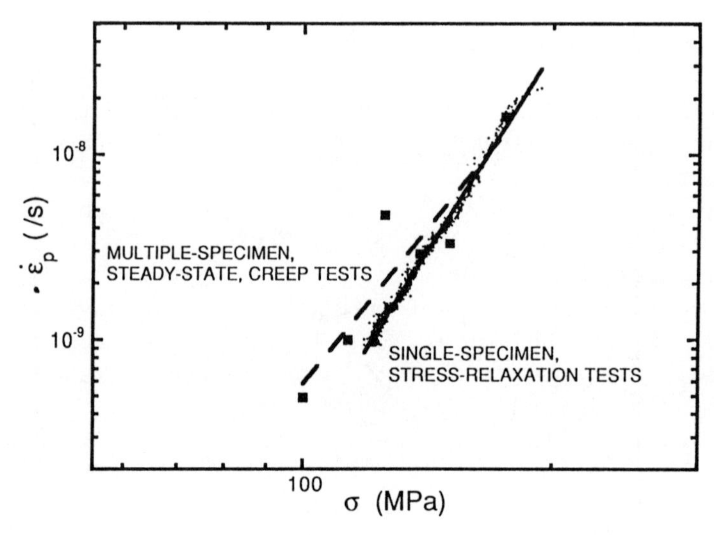

Figure 9. Comparisons of $\dot{\varepsilon}_p$ - σ relations for PY6 at 1260°C.

include either the onset or the end of GBS. For the materials of the present study, the long-term result of GBS is the formation of creep cavities which eventually coalesce to form cracks ultimately leading to the failure of the component. Strictly speaking, these discontinuities are also excluded under the application of the mechanical equation of state although such cavitation would not be expected during the relatively brief times of the stress-relaxation tests.

Another possible limitation of the application of mechanical equation of state to ceramic materials is the annealing effect on the intergranular phase after long-term exposure to elevated temperatures. In this case, annealing refers to the phenomenon [17,18] in ceramics in which initial amorphous intergranular phases crystallize into more refractory materials after exposures to elevated temperatures over periods of time. Given such refractory intergranular materials, the major creep mechanisms of these bulk ceramics, (i.e. intergranular deformation such as GBS) might no longer be effectively described by the mechanical equation of state since the state of strain hardening would no longer be constant.

Finally, it must be noted that the application of the equation of state is limited to the characterization of the macro-mechanical behaviour of the material. Microstructural effects of elevated-temperature exposure under stress will not be manifested using the stress-relaxation tests. The observation and analysis of such effects as creep cavitation, surface oxidation, slow crack growth are necessary for the understanding of material, microstructural behaviour and require long-term, constant-stress tests for these effects to develop.

CONCLUSIONS

Noteworthy conclusions are the following: 1) stress-relaxation tests, analyzed to give stress-strain rate response show promise in providing a unique state description for some structural ceramics under certain conditions; 2) the test method may provide further support for the general applicability of a mechanical equation of state for structurally stable materials; 3) the stress-relaxation test method provides a novel method for obtaining a phenomenological description of structural ceramics at elevated temperatures which heretofore had only been described on a limited basis using constant-stress creep tests; 4) additional investigations need to be conducted to separate statistical variations in the stress-relaxation results for structural ceramics from such phenomenological effects as limited plasticity and slow crack growth failure mechanisms.

ACKNOWLEDGEMENTS

Research sponsored by the U.S. Department of Energy, Assistant Secretary for Conservation and Renewable Energy, Office of Transportation Technologies, as part of the High Temperature Materials Laboratory Program, under contract DE-AC05-84OR21400 managed by Martin Marietta Energy Systems, Inc. The initiative of Professor David A. Woodford, Rensselaer Polytechnic Institute is gratefully acknowledged.

REFERENCES

1. A. Comfort, J. Cuccio, H. Fang, Life Prediction Methodology for Ceramic Components of Advanced Engines, WBS Element 3.2.2.3, in "Ceramic Technology for Advanced Heat Engines Project Semiannual Progress Report for October 1989 through March 1990," ORNL/TM-11586, Oak Ridge National Laboratory, Oak Ridge, Tennessee, pp. 402-411, (1990).
2. D.L. Vaccari, P.K. Khandelwal, Life Prediction Methodology, WBS Element 3.2.2.2, Ibid, pp. 397-401.
3. K.C. Liu, C.R. Brinkman, Cyclic Fatigue of Toughened Ceramics, W.B.S. Element 3.2.1.5, Ibid, pp. 359-372.
4. N.L. Hecht, Environmental Effects in Toughened Ceramics, W.B.S. Element 3.3.1.4, Ibid, pp. 412-437.
5. M.K. Ferber, M.G. Jenkins, Rotor Data Base Generation, WBS Element 3.2.1.6, in Ceramic Technology for Advanced Heat Engines Project Bimonthly Technical Progress Report for October 1990 through December 1990, ORNL/CF-90/385, Oak Ridge National Laboratory, Oak Ridge, Tennessee, pp. 40-46, (1990).
6. H.Riedell, "Fracture at High Temperatures," Springer-Verlag Berlin, Heidelberg, FRG (1987).
7. E.W. Hart, H.D. Solomon, Load Relaxation Studies of Polycrystalline High Purity Aluminium, Acta Metallurgica, 21:295-307 (1973).
8. H. Yamada, C-Y Li, Stress Relaxation and Mechanical Equation of State in B.C.C. Metals in Monotonic Loading, Acta Metallurgica, 22:249-253 (1974).
9. G.L. Wire, H. Yamada, C-Y Li, Mechanical Equation of State in Grain Boundary Sliding in Lead in Monotonic Loading, Acta Metallurgica, 22:505-512 (1974).
10. J.T.A. Roberts, Mechanical Equation of State and High-Temperature Deformation ($\geq 0.5\ T_m$) of Uranium Dioxide, Acta Metallurgica, 22:873-878 (1974).
11. D.A. Woodford, Measurement of the Mechanical State of a Low Alloy Steel at Elevated Temperature, Metallurgical Transactions A, 6A:1693-1697 (1975).
12. E.W. Hart, A Phenomenological Theory for Plastic Deformation of Polycrystalline Metals, Acta Metallurgica 18:599-610 (1970).

13. E.W. Hart, C.Y. Li, H. Yamada, G.L. Wire, Phenomenological Theory: A Guide to Constitutive Relations and Fundamental Deformation Properties, in "Constitutive Relations in Plasticity," A.S. Argon, ed, MIT Press, Cambridge, Massachusetts, pp. 149-197 (1975).
14. E.W. Hart, Constitutive Relations for the Nonelastic Deformation of Metals, J. Eng. Mater. Technol. 98:193-202 (1976).
15. J.-P. Poirier, "Creep of Crystals," Cambridge University Press, Cambridge, England, pp 29-33 (1985).
16. M.K. Ferber, M.G. Jenkins, V.J. Tennery, Comparison of Tension, Compression, and Flexure Creep for Alumina and Silicon Nitride Ceramics, Ceram. Eng. Scien. Proc., 11:1028-1046 (1990).
17. K.C. Liu, C.R. Brinkman, Tensile Cyclic Fatigue of Structural Ceramics, in "Proceedings of the 23rd Automotive Technology Development Contractors' Coordination Meeting," P-165, Society of Automotive Engineers, Warrendale, Pennsylvania, pp. 279-284 (1986).
18. D.R. Clarke, A Comparison of Reducing and Oxidizing Heat Treatments of Hot-Pressed Silicon Nitride, J. Amer. Ceram. Soc., 66:92-94 (1983).
19. S.M. Wiederhorn, B.J. Hockey, R.F. Krause, Jr, K. Jakus, Creep and Fracture of a Vitreous-Bonded Aluminum Oxide, J. Mater. Scien. 21:810-824 (1986).

PARTICLE IMPACT DAMAGE OF ENGINEERING CERAMICS

John E. Ritter

Mechanical Engineering Department
University of Massachusetts
Amherst, MA 01003

ABSTRACT

Sharp-particle impact of ceramics can lead to strength degradation and material loss through erosive wear. The strength degradation and erosive wear of a wide range of alumina and silicon nitride ceramics, varying in purity and grain size, as well as zirconia toughened ceramics was determined as a function of temperature up to 1000°C and impacting particle (SiC) kinetic energy (50 to 20,000 μJ). Since it is believed that sharp-particle impact damage of ceramics can be modeled by the idealized radial and lateral crack system produced by a Vickers indenter, static indentation damage in the same materials was studied using a Vickers microhardness indenter. Crack-microstructure interactions were characterized by SEM. The erosion results are compared to predictions based on an indentation fracture mechanics model that takes into account crack resistance toughening. Large grain size and toughened materials exhibit the strongest R-curve behavior and this results in strength degradation being less sensitive to the kinetic energy of the impacting particles. On the other hand, erosive wear was not significantly dependent on crack resistance toughening. Erosion temperatures up to 1000°C had little affect on erosive wear; however, post-erosion strength generally increased due to an annealing out of the localized residual stress around the impact flaw.

INTRODUCTION

Ceramics impacted by hard, sharp particles invariably suffer from surface cracking around the impact site.[1] This impact damage generally leads to strength degradation as well as material loss through erosive wear, causing a deterioration in performance, or even worse, catastrophic failure. Thus, the response of ceramics to particle impact must be understood before ceramics can be used reliably in structural applications.

It is generally believed that sharp-particle impact damage in ceramics can be modeled by the idealized flaw system produced by a sharp indenter such as a Vickers.[1-3] The sharp indenter produces two distinctive crack systems: radial cracks which are responsible for strength degradation and lateral cracks which are the source of material removal, see

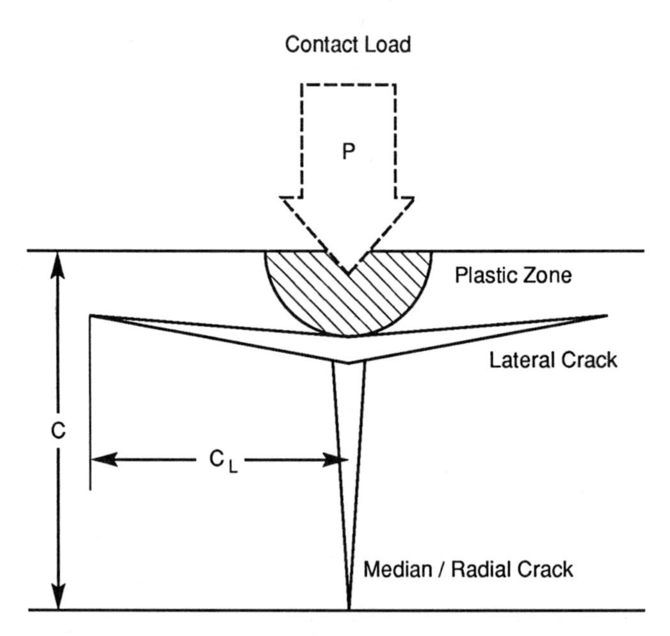

Figure 1. The indentation crack system.

figure 1. Both of these crack systems are driven to their equilibrium size by the residual stress about the indentation due to the mismatch between the plastic zone and the surrounding elastic matrix.[1-3] The indentation fracture mechanics model predicts the extent of cracking in terms of indentation load (or impact energy) and target material properties (toughness, hardness and elastic modulus). This model has recently been extended to include sharp-particle impact damage of ceramics that exhibit crack resistance toughening.[4]

The purpose of this paper is to discuss the impact damage (strength degradation and erosive wear) due to erosion with sharp particles (SiC) of a wide range of ceramics as a function of temperature and kinetic energy of the impacting particle. For comparison static indentation damage in the same materials was studied over a range of indentation loads. Some of these results on individual materials have been previously presented in Refs. 4-11. Note that impact damage by spherical particles is discussed in Refs. 12-15 and the erosive wear of numerous ceramics and ceramic composites in Refs. 16-25.

EXPERIMENTAL PROCEDURE

The materials that were studied are summarized in Table 1. The elastic moduli were given by the manufacturer, while the hardness and toughness values were measured as part of this study using conventional techniques. All materials had densities greater than 99% theoretical and were in the form of disks about 40 mm diameter and 2 mm thick. The samples were lapped and polished by the manufacturer. Note that the alumina CR materials contained sintering aids that made them fully dense and translucent. The coarse bimodal grain size of AS440 was developed to provide toughening.[11,26] For the zirconia based materials, the stress-induced phase transformation of zirconia from the tetragonal to monoclinic phase is thought to cause an enhanced toughening in these materials.[10]

Table I. Material Specifications

Material	Grain Size (μm)	Purity (%)	Elastic Modulus (GPa)	Hardness (GPa)	Toughness (MPa √m)
SLS Glass[a]			70	5.7	0.77
Alumina					
AD90[b]	4	90.0	276	11.3(2.0)[p]	2.9(0.3)
AD997[b]	2	99.7	386	16.1(2.0)	3.7(0.5)
AD995[b]	17	99.5	372	14.2(2.1)	4.4(0.3)
CR6[c]	22	99.9[l]	380	17.4(1.8)	4.2(0.5)
CR30[c]	8/60	99.9[m]	380	16.9(0.8)	4.5(0.6)
Silicon Nitride					
NC132[d]	3	97[n]	300	18.5(0.8)	5.5(0.4)
AS44[e]	1/6[i]	97[n]	304	15.6(0.6)	5.7(0.4)
AS440[e]	6/40[i]	97[o]	370	14.0(0.3)	8.3(0.6)
GN10[f]	3.5	97	306	15.0(0.5)	6.5(0.5)
GN10/30% SiC(w)[f]	j	-	336	18.8(0.4)	6-7
Zirconia					
TZP[g]	0.3	98[o]	198	13.9(0.9)	5.2(0.2)
ZTA[h]	1.0/2.5[k]	98	317	16.6(0.5)	4.3(0.2)

a. Soda-lime-silica glass, b. Coors Porcelain Co.; c. GTE Sylvania; d. Norton Co.; e. Allied-Signal Aerospace Co.; f. Garrette Ceramics Components, Allied Signal; g. tetragonal zirconia polycrystal [ref. 10], h. 10% zirconia (60% tetragonal) toughened alumina [ref. 9]; i. Grains are acicular in shape with the length (size given) being about 7 times diameter; j. SiC(w) 0.8 μ diameter with aspect ratio 10-20; k. ratio of zirconia to alumina grain size; l. Sintering aid 800 ppm MgO; m. Sintering aids 500 ppm MgO and 300 ppm Y_2O_3 ; n. Sintering aid MgO; o. Sintering aid Y_2O_3; p. Number in parenthesis represents one standard deviation.

Multiparticle impact (erosion) tests (up to 16 samples per test condition) at normal incidence were conducted in air at 25°C using a slinger-type apparatus as described in Refs. 4 and 5. In this apparatus commercial-grade SiC particles are fed into the center of a 50.8 cm diameter tubular rotor rotating in a horizontal plane, are accelerated to the end of the tube, and leave the rotor to strike specimens that are positioned around the perimeter of the apparatus. Impacting kinetic energies were varied over three orders of magnitude by varying the grit size and particle velocity. For the high temperature tests, the erosion cavity was heated by silicon carbide heating elements where the temperature was maintained at $\pm$ 10°C. The furnace was shut down immediately following the erosion test and the samples were allowed to cool slowly to room temperature.

Specimens were exposed to a fixed mass of erosion particles, ranging from 25 to 500 g depending on the grit size and particle velocity. The amount of grit used in each case was calculated to give the same cross-sectional area of particles impacting the specimens. Also, the particle quantities used was sufficient to reach steady-state erosion conditions. After erosion the mass loss by the samples was measured to at least 1% accuracy with an analytical balance. Assuming that the SiC particles are spherical, the number of particles impacting a given sample was estimated from the mass of SiC abrasive used, the particle size and density (3210 kg/m^{3}) of the SiC, and the geometry of the target sample relative to the slinger arm. The erosion rate, defined here as volume loss per particle impact, was calculated from the mass lost from the specimen per particle impact and the sample density. Note that for erosive wear the kinetic energy is based on the average particle size since wear is thought to represent a cumulative event. For strength degradation kinetic energy is based on maximum particle size since it is these particles that are responsible for the strength-controlling flaw.

Static indentation was carried out with a Vickers diamond indenter over a load range of 10 to 1000N. Indents for subsequent strength testing were placed in the center of the tensile surface of the specimen. Three to five samples per indentation load were used.

The post-indentation or -erosion strength was measured at room temperature in air at a fast crosshead speed and with the indent/eroded surface covered with mineral oil to minimize any fatigue effects. A ring-on-ring biaxial test fixture was used and the appropriate equation to calculate fracture strength is given in Ref. 27. After strength testing, fractographic analysis was carried out to ensure that the strength-controlling flaw was produced by indentation or impact.

STATIC VERSUS IMPACT DAMAGE

All materials with the exception of the CR aluminas exhibited the typical Vickers indentation pattern with radial cracks emanating from the four corners on the indent, see figure 2a and 2b. With the large grain size AD955 and AS440 the radial crack path was more irregular and tortuous, see figure 2b. Also with AD995, some intergranular chipping can be seen about the indent impression, indicating grain boundary cracking. For the two CR aluminas the indent impression was not well defined, see figure 2c, and was formed by extensive intergranular chipping underneath the indenter. Radial cracks emanate from the indent site in an irregular pattern. Fractographic analysis showed that the largest radial crack was found to be strength controlling.

Impact damage sites in comparison to indentation were more irregular and generally not reflective of the geometry of the impacting particle, see figure 3. This is probably due to the geometric irregularity of the impacting "sharp" particle that can cause it to hit on an edge or face

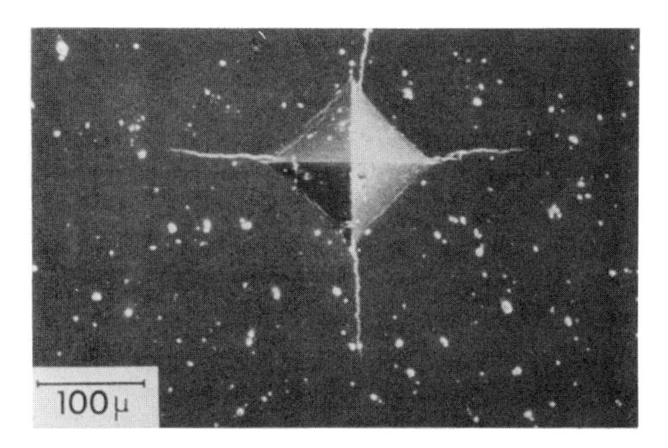

Figure 2a. Vickers indentation AS44 (P = 185 N)

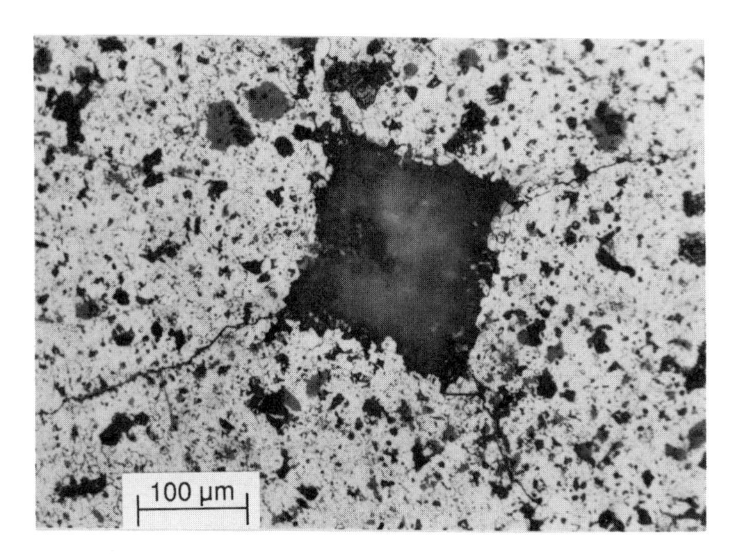

Figure 2b. Vickers indentation AD995 (P = 490 N)

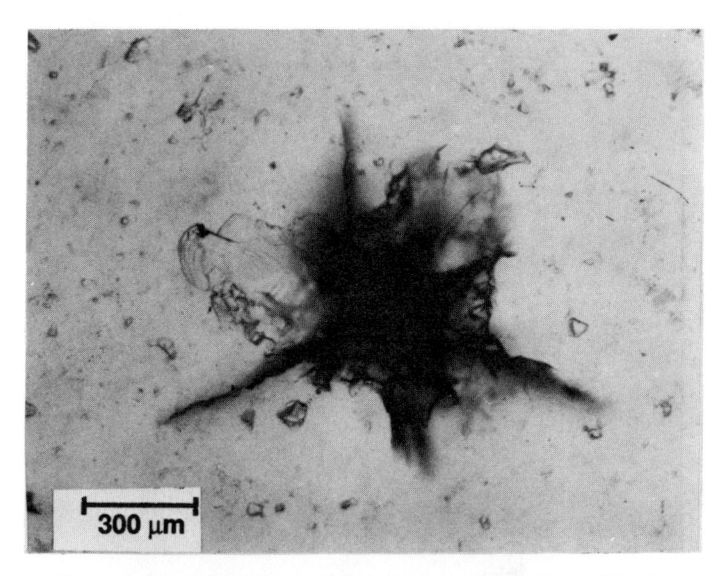

Figure 2c. Vickers indentation CR30 (P = 98 N)

rather than on a "sharp" point. It is important to note that material removal in all materials, especially at high impacting kinetic engeries, often involved surface pitting through grain boundary cracking (figure 3a and 3c), as well as through lateral cracking (figure 3b). Lateral crack chipping was more prevalent in the small grain size materials. However, with all samples radial cracks generally emanated from the larger impact sites and their size scaled with kinetic energy of the impacting particles - the larger the kinetic energy, the larger the radial cracks. Finally, no difference was seen in the impact damage of any of the samples tested up to erosion temperatures of $1000^{o}C$.

It is interesting to note that in contrast to impact by hard, sharp particles, lateral crack chipping was seldom observed in static indentation. However, recent research has shown that lateral crack chipping can be induced by repeated static indentations.[28] Thus, it appears that lateral crack chipping is promoted by the non-ideal shape of the impacting particle and by repeated impacts/indentations.

STRENGTH DEGRADATION

Post-impact strength can be derived from indentation fracture mechanics by assuming that the fracture resistance (K_R) depends on radial crack size (c) by[4,29,30]

$$K_R = k\ c^m \tag{1}$$

where k and m are fracture resistance constants. The limits on m are 0, corresponding to toughness being independent of crack length, and 0.5, corresponding to no catastrophic crack propagation. During a strength test of an indented flaw, the elastic-plastic residual stress about the indent supplements the applied tensile stress so that the net stress intensity factor (K) is[29,30]

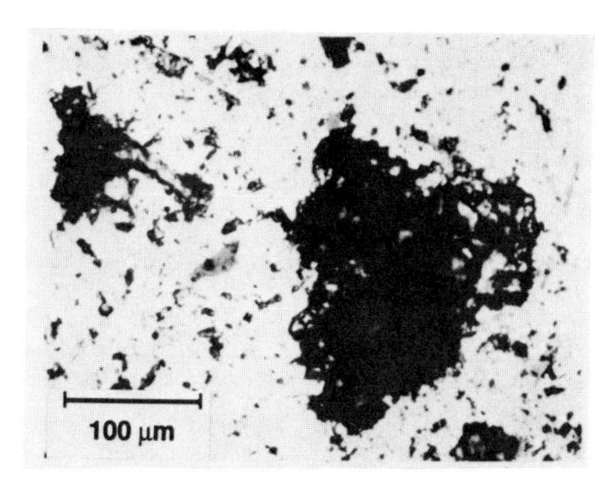

Figure 3a. Isolated impact site in AD995 eroded at U_k = 1220 μJ.

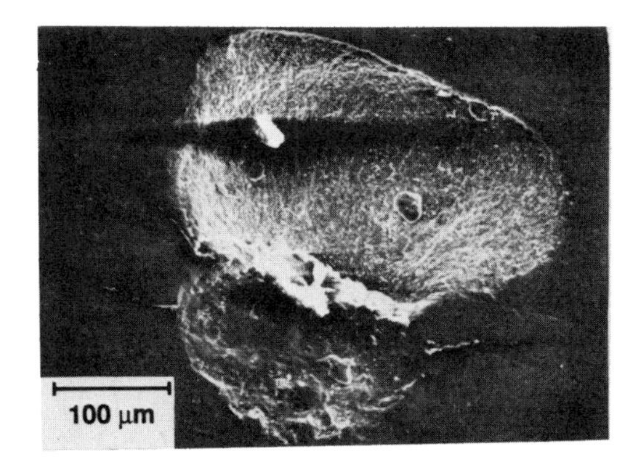

Figure 3b. Isolated impact site in AS44 eroded at U_k = 1942 μJ.

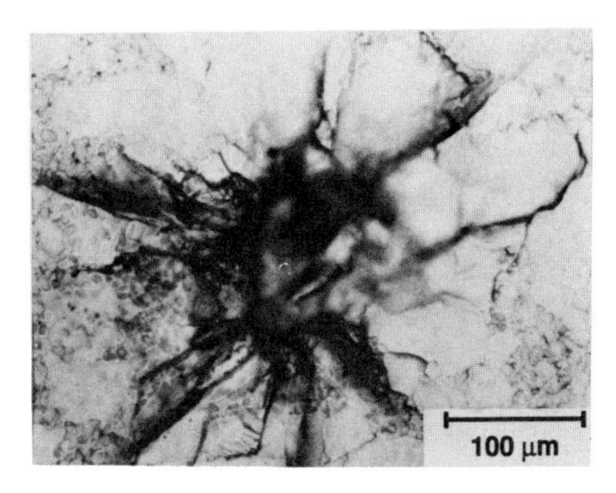

Figure 3c. Isolated impact site in CR30 eroded at U_k=3165 μJ

$$K = \frac{P\chi}{c^{3/2}} + Y\, \sigma_a c^{1/2} \quad (2)$$

where P is indentation load, χ is a residual stress constant, Y is the flaw/loading geometry constant (equal to 1.26 for a semicircular flaw in tension), and σ_a is the applied stress. The initial indentation crack size (c_o) is obtained from equation 2 by setting $K = K_R$ and taking $\sigma_a = 0$ to give

$$c_o = \left[P\left(\chi/k\right)\right]^{2/(2m+3)} \quad (3)$$

Crack extension under an applied stress occurs when the net stress intensity factor is equal to or greater than the fracture resistance. Therefore, by setting $K = K_R$ in equation 2 gives

$$\sigma_a = \frac{1}{Yc^{1/2}} \left(kc^m - \frac{\chi P}{c^{3/2}} \right) \quad (4)$$

A schematic plot of equation 4 is shown in figure 4 for non-zero and zero χ and for $m = 0$ and $m = 0.3$. For $\chi = 0$, spontaneous failure occurs with no crack extension when[29]

$$\sigma_{fo} = \frac{k}{Y} c_o^{(2m-1)/2} \quad (5)$$

where σ_{fo} is the failure stress for $\chi = 0$. For $\chi > 0$, the crack extends stably as the applied stress is increased up to the maximum at which failure occurs at σ_f. By taking the derivative of equation 4 it can be shown that this instability occurs at[29]

$$\sigma_f = \left(\frac{k}{Y}\right) \left(\frac{3+2m}{4}\right) \left[\frac{4\chi P}{k(1-2m)}\right]^{(2m-1)/(2m+3)} \quad (6)$$

$$c_m = \left[\frac{4\chi P}{k(1-2m)}\right]^{2/(2m+3)} \quad (7)$$

Equation 6 shows that for $m = 0$ the dependence of σ_f on P is to the negative one-third power and that this power decreases to 0 for $m = 0.5$, i.e. σ_f is not dependent on P, for $m = 0.5$. Thus, increasing the crack resistance toughening (m) makes a ceramic more resistant to indentation damage. Note that for $m \geq 0.5$, there is no maximum in the applied stress-equilibrium crack length curve, equation 4; thus, the applied stress can increase infinitely without reaching crack instability. Since this is physically impossible, m must be less than 0.5.

The reduction in strength caused by the residual stress about the indent is obtained by substituting equations 3 and 5 into 6 to give

$$\frac{\sigma_f}{\sigma_{fo}} = \left(\frac{2m+3}{4}\right) \left(\frac{4}{1-2m}\right)^{(2m-1)/(2m+3)} \quad (8)$$

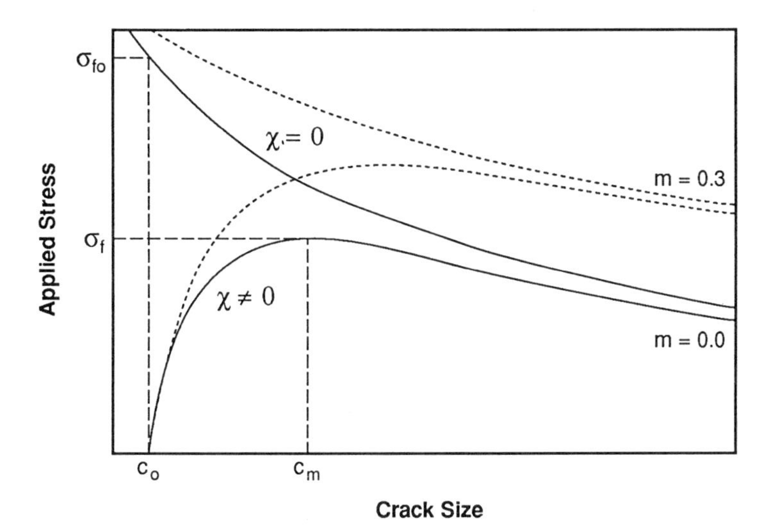

Figure 4. Schematic of applied stress as a function of crack size.

By taking the ratio of equations 3 and 7, the extent of stable crack growth is

$$\frac{c_m}{c_o} = \left(\frac{4}{1-2m}\right)^{2/(2m + 3)} \tag{9}$$

For m = 0, equations 8 and 9 show that the presence of residual stress can cause a strength reduction of about 53% and a relative crack extension of 2.5. For m > 0, the reduction in strength is less but the corresponding crack extension is greater. For example, for m = 0.30 the predicted strength reduction is 30% and the crack extension is 3.6.

Assuming that the particle impact event can be modeled as a quasi-static indentation, the relationship between impacting particle kinetic energy (U_k) and the equivalent indentation load is[31]

$$P = \alpha\ \xi\ H^{1/3}\ U_k^{\ 2/3} \tag{10}$$

where α is a constant related to the fraction of energy transferred to the target material, ξ is a geometric indenter constant (equal to 4.8 for the Vickers geometry) and H is hardness. Substituting equation 10 into equation 6 gives the post-impact (erosion) strength as a function of kinetic energy as

$$\sigma_f = \left(\frac{k}{Y}\right) \left(\frac{3+2m}{4}\right) \left[\frac{4\chi\alpha\xi H^{1/3}\ U_k^{2/3}}{k(1-2m)}\right]^{(2m-1)/(2m+3)} \tag{11}$$

For no crack resistance toughening ($m = 0$), equation 11 gives that the dependence of post-impact (erosion) strength on impacting kinetic energy is to the negative two-ninth power. For a non-zero m, equation 11 shows that crack resistance toughening makes strength less sensitive to impacting kinetic energy, hence more damage resistant. This is similar to the result above for the post-indentation strength.

Room Temperature Erosion

Figures 5, 6 and 7 summarize the post-indentation and erosion strength data for CR6, AS440 and AD995, respectively. Note that equation 10 was used to relate the U_k and P scales in figures 5, 6 and 7 assuming $\alpha = 1.0$ and $\xi = 4.8$. Based on the slopes (β) of the curves in figures 5, 6 and 7, m can be calculated using equation 6 for the indentation data ($\beta = (2m-1)/(2m+3)$) and equation 11 for the impact data ($\beta = 2/3(2m-1)/(2m+3)$). With both CR6 and AS440 m determined from the impact data is significantly greater than that determined from the indentation data. In addition, for CR6 and AD995 the post-indentation and impact strengths are similar; whereas, for AS440 the post-impact strength is considerably less than the post-indentation strength.

Table II summarizes the post-indentation and erosion strength data for all the materials studied. Given in the table are the m values calculated from the strength data and the strength values corresponding to $P = 100$ N and $U_k = 1000$ µJ. These results show that the materials with larger grain sizes have higher m values, i.e. exhibit stronger crack resistance toughening. It is believed that the coarser microstructures, especially with the bimodal grain sizes, lead to an irregular, torturous crack path and that this in turn enhances crack bridging by leaving unbroken grains behind the crack tip. The cumulative effect of these crack bridges setting up closure stresses behind the crack tip leads to crack resistance toughening.[32] However, m values determined from post-indentation strength

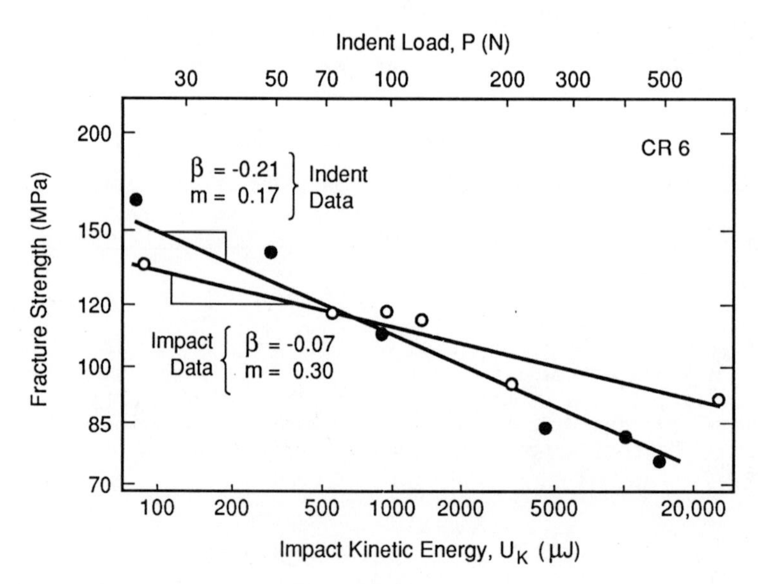

Figure 5. Comparison of post-indentation and erosion strength data for CR6. The solid lines represent the best-fit line with slope β.

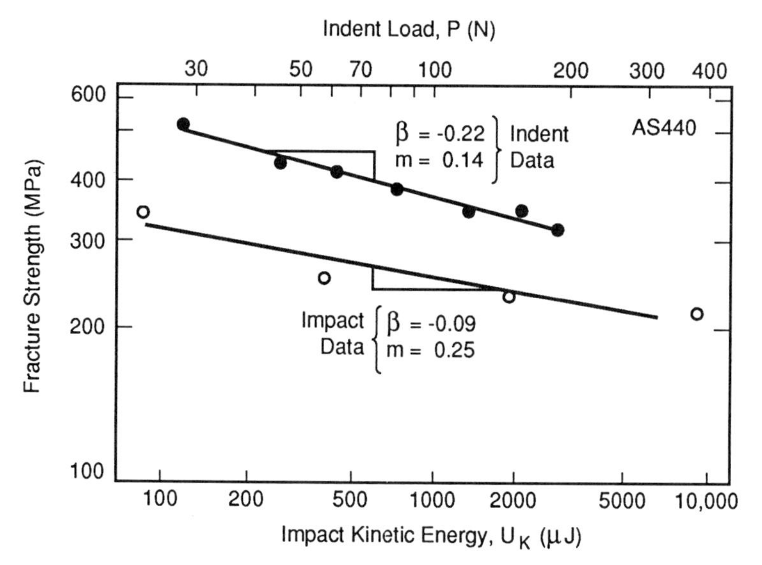

Figure 6. Comparison of post-indentation and erosion strength data for AS440. The solid lines represent the best-fit line with slope β.

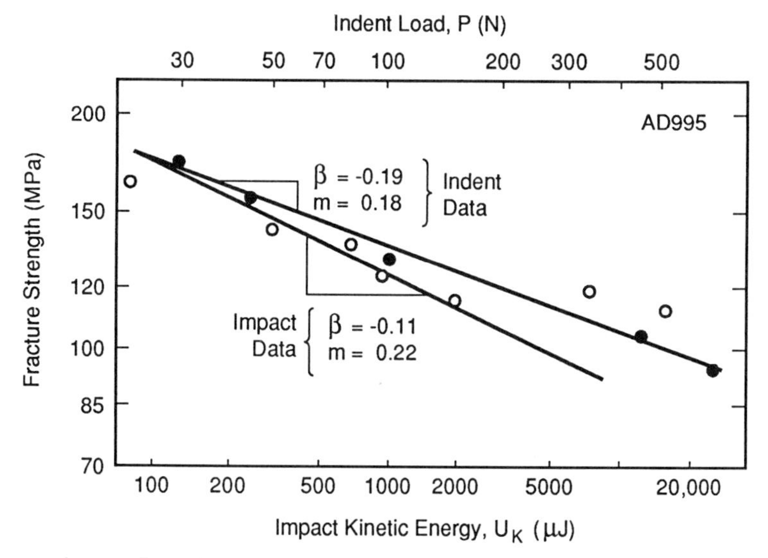

Figure 7. Comparison of post-indentation and erosion strength data for AD995. The solid lines represent the best-fit line with slope β.

Table II. Summary of Post-Indentation and Erosion Strength Data

Materials	Indentation m	Indentation σ_f (MPa) P = 100 N	Impact Energy m	Impact Energy σ_f (MPa) U_k= 1000 μJ
SLS Glass	0	28.7(0.5)	0	24.6(1.2)
Alumina				
AD90	0.05(.01)[a]	103(11)	0.03(.04)	86(12)
AD997	0.09(.01)	129(12)	0.20(.02)	114(11)
AD995	0.18(.01)	138(16)	0.22(.02)	124(14)
CR6	0.17(.02)	108(9)	0.30(.05)	112(8)
CR30	0.27(.02)	117(12)	0.35(.03)	116(14)
Silicon Nitride				
NC132	0.03(.01)	189(20)	0.06(.01)	190(20)
AS44	0.02(.01)	305(31)	0.15(.06)	196(12)
AS440	0.14(.02)	390(40)	0.25(.04)	254(16)
GN10	-	-	0.16(.02)	196(10)
GN10/30% SiC(w)	-	-	0.15(.04)	150(14)
Zirconia				
ZTP	-	-	0.18(.03)	210(24)
ZTA	-	-	0.21(.02)	146(15)

a. Number in parenthesis represents one standard deviation.

data are generally significantly lower than those determined from the post-erosion strength data, although the materials with low m values from indentation have low m values in impact, e.g. compare AD90 to AD995 and AS44 to AS440. The higher m values determined from the erosion data is probably related to the observation that at high impacting kinetic energies, extensive grain boundary cracking often resulted in chipping around the impact site to form a shallow pit, see figure 3a. This pit can reduce the stress intensity factor of the flaw by reducing the residual stress parameter χ since some (or all) of the plastic zone is removed and by lowering the stress intensity factor Y for the "pit/crack" flaw.[6,7] Both these effects would lead to higher post-erosion strengths at the high impacting kinetic energies, hence, higher m values. Importantly, at very high impacting energies post-erosion strengths sometimes show a tendency to level off (see figure 7). This is believed to be due to the pit radii approaching the radial crack size so that the stress intensity factor of the pit-crack flaw approaches an asymptotic value.[6,7]

Table II also shows that post-indentation strengths are either similar to the post-erosion strength or significantly greater. There are several complicating factors that make predicting post-erosion strength from indentation data difficult. Certainly, particles can rebound from the impact surface or fragment on impact.[33,34] Either of these effects would cause less energy to be transferred (lower α). Also the geometric irregularity of the impacting "sharp" particle can cause it to hit on an edge or face rather then a "sharp" point. This would affect the indenter constant ξ, the residual stress parameter χ, and the crack geometry parameter Y. The constant ξ increases for a more blunt indenter. A more blunt impact is expected to cause more grain boundary cracking. This leads to a shallow pit forming around the impact site and, as mentioned above, would be expected to reduce the parameters χ and Y. Since only the increase in ξ leads to a lower strength, post-erosion strengths would be expected to be generally higher than the post-indentation strengths. The fact that most of the materials in Table II have post-erosion strengths that are similar to post-indentation strengths implies that the parameters α, χ, ξ, and Y appropriate for impact damage balance out. However, the post-erosion strength data of the silicon nitride materials AS44 and AS440 and SLS glass are considerably less then the post-indentation strength data. This is believed to be due to multiple impact flaws linking up prior to final fracture, see figure 8. Examination of the fracture origins in these post-erosion strength samples generally showed that two or more impact flaws were adjacent to one another at the fracture origin. Thus, it is likely that these flaws linked up prior to final fracture and caused the much lower than expected strengths of the post-erosion data.[35]

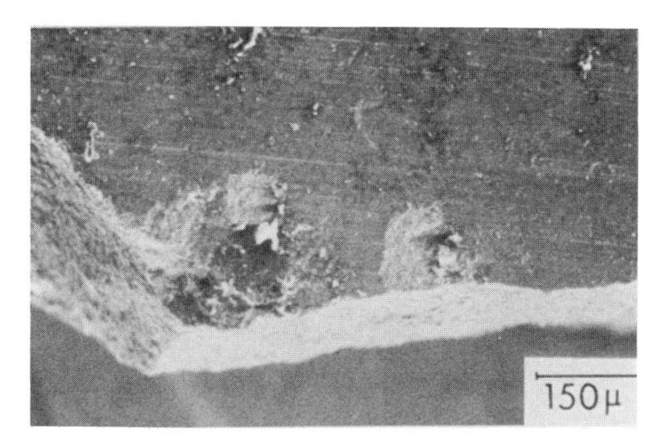

Figure 8a. Fracture origin in AS440 showing multiple impact flaws that linked up prior to final fracture.

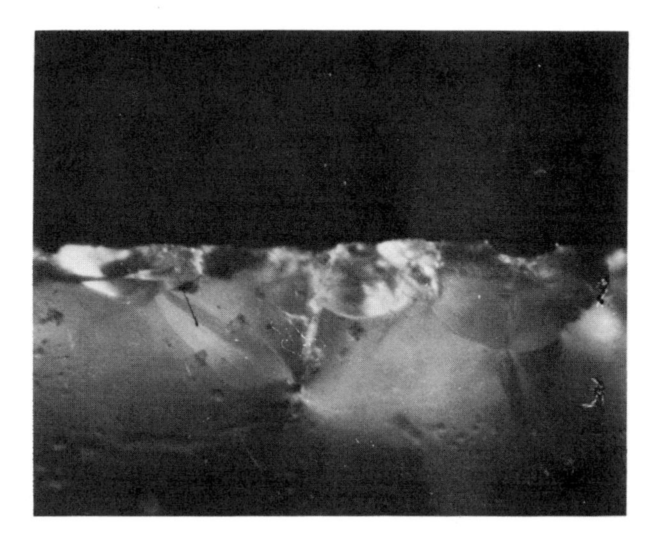

Figure 8b. Fracture origin in SLS glass showing multiple impact flaws linking up prior to final fracture.

In summary, the results in Table II show that the tougher silicon nitride and zirconia materials exhibit the least strength degradation on particle impact. AS440 is the toughest material and it exhibits the highest post-erosion strength. Thus, these results demonstrate that the materials with the best impact resistance are those with high toughness and a large crack resistance toughening.

Elevated Temperature Erosion

Annealing of eroded samples generally results in a modest increase in the post-erosion strength due to the relief of the elastic-plastic residual stress. Equation (8) predicts a maximum strength increase of 53% for m = 0 and 30% for m = 0.30. Annealing eroded SLS glass samples results in an increase of the post-erosion strength of about 20%[5] and annealing AD995 samples impacted at a moderate kinetic energy (345 μJ) increased the post-erosion strength about 40%.[7] The fact that the observed increase in post-erosion strength is significantly less than predicted suggests that the chipping about the impact site associated with multi-particle impact reduces the magnitude of the elastic-plastic residual stress acting on the flaw. Note that, as discussed in the next section, SLS glass samples are severely chipped due to overlapping of impact sites while ceramic samples exhibit chipping about the impact site due to both grain boundary cracking and lateral cracking. This explains why annealing increases the post-erosion strength of SLS glass less than AD995. On the other hand, annealing AD995 samples eroded at a high kinetic energy (5000 μJ) did not affect the post-erosion strength.[7] At very high impacting kinetic energies intergranular fracture in AD995 samples becomes so extensive that the resulting pit formation essentially relieves all of the elastic-plastic residual stress.

Figure 9 shows the dependence of post-erosion strength on erosion temperature for AS440. It can be seen that post-erosion strength increases with erosion temperature but that erosion temperature has little influence on crack resistance toughening. Figure 10 compares the post-erosion strengths of AS440 to AD995 as a function of erosion temperature. The relative insensitivity of post-erosion strength of AD995 to erosion temperature is thought to be due to grain boundary chipping relieving much of the elastic-plastic residual stress. Table III summarizes the post-erosion strength results of all materials tested as a function of temperature. It can be seen that post-erosion strength generally increases with increasing erosion temperature; however, the increases are less than that from annealing samples eroded at room temperature. This indicates that the elastic/plastic residual stress is only partially relieved by the high temperature erosion. This would be expected since the furnace was shut down and the samples began cooling immediately after erosion was completed. Thus, the impact damage that occurred just prior to the completion of erosion was exposed only momentarily to the high (annealing) temperature. For those materials that exhibit little sensitivity of post-erosion strength on erosion temperature, e.g. AD995, it is believed that chipping about the impact site relieves most of the elastic-plastic residual stress; hence, erosion temperature has little affect on strength. Note that ZTP is the only material that shows a decrease in strength after high temperature erosion. This is probably due to the tetragonal particles transforming back to monoclinic, thus providing less crack resistance toughening.

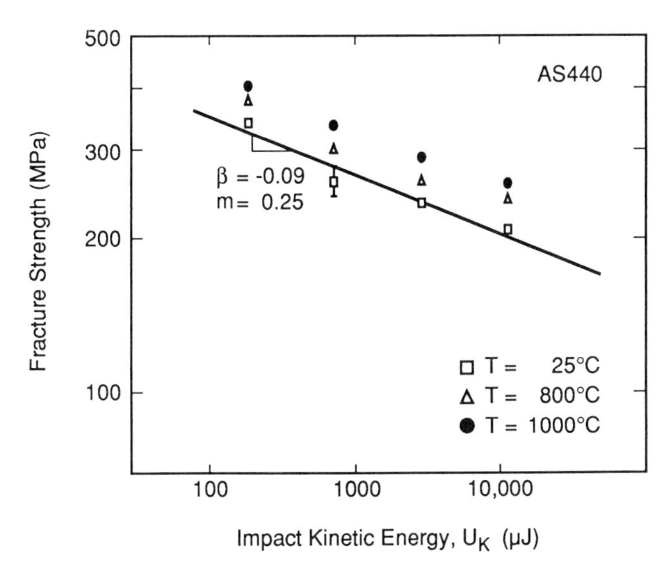

Figure 9. Post-erosion strength of AS440 as a function of erosion temperature.

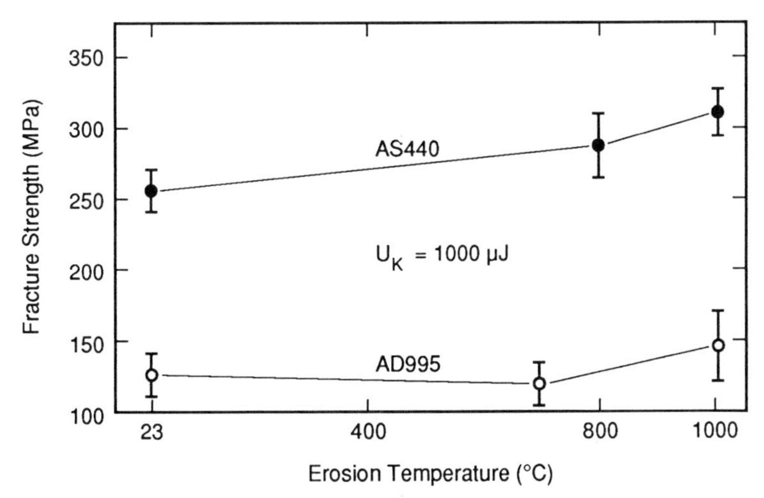

Figure 10. Comparison of post-erosion strengths of AS440 and AD995 at U_k = 1000 μJ as a function of erosion temperature.

Table III. Summary of the Post-Erosion Strength Results as a Function of Erosion Temperature

Material	23° C		800° C		1000° C	
	m	σ_f (MPa) U_k = 1000 μJ	m	σ_f (MPa) U_k = 1000 μJ	m	σ_f (MPa) U_k = 1000 μJ
SLS Glass	0.00	24.6(1.2) [a]	0.00	26.7(1.2) [b]	-	-
Alumina						
AD995	0.22(0.03)	124(14)	0.31(0.04)	119(10) [c]	0.35(0.05)	142(23)
Silicon Nitride						
AS44	0.15(0.06)	196(12)	0.24(0.04)	248(20)	0.20(0.03)	241(21)
AS440	0.25(0.04)	254(16)	0.26(0.03)	287(22)	0.26(0.01)	313(17)
GN10	0.16(0.02)	196(10)	0.22(0.04)	202(15)	0.15(0.04)	207(16)
GN10/30% SiC(w)	0.15(0.04)	150(14)	0.19(0.03)	157(12)	0.15(0.05)	157(14)
Zirconia						
ZTP	0.18(0.05)	210(24)	0.15(0.03)	176(18)	0.14(0.04)	172(15)

a. Number in parenthesis represents one standard deviation.

b. Erosion temperature was 520° C.

c. Erosion temperature was 700° C.

EROSIVE WEAR

Erosive wear can be derived from indentation fracture mechanics by assuming that material removal is by the lateral cracks curving upwards and intersecting the surface. Thus, volume removal per impact, i.e. erosive wear (V), is given by[2,5]

$$V \propto c_L^2 \, h \qquad (12)$$

where c_L is the size of the lateral crack and h is the depth of the lateral crack and is taken to be equal to the radius of the plastic zone[2,5]

$$h \propto E^{1/2} \, H^{-1} \, P^{1/2} \qquad (13)$$

where E is elastic modulus. Following the derivation in reference 2, except using equation (1) for toughness, it can be derived that

$$c_L^{(2+m)} \propto \frac{E^{3/4} \, P^{5/4}}{k \, H} \qquad (14)$$

Substituting equations 13 and 14 into equation 12 and utilizing equation 10 gives

$$V = \Gamma \, k^{-2/(m+2)} H^{-(m+1)/(3m+6)} \left(E/H\right)^{(m+5)/(2m+4)} \, U_k^{(m+7)/(3m+6)} \qquad (15)$$

where Γ is a constant. From equation 15 it can be seen that for m = 0, the predicted dependence of erosive wear on impacting kinetic energy is to the 7/6 power, and as m increases the slope decreases. However, the sensitivity of the slope to m is rather small (slope = 1.17 for m = 0 and slope = 1.05 for m = 0.33). This sensitivity is generally within experimental uncertainty.

As the photographs in figure 3 show, considerable material removal occurs by extensive grain boundary cracking and subsequent grain fallout to cause pitting. By assuming that the energy associated with the formation of a pit will be the grain boundary fracture energy γ multiplied by the product of the number of grains per pit and surface area per grain, it can be shown that[7,36]

$$V = \frac{\psi \, d \, U_k}{\gamma} \qquad (16)$$

where ψ is a constant and d is the grain diameter. Based on equation (16) the predicted dependence of erosive wear on impacting kinetic energy is linear. Thus, within experimental uncertainty this predicted dependency of grain boundary chipping on impacting energy is similar to that predicted on the basis of lateral crack chipping, equation (15).

Room Temperature Erosion

Figure 11 summarizes the room temperature erosive wear data for CR6 and AS440. The extensive grain boundary cracking exhibited by CR6, especially at the higher impacting kinetic energies, lead to significantly larger erosive wear for CR6 than for AS440 at the higher impacting kinetic energies.

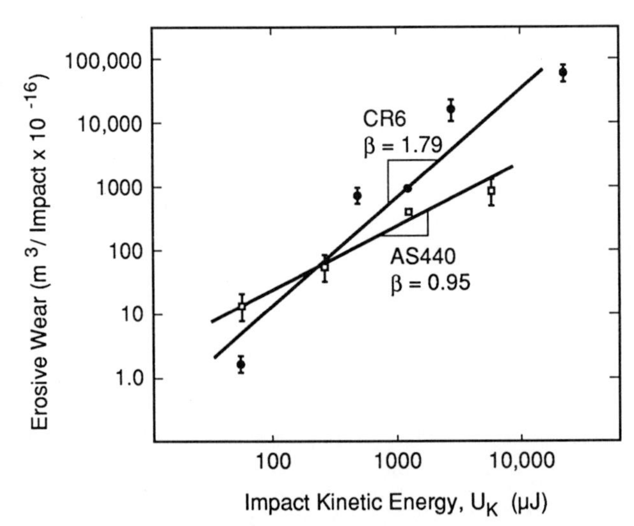

Figure 11. Comparison of erosive wear at room temperature of CR6 and AS440. The solid line represents the best-fit line with slope β.

Table IV. Summary of the Erosive Wear Results for Room Temperature Erosion

Materials	Slope, β	Erosive Wear, V[a]
SLS Glass	1.40(0.1)	90,900(11,400)
Alumina		
AD90	1.67 (.05)[b]	2154 (1487)
AD997	1.02 (.07)	912 (523)
AD995	1.27 (.03)	813 (327)
CR6	1.79 (.33)	1122 (125)
CR30	1.70 (.33)	1070 (153)
Silicon Nitride		
AS44	0.95 (.22)	432 (106)
AS440	0.95 (.11)	237 (125)
GN10	0.90(.18)	418 (66)
GN10/30% SiC(w)	0.89(.17)	569 (102)
Zirconia		
ZTP	0.86(.18)	159 (80)
ZTA	1.36(.13)	734 (353)

a. m^3/impact x 10^{-16} at U_k = 1000 μJ.

b. Number in parenthesis represents one standard deviation.

Table IV summarizes the erosive wear results for the various aluminas and silicon nitride eroded at room temperature. Given in the table are the regressed slope of the log V vs log U_k data and the value of V at U_k = 1000 μJ. Based on these results there does not appear to be any strong correlation between the slope and crack resistance toughening parameter m. In fact, within experimental scatter there appears to be little significant difference between the erosive wear of the various aluminas: however, the silicon nitride and zirconia materials at the higher impacting kinetic energies have a significantly lower erosive wear. Thus, crack resistance toughening, which has a strong influence on the post-indentation and post-erosion strength behavior, does not appear to have a significant effect on the wear behavior. This is probably related to the competing erosion mechanisms, grain boundary cracking/pitting and lateral crack chipping that are responsible for sharp-particle erosive wear in ceramic materials.

Another complication in trying to predict erosive wear results is that none of the models takes into account crack initiation.[1] Figure 12 compares eroded surfaces of ZTA with that of SLS glass. Although these samples were eroded with the same cross-sectional density of particles, it is evident that the SLS glass specimen is much more heavily eroded than the ZTA specimen. Thus, a large fraction of the impacts on ZTA must have been subcritical due to the non-overlapping of impact sites. Also with the ZTA specimen it can be seen that the impact damage is quite variable with a number of the impacts not causing either lateral or radial cracking. The influence of crack initiation on erosion damage would be a fruitful area for future research.

Because of these complexities in predicting erosive wear, dimensional analysis is an alternative method of obtaining relationships between the parameters that affect erosive wear.[37,38] Analysis of erosive wear behavior of a wide range of ceramic materials shows that erosive wear is proportional to fracture toughness to about the negative two power.[37,38] On the other hand, indentation fracture mechanics predicts from equation 15 a dependence of less than negative one power, depending on the value of m. The much stronger experimental dependency of erosive wear of ceramics on fracture toughness is believed to be primarily due to the dependency of the initiation of impact damage on toughness.

Elevated Temperature Erosion

Figure 13 summarizes the erosive wear results for AS440 for the three erosive temperatures studied. Within experimental scatter it can be seen that erosive wear is not strongly dependent on temperatures up to 1000° C. Figure 14 compares the erosive wear of AS440 to AD955 as a function of erosion temperature and Table V summarizes the erosive wear results (slope β, and wear, V) as a function of temperature. Although there is a general decrease in erosive wear at high temperature for all the materials tested, this trend is rather small in comparison to experimental scatter and the dependency of wear on impacting kinetic energy. This is undoubtedly related to the fact that the impact event is dynamic and that the controlling material parameters, primarily hardness and toughness, are little influenced by temperature under dynamic conditions. It is interesting to note that, consistent with the influence of erosion temperature on post-erosion strength, ZTP is the only material that showed a significant increase in erosive wear at high temperatures. This is again probably due to the tetragonal particles transforming back to monoclinic at high temperature, thus not providing any toughening.

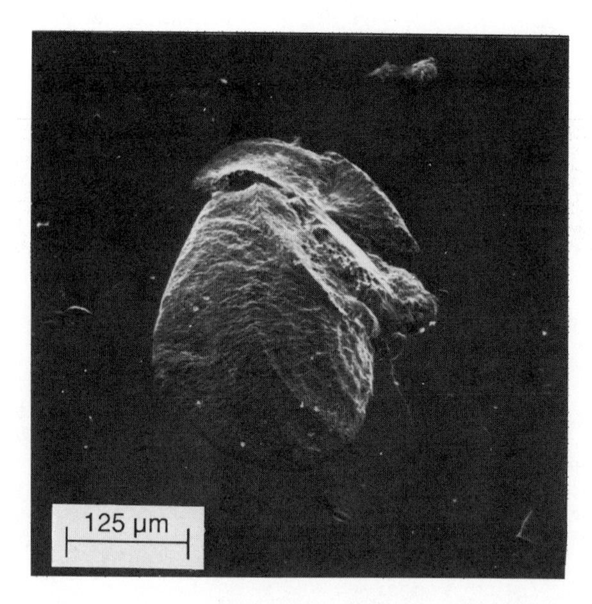

Figure 12a. SEM micrograph of eroded surface of ZTA, $U_k = 21248 \times 10^{-6}$ J.

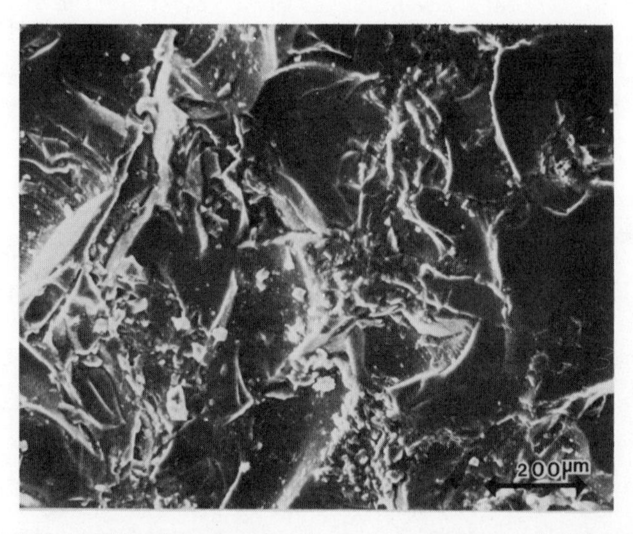

Figure 12b. SEM micrograph of eroded surface of SLS glass, $U_k = 2 \times 10^{-6}$ J.

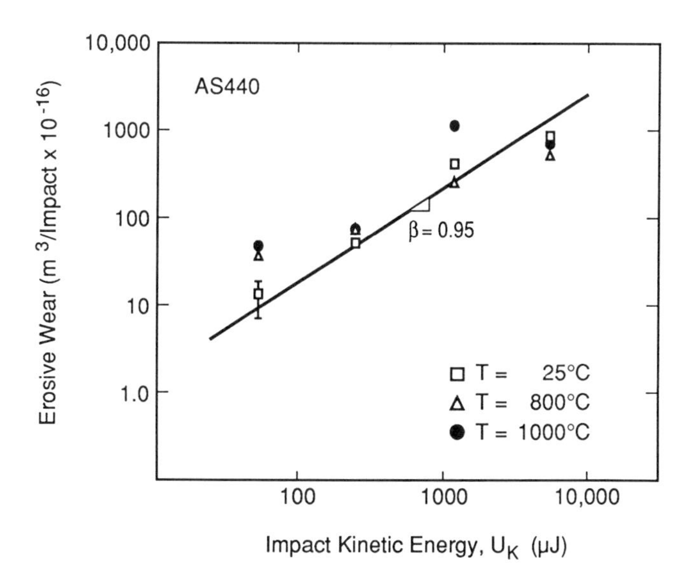

Figure 13. Erosive wear of AS440 as a function of erosion temperature.

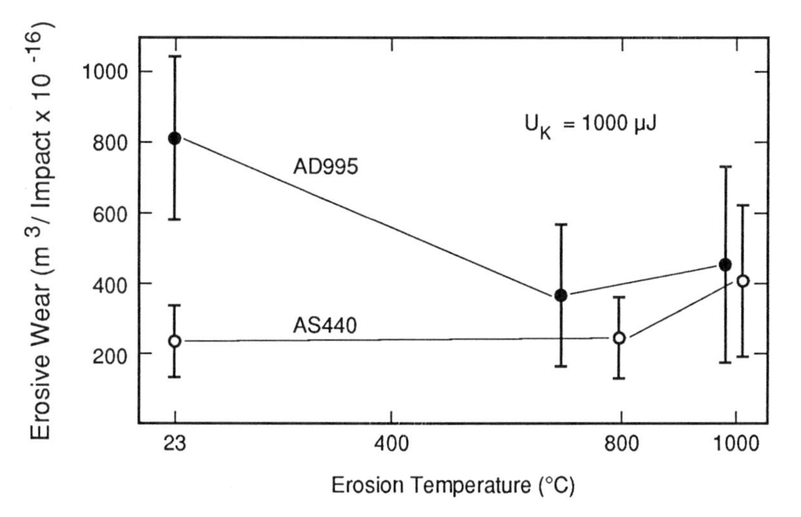

Figure 14. Comparison of erosive wear of AS440 and AD995 at U_k = 1000 μJ as a function of erosion temperature.

Table V. Summary of the Erosive Wear Results as a Function of Erosion Temperature

Material	23° C		800° C		1000° C	
	β	V [a]	β	V	β	V
SLS Glass	1.40(0.10)	90,000 (11,400) [b]	1.60(0.25)	63,100 (8,600) [c]	-	-
Alumina						
AD995	1.27(0.12)	813(327)	1.30(0.14)	363(215) [d]	1.29(0.15)	455(385)
Silicon Nitride						
AS44	0.95(0.22)	432(106)	0.78(0.15)	210(49)	0.72(0.29)	185(78)
AS440	0.95(0.11)	237(125)	0.61(0.04)	228(135)	0.72(0.26)	392(294)
GN10	0.90(0.18)	418(66)	0.71(0.17)	386(121)	0.82(0.24)	210(93)
GN10/30% SiC(w)	0.89(0.17)	569(102)	0.94(0.23)	384(79)	0.82(0.09)	227(138)
Zirconia						
ZTP	0.86(0.12)	150(80)	0.98(0.43)	228(54)	0.85(0.20)	449(157)

a. m^3/impact x 10^{-16} at $U_k = 1000$ μJ.

b. Numbers in parenthesis represents one standard deviation.

c. Erosion temperature was 520° C.

d. Erosion temperature was 700° C.

REFERENCES

1. R. F. Cook and G. M. Pharr, "Direct Observations and Analyses of Indentation Cracking in Glasses and Ceramics," J. Am. Ceram. Soc., 73(4) 787-817 (1990).
2. D.B. Marshall, "Surface Damage in Ceramics: Implications for Strength Degradation Erosion and Wear," pp. 635-56 in Nitrogen Ceramics, ed. F.L. Riley, Martinus Nijhoff, The Hague, Netherlands (1983). Netherlands(1983).
3. B. R. Lawn, B. J. Hockey and H. Richter, "Indentation Analysis: Applications in Strength and Wear of Brittle Materials," J. Microsc., 130 (3) 295-308 (1983).
4. K. Breder, J. E. Ritter and K. Jakus, "Strength Degradation in Polycrystalline Alumina Due to Sharp-Particle Impact Damage," J. Am. Ceram. Soc., 71 (12) 1154-58 (1988).
5. J. E. Ritter, P. Strzepa and K. Jakus, "Erosion and Strength Degradation in Soda-Lime Glass," Phys. Chem. Glasses, 25 (6) 159-62 (1984).
6. J. E. Ritter, P. Strzepa, K. Jakus, K. Buckman, and L. Rosenfeld, "Erosion of Glass and Alumina," J. Am. Ceram. Soc., 67 (11) 769-74 (1984).
7. J. E. Ritter, L. Rosenfeld, and K. Jakus, "Erosion and Strength Degradation in Alumina," Wear, 111 (4) 335-46 (1986).
8. J. E. Ritter, S. A. Carr, and K. Jakus, "Strength Degradation of Hot-Pressed Silicon Nitride Due to Erosion by Sharp Particle Impact," pp. 1363-70 in High Tech Ceramics, ed. P. Vincenzini, Elsevier Science Pub., Amsterdam (1987).
9. K. Breder, G. DePortu, J. E. Ritter, and D. D. Fabbriche, "Erosion Damage and Strength Degradation of Zirconia-Toughened Alumina," J. Am. Ceram. Soc., 71 (9) 770-75 (1988).
10. J. E. Ritter, G. DePortu, K. Breder, D. D. Fabbriche, "Erosion Damage in Zirconia and Zirconia Toughened Alumina," pp. 171-76 in Ceramic Developments, eds. C. C. Sorrell and B. Ben-Nissan, Mater. Sci. Forum, Vol. 34-35, Trans Tech. Pub. Ltd., Switzerland (1988).
11. J. E. Ritter, S. R. Choi, K. Jakus, P. J. Whalen, and R. G. Rateick, "Effect of Microstructure on the Erosion and Impact Damage of Sintered Silicon Nitride,"to be published in J. Mater. Sci., 1991.
12. Y. Akimune, "Impact Damage and Strength Degradation in a Silicon Carbide Reinforced Silicon Nitride Composite," J. Am. Ceram. Soc., 73 (10) 3019-25 (1990).
13. Y. Akimune, Y. Katano, and K. Matoba, "Spherical Impact Damage and Strength Degradation in Silicon Carbide Whisker/Silicon Nitride Composites," J. Am. Ceram. Soc., 72 (5) 791-98 (1989).
14. D. A. Shockey, D. J. Rowcliffe, K. C. Dao, and L. Seaman, "Particle Impact Damage in Silica Nitride," J. Am. Ceram. Soc., 73 (6) 1613-19 (1990).
15. S. M. Wiederhorn and B. R. Lawn, "Strength Degradation of Glass Resulting from Impact with Spheres," J. Am. Ceram. Soc., 60 (9-10) 451-58 (1977).
16. S. Wada, and N. Watanabe, "Solid Particle Erosion of Brittle Materials (part 1): The Relation Between Erosive Wear and Properties of Gas Pressure Sintered Si_3N_4," Yogyo Kyokai-Shi, 94 (11), 113 (1986).
17. S. Wada, and N. Watanabe, "Solid Particle Erosion of Brittle Materials (part 2): The Relation Between Erosive Wear and α or β Phase Content of Hot Pressed Si_3N_4," Yogyo Kyokai-Shi, 95 (5), 468 (1987).
18. S. Wada, and N. Watanabe, Solid Particle Erosion of Brittle Materials (part 3): The Interaction with Material Properties of Target and That of Impingement Particle on Erosive Wear Mechanism," Yogyo Kyokai-Shi, 95 (6), 574 (1987).

19. S. Wada and N. Watanabe, Solid Particle Erosion of Brittle Materials (part 4): The Erosive Wear of Thirteen Kinds of Commercial Al_2O_3 Ceramics," Yogyo Kyokai-Shi, 95 (9), 836 (1987).
20. S. Wada and N. Watanabe, "Solid Particle Erosion of Brittle Materials (part 5): The Effect of Impingement Angle," J. Ceram. Soc. Jpn. Inter. Ed., 95, 925 (1987).
21. S. Wada and N. Watanabe, and T. Toshihiko, "Solid Particle Erosion of Brittle Materials (part 6): The Erosive Wear of Al_2O_3-SiC Composites," J. Ceram. Soc. Jpn. Inter. Ed., 96, 113 (1988).
22. S. Wada and N. Watanabe, "Solid Particle Erosion of Brittle Materials (part 7): The Erosive Wear of Commercial Ceramic Tools," J. Ceram. Soc. Jpn. Inter. Ed., 96, 323 (1988).
23. S. Wada and N. Watanabe, "Solid Particle Erosion of Brittle Materials (part 8): The Erosion Rate of ZrO_2 Ceramics," J. Ceram. Soc. Jpn. Inter. Ed., 96, 587 (1988).
24. S. Wada, N. Watanabe, and T. Toshihiko, "Solid Particle Erosion of Brittle Materials (part 9): The Erosive Wear of Al_2O_3-SiC Particle Composites," J. Ceram. Soc. Jpn. Inter. Ed., 96, 737 (1988).
25. S. Wada, N. Watanabe, and H. Hasegawa, "Solid Particle Erosion of Brittle Materials (part 10): Strengthening by Erosion in Tetragonal ZrO_2 Ceramics," J. Ceram. Soc. Jpn. Inter. Ed., 97, 128 (1989).
26. C-W. Li and J. Yamanis, "Super-Tough Silicon Nitride with R-Curve Behavior," Ceram. Eng. Sci. Proc., 10 (7-8) 632-45 (1989).
27. D. H. Shetty, A. R. Rosenfeld, P. McGuire, G. K. Bansal, and W. H. Duckworth, "Biaxial Flexure Test for Ceramics," 59 (12) 1193-97 (1980).
28. M. Reece and F. Guiu, "Repeated Indentation Method for Studying Cyclic Fatigue in Ceramics," J. Am. Ceram. Soc., 73 (4) 1004-13 (1990).
29. J.E. Ritter, M. R. Lin, and T. J. Lardner, "Strength of Poly(methylmethacrylate) with Indentation Flaws," J. Mater. Sci., 23 2370-78 (1988).
30. R. F. Krause, "Rising Fracture Toughness from the Bending Strength of Indented Alumina Beams," J. Am. Ceram. Soc., 71 (5) 338-43 (1988).
31. S. M. Wiederhorn and B. R. Lawn, "Strength Degradation of Glass Impacted with Sharp Particles: I. Annealed Surfaces," J. Am. Ceram. Soc., 62 (1-2) 66-70 (1979).
32. Y. W. Mai and B. R. Lawn, "Crack Interface Grain Bridging as a Fracture Resistance Mechanism in Ceramics: II, Theoretical Fracture Mechanics Model," J. Am. Ceram. Soc., 70 (4) 289-94 (1987).
33. L. M. Murugesh and R. O. Scattergood, "Effect of Erodent Properties on the Erosion of Alumina," to be published J. Mater. Sci., 1991.
34. J. L. Routbort and R. O. Scattergood, "Solid Particle Erosion of Ceramics and Ceramic Composites," to be published Erosion of Ceramic Materials, ed. J. E. Ritter, Trans. Tech. Pub. Ltd., Switzerland, 1991.
35. J. E. Ritter and R. Davidge, "Strength and Its Variability in Ceramics with Particular Reference to Alumina," J. Am. Ceram. Soc., 67 (6) 432-37 (1984).
36. J. E. Ritter, "Erosion Damage in Structural Ceramics," Mater. Sci. and Engr., 71 195-201 (1985).
37. J. E. Ritter, K. Jakus, M. Viens, and K. Breder, "Effect of Microstructure on the Impact Damage of Polyerystalline Alumina," pp. 55.1-55.6 in Proc. 7th International Conference on Erosion by Liquid and Solid Impacts, eds. J. E. Field and J. P. Lear, Cambridge, England, Sept. 1987.
38. S. M. Wiederhorn and B. J. Hockey, "Effect of Material Parameters on the Erosion Resistance of Brittle Materials," J. Mater. Sci., 18, 766-80 (1983).

STRENGTH DEGRADATION OF CERAMICS CAUSED BY RESIDUAL STRESS AROUND IMPRESSION

Hiroaki Makino, Nobuo Kamiya and Shigetaka Wada

Toyota Central Research and Development Laboratories, Inc.
Nagakute, Aichi, 480-11, Japan

ABSTRACT

The surface of the specimens of three kinds of silicon nitrides and two kinds of sialons was indented with a conical diamond indenter. The flexural strengths of the specimens decreased over certain indentation loads, which were dependent on the material properties, especially on the hardness. It was thought from the observation of the fracture surfaces that the strength degradation was due to the residual stress formed around impressions. The residual stress was evaluated by the method based on a plasticity theory. With an increase in the hardness of the materials, the evaluated indentation load causing residual stress decreases. The evaluation of the residual stress indicates that the material with higher hardness has greater resistance against the localized contact stress. The experimental results coincided with the estimation.

INTRODUCTION

Ceramic components happen to fracture when a highly-localized stress is applied to the components by the contact with other ceramic components. Therefore, it is important to study the contact damage in ceramic materials.

The contact damage phenomena due to highly-localized contact stress have been investigated by means of indentation techniques. For example, local plastic deformation occurs in the case of indentation with a sharp indenter such as Vickers or Knoop indenter, which leads to median cracks[1-3]. In the case of indentation with a blunt indenter, Hertzian cracks are introduced into ceramics due to the elastic contact stress[4]. The relationships between the residual strength and the damages are analyzed based on the linear fracture mechanics[5-7].

We reported previously that in the case of indentation with a blunt indenter with a small tip radius of 0.2 to 1 mm, plastic deformation occurs without any macroscopic cracks, and many intergranuler as well as transgranuler cracks which are about the size of one or a few grains are generated in the plastic deformation area[8]. The area containing many microcracks of a few grain size is referred to as the damaged texture. It was also found that the region where the damaged texture extends is dependent especially on the grain size of the materials[9]. These studies revealed that the strength degradation occurs even if no contact damages are introduced into the materials[8-10]. The cause of strength degradation is supposed to be the residual stress around impressions.

In this paper, therefore three kinds of silicon nitrides and two kinds of sialons were indented with a conical diamond indenter, and then residual strengths were measured. The residual stress formed around impressions was evaluated based on a plasticity

Fracture Mechanics of Ceramics, Vol. 10
Edited by R.C. Bradt *et al.*, Plenum Press, New York, 1992

theory, and the relationship between the strength degradation and the residual stress was discussed.

EXPERIMENTAL PROCEDURE

Materials

Three kinds of silicon nitrides and two kinds of sialons were used. Mechanical properties of the materials tested are shown in table 1. The materials with various hardnesses were chosen. A, B and C are composed of β-silicon nitrides and, D and E are the composites of α'/β'-sialons.

Table 1. Mechanical properties of materials tested

Properties \ Materials	A	B	C	D	E
Young's modulus / GPa	307	308	315	337	338
Hardness / GPa	15.7	15.2	14.5	18.4	19.9
Fracture toughness / $MPam^{1/2}$	5.8	6.0	7.8	5.6	5.8
Flexural strength / MPa	1200	1320	730	1340	1070

Indentation and Strength Measurement

The surfaces (4×40 mm) of flexural specimens (3×4×40 mm) were polished into a surface roughness of Rmax 0.2 μm. The center of the polished surfaces was indented with a conical diamond indenter having a 0.2 mm tip radius and a 120° tip angle at the load of 29 to 147 N. An Instron universal testing machine model 4302 was used for the indentation. The loading and unloading rates were 0.1 mm/min.

Flexural strengths of the indented specimens, outer span 30 mm and inner span 10 mm, were measured so that the maximum tensile stress was applied to the indented place. Fracture surfaces were observed with a scanning electron microscope to identify fracture origins.

Evaluation of Residual Stress

A plasticity theory was employed to determine the residual stress around the impression. The model of the plastic deformation during indentation and the schematic view of residual stress distribution are shown in fig.1. The plastic deformation of an expanded spherical cavity in an infinite body[11] was applied to the plastic deformation beneath the conical indenter. The residual circumferential stress σ_θ^R after unloading becomes a maximum value at the boundary between plastic and elastic deformations shown in fig.1 and the form of the maximum residual stress σ_θ^R is[11],

$$\sigma_\theta{}^R = \frac{P_0}{2}\left\{1 - \frac{P}{P_0}\exp\left(1 - \frac{P}{P_0}\right)\right\},\quad P_0 = \frac{2}{3}Y, \tag{1}$$

where P_0, P and Y are the contact pressure between materials and the indenter at yielding, the maximum contact pressure and yield stress of the material, respectively. The yield stress Y of the ceramics is considered almost the same value as Vickers hardness Hv[12]. Eq.(1), then, is written as eq.(2).

$$\sigma_\theta{}^R = \frac{Hv}{3}\left\{1 - \frac{3P}{2Hv}\exp\left(1 - \frac{3P}{2Hv}\right)\right\} \tag{2}$$

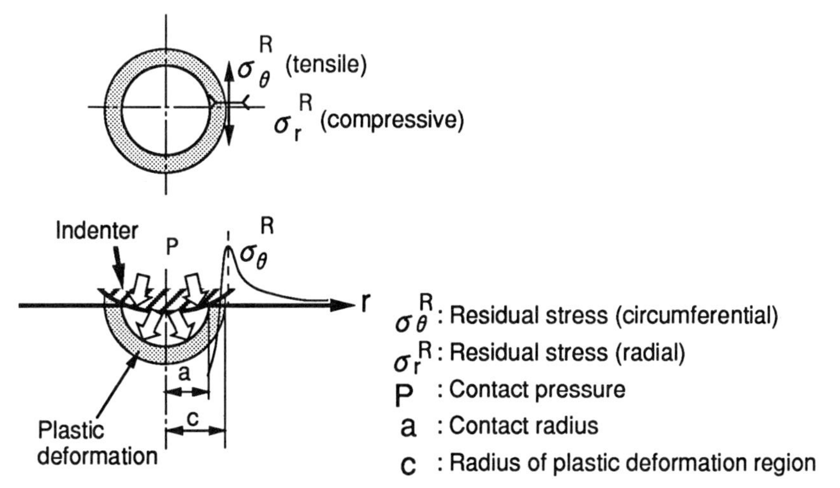

Fig. 1. Plastic deformation model during indentation and schematic view of residual stress distribution.

Therefore, the maximum residual stress σ_θ^R can be calculated by substituting the value of Vickers hardness Hv and the maximum contact pressure P, which is measured by the following experiments.

Measurement of Maximum Contact Pressure

Thin carbon coating was deposited on the polished surface of the specimens, then the surface was indented with the conical indenter. The carbon coating on the surface in the contact area was removed after the indentation, so the contact radius was measured with an optical microscope. The maximum contact pressure P was calculated from the maximum indentation load and the contact radius. The maximum residual stress was evaluated by substituting the value of Hv and P into the eq.(2).

RESULTS AND DISCUSSION

Residual Strength

The relationship between the indentation loads and the flexural strength of material A is shown in fig.2. The sites of the fracture origins switched from inherent flaws to the place around impressions above 88 N of indentation loads and the strengths were degraded above 103 N, which is the critical indentation load causing strength degradation, Pc, of material A. The critical indentation load, Pc, indicates the value of resistance against the localized contact stress; i.e. the higher the Pc of a material is, the superior the resistance against the localized contact stress of the material is.

The scanning electron micrograph of the fracture surface of material A, which was indented at the load of 118 N, is shown in fig.3. The fracture origin was identified to be adjacent to the impression near the surface. In that area, no cracks were observed which might degrade the flexural strength. It is known that the residual stress is formed near the impression[11]. The morphology of the fracture surface indicated the existence of the residual stress around the impression. To confirm the existence of the residual stress around the impression, the surface near the impression was indented with a microvickers indenter, as shown in fig.4. The crack extended from the corner of microvickers indentation confirms the existence of the residual stress. The fracture surfaces were similar at the indentation loads of 88, 103 and 118 N in material A, so the residual stress was considered to degrade the flexural strength at these indentation loads.

Radial cracks were observed around the impressions indented at the loads above 147 N, and these radial cracks were found to be the fracture origins from the observation of the fracture surface. Therefore, the radial cracks were considered to degrade the flexural strength above the indentation load.

The relationships between the indentation loads and the flexural strengths of the 5 kinds of materials are shown in fig.5. The strengths of each material were degraded above certain indentation loads, Pc, which was varied according to the material. The cause of the strength degradation slightly over the critical loads in all the materials

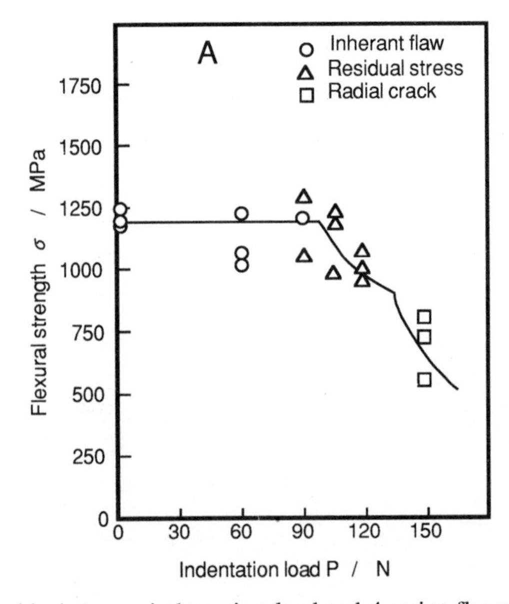

Fig.2. Relationship between indentation load and 4 point flexural strength of A.

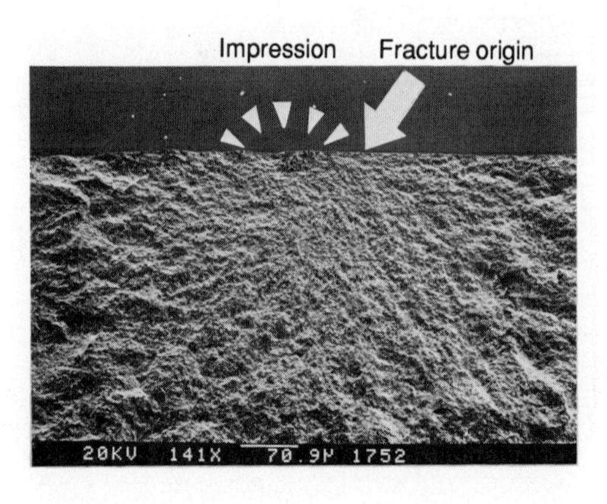

Fig.3. Scanning electron micrograph of fracture surface of A at indentation load of 118N.

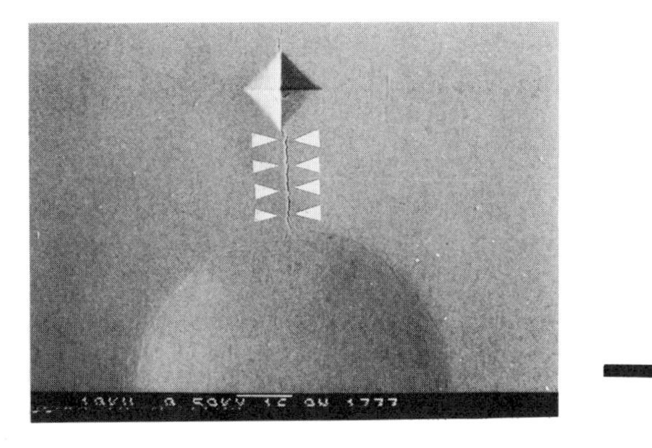

50 μm

Fig.4. Microvickers indentation around an impression indented at the load of 118 N. The crack extended from the corner of microvickers indentation indicates the existence of the residual stress around the impression.

were estimated to be the residual stress around impressions as well as in material A, because there were no traces of radial or Hertzian cracks in the fracture surfaces.

Radial cracks were observed at higher indentation loads, and these radial cracks caused the strength degradations at such higher indentation loads.

Evaluation of Residual Stress

The relationships between the indentation loads and the maximum residual tensile stresses, which were evaluated by eq.(2), are shown in fig.6. The residual stresses increased proportionally with increasing indentation loads except at high loads in A, B and C. It is supposed that the generation of the residual tensile stress causes the degradation of the flexural strengths. Therefore, the relationship between the critical

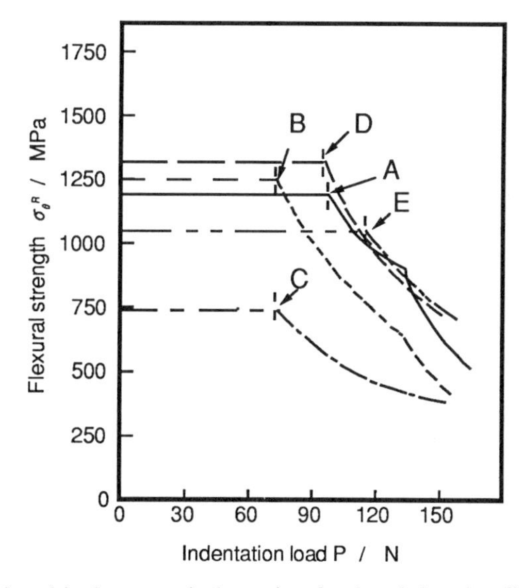

Fig.5. Relationship between indentation load and 4 point flexural strength of 5 kind of materials. Arrows indicates the critical indentation loads Pc causing strength degradation.

indentation load Lc at which the residual stress was generated, which is decided from the extrapolation of the relation between the residual stresses and indentation loads (fig.6), and the critical indentation load causing strength degradation Pc is shown in fig.7. The Pc became higher with increasing Lc. Therefore, it was thought that the material with high Lc has great resistance against the localized contact stress.

However, the two values did not coincide each other; i.e. the flexural strength did not degrade until the residual stress became a certain value. The reasons were considered as follows; ① the effective stressed volume of weibull statistics where the residual stress was generated was so small that the strength degradation did not occur, ② a small rise was observed around the impression, which might relax the residual stress, ③ the hypothesis that the yield stress is equal to the Vickers' hardness was over estimation. In general, the yield stress of metals are almost equal to one third of Vickers hardness of metals, empirically. Qualitative explanation could be obtained from the model, but further improvements are necessary for the model.

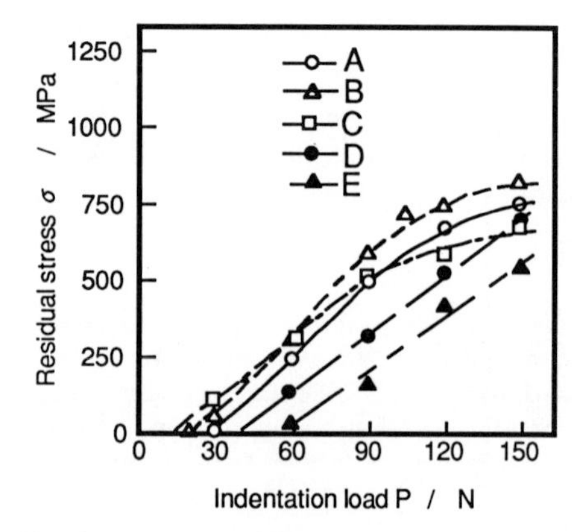

Fig.6. Residual tensile stress around impressions plotted against indentation loads.

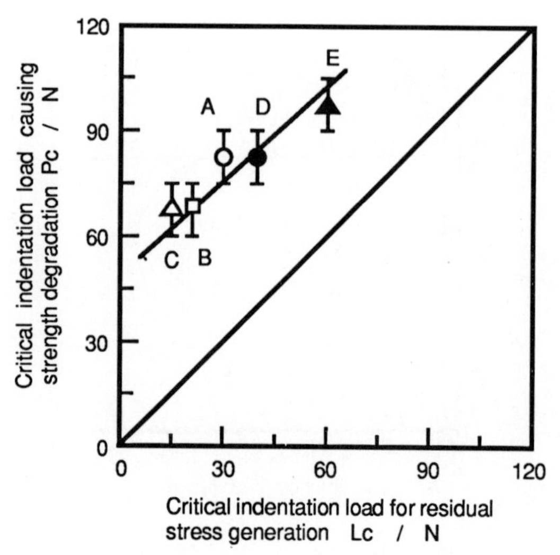

Fig.7. Critical indentation load causing strength degradation plotted against indentation load at which residual stress was generated.

Hardness and Strength Degradation

Fig.8 shows the relationship between the hardness of the materials and the indentation load at which the residual stress was generated. The higher the hardness was, the higher the indentation load of the residual stress formation was. The relationship between the material hardness Hv and the critical indentation load causing strength degradation Pc, which is derived from fig.7 and fig.8, is shown in fig.9. Fig.9 indicates that the higher the hardness is, the higher the critical indentation load causing strength degradation is; i.e. the higher the material hardness is, the superior the resistance against the localized contact stress becomes.

The conclusion derived from fig.9 might be limited in some special case. That is, the damage with a conical indenter was just plastic deformation for all materials at slightly over the Pc in this experiment. The Pc may not proportional to the Vickers hardness when different kinds of damages are generated in materials.

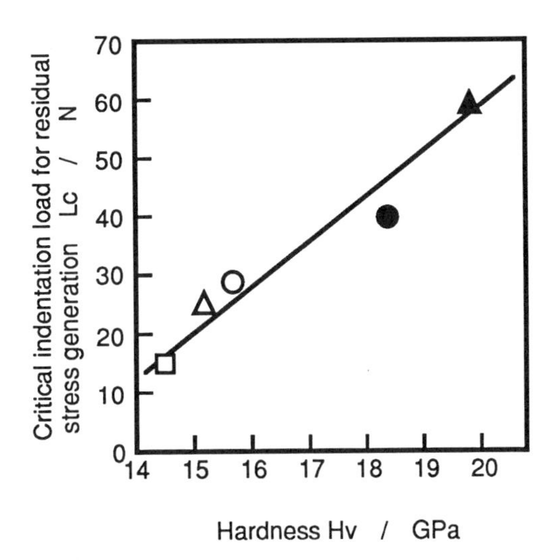

Fig.8. The relationship between the material hardness and the indentation load at which the residual stress was generated.

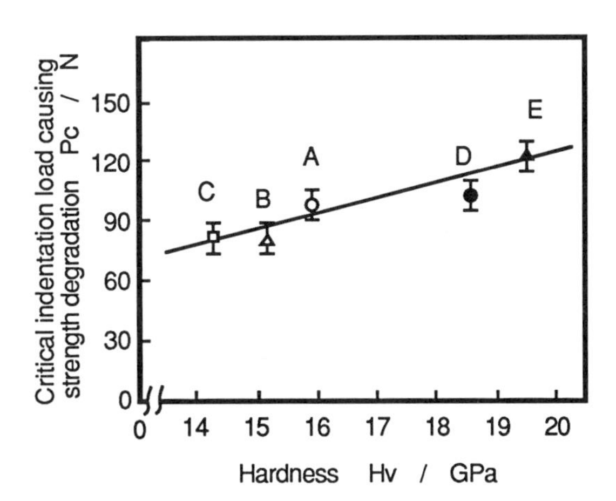

Fig.9. The relationship between the material hardnesses and the critical indentation load causing strength degradation.

SUMMARY

1 Flexural strengths of the material were degraded above certain indentation loads due to the residual stress formed around impressions.
2 The residual stress around the impression was evaluated based on a plasticity theory. The critical indentation load causing the residual stress was higher in accordance with the material hardness. The estimated results indicated that the material with high hardness has high resistance against the localized contact stress.
3 The critical indentation loads causing strength degradation became higher in accordance with the material hardnesses, as indicated from the residual stress evaluation.
4 These results indicate that the higher the material hardness is, the superior the resistance against the localized contact stress becomes.

REFERENCES

[1] B. R. Lawn and M. V. Swain, J. Mater. Sci., 10, 113-122 (1975)
[2] B. R. Lawn and E. R. Fuller, J. Mater. Sci., 10, 2016-2024 (1975)
[3] D. B. Marshall and B. R. Lawn, J. Mater. Sci. ,14,2001-2012 (1979)
[4] R. Warren, Acta. Metall. ,26, 1759-1769 (1978)
[5] B. R. Lawn, E. R. Fuller and S. M. Wiederhorn, J. Am. Ceram. Soc. ,59, 193-197 (1976)
[6] B. R. Lawn, P. Chantikul and G. R. Anstis, J. Aust. Ceram. Soc. ,16, 4-9 (1980)
[7] B. R. Lawn, J. Am. Ceram. Soc. ,58, 428-432 (1975)
[8] H. Makino, N. Kamiya and S. Wada, J. Mater. Sci. Lett. , 7, 475-476 (1988)
[9] will be published in J.Am.Ceram.Soc.
[10] H. Makino, N. Kamiya, S. Wada and O. Kamigaito , Proc. MRS Int. Mtg. on Adv. Mats. V5, 373-378 (1989)
[11] R. Hill , "The mathematical theory of plasticity", Oxford University Press, London (1950)
[12] Y. Enomoto , in Japanese, Science of Machine, 37, 31-36 (1985)
[13] A. G. Evans , ASTM. STP678, 112-134 (1979)

EVALUATION AND SIMULATION OF THERMAL SHOCK BEHAVIOR

BY LIQUID SOLDER QUENCHING

H.Uchimura, A.Kokaji and M.Kaji

Central Research Laboratory, Kyocera Corporation
1-4 Yamashita, Kokubu, Kagoshima, 899-43, Japan

ABSTRACT

Thermal shock test was studied using liquid solder as the cooling medium to obtain a constant heat transfer coefficient. Tested materials were a silicon nitride, a silicon carbide and an alumina. In the case of silicon nitride, no sudden degradation of the strength was observed. This phenomenon is caused by the statistical deviation of the strength. We have proposed to use the Weibull plot method to determine ΔTc. Applying this method to the experimental results, we have shown that the average ΔTc can be well defined, and this method can be consistently applied to many different ceramics. Considering the effective volume of specimen and the temperature dependence of the material properties, the critical temperature difference was calculated. The calculated value and the experimental data agree very well. We have shown that the judgment of materials by only ΔTc may lead us wrong conclusion.

INTRODUCTION

Recently, ceramics have been increasingly utilized in various commercial applications. Silicon nitrides and silicon carbides are particularly interesting high temperature structural materials, because of their light weight and stability at high temperature, compared with metals. However, they receive higher thermal shock fracture during rapid temperature changes, since their thermal conductivity is lower and their Young's modulus is higher compare with metals. For applications of ceramics at high temperature, it is necessary to understand the thermal (shock) stress behavior as well as the mechanical strength and other physical properties. Especially, the thermal shock properties must be carefully studied when ceramic parts are exposed to severe temperature changes such as starting up and shutting down of a heat engine system.

Many studies regarding the thermal shock properties of ceramics have been done by Hasselman[1-3] and other researchers[4-7]. Endurance to thermal shock has often been estimated by a quenching method. Thermal shock resistance is characterized by the critical temperature difference (ΔTc). ΔTc is defined as the amount of quench temperature difference (ΔT) where the strength starts to degrade. For a rapid quenching evaluation, water quench methods have been popular. However, it has been reported that the heat transfer coefficient varies greatly for different conditions. This comes from by boiling phenomena of water[8-10]. Midorikawa et al.[11] have

Fracture Mechanics of Ceramics, Vol. 10
Edited by R.C. Bradt *et al.*, Plenum Press, New York, 1992

reported that more than one ΔTc may be obtained when this phenomenon occurs. In addition, it is reported that specimens are covered by the air when they are dropped into the water[12]. These results demonstrate that water quench methods for thermal shock have several ambiguous features.

In this paper, thermal shock is studied by a quench method using liquid solder as a cooling medium. It is known that the liquid solder has a constant heat transfer coefficient, thus elimination the boiling effect and clear evaluation of ΔTc available. Test materials are commercial silicon nitride, silicon carbide and alumina. Furthermore, Weibull statistics[13] is used to obtain a well defined ΔTc from wxperimental data. In the next chapter, our experimental setup is explained. Then, experimental results and averaged critical temperature difference will be discussed.

EXPERIMENTAL PROCEDURES

Experimental Equipment

A schematic illustration of the experimental equipment for the sol der-quench thermal shock test is shown in Figure 1. Four specimens are fixed to the tip of the dropping bar. They are quenched into the molten solder bath directly from the furnace. A part of the specimens (70%) enter the solder. The rubber stem is attached to reduce the speed of specimen smoothly and to prevent the recoil of the bar. his setup made it possible to avoid specimens floating on the surface of the solder, and control the attitude and speed of the specimens. Figure 2 shows the track-like shaped solder bath. The solder is continually circulated by a rotating screw as shown in the figure. The temperature of the molten solder bath is controlled 200±1°C.

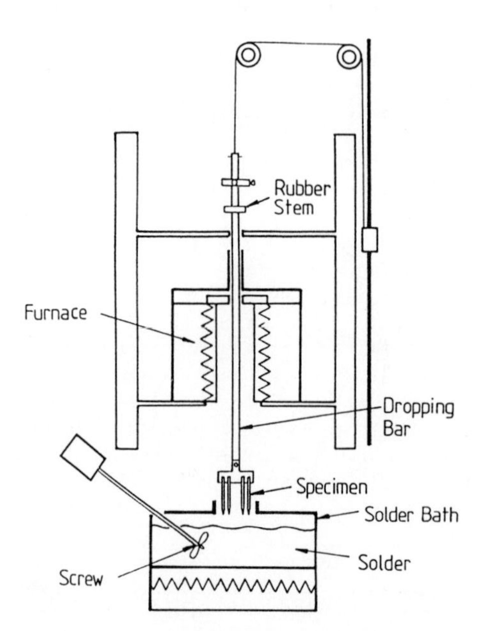

Figure 1. Experimental Equipment of the Solder-Quenching Thermal Shock Test

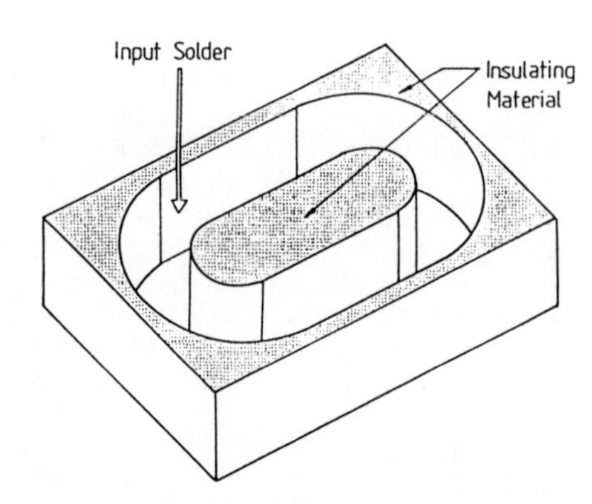

Figure 2. Solder Bath Configuration

Specimens

A rectangular bar is often used for the thermal shock quench test. According to finite element analyses, the maximum principal stress is tensile parallel to the direction of the specimen span at the corner, as is shown in Figure 3. Accordingly, the fracture origins are mainly located at corners of the specimen. Thus the fracture may be greatly affected by the machining of the corners.

In the case of a columnar bar, however, a uniform stress is generated on the surface. A pencil-shaped specimen (shown in Figure 4) reduces the solder splashing and the air drawing. These specimens were heat treated to reduce surface flaw effects. Silicon nitride and alumina specimens were kept at 1000°C for 2 hours and silicon carbide specimens were kept at 1200°C for 2 hours. Diameters of the specimens were 4, 8 and 12mm for the silicon nitride, and only 8mm for silicon carbide and alumina. Physical properties of these materials are summarized in Table 1 and the Appendix A. Four-point bending strength and Weibull modulus were determined at room temperature with displacement rate of 0.5mm/min. The inner and outer spans of the flexure fixture were 10 and 38mm, respectively. The strength of specimens (8mm diameter) are measured. 36 specimens were tested for each material.

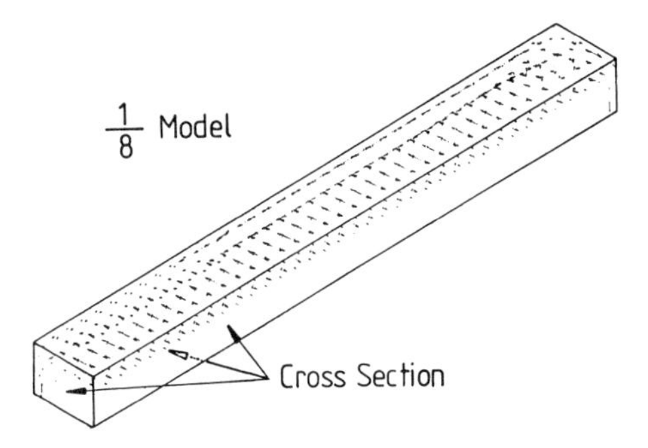

Figure 3. Result of the FEM Analysis for a Rectangular Bar

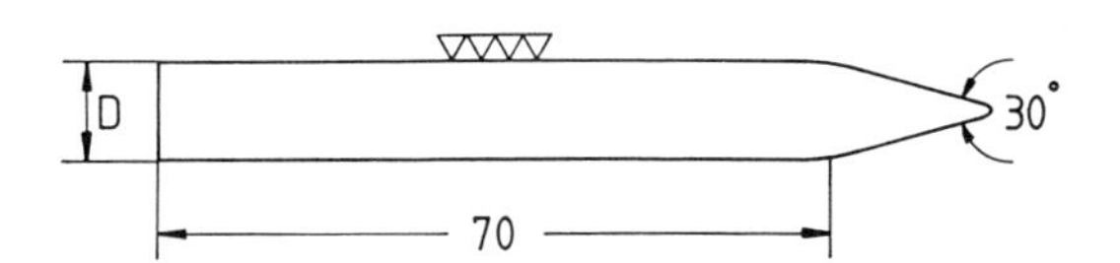

Figure 4. Pencil-Shaped Test Specimen

Table 1. Physical Properties of the Si_3N_4, SiC, Al_2O_3

	Si_3N_4	SiC	Al_2O_3
Fracture Strength (MPa)	676	490	323
Weibull Modulus	15.6	6.4	16.2

Heat Transfer Coefficients

To clarify the experimental condition, heat transfer coefficient was estimated by an empirical equation for liquid metals. The empirical equation for a columnar bar[14] in forced convection has been obtained as follows:

$$Nu = 1.125\ (Re*Pr)^{0.413} \quad , \qquad (1)$$

where Nu, Re and Pr are Knuselt's, Reinhold's and Prantle's numbers, respectively. The flow speed of the solder relative to the specimen is required to obtain the Re. The flow speed of the solder in the bath is meas ured, 0.3m/sec. The dropping speed of specimen was measured as a function of the time. This is shown in Figure 5. In the present experiment, a negative acceleration (almost constant) is applied by the rubber stem. According to this figure, the average speed of the specimen in the solder is 1.1m/sec. Using these values and the properties of the solder listed in Table 2, the heat transfer coefficient, h, is written as follows;

$$h = Nu*\lambda_s/R$$

$$= 2.1*10^3 * R^{-0.587} \quad . \qquad (2)$$

Where λ_s and R are the thermal conductivity of solder and the radius of specimen, respectively. For the case of the 8mm diameter specimens, the h is $5.4*10^4 W/m^2K$.

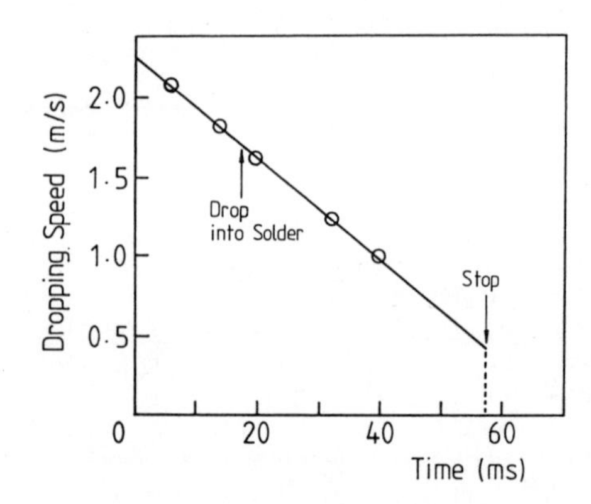

Figure 5. Relationship between the Dipping Speed and Time

Table 2. Properties of Solder at 200°C

Thermal Conductivity	(W/mK)	$2.63*10^1$
Fluid Viscosity	(m^2/sec)	$2.70*10^{-6}$
Specific Heat	(J/Kg·K)	$2.16*10^2$
Bulk Density	(Kg/m^3)	$8.51*10^3$

EXPERIMENTAL RESULTS AND DISCUSSION

Residual Strength

Results of thermal shock tests are listed in Table 3. Residual strength of each specimen was measured by the four-point bending test using the portion close to the cone. In the case of silicon nitride (8mm diameter) are shown in Figure 6. The averaged residual strength is gradually decreased as the ΔT increased from 600 to 700°C. In this case, it is difficult to estimate of ΔTc. Similar results have been reported by Coppola et al.[15] in the case of water quenching method and confirmed by Ashizuka et al.[16] Ashizuka et al. analyzed their result from a statistical point of view.

We consider as follows. The thermal stress, σ_t, which is generated by the temperature difference, ΔT, can be written as follows:

$$\sigma_t = \frac{\alpha E \sigma^*}{1-\nu} * \Delta T$$

$$\equiv \frac{1}{\beta} * \Delta T \quad , \qquad (3)$$

where α, E, ν and σ^* are thermal expansion coefficient, Young's modulus, Poisson's ratio and the dimensionless maximum thermal stress, respectively. Fracture initiates when the thermal stress exceeds the strength of the specimen. If ΔT is critical for that specimen, it is possible to define this ΔT as the ΔTc for the specimen as we can define the stress expressed by Equation (3) a critical stress. This means that ΔTc has a distribution similar to the material strength distribution, and that the distribution of ΔTc is enhanced by β. The residual strength of a specimen damaged by thermal shock exceeding its ΔTc experiences spontaneous degradation. However, the average residual strength appears to be gradually decreased, because of the statistical distribution of the strengths. Therefore, the range of changing residual strengths tends to be expanded when the Weibull modulus is low[17] and/or β is large. The spontaneous degradation of the residual strength observed for alumina may be explained by its small β.

Averaged Critical Temperature Difference

It is obviously difficult to estimate a specific ΔTc directly from the residual strength data, when it decreases gradually. Therefore, another approach must be considered.

In the present test, the heat transfer coefficient is considered to be constant, because liquid solder dose not boil within the test temperature range. In this case, β becomes constant and the thermal stress is linearly dependent on ΔT. Therefore, the ΔTc values for each specimen should have the same distribution as the strength distribution. Strictly speaking, thermal stress is not proportional to ΔT over the whole temperature range, because β consists of physical constants which are usually dependent on temperature. However, it is possible to consider that physical properties are almost constant for the range where the residual strength decreases. Using this approximation, the thermal stress can be considered proportional to ΔT.

Treating ΔT the same as the strength, it is considered to be appropriate to take the average of ΔTc value as a material constant. However, it is difficult to extract the ΔTc directly from the ΔT, which differs from the strength case. The reason is that ΔTc may be smaller than ΔT and is not always equal to ΔT even if a specimen is fractured (or cracked) at ΔT. The ΔTc determined by gradually increasing ΔT with the same specimen may not be exactly the ΔTc because of fatigue effect, such as slow crack growth.

The averaged ΔTc ($\langle \Delta Tc \rangle$) can be obtained as follows. When the

Table 3. Results of Thermal Shock Tests

Material	Diameter [mm]	ΔT [°C]	Residual Strength [MPa]	Tested Number	Cracked Number
Si_3N_4	4	600	652	5	0
		700	691	4	0
		725	627	8	1
		738	530	8	3
		750	539	8	3
		775	474	8	4
		800	527	8	8
	8	600	630	4	0
		620	598	8	0
		630	425	8	3
		640	386	8	4
		650	304	8	5
		670	242	8	7
		700	208	8	8
	12	600	555	9	0
		610	399	15	5
		620	338	15	7
		640	210	15	11
SiC	8	327	289	12	5
		344	241	10	5
		360	232	10	6
		381	212	10	7
Al_2O_3	8	130	341	12	1
		135	273	12	3
		140	222	12	7
		150	180	16	11

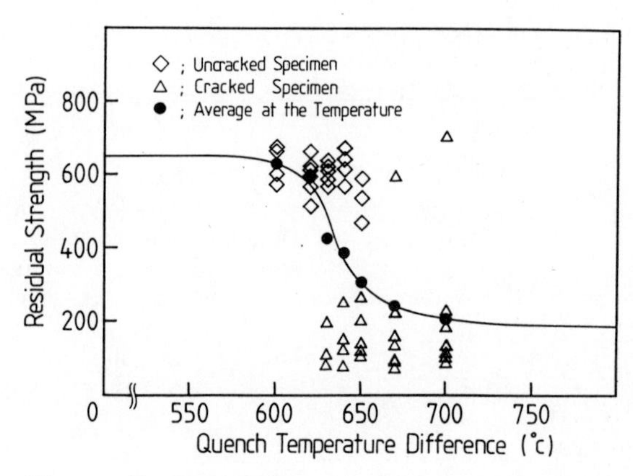

Figure 6. Transition of Residual Strength

strength follows two-parameters Weibull statistics, the thermal shock fracture probability follows the Weibull distribution as shown below, since ΔT and the generated thermal stress have a linear relationship.

$$F(\Delta T) = 1 - \exp[-\int (\Delta T/\Delta T_o)^m dv] \quad ; \Delta T>0$$

$$F(\Delta T) = 0 \quad ; \Delta T \leq 0 \qquad (4)$$

Where $\Delta T_o = \sigma_o * \beta$, and m is the Weibull modulus. When a specimen obtains macroscopic cracks at some ΔT, it is regarded as fractured and the fracture probabilities are plotted on the Weibull probability diagram. $\langle \Delta Tc \rangle$ is then defined as that temperature difference where the fracture probability is 0.5. The fracture probability must be assessed for at each ΔT, not by the rank order method used for the conventional Weibull plot of the fracture strength.

When the fracture probability follows the Weibull distribution, the value at the probability of 0.5 does not correspond to the statistical mean. Fracture probability at the averaged ΔT depends on the Weibull modulus, m, as follows.

$$F = 1 - \exp[-\{\Gamma(1+1/m)\}^m] \quad , \qquad (5)$$

where $\Gamma()$ is the gamma function. This value changes according to m as shown in Table 4. If we use this variant values depending on m, the definition becomes complicated. Therefore, the fracture probability which determines $\langle \Delta Tc \rangle$ is taken to be 0.5.

Table 4. Fracture Probabilities for the Statistical Means

Weibull Modulus	2	5	10	20	50	100
Fracture Probability	0.544	0.479	0.455	0.443	0.435	0.432

Weibull Plot of the Critical Temperature Difference

Figure 7-a shows Weibull plots of 3 different diameter specimens of the silicon nitride. Figure 7-b shows Weibull plots of the three different materials of 8mm specimens. Figure 7-a shows that the experimental values deviate from a line in the case of 4mm specimens. In cases of the 8 and 12mm diameter ones, good fitting lines are obtained. Figure 7-b indicates that other materials of 8mm specimens also show good linear dependence. For the case of alumina, the fracture probability changes very rapidly from about 0 to 1 within the range of only 20°C.

These results show that a ΔTc is well defined using Weibull statistics, and that a well defined $\langle \Delta Tc \rangle$ can be obtained from the Weibull plot drawn as a function of ΔT. It is also evident that this method can be consistently used for different materials from silicon nitride to alumina.

Crack Patterns and Fracture Origins

Crack patterns of silicon nitride, silicon carbide and alumina, which occur at about $\langle \Delta Tc \rangle$, are shown in Figures 8-a,b and c, respectively. These crack patterns are obtained by rolling a dye-penetrated cracked specimen on a sticky tape. From a careful observation of branch angles of these cracks, the crack propagation direction was analyzed. Only one fracture origin was found for each pattern. This confirms that the weakest link hypothesis, or Weibull statistics, is applicable to the thermal shock fracture phenomenon.

The exposed fracture surfaces have also been examined. In the case of silicon nitride, almost always a void was found on the crack line where the fracture origin is likely to be. The fracture surface is shown in Figure 9-a. When there is no obvious void at the surface, the fracture origin is usually located inside of the specimen (see Figure 9-b). In the case of alumina, abnormally large grains are often found at the surface and the inside of specimens. In the case of silicon carbide, crack propagates by an intra-granular fracture and high density cracks were generated. Therefore, it was not possible to investigate the fracture origin. For all cases, clearly fracture patterns of mirror, mist, and hackle, could not be observed near to the fracture origin.

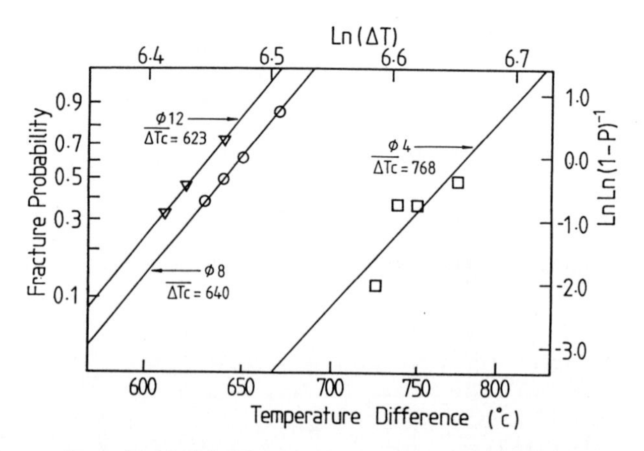

Figure 7-a. Weibull Plots of the Critical Temperature Difference for Different Diameters

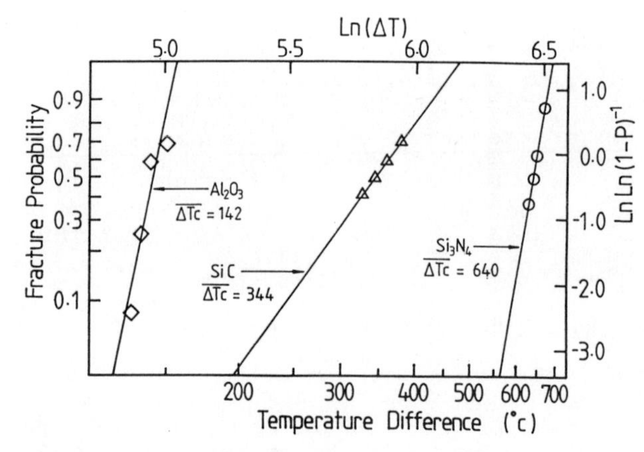

Figure 7-b. Weibull Plots of the Critical Temperature Difference for Different Materials

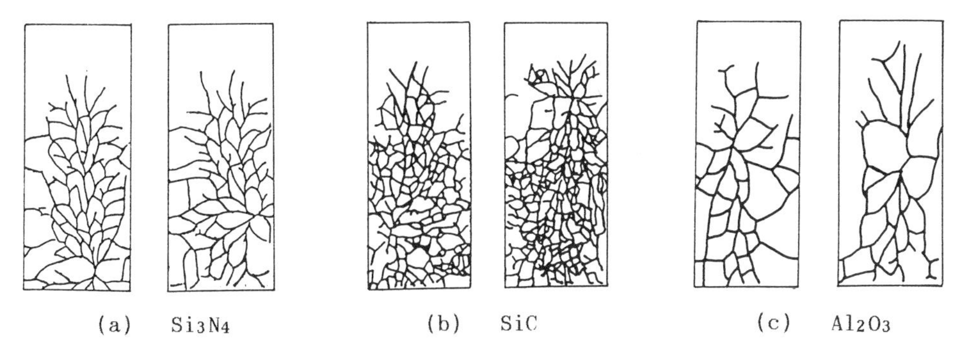

Figure 8. Crack Patterns

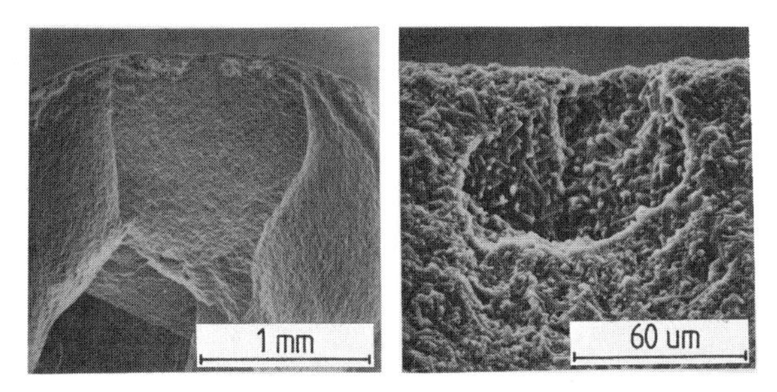

Figure 9-a. Surface Fracture Origin in Si_3N_4

Figure 9-b. Volume Fracture Origin in Si_3N_4

Generally, ΔTc is a function of the Biot's modulus, Bi, and written by the following equation:

$$\Delta T_C = \frac{S(1-\nu)}{\alpha E} \cdot \frac{1}{\sigma^*(Bi)} \quad . \tag{6}$$

In this equation, S is a specimen strength. The functional form of σ^* varies with specimen profile. Material properties (α, E, ν) depend on temperature. The S is calculate by using the effective volume or effective surface, since the S of ceramics obeys Weibull statistics. Approximate expression for the σ^* and the effective volume are obtained, restricting the shape to a columnar one. For the thermal stress calculations, analytical expressions of Jaeger[18] are used (see Appendix B).

Dimensionless Maximum Thermal Stress

The following approximate expression of σ^* has been often used for the case of the columnar shape[19].

$$\sigma^*(Bi) = \left[1.451\left(1 + \frac{3.41}{Bi}\right)\right]^{-1} \tag{7}$$

This equation, however, is available only for small values of Bi. The calculated value becomes lower than the actual one when Bi is large. Becher et al.[20] used a function:

$$\sigma^*(Bi) = \left[1.5 + \frac{B}{Bi} - 0.5\cdot\exp\left(-\frac{C}{Bi}\right)\right]^{-1} , \tag{8}$$

and obtained B=4.67 from Jaeger's solution. In the reference 20, the value of C was not written. We have fitted the expression (8) to analytically obtained data and obtained B=5.2 and C=29.3. Calculated σ^* by Equations (7) and (8) are shown in Figure 10, comparing with the analytic σ^* obtained form Jaeger's expression. σ^* calculated by Equation (8) show better agreement with the analytic σ^* in the wide range of Bi.

Effective Volume at Thermal Shock Stress

To evaluate the specimen strength, S, the effective volume or effective surface at the instant of damage must be calculated[21,22]. There is an argument that effective surface should be taken, since the maximum stress is generated on the surface during rapid quenching. However, we take the effective volume for estimating S, because the fracture origin has been observed inside the specimen, as well as on the surface as described previously. In the case of the columnar specimen, σ_{rr}, $\sigma_{\theta\theta}$ and σ_{zz} are the principal stresses. During the rapid quenching, σ_{rr} is compression, so that the effective volume can be calculated by the following equation. Here we use the effective volume ratio (V^*) defined by the ratio of the effective volume to the real volume (V_0).

$$V^* = \frac{1}{V_0}\int_V \left[\left(\frac{\sigma_{zz}}{\sigma_m}\right)^m + \left(\frac{\sigma_{\theta\theta}}{\sigma_m}\right)^m\right] dV \tag{9}$$

Where σ_{max} is the maximum thermal stress. In this calculation, $\sigma_{\theta\theta}$ and σ_{zz} are set to be zero when they are compressive. The calculated V^* is shown in Figure 11 as a function of time. This shows that the thermal stress is spread out in wider range as the time increase. The S decreases as V^* increases. We estimate V^* when the thermal stress reaches the maximum value. Accordingly to the Weibull theory of fracture, this may not

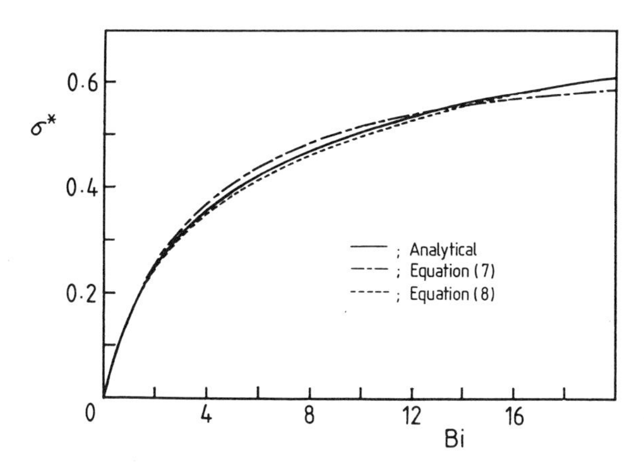

Figure 10. Dimensionless Maximum Thermal Stress

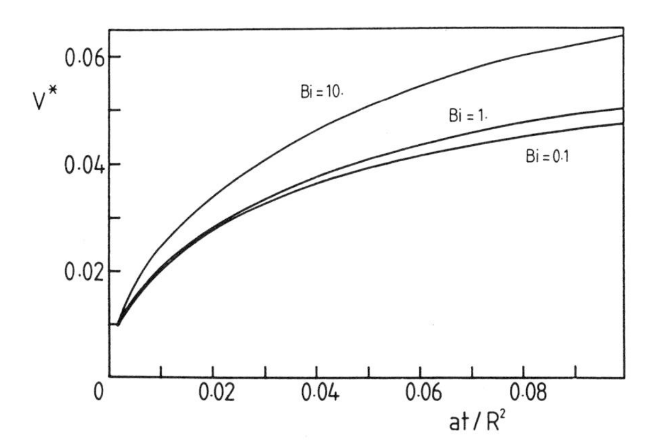

Figure 11. Effective Volume Ratio

coincide with the time when a crack is generated, since V^* keeps to increase as function of time. However, as is discussed in the next chapter, the final result of ΔTc is not much affected by this approximation for V^*.

An approximate expression for the V^* can be derived considering following conditions. Firstly, when the Weibull modulus, m, is zero or infinite, V^* is

$$\lim_{m \to 0} V^* \doteqdot 1 \quad , \quad \lim_{m \to \infty} V^* = 0 \tag{10}$$

Secondly, V^* is proportional to the 1/m when m is large, as shown in Figure 12. Thus, V^* can be written by a functional form as follows:

$$V^* = 1 - \exp\left[-\frac{1}{m} \cdot f(Bi)\right] . \tag{11}$$

V^* decreases as Biot's modulus, Bi, increase as can be seen Figure 11. We assume the function form of f(Bi) as follows:

$$f(Bi) = \frac{a}{Bi + b} + c . \tag{12}$$

The parameters, a, b and c are determined by fitting V^* of Equation 11 to the analytical V^*. Finally the following expression for V^* is obtained.

$$V^* = 1 - \exp\left[-\frac{1}{m}\left(\frac{8.3}{Bi + 9.9} + 0.26\right)\right] \tag{13}$$

The calculated V^* by this equation and the analytical value are shown in Figure 13.

Calculation of Critical Temperature Difference

The temperature dependence of material properties are taken into account using approximated equations listed in the Appendix A. The heat transfer coefficient, h, is given by Equation (2). The S, at the instant of thermal shock is calculated by the following equation and Equation (13), using the four-point bending test data listed in Table 1.

$$S = \left(\frac{V_0 \cdot V^*}{V_{4b}}\right)^{-1/m} \cdot \sigma_{4b} \tag{14}$$

Here σ_{4b} is the mean strength and V_{4b} is the effective volume in the four-point bending test. Finally the ΔTc can be calculated by equations (6) and (8). The obtained ΔTc is a function of temperature, since the material properties depends on the temperature. Therefore, the final results are calculated by an iteration technique.

The reference temperature for each material property was determined as follows. As for the E, the temperature of the specimen surface was used, since E is mainly related to the thermal stress and this stress is maximum at the surface. For the thermal conductivity, the temperature of the center of the specimen was used, since the average temperature of the whole conducting area may be closer to the central temperature of the specimen. As for the α, the mean temperature of the surface and the center of the specimen was applied. In the following calculations, the temperature of the specimen surface is set to 200°C which is the temperature of the molten solder, and the temperature of the specimen center is the temperature of the furnace.

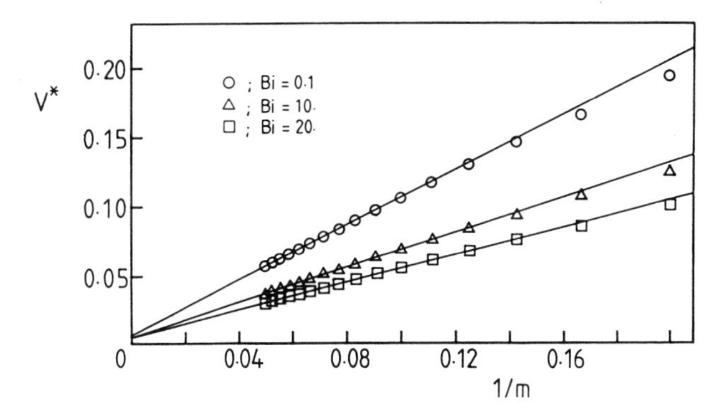

Figure 12. Effective Volume Ratio

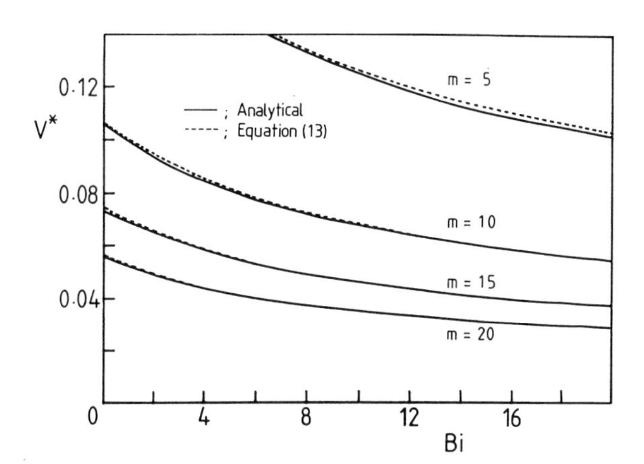

Figure 13. Effective Volume Ratio

NUMERICAL RESULTS AND DISCUSSIONS

Comparison between the Calculation and Experiments

Calculated ΔTc by Equation (6) and measured ΔTc values as shown in Figure 14 as functions of the specimen radius. This is also summarized in Table 5. In this table, $\Delta T(x)$ means the temperature difference at fracture probability of x. Accordingly, $\Delta T(0.5)$ corresponds to $\langle \Delta Tc \rangle$. ΔTc^e is the experimental data. Evaluated values agree with the experimental ones, within a range of 20°C. It is evident from Table 5 that silicon carbide is superior to alumina at the fracture probability of 0.5, but the situation is actually reversed below a fracture probability of 10^{-5}. This comes from that m of the silicon carbide is lower than that of the alumina.

In a sense of design, only $\langle \Delta Tc \rangle$ is not very useful, because it specifies the temperature difference for the fracture probability of 50%. A much lower fracture probability will be required for a practical design. This tells us that the judgment of materials from experimental data ($\langle \Delta Tc \rangle$) may lead us to wrong conclusions. Considerations with the m is very important as is the case of bending test.

Table 5. Comparison of Calculated ΔTc with Experimental Data

Materials	2R [mm]	Bi	S [MPa]	$\Delta Tc(10^{-5})$ [°C]	$\Delta Tc(0.5)$ [°C]	ΔTc^e [°C]
Si_3N_4	4	11.1	571	370	756	768
	8	14.0	525	326	666	640
	12	16.1	500	302	618	623
SiC	4	3.73	365	84	480	-
	8	4.61	295	64	364	344
	12	5.26	261	54	311	-
Al_2O_3	4	11.3	322	86	171	-
	8	14.6	298	74	148	142
	12	17.1	284	68	136	-

Maximum Thermal Stress and Material Strength

We have estimated V^* when the thermal stress reaches the maximum value. Practically, however, the fracture at the ΔTc will occur shortly after the maximum thermal stress is achieved[7], since V^* continues to increases with time. Then, the true ΔTc is the temperature difference at which the thermal stress curve contacts the specimen strength, S, curve (see Figure 15). The time deviation is the time difference between the contact point and the point when the thermal stress reaches the maximum value. The error for the estimation of strength is the strength difference of thermal stress and S when the thermal stress reaches the maximum value. The results of strength and time deviation are summarized in Table 6. The time deviation are big, but the strength deviation are small. Consequently, the assumption for the calculation of V^* has no problem.

Table 6. Time Deviation and Strength Deviation

Weibull Modulus	(m)	5	5	5	20	20	20
Biot's Modulus	(Bi)	1	5	20	1	5	20
Time Deviation	(Δt)	0.014	0.012	0.009	0.004	0.003	0.002
Strength Deviation	($\Delta\sigma$)	0.0034	0.0135	0.0368	0.0002	0.0008	0.0022

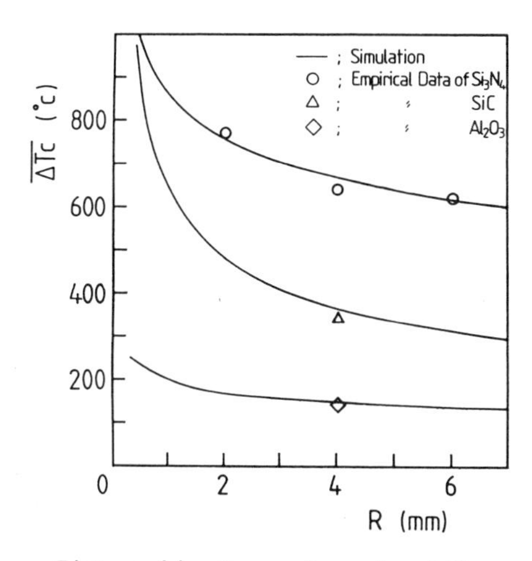

Figure 14. Comparison by ΔTc

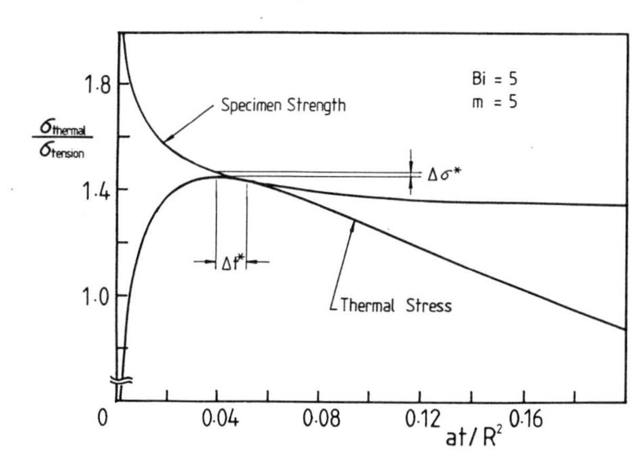

Figure 15. Thermal Stress and Specimen Strength

SUMMARY AND CONCLUSIONS

Thermal shock test was studied using liquid solder as the cooling medium to obtain a constant heat transfer coefficient. Test materials were a silicon nitride, a silicon carbide and an alumina in the form of pencil-shaped specimens to reduce the solder splashing and air drawing.

In the case of silicon nitride, no sudden degradation of the strength was observed. This phenomenon is caused by the statistical deviation of the strength. The degradation area becomes wider when the Weibull modulus is smaller and/or β is larger.

It is obviously difficult to estimate a specific critical temperature difference, ΔTc, when the residual strength decreases gradually. We have proposed to use the Weibull plot method to determine ΔTc. It is considered to be appropriate to take the average of ΔTc as the temperature difference whose fracture probability is 0.5 from a Weibull plot. Applying this method to the experimental results, we have shown that the average ΔTc can be well defined. It is also shown that this method can be consistently applied to many different ceramics.

From the observations of crack patterns of specimens, only a single fracture origin was found for each pattern. This confirms that Weibull statistics is applicable to thermal shock fracture phenomena.

Approximated expressions for the dimensionless maximum thermal stress and the effective volume of specimen were obtained for a columnar type model specimen. Using these equations and considering the temperature dependence of material properties, the ΔTc was calculated. The calculated value and the experimental data agree very well.

It was shown that the judgment of materials by only ΔTc may lead us wrong conclusion. Weibull modulus should be properly taken into account for thermal shock reliability analyses.

ACKNOWLEDGEMENT

The authors are greatful to acknowledge Professor R. C. Bradt of University of Nevada, Reno for his technical discussions. They also thank to our colleagues, Dr. H. Miyata and Dr. M. Nishimura for their advices and Mr. S. Nakagawa for his FEM analyses.

REFERENCES

1. D.H.P.Hasselman, "Unified Theory of Thermal Shock Fracture Initiation and Crack Propagation in Brittle Ceramics", J.Am.Cerm.Soc., 52 [11] 600-604(1969)
2. D.H.P.Hasselman, "Strength Behavior of Polycrystalline Alumina to thermal Shock", ibid., 53 [9] 490-95(1970)
3. D.H.P.Hasselman, "Role of Material Properties in the Thermal-Stress Fracture of Brittle Ceramics Subjected to Conductive Heat Transfer", ibid., 67 [6] 393-98(1984)
4. W.D.Kingry, "Factors Affecting Thermal Stress Resistance of Ceramic Materials", ibid., 38 [1] 3-15(1955)
5. R.W.Davidge and G.Tappin, "Thermal Shock and Fracture in Ceramics", Trans.Br.Cerm.Soc., 66, 405-22(1967)
6. T.K.Gupta, "Strength Degradation and Crack Propagation in Thermally Shocked Al_2O_3", J.Am.Cerm.Soc., 55 [5] 249-53(1972)
7. J.P.Singh, J.R.Thomas and D.P.H.Hasselman, "Analysis of Effect of Heat-Transfer Variables on Thermal Stress Resistance of Brittle Ceramics Measured by Quenching Experiments", ibid., 63 [3,4] 140-43(1980)
8. J.P.Singh, Y,Tree and D.P.H.Hasselman, "Effect of Bath and Specimen Temperature on the Thermal Stress Resistance of Brittle Ceramics Subjected to Thermal Quenching", J.Mater.Sci., 16 [8] 2109-18(1981)
9. W.P.Rogers, A.F.Emery, R.C.Bradt and A.S.Kobayashi, "Statistical Study

of Thermal Fracture of Ceramic Materials in the Water Quench Test", J.Am.Cerm.Soc. 70 [6] 406-12(1978)
10. M.Ishizuka, T.Sato, T.Endo and M.Shimada, "Thermal Shock Fracture Behavior of ZrO_2 Based Ceramics", J.Mat.Sci.Letter, 24, 4057-61(1989)
11. M.Midorikawa and T.Iseki, "Thermal Shock Testing of Dense SiC by Water-Quenching", Yogyo-Kyokai-Shi, 94 [1] 141-45(1986)
12. T.Sakuma, U.Iwata and K.Takaku, "Study Regarding Thermal Shock Resistance of Ceramics: Examination of Liquid Quenching Method", in Proceedings of the Sinetsu district, Nagano, Japan, on The Japan Society of Mechanical Engineers, pp.7-8(1990)
13. W.Weibull, "A Statistical Distribution Function of Wide Applicability", J.Appl.Mech., 18, 293-97(1951)
14. R.Ishiguro, K.Sugiyama and T.Kumada, "Heat Transfer around a Circular Cylinder in Liquid-Sodium Crossflow", Int.J.Heat Mass Transfer, 22[7] 1041-48(1979)
15. J.A.Coppola and R.C.Bradt, "Thermal-Sock Damage in SiC", J.Am.Cerm.Soc., 56 [4] 214-18(1973)
16. M.Ashizuka, T.E.Easler and R.C.Bradt, "Statistical Study of Thermal Shock Damage of a Borosilicate Glass", ibid., 66 [8] 542-50(1983)
17. P.Stanley and F.S.Chau, "Failure Probability Calculations for Brittle Components Subjected to Transient Thermal Stress", Proc.Br.Cerm.Soc., 32, 119-32(1982)
18. J.C.Jaeger, "On Thermal Stress in Circular Cylinders", Philos.Mag., 36 [257] 418-28(1945)
19. J.P.Singh, J.R.Thomas and D.P.H.Hasselman, "Analysis of Effect of Heat-Transfer Variables on Thermal Stress Resistance of Brittle Ceramics Measured by Quenching Experiments", J.Am.Cerm.Soc., 63,140-44(1980)
20. P.F.Becher, D.Lewis III, K.R.Carman and A.C.Gonzalez, "Thermal Shock Resistance of Ceramics: Size and Geometry Effects in Quench Tests", Am.Cerm.Soc.Bull., 59 [5] 542-48(1980)
21. S.S.Manson and R.W.Smith, "Theory of Thermal Shock Resistance of Brittle Materials Based on Weibull's Statistical Theory of Strength", J.Am.Cerm.Soc., 38 [1] 18-27(1955)
22. G.G.Trantina, "Statistical Fracture Analysis of Brittle Materials in Thermally Stressed Components", pp.229-44 in Thermal Stress in Severe Environments, Edited by D.P.H.Hasselman and R.A.Heller. Plenum, New York, 1980

APPENDIX

A. Material Properties

Young's Modulus

Si_3N_4 ; $-1.244*10^{12} * T^{-1} + 3.582*10^{11} * T^0 - 3.188*10^8 * T + 5.379*10^5 * T^2 - 2.973*10^2 * T^3$

SiC ; $-7.071*10^{12} * T^{-1} + 4.190*10^{11} * T^0 - 1.525*10^8 * T - 1.098*10^4 * T^2 - 5.811*10^0 * T^3$

Al_2O_3 ; $+1.131*10^{10} * T^{-1} + 3.730*10^{11} * T^0 - 4.014*10^7 * T + 1.099*10^3 * T^2 - 3.539*10^0 * T^3$

Thermal Expansion Coefficients (Expansion Ratio)

Si_3N_4 ; $+1.046*10^{-2} * T^{-1} - 3.417*10^{-4} * T^0 + 2.398*10^{-5} * T + 1.071*10^{-9} * T^2 - 1.421*10^{-13} * T^3$

SiC ; $+1.597*10^{-3} * T^{-1} - 1.496*10^{-4} * T^0 + 2.775*10^{-5} * T + 2.666*10^{-9} * T^2 - 1.029*10^{-12} * T^3$

Al_2O_3 ; $+2.178*10^{-2} * T^{-1} - 3.008*10^{-4} * T^0 + 6.136*10^{-5} * T + 3.338*10^{-9} * T^2 - 1.229*10^{-12} * T^3$

Thermal Conductivities

Si_3N_4 ; $+1.513*10^1 * T^{-1} + 2.505*10^1 * T^0 - 1.849*10^{-2} * T + 1.370*10^{-5} * T^2 - 6.345*10^{-9} * T^3$

SiC ; $- 8.806*10^{1} * T^{-1} + 7.604*10^{1} * T^{0} - 8.435*10^{-2} * T + 6.719*10^{-5} * T^{2} - 2.338*10^{-8} * T^{3}$

Al_2O_3 ; $+ 4.671*10^{2} * T^{-1} + 1.737*10^{1} * T^{0} - 1.425*10^{-2} * T + 4.640*10^{-6} * T^{2} + 1.235*10^{-10} * T^{3}$

Poisson's Ratios

Si_3N_4 ; 0.28 , SiC ; 0.16 , Al_2O_3 ; 0.23

B. Thermal Stress of the Infinite Columnar Specimen

Thermal stress of an infinite columnar specimen is as follows.

$$\sigma_{rr} = A \sum_{n=1}^{\infty} B_n \left[J_1(C_n R) - \frac{R}{r} J_1(C_n r) \right]$$

$$\sigma_{\theta\theta} = A \sum_{n=1}^{\infty} B_n \left[J_1(C_n R) + \frac{R}{r} J_1(C_n r) - C_n R J_0(C_n r) \right]$$

$$\sigma_{zz} = \sigma_{rr} + \sigma_{\theta\theta}$$

$$A \equiv \frac{\alpha E \Delta T}{1 - \nu}$$

$$B_n \equiv \frac{2 Bi \exp(-aC_n^2 t)}{C_n R [(C_n R)^2 + Bi] J_0(C_n R)}$$

where σ_{rr}, $\sigma_{\theta\theta}$ and σ_{zz} are the stresses in the r, θ and z directions. And a is the thermal diffusibility, J is the Bessel function and C_n is a solution of next equation.

$$C_n R J_1(C_n R) = Bi J_0(C_n R)$$

ANALYSIS FOR STRENGTH DEGRADATION OF INDENTED SPECIMENS DUE TO THERMAL SHOCK

J.H.Gong, Z.D.Guan, and D.C.Jiang
Department of Materials Science and Engineering
Tsinghua University, Beijing 100084, P. R. China

ABSTRACT

An analysis is proposed for the crack propagation and strength degradation in ceramics subjected to thermal shock with a Vickers indented specimen. In this analysis, fracture mechanical approach of indentation flaw, which is used to simulate the natural surface damage in materials, is applied to determine the driving force for crack propagation and the retained strength of specimens due to thermal shock. The analysis provides a functional relationship between the retained strength and the extent of crack propagation;the latter is a function of the temperature difference of thermal shock, ΔT. When ΔT is less than the critical value, ΔT_c, the extent of crack propagation is small and the associate retained strength is invariant and independent of ΔT. When $\Delta T \geqslant \Delta T_c$, however, the crack dimension will extend above a critical value, c_m, due to thermal shock, then causes an evident degradation in retained strength. It is confirmed in experiments that the proposed analysis seems to give a more realistic prediction than the conventional model.

INTRODUCTION

The thermal shock failure of ceramics has long been a problem in their application at elevated temperatures. Instantaneous decrease in strength of ceramics subjected to thermal shock has been investigated by many authors [1-4]. According to their investigations, strength behavior as a function of quenching temperature difference of thermal shock (ΔT) can be illustrated as Figure 1. It is shown that a critical temperature difference, ΔT_c, exists below which the retained strength of ceramics, σ_r, is unchanged and independent of ΔT and above which σ_r decreases suddenly as ΔT increases. Such a critical temperature difference is usually used as a "Quantitative" measure of thermal shock resistance of ceramics.

Using a fracture mechanical approach , Hasselman [1] has derived a model which describes strength degradation and crack propagation in brittle ceramics subjected to thermal shock. Although this model has been verified experimentally by several authors [2-4], there are some discrepancies existed between theoretically expected and experimentally observed thermal shock behavior. For instance, Hasselman's model indicates that the reason for the existence of a constant strength region (Figure 1) is that no crack propagation occurs for $\Delta T < \Delta T_c$. A continuous crack propagation, however, has been reported to occur at $\Delta T < \Delta T_c$ and has no influence on the constant strength region shown as Figure 1 [5].

The present paper try to analyze the relation between the retained strength after thermal shock and the extent of crack propagation. Since the thermal shock failure

Fracture Mechanics of Ceramics, Vol. 10
Edited by R.C. Bradt *et al*., Plenum Press, New York, 1992

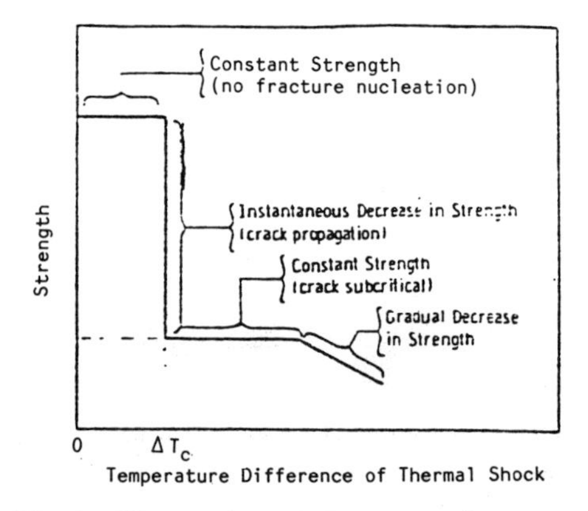

Fig.1 Illustration of Hasselman's Model

process is usually controlled by naturally surface flaws in materials, as reported by Faber et. al. [6], and the latter can be simulated with an indentation flaw [7], an indentation fracture mechanical approach is used as the background theory for this study.

THEORETICAL CONSIDERATION

According to indentation fracture mechanics, the stress intensity factor at the tip of a Vickers indentation flaw, which has cracked during indentation, at initial state, K_{I0}, arises from the residual stress about the crack tip [8]:

$$K_{I0} = K_{IC} = \frac{\chi_r P}{c_0^{3/2}} \tag{1}$$

where K_{IC} is the material fracture toughness; c_0 is the dimension of surface flaw after indentation; P is the contact load used to produce the surface flaw; and χ_r is a constant which characterizes the magnitude of the residual stress.

Eq.(1) shows that the surface flaw has been in a critical state before thermal shock. In this state, the crack propagation can be caused by any applied stresses.

The contributions from the thermal shock process to the driving force for propagation of surface flaw include two espects: the relief of residual stress due to temperature rises and the formation of surface thermal stress, σ_T, due to quenching. Both are functions of temperature difference of thermal shock, ΔT. The variation of σ_T with time in a quench test is shown schematically in Figure 2 [9]. These curves pass a peak value, σ_T^P, at $t=t_c$ and then decreases quickly with the time, for the temperature difference between the surface and the inside of specimen diminishes. Finally, σ_T becomes a smaller level, σ_T^U, at the end of the quench test. So during this process, the peak value of stress intensity about the crack tip can be formulated as:

$$K_T^P = \frac{\alpha \chi_r P}{c^{3/2}} + \sigma_T^P (\Omega \pi c)^{1/2} \qquad (c > c_0) \tag{2}$$

where Ω is a crack geometry constant; α is a relieve coefficient of residual stress and has a limit of 1 (no residual stress relieve) and 0 (complete residual stress relieve) [10]. In general, the variaty of the value of α is so small that K_T^P determined from Eq.(2) usually exceeds the critical value, K_{IC}, due to the supplement of σ_T^P, then causes the surface flaw to extend. Considering the decrease of surface thermal stress with quenching time (Figure 2), it can be expected that a continuous decrease in stress

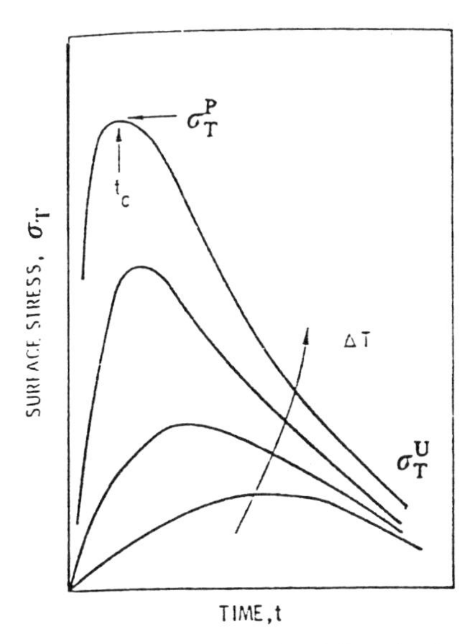

Fig.2 The Variation of σ_T with Time in Quench Test

intensity about the crack tip occurs during quench test. Specially, crack extension will stop when K_T^P drops below K_{IC}. Thus after the quence test, the ultimate stress intensity reduces as:

$$K_T^U = \frac{\alpha \chi_r P}{c_p^{3/2}} + \sigma_T^U (\Omega \pi c_p)^{1/2} \tag{3}$$

where c_p is the final crack dimension after quench test.

During the subsequent strength test, the stress intensity due to applied stress, σ_a, is given by [8] :

$$K_a = \sigma_a (\Omega \pi c_p)^{1/2} \tag{4}$$

Thus, the total stress intensity at crack tip, K_I, can be described as the superposition of Eq.(3) and (4):

$$K_I = \frac{\alpha \chi_r P}{c_p^{3/2}} + \left(\sigma_a + \sigma_T^U \right) (\Omega \pi c_p)^{1/2} \tag{5}$$

Eq.(5) shows an increasing tendency in K_I as the applied stress increases from zero. It can be expected that K_I will remain constant, K_{IC}, when σ_a exceeds a certain value, then according to indentation fracture mechanical analysis, [8] the surface flaw undergoes a stage of presuror stable growth. This is shown formally by inserting $K_I = K_{IC}$ into Eq.(5) to obtain the stress / equilibrium–crack relation:

$$\sigma_a + \sigma_T^U = \frac{K_{IC}}{(\Omega \pi c)^{1/2}} \left[1 - \frac{\alpha \chi_r P}{K_{IC} c^{3/2}} \right] \tag{6}$$

Inserting Eq.(1) into Eq.(6) and differentiating applied stress σ_a with respect to crack length, c, and put it equal to zero, we obtain from Eq.(6) that the function $\sigma_a(c)$ has a maximum value σ_m at:

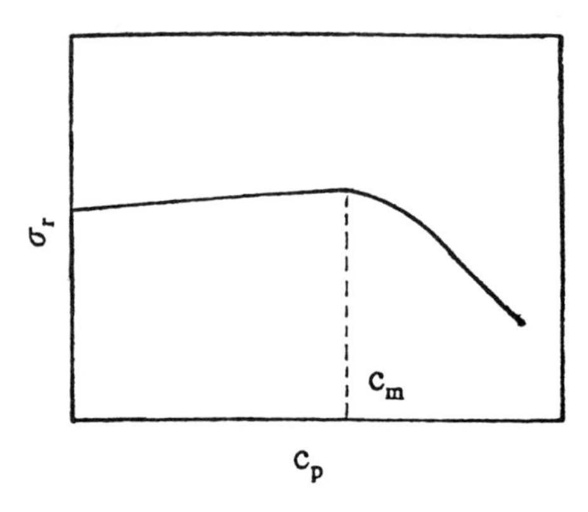

Fig.3 Illustration of the Presented Analysis

$$c_m = (4\alpha)^{2/3} c_0 \quad (7)$$

$$\sigma_m = \frac{0.47K_{IC}^{4/3}}{(\Omega\pi)^{1/2}(\alpha\chi_r P)^{1/3}} - \sigma_T^U \quad (8)$$

Eq.(7) and (8) predict that the indentation flaw with dimension, c_p, is smaller than a critical value, c_m, after thermal shock, will undergo a steady growth before failure occurs during the strength test, and the retained strength of specimen, σ_r, can be given by Eq.(8), i.e., $\sigma_r = \sigma_m$. In general, the surface thermal stressm, σ_T^U, is very small compared with the retained strength, σ_r, so Eq.(8) can be approximately treated as:

$$(\sigma_r)_{c_p < c_m} = \frac{0.47K_{IC}^{4/3}}{(\Omega\pi)^{1/2}(\alpha\chi_r P)^{1/3}} \quad (9)$$

Since K_{IC}, Ω, and χ_r are constants and the variaty of α is very small with ΔT, Eq.(9) indicates that σ_r is independent of ΔT approximately.

On the other hand, if the indentation crack has extended to a dimension equal to or larger than c_m during thermal shock, there will be no steady growth occured during the subsequent strength test and the retained strength of specimen, σ_r, is given by inserting $c = c_p$ into Eq.(6), i.e.:

$$(\sigma_r)_{c_p > c_m} = \frac{K_{IC}}{(\Omega\pi c_p)^{1/2}}\left(1 - \frac{\alpha\chi_r P}{K_{IC} c_p^{3/2}}\right) - \sigma_T^U \quad (10)$$

Eq.(10) shows that σ_r is smaller as c_p is larger in the range of $c_p > c_m$.

To illustrate, Eq.(9) and (10) are plotted in Figure 3. Since c_p is an increment function of temperature difference of thermal shock, ΔT, Figure 3 shows a similar tendency for retained strength variation as shown in Figure 2. But two different points exist between the two analysis. First, the experimentally observed crack propagation during thermal shock is incorporated into the presented analysis. It is predicted in the presented analysis that the extent of such a crack propagation increases with ΔT and reaches a critical value, c_m, when $\Delta T = \Delta T_c$. Second, the presented analysis predictes that the retained strength of thermal shocked specimen decreases continuously, not suddenly, in the range of $\Delta T > \Delta T_c$ while keeps a constant approximately in the range of $\Delta T < \Delta T_c$. Especially, a steady increase or decrease tendency will also appear sometimes in retained strength in the range of $\Delta T < \Delta T_c$, according to Eq.(8) and (9), for the values of α and σ_T^U decreases and increases as ΔT increases, respectively.

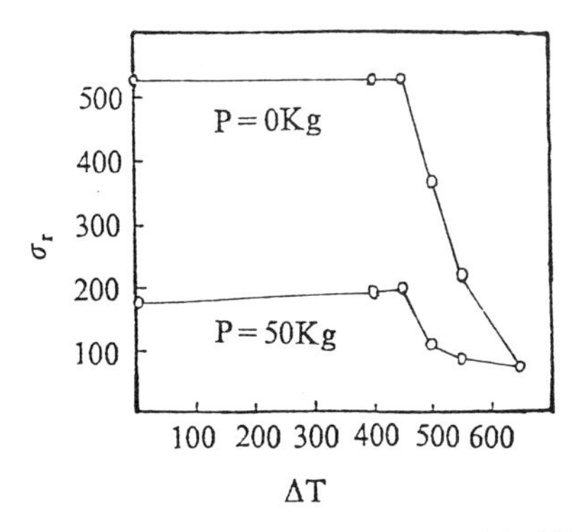

Fig.4 The Variation of σ_r with ΔT

EXPERIMENTS

The hot–pressed silicon nitride (HPSN) is used for this study. The preparation of flexure test specimens (approximately 80mm long × 6mm wide × 4mm thick) includes machining, grinding, and polishing. The final polishing is done with M2.5# diamond paste on the tensile surface. The edges are chamfered lengthwise to prevent notch effects. Then a indentation flaw is made with a Vickers indenter at the center of tensile surface with a diagonal perpendicular to the length direction.

Specimens are held for 25 min in a laboratory furnace and quenched into a bath of fluid water held at 34℃ . The severity of the quence is veried by adjusting the furance temperature. After the quence, room temperature strength is measured using three point bending at a crosshead speed of 0.1mm / min and a span of 40mm.

RESULTS AND DISCUSSIONS

The variation of retained strength, σ_r, with temperature difference of thermal shock, ΔT, of specimens with and without indentation flaw is shown in Figure 4. Two curves in Figure 4 exhibit decreasing tendency at same ΔT_c. This seems to say that the introduction of indentation flaws into specimens has no effect on the strength behavior in thermally shocked ceramics. Using the Hasselman's model and the presented analysis, the critical temperature difference, ΔT_c, can be determined to be 450℃ for HPSN.

Table I Retained Strength as a Function of Thermal Shock Temperature Difference

ΔT (℃)	0	400	450	500	550	650
σ_f (MPa)	538	540	543	374	226	36
No. of Samples	12	6	8	9	6	5

It is worth notice that a steady increase appears in retained strength as ΔT increases in the region of $\Delta T < \Delta T_c$ (Table I), as prediction as the present analysis. Similar tendency has been reported for glasses by Krohn et. al. [11] and Badaliance et. al. [5] , and for Zro_2 by Evans et.al. [4] . A constant strength region [1, 2] , even a decreasing tendency in retained strength [12] , at same region of ΔT have been also reported. These experimental phenomenon can be explained using the present analysis, concerning the effects of surface thermal stress, σ_T^U, and the relieve coefficient of residual stress, α, in Eq.(8).

The extent of crack propagation as a function of thermal shock temperature difference is given in Table 2. These results show an evident crck propagation at $\Delta T \leqslant \Delta T_c$, which has little effect on retained strength, as shown in Figure 4. This seems to be another evidence to support the present study.

Table II Crack Size as a Function of Thermal Shock Temperature Difference

P(Kg) \ ΔT(℃)	0	300	350	400	450
10	358	385	468	460	460
20	448		570	580	590
50	739	736	938	1070	1090

CONCLUSIONS

Using indentation fracture mechanical approach, an indentation fracture analysis is proposed for strength degradation and crack propagation in thermally shocked ceramics. Some predictions are made from this analysis:

(1) The surface flaw preexisted in materials may exhibit propagation under any given conditions of thermal shock. The extent of propagation is increases with ΔT.

(2) When $\Delta T \leqslant \Delta T_c$, the extent of crack propagation is small and has little effect on retained strength. So there is an approximately constant strength region existed in this case.

(3) When $\Delta T > \Delta T_c$, the extent of crack propagation exceeds a critical value, c_m, then cause an evident degradation in retained strength.

These predictions are confirmed in experiments.

REFERENCES

〔1〕 D.P.H.Hasselman, J. Am. Ceram. Soc., 52 (1969) 600–4
〔2〕 T.K.Gupta, J. Am. Ceram. Soc., 55 (1972) 249–53
〔3〕 K.Matsuskita, A.Kuratani, et. al., J. Mater. Sci. Lett., 3 (1984) 345–48
〔4〕 J.R.E.Evans, R.Stevens, and S.R.Tan, J. Mater. Sci., 19 (1984) 4068–76
〔5〕 R.Badaliance, D.A.Krohn, and D.P.H.Hasselman, J. Am. Ceram. Soc., 57 (1974) 432–36
〔6〕 K.T.Faber, M.D.Huang, and A.G.Evans, J. Am. Ceram. Soc., 64 (1981) 296–301
〔7〕 M.V.Swain; pp.257–72 in "Fracture Mechanics of Ceramics", Vol.3. Edited by R.C.Bradt, D.P.H.Hasselman, and F.F.Lange. Plenum, New York (1977)
〔8〕 D.B.Marshall, J. Am. Ceram. Soc., 66 (1983) 127–31
〔9〕 A.G.Evans, M.Linzer, et. al., J. Mater. Sci., 10 (1975) 1608–15
〔10〕 J.E.Ritter, K.Jakus, and P.Shi, J. Am. Ceram. Soc., 71 (1988) 426–29
〔11〕 D.A.Krohn and D.P.H.Hasselman, J. Am. Ceram. Soc., 56 (1973) 337–38
〔12〕 Y.W.Mai and A.G.Atkins, Ceram. Bull., 54 (1975) 593

STRENGTH ANALYSIS OF CERAMICS UNDER MECHANICAL AND THERMAL LOADING

Aalexander D. Izotov*, Vladislav B. Lazarev*, and
Vladimir Ya. Shevchenko**

*Kurnakov Institute of General and Inorganic Chemistry, Academy of Science of the USSR, 117907, GSP-1, Moscow B-71, Leninsky Pr.31

**High Tech Ceramic Res. Center, Academy of Sciences of the USSR 119361 Moscow Ozernaya str. 48

1 INTRODUCTION

In recent years increasing attention has been given to the strength analysis of ceramic materials. This is because until recently there were comparatively few theoretical studies of failure of `brittle' bodies such as ceramics, and the field of applications of ceramics has expanded significantly. Of special interest is the fact that ceramics are used under extreme loading conditions, at great energy densities in a dynamic regime, for example, in a gas turbine or in a special tool.

One of the basic problems in the study of the atomic mechanisms of deformation and fracture of solids is the evaluation of the very character of these processes. The elastic, brittle, or plastic behavior of materials under the effect of external loads is determined by the concrete conditions of loading of the solid and its physicochemical nature. The strength characteristics and stability of materials in extreme conditions (at high temperatures, high-speed impact deformations, etc.) are usually evaluated based on the calculation of the ideal strength of substances with a defect-free, ideal structure [1, 2]: values characterizing the absolute mechanical instability of the crystal lattice due to anharmonicity [3,4]. The effect of temperature.

2 IDEAL STRENGTH (OK)

The simplest and most common approach to the calculation of the ideal strength of solids is based on the use of two-particle interatomic potentials [1]; the ideal shear strength is usually estimated with consideration of the Frencel'-Kontorova linear model [5,6]. The dynamic properties of a linear chain of atoms was used for calculating the ideal breaking strength in [7-9]. The three-dimensional analog of a linear chain of atoms is a simple cubic structure, and for this reason, an examination of its strength properties is a natural extension of the model of a virtual isotropic crystal with interaction of only the nearest atoms.

Based on the two-particle interatomic Morse potential and the three-dimensional generalization of the Frenkel'-Kontorova model , the calculation of the ideal breaking and shear strength of a simple cubic structure as a function of the temperature was conducted by the self-consistent harmonic method in [10] , which was successfully used in examining the lattice instability of a loaded linear chain of atoms [11,12] and estimating the ideal breaking strength of cubic crystals [13,14]. The combined calculationn of the ideal breaking and shear strength in one model [15] permits estimating the relative stability of different substances as a function of the temperature and the possible change in the type of fracture during deformation, which is especially important in high-speed loading when the strength characteristics of the materials can approach their ideal values [16,17].

In conditions of high-speed deformation when the rate of this process is higher then the relaxation rate, it is possible to assume that all of the lading energy is consumed for occurrence of the elementary events of fracture (ruptures of interatomic bonds, elementary shear) without dissipation. i.e., adiabatically. The lower rate of the process of mechanical and thermal relaxation and dissipation of energy in comparison to the rate of deformation and fracture causes the quasi-microscopic character of fracture, which will not take place in the entire bulk of the material but in the narrow region of the fracture front. The strength characteristics of the materials will approach their ideal values, and the energy expenditures will be determined by the fundamental energetic physicochemical parameters of the substance (for examples, the heats of fusion, dissociation and atomization). In high-speed loading, the behavior of a solid can be examined in an adiabatic model of quasi-microscopic fracture, the relative stability of materials can be estimated based on their ideal breaking and shear strength, and this can be conveniently done with the two-particle interatomic potentials.

The ideal breaking strength of a single bond at 0 K (two-atom system) based on the Morse potential and dynamic properties of the system is [16]

$$\sigma^{o}_{max} = \frac{v_1}{r_o}\sqrt{\frac{D_b\bar{m}}{8}}\ \frac{1}{A_o} \qquad (1)$$

where D_b is the binding energy, r_o is the distance between atoms, $\bar{m}$ is the average mass of the atom, v_1 is the rate of transmission of longitudinal elastic vibrations (sound), and A_o is the area per bond.

In a linear chain of atoms (where each atom has two bonds), as in a two-atom system, the area per bond is set equal to r_o^2 . Then the ideal breaking strength is

$$\sigma^{o}_{max} = \sqrt{\frac{1}{8} D_{at}\rho_o v_t^2} \qquad (2)$$

where D_{at} is the volume density of the binding energy (atomization energy), i.e., the energy required for atomization (dissociation) of a substance of unitary volume; ρ_o is the density of the substance, equal to $\bar{m}/V_{at}$, where V_{at} is the volume per atom, in the given case equal to r_oA_o; $\rho_o v_1^2 \equiv L$ is the "longitudinal" elastic modulus of the substance.

In passing to the three-dimensional case, it is necessary to consider the average coordination number (average number of interatomic bonds per atom) n_b and to determine area A_o. We will use the rules for construction of a Wigner-Seitz cell whose volume corresponds to V_{at}. In this case, volume per interatomic bond, V_b, will be represented by a bipyramid, the distance between its apices will be equal to r_o, and the area of the base will be equal to A_o. Then

$$V_{at} = V_b \frac{n_b}{2} = \frac{r_oA_o}{3} \frac{n_b}{2} \tag{3}$$

The interatomic bond energy is determined from the relation

$$D_{g\text{-}atom} = D_b N_A n_b \frac{1}{2} \tag{4}$$

where $D_{g\text{-}atom}$ is the energy of atomization of a g-atom of substance, N_A is Avogardo's number, and the coefficient $\frac{1}{2}$ arises due to the fact that in summation of the energies of the interatomic bonds, each bond is considered twice. In consideration of (3) and (4), the idealbreaking strength of a three-dimensional solid is

$$\sigma^o_{max} = \sqrt{\frac{1}{8} D_{at} \rho_o v_t^2} \tag{5}$$

The ideal shear strength in the Frenkel'-Kontorova model is calculated from an examination of the interaction of two parallel linear atomic chains with the equation

$$\tau^o_{max} = G_o r_o / (2\pi d) \tag{6}$$

where $\rho_o v_t^2 \equiv G_o$ is the shear modulus, v_t is the transverse sound velocity in the substance, and d is the distance between atomic chains.

Before turning to an examination of the temperature dependence of the ideal strength, we will indicate the restrictions introduced by the approximations with respect to the substances examined. The use of the model of an isotropic virtual crystal implies that all bonds of atoms in the solid examined should be the same, and the structure should be highly symmetric. Consideration of the interaction only between neighboring atoms makes it necessary for the Wigner-Seitz cell to be electroneutral, which excludes purely ionic crystals from the examination. However, the lest approximation facilitates the determination of the parameters of the Morse potential, i.e., it permits using the atomization energy, density, interatomic distances, and elastic moduli of the crystals for this purpose. In addition, consideration of the interaction of the closest atoms legitimizes the use of the parameters of the crystals alone, and not the two-atom molecules, in the calculations, which was incidentally violated in [11]. As the calculations in [13] showed, going beyond the framework of this approximation can lead to significant numerical differences with the experimental data, which is in our opinion due to the inadequacy of the paired potential to the real interatomic interaction, especially for large distances in comparison to the interatomic distances, i.e., the "tail" of the potential.

3 IDEAL STRENGTH (TEMPERATURE DEPENDENCE)

The temperature dependence of the ideal breaking and shear strength was calculated by the method in [10,11] by modeling the material of a virtual crystal with s simple cubic structure (a_s is the unit cell parameter) consisting of atoms of mass $\bar{m}$ in positions r_i with the paired interatomic potential of the interaction $U(\vec{r}_{ij})$ (where $\vec{r}_{ij} = \vec{r}_j - \vec{r}_i$) [15]. The Hamiltonian of such a system, written with consideration of the Frenkel'-Kontorova model, is

$$H = \frac{1}{2\bar{m}} \sum_i \sum_s p_{is}^2 + \sum_{i<j} \sum_s \sum U(\vec{r}_{ij}) \Bigg|_{\langle \vec{r}_{ij} \rangle_{s \neq o}} + \sum_{i<j} \sum_{s \neq p} \sum \frac{G}{4\pi^2} \frac{a_s^4}{\langle \vec{r}_{ij} \rangle_p} \left\{ 1 - \cos \frac{2\pi r_s}{a_s} \right\} \Bigg|_{\langle \vec{r}_{ij} \rangle_{p \neq o}} \tag{7}$$

$$\vec{r}_i = \langle \vec{r}_i \rangle + \vec{u}_i \tag{8}$$

where p_{is} is the pulse coupled with the Cartesian coordinates (s) of the i-th atom $\vec{u}_i$ is the vector of movement of the i-th atom from the thermal equilibrium position $\langle \vec{r}_i \rangle$. The third terms in Eq. (7) is the generalization of the contribution of the "support" to the Hamiltonian to the three-dimensional case by analogy with the linear Frenkel'-Kontorova model written in potential form, were r_s is the running linear coordinate "along the support", and G is the shear modulus (for the one-dimensional case, see, e.g., [1,18]).

The approximation of interaction of only the nearest atoms in crystal with a simple cubic structure, as demonstrated in [4], results in a zero shear modulus. Use of the Frenkel'-Kontorova model, which is taken into consideration in the Hamiltonian (7), permits eliminating this difficulty and subsequently describing the shear behavior of the crystal. This is equivalent to selecting the angular dependence of the interatomic potential, which is a function of the interatomic distances [19].

It is possible to calculate the free energy of the dynamic system F with the Hamiltonian H using a model system based on the variation principle in statistical mechanics [20] and the Gibbs-Bogolyubov inequality [21] by minimizing the right part of the expression

$$F \leq F_o + \langle H-H_o \rangle \equiv F_m \tag{9}$$

where F_o is the free energy of the model system, and the averages ($\langle ... \rangle$) are taken within the limits of the canonical set characterized by the Hamiltonian H_o.

Following the self-consistent harmonic method in [10], the thermodynamic properties of the system examined can easily be calculated using harmonic oscillators (Einstein model) with variation force constants f_i and lattice constants a_s as the model system. In the high-temperature approximation in conditions of the effect of external forces on the atoms of the crystal, the minimum value of the free energy is determined by two systems of equations for self-consistent calculation of the force constants f_i and equilibrium positions of the atoms $\langle \vec{r_i} \rangle$. The paired interatomic Morse potential $U(r_o)$ was used with consideration of the dynamic properties of the linear chain of atoms, setting the condition

$$U''(r_o)=L_o r_o \tag{10}$$

and the expression for the shear modulus G at a finite temperature in the form

$$G = \frac{\langle U''(\langle \vec{r_{ij}} \rangle_p) \rangle}{\langle \vec{r_{ij}} \rangle_p} \frac{1-2\nu}{2(1-\nu)} \tag{11}$$

where ν is the Poisson ratio; averaging was conducted according to Gauss.

A simple cubic lattice under homogeneous tensile stress σ in direction [100], which results in an increase in the distance between two neighboring planes normals to [100] by a value of δr analog this axis, was examined for calculating the temperature dependence of the ideal breaking strength. The ideal strength was determined with the maximum of the function $\sigma[100](\varepsilon)$ for each temperature in the self-consistent calculation with the equations [15].

A simple cubic lattice under the effect of homogeneous shear stress τ in direction [100], resulting in shear of neighboring atomic planes of type (001) by value of δr, was examined for calculating the temperature dependence of the ideal shear strength. The equations for the calculation are [15].

It should be emphasized that the temperature dependence of the ideal strength in the model examined are determined (as follow from [15]) by only two parameters of material: ε^*_{max} and ν; ν is the Poisson ratio, and the physical meaning of the ε^*_{max} is the effective critical elastic deformation corresponding to the Hooke law $\sigma^o_{max} = \varepsilon^*_{max} L_o$.

The obvious relation: $\sigma_{max} = 2\sqrt{2\tau_{max}}$, is the condition of the transition from brittle (rupture of an interatomic bond) to plastic (due to shear) character of fracture (i.e., shear is realized when $2\sqrt{2\tau_{max}} < \sigma_{max}$). The results of the calculation are conveniently represented in the coordinates: ideal strength normalized for the modulus $L(2\sqrt{2\tau^*_{max}}$ and $\sigma^*_{max})$ - reduced in the temperature T/T_c, where T_c is the critical temperature of stability of the solid state of the substance, calculated from the equation [15].

As the calculations in the range of numerical values of the parameters of $0.1 \le \nu \le 0.4$ and $0.05 \le \varepsilon^*_{max} \le 0.2$, which covers most known inorganic substances, showed, there are three types of temperature dependencies of τ^*_{max} and σ^*_{max} (Fig.1). They characterize the brittle character of fracture (Fig.1a) when the breaking strength is lower than the shear strength and plastic fracture (Fig.1c) in the reverse sequence of the ideal strengths (both in the entire temperature range), and the case of the transition from the brittle to the plastic type with an increase in the temperature (Fig.1b). The concrete relations of parameters ν and ε^*_{max} corresponding to the these cases are shown in Fig.2.

The diagram makes it possible to evaluate the type of fracture of materials based on the adiabatic model of quasi-microscopic fracture. The results obtained also permit establishing the physical criterion of the "brittleness-plasticity" of substances, actually equal to the renormalized value of the effective critical elastic deformation ε^*_{max}:

$$R = \varepsilon^{*o}_{max} / \varepsilon^*_{max} \tag{12}$$

where ε^{*o}_{max} is the found from the condition $\sigma^o_{max} = 2\sqrt{2\tau^o_{max}}$.

The elastic and strength characteristics of elementary substances and some ceramic materials are reported in Table 1. The experimental data required for the calculations were taken from [22-26]. Within the adiabatic model of quasi-microscopic fracture, it is possible to state that substances with a value of $R \le 1$ undergo plastic fracture (see region c in Fig.2), and fracture is primarily brittle when $R \le 1.5$-1.6 (region a in Fig.2). A value of R between the indicated values indicates the possibility of the transition from brittle to plastic fracture as the temperature increases (region b in Fig.2).

The quantitative value of ε^*_{max} (effective critical elastic deformation) is determined by the ratio at two values: the volume density of the energy of dissociation (atomization) of the substance and the "longitudinal" or "linear" elastic modulus $L = \rho_o V_1^2$, where ρ_o is the density of the substance and V_1 is the density of the substance and V_1 is the longitudinal sound velocity in the substance. It is necessary to note that the product of these values, as demonstrates in [9,16], characterizes the dynamic stability of the substances in the adiabatic model of brittle and elastoplastic fracture of solids in high-speed deformations.

TABLE 1. Elastic and Strength Characteristics of Substances

Substance	G_o,GPa	ν,rel. units	L_o,GPa	D_{at}, GPa	σ^o_{max},GPa	R,rel.units
Be	152.4	0.042	317.7	66.22	41.87	1.638
Mg	19.3	0.277	62.6	10.92	7.548	1.151
Ca	8.8	0.180	22.4	6.768	3.554	1.115
Sr	6.08	0.280	20.3	4.921	2.885	0.949
Ba	4.91	0.28	16.5	4.681	2.071	1.067
Ti	43.3	0.321	165.0	44.02	24.60	0.792
Zr	34.34	0.38	170.0	43.51	24.83	0.623
Hf	30.4	0.38	149	45.50	23.80	0.575
V	48.3	0.358	218.7	60.46	27.10	0.802
Nb	58.80	0.35	250.0	66.46	20.38	0.872
Ta	70.7	0.337	287.6	71.88	33.89	0.939
Cr	116	0.207	314	54.67	30.88	1.691
Mo	125.8	0.294	431.4	69.88	40.92	1.384
W	160	0.280	523	87.81	50.51	1.426
Fe	94.1	0.27	298.6	58.31	38.09	1.112
Ru	172.7	0.25	514.7	76.18	57.16	1.360
Os	223.7	0.25	671.0	92.56	71.94	1.400
Co	93.2	0.287	311.6	63.95	40.75	1.030
Rh	154.2	0.258	472.6	67.11	51.41	1.350
Ir	230	0.244	680	78.54	66.71	1.552
Ni	101.1	0.272	322.4	65.44	41.93	1.085
Pd	52.48	0.372	257.7	43.50	30.56	0.773
Pt	65.1	0.393	369.8	62.15	43.76	0.670
B	179.2	0.227	507.8	124.40	46.83	1.722
Al	28.6	0.353	126	32.60	18.50	0.696
In	7.3	9.425	55.9	15.69	7.60	0.43
Tl	7.6	0.406	42.1	10.22	5.99	0.571
C diamond	533.4	0.07	1153.3	208.15	81.66	2.940
Si	66.5	0.223	186.5	37.15	13.87	2.158
Ge	56.46	0.199	150.2	27.46	10.70	2.375
β-Sn	19.9	0.352	87.1	18.47	8.19	1.094
Pb	10.39	0.391	57.68	10.77	7.20	0.650
TiB_2	215	0.28	704	123.13	60.05	1.620
ZrB_2	221.2	0.121	512.9	111.24	68.95	1.444
UB_2	130	0.3	460	138.09	72.73	0.805
HfB_2	228	0.118	526	117.37	71.73	1.431
B_4C	200	0.182	515.9	144.13	48.20	1.874
α-SiC	200	0.168	501	99.20	37.14	2.424
TiC	193.1	0.184	498	112.76	48.37	1.794
ZrC	164.5	0.192	429.3	99.61	42.21	1.754

TABLE 1. (continued)
Elastic and Strength Characteristics of Substances

Substance	G_o,GPa	ν,rel. units	L_o,GPa	D_{at}, GPa	σ^o_{max},GPa	R,rel.units
α-Si_3N_4	86.7	0.257	265.1	96.25	24.65	1.583
TiN	176.2	0.25	526.9	112.48	49.69	1.602
ZrN	160	0.25	480.3	101.29	45.02	1.600
UN	140.2	0.25	420.4	113.08	44.50	1.418
SiO_2	41.8	0.203	112.2	81.37	13.00	1.450
BeO	159.2	0.237	461.8	142.06	42.69	1.679
MgO	131.5	0.187	341.3	90.37	35.85	1.651
Al_2O_3	163.5	0.232	620.8	119.94	49.82	1.477
HfC	221.3	0.14	528.6	103.60	47.77	2.085
UC	155.98	0.346	598.4	115.83	53.74	1.306
NbC	217	0.21	590.3	117.61	53.78	1.816
TaC	221	0.24	646.7	122.54	57.46	1.731
WC	280	0.31	1005	129.15	73.50	1.717
UC	91.55	0.285	285.9	72.49	29.39	1.402
AlN	130.8	0.179	334.3	88.89	28.73	2.050

The calculations of the temperature dependence of the ideal strength were conducted in the approximations used in describing loading of materials and their physicochemical properties and in the assumption of homogeneous deformation. This must be taken into consideration in examining concrete problems.

4 THERMAL LOADING

Shevchenko at al [7] were the first to call attention to some specific details of the process of ceramic failure under high-speed mechanical loading, and to propose a model of the so-called `dissociative' failure. The model satisfactory describes the behavior of ceramics at loading speeds of the order of the speed of sound in these ceramics, as well as their conduct in a region of `weak' loading where the failure parameters are functions of the ratio of kinetic loading energy to elastic impedance [27].

The analysis of ceramic failure under great thermal loads (thermal shocks) is more complicated. In this paper we develop a theoretical model for a solid with thermal loading of one of its surfaces, to study the thermal endurance of ceramics.

When the temperature near one of the boundaries of a solid changes rapidly, thermal stresses develop in its bulk and lead, when the ultimate strength is exceeded, to catastrophic failure of the body. Numerous estimates of the thermal endurance criterion are based on these notions. The criterion is commonly understood to be the difference, ΔT, between the mean temperature of a body and the temperature on its boundary when failure occurs. Hasselman [28] gives an analytical form of the criterion as

$$\Delta T = \frac{\sigma_1(1-\mu)}{\alpha E} \tag{13}$$

where μ is the Poisson coefficient, α is the linear expansivity, and E is Young's modulus.

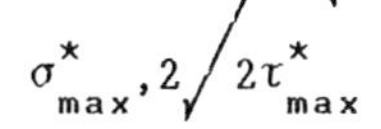

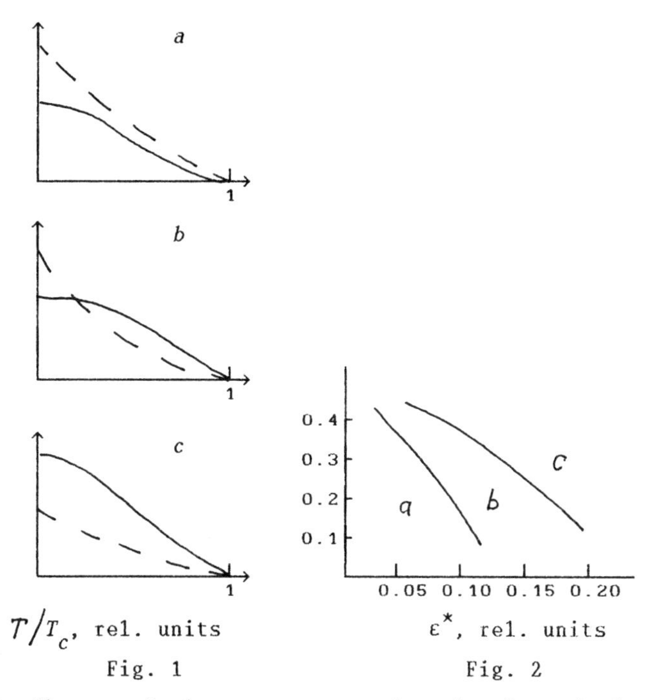

Fig. 1 Fig. 2

Fig.1. Shape of the temperature dependencies of the ideal breaking (solid lines) and shear strength (broken line) of materials undergoing brittle (*a*) and plastic (*c*) fracture in the entire temperature range and undergoing the transition from brittle to plastic fracture with an increase in the temperature (*b*).

Fig. 2. Diagram of the types of fracture based on the the adiabatic model: *a*) brittle; *b*) in going from brittle to plastic with an increase in the temperature; *c*) plastic.

Hasselman also gives some modifications to the equation for a number of particular cases, but the comparison of experimental values of the thermal endurance criterion with values calculated from equation (1) reveals systematic differences; the experimental values are as a rule larger [27]. Despite some difficulties in obtaining these estimates — data for same specimens are not always available and thermal lading conditions differ in different papers — the differences discovered show that the criterion of equation (1) some aspects of energy dissipation under thermal loading have not been taken into account, for example the development of wave processes.

Consider the simple case of a homogeneous rod of length *1* at a temperature T_0 with one its ends in contact with a medium at a higher temperature T_1. The heat flow through the rod's boundary is proportional to the difference in temperature:

$$- k \frac{dT}{dx} \bigg|_{x=+0} = h \, (T_1 - T|_{x=0}),$$

where k is the thermal conductivity, h is the heat transfer coefficient, and $x = 0$ refers to system of coordinates with its origin on the boundary of the rod. The initial conditions are

$$T|_{t=+0} = T_0 \quad \text{and} \quad \frac{dT}{dt}\bigg|_{t=+0} = 0 \, .$$

The solution of the heat equation for this case is

$$T = T_0 + \gamma(T_1 - T_0)\exp(\pi\gamma^2 t) \int_0^t z^{-1/2}\exp(-\pi\gamma^2 z - x^2/4\chi z)\,\mathrm{erf}[\gamma\pi^{1/2}(t-z)^{1/2}]dz, \tag{14}$$

where $\chi = k/c_p\rho$ is the thermal difficulty, c_p is the specific heat, ρ is the density, and $\gamma = h/(\pi k c_p \rho)^{1/2}$.

We assume that thermal loading will be accompanied by the propagation of elastic compression waves along the rod.

By using the relations [29] we obtain

$$\Delta T_{c\tau}(x, t) = \frac{\sigma_1 \nu^{3\backslash 2}(\nu t - x)^{1/2}}{E\alpha\gamma\chi} \, . \tag{15}$$

or

$$\Delta T_{c\tau} = \frac{\sigma_1 c_p}{h'\alpha\nu} \, ,$$

where $\nu_T = (\chi/t)^{1/2}$ is the speed of the heat wave, $h' = h/\sqrt{\pi}$.

It is interesting to note that the results transform in the limit as $\nu \longrightarrow 0$ into expression (13).

In spite of some simplifications assumed in the calculations [29], equations (15) and (16) obtained in the frame of our model allow us to estimate the value of the thermal endurance criterion and its variation with coordinate. This can be done for different materials (besides ceramics) under thermal lading, and other thermal and mechanical properties can also be estimated.

5. CONCLUSIONS

The temperature dependences of the ideal breaking and shear strength of substances of different physicochemical natures were calculated with a self-consistent harmonic method within the framework of the adiabatic model of quasi-microscopic fracture.

The physical criterion of the transition from brittle to plastic fracture with an increase in the temperature was substantiated.

It was shown that the ratio of two values: The volume density of the energy of dissociation (atomization) to the "longitudinal" elastic modulus, characterizes the degree of "brittleness-plasticity," while the product of the same values determines the dynamic stability of the substances in the adiabatic model of quasi-microscopic brittle and elastoplastic fracture of solids in high-speed loading.

A theoretical study of the thermal endurance of ceramics has been conducted. A model based on the assumption of the simultaneous development of heat and elastic waves in a solid under thermal loading of its surfaces has been considered.

6 REFERENCES

1. N.H.Macmillan, "Ideal strength of solids", in: Atomic Theory of Fracture [Russian translation], Mir, Moscow (1987), pp. 35-103.
2. A.Kelly and N.H.Macmillan, Strong Solids, Clarendon, Oxford (1986).3. M.Born, "Thermodynamics of crystals and melting", J. Chem. Phys., 7, No.8, 591-603 (1939).
4. M.Born and K.Huang, Dynamical Theory of Crystal Lattices, Clarion, Oxford (1954).
5. Ya.I.Frenkel'(J.Frenkel) Z.Phys., B37, No.7-8, 572-609 (1926).
6. T.A.Kontorova and Ya.I.Frenkel' Zh. Eksp. Teor. Fiz., 8, No. 1, 89-99.
7. V.Ya.Shevchenko, A.D.Izotov, V.B.Lazarev, N.M.Zhavoronkov, Izv. Acad. Sci. USSR, Inorg. Mater. 20, No.6, 1047-1052 (1984).
8. A.D.Izotov, V.B.Lazarev, Izv. Acad. Sci. USSR, Inorg. Mater. 21, No.5, 706-711 (1984).
9. V.Ya.Shevchenko, A.D.Izotov, V.B.Lazarev, N.M.Zhavoronkov, Dokl. Acad. Sci. USSR, 285, No.6, 1401-1405 (1984).
10. T.Matsubara and K.Kamiya, Progr. Theor. Phys., 58, No. 3, 767-776 (1977).
11. K.Nishioka, S.Nakamura, T.Shimamoto, and H.Fujiwara,Scr. Metall., 14, No.5, 497-502 (1980).
12. V.L.Gilyarov and V.A.Petrov, Fiz. Tverd. Tela, 25, No.2, 472-477 (1983).
13. K. Nishioka and J.K.Lee, Philos. Mag.A, 44, 779-798 (1981).
14. V.L.Gilyarov, Fiz. Tverd. Tela, 25, No.3, 944-946 (1983).
15. A.D.Izotov and V.B.Lazarev, Dokl. Akad. Nauk SSSR, 305, No.5, 1155-1159 (1989).
16. A.D.Izotov, V.B.Lazarev, Zh. Fiz. Khim. 60, No.1, 192-196 (1986).
17. A.A.Kozhushko, I.I.Rykova, A.D.Izotov, V.B.Lazarev, Izv. Akad. Nauk SSSR, Neorg. Mater., 23, No.12, 2078-2082 (1987).
18. V.L.Gilyarov, Fiz. Tverd. Tela, 29, No.4, 1055-1059 (1987).
19. M.Pfuff, GKSS Rep. E, No.25, 1-35 (1986).
20. R.P.Feynman, Statistical Mechanics, Benjamin-Cummings, Menlo Park,California (1972).
21. A.Münster, Statistical Thermodynamics, Vol.2, Academic, New York(1974), p.744.
22. C.Kittel, Introduction to Solid State Physics, Wiley, New York (1968).
23. O.H.Krikorian, Science of Hard Materials, Institute of PhysicsConference Series, No.75, Adam Hilder, Bristol-Boston (1986),pp.137-154.

24. I.N.Frantsevich, F.F.Voronov, and S.A.Bakuta, Handbook of ElasticConstants and Elastic Moduli of Metals and Nonmetals [in Russian],Naukova Dumka, Kiev (1982).
25. Landolt-Börnstein Numerical Data and Functional Relationship inScience and Technology [in German], New Series, Group 3, Cryctaland Solid State Physics, Vols.11, 18, Springer-Verlag, Berlin(1979, 1984).
26. Handbook of Thermal Constants of Substances [in Russian], V.P.Glushko (ed.), Nauka, Moscow (1978-1982).
27. A.G.Evans, T.G.Langton, Structural Ceramics [translated into Russian], Metallurgiya, Moscow (1980).
28. D.P.H.Hasseman, Bull. Am. Ceram. Soc., 49, 1033 (1970).
29. V.Ya.Shevchenko, I.E.Smolyarenko, N.M.Zhavoronkov, High Temp.-HighPress., 20, No.3, 325-329 (1987).

DESIGN METHODOLOGY FOR MANUFACTURING GLASS CATHODE RAY TUBES

Asish Ghosh, Chin Y. Cha and Shridhar Vaidyanathan

Philips Display Components Co., Ann Arbor MI-48106, USA

INTRODUCTION

Analytical determination of manufacturability of brittle material components used to be difficult due to the lack of available data and statistical techniques to evaluate the structural reliability of the components during manufacture. However, the understanding of the mechanical behavior of glass and ceramic components has increased to an extent, where products can now be designed and the feasibility of manufacturing analyzed theoretically. This is possible because of studies on reliability analysis of brittle materials[1-8]. Economics work in favor of analyzing the feasibility of manufacturing theoretically, prior to making prototypes.

Brittle materials, unlike metals, exhibit a distribution of strength values, which is dependent on the distribution of inherent flaws in the material[9-11]. Because strength is not a single value parameter, deterministic approach to evaluate reliability of brittle materials is not the best alternative. However, a probabilistic approach is better applicable for brittle materials[1,3,7,11].

The structural reliability of brittle materials can be determined by many techniques, some of these are the Weibull method, strain energy release rate technique and the energy density approach[2]. The techniques listed above have been used extensively for heat engine components such as turbine rotors[3,7]. However, to our knowledge, these techniques have not been applied for determining the structural reliability of cathode ray tubes.

This work proposes a step by step design procedure for fabricating glass components that are not only structurally feasible but are also manufacturable. This procedure can be followed by product development engineers to theoretically determine the manufacturability of products made of brittle materials at the fastest process speeds possible. As an example, manufacture of cathode ray tubes (CRTs) is considered. In addition, it is suggested to use a probabilistic approach instead of a deterministic approach to evaluate new designs of cathode ray tubes (CRTs).

DESIGN STEPS

Figure 1 illustrates the proposed flow chart to determine the manufacturability of brittle material components. The product identification stage includes the design model of the product, defining the shape, dimensions and the materials. The model is then divided into various sub-component or sub-assemblies. Using design for assembly (DFA)[12] methodology the minimum number of parts and the least number of process steps, that are required to manufacture the product, is determined. As a consequence, a list of process steps, in terms of thermo-mechanical loads, to assemble the sub-parts in the most economical way, is obtained. This drawing becomes the initial design model for further analysis.

Finite element analysis (FEA) is done for the product design model for each thermo-mechanical load required to assemble the various components. Stress values are generated for the different regions of the product during and after each thermo-mechanical load. The mechanical properties of the materials used are determined either by actual testing or using available values in the literature. At this point the stress values using FEA are used as input data for a reliability program, for example CARES (Ceramics Analysis and Reliability Evaluation of Structures)[7], to determine the probability of failure of the different regions of the product. Based on the reliability analysis, alterations in materials and/or dimensions are attempted in regions of high stresses. An iterative cycle continues until an optimal design, which is structurally stable as well as manufacturable, is obtained.

At this point an analysis of the process equipment being used (for assembly) is considered to obtain the magnitude of thermal loadings. Finite element analysis of the optimal design model is done to obtain stress values for the existing process speed with the existing equipment. This is followed by a probabilistic reliability analysis of the structure due to transient thermal stresses. This analysis indicates if faster manufacturing speeds could be considered for processing of this design model using the existing equipment.

To maximize the manufacturing speed, stress values for various process speeds with the existing process equipment is calculated using FEA. Probabilistic reliability analysis is done for these stress values. If any changes in the equipment design are feasible, it is incorporated in the thermal stress calculations in order to reduce the probability of failure. Subsequently changes are made to the material and/or dimensions in conjunction with equipment

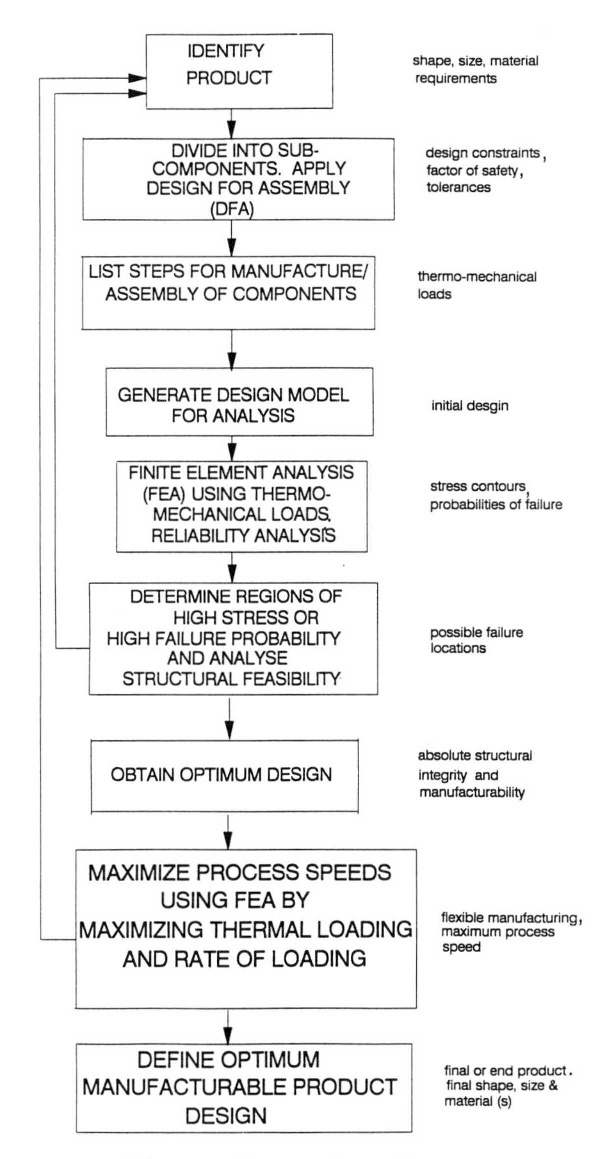

Figure 1. Design Steps

alterations, iterating, to obtain the most manufacturable design.

EXAMPLE: DESIGN OF CATHODE RAY TUBES

In this example, a 68.58 mm (27 inch) diagonal cathode ray tube was chosen. The first step was to identify the various dimensions of the tube. The dimensions of the glass envelope were based on optical performance required, structural stability and the limits of X-ray radiations. The dimensions of the inside metal components were also based on the optical performance required.

Design for assembly (DFA)[12] was used to physically divide the CRT into sub-components or sub-assemblies. As a result of the analysis the different process steps used for the assembly of a cathode ray tube is obtained. Figure 2 lists the various steps required for assembly of a CRT along with the type of loading (process conditions) applied. Mechanical loadings are permanent stress conditions but thermal loadings during manufacturing are transient stress conditions. Only thermal loading is applied in the frit seal stage because the glass is heated to greater than 400°C. Both thermal as well as mechanical loading is applied in the evacuation step (exhaust process) because the envelope is heated to 400°C during evacuation. The vacuum in the envelope is about 10^{-5} torr almost as soon as the evacuation process starts. The final pressure is about 10^{-8} torr. The tube is subjected to permanent mechanical loading at the final step due to the application of the implosion protection band for ensuring safety of tubes.

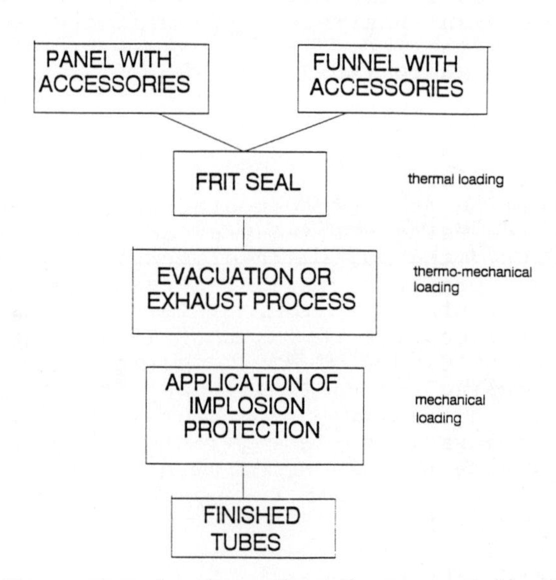

Figure 2. Steps Required to Manufacture Cathode Ray Tubes

Frit sealing is done to seal the front face (panel) to the rear cone (funnel). Although the thermal stresses generated during the frit cycle was determined, process rates are governed more by the type of frit used, rather than the structural design of the tube. Therefore, process rate of the frit sealing step cannot be maximized unless and until the frit type is changed. Hence, maximizing the frit sealing step was not considered for this analysis. However, in the evacuation step the process rates are governed by the dimensions of the tube as well as the type of heating equipment used. Therefore, process rate maximization is addressed.

Experimental Procedures

Since the envelope of the CRTs is made of glass, the material properties required, for determining structural stability of the design, are the elastic modulus, Poisson's ratio, thermal expansion coefficient, thermal conductivity, specific heat and the modulus of rupture or strength.

Modulus of rupture (MOR) of panel glass (Corning 9068) was measured using a four point bend test setup on specimens of dimensions 6.35 X 25.4 X 101.6 mm. The upper and lower span dimensions were 19.05 mm and 95.25 mm respectively. All specimens were tested under ambient conditions. All edges were beveled and etched with HF acid to ensure rupture to occur from the center of the specimen. Weibull statistics[14] were used to determine the characteristic strength and the Weibull moduli.

Finite element analysis (FEA) modeling was done to determine the optimal design. ANSYS 4.4[1] was used as the finite element analysis code for stress analysis. Table 1 lists the description of the number and type of elements used for finite element analysis. Predominantly two layered 8-node, 3-D, isoparametric solid elements (Stif 45) were used for structural analysis.

To better understand the stresses generated during thermal loading in the existing evacuation equipment, actual temperatures were measured at discrete locations on the tube during the evacuation process. Temperature profiles on the tube were generated using the measured temperatures by using 8-node isoparametric thermal solid elements (Stif 70). Based on the temperature profiles, thermal stresses using static conditions were calculated at four different temperatures of the inside center face (ICF) using Stif 45 elements. The temperatures were 50°C, 300°C, 396°C during the heating cycle and 275°C during the cooling cycle. To determine the maximum manufacturing or evacuation speeds thermal stress were calculated for various heating rates using FEA. The thermal stresses were calculated by using a sandwich

1 Swanson Analysis Systems Inc., Houston, PA

Table 1. Summary of Element Types Used

	Vacuum Stress	Measured Thermal Stresses Temperature	Stress	Calculated Thermal Stress Middle	Outside & Inside
Element Type	Stif 45	Stif 70	Stif 45	Stif 95	Stif 93
No. of Elements Through Thickness	2	1	1	1	2
Total No. of Elements	3968	936	936	454	908
Boundary Conditions	0 displacement at neck in all three directions				

Material Parameters:

E = Elastic Modulus = 69.4 GPa
ν = Poisson's Ratio = 0.23
α = Thermal Expansion = 9.9×10^{-6} mm/mm ^{o}C
K = Thermal Conductivity = 1.047×10^{-3} W/mm ^{o}C
W = Convection Coefficient = 75 W/m^{2} ^{o}C
s = Specific Heat = 865 J/gm^{o}C

structure of elements through the thickness of glass. 3-D 20-node isoparametric solid elements (Stif 95) were sandwiched between 8-node isoparametric shell elements (Stif 93).

Results and Discussion

Figure 3 shows the Weibull plot for the modulus of rupture (MOR) for the panel glass. The Weibull modulus is 5.48 and the characteristic strength is 85.71 MPa. The characteristic strength is defined as a stress value which has a probability of failure of 0.632 for the surface area/volume of the MOR sample under maximum bending moment.

Figure 4 shows the stress contour when the tube is evacuated. The maximum principal stress is 7.26 MPa and is located on the outside surface at the end of the minor axis (short axis) in the skirt area of the panel. Upon application of the implosion protection band the maximum principal stress reduces to 5.90 MPa but the location remains the same[13]. These are permanent mechanical stresses which exist in the finished product. The ANSYS stress file can be used as the input data file for CARES[7] and the probability of failure of the tube determined for the permanent mechanical loads, namely evacuation and application of the implosion protection band.

The probability of failure or reliability is calculated by CARES[7] by using the Weibull modulus and the characteristic strength determined from MOR tests. The

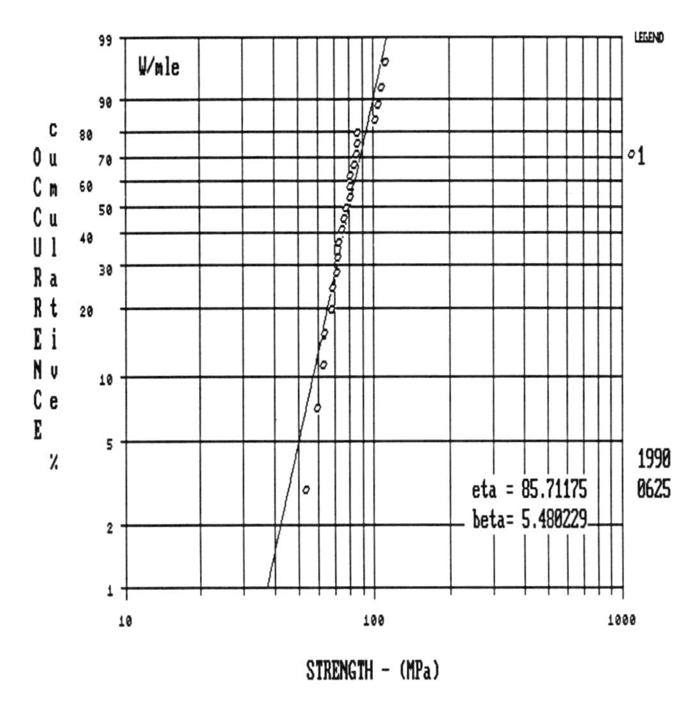

Figure 3. Weibull Plot of Modulus of Rupture of Panel Glass Bars.

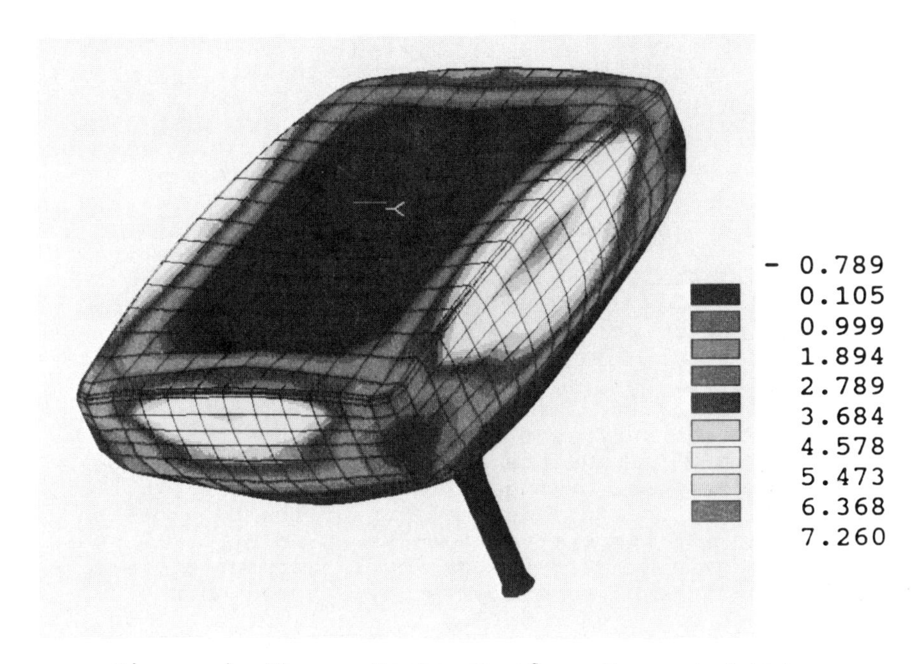

Figure 4. Stress Contours of an Evacuated Tube

Table 2. Summary of Calculated Stresses from Experimentally Measured Temperatures

Inside Center Face Temperature (°C)	Thermal Stress (MPa)		Thermo-mechanical Stress (MPa)	
	Inside	Outside	Inside	Outside
50	26.53	19.42	30.66	23.08
300	34.06	24.10	38.18	27.69
396	4.04	14.03	17.54	24.88
275	2.46	8.89	7.90	14.26

Weibull modulus is taken to be constant for the material (5.48 for this panel glass) but the probability of failure for the same stress values will depend on the surface area/volume of material which is subjected to that stress. Larger surface area/volume will have higher probabilities of failure for identical stresses. Results from this analysis will be addressed in future work.

Existing Process Conditions

Temperatures at different locations were measured for the CRT as it was processed in the existing evacuation equipment. The temperature profile for air and the inside center face (ICF) as a function of time is shown in Figure 5. The maximum air temperature is 400°C. The ICF temperature lags behind the air temperature during the heating cycle with the maximum difference being at 33 minutes. The air temperature is about 370°C while the temperature of the ICF is 300°C at 33 min. into the evacuation process. Maximum ICF temperature is 396°C which is attained at 41 minutes but by this time the air temperature has already started cooling. During the cooling cycle the ICF temperature is higher than the air temperature with the maximum temperature difference being at 75 minutes. The ICF temperature is 275°C while the air temperature is about 225°C.

Thermal stresses in the actual evacuation cycle were calculated using FEA under static conditions at inside center face temperatures of 50°C, 300°C and 396°C during the heating cycle and 275°C during the cooling cycle using the temperature profiles measured above (Figure 5). Principal stress contours for the inside and the outside surface was calculated. Figure 6 shows the stress contours on the inside surface of a quarter model of the tube during the heating cycle at ICF temperature of 300°C. Glass will fail predominantly from surface flaws therefore the surface stress contours will indicate the stress values that the flaws will experience. Maximum principal stresses (Thermal) on the inside and the outside surfaces of the tube at the four different ICF temperatures are shown in Figure 7 and tabulated in Table 2. Stresses on the inside surface are higher during the heating cycle while stresses on the outside surface are higher during the cooling cycle. This is expected because during the heating cycle the outside surface expands faster than the inside surface because the heating is from outside. During the cooling cycle the outside air temperature is lower than the tube temperature. The maximum thermal stress is 34.06 MPa (inside surface)

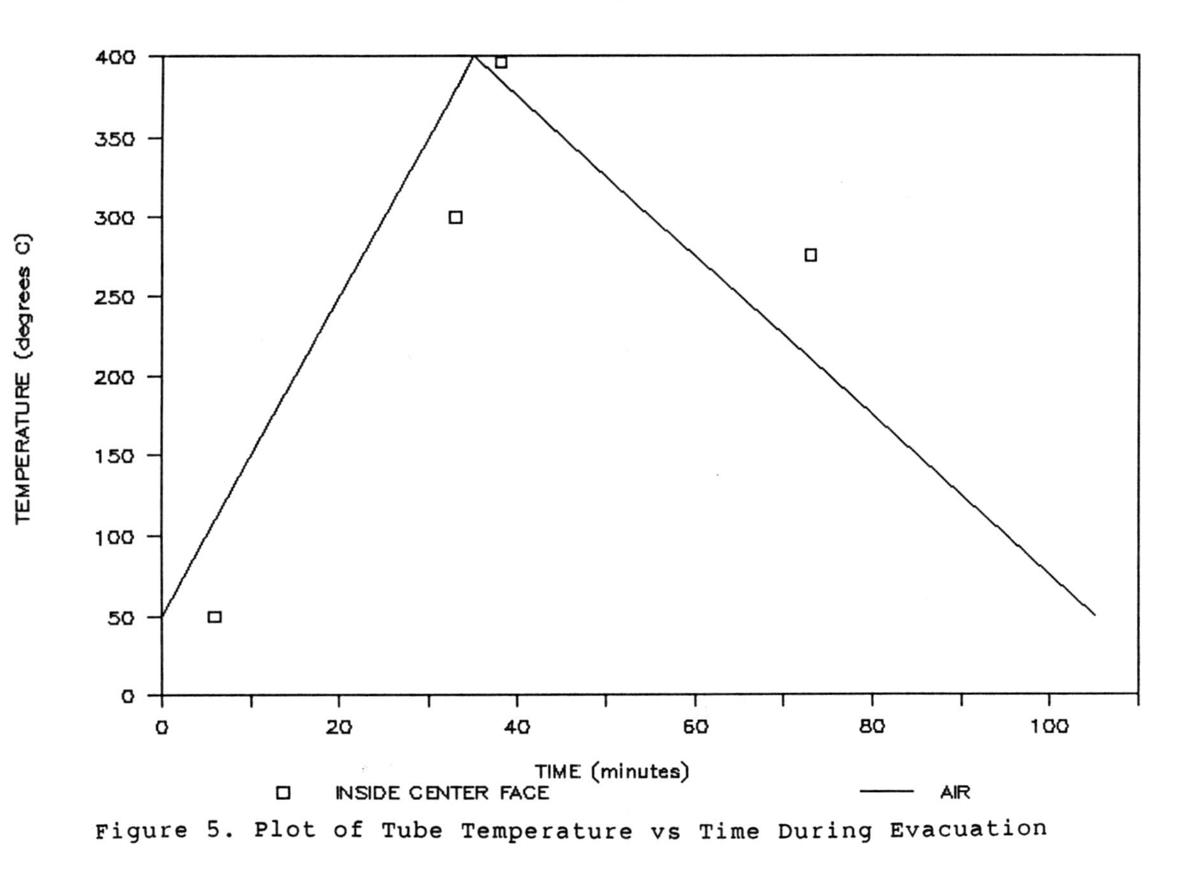

Figure 5. Plot of Tube Temperature vs Time During Evacuation

during the heating cycle when the ICF is at 300°C while during the cooling cycle the maximum principal stress (thermal) is 8.89 MPa which is on the outside surface. An important observation is that during the heating cycle the temperature gradient between the outside and the inside surface is greater than what it is during the cooling cycle, therefore, absolute maximum principal stresses are reached during the heating cycle. The reason is that the cool down cycle is done at a controlled rate in order to avoid the generation of residual strain in glass.

The mechanical load due to evacuation is added to the transient stresses to yield total thermo-mechanical stresses. Maximum principal thermo-mechanical stresses at the four different ICF for the inside and the outside surfaces are shown in Figure 8. The trends of thermo-mechanical stresses as a function of time during the evacuation process is similar to the thermal stresses but is greater in magnitude. The maximum thermo-mechanical stress

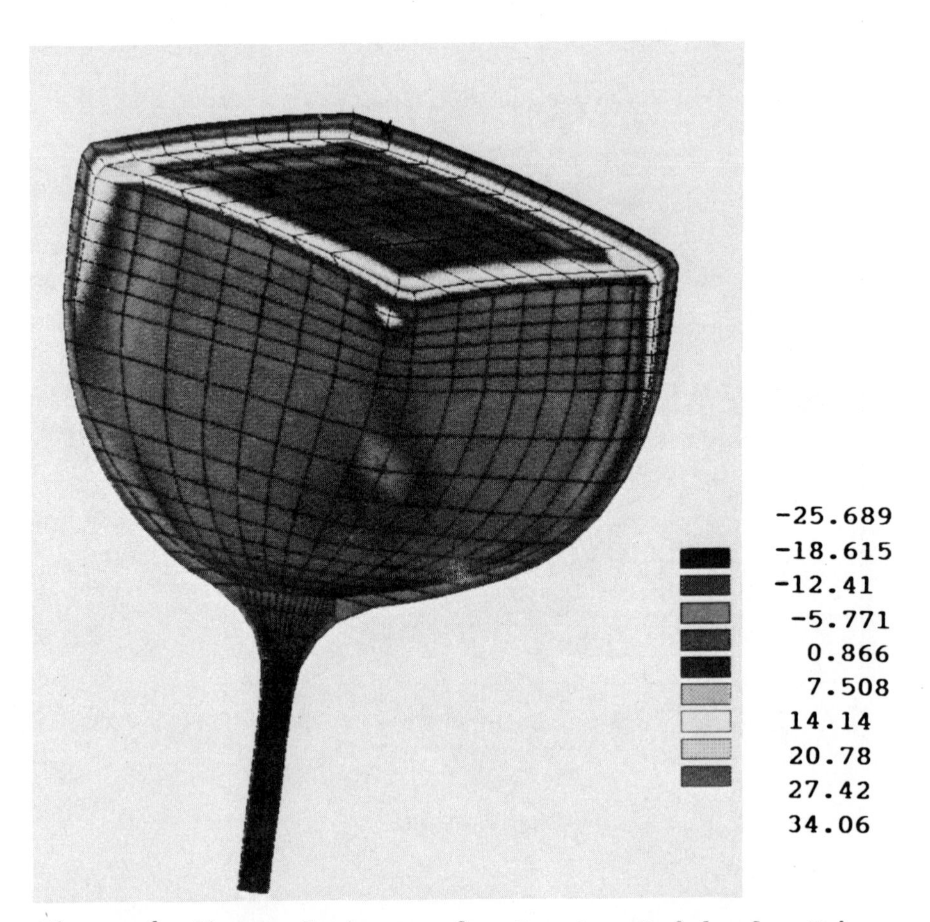

Figure 6. Stress Contours of a Quarter Model of a Tube on the Inside Surface at 300°C During Heating

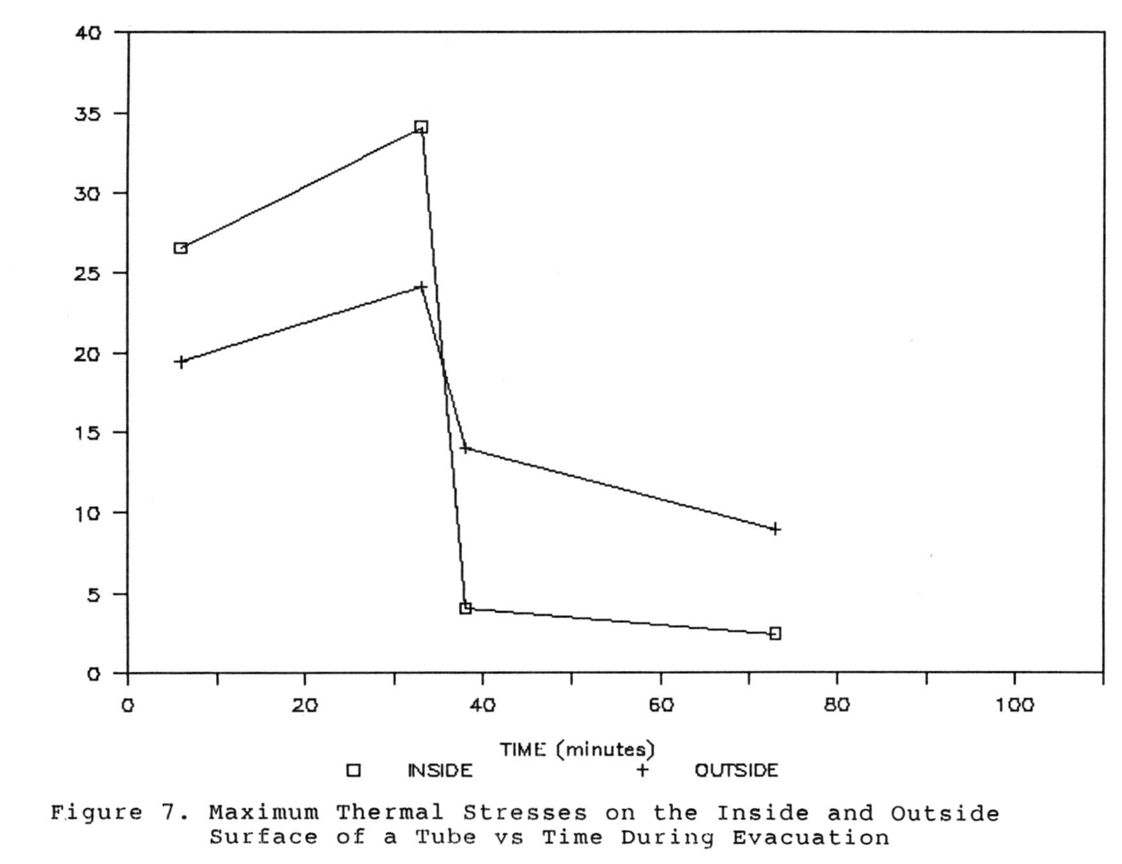

Figure 7. Maximum Thermal Stresses on the Inside and Outside Surface of a Tube vs Time During Evacuation

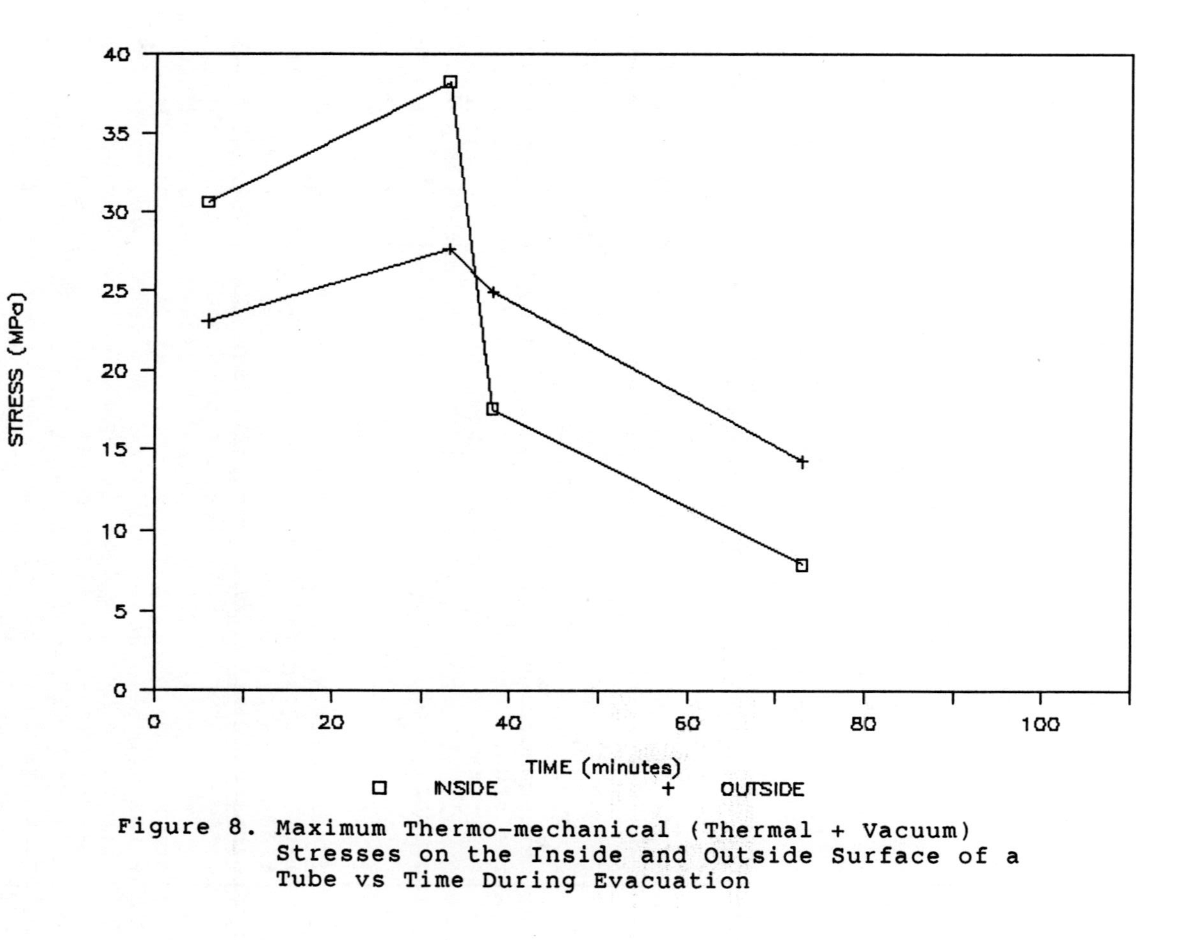

Figure 8. Maximum Thermo-mechanical (Thermal + Vacuum) Stresses on the Inside and Outside Surface of a Tube vs Time During Evacuation

Table 3. Summary of Maximum Stresses and Locations

Analysis	Stress (MPa)	Location
Vacuum Stress	7.26	End of minor axis, Skirt area of panel
With Implosion Protection	5.90	End of minor axis Skirt area of panel
Thermal Stress Heating Rate: 7°C/min	16.46	Diagonal of panel Inside surface
Thermal Stress Heating Rate: 10°C/min	23.51	Diagonal of panel Inside surface
Thermal Stress Heating Rate: 13°C/min	30.56	Diagonal of panel Inside surface

is 38.18 MPa on the inside surface during the heating cycle when the ICF is at 300°C. On the outside surface the maximum thermo-mechanical stress is 27.69 MPa again at ICF of 300°C, during the cooling cycle.

The thermal and the thermo-mechanical stress values at the four different ICF temperatures can be the input data for the program CARES[7] to yield probabilities of failure for the different regions of the tube under current equipment conditions. This analysis will be addressed in a separate study. Based on the magnitude and distribution of stresses, changes in the glass dimensions are made while meeting all the other design constraints to obtain the optimum design of the CRT.

Process Speed Maximization

To maximize the process speeds thermal stresses were calculated at three different heating rates, namely 7°C/min, 10°C/min and 13°C/min. Since maximum principal stresses, due to thermal gradients, are generated when the ICF is at 300°C, static thermal calculations were made for the three heating rates at this ICF temperature. Maximum principal stresses for the different ramp up rates are given in Table 3. The maximum stress increases from 16.46 MPa at 7°C/min to 23.51 MPa at 10°C/min and is 30.56 MPa at 13°C/min. Principal stress contours from FEA can be used as input data for CARES to obtain reliability of the structure at the different heating rates. Decision to increase the heating rate can be made based on the reliability analysis such that the probability of failure of the product is less than 10^{-6}.

The thermal stresses are dependent on the thickness of the glass, specific heat, thermal conductivity and thermal expansion coefficient. In addition, the stresses are dependent on the type of heating such as radiant heating or convection heating and the heating rate. Since the thermal expansion, specific heat and the thermal conductivity are material properties the only parameters that can be changed are the type of heating and the thickness of glass which changes the basic design of the tube. Convection heating is more uniform and generates less stress while radiant heat is more direct and could generate localized stresses. Depending on the type of application either type of heating could be preferred over the other; discussion of which will not be considered here. It would suffice to say that the actual thermal stresses can be altered by changing the type of heating. This analysis is done to consider the possibility of changes in the equipment in terms of heat shields or additional heaters at different locations of the oven to obtain reduced thermal stresses. This change could increase process speeds without changing the design of the product.

Changing the thickness of the glass required using the design optimization package and the thermo-mechanical stress analysis, in an iterative manner. As a result the optimal manufacturable design was obtained.

NEED FOR USING PROBABILISTIC DESIGN METHODOLOGY FOR DETERMINING MANUFACTURABILITY OF CATHODE RAY TUBES

Analytical stress analysis for determining the manufacturability of CRTs should be done extensively because of our increased understanding of mechanical behavior of glass as a material as well as the availability of various tools for reliability calculations.

Since glass is a perfect brittle material exhibiting a distribution of flaws it is practical to follow the design approach which uses probabilistic failure analysis, because there is no single strength value that can be used for design of CRTs. The basic advantage of using probabilistic design methodology is that there is little overdesigning of the product. Deterministic analysis which is being used by most cathode ray tube manufacturers results in overdesigning. Probabilistic reliability analysis ensures that the product is designed optimally to be processed at the fastest possible speed. Regions of high transient stresses can be determined analytically and alterations in the equipment can be made to reduce the stresses.

CONCLUSIONS

Step by step procedure to design manufacturable products using probabilistic analysis is given. The structural stability is determined using FEA followed by a reliability analysis to obtain regions of high failure probabilities. Changes in material and/or dimensions in conjunction with alterations in process equipment are done iteratively to obtain the optimum manufacturable product design.

The usefulness of the proposed design steps for manufacturability is indicated by following it for the design for manufacturability of a 68.58 mm diagonal CRT. The advantages of using a probabilistic approach rather than a deterministic approach to obtain proper design of CRTs is listed.

The permanent mechanical stresses generated in a cathode ray tube during mechanical and thermal loading of the tube is determined to be 7.26 MPa due to evacuation. This is reduced to 5.90 MPa upon application of the implosion protection band.

The Weibull modulus for the panel glass is 5.48 and the characteristic strength is 85.71 MPa measured by using four point bend tests.

During the heating cycle, inside surfaces have higher thermal and thermo-mechanical stresses while during the cooling cycle outside surface experiences higher stresses. The maximum principal thermal stress is 34.06 MPa on the inside surface during the heating cycle when the ICF is at 300°C, while the maximum principal thermo-mechanical stress is 38.18 MPa also at the same point of time.

ACKNOWLEDGMENTS

The assistance of D. Welch, J. Calomeni-Michels and J. Guenther in preparing the manuscript is acknowledged. Useful discussions with K. Fulton, R. Corson, J. Salem and M. Jenkins is greatly appreciated.

REFERENCES

1) J.A. Salem, J.M. Manderscheid, M.F. Freedman and J.P. Gyekenyesi, "Reliability Analysis of a Structural Ceramic Combustion Chamber," Presented at Int. Gas Turbine and Aeroengine Congress and Expo., Orlando, FL, June 3-6, 1991
2) L.Y. Chao and D.K. Shetty, "Equivalence of Physically Based Statistical Fracture Theories for Reliability Analysis of Ceramics in Multiaxial Loading," J. Amer. Cer. Soc., 73, [7] 1917-21 (1990)
3) M.C. Shiao, V.K. Nagpal and C.C. Chamis, "Probabilistic Structural Analysis of Aerospace Components Using Nessus "
4) J.A.M. Boulet, "An Assessment of the State of the Art in Predicting the Failure of Ceramics," Final Report to DOE Contract No. DE-AC05-85OR21400, March 1988
5) J.P. Gyekenyesi and N.N. Nemeth, "Surface Flaw Reliability Analysis of Ceramic Components with the SCARE Finite Element Postprocessor Program," J. of Eng. for Gas Turbines and Power, Trans. of ASME, 109, [7] 274-281 (1987)
6) S.S. Pai and J.P. Gyekenyesi, "Calculation of Weibull Strength Parameters and Batdorf Flaw-Density Constants for Volume- and Surface-Flaw Induced Fracture in Ceramics," NASA TM 100890, October 1988
7) N.N. Nemeth and J.M. Manderscheid and J.P. Gyekenyesi, "Design of Ceramic Components With the NASA/CARES Computer Program," NASA TM 102369, April 1990
8) S.A. Szatmary, J.P. Gyekenyesi and N.N. Nemeth, "Calculation of Weibull Strength Parameters, Batdorf

Flaw Density Constants and Related Statistical Quantities Using PC-CARES," NASA TM 103247, October 1990

9) S.B. Batdorf and J.G. Crose, "A Statistical Theory for the Fracture of Brittle Structures Subjected to Non-uniform Polyaxial Stresses," J. App. Mech., 41, 459-65 (1974)

10) S.B. Batdorf and H.L. Heinisch, Jr., "Weakest Link Theory Reformulated for Arbitrary Fracture Criterion," J. Amer. Cer. Soc., 61 [7-8] 355-58 (1978)

11) A.G. Evans, "A General Approach for the Statistical Analysis of Multiaxial Fracture," J. Amer. Cer. Soc., 61 [7-8] 302-308 (1978)

12) G. Boothroyd and P. Dewhurst, "Handbook Design for Assembly," University of Massachusetts, Amherst, MA 1984

13) A. Ghosh, C.Y. Cha, S. Vaidyanathan and R.C. Bradt, "Finite Element Stress Analysis and Crack Path Prediction of Imploding CRT," pp 1-24 in Ceramic Transactions, Volume 17, Ed. V.D. Frechette & J.R. Varner, Amer. Cer. Soc,. Westerville, OH 1991

14) W. Weibull, "A Statistical Theory of the Strength of Materials," Ingniors Vetenskaps Akademien, Handlinger, #151

Session Chairs

K. Ahlborn, Central Research and Deveropment, Carl Freudenberg, 6940 Weinheim GERMANY

Y. Akimune, Central Engineering Lab., Nissan Motor Co. Ltd., Yokosuka 237, JAPAN

R.C. Bradt, Mackay School of Mines, University of Nevada-Reno, Reno, Nevada 89557-0047 USA

J.T. Dickinson, Dept. of Physics, Washington State University, Pullman, WA 99164-2814 USA

G. Fantozzi, INSA, GEMPPM, U.A. 341, Lyon, FRANCE

K. Funatani, Japan Fine Ceramics Centre, Nagoya 456, JAPAN

D.P.H. Hasselman, Dept. of Materials Engineering, Virginia Polytechnic Institute and State University, Blacksburg, Virginia 24061, USA

K. Hirao, Dept. of Industrial Chemistry, Kyoto University, Kyoto 606, JAPAN

Y. Inomata, National Institute for Research in Inorganic Materials, Tsukuba 305, JAPAN

M.G. Jenkins, Oak Ridge National Laboratory, Oak Ridge, Tennessee 37831-6064, USA

Y. Kagawa, Institute of Industrial Science, The University of Tokyo, Minato-ku 106, JAPAN

M. Kaji, Central Research Lab., Kyocera Corporation, Kokubu 899-43, JAPAN

T. Kishi, Research Centre for Advanced Science and Technology, The University of Tokyo Megro-ku 153, JAPAN

Y-W. Mai, Center fo Advanced Materials Technology, University of Sydney Sydney, N.S.W. 2006, AUSTRALIA

M. Matsui, Analysis and Properties Research Lab., NGK Insulators, Ltd., Nagoya 467 JAPAN

Y. Matsuo, Dept. of Inorganic Materials and Engineering, Tokyo Institute of Technology Tokyo 152, JAPAN

N. Miyata, Dept. of Industrial Chemistry, Facal. of Engineering, Kyoto University Sakyo-ku 606, JAPAN

D. Munz, Institut für Materialforschung II, Kernforschungszentrum Karlsruhe 75 Karlsruhe 1, GERMANY

K. Niihara, The Institute of Scientific and Industrial Research, Osaka University Ibaraki 567, JAPAN

T. Nishida, Dept. of Chemistry and Materials Technology, Kyoto Institute of Technology Sakyo-ku 606, JAPAN

J.E. Ritter, Dept. of Mechanical Engineering, University of Masachusetts Amherst, MA 01003, USA

J.A. Salem, NASA Lewis Research Center, Cleveland, Ohio 44135, USA

V.Ya. Shevchenko, High-Tech Ceramics Research Center, Academy of Science of The USSR
Ozernaya 48, Moscow 119361, USSR

R.W. Steinbrech, Institute for Reactor Materials, Research Center Jülich
5170 Jülich, GERMANY

M.V. Swain, Dept. of Mechanical Engineering, University of Sydney
N.S.W., Sydney 2006, AUSTRALIA

Y. Tajima, R&D Center, NGK Spark Plug Co., Ltd., Komaki 485, JAPAN

K.W. White, Dept. of Mechanical Engineering, University of Houston
Houston, TX 77204-4792, USA

D.S. Wilkinson, Dept. of Materials Science and Engineering, McMaster University
Hamilton, Ontario L8S 4L7, CANADA

Authors

W.F. Adler, General Research Corporation, Santa Barbara, CA 93111, USA

K. Ahlborn, Central Research and Deveropment, Carl Freudenberg, 6940 Weinheim GERMANY

S. Akada, Asahi Glass Co., Ltd., JAPAN

T. Akiba, Central Engineering Lab., Nissan Motor Co. Ltd., Yokosuka 237, JAPAN

Y. Akimune, Central Engineering Lab., Nissan Motor Co. Ltd., Yokosuka 237, JAPAN

H. Awaji, Japan Fine Ceramics Centre, Nagoya 456, JAPAN

S. Baik, Dept. of Materials Science and Engineering, Pohang Institute of Science and Technology Pohang 790-600, KOREA

S.M. Barinov, High-Tech Ceramics Scientific Research Center, Academy of Science of The USSR, Ozernaya 48, Moscow 119361, USSR

R.C. Bradt, Mackay School of Mines, University of Nevada-Reno, Reno, Nevada 89557-0047, USA

J.H.A. Bressers, Institute for Advanced Materials, Joint Research Center - CEC 1755 ZG Petten, THE NETHERLANDS

C.Y. Cha, Philips Display Components Company, Ann Arbor, MI 48106-0963, USA

D. Chakraborty, Central Glass and Ceramic Research Institute, Calcutta 700032 INDIA

S. Chandrasekar, School of Industrial Engineering, Purdue University West Lafayette, IN 47907, USA

Y.M. Chen, School of Industrial Engineering, Purdue University, West Lafayette IN 47907, USA

J-L. Chermant, Laboratoire D'etudes et de Recherches Sur les Materiaux, LERMAT URA CNRS 1317, ISMRa, 14050 CAEN Cedex, FRANCE

S.R. Choi, Cleveland State University, Cleveland, Ohio 44115, USA

T-J. Chuang, Ceramics Division, National Institute of Standards and Technology Gaithersburg, MD 20899, USA

A. Chulya, Cleveland state University, Cleveland, Ohio 44115, USA

B. Cotterell, Center for Advanced Materials Technology, University of Sydney
Sydney, N.S.W. 2006, AUSTRALIA

S.K. Datta, Central Glass and Ceramic Research Institute, Calcutta 700032, INDIA

J.T. Dickinson, Dept. of Physics, Washington State University, Pullman
WA 99164-2814, USA

K.Y. Donaldson, Dept. of Materials Engineering, Virginia Polytechnic Institute
and State University, Blacksburg, Virginia 24061, USA

K. Duan, Center for Advanced Materials Technology, University of Sydney, Sydney
N.S.W. 2006, AUSTRALIA

S.N. Dub, Institute for Superhard Materials of The Ukrainian Academy of Science
2 Avtozavodskaya St., 254153 Kiev, USSR

M. Elices, Dept. of Materials Science, Polytechnic University of Madrid
28040- Madrid, SPAIN

G. Fantozzi, INSA, GEMPPM, U.A. 341, Lyon, FRANCE

T.N. Farris, School of Aeronautics and Astronautics, Purdue University
West Lafayette, IN 47907, USA

M.K. Ferber, Oak Ridge National Laboratory, Oak Ridge, Tennessee 37831-6064, USA

T. Fett, Institut für Materialforschung II, Kernforschungszentrum Karlsruhe
75 Karlsruhe 1, GERMANY

P. Fourvel, SEP, VERNON 2700, FRANCE

S. Fukui, Faculty of Mechanical Engineering, Technical college of Takamatsu
Takamatsu 761 JAPAN

K. Funatani, Japan Fine Ceramics Centre, Nagoya 456, JAPAN

Y. Furuse, Engineering Research Center, Tokyo Electric Power Co., Inc.
Chofu 182, JAPAN

Y-C. Gao, Center fo Advanced Materials Technology, University of Sydney
Sydney, N.S.W. 2006, AUSTRALIA

A. Ghosh, Philips Display Components Company, Ann Arbor, MI 48106-0963, USA

M. Gomina, Laboratoire D'etudes et de Recherches Sur les Materiaux, LERMAT
URA CNRS 1317, ISMRa, 14050 CAEN Cedex, FRANCE

J.H. Gong, Dept. of Materials Science and Engineering, Tsinghua University
Beijing 100084, CHINA

Z.D. Guan, Dept. of Materials Science and Engineering, Tsinghua University
Beijing 100084, CHINA

L. Guazzone, Dept. of Mechanical Engineering, University of Houston
Houston, TX 77204-4792, USA

D.P.H. Hasselman, Dept. of Materials Engineering, Virginia Polytechnic Institute
and State University, Blacksburg, Virginia 24061, USA

H. Hatta, The Institute of Space and Astrnautical Science, Sagamihara 229, JAPAN

A. Hattori, Dept. of Materials Science and Engineering, Nagoya Institute of
Technology, Nagoya 466, JAPAN

J.C. Hay, Dept. of Mechanical Engineering, University of Houston
Houston, TX 77204-4792, USA

S. Hayashi, Structure and Strength Dept., Research Institute, Ishikawajima-Harima
Heavy Industries Co., Ltd., Koto-ku 135, JAPAN

K. Hirao, Dept. of Industrial Chemistry, Kyoto University, Kyoto 606, JAPAN

W.S. Hong, Institute for Defense Analyses, Alexandria, Virginia 22311-1772, USA

X. Hu, Center for Advanced Materials Technology, University of Sydney
Sydney, N.S.W. 2006, AUSTRALIA

Y. Ikuhara, Japan Fine Ceramics Centre, Nagoya 456, JAPAN

S. Ito, Government Industrial Research Institute- Nagoya, Nagoya 462, JAPAN

A.D. Izotov, Kurnakov Institute of General and Inorganic Chemistry, Academy of
Science of The USSR, 117907, GSP-1 Moscow B-71, Leninsky Pr.31, USSR

M.G. Jenkins, Oak Ridge National Laboratory, Oak Ridge, Tennessee 37831-6064
USA

L.C. Jensen, Dept. of Physics, Washington State University, Pullman
WA 99164-2814, USA

D.C. Jiang, Dept. of Materials Science and Engineering, Tsinghua University
Beijing 100084, CHINA

H. Jinno, Dept. of Chemistry, Sophia University, Tokyo, JAPAN

Y. Kagawa, Institute of Industrial Science, The University of Tokyo
Minato-ku 106, JAPAN

M. Kaji, Central Research Lab., Kyocera Corporation, kokubu 899-43, JAPAN

I. Kameyama, Dept. of Chemistry and Materials Technology, Kyoto Institute of
Technology, Sakyo-ku 606, JAPAN

A. Kamiya, Government Industrial Research Institute- Nagoya, Nagoya 462, JAPAN

N. Kamiya, Toyota CRDL, Aichi-Gun 480-11, JAPAN

W. Kanematsu, Government Industrial Research Institute- Nagoya, Nagoya 462, JAPAN

K. Kawaguchi, Research and Development Division, Sumitomo Metal Industries, Ltd. Amagasaki 660, JAPAN

H. Kawamoto, Higashifuji Technical Center, Toyota Motor Corp., Susono 410-11 JAPAN

D. Kervades, Laboratoire D'etudes et de Recherches Sur les Materiaux LERMAT, URA CNRS 1317, ISMRa, 14050 CAEN Cedex, FRANCE

S. Kimura, Dept. of Inorganic Materials, Tokyo Institute of Technology Megro-ku 152, JAPAN

M. Kinoshita, Government Industrial Research Institute- Osaka, Osaka 563, JAPAN

T. Kishi, Research Centre for Advanced Science and Technology, The University of Tokyo, Megro-ku 153, JAPAN

H. Kishimoto, Dept. of Mechanical System Engineering, Toyota Technological Institute, Nagoya 468, JAPAN

K. Kitakami, Canon Research Center, Atsugi, JAPAN

H.N. Ko, Nakanihon Automotive College, Kamo-gun, Gifu 505, JAPAN

A.S. Kobayashi, Dept. of Mechanical Engineering, University of Washington Seattle, Washington 98195, USA

T. Kobayashi, Dept. of Production Systems Engineering, Toyohashi University of Technology, Toyohashi 441, JAPAN

Y. Kodama, Government Industorial Research Institute- Nagoya, Nagoya 462, JAPAN

Y. Kogo, Materials and Electric device Laboratory, Mitsubishi Electric Corporation, Sagamihara 229, JAPAN

N. Kohler, Fachhochschule Rhineland-Pfalz, Koblenz, GERMANY

A. Kokaji, Central Research Lab., Kyocera Corporation, kokubu 899-43, JAPAN

K. Kubo, Government Industrial Research Institute- Nagoya, Nagoya 462, JAPAN

S.C. Langford, Dept. of Physics, Washington State University, Pullman WA 99164-2814, USA

V.B. Lazarev, Kurnakov Institute of General and Inorganic Chemistry, Academy of Science of The USSR, 117907, GSP-1 Moscow B-71, Leninsky Pr.31, USSR

S.M. Lee, Dept. of Materials Science and Engineering, Pohang Institute of Science and Technology, Pohang 790-600, KOREA

J. Llorca, Dept. of Materials Science, Polytechnic University of Madrid
28040 - Madrid, SPAIN

E.H. Lutz, Kelterbergstr. 15, W-7562 Gerns Bach, GERMANY

Y-W. Mai, Center fo Advanced Materials Technology, University of Sydney
Sydney, N.S.W. 2006, AUSTRALIA

A.L. Maistrenko, Institute for Superhard Materials of The Ukrainian, Academy of
Sciencec ,2 Avtozavodskaya St., 254153 Kiev, USSR

H. Makino, Toyota CRDL, Aichi-Gun 480-11, JAPAN

T. Makino, NGK Insulators, Ltd., Nagoya 467, JAPAN

C.S. Martins, Institute for Advanced Materials, Joint Research Center - CEC
1755 ZG Petten, THE NETHERLANDS

M. Masuda, Analysis and Properties Research Lab., NGK Insulators, Ltd.
Nagoya 467, JAPAN

M. Matsui, Analysis and Properties Research Lab., NGK Insulators, Ltd.
Nagoya 467, JAPAN

R.L.K. Matsumoto, Hercules Research Center, Hercules Incorporated
Wilmington, DE 19894-0001, USA

Y. Matsuo, Dept. of Inorganic Materials and Engineering, Tokyo Institute of
Technology, Tokyo 152, JAPAN

J. Matsuoka, Dept. of Chemistry for Materials, Mie University, Tsu 514, JAPAN

D.J. Mihora, General Research Corporation, Santa Barbara, CA 93111, USA

B.S. Min, Samsung Electrics Suwon, KOREA

N. Miyahara, Dept. of Mechanical Engineering, Nagaoka University of Technology
Nagaoka 940-21, JAPAN

T. Miyajima, Government Industorial Research Institute- Nagoya, Nagoya 462, JAPAN

N. Miyata, Dept. of Industrial Chemistry, Facal. of Engineering, Kyoto University
Sakyo-ku 606, JAPAN

A.K. Mukhopadhyay, Dept. of Mechanical Engineering, University of Sydney
Sydney, N.S.W. 2006, AUSTRALIA

D. Munz, Institut für Materialforschung II, Kernforschungszentrum Karlsruhe
75 Karlsruhe 1, GERMANY

N. Murayama, Government Industorial Research Institute- Nagoya, Nagoya 462, JAPAN

Y. Mutoh, Dept. of Mechanical Engineering, Nagaoka University of Technology
Nagaoka 940-21, JAPAN

S. Nakahara, Dept. of Mechanical Engineering, Kansai University, Suita 564, JAPAN

A. Nakahira, The Institute of Scientific and Industrial Research, Osaka University
Ibaraki 567, JAPAN

K. Nakano, Government Industrial Research Institute- Nagoya, Nagoya 462, JAPAN

Y. Nakasuji, Analysis and Properties Research Lab., NGK Insulators, Ltd.
Nagoya 467, JAPAN

H. Nakayama, Faculty of Junior College of Automobile Industory
Osaka Sangyo University, Daito 574, JAPAN

K. Niihara, The Institute of Scientific and Industrial Research, Osaka University
Ibaraki 567, JAPAN

T. Nishida, Dept. of Chemistry and Materials Technology, Kyoto Institute of
Technology, Sakyo-ku 606, JAPAN

T. Nishikawa, Dept. of Materials Science and Engineering, Nagoya Institute of
Technology, Nagoya 466, JAPAN

H. Nishimura, Dept. of Mechanical Engineering, Tokyo Metropolitan University
Hachioji 192-03, JAPAN

R. Nowak, Government Industrial Research Institute- Osaka, Osaka 563, JAPAN

T. Ogasawara, Central Engineering Lab., Nissan Motor Co., Ltd., Yokosuka 237
JAPAN

T. Ogawa, Dept. of Mechanical Engineering, Gifu University, Gifu 501-11, JAPAN

T. Ohji, Government Industrial Research Institute- Nagoya, Nagoya 462, JAPAN

T. Oikawa, Advanced Technology Research Center, NKK Corporation, Kawasaki 210
JAPAN

N. Okabe, Heavy Apparatus Engng. Lab., Toshiba Corporation, Kanagawa 230, JAPAN

A. Okada, Central Engineering Lab., Nissan Motor Co., Ltd., Yokosuka 237, JAPAN

T. Okada, ISIR, Osaka University, Ibaraki 567, JAPAN

A. Okura, IIS, The University of Tokyo, Tokyo 108, JAPAN

C. Olagnon, INSA, GEMPPM, U.A. 341, Lyon, FRANCE

H. Omura, Matsushita Electric Industrial Co., Ltd., JAPAN

S. Owaki, ISIR, Osaka University, Ibaraki 567, JAPAN

F. Pernot, Laboratoire de Science des Matériaux Vitreux- CNRS, URA, USTL.
Place E. Bataillon 34095 Montpellier Cédex 05, FRANCE

J. Peussa, Federation of Finnish Metal, Engineering and Electro-Technical Industries, SF00130, Helsinki, FINLAND

G. Pezzotti, The Institute of Scientific and Industrial Research, Osaka University Ibaraki 567, JAPAN

J. Phalippou, Laboratoire de Science des Matériaux Vitreux- CNRS, URA, USTL. Place E. Bataillon, 34095 Montpellier Cédex 05, FRANCE

J.W. Pier, Hercules Research Center, Wilmington, DE 19894, USA

M.A. Rigdon, Institute for Defense Analyses, Alexandria, Virginia 22311-1772, USA

J.E. Ritter, Dept. of Mechanical Engineering, University of Masachusetts Amherst, MA 01003, USA

L.G. Rosa, Institute Superior Técnico, Lisbon, PORTUGAL

M-H. Rouillon, , Laboratoire D'etudes et de Recherches Sur les Materiaux LERMAT, URA CNRS 1317, ISMRa, 14050 CAEN Cedex, FRANCE

M. Saadaoui, Ecole Mohammadia D'ingénieurs, Rabat, MAROCCO

S. Sakaguchi, Government Industorial Research Institute- Nagoya, Nagoya 462, JAPAN

M. Sakai, Dept. of Materials Science, Toyohashi University of Technology Toyohashi 441, JAPAN

T. Sakai, Faculty of Science and Engineering, Ritsumeikan University Kita-ku 603, JAPAN

J.A. Salem, NASA Lewis Research Center, Cleveland, Ohio 44135, USA

J.M. Sater, Institute for Defense Analyses, Alexandria, Virginia 22311-1772, USA

R.A. Schultz, Mackay School of Mines, University of Nevada-Reno Reno, Nevada 89557-0047, USA

K. Sekine, Institute of Industrial Science, The University of Tokyo Minato-ku 106, JAPAN

V.Ya. Shevchenko, High-Tech Ceramics Research Center, Academy of Science of The USSR Ozernaya 48, Moscow 119361, USSR

G. Sines, Dept. of Materials Science and Engineering, University of California Los Angeles, CA 90024, USA

N. Soga, Dept. of Industrial Chemistry, Kyoto University, Kyoto 606, JAPAN

M. Steen, Institute for Advanced Materials, Joint Research Center - CEC 1755 ZG Petten, THE NETHERLANDS

R.W. Steinbrech, Institute for Reactor Materials, Research Center Jülich 5170 Jülich, GERMANY

K. Sugihara, Dept. of Mechanical Engineering, Kansai University, Suita 564, JAPAN

A. Suzuki, Structure and Strength Dept., Research Institute, Ishikawajima-Harima Heavy Industries Co., Ltd., Kotoku 135, JAPAN

K. Suzuki, Faculty of Education, Niigata University, Niigata 950-21, JAPAN

M.V. Swain, Dept. of Mechanical Engineering, University of Sydney N.S.W., Sydney 2006, AUSTRALIA

Y. Tajima, R&D Center, NGK Spark Plug Co., Ltd., Komaki 485, JAPAN

Y. Takagi, Advanced Technology Research Center, NKK Corporation, Kawasaki 210 JAPAN

J. Takahashi, Dept. of Materials Science and Engineering, Nagoya Institute of Technology Nagoya 466, JAPAN

M. Takatsu, Dept. of Materials Science and Engineering, Nagoya Institute of Technology Nagoya 466, JAPAN

N. Takeda, Research Center for Advanced Science and Technology, The University of Tokyo Meguro-ku 153, JAPAN

S. Takeda, Dept. of Applied Chemistry, Okayama University, Okayama 700, JAPAN

H. Tanaka, Dept. of Mechanical Engineering, Nagoya University, Nagoya 464-01, JAPAN

I. Tanaka, The Institute of Scientific and Industrial Research, Osaka University Ibaraki 567, JAPAN

K. Tanaka, Dept. of Mechanical Engineering, Nagoya University, Nagoya 464-01, JAPAN

T. Tanaka, Faculty of Science and Engineering, Ritsumeikan University, Kyoto 603, JAPAN

I. Tari, Dept. of Applied Chemistry, Okayama University, Okayama 700, JAPAN

H. Tsuruta, Engineering Research Center, Tokyo Electric Power Company, Chofu 182 JAPAN

H. Tsuruta, Analysis and Properties Research Lab., NGK Insulators, Ltd., Nagoya 467 JAPAN

H. Uchimura, Central Research Lab., Kyocera Corporation, kokubu 899-43, JAPAN

A. Ueno, Dept. of Mechanical System Engineering, Toyota Technological Institute Nagoya 468, JAPAN

K. Ueno, Government Industrial Research Institute- Osaka, Osaka 563, JAPAN

K. Urashima, R&D Center, NGK Spark Plug Co., Ltd., Komaki 485, JAPAN

S. Vaidyanathan, Philips Display Components Company, Ann Arbor, MI 48106-0963, USA

A. Venkateswaran, Dept. of Materials Engineering, Virginia Polytechnic Institute and State University, Blacksburg, Virginia 24061, USA

S. Wada, Toyota CRDL, Aichi-Gun 480-11, JAPAN

F. Wakai, Government Industorial Research Institute- Nagoya, Nagoya 462, JAPAN

S. Wakayama, Dept. of Mechanical Engineering, Tokyo Metropolitan University Hachioji 192-03, JAPAN

M. Watanabe, R&D Center, NGK Spark Plug Co., Ltd., Komaki 485, JAPAN

K.W. White, Dept. of Mechanical Engineering, University of Houston Houston, TX 77204-4792, USA

D.S. Wilkinson, Dept. of Materials Science and Engineering, McMaster University Hamilton, Ontario L8S 4L7, CANADA

T. Woignier, Laboratoire de Science des Matériaux Vitreux- CNRS, URA, USTL. Place E. Bataillon 34095 Montpellier Cédex 05, FRANCE

N. Yamada, Analysis and Properties Research Lab., NGK Insulators, Ltd., Nagoya 467 JAPAN

Y. Yamade, Research and Development Division, Sumitomo Metal Industries, Ltd. Amagasaki 660, JAPAN

K. Yamaishi, Dept. of Mechanical Engineering, Nagaoka University of Technology Nagaoka 940-21, JAPAN

S. Yamamoto, Faculty of Science and Engineering, Ritsumeikan University, Kyoto 603 JAPAN

S. Yamauchi, Dept. of Production Systems Engineering, Toyohashi University of Technology Toyohashi 441, JAPAN

Y. Yamauchi, Government Industrial Research Institute- Nagoya, Nagoya 462, JAPAN

Y-N. Yan, Tsinghua University, Beijng, CHINA

L.X. Zeng, Dept. of Materials Science and Engineering, Tsinghua University Beijing 100084, CHINA

X.F. Zhao, Dept. of Materials Science and Engineering, Tsinghua University Beijing 100084, CHINA

L. Zhou, Center for Advanced Materials Technology, University of Sydney Sydney, N.S.W., 2006, AUSTRALIA

INDEX